AF619035

I. Roth H. Lindorf
South American Medicinal Plants

Springer-Verlag Berlin Heidelberg GmbH

I. Roth H. Lindorf

South American Medicinal Plants

Botany, Remedial Properties and General Use

With 313 Figures and 9 Tables

Professor Dr. INGRID ROTH
Moosweg 2
79244 Münstertal
Germany

Professor Dr. HELGA LINDORF
Universidad Central de Venezuela
Biología Tropical
Aptdo. 21201
Caracas 1020
Venezuela

DOI 10.1007/978-3-662-04698-2

Library of Congress Cataloging in Publication Data South American medicinal plants: botany, remedial properties and general use/Ingrid Roth, Helga Lindorf (eds). p .; cm

1. Medicinal plants--South America--Encyclopedias. Roth, Ingrid, 1920- II. Lindorf, Helga.

http://www.springer.de

Originally published by Springer-Verlag Berlin Heidelberg New York in 2002
MyCopy version of the original edition 2002

Production: PRO EDIT GmbH, Heidelberg, Germany
Cover design: design & production, Heidelberg, Germany
Typesetting: Mitterweger & Partner, Plankstadt
Printed on acid-free paper SPIN 10797324 14/3130/Re - 5 4 3 2 1 0
www.springer.com/mycopy

Motto

A herb is growing for every sort of disease

Gegen jede Krankheit ist ein Kraut gewachsen

Dedications

Dedicated to my dear friend Stefan Fankhauser
INGRID ROTH

Dedicated to my parents, and my brothers Willie, Carlos and his family
HELGA LINDORF

Preface

Books dealing with medicinal herbs have been much in demand in the last decade, since interest in the use of 'healing plants' instead of chemistry wrapped up in the form of pills has gradually increased. Interest in natural methods of healing by medicinal plants advanced with the ecological conscience of man. European officinal plants are well known and have already been applied for hundreds of years or even more. However, as the tropical forests attracted more attention, the great wealth of useful plants stored in tropical regions became obvious. India is probably a leading country in the knowledge of indigenous tropical medicinal plants, some of which are also known or are even common in Europe. Further useful and medicinal plants come from China and other countries of the Far East (eg. *Ginkgo*, Ginseng) or from Africa.

Although Latin America has supplied the world with quite a few very important useful plant products, such as quinine and rubber, the majority of useful plants has remained unknown to the rest of the world. Only very recently, North American pharmaceutical companies have invaded the tropical forests, principally of Central America, exploring useful, mainly medicinal plants (see also the preface in Roth 1992 'Coastal vegetation and mangroves of Venezuela'). These companies send science-brokers to the forests in search of the still unknown chemical compounds. Within a few years – they promise – a pharmaceutical preparation will appear on the market.

However, thousands of years ago, the Indians of South and Central America had collected, named and used many of those plants and had also cultivated quite a few of them (Brücher 1989). Their knowledge has partly been transmitted to the present population and local books on popular indigenous medicinal plants are available in most of the South and Central American counties, but are apparently little known or seldom used by the rest of the world (see the series SECAB and the bibliography in this book).

This book has mainly been written to show what an enormous wealth of useful and medicinal plants is hidden in the Neotropics which, however, will soon be lost by constant destruction of the tropical forests by man. Furthermore, the knowledge of useful and medicinal plants is lost as soon as tradition vanishes through the import of new products from industrialized countries. Already in 1926, Pittier regreted that the use as well as the production of indigenous medicinal-herbal products, foodstuffs and other local industrial vegetable products, such as special oils, waxes, fibers, tannic substances, cork etc., diminished alarmingly, as importation of artificial or foreign products from the USA and Europe increased. This decrease in natural indigenous production continues up to the present day such that also the knowledge of these useful plants is being gradually forgotten. As local industries disappeared, unemployment consequently increased.

However, there is an enormous variety of actually and possibly useful officinal plants in South America and particularly in Venezuela with its very variable geographic and climatic conditions and biotops that have never been studied exhaustively. It would be easy for the authors to enumerate a thousand of them. But how can we get to know these plants and their uses? First of all it is necessary to study the native bibliography. Most of the South American countries have their own special books dealing with medicinal or useful plants. The collection of SECAB is one of the most valuable compilations of this kind and includes the flora of Bolivia, Colombia, Chile, Ecuador, Panama, Peru, and Venezuela. Pittier (1926) contributed a great deal to the knowledge of the useful Venezuelan plants. Gupta (1995) studied 270 Ibero-American medicinal plants. Schultes & Raffauf (1990) dealt with the 'healing forest' of the northwest Amazon region. Brücher (1989) covered the genetics of neotropical useful plants. More recently, the interest in useful tropical plants has grown steadily such that several books dealing with taxonomic description, chemical contents, pharmacology, medicinal use (including recipes), ecological requirements and cultivation of medicinal plants have already appeared, written by native botanists, pharmacists, experts on human nutrition, chemists etc. Experiments were carried out and quite a few neotropical herbs and trees are already recognized as officinal medicinal plants. A good survey of useful and medicinal plants up to 1968 is given by UPHOF.

However, there are numerous possibilities of utilization of tropical plants. There are not only remedies against the common sicknesses, but there is also hope to cure cancer and AIDS with the help of plants. Time is thus growing short to get to know the plants, to recognize them in the field and to maintain and cultivate them. To identify a certain species in the great wealth of tropical plants, one must know their structure. Although taxonomy, morphology and anatomy have long since been devaluated as old-fashioned disciplines, they now acquire a new meaning, when the tropical flora is concerned. As a consequence of the depreciation of descriptive botany and because of the overestimation of the Northern flora, there is great need of taxonomists, morphologists and anatomists and their experience in the tropical flora. Many of the plant species collected cannot be identified at present or have to wait for identification for many years, as there are not enough specialists to identify them. There are also certain plant families such as the Myrtaceae or the Sapotaceae, for example, which are difficult to identify because the species are similar. Other species flower very seldomly and can only be identified by outstanding features of the vegetative organs. This was the main reason why we started to study the microscopic structure of useful and medicinal tropical plants.

The European medicinal plants are usually called officinal herbs. Historically, medicine was born in Europe with the use of biologically active herbs. We must emphasize, however, that a larger part of the useful and medicinal tropical plants consists of shrubs, trees or lianas. This is another difference between the northern and the tropical flora. Herbs often represent adaptations to the adverse northern climate, while trees can better develop in the more favourable tropical environment.

Only angiospermous plants are considered in this volume. Gymnosperms are however almost absent from tropical America – besides a few species of *Pinus, Podocarpus, Ephedra*, and *Gnetum*. But the healing properties of the enormous diversity of cryptogams: ferns, mosses, fungi, lichens and algae, still remain to be investigated.

The next step in finding information on promising plants would be to visit the so-called 'herbolarios', open stands where healing plants are offered. They are scattered all over the country along highways or on local markets and are provided by native experts of healing plants. These people still maintain the knowledge of useful plants that has been transmitted to them by their ancestors. They know exactly the right plant species, the organs which have to be used, the methods to prepare the drug, the way of administrating the medicine; they also are informed whether the plant has been collected wild or whether it was cultivated and they know the vernacular and local name of the plant. As far as the local name ist concerned it is very important to know the site where the plant comes from because vernacular names may change from site to site and the same names may signify different species in different places.

The final – and perhaps most authenic – information is found among the living Indian tribes who not only know healing plants, but also nacrotics (SCHULTES & RAFFAUF 1990). However, there are only few Indian tribes still living in Venezuela.

The reader may, however, take into account that the knowledge accumulated by the Indians is empiric, it does not proceed from exact scientific investigation. Some information about sicknesses or recipes may therefore appear unusual, e.g. when a medicine is used for good luck or when recipes are inaccurate, e.g. 'a handful leaves'.

In most cases, healing substances are only found in one or the other organ, or are enriched in determined organs. When buying drugs at the 'pharmacological' stalls, only plant parts or organs are handed to the buyer, who then has to find out which plant species is the 'donor' and which taxonomic name lies behind it as plants are usually sold by their vernacular names. In many cases it is consequently impossible to discover the botanic identity of the drug, particularly when only roots are presented. A good anatomical description may be very helpful in this situation! It should therefore be attempted to compose an entire pharmacognostic work of South American (Venezuelan) officinal plants based on the thorough study of plant anatomy. Up to the present, the inner structure of the organs, shape and size of starch grains, type of laticifers, formation of secretory canals or glands, shape of fibers and stone cells, form of crystals and of other inclusions etc. have been studied only of very well known plants. But with these anatomical data it would be possible to identify plant species. We therefore noted – at the end of each plant description – a few characteristics which facilitate the determination of species. Only a few species are easily identified when only a few anatomical properties are known: The venation pattern of the leaves of *Touroulia guianensis* which is sufficient to identify the species (ROTH 1996), is a unique case.

Each species is studied in the following way: First a short characterization of the family is given. The vernacular name or names of each species are mentioned. The taxonomic description is followed by indicating the probable species origin, the historical background and the actual occurrence of the species in the country. In the anatomical description we either mention all parts of the plant or only the useful organs. As far as possible, economical utilization of the species, the names of the drugs, the method of use, the healing properties, the chemical contents (as far as studied), toxicity, varieties and related species, possible cultivation, special characteristics which facilitate identification are referred to. It is obvious that the above mentioned information was not available for ALL species studied and that certain information is missing here and there.

We gave weight to the study of only little known plants of medicinal or economical use, but we also included some better known species. We laid emphasis on the South American or Venezuelan origin of plants. But also we thought that some plants not indigenous to South America, though cultivated in this region, should be added.

Most specimens were recently collected and studied, some were however adopted from our earlier publications.

The recently studied plants that were collected at different sites, are all preserved in the form of a professional herbarium which will be incorporated in the National Herbarium of the University of Caracas, Instituto Botánico, Venezuela.

The recipes for the preparation of drugs were adopted from other books, but no guarantee is given that the mentioned drug has the desired effects nor that all undesired secondary effects are included. Our contribution must be understood as an attempt to show the many possibilities of applying neotropical plants medicinally or economically. It is suggested that their chemical structure should first be explored thoroughly and that healing effects, possible toxicity and side effects should thereafter be studied experimentally. Certainly, Indians carried out experiments on themselves for possibly hundreds of years, but they may have already been accustomed to certain poisonous substances or may have been or still are more resistant to drugs than the so-called civilized people.

Nevertheless, the experimental and standardizing in-vitro and in-vivo methods to prove the effectiveness of phytopharmaceuticals and of medicines in general depend both on objective and subjective factors, such as the state of health, psychical disposition, spiritual state and social competence of the test persons or patients, and consequently their value is limited. Most effectiveness must probably be ascribed to 'multi-purpose' phytochemicals which contain a mixture of numerous active substances. When the active substance or substances are isolated, a modified and perhaps improved medicine can be developed. But monotherapeutica can possibly lead to 'over-therapy'; furthermore, other active substances are eliminated in this way.

Legislation of medicines is very different in the various countries. In Europe, quality, safety, efficacy and standardization are most important. In the USA, on the contrary, many phytopharmaca are considered 'nutriceuticals', as for example, chips impregnated with *Ginkgo* extract.

As to the chemical contents of already (partly or completely) studied medicinal plants, we could not mention all the secondary compounds found in each species, so we only mentioned the most common ones. Readers who are interested to know all the results in detail are referred to the cited bibliography and/or may consult the series of SECAB and GUPTA (1995).

In this volume we are dealing explicitly with useful and medicinal plans at the same time because the majority of the plants studied have only rarely a single use, but serve several purposes at the same time. They may have edible organs (fruits, seeds, leaves) or may contain oil, fibers, resin etc, on the one hand, but on the other may still have medicinal properties. Even the medicinal applications may be diverse and manifold. Healing properties of a sinlge plant species may encompass sicknesses as different as diarrhoea, cough, dysentery, rheumatism, allergies, and menstrual problems. These applications which are transmitted verbally by the indigenous people or by special books which originally are also based on tradition have to be proved first chemically and experimentally. In quite a few cases these requirements have already been met. In most examples, however, these investigations are still lacking and will require many years of work. For this reason and possibly because certain healing properties ascribed to determined plant species are unrealistic, we considerd it inadequate to present a list of sicknesses in alphabetical order coordinated with plant species that could possibly cure these diseases. Furthermore, we do not recommend the application of the recipes mentioned in our text, because a plant is not a verified pharmaceutical product: chemical contents in plants not only vary in the distinct organs (leaf, stem, root etc), but also according to the sub-species and to the environmental conditions of the individual plants. Moreover, native people may be accustomed to certain chemical compounds in plants which they use repeatedly, while the same compounds may be harmful to other persons. Chemical compounds have to become standardized before the use of the corresponding healing plants may be considered safe.

At present, Venezuela can be considered a leading country in the investigation of medicinal and useful plants in South America. The ethnobotanical congress in Caracas (1999) evidenced a number of new activities in this respect. PILAR RODRIGUEZ, a former student of the senior author, gave a great impulse to the investigation of medicinal plants in Venezuela (1983). The senior author spent 21 years at the Universidad Central de Venezuela in Caracas in research and teaching. The junior author, Helga Lindorf, a native Venezuelan, is a former student and collaborator of the senior author, Ingrid Roth.

In writing this book, the collaboration with the Central University of Venezuela in Caracas was most important. Prof. Helga Lindorf, has accomplished the most important part. She collected and identified the plants with the help of her assistants. Herbarium specimens were prepared and can be consulted in the National herbarium of the Botanical Institute of the Central University of Caracas. A large part of the bibliography used was provided by Prof. Helga Lindorf who directed the scientific activities in Venezuela.

The microscopic slides were prepared by the assistants of Prof. Lindorf at the Laboratory of Tropical Botany, Central University of Venezuela, Caracas. Our particular thanks go to Carlos Blanco, Aura Castillo, Mercedes Castro, Alexandría Jiménez, Silvia Llamozas, and Silvia Pérez.

Habitus photographs were in part made at the collection site by Bruno Manara who also collected part of the plants and to whom we express our sincere thanks.

Most of the scanning microscope photographs were made by Prof. Helga Lindorf at the University of Caracas.

Most of the microscopic photographs are taken with a Leitz light microscope ('Das Mikroskop'), a loan of the DFG in Bonn, an organization we owe our gratitude.

Drawings are originals made by the senior author Ingrid Roth, if not otherwise indicated.

We are grateful to Dipl. Interpreter Assia Boutskoy for the correction of the English text.

Our sincere thanks are due to the European Commission for Science, Research and Development (Cooperation with Emerging Countries and International Organizations) in Brussels, for the generous financial support of our research.

Contents

PART I

Introduction to South American Medicinal Plants

Introduction

As this book will be of use for experts of very different branches of science, such as plant anatomists, taxonomists, ecologists, forestry engineers, pharmacists, medical doctors, ethnologists, teachers and the like, it seemed advantageous to the authors to give some introductory explanations as to the effects of phytopharmaceuticals as well as of the basic knowledge concerning therapy and toxicity. We also mention the possibilities of finding new promising and already known plant species and the difficulties of their identification, as the tropical forest is not a Botanical Garden in which plants are labelled. Even experts can easily be perplexed, when flowers are not available which is very often the case. Nor may all specialists be familiar with the anatomy of tropical plants and particularly of tropical barks, product of waste in the utilization of valuable tropical timber which can and should be applied in manifold ways. Besides a number of doctors are not yet aware of the importance of a vegetable diet in medicine. As the large pharmaceutical companies of the USA are now earnestly interested in the search for new vegetable drugs to combat AIDS, cancer, heart and infectious diseases, we also include instructions, on how to find new plant species which may supply important drugs.

Phytochemicals and their curative properties

In the past, the application of herbal drugs was often connected with religious ritual ceremonies, with the use of narcotics and hallucinogenics. It is possible that some medicinal effect was due to the faith in the healing powers of a charismatic magian.

This is perhaps the reason, why even today people still believe that witchcraft is also involved in the healing effects of plants. But in fact, today we are in the fortunate situation of knowing the healing substances by their chemical formula and can also explain most of their actions on the human body. They are commonly called secondary vegetable compounds in contrast to the primary compounds which are generally used as foodstuff by man (e. g. proteins, fat, carbohydrates and roughage). The secondary substances consist of a large variety of chemically different compounds which only occur in small quantities in the plants and commonly have pharmacological effects on man. However, criteria for the distinction of primary and secondary vegetable compounds are still dubious. The expression secondary compounds was therefore recently substituted by the term phytochemicals.

Some of these secondary compounds or phytochemicals which play their part in medicine or in foodstuff are mentioned below.

Etheric oils

Etheric oils have a characteristic aromatic smell or taste and occur externally on the plant in glandular hairs or scales, and internally in oil cells, cavities or canals. They are frequently found in such families as the Lauraceae, Myrtaceae, Rutaceae, Zingiberaceae, Apiaceae etc. Etheric oils correspond to a mixture of several substances, comprising terpenes, (mono-, sesqui-, diterpenes), phenols, acids, sulphur containing and N-containing substances.

According to their smell and taste, they may be classified as aromatica, aromatica-amara (with a bitter taste) and aromatica-acria (with a pungent taste). They may be irritant, rubefacient and vesicant, and promote a better blood supply to the skin. Some volatile oils are antiphlogistic and cure inflammations. Volatile oils are also disinfectant, antiseptic and antibiotic. The secretolytic effect of expectorants such as of *Foeniculi* fruct. or *Eucalypti* fol. is well known. Volatile oils used as diuretics promote the blood supply to the kidneys. Etheric oils as spasmolytics mainly exercise an effect on the smooth muscles of the intestine, gall bladder, bronchial tubes and blood vessels. Etheric oils may further be used as carminatives, stomachics, choleretics and may regulate secretion and motility of the stomach. Etheric oils used as sedatives are mostly applied internally. Analeptic etheric oils exercise a stimulating effect on heart and blood circulation. Emmenagogues obtained from volatile oils have a positive influence on amenorrhoea and dismenorrhoea, when given in small doses, but may have an abortive effect in high doses. Other volatile oils are anthelmintics. Some are used as insecticides and as repellents. Undesired side effects may arise when volatile oils are used in excessive quantities or over a prolonged period of time. For example, cancerogenic, abortive, allergic, narcotic, hepatotoxic effects have been reported.

Certain drugs made of etheric oils are used as pungent tasting condiments so that the pleasant flavour is increased and appetite and digestion are stimulated. Chillies (*Capsicum frutescens*), nutmeg (*Myristica fragrans*), cloves (*Syzygium aromaticum*) and *Vanilla* may be mentioned in this connection. Flowers, leaves and fruits of *Citrus* species are also used pharmaceutically.

Resins (and balsams)

Resins (and balsams) are mainly found in large quantities in the bark and wood of tropical trees. Solutions of resins in etheric oils are called balsams. They are distinguished according to their chemical composition, into terpene resins, benzene resins and gum resins. Elemi is a terpene resin, Peru balsam and guaiac resin are benzene resins.

Terpenes are synthesized only by plants and some microorganisms. For man, they have a particular value as aroma substances, e.g. menthol or limonene. Ninety percent of the *Citrus* oil consists of limonene. Both menthol and limonene have anticarcinogenic effects. Limonene increases the activity of detoxicating enzymes in liver and intestine. In the future, this substance may possibly play an important part in the prophylactic treatment of cancer.

Drugs with bitter principles

Drugs with bitter principles or 'amara' are applied to stimulate saliva, stomach and gall secretion. Terpenoid and non-terpenoid bitter principles are distinguished. The bitter taste is produced by stimulation (or irritation) of the bitter-receptors in the taste buds on the bottom of the tongue. Monoterpenes, sesquiterpenes, diterpenes, tri and secotri terpenes are known. Within the non-terpenoids, the flavanone glycosides, the phloroglucinol derivatives, and the alkaloid quinine may be mentioned.

Vegetable sweeteners

Among the vegetable sweeteners, calorie-rich and calorie-free sweeteners (non-caloric) or compounds poor in calories may be distinguished.

Caloric sweeteners are saccharose, glucose and the so-called sugar substitutes, such as fructose, sorbite, xylite, mannitol and sorbose; they have about the same nutritive value as saccharose, but are less readily or not at all resorbed by the intestine. As fructose, sorbite, xylite are degraded wether insulin is present or not and do not change the blood sugar level, they are very suitable for diabetics. The calorie-free sweeteners are extracted from higher plants and contain a much higher sweetening power than saccharose.

Xylite, sorbite and sorbose are not cariogeneous since they are not degraded by oral bacteria, leading to caries. Fructose exercises a decreasing effect on blood alcohol. With the exception of the glycosides, sugar substitutes do not have any nutritional value and are therefore appropriate for weight reduction diets.

Due to the suspected carcinogenic action of the synthetic sweeteners, the herbal sweeteners increasingly gain importance. Herbal sweeteners belong to very different chemical categories, such as peptides, glycoproteins, aldehydes, sesquiterpenes, or such compounds as neohesperidine, dihydrochalcone and aspartame. They have sweetening values 50–3000 times higher than saccharose.

The structural prerequisite of the sensation 'sweet' is, according to SHALLENBERGER & ACREE (1967), due to the fact that the molecule possesses a proton donor group AH and a proton acceptor group B at a distance of 2.5–4.0 Å. The receptor of the sensation 'sweet' therefore should have a complementary arrangement of proton acceptor and proton donor at the tip of the tongue, so that hydrogen bonds are developed between the 'sweet' molecule and the receptor (for further details see WAGNER 1993).

Glycyrrhicic acid is a sweet principle of Liquiritiae radix, which has a sweetening capacity about 50 times stronger than that of saccharose. In pharmacy, the ammonium salt is used as a taste corrigent, for aromatization of chewing tobacco and for the production of the true Porter beer. It tastes of licorice and has an effect similar to corticoid so that it is not applied as a pure sweetening agent. Secondary effects may be noticed in long therm use.

Stevioside is a diterpeneacid-glucosyl-sophoroside which has a sweetening effect 150–300 times greater than that of saccharose. It is permitted as a sweetening agent only in Japan and in South America.

Osladin is a steroid saponin extracted from the rhizome of *Polypodium vulgare*. With a sweetening capacity 3000 times higher than saccharose it is one of the sweetest natural agents. Thaumatin I and II are peptides with a sweetening capacity exceeding 1600 times that of saccharose. The sweetening effect is however lost by the action of proteolytic enzymes or through heat and treatment with acids.

Monelline is a glycoprotein; the sweetening effect is bound to the carbohydrate and amounts to 3000 times that of saccharose.

Miraculine of *Synsepalum dulcificum* is a glycoproteid which has no proper sweetening power, but makes sour or unsweetened food taste sweet after addition of sour principles. 'Miracle-fruit-drops' for diabetics are on the market in the USA.

Perilla-aldehyde – aldoxime is found in the etheric oil of *Perilla* and has a sweetening power about 2000 times that of saccharose. It is used in Japan as a substitute of honey and licorice.

The flavanon-7-0-glycosides naringin and neohesperidin of *Citrus* fruits may be converted into sweet-tasting dihydrochalcone-O-glycosides. Neohesperidin-dihydrochalcone is 1000–3000 times as sweet as saccharose, it is heat and acid resistant.

Hernandulcin of *Lippia dulcis* is a sesquiterpene, 100 times sweeter than cane sugar.

Dipeptidester of asparaginic acid, such as aspartame, exceeds the sweetening power of saccharose by 180–200 times.

Considenng all these vegetable sweeteners and all these sugar substitutes which have not the pernicious effects of refined sugar, as they do neither provoke caries on teeth nor do they promote dangerous diabetes and, additionally, have little or no caloric value, it is not understandable that an alternative to refined sugar (Saccharose) to date, could not be found. But the solution of the mystery is the agricutural market in which sugar is extremely protected. Although production surpasses the domestic need (in Germany by about 40 %), a subsidy of 300 million Euro is yearly paid for storage. The high import duties and the rigid system of contingents in sugar production burden the consumers of the European Union with 6 million Euro yearly.

A reform of the EU sugar market regulations would thus be very much appreciated. The sugar lobby should be better controlled in the future and new products with innocuous sweeteners or sugar substitutes could then gradually replace refined sugar.

Glycosides

Glycosides play an important part in heart therapy. Up to date, about 15 plant families which supply 'heart glycosides' are know, such as Apocynaceae, Asclepiadaceae, Euphorbiaceae, Sterculiaceae, Tiliaceae, Meliaceae and others. In moderate doses, these glycosides cause an improvement of the economy of the heart function. The mechanism rests on a rhythmic intracellular release of cations. Heart glycosides of plants growing in tropical South America are less known than those of European plants.

Saponins

Saponins are well known for their surface activity (foam production) and their bitter taste. Saponin drugs mainly contain triterpene and steroid glycosides as principal active agents. Saponins correspond to glycosides which are split into sapogenine and sugar by acid or enzymatic hydrolysis.

The larger part of saponins occurring in nature are triterpene saponins; more than 120 different triterpenes are known up to date.

Saponins have a very wide-spread occurrence in higher plants. Their fungicidal properties are probably the reason of this phenomenon. Triterpene saponins are principally found in dicotyledons. Species particularly rich in saponins occur, for example, in Sapindaceae, Sapotaceae, Araliaceae, Fabaceae etc. Steroid saponins, on the other hand, are mainly found in monocotyledons, e.g. in Liliaceae, Dioscoreaceae, Agavaceae (*Agave* and *Yucca*).

Saponins have the property of reducing the interfacial surface tension of heterogeneous systems; the presence of a hydrophilic and a hydrophobic molecular part is responsible for this phenomenon. In this way, formation of foam (with liquid-gaseous phases), an emulgator effect (with liquid-liquid phases) and a dispersing effect (with liquid-solid phases) are achieved.

Saponins are not used as detergents any more, though they are used in food manufacture as additives for foaming drinks (like ale) and as taste additives. In pharmacy, they are used as emulgators and as dispersants, in cosmetics as dispersants for tooth paste, shampoos, mouth wash and gargles.

The haemolytic effect of many saponins is due to their abilily to form complex bonds with sterols and proteins. The haemolytic action helps to evaluate saponin drugs. The complex formation with cholesterol is used for the quantitative determination and isolation of saponins.

The pharmacological effects of saponins are only partly explicable by their physical-chemical properties. The expectorant effect is due to the increasing surface activity, the liquefaction of slime and the local irritation of the mucosa. The diuretic effect of saponins is achieved through osmosis or by direct irritation of the kidney epithelia. Antibiotic, antimycotic, molluscicidal and antiviral activities are inherent in quite a few saponins. The antimycotic activity is caused by a complex formation with sterols. Leprosy and tuberculosis can be cured by certain saponins.

The diuretic effect is often activated by flavones and etheric oils. This is one of the many examples that the complete herbal medicine often exercises a better effect on the sick person than the synthetic pill.

The resistance of tropical timber to termites is due to the saponin content. Saponins with molluscicidal properties could be applied to destroy the snail which transmits Schistosomiasis (*Bilharzia*). The better resorption of certain vegetable extracts – in comparison with the pure chemical substances – is attributed to the presence of saponins. Further – more saponins are beneficial because they have anticarcinogenic, anti-inflammatory, antimicrobial, immunomodulating, and cholesterol reducing properties. Most saponins are tissue irritants and toxic so that intramuscular or subcutaneous injections could prove dangerous.

Alkaloids

Alkaloids are nitrogenous plant bases. Acid amides such as capsaicin belong to the protoalkaloids. True alkaloids always bind nitrogen cyclically. Primary, secondary, tertiary and quarternary ammonium bases are distinguished; most alkaloids are however derived from tertiary amines.

In the plant, alkaloids are found as water soluble salts of organic plant acids, e.g. of acetic acid, oxalic acid, lactic acid, tartaric acid, malic acid etc. Some alkaloids are bound to tanning substances. Numerous alkaloids, such as strychnine or quinine, have a bitter taste.

Typical alkaloid families are: the Annonaceae, Lauraceae, Menispermaceae, Punicaceae, Caricaceae, Cactaceae, Rhamnaceae, Rutaceae, Apocynaceae, Rubiaceae, Solanaceae and many other dicotyledonous families. Some alkaloids are also found in the animal kingdom. Certain alkaloids, such as nicotine, ephedrine or caffeine occur frequently, others are confined to certain families or some representatives of these families, such as morphine, emetine and strychnine. Roots and shoots are preferred sites of alkaloid biosynthesis; however alkaloids are also transported from one organ or plant part to the other. Possibly, the function of alkaloids in the plant is to transfer and store nitrogen. Furthermore, they may have a protective function against bacteria, fungi and viruses and against predation. Some important alkaloid drugs are: *Cacao* semen, Curare, *Colae* semen, *Granati* cortex, Guarana, *Ipecacuanhae* radix, *Yohimbe* cortex, *Nicotianae* folium, *Cinchonae* radix, Opium etc.

Opium has a constipating effect, noscapine exerts an antitussive action; *Ipecacuanha* alkaloids have emetic properties.

Cinchona is found in the Andes at altitudes between 1000 and 3500 meters.

Quinine has antipyretic and analgesic effects. As a protoplasm poison it combats malaria. Yohimbine acts as a sympatholytic and reduces blood pressure. Curare (a complex concept) is an arrow poison, extracted from the bark with boiling water. Purin (caffeine), present in *Theobroma cacao, Cola nitida* and *C. acuminata*, increases the contractile power of the heart. Guarana is the drug richest in caffeine, containing 2.5–8 % caffeine and up to 25 % catechin tannin. The drug combats headache and is also used as a stimulant. The alkaloids of *Granati* cortex paralyze the muscles of *taenia* (*tapeworm*). Some alkaloids have bactericidal, fungicidal, and insecticidal properties.

Euphorics and hallucinogenes are not specially treated in our book, because there already exists enough bibliography on this subject (see SCHULTES & HOFFMANN 1979). Well-known drugs are opium, morphine and derivatives, marihuana, mescaline etc.

Laxatives

Laxatives or purgative substances are mainly peristaltics which have a stimulating influence on the smooth muscles of the intestine, such as anthrachinones, oil of *Ricinus*, oil of *Croton* and the so-called resinous drugs. The effect probably derives from a release of or an increased production of histamine or other mediators. Hydragoga or filling peristaltica, on the other hand, act by water resorption through swelling, by osmotic retention of water, by secondary inhibition of water resorption or by increased water secretion into the intestine and the subsequent pressure on the intestinal wall. Laxative substances are often found in tropical plants.

Flavonoids

Flavonoid drugs usually have a content of 0.5–3 % flavonoids. Up to date, more than 300 different flavonoids have been isolated from plants. Flavone glycosides, flavonoid aglyca and anthocyanines are found dissolved in the cell sap. They mainly occur in aerial plant organs. Of all secondary compounds, the flavonoids show the most extensive distribution in the plant kingdom. They are found everywhere in Angiosperms and Gymnosperms. The frequent occurrence of glycosides of the flavanone series is striking in *Citrus* and *Pinaceae*.

Leguminosae and Asteraceae are distinguished by their richness in flavonoids. The biological function of flavonoids in the plant is to attract insects for pollination, to protect it from viral and fungal infections, and from predation by insects. Further tasks are to control growth in connection with growth hormones, to inhibit enzyme effects and to participate in cell oxidoreduction systems.

Technically, flavonoids are applied to dye wool, and to improve tea mixtures. Flavonoids are chelating agents and are used for the identification of metallic ions. Dehydrochalconeglycosides which are obtained from *Citrus* flavanone glycosides have a sweetening power 1000–3000 times superior to that of saccharose. Polyhydroxyflavones, such as quercetin, have antioxidative properties.

Pharmacologically, flavonoids exert antihaemorrhagic, antisclerotic, antiphlogistic, antioedematic, spasmolytic, choleretic, diaphoretic, diuretic and antiallergic activities on the human body.

Coumarins

The name coumarin is derived from the Indian name 'coumarouna', given to *Dipteryx odorata*, a tree endemic of Venezuelan Guiana. The first non-substituted benzo-α-pyron was extracted in crystalline form from the seeds of this tree. Today we know about 1000 coumarin compounds. Some coumarins inhibit seed germination and promote growth of roots; they may thus be of some importance for the plant.

Therapeutically, some coumarins are antiphlogistic, spasmolytic, bactericidal, promote circulation or suppress oedema, but in higher doses they have narcotic effects. The coumarins umbelliferone, esculetin and herniarin are suitable additives to skin creams and lotions as sunblockers. Since they absorb UVB rays which produce erythemas.

Aflatoxins (cyclopentanocoumarins) are metabolic byproducts of the fungus *Aspergillus flavus* which develops on such foodstuffs as cereals, nuts (peanuts), legumes, coconuts, almonds fishmeal etc. They cause damages to the liver. Additionnally aflatoxin B1 has carcinogenic properties.

Novobiocin has been isolated from *Streptomyces sphaeroides* and *S. niveus*. It is mostly effective against gram-positive bacteria and is indicated in infections with *Staphylococcus*, when penicillin and other antibiotics do not help.

Tannins

Tanning substances are well-known for their astringent and antidiarrhoeic properties. By oxidation, enzymatic polymerisation or acid condensation, the light brown or colourless 'tannins' are converted to dark brown, black or red, water insoluble and physiologically ineffective products (phlobaphenes, tannin reds). To distinguish hydrolytic from condensed tannins, tests with formaldehyde and HCl (red precipitation) or iron(III)chloride (blue or green colour) can be used in microtechniques.

The tanning properties are due to the mutual reactions with the collagen fibers (protein) of the skin. The tanning process of the fiber is completed, when the collagen has absorbed about half of the weight of the fiber contained in the tanning substance.

By forming a coagulating membrane on the surface of the mucosa or the connective tissue, tannins soothe irritations, are antiinflammatory, act as local anaesthetics, inhibit secretions of the skin and wounds and are bactericidal.

The antiphlogistic effect might be due to repression of certain inflammatory mediators (5-lipoxygenase).

The hypoglycaemic effect possibly depends on the repression of liver enzyms (glucosidases).

Tannins are applied externally for wounds, burns, frostbites, bleedings, haemorrhoids and inflammations of the mouth and the thorax (angina, gingivitis, stomatitis, periodontitis.

Internally tannins cure gastritis, diarrhoea, irritations of the stomach and intestine. Tannins are also used as antidotes for alkaloid and heavy metal poisoning (black coffee or tea).

A well-known drug is Ratanhiae radix of *Krameria triandra*. A non officinal drug is obtained from *Uncaria gambir* (Gambir catechu). Both species are studied in this book.

Carbohydrates

Many drugs are obtained from carbohydrates. Fructose is present in many fruits and in honey. It is used in the therapy of liver intoxications, in cases of acute alcohol poisoning and as an infusion in hypoglycaemia.

Pure honey consists of 1–10 % saccharose, some protein, organic acids, acetylcholine, etheric oil, mineral salts, colourants and 20 % water. It has a bactericidal and secretolytic effect (osmosis) and is therefore applied internally for colds and externally for wound treatment. Natural honey also contains pollen grains. In tropical natural honeys, hundreds of distinct pollen grains coming from different plant species may be found (observation of the senior author).

Figs (Caricae) are mild, osmotically acting laxatives and are used for osmodiuresis and as a filling material for chewing tablets. Figs contain 50 % invert sugar, slime, pectin and organic acid.

Sorbitol occurs in the fruits of Rosaceae (*Sorbus aucuparia*). It serves as a mild laxtive and as diabetic sugar.

Xylite is also a well-known sugar substitute for diabetics (see above).

Amylum (starch) occurs in large quantities in the storage organs of plants, such as tubers, bulbs etc. It is a 2-component mixture of amylose and amylopectin. Starch has a strong water-absorbing power and is applied for its absorptive capacity of wound secretions, and as a powder substrate. It also has a cooling effect and is suitable as a lubricant.

Because of its swelling capacity, starch is used as a tablet explosive. It is also applied as a dietetic and a mucilage.

Industrially, starch is suitable for manufacture of ethanol, dextrin, and grape sugar, as a thickening medium, a glue and for dressing and finishing and even as reagent in iodometry.

Pectins

True pectins have the ability to pass from the state of a sol into a gel in a watery solution. Frequently, pectins gelatinize only when sugars, acids and salts are added.

Pectins are appropriate for diarrhoea, gastroenteritis and ulcers. For their haemostyptic action, pectins are applied in the treatment of wounds. They are also used as liquid blood substitute, as emulgators and as thickening agents, for the manufacture of tablets and for the partial synthesis of heparinoids.

Polysaccharide drugs are obtained from sea algae; agar-agar is extracted from Rhodophyceae.

Polysaccharides of fungi which non-specifically stimulate the immune system are used together with chemosynthetics for tumor therapy.

Plant gums, such as gummi arabicum, are heteropolysaccarides, being used internally as a mucilage, as stabilizers and emulgators. Technically, the gums are applied for glues, adhesives, as thickening and binding agents.

Slimy drugs with antiphlogistic properties are applied externally for inflammations, tumors, glandular swellings, and internally as laxatives, antidiarrhoeals, and for the treatment of intestinal or gastric inflammations (e.g. *Psyllii* semen, *Malvae* flos). Mucilages are non-sticky heteropolysaccharides which can be extracted with hot or cold water from the drug.

Lipoids

Fat, oil, wax and other lipoids are storage substances in plants which are often found in large quantities and are therefore suitable for the preparation of drugs. Fats with specific activities are, for example, *Ricini* oleum, laxative, and *Crotonis* oleum, likewise a laxative. However, 20 drops or 5 seeds can be deadly.

Fats and oil occur in all parts of the plants, but may be accumulated in seeds and fruits. The oil content of seeds and fruits may reach as much as 50 % of the total weight. About 3/4 of all higher plants contain fat as their main storage product in seeds. Only in 1/4 of the higher plants do starch and proteins prevail. *Arachidis* oleum, *Cacao* oleum, *Cocos* oleum are South American products.

Vegetable fats and oils are obtained by pressing or extraction. Fats and oils are indispensable as energy providers of the human and animal body. As storage fats they can be stored indefinitely in living beings, but may be mobilized at any time. From the physiolgical point of view, saturated and unsaturated fatty acids have to be distinguished. Some unsaturated fatty acids, such as linolic acid, are indispensable for mammals: Since they can not synthesize them, these fats are essential for them. The highly unsaturated fatty acids have the ability of reducing the cholesterol level in the blood and are consequently used against high blood pressure and to prevent arteriosclerosis.

Less valuable fats and oils are used industrially for the manufacture of soap; drying oils are applied for the fabrication of boiled oil, siccatives, stand oil, thick oil or as basis of linoleum. Oil of ricinus (castor oil), is a favoured additive in cosmetics, due to its high viscosity e.g. 'brillantine' (for hair care).

Lipoids in an ample sense are waxes (cera). They occur on the surface (epidermis) of plants and find application in pharmaceutical technology as indifferent auxiliary substances, as additives for creams and ointments to increase their consistency, while in industry, they are used for polishes and for transparent paper.

Tars

Tars are obtained by dry distillation from certain timbers and coals. They are composed of a mixture of aliphatic and aromatic carbohydrates, phenols, aldehydes and resinic acids. In combination with other substances they are used for oil baths, lotions, tinctures and ointments for the treatment of skin diseases, such as psoriasis, eczema, pruritus, seborrhoea.

Other substances

Other commercial drugs are only briefly dealt with here. Artichockes (*Cynara scolymus*) contain cynarine which has choleretic effects and acts as an antihepatotoxic.

Ginkgo folium is used for deficient cerebral and peripheral arterial blood circulation due to arteriosclerotic, postthrombotic or diabetic vessel degenerations.

Passiflora incarnata, indigenous to Central and South America, contains flavonoids, coumarins and maltol. Its sedative and antidepressive effects are probably due to the presence of maltol and of flavonoids.

Antidotes

Antidotes for snake and spider poisons are frequently found in the native bibliographies; a large number of plant species used as antidotes to snake bites and scorpion stings etc. are mentioned in them. We just mention a few in the following. *Nicotiana tabacum, Heliconia, Urera, Calathea, Abuta grandifolia, Potalia amara, Hibiscus moschatus, Cyclanthus, Besleria drymophila, Bombax, Hedyosmum, Quiina, Brunfelsia* etc. However, the chemistry of all these plants is not known and consequently we do not know the effective chemical compounds of the antidotes. In some plants, alkaloids were found, in others volatile oils. Furthermore we do not know whether the applied medicines are real antidotes or whether they have only antiinflammatory, bactericidal or detumescent properties.

Enzymes

Enzymes are likewise secondary compounds used in human medicine. Papain is obtained from *Carica papaya*, bromelain from *Ananas comosus* and ficin from *Ficus* species. All three enzymes are used for digestive problems of stomach and intestine, for wound healing, for the treatment of oedema and of inflammations and to promote blood coagulation.

Papain is a proteolytic enzyme extracted from the latex of unripe fleshy fruits of *Carica papaya* L. which breaks down proteins into amino acids. By drying the milk sap, the industrial product papayotin is obtained. It is used to help enzymatic digestion, to enzymatically clean wounds and as an additive to cleansing agents. Recently, papain and its derivatives are applied in injections to cure damage of intervertebral discs by chemonucleolysis. In earlier times, papain also served as an anthelminticum. In the food industry it is used as a meat tenderizer.

The proteolytic enzyme bromelain is extracted from the press-sap of *Ananas comosus*. Its activity is similar to that of papain. It is applied for digestive disturbances, prophylactically for postoperative oedema, for the treatment of inflammation and as a reagent for testing erythrocytic antigens and antibodies. Ficin extracted from the fresh milk sap of several *Ficus* species has effects similar to those of papain and bromelain.

Lectins

Lectins occur in roots, leaves, barks and principally in seeds. Most of them are glycoproteins. The best known lectins are ricin of *Ricinus communis*, abrin of *Abrus precatorium* and concanavalin of *Canavalia ensiformis*. They do agglutinate erythrocytes, lymphocytes, fibrocytes, spermia and malignant cells. Lectins are applied for the distinction of different blood groups, for the diagnosis of hereditary immune defects, the detection of mallignant cells (lectin-bound cytostatica) and for the isolation of glycoproteins. The *Viscum* lectin activates nonspecific immune defense mechanisms.

Vitamins

Vitamins are almost exclusively formed by microorganisms and higher plants with the exception of vitamins A and D_2 which are also produced by animals. Vitamins are necessary for growth and sustenance of life functions of animals and humans, though only in small amounts.

Vitamins and enzymes are of great importance in cosmetics, e.g Vitamin A for regeneration processes of the epithehum cells. β-carotene regulates the oxidative stress of the skin. Vitamin C protects the skin from free radicals, stimulates neoformation of cutaneous cells and of collagen, improves the elasticity of the skin. Vitamin E (α-tocopherol-acetate) diminishes the risk of premature aging of the skin by UV radiation, maintains the elasticity of the vessel walls, improves the microperfusion of the skin as weil as the amount of humidity in the cornea, reduces wrinkles and renders rough spots elastic and sleek. Vitamin F controls dry skin and disturbances of keratinization and alleviates pruritus. Vitamin B5 improves the cutaneous metabolism, promotes the moisture retention capacity of the cornea and acts as a sedative and an antiinflammatory. Coenzyme Q10 (ubiquinone) activates stressed, exhausted skin, improves cell function and cell renewal, protects from noxious free radicals, smoothes wrinlkes and exercises a positive influence on dry and sensitive skin.

Carotenoids in general have antioxidative effects, stimulate the immune system, reduce the frequency of light-induced tumors, inhibit mutagenesis, and prevent damage to the cell nucleus.

Antibiotics

Antibiotics are metabolic products of lower or higher organisms which - in small doses - inhibit the multiplication of pathogens or even destroy them. Accordingly, they can be distinguished into bacterio-, viro- and fungicides as well as amoebicides.

Of the 2500 antibiotics known up to date, most come from the actinomycetales (*bacteria*). The second place is occupied by fungi, followed by the eubacteriales. Antibiotic substances of higher plants which can be applied therapeutically, are rare (2 %). The 25–30 antibiotics important in therapy are obtained from *Streptomyces, Bacillus, Penicillium* and *Cephalosporium* species. The reason for their specific usefulness rests on their primitive metabolism, the high rate of mutations and their saprophytic habit.

Antibiotics are mainly applied in the control of bacteria and, only to a lesser extent in that of fungi, trichomonas, amoebas and viruses.

Cytostatics and immunostimulants

Cytostatics (antineoplastica) which impede the growth of cancerous cells, interfere directly, attacking the cellular cycle of proliferating cells. Examples are: oncolytic antibiotics extracted from *Streptomyces*, certain alkaloids (e.g. *colchicum* alkaloids), lignans (e.g. podophyllin of *Podophyllum*), extracts of *Viscum* (with lectins, viscotoxins), and immune therapeutics non-specifically stimulating the defense mechanisms of the human body.

Vegetable virostatic or virucidal substances are some alkaloids, proteins, etheric oils, terpenes, polyphenols, lignans and compounds of phenylpropan.

Substances stimulantig the immune system are, for example, antigens which cause the formation of antibodies; they may be bacteria or viruses, toxins of bacteria, alien proteins or blood cells, polysaccharides of fungi and higher plants, mucopolysaccharides or lipopolysaccharides. However most of the antigens correspond to proteins or polysaccharides.

Immuneprophylactics and therapeutics such as cytokinins, which are mediators of cellular interactions, exercising regulating functions on the entire immune system, could become key substances in the future for a new type of medical therapy. They comprise interferons, interleukins, tumor necrotic factors and haematopoietic growth factors.

Echinaceae angustifoliae radix is a drug used to increase the immune defense against infections, chronic inflammations and as an aid to tumor therapy. Apparently, granulocytes and macrophages are stimulated by this drug. Polysaccharides, acid esters, acid amides and/or poly-yns seem to be responsible for this reaction.

Other immune stimulants are *Echinaceae purpureae* radix, *Baptisiae tinctoriae* radix, certain substances extracted form *Eupatorium perforatum, Thuja occidentalis, Arnica montana, Viscum album, Calendula officinalis* etc.

Immunosuppressives

Immunosuppressives are indicated in the case of organ transplantation and in autoimmune diseases (e.g. cyclosporin extracted from a fungus). Vegetable antirheumatics and antiarthritics with corticomimetic properties have immune suppressant effects (extracts of *Glycyrrhiza, Harpagophytum, Lycium, Solanum* species).

These effects are probably due to steroid saponins, steroid alkaloids or iridoids.

Antiallergenics

Antiallergenics are specific substances for the hypo- or desensitization of the human body. According to the route of contact, allergens may be distinguished into the following categories: inhalation allergens, food allergens, injected allergens and contact allergens. Many plants can provoke allergies. The strongest contact allergens are phenolic substances of *Toxicodendron radicans*. Allergies can furthermore be generated by gallic acid, salicylic acid, cinnamic acid and anthranilic acid, by sesquiterpene lactones; amoung the volatile oils, cinnamon oil, oil of cloves as weil as Peru balsam are potential allergens or allergen producers.

For a hyposensibilisation of the patient, therapy may be carried out by injecting increasing doses of the allergy generating plant material in diluted extracts at certain intervals of time.

Inhalation allergens are, for example, found in pollen grains. For the treatment of pollen allergies, pollen extracts are standardized in the form of pollen units (noon units) which correspond to the activity of 0.001 mg pollen. Likewise, mixed preparations for polyvalent treatment, are available.

However, treatments with phytopreparations, such as of *Luffa operculata* (hay fever) or *Galphimia glauca* and *Allium cepa* (pollinosis and asthma) also exist. For the treatment of allergic asthma atropin, theophyllin and ephedrin are available.

For more detailed information on pharmaceutical biology we recommend WAGNER (1993) and ROEMPP (1997).

Bioactive substances - mainly secondary plant compounds - in food

Numerous substances found in food have carcinogenic effects when used in high doses, at least in animal tests. Most toxic is aflatoxin B, produced by certain fungi which contaminate foodstuff. However, some fruits and vegetables have anticarcinogenic effects, probably due to the antioxidative action of vitamins C and E and the provitamin A (β-carotene). Other substances with protective anticarcinogenic effects are roughages (fibers), phenolic acids, saponins, phytosterols, carotenoids, flavonoids, indoles, thiocyanates, sulphides, monoterpenes, phytoestrogens, antioxidants, protease inhibitors, minerals and substances of fermented footstuff, e.g. yoghurt, curdled milk, pickled cabbage, fermented red beets (WATZL & LEITZMANN 1995).

Antimicrobial substances

For thousands of years, certain foods and condiments were used by the natives to combat infections with bacteria, fungi and viruses. Sulphides, thiocyanates, phenolic acid, flavonoids and saponins (tomato) are active substances with antimicrobial effects which can be applied prophylactically to prevent infectious colds.

Antioxidants

To keep within limits the noxious effects of free radicals and oxidants, the human body develops certain protective measures. The following antioxidants have been found in human blood: vitamin C, glutathione, bilirubin, uric acid, vitamin E, and carotene, lycopene, zeaxanthin. Bilirubin impedes lipid peroxidation and as an antioxidant is almost as effective as vitamin E. Antioxidative food substances have protective effects against numerous diseases, such as cancer, heart and circulatory diseases. Vitamin C specifically protects the gastro-intestinal tract from cancer. Probably, selenium also has a preventive effect from cancer.

Other effects caused by plant compounds

- *Antithrombotic activities* are ascribed to other secondary plant compounds, for example, sulphides, flavonoids, adenosine, gingerol.
- *Blood pressure regulating effects* are ascribed to certain alkaloids of subtropical and tropical plants, e.g. caffeine in coffee, tea, cola, cocoa, adenosine and allicin (in the onion),
- *Cholesterol reducing effects* are also found in secondary plant compounds. Food prepared from animals generally increases the cholesterol content in the blood, while vegetable food is free from cholesterol. Vegetable roughage and secondary plant compounds, such as saponins, phytosterols, tocotrienoles, indole, anthocyanins and allicin decrease the cholesterol content of the human body.
- *Glucose regulating effects* of phytochemicals are not uncommon. Diabetes mellitus, obesity and disturbances of the lipid metabolism are related to alimentation rich in energy e.g. a high consumption of sweets and sweet drinks and a low consumption of roughage or dietary fibre. Certain phytochemicals reduce digestion and absorption of starch; they are mainly found in cereals and legumes in the form of phytic acid, tannins etc.
- *Immunomodulating effects* of secondary plant compounds are further beneficial properties found in higher plants. The human immune system participates in pathophysiological mechanisms of numerous diseases, such as AIDS, cancer, heart and circulatiory or rheumatic diseases. By modulating the immune system, the origin and development of certain diseases can be influenced so that the state of health is improved. The essential constituents of the immune system are macrophages and lymphocytes. Certain secondary plant compounds can however act as immune modulators, e.g. carotenoids stimulate the immune system against cancer of lungs and stomach. Likewise saponins augment the immune increasing effects on the human body. Flavonoids particularly quercetin possibly influence the immunologic balance and exert an immune suppressive action on the body. Furthermore sulphides have anticarcinogenic effects. Phytic acid stimulates the cytotoxicity of the natural killer cells.
- *Antiinflammatory effects* of sulphides. Flavonoids can reduce several symptoms of inflammatory reactions.
- *Antiphlogistic effects* of fungi (Ganoderma and others) are used in Chinese popular medicine.

Dietary fibres and fermented food

Dietary fibre or roughage are parts of food which cannot be digested by man. They only occur in plants in the form of fibres and sclereids, as mechanical supporting elements and as filling and protective material. They consist mainly of cellulose, lignin etc. Water-soluble roughage forms viscous solutions, while water-insoluble dietary fibre has a high swelling capacity. Roughage can further be classified according to the position in the plant: either as structural components of the plant (cellulose, hemicellulose, lignin and some pectins) or non-structural components (e.g. slime and gums). Chemically, roughage corresponds mainly to polysaccharides with the exception of lignins. As roughage can have very distinct properties, the content of roughage in food is very difficult to identify (e.g. the test of raw fibre). The most modern test is the enzymatic identification. However, a chemical analysis does not reveal the biological effects. These mainly depend on the physical properties of the dietary fibre. Consumption of roughage and diseases of civilization seem to be related to one another; in countries where roughage consumption is large, these diseases are very rare.

WATZL & LEITZMANN (1995) emphasize that the following diseases are favoured by a roughage deficit: obesity, high blood pressure, hypercholesterolaemia, hardened arteries, gallstones, diabetes mellitus II, haemorrhoids, diverticulitis, appendicitis, hernia, varicose veins, cancer of the colon. It is recommended that 30 g of roughage be consumed daily. Cereals, legumes, fruit and vegetables provide sufficient roughage. The most frequent types of cancer due to a roughage deficit are: cancer of lungs, colon (large intestine), breast, prostate, ovary, endometrium and pancreas.

Part of the anticancerous effect of roughage is due to the binding of carcinogenic substances in the gastrointestinal tract and to the acceleration of their elimination from the human body. Roughage can also lead to a change of the enzymatic activity in the intestinal flora. Some roughages become water-soluble in this way and therefore can be easily eliminated.

Roughage with its high swelling capacity dilutes the concentration of carcinogens, on the one hand, and shortens the transit time and contact of carcinogenic substances in the intestine.

Roughage has also blood-glucose regulating effects. A high consumption of dietary fibre coincides with a low frequency of diabetes mellitus II, in epidemologic studies. With an increase of less than 5 g roughage a day in the food, positive effects can be achieved. Dietary fibre diminishes the enzymatic action of the pancreas, probably by adsorption, and increases the volume and the viscosity of the food chyme so that resorption of glucose is retarded. The blood-glucose level thus increases more slowly and reaches lower maximal values than with a food poor in roughage. Effectiveness of dietary fibre increases with its viscosity; fibre may be able to form gels and to slow down evacuation of the stomach.

A further benefit of roughage is its cholesterol reducing effect. A high cholesterol level is a risk factor for heart and circulatory diseases. Apart from the many cholesterol reducing secondary plant compounds already mentioned, dietary fibre furthermore exercises a series of physiological and chemical effects on the serum cholesterol. To date, four distinct mechanisms of dietary fibre acting on blood lipids have been described: a reduction of the lipase activty, binding of primary bile acids, reduction of the time of transit, and inhibition of cholesterol synthesis. Water-soluble dietary fibre, such as pectin, guar and hemicelluloses as well as lignin (water-insoluble) bind bile acids more readily than certain waterinsoluble roughage.

There are also immune modulating effects of dietary fibre. Hemicelluloses, pectins, glucans and gums exercise immunologic effects on nonspecific and specific defense mechanisms (WALDRON & SELVENDRAN 1993). Immune modulating contents could be found accumulated in the Asteraceae and Poaceae. Among the Poaceae, principally the cereals are important for human food. The immune modulating substances in cereals are mainly indigestible polysaccharides. The number of possible polysaccharide structures which can react with surface structures of immune competent cells is almost unlimited (WATZL & LEITZMANN 1995). Structural types of immune-stimulating polysaccharides can be classified as homoglycans, glucans and mannans, on the one hand, and as heteroglycans, hemicelluloses and plant gums, on the other. They exert tumor suppressant effects. Polysaccharides with anti tumor effects can activate cells of the immune system. They increase, for example, the cytotoxicity of macrophages against tumor cells and stimulate their interleukin I production. They also stimulate the cytotoxicity of natural killer cells as well as of cytotoxic T-lymphocytes. The polysacharide concentrations which show antitumor effects in animal tests amount to 1–200 mg/kg body weight (WALDRON & SELVENDRAN 1993).

Cereal polysaccharides, such as the glucans, do not only have an effect on human immune cells, but also activate the defense mechanism of other mammals, invertebrates and even of plants.

Although the following paragraph deals not only with vegetable substances, we think it adequate to add a few words on the healing substances in fermented food. However, plants are also involved in fermentation, e.g. yeast; formerly, even bacteria were attributed to the plant kingdom. And many plants are used for fermentation processes of our food. Lactic acid fermentation is a process in which food is chemically converted by microorganisms, such as bacteria and yeast, so that lactic acid is produced. Fermentation is one of the most ancient preservation methods used by man. It is a very complex process in which several microorganisms are involved: in yoghurt production, for example, *Streptococcus thermophilus* and *Lactobacillus bulgaricus*; in pickled cabbage *Leuconostoc mesenteroides, Lactobacillus plantarum, L. brevis, Pediococcus cervisiae, Streptococcus faecalis*. Reduction of the pH value as well as degradation of easily available carbohydrates are the main reasons for the preserving effects of fermentation.

Vegetables, legumes and cereals are principally suitable for fermentation as well as milk, meat and fish.

The most important end product of fermentation is lactic acid, i.e. laevorotatory and dextrorotatory. Lactic acid bacteria use lactose as their energy source; it is intracellularily split into galactose and glucose; from glucose, lactic acid is finally produced.

For hundreds of years, yoghurt has been considered good for human health. The high life expectancy of Bulgarians is due to the consumption of fermented milk products. Elie Metchnikoff, Nobelprize winner, understood that lactic acid bacteria control the undesired putrefacient bacteria in the intestine.

Persons with lactose intolerance do not have lactase activity in the epithelium of the small intestine; gastrointestinal problems are the consequence, possibly in the form of diarrhoea. A lactase deficit is normal for most of the world population, only white north and middle Europeans show intestinal lactase activity also in the adult stage.

But lactose intolerant persons digest lactose in fermented milk products relatively well, although usually less than 30 % of the lactose is broken down. Probably vital lactic acid bacteria produce this effect which is much reduced in heated yoghurt.

The cholesterol reducing effect of fermented milk products depends on the activity of the special stock of lactic bacteria. Lactobacillus LC 1 of Nestle is recommended. The reduction possibly takes place by assimilation of cholesterol in the intestine through lactic acid bacteria.

Lactic acid bacteria also exert antimicrobial effects. The human digestive tract contains about 10^{14} microorganisms composed of 400–500 distinct species, including Lactobacillus which exercises diverse protective effects on the intestine, e.g. an antimicrobial effect on undesired microorganisms; it hinders the attachment and multiplication of unphysiological microorganisms on the intestinal epithelium. This activity is called colonisation resistance; it is due perhaps to the competition for foods and for epithelial points of contact. The production of bacteriocines plays a further part; these substances are antimicrobial peptides which are effective against related species of bacteria, such as *Staphylococcus aureus, Salmonella thyphimurium, Clostridium perfringens, Escherichia coli, Vibrio* spp. and *Shigella* spp. Some of these bacteriocines, such as nisin are already used for preservation of food.

The therapeutic effect of lactic acid bacteria is used to cure diarrhoea and intestinal inflammations, gastroenteritis etc. (*Lactobacillus casei* GG).

Similarly, fermented milk products have a positive influence on vaginal infections with *Candida albicans*. Certainly, lactic acid bacteria also protect fermented food from microbial damage and decay.

Finally, the anticarcinogenic effects of lactic acid bacteria should be mentioned. Lactic acid bacteria activate the immune system and protect from tumors, inhibit faecal enzymes which take part in the activation of procarcinogens, bind mutagenic substances in the intestine and thus inactivate them. It could be proved that yoghurt exercises a tumor inhibitory effect in the intestine.

Lactic acid bacteria are able to modulate the immune reaction in the intestinal microflora; they augment the concentration of immunoglobulins.

Bacteria present in yoghurt can increase the production and secretion of interferon-γ and of interleukin-I β as well as of tumor necrosis factor α (with the aid of lymphocytes.) *Lactobacillus bulgaricus* and *L. casei* can stimulate macrophage activity for several days.

Nitrite can be converted into carcinogenic nitrosamines in the gastrointestinal tract. Several *Lactobacillus acidophilus* stocks are able to withdraw nitrite from nitrosamine formation, by absorbing it in their cells.

Gallic acids play an important part in the development of cancer in the intestine. *Lactobacillus acidophilus* however, reduces the rate of conversion of primary into secondary gallic acids, thereby lowering the risk of cancer formation.

Patients with intestinal cancer have lower amounts of *Lactobacillus acidophilus* in the stool than healthy persons.

The consumption of heterocyclic aromatic amines in the form of fried meat (twice a day for 3 days) increases the mutageneity of stools and urine, while the presence of *Lactobacillus acidophilus* significantly reduces mutageneity. Possibly, the mutagens are bound by *Lactobacillus* so that they do not come in to contact with the intestinal epithelium.

In epidemological studies, it could be proved that a high consumption of yoghurt protects from breast cancer.

However, not only fermented milk products, but also vegetables prepared by lactic acid fermentation can impede tumor growth, e.g. fermented red beets.

For a more detailed information we recommend the book of WATZL & LEITZMANN (1995).

The scientific data prove very convincingly the medicinal potential of secondary plant compounds, although part of the cited data was elaborated by animal tests. It could however, be proved, that a principally vegetarian diet not only protects from cancer, but also from heart and circulatory diseases; apparently, bioactive substances have prophylactic effects on several diseases which depend on alimentation.

To date, the medicinal effects of primarily individually isolated bioactive substances are known. In each meal, usually a complex mixture of many bioactive substances is present – and this is usually also the case for each plant species. The cholesterol level in the blood is, for example, not only affected by dietary fibre, but also by lactic acid bacteria in fermented food (yoghurt, fermented red beets and pickled cabbage), by saponins, sulphides, and phytosterins. The additive, synergistic or antagonistic action of these bioactive substances can not as yet be quantified. The vast field of the complex interactions of all these compounds still lies ahead of us to be studied.

Furthermore we can derive from the bibliography that secondary plant compounds as natural components of a mixed diet do not pose any risk to human health. On the contrary, the consumption of a principally vegetarian diet remains the most secure way to acquire the protective effects of phytochemicals and of other bioactive substances for the promotion of human health (WATZL & LEITZMANN 1995)

Therapy

Medicinal plants represent the most ancient therapy in the world. Actually two thirds of the worlds population use phytotherapy. That classic medicine already makes use of officinal herbs is often overlooked. Numerous modern medicines are obtained directly or in some altered form from plants. Worldwide, more than 20 000 different plant species are applied in herbal medicines. The therapeutic use of herbs is based on a long-lasting tradition of folk medicine. In many cases, traditional herbal medicines are much better founded and experimentally proved (by self medication of the ancestral generations) than modern medicines the effects of which are only tested on white mice or rats. It has been documented that the real efficiency of about 16 000 medicines which are daily used and prescribed by German doctors, have not yet been tested. Frequently, natural medicines are included in this category. However, doctors as well as patients for many years have had good experience with these milder medicines that often are less expensive.

Actually, the healing effects of plants and their extracts are the base of modern medicine development. According to the opinion of leading pharmacists, numerous side effects of synthetic medicines can be avoided, when the natural herbal substances are used. Some of the most active and most popular herbal compounds are digitalis glycosides, atropine and morphine.

Herbal medicines usually stimulate, mobilize and strengthen the natural power of the body to resist diseases. Synthetic medicines, on the other hand, often reduce this power and frequently cause undesired side effects.

Although herbal medicines do usually not produce undesired side effects, this is not always true. Nevertheless, risks are – as a rule – less common in properly applied natural medicine than in synthetic medicaments. Phytopharmaceuticals however have to respond to the rules of quality, effectiveness and harmlessness.

Definition of the term 'drug'

We define the term drug in the sense of a chemical substance (either natural or synthetic) which is used in medicine and affects the function of organs or tissues of the human body.

However, the word drug is also used in the sense of narcotics or hallucinogenics. Drugs are also defined as dried (or fresh) substances mainly of vegetable origin, but also derived from animals, including certain raw products such as fats, essential oils, resins, balsams or gums, as well as metabolic products from microorganisms (antibiotics).

The term officinal drug refers to those medicines that have been included in pharmacopoeias, as well as to the pure substances isolated from them. Their quality is subject to strict control of identity, testing, quality, denomination, preparation, dosage, storage and sale. In our book, only a few real officinal drugs are included, on the contrary, our intention was to present new vegetable medicines or drugs which are mainly known to the native population and are so far not on the market, but could be promising medicines for the whole world in the future.

Most of the drugs mentioned in our book are therefore non-officinal drugs, not covered by phar macopoeias and not subject to legally binding tests. Nevertheless, in the drug trade, these drugs often represent a considerable share of the market. Most of them are used for tea assortments and for extract production, while others serve technical purposes.

Vegetable drugs are required and processsed in many different industrial fields, among them tanning and dyeing industries, pharmaceutical industries, in confectionery and liqueur industries and cosmetics etc. Recently, vegetable drugs have also been used in household products, a development which was initiated by the increasing awareness of the environmental problems of our world ('Green Movement'). Vegetable drugs are found in cleansing agents, detergents, in fabric softeners, in floor and surface cleaning products and in furniture care. Examples of vegetable sources are – *Aloe vera*, jojoba, avocado, soapwort etc. Vegetable drugs are also found in varnishes, insecticides and fungicides, in fireworks, perfumes and as aromatics in tobacco and other products.

Toxicity

'Ein Kraut das heilt, kann auch töten' (A herb which heals can also kill).

Toxicity of medicine depends on quantitiy is a rule given by Paracelsus. Phytopharmaca are not always harmless. Dose and period of time of administration are of great importance for the effect of a medicine.

As a rule, effects and side effects are difficult to calculate in medicinal plants.

Nevertheless, about 50 % of the top medicines come from the tropical forests – money which grows on the trees – a six milliard dollar market up to date.

Poisonous plants of Venezuela were studied by BLOHM (1962). He mentions altogether 35 families of Angiosperms. He defines poisonous plants as those which (under natural conditions and upon contact with or introduction of relatively small amounts into an average individual, human or animal, of normal health) are chemically capable of inducing disease or death. The symtoms of poisoning may either appear immediately or only after a lapse of time. Some plants which are ordinarily non-toxic when ingested in relatively small quantities, may, when minute amounts are repeatedly ingested over a period of time, ultimately produce harmful effects. This is due to the gradual accumulation of one or more poisons in the body until the threshold of toxicity is reached.

A plant may be poisonous at certain stages of its life cycle, yet be innocuous at others. Environmental conditions may influence the toxicity of a plant to a greater or lesser extent. Not all parts of a plant are necessarily toxic, and if they are, they may not all show the same degree of toxicity. Furthermore, there may be an individual variation in the toxicity of plants of the same species. The quantity of a poisonous plant ingested or the extent and intesity of contact with such a plant are obviously very important in determining the degree and seriousness of poisoning.

On the other hand, individuals affected by a poisonous plant react differently, according to the weight, age and overall physiological condition. In animals, sex and breeding may influence the susceptibility of an organism to a particular poison. Different individuals of the same species may thus show varying reactions to the same poisonous plant.

Classification of poisonous plants can be carried out according to the chemical nature of the toxic principles involved, in conformity with the symptoms which poisonous plants produce on human and animal organisms or in persuance to the taxonomic principles of plants. BLOHM 1962 preferred an arrangement according to taxonomical points of view.

Each cited plant species is given a brief taxonomic description, its habitat is mentioned, the poisonous principles are noted and the conditions of poisoning are thoroughly decsribed. Coloured or black and white photographs accompany the text. A glossary is annexed. The bibliography comprises 394 citations.

Although our knowledge of poisonous plants has expanded enormously since 1962, the descriptions of conditions of poisoning are still valid and of great value.

Allergy

Plants may not only be toxic, but cause allergies.

A comprehensive survey of allgery provoking plants and plant allergens has been carried out by HAUSEN & VIELUF (1997). Plants are present everywhere: in our food, in cosmetics, in pytopharmaca, in the form of ornamental plants etc. In some sensitive individuals, certain secondary compounds of plants may provoke irritation of the skin and the mucous membranes or may exert an influence on the general state of health. The content of secondary compounds of plants is however subject to fluctuations, according to the area where the plant grows, to soil and climate, to the time of harvest and so on. Recently, people tend to avoid chemical medicines and to prefer natural products with the prefix 'bio', such as teas, extracts, medicinal herbs; alternative medicine is in this way reviving. However, this trend may provoke allergies. With the construction of green houses, an enormous variety of exotic ornamental plants can be cultivated. People engaged in this activity are particularly exposed to the danger of allergies. Furthermore, exotic plants are imported from everywhere in the world; but with new exotic plants, the risk of allergic diseases has increased in Europe and in other temperate regions. In this situation it is difficult for the doctor to identify the plants which affect the patients and which are brought to the consultation offices to be shown to the doctor. The many trivial names given to the plants are also confusing, so that finally a botanist has to be consulted to solve the problem.

When a contact allergy has been diagnosed, the exact botanical identification of the plant is indispensable in order to detect the secondary compounds responsible for the disease. The undesired effects of plants on the human skin may be mechanically, irritative, phototoxic, photoallergic or allergic. Mechanical irritations may be generated by thorns, stings, hairs, or raphides which perforate the skin. Irritative contact allergies may provoke inflammations. Acetylcholine, histamine, serotonin and latices are some substances found among others in the Euphorbiaceae and Urticaceae. Contact with the plant sap can generate oedema, necrosis, ulcers or blisters and redness. Even some vegetable enzymes can provoke relatively mild irritations.

Phototoxic reactions are due to furocoumarin occurring, for example, in Rutaceae, Umbelliferae, Moraceae.

Allergenic effects of plants on humans can be produced by a great variety of secondary compounds. The plant develops these compounds to combat infections by fungi, bacteria, viruses or to repel insects and to prevent animal predation. Secondary compounds with allergenic effects on man are, for example, glucosinolates, allylsulfides, coumarins, quinones, flavonoids, stilbenes, terpenes, lactones and many others. Some allergenic tropical and subtropical plants are *Cinnamomum zeylanicum, Citrus sinensis, Gossypium hirsutum, Helianthus annuus, Krameria tridentata, Philodendron scandens, Syzygium aromaticum,* and *Vanilla planifolia.* HAUSEN & VIELUF (1997) gave a complete list of all chemical compounds found in plants which were identified as contact allergenes.

Vegetable antigens may be the cause of instantaneous allergic reactions, particularly certain substances of pollen grains and of food (mainly leaves and fruits). Rhinitis, conjunctivitis or asthma may be the result. Many of the allergens are watersoluble glycoproteins (with a molecular weight of 10–60 kD), e.g. in pollen grains. Proteins in the latex of many plant species are suspected allergens.

Pseudoallergic reactions can be generated by plant species, such as *Ananas comosus, Capsicum annuum* and *C. frutescens, Carica papaya, Ceiba pentandra, Ficus carica, Gossypium hirsutum, Nicotiana tabacum.*

Plant species with phytoallergens are, for example, *Ananas comosus, Arachis hypogaea, Hevea brasiliensis* (latex), *Lycopersicon esculentum, Phaseolus vulgaris, Solanum tuberosum.*

It is however obvious that as far as allergies are concerned, plants of temperate regions, especially of Europe, are much better studied than tropical and subtropical plants, particularly plant species from the Neotropics.

The allergic effects of only a few of the plants mentioned in this book have been studied.

How can useful and medicinal plants be found?

Venezuela, as a tropical country, possesses an enormous wealth of useful and medicinal plants. It would be easy for the authors to enumerate a thousand of them. But how can we get to know these plants and their uses? First of all it seems to be necessary to consult the native bibliography. Most South American countries have their own special books dealing with medicinal or useful plants. The SECAB collection is one of the most valuable compilations of this kind which includes the flora of Bolivia, Columbia, Chile, Ecuador, Panama, Peru and

Venezuela. PITTIER (1926) contributed a great deal to the knowledge of useful Venezuelan plants. GUPTA (1995) studied 270 Ibero-american medicinal plants. SCHULTES & RAFFAUF (1990) deal with the „healing forest‘ of the northwest Amazon region. BRÜCHER (1989) treated the genetics of neotropical useful plants. More recently, the interest in useful tropical plants has been growing steadily so that several books dealing with taxonomic description, chemical contents, pharmacology, use (including recipes), ecological requirements and cultivation of medicinal plants have appeared written by native botanists, pharmacists, experts on human nutrition, chemists etc. (see the bibliography of the respective paragraphs). Experiments were carried out and quite a few neotropical herbs and trees have already been recognized as officinal medicinal plants. A good survey of useful and medicinal plants up to 1968 is given by UPHOF.

Information about medicinal plants is scarce on the Internet. The web pages of the University of Greifswald (in German language) most informative deal with the distribution of the species, description of the drug, origin, contents, dose and route of administration, undesired side effects and toxicity.

The richest English page is called, 'Plants for a future -data base', citing 7000 plant species of economic use including medicinal plants.

A further source is the 'Health link data bank' of a commercial undertaking which gives the definition of a drug, a description and occurrence of the plant species, a microscopic description, indication and action of the drug, contraindications, combination possibilities with other drugs, preparation and dosage.

As internet pages have already reached a gigantic volume, it becomes more difficult every time to find out and select the most interesting information, particularly when new and attractive offers are concerned. However, the information on medicinal plants is so scarce that it is necessany to study books. Furthermore, the production of really good pages is connected with a great expenditure of time and energy and it is therefore supposed that these will be available in the future only through payment so that the net will increaslngly be filled with useless data i.e. with copies of pupular publications.

The next step in the acquisition of information on officinal plants would be to visit the so-called herbolarios, open stands where healing plants are offered. The 'herbolarios' are often found along highways or on local markets. They are scattered all over the country and provisioned by native experts of healing plants. They still preserve the knowledge of useful plants transmitted to them by their ancestors. They know exactly the right plant species and the organs which have to be used, the methods of preparing the drug, the route of administration of the medicine; they are also informed whether the plant has been collected wild or whether it is cultivated and they know the vernacular and local names of the plant. With reference to the local name, it is very important to know the site from which the plant comes because vernacular names change from site to site and the same names may signify different species at different places. Drugs of this kind however have to be handled with care. We may take into consideration that the native people are accustomed to these drugs for even hundreds of years while other people may react in very different ways. It is not therefore recommended that the plants are used the same way, and coution is advisable.

The final - and perhaps most authenic - information is obtained from the living Indian tribes who not only know healing plants, but also narcotics (SCHULTES & RAFFAUF 1990). However, there are only few Indian tribes still living in Venezuela and scientists may have trouble, when trying to communicate with the natives, as there exist about a thousand different languages and dialects in South America (KENNETH KATZNER 1996).

For all these reasons, it seems to be very urgent to register all the healing plants which are still known. Besides, the plants have to be described taxonomically so that they may be identified later on; which parts of the plants have the healing properties should also be noted. In most of the cases, healing substances are only found in the one or the other organ, or are enriched in particular organs. When buying drugs at the „pharmacological‘ stalls, only plant parts or organs are handed to the buyer, who then has to find out which plant species is the donor and which scientific name it has, as plants are usually sold by their vernacular names. Consequently it is in many cases impossible to guess the latin names, particularly when only roots are presented. A good anatomical description may be very helpful in this situation. An entire pharmacognostic work of South American Venezuelan officinal plants based on the thorough study of plant anatomy should be attempted. To date only very well known plants are characterized according to the inner structure of the organs, shape and size of starch grains, type of laticifers, formation of secretory canals or glands, shape of fibers and stone cells, form of crystals and other inclusions etc. But with these anatomical data it would be possible to identify plant species. We therefore noted at the end of each plant description a few characteristics which facilitate the determination of species. Only a few species are easily identified when only a few anatomical properties are known. That the venation pattern of the leaves of *Touroulia guianensis* (Fig. 224) is sufficient to identify the species (ROTH 1996) is for instance a unique case.

Search for active substances

Nearly 80 % of the food demand throughout the world is provided by the tropics, in which about 8000 edible plant species, mostly used by the native population, are found. Of the estimated 400–500 000 plant species around the globe only a small percentage have been phytochemically tested. In comparison with the abundant occurrence and diversity of tropical medicinal plants, our knowledge in this sphere is very restricted, although these plants are a valuable resource of the public health system in developing countries. Less than 10 % of the worldwide existing Angiosperms are being studied yet concerning their pharmaceutical properties (BERMUDEZ 1999).

However, phytopharmaceuticals are much used throughout the world. Approximately 50 % of all medicines used in Germany originate from plants. A list of more than 21 000 plant species which are globally used for medicinal purposes has been compiled by the WHO (GROOMBRIDGE 1992). It is frequently far more economical to use plants as raw material rather than chemical substances. Nevertheless, to date pharmacological properties of only about 10 % of the assumed total plant species existing world-wide have been investigated.

Almost every plant is useful for the native people, the Indians. Additionally most plant species serve not only one but several purposes. In our modern times, about 80 % of the medicines used worldwide are of vegetable origin. Especially the National Cancer Institute (NCI) of the USA is interested in cytostatica extracted from plants. Between 1960 and 1982, thousands of plant extracts were tested for their antitumour properties. To date, seven substances for cancer therapy were obtained from that programme. The number of active substances present in a single plant species is overwhelming. *Catharanthus oseus*, for example, contains 70 different alkaloids. Some plant species contain dozens of active substances with similar effects.

Tests are constantly improved by increasing technical perfection, which allows an enormous quantity of tests to be conducted in a relatively short time. Since 1986, the NCI has concentrated on tropical and subtropical regions with their immense richness of species which exceeds European regions, at least 10-fold. The authors however consider this an understatement.

The New York Botanical Garden in USA specializes in Central and South American plants. The plant collectors visit healers and shamans who live in regions rich in plant species and to whom the knowledge of medicinal plants has passed down from many former generations and who know now to prepare the medicines. This knowledge is however also transmitted in many books dealing with medicinal plants of the respective regions.

The chemical laboratories, on the other hand, need up to a kilogramm of the plant material for tests and the preparation of extracts. Dried and pressed plant material is in the meantime sent to Herbaria for identification, but this becomes more difficult every time because there are insufficient numbers of taxonomists and specialists dealing with certain plant families in the world.

Sometimes, the analysts need 50–100 kg of a single plant species to isolate the active substances and to identify their chemical structure, mainly using spectroscopic methods.

Nowadays not only are national institutions interested in tropical plant species, but also North American pharmaceutical companies, such as Merck, Glaxo and Shaman Pharmaceuticals. Merck Pharmaceuticals invest a million dollars in cataloguing the plant species of a part of the rain forest in Costa Rica. In Belize, Central America, the first ethnobiomedical forest reserve was founded by the Association of Traditional Healers with the aid of scientists from the New York Botanical Garden.

In Berlin, Germany, the company AnalytiCon has specialized in the characterization of natural substances. In the meantime, North American Phamaceutical Companies are already the treasure hunters of new chemical compounds in tropical plants. Science brokers search the tropical forest for wild medicinal plants to supply the industry with new products.

It is now the task of pharmaceutical companies to study the chemical composition of all plants used in popular medicine in order to find out, whether they are of real therapeutic value. But analytical laboratories of this kind should become established in the South American countries themselves to help industrialize the developing countries and not in the USA or in Europe.

Finally, we should not forget that popular medicine which uses plants with curative properties, is the medication of the poor class in underdeveloped countries and that many people living in the country far away from cities and doctors have no other choice!

International agreements

The access to tropical plants is controlled by the Agreement of Species Protection in Washington and by the UN Convention of Biological Diversity, which was signed in Rio de Janeiro by 150 countries in 1992. However, the regulations are contradictory.

Particularly, the proceedings of North American scientists do not always please the developing countries. The patenting of native plants such as *Curcuma*, and of medicinal knowledge by North American investigators caused indignation in India, when it became obvious that these plants are well-known in traditional Indian medicine and were documented by scientific investigation.

Scientists at AnalytiCon in Berlin consider the ethnobotanical search for new useful plant species as too wasteful, too time-consuming and too expensive. This is certainly correct, when the quantities of plant material and the possible forest destruction are considered. Many extracts do not show the effects indicated by the healers, and only have nonspecific effects. In this connection the authors do not unterstand why the native and international bibliography on the subject is neglected. Chemical contents of most of the important plant genera, are already known and related species may produce similar substances.

The search for useful and medicinal plants should start with the bibliography and concentrate on 'promising' plants, hundreds of which are, for example, mentioned in the SECAB series, in UPHOF, SCHULTES & RAFFAUF 1990 etc. These include descriptions of chemical compounds.

The 'Junta del Acuerdo de Cartagena' was founded in Lima, Peru in 1996 to protect the biodiversity of the Neotropics. This agreement was the first answer to the conflict between developing and the industrialized countries. It was proposed that living creatures have to be protected and patented, the rights held by the countries from which they originated (SILVIA 1999).

Isolation and standardization of phytopharmaca

The isolation of the pure substance in herbal extracts is very difficult and time consuming. An extract can be composed of a hundred different compounds in varying concentrations; it has therefore to be broken up into fractions by diverse chemical separation processes. Each of these fractions has to be tested for its activity. The active fraction then has to be further refined and processing is continued in the same way until the pure substance is obtained. The weaker the concentration of a pharmacological substance, the more fractions have to be processed and the more lengthey the procedure. The isolation process may even take a whole year and another year may pass to identify the exact chemical structure of the substance. However many structures particularly in marine organisms, are so complex that it is not financially viable to prepare them synthetically.

Only the largest and most modern North American companies have the capacity to systematically search for new phytopharmaceuticals.

The so-called 'leading substances' ('Leitsubstanzen') are explored with X-ray crystallography or nuclear magnetic resonance (NMR) spectroscopy. Modern pharmaceutical companies, such as Glaxo, are able to test more than 50 000 substances a month. Colourimetric assays indicate, for example, whether the test substances can inhibit tumor cells. In 10 000 tests about one lucky strike is found; and then, it takes still 10 more years until the medicine is fully developed for the market, while processing amounts to several hundred millions of dollars.

The Merck, Sharp & Dome Company in Costa Rica runs the best known model of commercial utilization of the tropical forests on an area of 50 000 km^2. The Instituto Nacional de Biodiversidad of Costa Rica has a share in the Company's profits.

Shaman Pharmaceuticals, on the other hand, searches for plants which are used by the native curanderos (healers) to cure diseases. However the general trend is away from the gathering ethnobotany and also away from the empiric search for active substances, through a well-directed chemical modification of natural substances leading to the creation of new medicines (designer drugs).

There has been a great world-wide boom of phytopharmaca and an increase in the phyto market. The most rapid (6 %) growth of the market at the moment is observed in England. Top herbs are - *Echinacea*, garlic and ginseng. In developing countries, about 90 % of the medicines still come from the native flora.

The development of therapeutically and qualitatively uniform and homogeneous medicines which can be pharmacologically and clinically reproduced is the most important challenge of phytopharmacology. However the assessment of the correct dose for phytopharmaca still remains difficult due to their complexity. Therefore additive or synergistic effects freqeuntly result. Furthermore, the content and quantity of effective substances are subject to variation in the plants. As about 70–80 % of the medicines used world-wide are still obtained by uncontrolled collections and ruinous exploitation and given that phytopharmaca for self medication become more popular, the cultivation of tropical plants is suggested; carried out systematically, it has to be intensified and the quality criteria have to be adapted to meet the international standards.

Historical considerations

Not even 0.5 % of the 250 000 angiospermous plant species world-wide have been thoroughly studied as to their possible pharmacological effects. However, according to the WHO, more than 3.5 million people are dependent on plants as their basic medicinal supply. Traditional medicine is cheaper and easier to obtain. The traditional medicines mostly cure stomach and intestinal problems, inflammations, skin diseases and gynaecological disorders. These are also the diseases which most frequently occur in tropical regions; probably the native population searched principally for plants to cure these sicknesses. Western chemical medicines, on the other hand, are more frequently used for heart and circulatory diseases, nervous problems, cancer and infections, disorders which are apparently a consequence of civilisation! The native people experimented over hundreds of generations to obtain the best phytopharmaca. Moreover, the native people obtain food, clothes, building materials, means of transportation and accessories for rituals and body decoration from plants. The knowledge of useful plants from the New World has not only changed the food habits in the Old World, but also its culture. Potatoes, tomatoes, maize, peanuts, maniok, cucurbits, chilli pepper etc. were not known in the Old World before the conquest of the New World by the Spaniards. The trade with opium, quinine, cotton, rubber and sugar cane, has changed the destiny of entire nations.

For about 5000 years, the Indians were in search of useful natural products in their country and cultivated a large number of useful plants. Only 500 years ago through the conquest by Columbus, these plants became known to the Western World. Plants such as beans, sweet potatos, paprika, sunflower, pineapple, avocado, guavas, vanilla, cocoa, mate, tobacco etc. resulted from Indian breeding. Rubber, cashew nuts, brazil nuts, china bark, peru balsam, ratanhia root, guajak, *Echinacea*, were used by the Indians as wild-grown plants.

The first 'Americans' came from Asia some 20 000 years ago. The most ancient archeological findings of used plants are cucurbit seeds, about 14 000 years old, and avocados, 10 000 years old. Calabash fruits were already used as vessels about 9000 years ago, cotton was collected as early as 7800 years ago. Cultivation of plants and agriculture started at the latest, about 5000 years ago. The words maize, tomato, chocolate, ananas, are Indian names: 'maize' is derived from Aruak, 'tomatl' and 'chocoatl' from Aztec, ananas from the Tupi 'nana'. Vanilla, an orchid, has been propagated vegetatively since time immemorial. The milky sap latex of *Manilkara zapota*, a Sapotacea, has been manufactured in Chile before the conquest by the Spaniards. Fiber plants such as cotton (*Gossypium hirsutum* and *G. barbadense*) as well as *Agave sisalana* and palms like *Leopoldina piassaba* and *Attalea funifera*, besides many others were used as fiber plants by the Indians.

Timbers of hot and wet tropical regions become durable due to their deposits of antiseptic substances, such as tannins or other polyphenols. They are therefore of special value for the native population. The Caesalpiniaceae, for example, supply many durable timbers for construction, furniture and for dyeing. Timber that is resistant against termites is very valuable for ship-and-house-building. These timbers were always preferred by the Indians, because they contain substances that repel insects. However the insecticides are of very different chemical origin and also exercise very different effects on insects.

As compared with the old world, in tropical America there exist relatively few medicinal plants which contain volatile oils, balsams or resin and only few cardenolide glycosides and anthracene derivates are known. On the contrary – the alkaloid drugs are numerous. In the South American tropics and subtropics, there are still many medicines in use which combat fever, inflammations, parasites, snake bites; equally, insect repellents are frequently employed for protection against mosquitoes and other insects, particularly in hot river lowlands, e.g. in the Orinoco and Amazon regions. It is not surprising that quite a few of these Indian drugs are now used in homoeopathics. WOLTERS (1992) cites 75 well known medicinal plants from America which are mentioned in the DAB, HAB and DAC (Deutsches Arzneibuch, Homoeopathisches Arzneibuch, Deutscher Arzneimittel Codex) and which are apparently much in use. And again, it has to be emphasized that natural medicines have been tested much longer (often over hundreds or even thousands of years) than the chemical clubs.

While many of the condiments used in Europe came from the Far East, already 2000 years B.C.

South American condiments which were used by the Indians for thousands of years, became known in Europe only after the conquest of Columbus.

Central and South America

It is estimated that more than 20 alimentary plant species were cultivated on the continent before its discovery in 1492. The Aztecs in Mexico knew how to use more than 1200 plants. Hernan Cortes was cured from a wound on the head by native medicine. The Incas trained 'pharmacognosts' specialized in the search for new natural drugs. However, in the primitive medicine, a marked influence of religious mysticism existed. The healing was accompanied by ritual dances, exorcism, faith healing and the use of amulets.

In 1884, the first school of pharmaceutic studies was founded in Venezuela by Vicente Marcano. However, in the 20th Century, interest in natural products sharply decreased owing to the development of chemotherapy and the synthetic preparation of medicines.

As a consequence of the discovery of America, the European standard of life was improved by the introduction of products such as maize, cocoa, potatoes, tobacco and such beautiful flowers as orchids. The Aztecs, Mayas and Incas cultivated maize. The name potato is derived from the Ahitian Indian 'batata' changed to 'patata' in Spanish. The species comes from the Peruvian 'altiplano' and is one of the few plants that can be cultivated above 3000 m. In 1534, the potato was first taken to Spain and sold at enormously high prizes. A medicinal plant of great importance (due to the febrifugal effect of quinine) was *Cinchona* named after the countess Chinchon of Peru.

Nicotiana tabacum originally from Peru and Ecuador and *N. rustica* from Yucatan were used as stimulants and narcotics long before the discovery of America.

Another plant which influenced the world economy is *Hevea brasiliensis* with its latex called caoutchouc. Plants supplied arrow poisons, fish poisons, narcotics and hallucinogenics for the native people. Amerinidian tribes of Central and South America domesticated and improved the yield and quality of numerous crop plants which today enrich the daily diet of the highly developed industrial nations (BRÜCHER 1989).

Columbus and his crew were amazed by the diversity of exotic fruits, vegetables and spices and other useful plants, such as sisal, cotton, rubber etc. they saw in South America.

Plenty of vegetable food reserves still 'unused' by industrial nations exist in the Neotropics and are only known to the natives. From Amerindian languages come such familiar names as tobacco, avocado, cocoa, caoutchouc (from 'cau-uchu') or chicle. See also the explanation of the origin of Venezuelan plant names in ROTH 1981, p. 7–10.

At the time of colonization by the Spaniards, three great cultures were found in Latinamerica: Aztec, Maya and Inca. In a manuscript entitled 'Libellus de medicinalibus Indorum' the doctor M. de la Cruz described 250 'plantas curativas' already in 1552. A scientific expedition sent by Phillip II to Iberoamerica between 1571 and 1577, collected not less than four thousand medicinal plants which were described in the 'Tesoro de las cosas medicinales de Nueva Espana'. Several important publications on medicinal plants followed in 1565, 1579, 1580, 1589, published by different authors. Bibliography which may be of interest in this connection includes Grosourdy's 'The Creole Botanical Doctor', Figueroa's 'Diseases of the Conquistadores', and Picaza's 'Medicine of Vegetable Origin during the Cuban Wars of Liberation'.

But already the Mayas elaborated illustrated books known as Codices. The oldest is the 'Codex Dresdensis' dating from the 10th century; some of its chapters were dedicated to the preparations obtained from medicinal plants. Occultism, astrology and religion were integrated in their healing methods.

The Inca culture was superior to the Aztec cuture in its technology and social organisation, but the Incas completely ignored all pictorial representation. Inca medicine was supported by a magic-religious intervention of the priest and the role of the medicine man was empirical. The medicine men who collected roots and other vegetable parts were known as 'hampi kamayok', because songs and prayers accompanied their labour. They applied the 'Theory of Signatures' – in a similar way as Paracelsus in the 16th century supposing that each plant resembles the part of the body which it can cure. Likewise they applied the 'Theory of Affinities' between plants and human individuals, due to astrological influence. Readers who want to know more about these details are referred to A.M. ALBORNOZ 1993 and the bibliography cited in his publications.

Ethnobotany in Venezuela

Ethnobotanical investigation started in Venezuela with the first descriptions of useful plants by HUMBOLDT and BONPLAN in their famous work

'Viaje a las regiones equinocciales del Nuevo Continente' (1799–1804). ERNST (collected works 1986) continued from the end of the 19th and beginning of the 20st century these studies, which were crowned by the 'Manual de las plantas usuales de Venezuela' of PITTIER (1926) and 'Plantas communes de Venezuela' by SCHNEE (1960).

More recently, ethnobotanical studies were cariied out principally in the southern Orinoco region where native people live (TILLET & FUENTES 1983; DELASCIO 1984; BALICK 1985; PETERS et al. 1989; HERNANDEZ et al. 1994; GASTILLO 1995, 1998, 1999; FERNANDEZ DEL VALLE et al. 1999; WILBERT 1995, 1996; GUANCHEZ 1996; NARVAEZ & STAUFFER 1999, studies which cover the entire Amazon region).

Today, much importance is laid on ethnobotanical studies in Venezuela. The first 'Ethnobotanical Symposium' was held in Venezuela in 1999. About 100 scientists took part, 7 conferences were held and 46 papers were presented.

SILVA (1999) recently carried out ethnobotanical studies on the Paria peninsula. Two ethnic groups were living there before the conquest by the Spaniards: the Acios who possibly pertain to the Arahuaca and the Pariagotos who are part of the Caribic population. HUMBOLDT & BONPLAND (1799–1800) observed in the 18th century that these ethnic groups still maintained holy places of worship, such as forests and caves (MANARA 1996).

Women play an important part in the beneficial use of plants on the Paria peninsula and in other parts of South America. They know how to use and to prepare the plants. Many useful plants are cultivated by women around the house in a special house garden. Women likewise play their part in agriculture and in the knowledge of the biological diversity. Furthermore they maintain the agricultural system called, 'conuco' carried out in Venezuela in the form of small plantations (ROYERO ET AL. 1999).

Today, many ethnobotanical studies are carried out not only in Venezuela, but also in other South Amercan countries, such as Colombia, Peru, Argentina, Chile and so on, but ethnobotany is not the subject matter of this book.

South American plants and their uses

Useful plant species may be classified according to the organs or plant parts which supply the economically used products, such as roots, tubers (arising from roots or stems), rhizomes, shoots, stems, bark, wood, buds, branches, leaves, flowers, fruits, seeds – or according to the products themselves, e.g. fiber, cork, oil, starch, resin, rubber, alkaloids, saponins, tannins etc. In our classification we subdivided the plants according to their usefulness into: Nutritive plants with edible parts, plants of economic use which may be of interest for industrial purposes, and medicinal plants. However, a sharp distinction is impossible as nutritive plants can also be of interest for industrial production and because a single plant species may supply several different products.

Alimentary plants

Of the nutritive or alimentary plants, the fruits and/or seeds may be edible or the roots and tubers may be used as a vegetable. In other cases, the leaves and/or shoots (shoot tips or 'hearts') or even the flowers are eaten. Some edible parts are eaten raw, e.g. in salads, others are boiled as vegetables or cooked for the preparation of jelly, jam, and marmalade. Drinks are mainly made of fresh fruit, of seeds (coffee, cocoa) or of leaves (tea). Fruits and seeds, but even rhizomes are prepared as condiments.

Edible fruits

The number of edible fruits in Venezuela is large (Roth 1977, Roth & Lindorf, several papers) and most of them unkown in Europe or elsewhere, because their transportation is very difficult or because they are not cultivated for export. Well-known indigenous fruits (and infrutescences) are the delicious cherimoyas, species of Annona, the pinapple, the zapote (*Achras zapota*), the guayabas which resemble the quince in taste and consistence. But also many non-cultivated wild plants supply tasty edible fruits which could be cultivated for export, such as the beach grape, the mammey, several Sapotaceae (besides *Achras zapota*), Rosaceae and many others which could also be improved by cultivation (BRÜCHER 1989). In these fruits, the parenchymatous tissue surrounding the seeds is fleshy, juicy, sweet and edible. It contains sugars, vitamins and minerals. Other fruits, however, have a hard pericarp, but an edible embryo (stuffed with proteins, starch or oily substances) such as the brazil nut and many other species of the same family, the Lecythidaceae, not to mention the other numerous seeds which supply nuts.

Inedible fruits

Some inedible fruits which have a very hard shell are used as vessels, such as *Lagenaria vulgaris* (siceraria), the bottle gourd, or *Crecentia cujete*, the calabash tree, and of other species of *Crescentia*. Species of *Luffa*, the sponge gourd, have fruits which are penetrated by a dense network of fibers; this fibrous skeleton is used as a sponge or a filtering material and a shock absorber. Nonbitter selections of *Luffa* are however eaten fresh or cooked as a vegetable. Quite a few fruits of wild plants eaten by the Indians are poisonous and have to be prepared first to become edible, e.g. the cashew nut has to be roasted before it is eaten.

Tubers

Certain subterraneous tubers, such as of the potato plant are indigenous to the South American Andes; they correspond to thickened shoots with restricted growth in length. A true edible root is that of *Arracacia xanthorrhiza*, a relative of the celery. It is also native to the South American Andes and is used as a substitute for the potato, but has a stronger and very peculiar taste. Root tubers are also produced by several species of *Dioscorea*. The rhizomes of *Xanthosoma sagittifolia* (Araceae), are similar in their use and aspect to those of *Alocasia* and *Colocasia* of the Old World.

Vegetables

Plants used as vegetables often have edible shoots or leaves, e.g. some Araceae such as *Monstera* and *Philodendron*. Certain young palm shoots (palm hearts) are vegetable delicacies.

Fleshy plant parts mainly contain water and starch or sugar (shoots of the sugar cane). Starch is abundant in many tubers and rhizomes or in grains e.g. in maize. The avocado fruit is rich in oil. The fruit of the oil palm, *Elaeis oleifera*, is the fruit richest in oil. However, protein plants, such as the Leguminosae are less common. Proteins are preferably found as storage substances in seeds.

Drinks

Drinks squeezed out from fruits are particularly rich in the tropics. A large variety of fruits are available, such as species of *Passiflora* which have a very aromatic sweet-sour pulp. *Ananas sativus*, species of *Annona* (cherimoyas and guanabanas), species of *Eugenia*, *Fragaria ananassa* (of South-American origin!), *Mammea americana* tasting like an apricot, *Achras zapota*, *Carica papaya* and many others which are only known and used locally by the natives, supply delicious drinks.

We may also refer to drinks already in use in precolombian times obtained by fermentation from the juice of palm trees, such as *Acroconia sclerocarpa*, from which an alcoholic drink is made. Beneath the leaves there is a cavity which contains a sweet and clear liquid tasting like champagne. From the liquid of *Attalea speciosa*, another palm tree, a very much appreciated wine is obtained by fermentation. Another palm tree, *Oenocarpus bataua*, which reaches a height of up to 12 m, has 2 cm long ovoid fruits from which a fermented drink is prepared by the Indians (Caribs, Araucos and Guaraunos). Cocoa is obtained from fermented beans since time immemorial.

Even maize and pineapple are used for the preparation of refreshing alcoholic drinks. Less dangerous in its effect is the juice obtained from *Vitis caribaea*, the water climber, which releases large quantities of clean water when cut rapidly. It may have helped many persons to survive when lost in the woods.

Condiments

A special plant group supplies condiments. The spices are used in small quantities for their peculiar taste. The aromatic flavour of the vanilla fruit, an orchid, is used in chocolate and in many other sweets. Besides the nutmeg, other Myristicaceae, such as *Virola sebifera*, have an aril with a pungent taste. There are quite a few species of *Capsicum* which supply peppers, (chilli pepper, Spanish pepper, hot cayenne and paprika). They already played a notable part, when Columbus discovered the New World. Products such as chocolate made of the fruits of *Theobroma cacao* have rather stimulant effects akin to tobacco.

Fodder plants

There are large numbers of pasture and fodder plants for animals, but they are only little documented in Venezuela; included are many species of Leguminosae and Gramineae (e.g. *Paspalum*). Besides the leaves, also fruits and seeds, flowers, barks and even roots are eaten by cattle. Farmers feed their livestock not only with plantains and bananas, but also with grains of maize or with the fruits of *Guazuma ulmifolia*, Sterculiaceae, of *Enterolobium cyclocarpum*, Mimosaceae, *Samanea saman*, Mimosaceae, and of *Leucaenia trichodes*, Mimosaceae. The tubers of *Ipomoea batatas* corresponding to adventitious roots are not only eaten by men, but are also used as a cattle fodder.

Ornamental plants

There are thousands of ornamental plants originating from neotropical countries. There are quite a few well illustrated and voluminous books dealing with the flowering trees of Caracas or with the Venezuelan ornamental flora (e.g. written by JESUS HOYOS, see the bibliography). HOYOS cites more than 300 ornamental trees for Caracas, most of which are indigenous to Venezuela. BRAUN (see bibliography) published several books on Venezuelan palm trees most of which are ornamental. We may also refer here to the beautiful orchids (DUNSTERVILLE, FOLDATS) (1959–1976) which are cultivated in European countries. *Canna indica, Petraea volubilis*, species of *Lantana, Datura, Ixora, Calliandra, Acacia, Malpighia, Euphorbia* (Christmas star), *Pyrostegia, Allamanda, Petunia* and *Eichhornia crassipes*, a beautiful water plant found in green houses, are all planted in European gardens, houses and green houses. We should also mention the numerous species of the very ornamental *Heliconia* (ARISTEGUIETA), *Calathea, Ctenanthe, Salvia splendens, Zebrina pendula, Rhoeo discolor*, many Bromelias, such as *Aechmea* and *Bilbergia*, many cacti, Araceae (*Philodendron, Dieffenbachia, Anthurium, Monstera*) etc. which come from South America and which adorn our European rooms and gardens.

Fibers

Fibers are found in the phloem as well as in the xylem. They are usually abundant in the secondary phloem of the bark. However, the bark usually contains less fiber than the wood, as wood is mainly composed of fibers. The fibers of the stem in the central cylinder are strong, long and flexible. Strong fibers are also present in certain leaves. Fibers from seeds or fruits, on the other hand, are thinner and less resistant to mechanical forces.

Straw hats, baskets, table mats and other utensils are weaved from the leaves of several palm species. Baskets are also made from the leaves of some Gramineae (species of *Guadua, Panicum*).

Shoots of *Chusquea scandens* and of other climbing plants are used for twisting ropes. Fibers of the stem (cortex) are mostly used in ropery. Branches of lianas serve as cables and ropes. Many fibers obtained from bark, wood and leaves are fine enough to be woven as fabrics for textiles. Although the number of species used in this way is large, the leaves of *Ananas sativus*, Bromeliaceae, and the cortex of *Pouzolzia occidentalis*, Urticaceae, may be mentioned. There are, of course, several other representatives of Urticaceae which supply fibers for textiles.

Even the cottons are partly of neotropical origin. Cotton has been cultivated by the Incas of Peru since 3600 B.C. Cotton is possibly the world's most important and oldest fiber plant. There are 5 different species known in the New World. The useful fiber originates from the outer seed epidermis and reaches 2–4 cm in length. The seeds also provide valuable forage, because they contain oil and protein. However, there are also many other vegetable wools. Commercial kapok comes from several species belonging to the Bombacaceae family. The original and best known kapok is the product of *Ceiba pentandra*, an enormous tree about 30–50 m high. The fibers arise in this case from the inner epidermis of the fruit pericarp (endocarp) in which the seeds are embedded. The fibers are valuable as filling material for jackets, cushions, mattresses etc. However, other genera of Bombacaceae, such as *Bombax, Bombacopsis*, and *Ochroma* have similar hairs which could likewise be used industrially. It is even said that the fibrous hairs of *Ochroma* are superior in quality to those of *Ceiba*!

Furthermore any plant parts with sufficiently long fibers may be used for paper fabrication. Straw from cereals, Esparto grass, and woody plants may equally supply paper pulp. There is a great number of species which could be used for many purposes, but the problem of the tropics is that none of these species occur in the same large quantities as comparable plants in the northern hemisphere, where relatively few species occur gregariously and in large numbers.

Cork

Besides *Quercus suber*, the cork oak, and *Quercus occidentalis* which are cultivated for cork yield in Europe, many tropical tree species could supply cork. Some American trees have comparable common names, such as black cork ('corcho negro', *Guatteria anomala*), coloured cork ('corcho colorado', *Beliota mexicana*, Tiliaceae) or cork ('corcho', *Heliocarpus donnell-smithii*, Tiliaceae). Another cork tree is *Thespesia populnea*, Malvaceae, a Venezuelan plant. Plant families which develop a thick bark and rhytidome are the Mimosaceea, Papilionaceae, Flacourtiaceae and Bombacaceae. The thickest cork formations were found in Opiliaceae, Mimosaceae, Papilionaceae and Simaroubaceae by ROTH (1981). A cork width of 10 mm is however a maximum in the humid tropical forest. Possibly the cork may become thicker when frequently cut. Apparently, the cork material is not very much in demand in Venezuela or is simply imported from other countries instead of being produced in the country itself. Surprisingly, no mention of cork is made, when useful Venezuelan plants are considered (PITTIER 1926).

Oil

Oil in commercially adequate quantities is mostly found in seeds, where it is deposited for the developing embryo. Several palm trees are known for their oily seeds, such as *Attalea humboldtiana* which produces an excellent oil in the seeds as well as in the fruits. Further oil palms are *Oenocarpus bataua*, *Acrocomia sclerocarpus*, and species of *Bactris*. From the fruit and the seed of *Elaeis oleifera*, the best known oil producing palm, two different kinds of very fine oil are extracted. Likewise, the fruit of *Jessenia polycarpa* contains an excellent oil which (mixed with water) is used by the natives as a substitute for milk. Another palm species with oily fruits is *Cocos orinocensis*. The fruit mesocarp and the seed of *Mauritia flexuosa*, Palmae, contain an oil of high quality. *Ceiba pentandra* of the Bombacaceae family supplying the kapok of commerce, has seeds which contain an excellent oil that is used industrially in certain parts of Venezuela. The peanut, native of South America, is a further example of oily seeds. The two cotyledons of the embryo are rich in oil. Besides, the peanut is also a protein source. The bitter-tasting oil extracted from the seeds of *Carapa guianensis* is more often used in medicine than for culinary purposes. Other oily seeds are the brazil nut (*Bertholletia excelsa*, Lecythidaceae), the seeds of *Pachira insignis* (Bombacaceae), of the species of *Bombacopsis* (Bombacaceae), *Feuillea cordifolia* (Cucurbitaceae), *Virola venezuelensis* (Myristjcaceae), the Venezuelan nutmeg, *Virola sebifera* (Myristicaceae), and *Dialyanthera otoba* (Myristicaceae). The seeds of *Jatropha curcas* (Euphorbiaceae) contain an oil which is used as a purgative and as a very drastic vermifuge, but is also employed for illuminants and for soap production. The bark of *Bursera graveolens* (Burseraceae) exudes an oil which is medicinally applied. A special oil used for medicinal purposes is extracted from *Amyris balsamifera* (Rutaceae).

A fruit very rich in oil is the avocado which is usually eaten raw (ROTH 1977). The well-known jojoba oil, frequently used in cosmetics, is a product of the American species *Simmondsia chilensis* (synonym *S. californica*). The oil is very resistant to bacterial degradation and does not become rancid. It is very beneficial for the skin and contains vitamin A.

Gums and rubber

The Moraceae and the Euphorbiaceae as well as other plant families supply very important gums and caout-chouc. The Maya Indians used the latex of *Castilla elastica*, Moraceae, in pre-conquest times to make rubber balls for their games and to manufacture water-proof clothing. The laticifers are found in the primary cortex, the pericycle, the primary and secondary phloem and in the pith. Latex is however also found in species of *Artocarpus*, *Brosimum*, *Cecropia*, *Chlorophora*, *Clarisia*, *Ficus*, *Morus*, *Pourouma*, *Sapium* and in many other genera of the Moraceae. Most species of *Euphorbia* together with some other Euphorbiaceae contain laticifers. From an economic point of view, the most important species is however *Hevea brasiliensis*, native to the Amazon region. The latex is chiefly situated in the bark of the stem. Laticifers are also present in the pith and in the xylem rays. High-yielding plants have latex elements with a larger diameter. *Manihot glaziovii* is another latex-yielding Euphorbiacea from South America. It supplies the well-known 'cera' rubber. The latex of the succulent species of *Euphorbia* and of other Euphorbiaceae has not been used commercially as a source of rubber up to the present day, because of its high resin content. Other Euphorbiaceae are either oil or resin producers or have medicinal properties.

Another important familiy that supplies rubber is the Sapotaceae. Several species of *Calocarpum* are used for their latex. The famous 'chicle' which is the basic substance of chewing gum production is obtained from *Calocarpum mammosum*. There are also many species of *Chrysophyllum* and *Manilkara*, of *Mimusops* and *Pouteria* and of other Sapotaceae containing latex. Many are used for their tasty fruits, such as *Achras zapota*, species of *Manilkara*, *Chrysophyllum* and *Pouteria*. *Mimusops balata* was one of the principal sources of caoutchouc and guttapercha and occurred in the Guianas and the Orinoco region, but has been exploited industrially to such an extent that the tree has almost disappeared.

Resin

Stem, branches and fruits of *Protium heptaphyllum* (Tacamahaco, Burseraceae) contain a perfumed oleoresin. Additionally, other species of *Protium* supply resin. The bark of *Cercidium praecox*, Caesalpiniaceae, is constantly covered with a sweet tasting yellow-greenish resin from which very fine soap is prepared. *Symphonia globulifera*, Guttiferae, is also resiniferous. The American copal, a resin, is found in the stem and roots of *Hymenaea courbaril*. Varnish and laquer are obtained from the resin which is gathered from the soil beneath the tree. The most important oleoresin is the copaiba balsam extracted from *Copaifera officinalis*, Caesalpiniaceae. It is found in the secretory canals of the stem and of the main branches; the canals sometimes reach a diameter of several centimeters. Some trees even supply up to 50 kg of balsam. The pressure of the liquid within the canals sometimes becomes so high as to make the stem explode with a sharp detonation. Although *Myroxylon balsamum*, Papilionaceae, supplies the Tolu balsam, this is not used industrially in Venezuela; it is however known in perfumery and in medicine.

Wax

Vegetable waxes may be obtained from species of *Ceroxylon*, Palmae, e.g. from *Ceroxylon andicola*, the stem of which is covered with wax. Quite a few plants supply dyes for foodstuff, for body painting, for microscopic use as colorants, or to paint clay vessels and for woven straw and wicker work.

Timber

About half of Venezuela is covered with forest in which hundreds of species representing large trees grow. Quite a few of them supply very valuable timber, such as *Swietenia mahagoni* (Caoba), *Cedrela odorata* (cedro), *Casearia praecox* (zapatero de Maracaibo), *Toluifera balsamum* (*Myroxylon toluiferum*, Balsamum de Tolu), *Bulnesia arborea* (vera), *Guaiacum officinale* (Guayacan), *Tecoma pentaphylla* (apamate), *Hura crepitans* (Habillo), *Caesalpinia granadillo* (ebano), and many others. The most important timber trees of the New World are cited by RECCORD & HESS (1943).

However, not only the very valuable timbers are used, e.g. for furniture, but wood may be used for many purposes, such as cabinet work, carpentery, or for pales, scaffolds, fences, trestlework, tools, handles, boxes and cases, construction in general etc. furthermore it is used as firewood, fuel and for charcoal production. Any wood, even of minor quality, is used by the native population. The valuable timbers of the world are generally very well known in their structure and uses, so that we do not have to describe them again here. Besides, many timbers are only locally used and are not of world-wide importance. PITTIER (1926) calculated that there are about 250–300 important timber tree species in Venezuela.

The most important characteristics of economically interesting wood are: specific weight, density, durability and resistance to decay, working properties, finish, lustre, texture and consistency, odor, colour, taste, veining and grain, and combustibility. Resistance to ants and termites is highly appreciated and not infrequently found in tropical wood.

In our studies, we did not consider the trees which only supply timber, because there are many books available dealing with this subject. But trees often have further uses supplying, for example, tannins, resins, oils, dyes or other products besides timber, and these are included in our descriptions.

Medicinal plants

Medicinal plants are usually classified according to the sickness they cure. The number of remedies obtained from plants is very large: there are antiasthmatics, anticancer drugs, antihaemorrhoidals, antileprous medicines, antiveneric, antidiarrhetics, antidysenterics, antiscorbutics, antispastics, anthelmintics, antihydropsics, antiotalgics, antirheumatics, astringents, tranquillizers, remedies for the liver and cholagogues, sudorifics, diuretics, emetics, remedies against women's diseases, or for skin diseases, antipyretics, expectorants, haemostatics, even hypnotics and aphrodisiacs which are mentioned in popular medicine. Also ophthalmics, odontologics, medicines against parasites, laxatives, tonics and stimulants, remedies for the treatment of wounds, and even against AIDS and other new diseases have been found,

PITTIER (1926) cites more than 300 medicinal plants from Venezuela. The popular medicine in South America has the same origin as in other countries: at the beginning superstition was involved in the experience of real effects. However, the knowledge transmitted to the present population is based on the experience and selfexperimentation of Indians for hundreds or even thousands of years. Many plants contain compounds which activate the immune system, others are pain soothing, have narcotic effects, have disinfectant or healing powers, contain certain alkaloids or glycosides, are tanniferous or contain certain enzymes and vitamins, etheric oils, aflatoxins, bitter substances, coumarins, quinones, phytoalexines, phenolic compounds, saponins, mucilage, and many other substances, all of which we can not enumerate here.

The angiospermous medicinal plant species of Venezuela actually used are estimated to about 500 by Rodriguez (personal communication). However, this is certainly an underestimation of the actual wealth of Venezuelan medicinal plants. Unfortunately, the knowledge of medicinal plants acquired by the Indians during hundreds or even thousands of years of experimentation is being lost gradually, as the traditions are not practiced any more and because the Indians themselves are continuously decimated or wiped out. SCHULTES & RAFFAUF 1990 studied more than 1500 species in northwest Amazonia, 50 % of which were pharmacologically studied, but only a part of the plants is used by the natives. It may therefore be expected that there is a much larger number of useful angiospermous plant species in Venezuela, not counting the great mass of cryptogamous plants which are far from being studied completely.

Identification of medicinal and economically useful plants

Besides the already well-known vegetable drugs, there is an overwhelming number of South American/Venezuelan plants which were already used for hundreds or even thousands of years by the Indians and which could be used now, if they were only known to us. Large North American companies are continously searching for these promising plants in order to find new medicines to cure cancer, AIDS, Alzheimers disease etc. However, to recognize these ,promising plants in the wilderness, forest, desert, mountains etc., we must be able to identify them unmistakeably.

Taxonomists principally need the flowers to identify higber plants.This may however prove difficult, when plants can flower only every 10–20 (or even 30) years or only once in their lifetime. Species of the bamboo family are weil known for their very long vegetative phases of up to 30–35 years until they flower. Consequently there exists a great confusion in the identification of species in this family.

A correct taxonomic description of a new plant species should include as many peculiarities as possible. The taxonomist mainly describes the outer appearance of the plant, i.e. its gross morphology.

There is not only the distinction between herbs (whether erect or prostrate) - shrubs and trees, there are also the many life forms, such as climbers, epiphytes, parasites, saprophytes, geophytes with subterraneous organs (rhizomes, stolones, storage tubers, scaly bulbs and onions), marsh plants, hygro and hydrophytes, mesophytes, xerophytes, as well as the special growth forms in the high mountains, such as cushion plants, dwarf shrubs, the trellis forms, clusters or wisps of grasses, the rosettes and the caulirosellas or caulescent rosettes of *Espeletia* (ROTH 1995), the therophytes which endure adverse periods in the seed form, and finally the giant herbs, such as *Carica papaya* (see also VARESCHI 1970). Furthermore it should be taken into consideration that the German flora, for example, possesses only about 15 % of woody plants, while the Amazon region contains 88 % of tree species (VARESCHI 1980).

The vegetative structures, such as leaves, twigs, stem, crown, growth habit and buds are readliy apparent to the eye. As the number of herbaceous species is higher in temperate regions than in the tropics, keys to woody European plants are easier to elaborate. However, vegetative characteristics lack necessary variability of type, providing stability and constancy within the species, (LAWRENCE 1951).

In refering to the vegetative properties of the plant, the size of the plant (particularly of trees), the formation of the stem, the type of ramification (monopodial, sympodial), distinction of makro and brachyblasts, epi and hypotony of lateral twigs, the appearance of the crown, presence of spines, outer structure, colour and exudation of the bark are of interest (ROLLET 1980, 1982).

Further important characteristics are: The position of the leaves on the twig whether alternate or opposite, in distichy or decussation, in the form of whorls of more than 2 leaves; the shape of the leaves; whether simple or compound, size of leaves, leaf base with a sheath or stipules, shape of leaf base and leaf apex, formation of margins: whether undulate, crenate, entire, dentate, serrate etc. (LAWRENCE 1951), form of leaves; needle-shaped, scaly or laminate, types of leaf shape (VARESCHI 1980), colour of upper and lower leaf side, pubescence (hair types: LAWRENCE 1951), length of the petiole or sessile leaves, persistent or deciduous leaves, consistency of the leaves (ROTH 1984, 1977A), finally venation patterns: whether parallel or reticulate (ROTH 1996).

In the leaf, macro and microvenation are helpful characteristics for diagnosis. KLUCKING (1986–1995) systematically studied the macrovenation of the following families: Annonaceae, Lauraceae, Myrtaceae, Melastomaceae, Combretaceae, Flacourtiaceae. In many cases, the variation of patterns is not very remarkable; there is a midrib and a certain number of side ribs. Von ETTINGHAUSEN (1854/1861) has first studied the regularities of vein pattern formation in leaves, adding emphasis to the identification of fossil species; certainly, the venation skeleton is the best preserved part in fossile leaves.

Many more details are however supplied by the micro or minor venation of leaves (Figs. 1, 2, 3). ROTH (1996) presented a classification of the meshwork of the minor venation in leaves from representatives of the Caesalpiniaceae, Euphorbiaceae, Flacourtiaceae, Guttiferae, Lauraceae, Lecythidaceae, Melastomaceae, Meliaceae, Mimosaceae, Moraceae, Myristicaceae, Myrtaceae, Ochnaceae, Papilionaceae, Polygonaceae, Quiinaceae, Rosaceae, Rubiaceae, Sapindaceae, Sapotaceae, Simaroubaceae, Sterculiaceae, Tiliaceae, Verbenaceae and Vochysiaceae. The minor venation is visible only under the microscope. The species with the most characteristic pattern of the minor venation is *Touroulia guianensis*, Quiinaceae (ROTH 1996, P. 134). No mesh formation is observed here, but instead long conspicuously undulated anastomoses occur. *Touroulia guianensis* is possibly the only species in the world – as far as we know – which can be immediately identified by its minor leaf venation.

Besides the venation pattern of the leaf blade, other venation patterns, such as that of the leaf petiole, as seen in a transverse section, and that of the rhachis or midrib in transection can be useful for taxonomic diagnosis. In quite a few tropical plants, even veins of secondary order show interesting patterns, as seen in a transverse section (ROTH 1996).

Leaf anatomy of neo tropical trees has been studied by ROTH (1984, and the bibliography cited there), 48 different families with 232 species were anatomically studied. Barks and leaves mainly came from Venezuelan Guiana. Leaf structure of plants of the coastal regions, mountainous regions (Andes, Cordillera de la Costa) and the cloud forest was also studied. We are used to identifying wood in transverse and longitudinal radial and tangential sections, but also the bark supplies interesting details. Classification of bark patterns in neo tropical trees was presented by ROTH (1981). The hardbast in the bark in the form of fibers or sclereids is arranged in very variable and characteristic groups, plates, rings or in other ways, as seen in a transverse sec-

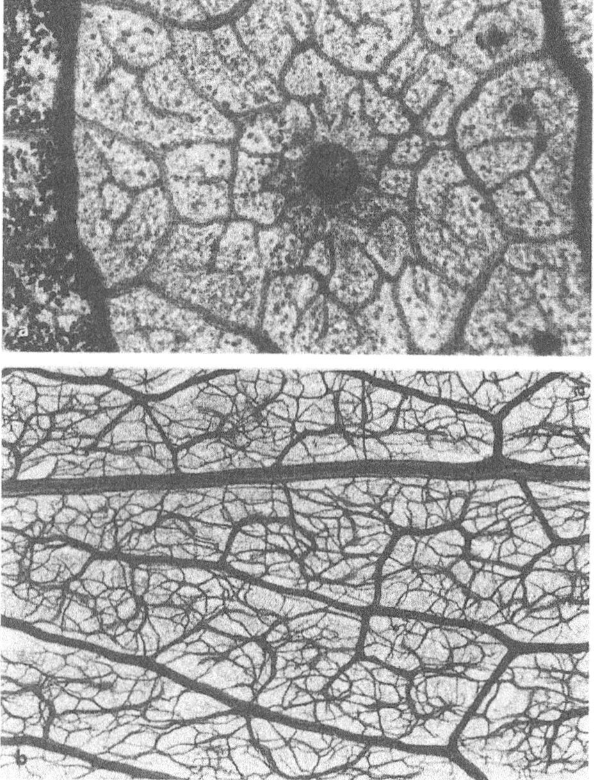

Fig. 1. Venation pattern of leaves. **a** *Vismia guianensis*, Guttiferae, with secretory cavity in the mesophyll (Roth 1996). **b** *Aspidosperma megalocarpon*, Apocynaceae, with large and small meshes and few free endings (ROTH 1996).

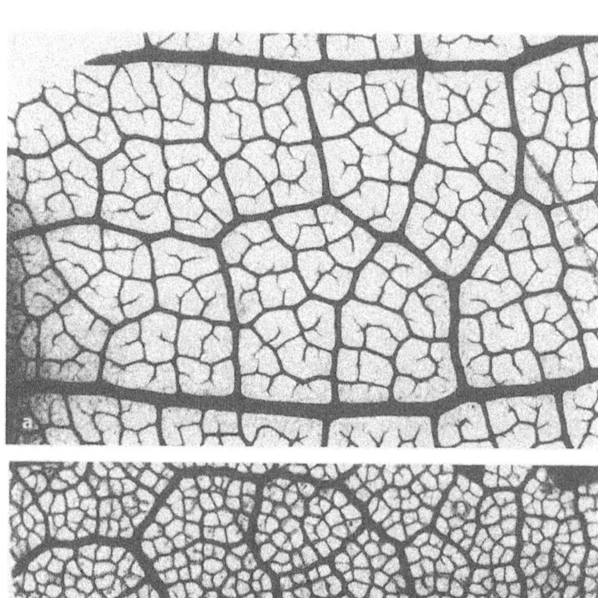

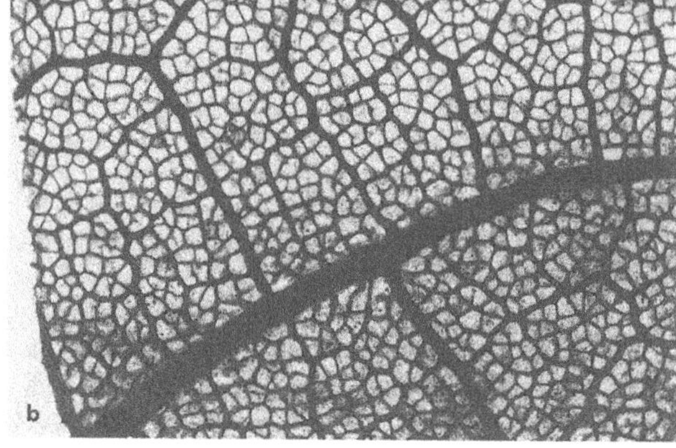

Fig. 2. Venation pattern of leaves. **a** *Protium neglectum*, Burseraceae. Larger meshes are subdivided into smaller ones; ramifications and endings are rare. **b** *Sterculia pruriens*, Sterculiaceae. The meshes are square-shaped or polyedric, they are mostly closed and have none or only a single free ending (ROTH 1996).

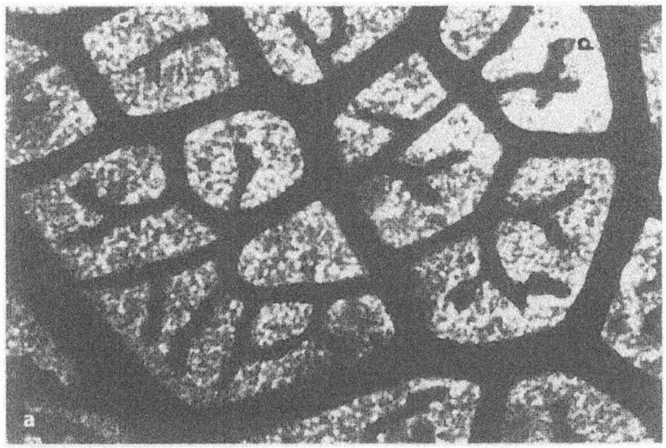

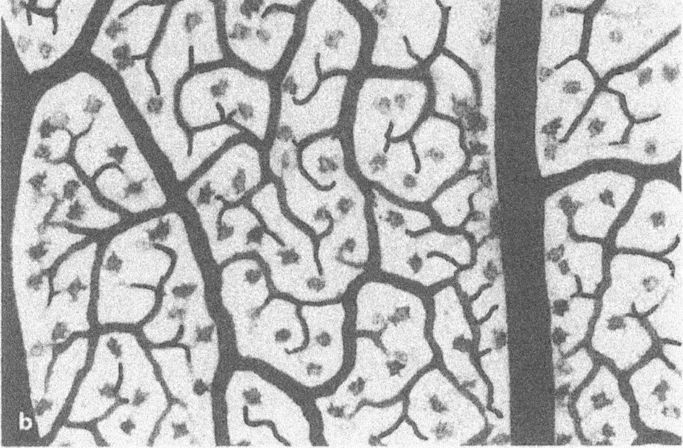

Fig. 3. Venetaion pattern of leaves. **a** *Eschweilera odora*, Lecythidaceae, with very thick mesh walls and strong and short branches. **b** *Mouriria sideroxylon*, Melastomaceae. Close to the endings lie isodiametric brachysclereids with short arms (ROTH 1996).

tion, so that keys of sclerenchyma distribution could be worked out (ROTH 1981, P. 422 FF). The hardbast is also the most durable tissue of the bark and is best preserved in fossile barks. Further helpful diagnostic features in the bark are rays, secretory cells, cavities and canals, the sheath parenchyma around the canals, crystals, and lenticels on the bark surface. Distribution of these tissues is very variable and therefore useful for identification. However, in some species, the bark of the young tree differs from that of the adult tree, particularly in the arrangement of the hardbast. Fluctuations in shape and frequency of crystals may occur here and there and are often due to ecological/environmental conditions. Certain plant families have a very characteristic bark pattern, while sclerenchyma distribution and arrangement in the bark in others are very heterogeneous. ROTH (1981) studied the bark structure of 280 plant species of Venezuelan Guiana belonging to the following plant families: Annonaceae, Capparidaceae, Flacourtiaceae, Vochysiaceae, Guttiferae, Quiinaceae, Caryocaraceae, Bombacaceae, Sterculiaceae, Tiliaceae, Elaeocarpaceae, Humiriaceae, Rutaceae, Simaroubaceae, Burseraceae, Meliaceae, Dichapetalaceae, Olacaceae, Opiliaceae, Celastraceae, Rhamnaceae, Sabiaceae, Sapindaceae, Anacardiaceae, Mimosaceae, Caesalpiniaceae, Papilionaceae, Rosaceae, Rhizophoraceae, Combretaceae, Myrtaceae, Lecythidaceae, Melastomaceae, Araliaceae, Rubiaceae, Sapotaceae, Apocynaceae, Boraginaceae, Bignoniaceae, Verbenaceae, Nyctaginaceae, Polygonaceae, Myristicaceae, Hernandiaceae, Lauraceae, Euphorbiaceae and Moraceae.

Certain cell contents, such as crystals, proteins, starch, oil, latex, secretory and other substances are important criteria for diagnosis. Some substances such as lignins, cellulose, starch, proteins, oils and fatty substances, cutins and suberins etc. can easily be tested for by specific reagents.

Hairs which may be present on all organs are very variable. They can be simple and unicellular, multicellular, ramified, curled, star-shaped, twin-shaped, 2-armed (Sapotaceae), scaly, bristleshaped, in the form of cystoliths (Boraginaceae).

Most peculiar are the glandular hairs with a foot, a middle and a head cell or those composed of many cells with a multicellular (4–8–16 celled) head or emergences with a multicellular foot and a syringe-like opening mechanism (Urticaceae). The digestive glands with a secreting epithehum (Drosera, Nepenthes: ROTH 1954) are of special interest.

For an exhausting pharmacognosy, knowledge of the complete anatomy of the medically used organ or plant part is indispensable. Particularly, fruits and seeds often have a very characteristic inner structure, e.g. with a sclerenchyma, special palisade cells or layers, crystal layers, light lines (legumes) etc. which facilitate an unmistakable identification.

The reproductive characteristics are generally more constant and the number of components and mathematical combinations are much greater.

Classification of inflorescences, (their ramification, position on the plant) is necessary for identification (TROLL 1964/69).

In the flower, spiral or cyclic position of members, number, shape and colour of sepals and petals, whether members are separate or united (gamopetalous), shape of the crown, aestivation of petals, number and shape of stamens, whether free or united with petals or gynaecium, way of opening of anthers, apo or syncarpous gynaecium, number of locules, number of styles and stigmas, and their shape, superior or inferior ovary position, arrangement of ovules, development of the fruit, fruit types, whether dehiscent or indehicent, colour of fruit, surface formation, presence of spines or hairs and protuberances, glabrous or viscid surface, way of dispersal (ROTH 1977), all these characteristics are of taxonomical importance; furthermore, shape and outer aspect of the seeds, size, length, width, diameter, colour, sculpturing, weight, number of seeds per locule, way of dispersal of seeds etc. play their part in identification.

TILLET & HAJEK (1999) emphatically stress the importance and the difficulties arising in the identification of tropical medicinal plants. Tillet is the founder of the herbarium 'Dr. Victor Manuel Ovalles (MYF)' at the Faculty of Pharmacy, Universidad Central de Venezuela, in Caracas, and is currently its curator. The herbarium which comprises 17 000 already processed samples and a collection of 15 000 samples which still have to undergo their final processing, is exclusively dedicated to medcinal, useful and toxic plants of Venezuela. The main challenge is to find new chemical compounds which may be interesting for the development of new medicines. Additionally chemo-taxonomical studies are carried out in the Faculty. The herbarium lends itself for ethnobotanical investigations so that there is not only a collaboration with the Chair of Ethnobotany, but also with biologists, anthropologists, geographers and sociologists from all over the world. These are mainly concerned with the use of the plants, with the ethnic groups which employ the plants and with the languages or dialects spoken by these groups. In the Faculty of Pharmacy, pharmacological, pharmacognostical and chemical studies are carried out. Some members of staff are also working with the analysis and chemistry of foodstuff as well as with vegetable substances utilized in cosmetics. A further important field is the investigation of toxic plants. Indeed, quite a few toxic plants supply medicines. Also, a special botanical garden has been created in which only medicinal, toxic and economically useful plants are cultivated.

However, one of the most important activities of the members of the herbarium is the correct collection of the material which has been and is still being assembled through large excursions of specialists in the whole world. Furthermore, new methods of plant preservation, drying and mounting of samples have been elaborated. For the preservation of samples in the field, alcohol (75 %–15 %) is applied instead of formalin so that the samples are better preserved. With a solution of 15 % alcohol in water, even the plant colours are preserved. For mounting of dried plants, a completely cellulosic glue was invented. The plants sampled in the museum up to date, mainly come from the Amazon region, from the Bolivar State and from Guiana. A library of scientific journals and books mainly concerned with plant systematics and ethnobotany is available to the investigators and students. There is no doubt that the collection of more than 30 000 samples of mainly tropical Venezuelan plants in the herbarium Ovalles of international repute, is of immense importance for the whole world. The engagement and the financial investment in this enterprise demonstrates what an important part investigation of tropical medicinal plants already plays in the world.

Identification of species is particularly difficult in the tropics, because of the overwhelming presence of species. For each plant family, another specialist has to be consulted. Additionally, in quite a few genera, aggregate species occur which can hardly be distinguished form one another due to their many congruent characteristics and similarities. There are also species which flower very seldom, bambus, for example, with a rhythm of every 30–35 years! Unusual peculiarities have then to be found and plant anatomy may be consulted successfully in each case.

The search for new medicinal plants in the tropics has initiated a revival of plant taxonomy. Unfortunately, there are not enough trained taxonomists in the world, so that some very difficult plant species have to wait years or even decennia to be identified. Such slow progress can not be accepted for the investigation of medicinal plants.

Fortunately the native South American population generally knows its own flora very well. There are always experts (peritos) who may help the scientist to find the right species. Though another obstacle are the many folkloric vernacular names given by the natives to the plants. The vernacular names are strictly regional so that a species called alcornoque in Guiana may have a completely different name in other parts of Venezuela; and vice versa, alcornoque possibly designates 2 very different plant species in the Cordillera de la costa and in the Andes, ROTH (1981, P. 7–10) explains some of these vernacular and pseudolatin plant names used in Guiana, their origin and ethymology.

Preparation of drugs

Herbal drugs may be prepared and used in very different ways. Natives frequently use the crude herbs, e.g. putting fresh leaves on wounds to disinfect and to cure them, or on the forehead to soothe headache. In other cases, fresh buds, flowers or fruits and seeds etc. are eaten raw. More frequently, however, plant parts are cut to pieces, broken, shredded, crushed, ground or pulverized. In this way, the healing substances come in direct contact with the skin, wound etc. or can be eaten more easily so that the effect of the healing substances is greater, as large amounts of cell sap and cell contents are released. For external use, cataplasms are prepared by crushing and grinding. For internal use, frequently a tea, decoction or infusion is recommended. For a tea, boiling water is poured on dry leaves, flowers or fruits, letting them steep for 5 minutes. But there are also teas which have to be boiled for several minutes or longer. The tea is then strained. An exception is made for all slimy drugs (e.g. Radix *Altheae*) which have to be prepared and taken cold.

Infusions are watery preparations of crushed pieces of plants on which boiling water is poured; they are then boiled in bain-marie for 5 minutes and repeatedly stirred and, after cooling are squeezed and filtered. They always have to be prepared fresh. When no quantities are indicated, a ratio of 1 (drug) : 10 (water) is used.

Decoctions, on the contrary, are watery preparations of crushed pieces of plants over which cold water is poured; afterwards they are boiled in bain-marie for 1/2 hours, squeezed, while still warm, and strained. Decoctions for which no exact quantities are prescribed, are prepared in the proportion 1:10 (drugs : water). They always have to be prepared fresh.

The time of boiling is commonly indicated in the recipes and depends on the substances that have to be released. By boiling them, certain poisonous substances may become chemically converted, but also beneficial substances, such as vitamins, may be destroyed.

Drugs with an enriched content of active substances are extracts, tinctures (substances dissolved in alcohol), percolates (filtered material), macerates (soaked and softened material), destillates, pressed juices.

Milksap, latex and resins are often obtained by the natives through incisions in the bark or in other plant parts.

Besides the therapeutically active substances in the herbal drugs, additional by-products and roughage, such as cellulose, pectins, lignin, wax, fibers and sclereids, etc. constitute the greater part of the phytopharmaca. According to its solubility, the roughage can be classified as water soluble substances which form viscous solutions and as water insoluble substances which have a high swelling capacity. Chemically, most of the roughage are polysaccharides, such as cellulose and hemicellulose. But the roughage also has physical properties with biological effects, such as a protective function on the intestine so that the risk of cancer is diminished; while a normal function of the intestine is secured by the swelling capacity of the fibers. For more details see WATZL & LEITZMANN 1995. Roughage is commonly defined as components of the human food which cannot be digested by the enzymes naturally occuring in the small intestine.

The plant organs or parts of organs used in popular medicine include leaves, stem, bark and wood, flowers, fruits, seeds, roots, rhizomes and all modifications of plant organs, such as tubers, bulbs etc., but even the entire plant may be used. The expression medicinal *herb* however, is misleading, particularly in the tropics, because the majority of medicinal plants corresponds to shrubs and trees. The nomenclature of the used plant parts comprises:

folium (green foliage), *herba* (green overground parts of a herb), *radix* (root), *rhizome* (rhizome, stem developing underground), *tuber, bulb, fructus* (fruit), *semen* (seed), *cortex* (bark of stem or root), *lignum* (wood).

Officinal and non-officinal drugs are distinguished in Europe. The officinal drugs supply pure substances by chemical isolation and are mentioned in the pharmacopeias. It should be quite clear that most of the herbal medicines mentioned in this book are not officinal drugs, but were detected by the native population through self experiments, perhaps for hundreds or thousands of years. The raw material usually comes from wild plants which are, with some exceptions, not cultivated. The content of active substances of the plants is therefore subject to variation, due to local races, to differences in soil and climate, to the time of harvesting and last but not least to differences in the diverse plant parts. Cultivation of herbal plants would thus be preferable, because the wild plants would not be harvested, confusion of plants would be avoided, the content in active substances would remain more stable, and cultivation of better races, e.g with a more profitable harvest would be possible.

However, vegetable drugs are not pure chemical compounds, but often contain a large variety of secondary compounds, some of which may react neutral, but others may counteract the desired effects or may simply be poisonous. As indicated above, poisonous effects may possibly be eliminated by a certain pre-treatment of the drug, for example, by heat, e.g. boiling or roasting the cashew nuts before eating. In many cases however, other secondary compounds may assist the healing substances in their beneficial effects so that a complete herbal drug may have a better effect than the pure chemical compound. These are the advantages and disadvantages of herbal drugs.

Methods of investigation

Studies of useful and medicinal plants may be carried out according to different points of view: e.g. particular plant parts, such as wood, bark, fruits or seeds may be the object of investigation. Other students of useful plants select certain contents of plants in which they are interested, e.g. alkaloids, mucilage and gums, etheric oils, resins and balsams etc. A further method to study economically important plants is to select a distinct region where the studies are carried out. This has the advantage that natives, special healers, shamans and so on can be consulted. SCHULTES & RAFFAUF 1990 preferred this method. Regional studies are presented by several Venezuelan authors. Some authors however focus their studies on certain plant familes emphasizing their taxonomic relationship. In this book we followed this method, observing the natural relationships of plants. Quite a few plant families are well-known for their particular contents, such as latex, tannins, resins etc.

Special plant parts and their uses

Bark and cork

When trees are mainly used for their valuable wood, a large amount of bark waste arises. As costs for cutting and transportation of valuable tropical wood are very high, an economic utilization of bark is necessary, the more so as the product is free of costs. However certain prerequisites have to be fulfilled. First, a continuous rate of use of large amounts of the same tree species has to be warranted. Furthermore, the method of decortication is important for further use; and finally, the necesary marketing conditions have to be insured (SCHNEIDER-BAUMS 1970).

Surplus bark is a serious residue problem which wood conversion industries must adress. The volume of bark residue is so great that it is receiving increased attention. Although bark supplies useful byproducts, there has never been a greater demand for the item on the market. However, air pollution regulations and high stumpage prices make maximum utilization inevitable. Development of new technologies make a valuable asset out of a costly waste. Wood processing companies are currently investing huge sums in the construction of bark processing plants.

Bark product manufacturers have to widely advertise their wares and strive to educate the public in their use. The Bark Utilization Committee of the Forest Products Research Society sponsors an annual 'Bark Products Promotion Award' for the best material any U.S. or Canadian Company, firm or dealer uses, so as to help advertise and sell any product made mainly from bark (HARKIN & ROWE 1971).

No other vegetable product should consequently be regarded with more care and interest than the bark. Otherwise, millions of tons of bark waste are thrown away every year. And this enormous waste mainly concerns the tropical forests. If we really care about tropical forests, we have to make use of the barks of valuable trees.

Bark structure of tropical trees is more complicated and more variable than that of trees from temperate regions.

Utilization of bark is partly determined by the structure of the bark, e.g. presence of fibers, cork, exudation, phloem parenchyma with contents such as starch, oil, fat, tannin, phenols, crystals. etc. The proportion of fibers is generally higher in tropical than in European barks. Frequency and length of fibers is very important for their utilization. Fibers and sclereids store lignin in their walls that has the same structure as wood lignin and can be used in the same way. The cork cells of the periderm usually contain suberin and wax, but may also have lignified walls.

The variability of bark structure of tropical trees is specific to the species so that an identification of the tree may be possible by the precise knowledge of the bark structure. This is of great importance for the forest engineer who has to identify tree species in the field mainly through bark morphology and anatomy (PARAMESWARAN 1971, ROTH 1981, ROLLET 1980 and 1982).

A prerequisite for the utilization of bark is therefore the knowledge of its most important anatomical, physical and chemical properties.

The bark has a highly complex and heterogeneous anatomical structure due to its manyfold functions. The chemical composition of bark also exhibits a great diversity so that common analytical data on bark samples are difficult to obtain and are often not very significant.

Vast differences in the nature and the amount of various chemical components and of diverse extraneous materials in the bark can be found even within a single species, depending on the age and the growth site of the trees sampled and on the fraction of bark examined. Differences in composition and variation in amounts of common constituents can, of course, be much greater between species and for these reasons, no good standarized methods of bark analysis exist (HARKIN & ROWE 1971).

Inner and outer structure of bark

According to ESAU (1977), the term bark may be used to designate all tissues outside the vascular cambium. In its secondary state, bark includes the secondary phloem, the primary tissue that may still be present outside the secondary phloem, the periderms, and the dead tissues outside the periderm. Ethymologically, the English word bark and the German word Borke have the same root, but the German word commonly designates the rhytidome, i.e. the dead tissue only.

Various tissue types of different origin may thus be included in the bark, particularly when primary cortex, not formed by the vascular cambium, still adheres to the outside of the secondary phloem. The vascular tissue of the young axis is still surrounded by the primary cortex which is directly derived from the apical meristem of the shoot, being composed of parenchyma with a photosynthetic layer beneath the epidermis and occasionally with peripherally situated collenchyma. A continuous cylinder of fibers may occur on the periphery of the primary vascular cylinder, either corresponding to primary phloem fiber or, more seldom, to perivascular fibers. This continuous pericycle cylinder is ruptured during growth of the stem in girth and is often restored by insertion of stone cells which develop from parenchyma cells. The crushed thin-walled cells of the primary phloem as well as some fibers of the protophloem remain on the inside of the primary cortex.

The secondary cortex, on the other hand, originates completely from the vascular cambium and all tissue types formed by this meristem towards the outside correspond to the secondary phloem, including sieve elements with their companion cells, phloem parenchyma, mechanical tissue and phloem rays. In this definition, the secondary cortex differs from the concept bark, as the latter may also include parts of the primary cortex (see above).

The primary tissues of the stem are usually separated at an early stage from the inner tissues by periderm formation. However, when no periderm formation takes place, the epidermis together with the primary cortex is maintained, e.g. on trees with a so-called green bark, such as species of *Cercidium*, *Pseudobombax* etc. The term bark is misleading in this case, because it includes a large part of primary tissues (ROTH & DELGADO 1968 A & B, ROTH 1981, ROTH 1963A). The bark seems to be green, because a photosynthetic green layer occurs beneath the transparent epidermis (ROTH 1981).

For forestry engineers who mainly work in the field, the ability to rapidly interpret a crude anatomical cross section of the bark is very advantageous, particularly in the tropics, when identification of trees becomes difficult. During the coordinated studies of the senior author with B. Rollet, an identification key for crude bark sections was elaborated by Roth and was successfully used by Rollet for rapid indentification of bark elements, such as rays, hard bast, cork and rhytidome. In this way a large number of species could be compared by means of their anatomical bark structure.

In this connection the senior author found it useful to make a distinction between 3 different regions of the bark when cross sections were concerned. The inner bark is defined as the conducting phloem. This region is often well distinguished from the middle bark – particularly in fresh material – by its different colour (frequently dark-brown in preserved material) and possibly by a layer of crushed sieve elements at the border line towards the middle bark. The inner bark may also differ from the middle bark by a very regular arrangement of cells in radial rows. The middle bark may be distinguished by a larger amount of mechanical tissue (fibers and sclereids), specialization and enlargement of parenchyma cells, obliterated sieve elements, crystal formation, formation of idioblasts, storage of starch or oil, partial sclerization of cells and other transformations or modifications. Due to a rhythmical sieve tube collapse, the phloem rays may adopt a very characteristic undulation. A very peculiar arrangement of the hard bast, either in the form of fibers or in the form of sclereids, may arise in the middle bark which greatly facilitates identification of families or even of genera. In the outer bark, on the other hand, the regular arrangement of cells and especially the pattern produced by the hard bast, is disturbed by a more or less vigorous dilatation of phloem rays and/or phloem parenchyma. Principally the rays expand in a tangential direction, though less commonly the phloem parenchyma also extends in this direction. Parenchyma cells may additionally differentiate into secondarily formed sclereids. Even secretory cells or canals and other idioblasts may become expanded tangentially. In this way, the outer bark adopts a very irregular aspect. Rays may become torn in different directions, mechanical rupture of tissue may occur, and secondarily formed sclereids may increase in amount so that more sclerenchyma may become accumulated in the outer bark than was originally present in the middle bark. There are however barks which also show little or no dilatation, particularly barks of very old trees where dilatation growth has almost ceased (ROTH 1981).

Finally we may draw attention to the fact that the above definition of inner-middle-outer bark was definitely created for field work and gross anatomy. In spite of the fact that certain anatomists dislike this distinction of three regions in the bark, these three zones can be recognized in most of the barks studied and definitely help in field work and identification.

The rhytidome develops in the bark by formation of successive periderm layers. The outermost and first periderm may either form in the epidermis itself which is a rare case, in subepidermal relatively superficial layers or, finally, in deeper lying tissues of the bark. In any case, a dedifferentiation, usually of the parenchyma cells, takes place so that a phellogen may develop. Not infrequently, the first periderm formation starts discontinuously, beginning around lenticels or small fissures and spreading from there in different directions. When the following deeper lying periderms are parallel to the first one, constituting concentric cylinders, a ringbark results. However this type of rhytidome is an exception. In most cases, the periderms cut out shells from the tissue so that the following periderms remain in contact with the former, penetrating into deeper tissue parts, but reemerging to meet with further outward-lying periderms, so that shell-shaped scales arise. As seen in cross section, the periderms seem to ramify. A scale bark is produced in this way. The common expressions ringbark and scale bark are misleading and the designations ringrhytidome and scalerhytidome would be more correct. Moreover, the distinction between a scaly and an annular rhytidome seems almost useless for identification of barks, as most of them are scaly or fissured. The outer aspect of the bark is very helpful in the identification of trees: colour of bark, bark surface – whether smooth or rough – presence of lenticels, exudations and their colour, consistency and smell, type of scales and their outlines, size, width, consistency and coulor, type of fissures (length and width, depth, shape and course of fissures; whether straight or wavy) are of great importance.

Periderms arise in the bark through formation of a so-called phellogen, a meristem which forms phellem or cork towards the outside and phelloderm towards the inside. Most frequently, the phellem differentiates into a corky tissue with suberized walls. The mature suberized cork cells are dead and contain air. Cork is consequently a very good insulating material either as a thermal barrier or as a barrier to humidity.

Fiber length and thickness, degree of lignification and wall thickness as well as flexibility are important for the utilization of bark. ESAU (1969) could show that phloem fibers are often longer than xylem fibers. Fiber length in the bark may vary between 300 and 4000 microns. Esau mentions a fiber length of 25 cm for *Boehmeria nivea*. Metcalfe & Chalk (1950) noted an arithmetic mean value of 1317 microns for fiber length in the wood of dicotyledons. A comparison of the primary and secondary phloem shows that primary fibers are considerably longer than secondary fibers. In *Cannabis*, fibers of the primary phloem reach an average length of 13 mm, while fibers of the secondary phloem only attain a length of 2 mm (ESAU 1969). However LIESE AND PARAMESWARAN (1972) reported a decrease of fiber length from the cambium in centrifugal as well as in centripetal directions which was interpreted by them as an increase in length of cambial initials with the increasing age of the tree. The mean diameter of fibers amounts to about 20 microns. The shape of the fibers usually varies less in the bark than it does in other plant parts. Septate fibers could be recorded in *Laetia procera* and *Cecropia sciadophylla*, referring to the material collected in Venezuelan Guiana (ROTH 1981). Some fibers retain their protoplasmatic content for a long time and may even contain chloroplasts (FAHN & LESHEM 1963).

Fiber pits are generally simple and slit-shaped, extending parallel to the micellary structure. Bordered pits occur very rarely. Pits are most abundant in the middle portion of the fiber, diminishing in number towards both ends. Within the fiber plates, the fibers are usually arranged in radial rows. Not infrequently, the fiber bundles are accompanied by septate crystal strands.

The phelloderm cells are often parencbymatous and thin-walled. However cell shape, type of wall thickenings and impregnation are subject to many variations. Due to their origin the cells of the phellem and the phelloderm are commonly arranged in radial rows. The cells may have the shape of bricks, as seen in a transverse section but may even adopt a spherical shape or the shape of the cells may rhythmically change so that for instance radially elongated cells alternate regularly with brick-shaped cells. An aerenchyma may arise in this way. Wall thickenings may be homogeneous or may become U-shaped, strengthening only the outer or inner tangential walls and part of the side walls. Phellem cells as well as phelloderm cells may transform into stone cells becoming lignified (see also ROTH

1981, p. 109). Roth mentions 26 different characteristics of phellem and phelloderm structure which allow 169 different combinations between phellem and phelloderm elements. The periderm is therefore considered an excellent diagnostic resource. Even the course of the periderms is very characteristic; in some species it is undulated, in others straight, and sometimes it produces very particular patterns in the bark. The pattern of the peridem within the bark is also dependent on the presence of mechanical tissue which forms an obstacle, since only living thin-walled cells can be transformed into a phellogen.

The mechanical tissue, whether in the form of fibers or in the form of sclereids, imposes very characteristic patterns on the bark, when a transverse section is considered. The arrangement of the mechanical tissue is most important for diagnostic purposes. Studying the material of approximately 280 species belonging to 48 different plant families, ROTH (1981) elaborated a key for the indentification of sclerenchyma patterns occurring in the barks of tropical trees (ROTH 1981, p. 420 ff.). Sclerenchyma cells may be solitarily dispersed in the soft tissue or, more commonly, may be arranged in the form of irregular or regular units or groups. Very often, the sclerenchyma appears in the form of plates. As tangential length, radial width and outlines of the plates (whether straight, undulated, ovalate, U-shaped, uniting in concentric rings, in superposed or alternating position with the preceding and the following plates, etc.) may vary, a wide range of combinational possibilities arises. Moreover, of course width and dilatation of rays may also impose a certain pattern on the bark.

Certain idioblasts, such as secretory cells, cavities or canals and crystals, e.g. septate crystal strands or large druses, may be useful for diagnostic purposes (ROTH 1981).

The outer morphology of the bark is very variable. The bark may be smooth, fissured or scaly. A ring bark is very seldom observed. In the tropics it is difficult to indentifiy trees by their outer bark aspect. Barks of young trees may differ from those of old trees. The similarities between species of the same genus, as well as the great number of different species in the forest often impede identification through the outer bark aspect. However, there exist relationships between the outer aspect and the inner structure of bark.

Characteristics which may be used for bark identification are bark colour, presence of lenticels, prickles or special marks on the bark and other outer bark patterns, such as a reticulate pattern, fissures, scales, furrows, and even exudations. The consistency of the bark – whether mealy, spongy, crumbly, brittle or hard, with a metallic sound when touched – is another valuable characteristic for identification. Consistency and inner structure of bark are certainly related to each other. With a cut of the machete it becomes obvious whether or not the bark has an exudation. Exudations may be white, yellow, orange or red; some are completely transparent.

The average thickness of bark amounts to 3–10 mm in European trees. The proportion of bark: wood varies according to the species, stem diameter, age and height of the tree, and is also dependent on the stand. The bark proportion increases with the age of the tree, with increasing altitude above sea level and northern exposure, with an unfavorable stand and with exposure to the weather side when the individual tree is concerned (ROTH 1981). The average proportion bark: wood is estimated at about 10% (SCHNEIDER-BAUMS 1970). The thickest bark found by ROTH (1981) in trees of Venezuelan Guiana was that of *Hieronyma laxiflora* with up to 46 mm in width.

For production purposes bark volume, specific weight, energy of combustion, bending and tensile strength, flexibility, swelling capacity, fire restistance, and extracts of bark are important.

Utilization

Bark has a long tradition of application in South America, beginning with the use by Indian tribes, some hundreds or thousands of years ago, but has gradually lost its importance due to the invention of new artificial products and because of its complexity. Cork, fiber, tannins, dyes, gums, resins, latex, fish and arrow poisons, foodsuff, flavourings, aphrodisiacs, antibiotics and other medicine can be obtained from the bark. Cinnamon for flavouring, quinine for malaria, Angostura bitters for digestive cocktails are well-known examples. However, barks with these properties are mainly limited to tropical plant species. The heterogeneity of the bark and the extreme differences between barks of distinct species render their use on a large scale difficult. Furthermore, physical properties of tropical barks such as density, thermal and mechanical qualities, and moisture proportions are only known of comparatively few barks (MARTIN, MARTIN et al. 1963–1971).

Any high grade utilization of bark will require large amounts of clean, dry bark from a single species. In the tropics where a large variety of species exists – each with only a few representatives –, available amounts of bark are insufficient for commercial processing so that often a bark mixtur from different species has to be used.

Factors such as tree size, accessibility of the stand, number of individuals per square kilometer, site of production, ability of the market to absorb products at a profitable price, competition from artificial products, and maintenance of a proper balance of products need careful evaluation. Although many pure chemicals could be isolated from barks, most of the pure organic compounds could not achieve a profitable large-volume market. This is particulary true of the neotropics.

Crude fractions, on the other hand, have definitely found markets. The largest market is for tannins which are commonly condensed, polymerized polyphenols. In South America, *Schinopsis quebracho colorado* (wood and bark) and *Rhizophora* have been processed successfully for tannin, *Schinopsis* was exploited so much in Argentina that it was almost eradicated. Availability of large quantities at a low price and the uniform and reproducible quality were significant factors in capturing the leather tanning market.

Processing of bark is carried out by physical or chemical upgrading. The cellulose and hemicellulose in the fiber portions of bark are largely similar to the corresponding materials from wood. Bark constituents are generally examined by extracting comminuted bark samples with various solvents. Chemical upgrading of bark is however difficult, due to the heterogeneity of the material. Barks are generally much richer in both quantity and complexity of extracts than the corresponding woods. A large number of pure organic chemicals can be isolated from the bark, such as tannins, flavonoids, alkaloids, carbohydrates, inositols, terpenoids, glycosides, saponins, esters, steroids, fats, lignans, and complex phenols. Tannins, waxes, balsams, essential oils, gums, mucilages, resins, latices and dyestuff extracted from bark may or may not represent relatively pure chemical entities.

The use of bast fibers

Bast fibers are obtained by a 6–8 week fermentation process so that the fiber bundles can be separated mechanically. Bast can be used as binding material or for weaving of baskets, hats, table sets or ropes and for matting. Vonnel textiles from Japan contain a large amount of fibers of *Rhizophora*, a mangrove.

The most antique paper of the Maya cultures was prepared from bast of *Ficus* species.

While natural bast has lost its importance in Europe, it is still very much in use in developing countries. It is appreciated for its length, elasticity and brilliance. The fibers of *Broussonetia* are manufactured into paper, umbrellas, cloth and deco foils. The bark of *Ficus* is still used in Mexico today to manufacture (by pressing) a type of paper used for folkloric painting (see frontispiece).

Some fibers of tropical trees reach a considerable length, e.g. those of *Sterculia ferruginea* (5 mm), of *Artocarpus rubrovenia* (6.8 mm), of *Ficus variegata* (8.95 mm), of *Cannabis* (13 mm), of *Boehmeria nivea* (50–250 mm).

The proportion of fibers is usually higher in tropical barks than in those of temperate regions. Frequency and length of fibers determine their use for particular purposes. Besides the fibers, lignified sclereids are often present in tropical barks. According to PARAMESWARAN (1971), the lignin of bark is identical with that of wood. Sclereids in the outer bark portions exclude their use in paper processing, because they would be visible in the form of granulated particles in paper. Barks rich in lignin, e.g. that of *Melaleuca* and *Acacia* species, can be used as extenders and as a carrier for phenolformaldehyde glues.

Experimental work has been carried out to incorporate fibers into plastics as a reinforcement for molded products. Minor amounts of bark can be tolerated in certain types of paper. Today, natural fibers are not only used as a substitute of plastics, those of flax and hemp can possibly even compete with glass fiber. Fibers of *Urtica* can strengthen glass fibers and plastic material, such as the interior outfit of cars or door frames. The advantage of nettle fibers is that they have a much lighter weight than glass fibers.

Certain bast fibers are even used as filling material for upholstery and for the fabrication of strainers and felt. Filter material made of barks can be applied for aeration of sewerages, as they absorb bacteria and other small particles, thus cleaning the drainages without exerting toxic effects on the aquatic flora and fauna.

The use of cork

The phellogen, e.g. of cork oaks, reacts to the artificial removal of cork by forming new larger cork layers. Cork has several favourable peculiarities: it has a low specific weight ($0.24\ g/cm^3$), is soft and elastic, is very durable and impermeable to gases and humidity, has a low conductivity of temperature and electricity and is easy to work with. Common cork contains suberin (about 58 %), cellulose (22 %), lignin (12 %), water (5 %), cerin (2 %), phellon and decacrylic acid (1 %). Cork therefore has manifold possibilities of utilization.

In the past, cork was mainly used as a plug for bottles. But nowadays, cork has many applications, e.g. as an insulating material, for floaters, bumpers, and cigarette tips. The residues are manufactured into insulating boards, acoustic boards, flooring, life belts, sealing and for cork helmets.

The forests of cork oaks occupy about 2 millions hectares throughout the world, centers of cork production are located in the western Mediterranean regions, the South-west of Spain, Portugal, North Africa and Greece.

Cork is an ideal insulating material due to its low specific weight, its high impermeability to liquids and its insulating effect for temperature.

New methods of bark processing allow a separation of cork from fibers. Products which mainly contain fibers are used as an additve to plastic products to improve the strength of the material. A mixture of cork and fibers serves as a mechanical stabilizer and as a filling material. Products which contain a surplus of lignin (65 %) provide an excellent carrier for atomizing agents in pesticides.

A mixture of cork particles, fiber and pulverized material is suitable as a carrying material for certain types of adhesives.

Waxes

Waxes are another product which can be obtained from bark (e.g. from species of Cercidium).

The corky fraction of many softwood barks is rich in wax. Wax polishes and carbon paper are 2 commercial possibilities. However, synthetic waxes now largely dominate the market.

Tanning bark

Bark is possibly most used for the extraction of tannins. By oxidizing agents or acids, the tannins can be condensed to phlobaphenes.

The bark can be manufactured into tannin extracts for tanning (e.g. the Argentine Quebracho). The tanning effect is due to secondary valency bonds between phenolic hydroxyl and the proteins of the animal skin which stabilize and strengthen the collagen.

The Chinese tanned animal skins 3000 years ago, but tanning was also practised by Romans, Germans and the Indians. Around 1875, importation of cheap tropical tannins had started. Imported tropical tanning material has a much higher tannin content than the European barks, such as that of spruce for example. The average content of extractable tannins in tropical barks amounts to 35–48 % (species of *Rhizophra* and *Acacia*), whereas the bark of oak contains only 10 %. On the other hand, a number of synthetic tannins which are relatively cheap have been developed and, additionally, have the advantage of standardization.

Synthetic tanning agents have not yet completely replaced the tropical vegetable tannins which are preferred for their valuable effect on the quality of leather. The surface of leather tanned with tannins of mangroves is smoother than that tanned with synthetic tannins.

Some tanniferous barks are distinguished by a high formaldehyde precipitation rate and are therefore suitable for manufacture of binding agents for adhesives. As phenol is still a costly additive in tropical countries, the utilization of tanniferous barks is interesting from the economic point of view.

Bark tannins find a large market, e.g. in oil-well drilling; large quantities are needed to thin the muds. The tannins act as clay deflocculants and control the viscosity and gel strength of drilling muds.

Another outlet for bark tannins is their use in adhesives, particularly for plywood and particle boards. A cold-setting, waterproof adhesive is developed from bark. The polyphenolic fractions can be extracted from bark in a particularly high yield with alkaline reagents. Tree barks are remarkably soluble in alkali; often over 50 % of bark will go into solution. By varying the extraction conditions as well as the purification and treatment steps a large variety of different polyphenolic products can be prepared.

The polyphenolic extracts can be used as dispersants, binders, deflocculants in ceramic clays, antioxidants, sequestering agents in boiler feed water, floatation agents in ore purification, as well as for vat dyeing of nylon and for desulfurization of gasoline.

Energy production and manufacture of boards

Since hundreds or thousands of years, native people of South America used the bark for energy production. This energy source has only recently been rediscovered. However, there are profound differences in the value of the distinct bark types, which depend on their composition.

The heat value of fats and resins amounts to about 8500–9100 kcal/kg; that of cellulose to 4150–4350 kcal/kg and that of lignin to 6100 kcal/kg.

The water content of the bark has a great influence on its heat value; it depends on the season, the climate and soil, on the method of transportation and the methods of decortication.

Bark humidity amounts to 50–65 % when dry decortication is used but can increase to 70–80 %, when the stems are stored in water. The drier the bark, the better the heat value. A completely dry bark has a calorific value of 4200–4600 kcal/kg.

Bark has a lower ignition effect than oil fuel, due to the presence of free water in the bark. The ash content of bark is higher than that of wood. The fresh bark has first to be dried and crushed. Dehydration may be carried out by a flue gas drying process or by a moisture expeller. Afterwards, the bark is compressed and shaped into tablets. A 3 month ageing process during which fermentation takes place, improves the strength of the boards. By higher temperatures and higher pressure, the bark can be briquetted for use as fire place fuel. Another major outlet is charcoal, e.g. for recreational use, for home barbecues etc.

The remnants of decortication can be used for the fabrication of boards of different types, e.g. insulating boards, hard boards, fiber boards or particle boards. However, the anatomical structure as well as the chemical composition of the bark is very different from that of wood, so that also the end products such as boards and tables give differemt results. The tensile strength of fir wood fibers exceeds that of bark fibers about 24 times. Improvement or refinement of boards is possible by addition of glue. The bending strength increases with increasing proportion of glue. To increase the resistance to humidity, hydrophobing agents, such as paraffin powder, are added.

Boards and tables made of bark are mainly used in construction, e.g. for roof insulating boards, possibly with diffusion canals which promote aeration to eliminate fungal attacks, or for hard boards. Fibrous barks are most favourable for board processing, but the fact that not all tropical barks are fibrous has to be taken into account. Mixed with wood, fiber boards consisting of up to 75 % bark may be processed.

For the employment of bark for scale boards, it is necessary to know the pH value, as the storage possibilities and some other peculiarities of the boards depend on this value. pH values of wood have been known for a long time, whereas pH values of bark have only been investigated sporadically. The pH values depend on the age and stand of the tree, the time of cutting and the height of the tree. VOLZ (1971) studied the pH value of some European trees and found that the pH value of the bark is lower than that of the respective wood. Furthermore, the pH value decreases with the age of the tree: the inner bark of the tree shows a higher value than the outer bark

Bark has generally less fiber than wood, so that it is slightly less strong than in wood.

Fiber boards made of tropical barks are distinguished by their near perfect thermic insulation properties. Since bark conducts heat less readily than wood, its use in insulation boards seems particularly attractive.

Many softwood barks are relatively rich in resins and waxes so that sizing of boards becomes unnecessary. A high extractive content of bark aids in binding the particles together.

Bark powder can be effective in absorbing oil slicks from water. Bark and saw dust may be used as an oil scavenger. Shredded bark is applied on playgrounds, bridle paths, golf cart tracks and as a base for practice ski slopes.

Bark for soil improvement and as a fertilizer

Bark is now used for mulch, humus and as a ground cover. Bark powder can be used for mulching on the soil surface or can be mixed with the soil. However it has mainly physical effects on the soil improvement, as it decomposes very slowly and because it is able to absorb water. But it has no fertilizing effects in this state. Bark decomposition takes much time, due to the unfavourable proportion of carbon to nitrogen in the bark. However, a new method of preparing a fertilizer from bark was invented by an Austrian biologist. Employing the hot composting of bark and with the aid of certain bacteria, the bark is transformed into manure within 8 weeks. Composting is additionally accelerated by a high water content of the bark (70–75 %) and by employing small bark particles.

Bark is used in mulching and soil improvement. Bark finds its most attractive low-grade outlets as a soil conditioner or mulch. The decomposition rate of bark is considerably lower than that of wood, hence it will last longer than mulch and will have a lower nitrogen consumption when incorporated into the soil. The pH values of some North American barks range between 3.5 and 8 (HARKIN & ROWE 1971). Bark lends body to sandy or silty soils, yet losens up clay soils, it can improve the tilth, structure and aeration of heavy soils and increases water absorption and penetration. It has a high ion-exchange capacity and, contains all the nutrients necessary for a good organic soil, with the exeption of nitrogen.

As mulch, it conserves moisture through weed control and reduces evaporation; it maintains uniformity of the soil and improves granulation of surface soils; it reduces top soil erosion and builds up organic matter and humus in soil with concurrent benefits to the soil microflora. Bark does not attract and does not provide nutrients for termites; this can be a major advantage in tropical South American countries. Bark tannins will form chelates with heavy metal cations and help retain important minerals in the soil; they also complex with soil nitrogen compounds and prevent their rapid breakdown or elution.

Beneficial effects in depressing soil-borne plant diseases have also been observed (possibly through phenolic compounds, lignins, flavonoids, vanillin, fatty acids etc.)

Inhibitory effects of bark on germination and early growth of other plant species have occasionally been observed. The effects that plant extracts may have on the soil microflora are difficult to control. Most plant extracts are rapidly decomposed in soil, and toxic effects are seldom observed beyond a short time after application.

Composting of wet bark obtained from hydraulic debarking is most attractive.

Both, hard wood and softwood barks can be mixed with soil to give media for packing root balls of young trees, ornamental shrubs etc.

Bark may also be applied as a potting medium for container stock and for rooting plant cuttings.

Many farmers now use bark as a substantial portion of their feedstock.

Hardwood bark can also successfully be used for animal bedding and as litter for poultry.

Bark as insulating material

Bark can be used as insulating material against frost. Railway platforms (embankments) are exposed to freezing in Scandinavian countries, Canada and other countries with winter frost. This can be avoided by the employment of compressed bark. The bark has an insulating effect and, at the same time, prevents the freezing of the soil by water absorption. Bark as an insulating material in the soil prevents frost-heaving of railroad tracks. Bark can also be used as an insulating material in horticulture.

Bark is widely used in the production of insulating boards that have various applications, for instance in construction. Insulation boards can even be made from unbarked wood, a method which considerably lowers the production costs.

Bark as indicator of air pollution

More recently, bark has been used as an indicator of air pollution, mainly with $S0_2$ emanating from industrial buildings. To study the degree of acidity emmission, aged trees with sufficiently thick barks are used. The pH value of the bark is the indicator of air pollution through acidity emmission. Comparing the average values of bark in the center of a large city with those in its periphery, an increase of the pH value, i.e. a decrease of the acidity content of bark is evident with increasing distance from the city. Indeed, each tree species has its particular pH value of bark, so that only comparisons of the pH values of one and the same species can give reliable results.

Simultaneously, the sulphur content in the bark decreases from the city center towards the periphery. Additionally, the sulphur content decreases from the bark surface (0–3 mm) towards the deeper lying parts (3–9 mm). The issued sulphur is absorbed by the bark surface. There is a linear interdependence between sulphur content and pH value. The pH value of the studied barks fluctuates between 2.7 and 4.35, the sulphur content varies between 0 and 250 $\mu g/cm^2$ (VOLZ 1971).

PARAMESWARAN (1974) found a weak acidity in tropical tree barks. pH values are generally lower in barks than in the corresponding wood. Barks of hardwood trees are less acid than those of softwood trees. The inner phloem (inner bark) has a slightly higher pH than the rhytidome of softwood trees. Buffering capacity of barks is greater than that of the respective woods. pH and buffering capacity are important for the manufacture of wood particle boards. The same data are necessary for the production of boards from whole tree barks. When bark is used as soil conditioner or mulch, its pH may affect the ion exchange capacity with the soil.

Bark may even be used for protection from water pollution. A Swedish paper plant invented a new method to clean polluted water from oil, phenols and similar substances. A bark powder absorbs the oil like blotting-paper, giving rise to lumps which can be burned off.

Plants (barks) which supply resin

A very large number of plant species contains gums, resins and gumresins.

NAIR (1995) cites almost 200 such plants mainly of tropical origin. Resins are widely distributed in the plant kingdom and several plant families are particularly known for resin production, such as Pinaceae, Leguminosae, Guttiferae, Burseraceae,

Anacardiaceae etc. Resins, gumresins and gums may be present in all organs of the plant, but occur in large quantities mainly in the bark of trees.

Hymenaea courbaril, Leguminosae, native of Venezuelan Guiana, exudes a resin of yellow to orange colour, which dries on the soil and hardens. It is called copal and can be used for laquers and for mastic in flooring. Furthermore, the tannin content of the bark of this species can be used commercially.

Symphonia globulifera, Guttiferae, has a bark which exudes a yellow resin. Drying in the air it assumes a red colour. The resin is used by the natives of Venezuelan Guiana.

The reader will find in this book many little-known examples of resinous barks.

Plants (barks) which supply caoutchouc

Quite a few plant families contain latex, a milky sap, mainly in their bark. *Hevea brasiliensis*, Euphorbiaceae, is the most famous example. Other plant species that supply caoutchouc are *Manihot glaziovii*, Euphorbiaceae, *Ficus elastica*, Moraceae, *Kickxia elastica*, Scrophulariaceae and many more species.

Caoutchouc is the polymerization product of the unsaturated hydrocarbon Isopren. It was an important product in the past, but is now replaced by synthetics. However, latex contains many other substances besides caoutchouc, e.g. resins, minerals, proteins, starch.

Hura crepitans, Euphorbiaceae, native of Venezuela, contains a toxic milk sap, which is used medicinally and as a fish poison.The Latex is often used as a remedy for man and animals by the natives.

Other uses of bark

Bark is used in field drainings. Bark is known as filling material, additive and extender. Bark is applied in herbicides or in a combination of fertilizer and herbicide. Bark extracts are components of adhesives. Extracts of certain barks can be utilized as cheap dyes for the textile industry, as well as for the production of oxalic acid. Even substitutes of cork can be prepared from bark. The pulverized fraction of the bark which is rich in extractive substances may be applied as a dispersion, emulsion or floatation agent. Charcoal production and extraction of cellulose are other possibilities of bark utilization.

The use of bark in the manufacture of drugs from bark extracts is very important. Analysis of the chemical substances present in bark extracts is, however, difficult and time consuming, due to the presence of many distinct substances, particularly in tropical barks. Nevertheless, bark contains a large variety of valuable products which can be used medicinally. This field of bioactive substances is now expanding rapidly.

Medicinal plants of selected regions

Venezuela

VELE, MILANO, FERNANDEZ, WILLIAMS & MICHELANGELI (1999) elaborated a list of medicinal plant species known in Venezuelan Ethnobotany. 1384 species belonging to 175 families have been found. To date about 626, 700 plant species were collected in Venezuela, so that the percentage of medicinal plants in the Venezuelan flora seems to be very low. It is however probable that a great number of medicinal plants used in former times have already been forgotten, due to the loss of tradition. The authors composed a list of the families in which these medicinal plants occur (see Table 1).

The authors also presented a diagram of the families in which most medical plants are found, (Diagram 1). About 40% of all medicinal plants registered in Venezuela are concentrated in these families. 51 of these species are the most popular ones in Venezuela.

A further diagram (Diagram 2) shows which diseases are most frequently cured by the medicinal plants.

In the regions where most natives live, such, as Guiana, the Orinoco Delta and the northeastern coastal region, the highest percentages of utilization of medicinal plants occur. Even in regions where a dense rural population lives, e.g. in the 'Llanos', the percentage of medicinal plants used by the people is high. Urban people however, do not use many medicinal plants. These circumstances are ascribed to the social rural conditions or to the conditions in which the indigenous, i.e. Indian tribes, live. The frequent use of medicinal plants is either due to tradition or to the poverty of people or to the difficult access to modern medicine. Indian tribes, for example, live too far away from any civilization.

The authors give a survey of the frequency percentage of plant parts preferably used in indigenous medicine. In general, they observed the following gradation: leaf, root, fruit, bark, flower, entire plant, seed, branches, latex, resin and sap.

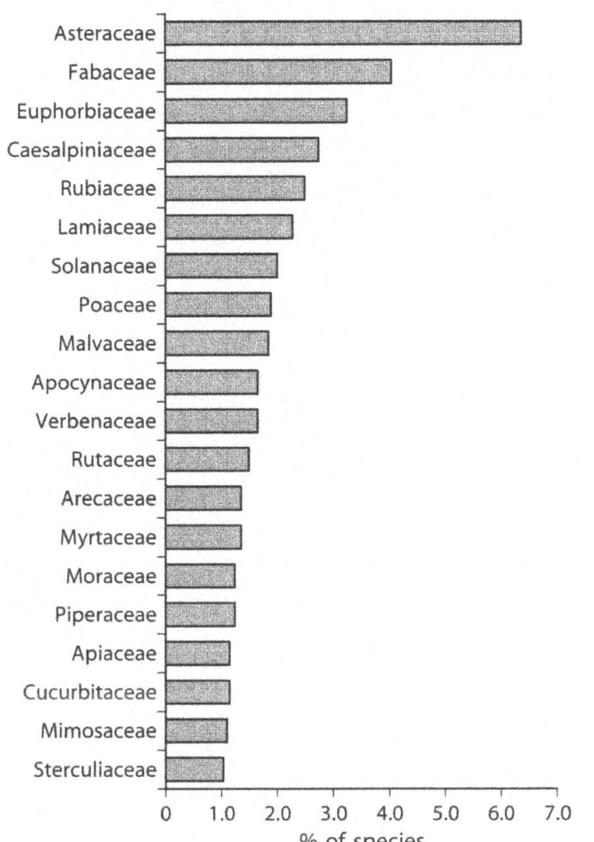

Diagram 1. Plant families in which medicinal plants are found.

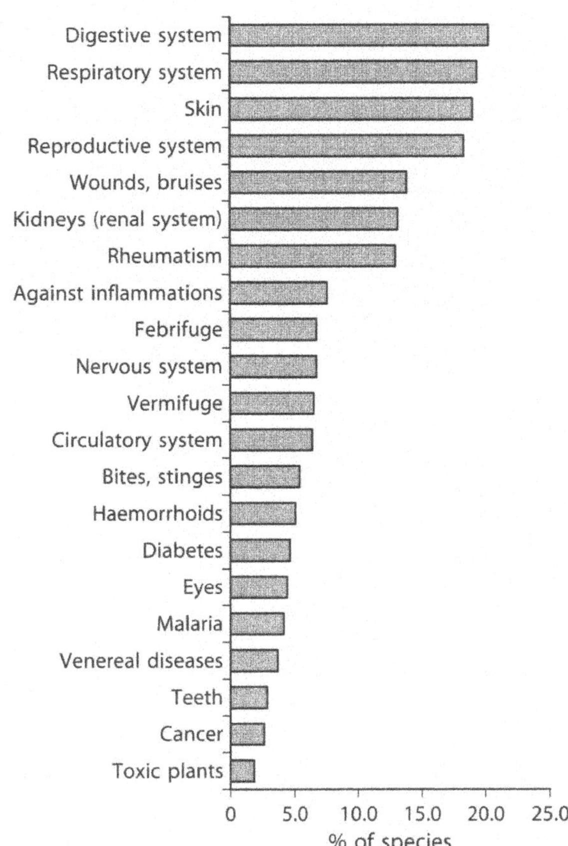

Diagram 2. Diseases cured by medicinal plants.

Table 1. List of medicinal plant species mentioned in Venezuelan ethnobotany

Family	No. of species
Acanthaceae	11
Aizoaceae	1
Alismataceae	1
Amaranthaceae	16
Amaryllidaceae	10
Ancardiaceae	7
Annonaceae	14
Apiaceae	20
Apocynaceae	29
Aquifoliaceae	2
Araceae	15
Araliaceae	3
Arecaceae	24
Aristolochiaceae	9
Asclepiadaceae	8
Asteraceae	113
Avicenniaceae	1
Balsaminaceae	1
Batidaceae	1
Begoniaceae	2
Bignoniaceae	12
Bixacae	1
Bombacaceae	3
Boraginaceae	17
Brassicaceae	10
Bromeliaceae	10
Burseraceae	16
Cactaceae	10
Caesalpiniaceae	48
Campanulaceae	3
Cannabinaceae	2
Cannaceae	2
Capparidaceae	7
Caprifoliaceae	4
Caricaceae	2
Caryocaraceae	2
Caryophyllaceae	2
Cecropiaceae	2
Celastraceae	3
Chenopodiaceae	4
Chloranthaceae	4
Chrysobalanaceae	3
Clethraceae	1
Clusiaceae	13
Cochlospermaceae	4
Combretaceae	6
Commelinaceae	10
Connaraceae	1
Convulvulaceae	7
Coriariaceae	1
Crassulaceae	5
Cucurbitaceae	20
Cupressaceae	2
Cycadaceae	1
Cyperaceae	10
Cyrillaceae	1
Dennstaedtiaceae	1
Dilleniaceae	5
Dioscoreaceae	2
Droseraceae	1
Elaeocarpaceae	1
Equisetaceae	4
Ericaceae	7
Eriocaulaceae	1
Erythroxylaceae	2
Euphorbiaceae	57
Fabaceae	72
Ficoidaceae	1
Flacourtiaceae	6
Gentianaceae	3
Geraniaceae	7
Gesneriaceae	3
Gnetaceae	1
Haemodoraceae	2
Hamamelidaceae	1
Hernandiaceae	1
Hippocastanaceae	1
Humiriaceae	2
Hydrophyllaceae	2
Hymenophyllaceae	1
Hypoxidaceae	1
Ilicineae	1
Iridaceae	5
Juglandaceae	1
Krameriaceae	2
Lamiaceae	39
Lauraceae	12
Lecythidaceae	7
Lemnaceae	1
Liliaceae	7
Linaceae	1
Loganiaceae	5
Loranthaceae	5
Lycopodiaceae	3
Lythraceae	2
Malphigiaceae	8
Malvaceae	32
Maranthaceae	2
Marcgraviaceae	1
Melastomataceae	6
Meliaceae	7
Menispermaceae	4
Mimosaceae	19
Monimiaceae	2
Moraceae	22
Moringaceae	1
Musaceae	4
Myristicaceae	10
Myrtaceae	24
Nyctaginaceae	8
Ochnaceae	1
Olacaceae	3
Oleaceae	4
Onagraceae	7
Ophioglossaceae	1
Orchidaceae	5
Oxalidaceae	3
Papaveraceae	4
Passifloraceae	7
Pedaliaceae	2
Petiveriaceae	1
Phytolaccaceae	5
Pinaceae	1
Piperaceae	22
Plantaginaceae	3
Plumbaginaceae	1
Poaceae	33
Polygalaceae	5
Polygonaceae	14
Polypodiaceae	14
Portulacaceae	3
Proteaceae	1
Pteridaceae	2
Punicaceae	1
Quiinaceae	1
Ranunculaceae	1
Rhamnaceae	8
Rhizophoraceae	1

Table 1. (Continued) List of medicinal plant species mentioned in Venezuelan ethnobotany

Family	No. of species	Family	No. of species
Rosaceae	17	Theophrastaceae	3
Rubiaceae	43	Tiliaceae	7
Rutaceae	26	Tropaeolacaeae	1
Salicaceae	1	Turneraceae	1
Sapindaceae	12	Typhaceae	1
Sapotaceae	5	Umbelliferae	1
Schizaeaceae	1	Urticaceae	8
Scrophulariaceae	5	Valerianaceae	2
Selaginellaceae	2	Verbenaceae	29
Simaroubaceae	5	Violaceae	2
Smilacaceae	5	Vitaceae	5
Solanaceae	35	Vochysiaceae	2
Sphenocleaceae	1	Winteraceae	1
Sterculiaceae	18	Zingiberaceae	9
Strelitziaceae	1	Zygophyllaceae	5

Northern Amazonas

CASTILLO SUAREZ (1999) studied the medicinal utilization of plant species of the northern Amazonas. The region is dominated by the rivers Cataniapo, Cuao, Sipapo and the middle Orinoco, and is covered by forest communities. It is located in the Departments Atures, Edo. Amazonas, Venezuela. It extends over a length of 120 km. The following plant families and species were studied.

Anacardiaceae: *Tapirira guianensis* AUBLET. The pericarp of the fruit is used to cauterize wounds on the sole of the foot and the caustic sap of the fruit is applied to remove warts.

Asteraceae: *Acanthospermum australe* (LOEFL) KUNTZE. The entire plant in decoction is used in the form of a cataplasm for cancer and ulcers.

Bignoniaceae: *Pleonotoma jasminifolia* (KUNTH) MIERS. The crushed roots are applied as an emetic in the case of biliar disorders.

Boraginaceae: *Cordia nodosa* LAM. The crushed and macerated bark prepared as a drink is used to treat snake bites.

Burseraceae: *Protium heptaphyllum* (AUBL) MARCHAND. The resin of the bark is used to treat stomachic diseases.

Caesalpiniaceae: *Campsiandra guayanensis* STERGIOS.The bark is utilized to control diarrhoea and vomiting. The bark of *Swartzia argentea* SPRUCE EX BENTH, scraped and put on the bite of a ray to alliviate the pain. *Tachigalia chrysophylla.* (POEPP.) ZARUCHI & HEREND. The scraped, hot bark in the form of a cataplasm is used for Leishmaniasis.

Celastraceae: *Maytenus laevis* REISS. The bark in decoction serves to cure rheumatism and arthritis. The scraped bark soaked in water is taken to control diarrhoea, stomach aches and general discomfort.

Clusiaceae: *Caraipa densifolia* MARTIUS. The dry bark is boiled in water for 5 minutes and applied on the skin for lice and scabs. *Clusia amazonica* PL. & TR. ROOT and bark in the form of a bath are applied in cases of leprosy. *Clusia renggerioides* PLANCH & TRIANA. The bark is applied to set fractured bones. *Vismia macrophylla* KUNTH. The boiled bark is applied in the treatment of ringworm and of white spots on the skin produced by fungi.

Cucurbitaceae: *Gurania spinulosa* (POEPPIG & ENDL.) COGN. The plant is used against stomach ache.

Euphorbiaceae: *Alchornea castaneifolia* (WILLD.) JUSS. The scraped bark in the form of a cataplasm is put on bites of a ray. *Alchornea schomburgkii* KLOTZSCH. The bark in decoction is used as a postmenstrual tonic.

Fabaceae: *Centrosema pubescens* BENTH. The roots are an emenagogue, they activate menstruation, promote uterine contractions, facilitate discharge from the vagina after childbirth or abortion and are also used for vaginal douches. Medicines are prepared with the flowers and roots for obstructions of the liver and the spleen. The seeds act as purgatives and are effective against worms. *Mucuma urens* (L.) DC. The hairs of the fruit in syrup or honey are purgative. The fruits without spines are prepared in an infusion to cure oedema. A decoction of the roots is used in cases of cholera. The aereal parts of the plant in infusion are applied for a hipbath and extracts taken in drops are effective against haemorrhoids. The seeds, rich in tannins, slime, resinous and pectic substances, in infusion act as a diuretic and against haemorrhoids. *Pterocarpus santalinoides* L'HER EX DC. The resin is used as an antidiarrhoeic, antihaemorrhagic and for cicatrizing. An infusion of the bark and the red resin is applied for the treatment of amygdalitis and to strengthen the gums.

Hippocrateaceae: *Hippocratea volubilis* L. An emulsion from the seeds which is useful in inflammation of the breast and the larynx is prepared.

Lecythidaceae: *Allantoma lineata* (MART. EX BERG) MIERS. The seeds are hallucinogenic.

Menispermaceae: *Abuta grandifolia* (MARTIUS) SANDW. The leaves in infusion act as an antipyretic.

Mimosaceae: *Enterolobium schomburgkii* BENTH. The plant is used to control pediculosis. *Inga thibaudiana* DC. The hot bark is used against pains caused by the bite of bagrewels.

Moraceae: *Brosimum utile* (KUNTH.) PITTIER SUBSP. *ovatifolium* (DUCKE) BERG. The latex cures asthma, some pulmonary diseases and bronchitis.

Myristicaceae: *Compsoneura sprucei* (DC.) WARB. The bark alleviates buccal ulcers.

Myrtaceae: *Myrcia citrifolia* (AUBLET) URBAN. The leaves in decoction have antiinflammatory and diuretic effects and are used in the treatment of diabetes, renal disturbances, haemorrhoids and inflammations of the uterus and the ovaries.

Olacaceae: *Minquartia guianensis* AUBLET. The bark is anthelmintic and a vermifuge.

Rubiaceae: *Psychotria poeppigiana* MÜLL.-ARG. The plant is useful as a remedy for ocular diseases.

Sapotaceae: *Pouteria caimito* (RUIZ & PAVON) RADLKOFER. The fruits are astringent.

Viscaceae: *Phoradendron piperoides* (KUNTH.) Trelease. Leaves and shoots in decoction are used for female diseases and menstruation.

Vitaceae: *Cissus erosa* RICH. The roots are used to cure oedema.

Cuyagua and Cata (Aragua)

Medicinal plants in Cuyagua and Cata (Edo. Aragua), Venezuela, were studied by SERRA GAY (1999). She gives a list of the commonly used plants in this region, also including species which are not indigenous to Venezuela. She counted 92 species altogether. Most of the plants are indicated below.

Acanthaceae: *Hygrophylla* sp., good for the kidneys. *Justicia pectoralis* JACQ., for the diseases of the breast and for flu. *Justicia lindaviana* LEONARD, good for blood circulation. *Justicia* sp to expel worms in dogs. *Ruellia tuberosa* L. for blood pressure, to cure bruises and coughs.

Amaranthaceae: *Achyranthes indica* MILL. for flu. *Gomphrena globulosa* L., for the nerves and the heart. *Iresine angustifolia* EUPHR., for haemorrhagia.

Apocynaceae: *Rauvolfia tetraphylla* L. for eruptions and for toothache and tumors of the gums.

Asteraceae: *Ageratum conyzoides* L. for eruptions in the vicinity of the eyes and for hordeolum. *Clibadium surinamense*, for scabies in dogs. *Helenium tenuifolium* NUTT: Antispasmodic and to cure stomach problems. *Mikania guaco* HUMB. & BONPL. for the stomach and against bad luck. *Parthenium hysterophorus* L. for diabetes. *Pluchea carolinensis* (JACQ.) SWEET for headache, spasms and to conserve youthfulness. *Synedrella nodiflora* (L.) GAERT. for fever and inflammations. *Tithonia diversifolia* (HEMSL.) GRAY, for bruises.

Begoniaceae: *Begonia humilis* AIT., for fever.

Bixaceae: *Bixa orellana* L. against hepatitis.

Boraginaceae: *Heliotropium indicium* L. for haemorrhagia of the uterus.

Capparidaceae: *Cleome gynandra* L., to clean the uterus.

Caprifoliaceae: *Sambucus mexicana* PRESL., for flu.

Caryophyllaceae: *Drymaria cordata* (L.,) WILLD., for varices.

Chenopodiaceae: *Chenopodium ambrosioides* L., against worms.

Commelinaceae: *Rhoeo discolor* (L'HER) HANCE EX WALP, for the kidneys.

Cucurbitaceae: *Momordica charantia* L. for haemorrhoids.

Euphorbiaceae: *Chamaesyce hirta* (L.) MILL., for aphtous ulcer in the mouth. *Chamaesyce cf. thymifolia* (L.) MILL., for the breast and the blood. *Jatropha gossypifolia* L. for bruises. *Phyllanthus niruri* L. for stones in the kidneys.

Fabaceae: *Brownea coccinea* JACQ. for haemorrhage in women. *Brownea grandiceps* JACQ., for haemorrhage of the uterus. *Cassia occidentalis* L., for the breast.

Haemodoraceae: *Xiphidium coeruleum* AUBL. for the ovaries, for the purification of the liver.

Lamiaceae: *Coleus amboinicus* LOUR., for the kidneys. *Hyptis* sp. for prayers for the health and incenses. *Ocimum* aff. *campechianum* MILL., for stomach ache. *Ocimum campechianum* MILL. for intestinal gases. *Ocimum* SP., for the nerves, fever, bruises and for the stomach. *Scutellaria purpurascens* SW. for good luck.

Loganiaceae: *Spigelia anthelmia* L., against worms.

Malvaceae: *Gossypium* cf. *hirsutum* for cotton. *Gossypium* aff. *purpurascens* POIR. for the thyroid gland. *Malachra alceifolia* JACQ. as a purgative for children. *Pavonia fruticosa* FAWCETT & RENDL. for the kidneys. *Urena sinuata* L. for the kidneys.

Myrtaceae: *Pimenta* aff. *Racemosa*, for brainwashing.

Oxalidaceae: *Oxalis* aff. *corniculata* L., for fever.

Phytolaccaceae: 'Tigua' (only vernacular name known) against blindness.

Piperaceae: *Peperomia pellucida* (L.) H.B.K. for the kidneys, the blood, zabañones (translation not known) and fungal infections of the skin. *Peperomia* sp. for bruises. *Potomorphe peltata* (L.) MIQ. for headache.

Poaceae: *Vetiveria zizanioides* (L.) NASH., against diarrhoea.

Rubiaceae: *Spermacoce cf. confusa,* used in witchcraft.

Scrophulariaceae: *Capraria biflora* L., for blood pressure and for bruises.

Sterculiaceae: *Guazuma ulmifolia* LAM., for pain in the kindneys and for the blood. *Melochia tomentosa* L. against fever and flu.

Verbenaceae: *Lippia micromera* SCHAU. to purify the ovaries.

Zingiberaceae: *Alpinia speciosa* SCHUM. for flu. *Costus* aff. *scabber* R. & P., for the kidneys. *Costus guanaiensis* RUSBY, for the kidneys and the liver.

Only those species which are not indigenous and are already very well known have been left out. Within the 92 species studied, the Asteraceae, Euphorbiaceae, Lamiaceae, Acanthaceae, Piperaceae, Amaranthaceae, Fabaceae and Zingiberaceae predominate. The remedies which predominate are for snake bites, for the kidneys, bruises, purification of the blood, flu and fever, and stomach disorders. Species used in superstition and/or ritual are also included.

Yutajé

FERNANDEZ DEL VALLE et al. (1999) found 145 plant species with medicinal properties pertaining to 72 plant families in the region of Yutajé (northern part of the Amazonas State, Municip. Manapiare). The plant families which show the highest numbers of medicinal plants are: Rubiaceae, Piperaceae, Asteraceae, Fabaceae, Mimosaceae, Caesalpiniaceae, Apocynaceae and Poaceae.

The ethnic groups who live in this area and use the medicinal plants are: the Piaroa, Hoti, Guahibo, Yabarana and Ye'kwana.

The plant species with medicinal properties are listed in Table 2.

MUNOZ et al. (1999) studied 24 medicinal plants belonging to 18 different families in Yutajé, State of Amazonas. The region lies in the 0rinoco/Amazonas river basin and is distinguished by the immense variety of habitats including Tepuyes, savannas, tropophyllous and ombrophilous forests and extense Morichales. Up to date, 230 plant families represented by 1.786 genera and 9.411 species, 22 % of which are endemic, were found in the Orinoco/ Amazonas basin. However, only 1 % of these plants has been investigated with reference to their secondary compounds. Due to the high incidence of diarrhoea in the population, the authors studied the antienterobacterial effects of the plants. *Salmonella* sp. was the one most resistant to plant extracts. The foliar extract of *Eschweilera* sp., on the other hand, showed the greatest activity against enterobacteria. The following species showed antienterobacterial activities in extracts:

Campsiandra implexicaulis, Clidemia hirta, Clusia sp., *Geonoma deversa, Gnetum nodiflorum, Helicostylis scabra, Jacaranda copaia, Piper hostmannianum, Protium heptaphyllum, Swartzia* sp., *Uncaria guianensis, Tococa guianensis, Phenakospermum guyanense, Costus spiralis, Curatella americana,* and *Eschweilera* sp.

The authors mentioned some of. the species studied with their applications (according to the bibliography).

***Campsiandra implexicaulis*, Caesalpiniaceae:** for diarrhoea and as a food.

***Clidemia hirta*, Melastomaceae:** for food, ulcer due to leishmaniasis.

***Tococa guianensis*, Melastomaceae:** for insect bites.

***Clusia* sp., Clusiaceae:** for muscular distensions and fractures.

***Vismia baccifera*, Clusiaceae:** for cutaneous fungi.

***Geonoma* sp., Arecaceae:** as perfume.

***Gnetum nodiflorum*, Gnetaceae:** as a food, for inflammations due to muscular distensions.

***Curatella americana*, Dilleniaceae:** for herpes and inflammations of the throat.

***Helicostylis scabra*, Moraceae:** for intestinal parasites and as a fungicide.

***Jacaranda copaia*, Bignoniaceae:** for colds and pneumonia, infections of the skin; it is cicatrizing and anticancerous.

***Eschweilera* sp., Lecythidaceae:** against parasites and as a fungicide.

***Mimosa pudica*, Mimosaceae:** relaxing, soporific, anticholeric, anthelmintic.

***Piper hostmannianum*, Piperaceae:** for gonorrhoea and amoebas.

***Protium heptaphyllum*, Burseraceae:** for colds, cercaricidal.

***Siparuna guianensis*, Monimiaceae:** for indigestion.

***Phenakospermum guyanense*, Strelitziaceae:** for snake bites, cicatrizing, burns, for food.

***Swartzia* sp., Caesalpiniaceae:** medicinal.

***Carica papaya*, Caricaceae:** abortive, anticonceptive, for intestinal parasites, for food.

***Costus spiralis*, Costaceae:** for food.

***Aristolochia disticha*, Aristolochiaceae:** for diarrhoea, contraceptive

***Uncaria guianensis*, Rubiaceae:** for dysentery.

The crude extracts were prepared with fresh plant material which was macerated using ethanol 70%, then filtered and evaporated at 40 °C. Extracts were made from leaves (64%), shoot (11%), bark (8%), fruit (6%), flower (8%) and seed (3%).

The following plant species clearly showed cytotoxic activities: *Eschweilera* sp., *Uncaria guianensis*, *Siparuna guianensis*, *Mimosa pudica*, *Jacaranda copaia*, *Helicostylis scabra*, *Clusia* sp.

Table 2. List of medicinal plants mentioned in the Venezuelan bibliography and found in Yutajé.

Species	Family	Uses (described in the literature)	Uses in Yutajé
Abelmoschus moschatus	Malvaceae	Antidote to poison, antispasmodic, against snakes	
Acantospermum australe	Asteraceae	Emenagogue	
Achyranthes aspera	Amaranthaceae	Antiseptic, haemostatic, against flu, antiinflammatory	
Amarnanthus dubius	Amaranthaceae	Febrifuge, vermifuge, antidiarrhoeic	
Amaranthus viridis	Amaranthaceae	Anthelmintic, febrifuge, pectoral	
Annona muricata	Annonaceae	Antihypertensive, emetic, sedative	
Aphelandra scabra	Acanthaceae	Antiasthmatic	
Astrocaryum gynacanthum	Arecaceae (Palmae)	Vermifuge	
Bauhinia guianensis	Caesalpiniaceae	Astringent, depurative, against syphilis	Against malaria
Bidens cynapiifolia	Asteraceae	Antiseptic, fungicid	
Bidens pilosa	Asteraceae	Emenagogue, vulnerary, antidiabetic	
Boerhaavia diffusa	Nyctaginaceae	Antihaemorrhagic, abortive	
Borreria capitata	Rubiaceae	Digestive	Herpes in the mouth (aphtae)
Borreria verticillata	Rubiaceae	Against flu	
Bowdichia virgilioides	Fabaceae	Antihaemorrhagic, astringent, depurative	
Bromelia plumieri	Bromeliaceae	Diuretic, febrifuge, laxative, vermifuge	
Brownea coccinea	Caesalpiniaceae	Analgesic, haemostatic	Contraceptive
Byrsonima crassifolia	Malpighiaceae	Antiinflammatory, diuretic, febrifuge, dilating the bronchi (against bronchostenosis)	
Byttneria scabra	Sterculiaceae	Antidote, desinfectant, for the genitourinary tract	
Caladium bicolor	Araceae	Antirheumatic	For snake bites
Cardiospermum halicacabum	Sapindaceae	Antiinflammatory, febrifuge, insect repellent	
Carica papaya	Caricaceae	Anthelmintic, abortive, digestive	
Caryocar glabrum	Caryocaraceae	Emenagogue, fish poison, vesicant	
Caryocar microcarpum	Caryocaraceae	Fish poison	Fish poison
Cassia moschata	Caesalpiniaceae	Anthelmintic, against venereal diseases, astringent	
Cecropia peltata	Moraceae	Antidiabetic, cicatrizing, vulnerary	
Cipura paludosa	Iridaceae	Antidiarrhoeic, against venereal diseases	

Table 2. (Continued) List of medicinal plants mentioned in the Venezuelan bibliography and found in Yutajé.

Species	Family	Uses (described in the literature)	Uses in Yutajé
Cissampelos ovalifolia	Menispermaceae	Antiinflammatory, abortive, expectorant	
Cissus erosa	Vitaceae	Against malaria	
Clusia microstemon	Clusaiaceae/Guttiferae	Analgesic	
Cochlospermum orinocense	Bixaceae	Febrifuge, cicatrizing	Against malaria
Combretum fruticosum	Combretaceae	Depurative	
Conceveiba guianensis	Euphorbiaceae	Fertility stimulating	
Connarus ruber	Connaraceae	Vulnerary	
Cordia nodosa	Boraginaceae	Against parasites of the skin	
Costus arabicus	Costoideae/Zingiberaceae	Antidiabetic, diuretic	Contraceptive
Costus scaber	Costoideae/Zingiberaceae	Antidiabetic, antiinflammatory, antiseptic	Edible arillus, eaten by children
Costus spiralis	Costoideae/Zingiberaceae	Abortive, styptic	Contraceptive
Couma macrocarpa	Apocynaceae	Purgative	
Coutoubea ramosa	Gentianaceae	Toxic	
Crotalaria incana	Fabaceae	Febrifuge, vulnerary	
Croton cuneatus	Euphorbiaceae	Skin rejuventating	
Croton trinitatis	Euphorbiaceae	Antidiarrhoeic, antiinflammatory, sedative	
Cupania scrobiculata	Sapindaceae	Febrifuge, vomitive	
Curatella americana	Dilleniaceae	Antidiabetic, antidiarrhoeic, astringent, generates hypotension	For asthma
Cyperus articulatus	Cyperaceae	Against hypertension, diuretic, emenagogue	
Cyperus luzulae	Cyperaceae	Emenagogue	
Cyrilla racemiflora	Cyrillaceae		
Dalbergia hygrophilia	Fabaceae	Against ulcers	
Davilla kunthii	Dilleniaceae	Antidiarrhoeic	
Desmodium adscendens	Fabaceae	Against aches and impotence	
Duroia genipoides	Rubiaceae	Against aches, toxic	
Elephantopus mollis	Asteraceae	Medicinal, astringent	
Euceraea nitida	Flaccourtiaceae	Toxic	
Galactia jussiaeana	Fabaceae	Antidiarrhoeic, antispasmodic, against malaria	
Gnetum nodiflorum	Gnetaceae	Antiinflammatory	
Goupia glabra	Celastraceae	To treat cataract	
Guazuma ulmifolia	Sterculiaceae	Analgesic, astringent, diuretic, against syphillis	
Gurania spinulosa	Cucurbitaceae	Emenagogue, vulnerary	Food
Hamelia patens	Rubiaceae	Diuretic, vermifuge	
Heliconia psittacorum	Heliconieae/Musaceae	Against ulcers of the scalp	
Helicostylis scabra	Moraceae	Fungicide, vermifuge	
Helicteres guazumaefolia	Sterculiaceae	Contraceptive, haemostatic	
Heliotropium indicum	Boraginaceae	Abortive, stomachic, sudorific	
Himatanthus bracteatus	Apocynaceae	Antidote, febrifuge	
Hirtella racemosa	Chrysobalanaceae	Against ear and throat ache	
Humiria balsamifera	Humiriaceae	Vulnerary	
Hymenachne amplexicaulis	Poaceae	Antidote to poison	
Hyptis capitata	Lamiaceae/Labiatae	Antiinflammatory, cicatrizing, antiseptic	
Ichthyothere termimalis	Asteraceae	Against venereal diseases, sedative, galactogogue	
Inga edulis	Fabaceae	Antidiarrhoeic	Medicinal, against snakes
Inga pilosula	Mimosaceae	Ophthalmic	
Inga spuria	Mimosaceae	Against dermatitis	
Ipomoea carnea	Convolvulaceae	Antirheumatic, purgative	
Jacaranda copaia	Bignoniaceae	Vulnerary, cicatrizing, against infections	
Jacaranda obtusifolia	Bignoniaceaea	Depurative, desinfectant, vulnerary	
Justicia pectoralis	Acanthaceae	Aphrodisiac, cicatrizing, pectoral	
Laetia procera	Flacourtiaceae	Medicinal	
Lantana camara	Lamiaceae	Febrifuge, mucolytic, against leprosy, against measles	
Licania heteromorpha	Chrysobalanaceae	Against toothache, purgative	Expectorant
Luffa cylindrica	Cucurbitaceae	Antiinflammatory, antirheumatic, purgative	
Lycopodiella cernua	Lycopodiaceae	Against constipation, fever	

Table 2. (Continued) List of medicinal plants mentioned in the Venezuelan bibliography and found in Yutajé.

Species	Family	Uses (described in the literature)	Uses in Yutajé
Lygodium venustum	Schizaeaceae/ferns	Fungicide	
Macoubea guianinsis	Apocynaceae	Pectoral	
Malvabiscus arboreus	Malvaceae	Emollient	
Mandevilla scabra	Apocynaceae	Depilatory, to treat spines sticking to the skin	
Maranta arundinacea	Marantaceae	Analgesic, cholagogue, antidote to poisons	
Miconia tomentosa	Melastomaceae		
Mimosa pudica	Mimosaceae	Emetic, sedative, tonic, toxic	
Momordica charantia	Cucurbitaceae	Antidiabetic, against malaria, vulnerary	Against malaria
Oenocarpus bataua	Arecaceae	Antidote, antiinflammatory, toxic	Against asthma
Pachira aquatica	Bombacaceae	Abortive, febrifuge	
Paepalanthus fasciculatus	Eriocaulaceae	Astringent	
Palicourea rigida	Rubiaceae	Antiinflammatory	
Palicourea triphylla	Rubiaceae	Fish poison	
Panicum pilosum	Poaceae	Against flu	
Parkia pendula	Mimosaceae	Sadative, for haemorrhagia	
Paspalum conjugatum	Poaceae	For inflammations of the eye	
Paspalum repens	Poaceae	Abortive	
Pectis elongata	Asteraceae	Carminative, soothing, vulnerary	
Perama galioides	Rubiaceae	Astringent	
Petiveria alliacea	Phytolaccaceae	Anticancerous, antispasmodic	
Phoradendron crassifolium	Loranthaceae	Vulnerary	For fractures
Piper arboreum	Piperaceae	For dysentery	
Piper dilatatum	Piperaceae	Poison	
Piper hispidum	Piperaceae	Diuretic, fish poison	
Piper hostmannianum	Piperaceae	To treat spines sticking to the skin	
Piper marginatum	Piperaceae	Analgesic, digestive, febrifuge	For stings and bites, renal remedy
Piper tuberculatum	Piperaceae	Anthelmintic, emenagogue, insecticide	
Polypodium aureum	Polypodiaceae	Against blenorrhagia and intoxicaton, vermifuge	Medicinal, for beats, knocks and bruises
Portulaca oleracea	Portulaccaceae	Anthelmintic, ascaricide, laxative	
Posoqueria latifolia	Rubiaceae	Insect repellent	
Pothomorphe peltata	Piperaceae	Antiinflammatory, antidiarrhoeic, emetic, febrifuge	
Protium heptaphyllum	Burseraceae	Abortive, cicatrizing, fragrant	Medicinal
Protium unifoliolatum	Burseraceae	Analgesic, decongestant	
Psidium guineense	Myrtaceae	Antidiarrhoeic, astringent	
Psychotria capitata	Rubiaceae	Decongestant	
Psychotria microdon	Rubiaceae	Against earache	
Psychotria poeppigiana	Rubiaceae	Pectoral	Complementory hallucinogen
Psychotia racemosa	Rubiaceae	Toxic, rat poison	
Qualea acuminata	Vochysiaceae	Vermicide	
Ruellia tuberosa	Acanthaceae	Depurative, emetic, vomitive	
Sabicea villosa	Rubiaceae	Against malaria	
Sauvagesia erecta	Ochnaceae	For stomach ache	
Schefflera morototoni	Araliaceae	Stomachic, cardiac tonic	
Scoparia dulcis	Scrophulariaceae	Astringent, diuretic, vomitive, stomachic	
Senna occidentalis	Caesalpiniaceae	Abortive, diuretic, emenagogue	
Sida acuta	Malvaceae	Antiinflammatory, expectorant	
Siparuna guianensis	Monimiaceae	Antiinflammatory, digestive, febrifuge, sedative	Febrifuge, for bites and stings
Smilax maypurensis	Smilacaceae/Liliaceae	Depurative	
Souroubea guianensis	Marcgraviaceae	Sedative	
Strychnos panurensis	Loganiaceae	Medicinal	
Syngonium vellozianum	Araceae		
Tabebuia barbata	Bignoniaceae	Antidote, carminative, cicatrizing	Diuretic
Tapirira guianensis	Anacardiaceae	Againsts measles	
Theobroma grandiflorum	Sterculiaceae	For abdominal aches	
Tococa guianensis	Melastomaceae	For insect bites	
Uncaria guianensis	Rubiaceae	For dysentery	Stimulator of the immune system, antiinflammatory
Vanilla planifolia	Orchideae	Aphrodisiac, stomachic, febrifuge	

Table 2. (Continued) List of medicinal plants mentioned in the Venezuelan bibliography and found in Yutajé.

Species	Family	Uses (described in the literature)	Uses in Yutajé
Virola sebifera	Myristicaceae	Febrifuge, tobacco surrogate	Complementory hallucinogen
Virola surinamemsis	Myristicaceae	Digestive, cicatrizing, vulnerary	
Vismia baccifera	Clusiaceae/Guttiferae	Cicatrizing	
Vitex capitata	Verbenaceae	Against aches, haemostatic, vulnerary	
Vochysia ferruginea	Vochysiaceae	Febrifuge	
Xylopia aromatica	Annonaceae	Against snake bites, calmative, antidiarrhoeic	

Trujillo

A list of plants sold in the 'herbolarios' of the popular markets of the State of Trujillo, Venezuela, and their utilization was elaborated by BERMUDEZ & VELAZQUEZ (1999) (Table 3).

Tachira

An inventory of the medicinal plants of the State of Tachira, Venezuela, was elaborated by VERA & PABON PINTO (1999). Altogether 114 species were identified. The plant families which are best represented are the Compositae with 21 species, the Leguminosae with 7 species, the Euphorbiaceae with 7 species, the Gramineae with 6 species, the Amarantaceae and Labiatae each with 5 species, Pteridophytes, Malvaceae and Verbenaceae with 4 species each; Commelinaceae, Cyperaceae, Cruciferae, Acanthaceae, Solanaceae, Rubiaceae, Loranthaceae, Polygonaceae, Portulaccaceae, Piperaceae, Oxalidaceae, Lythraceae with 2 species each and 19 families with only one species.

Medicinal plants of the State Tachira and the diseases which they cure are indicated in the following tables (Tables 4 and 5).

In Table 5, a list of diseases and of the plants which cure them is presented, using the numbers of Table 4 for indentification

Many of the above named plant species are used for several diseases. 27 % of the species studied have diuretic properties. 60 % of the species are useful for the most frequently occurring diseases.

GONZALEZ DE COLMENARES, VERA & MEZA (1999) studied the aromatic and medicinal plants of the State of Tachira/Venezuela. 33 species belonging to 18 different families (Burseraceae, Cactaceae, Capparidaceae, Cucurbitaceae, Compositae, Dilleniaceae, Gramineae, Guttiferae, Labiatae, Lauraceae, Leguminosae, Loganiaceae, Myrtaceae, Orchidadeae, Rutaceae, Solanaceae, Verbenaceae and Zingiberaceae) and which mainly grow in mountainous regions, have been investigated. The plants are used in popular medicine, as spices, perfumes, insecticides and as ornamentals. Principally, the socioeconomically discriminated population which has no access to orthodox medicine, is dependant on the use of medicinal plants, Nevertheless, a lack of concrete information on the subject is evident in the region. The authors therefore found it necessary to study the ecological demands of the plants, their cultivation possibilities, their profitability, as well as the composition and the applicability of their extracts, such as essential oils and the like. The authors rightly emphasize that Venezuela invests large sums of foreign currency to import essences and extracts of essential oils, necessary in the production of foodstuff, for pharmacology and cosmetology. The continuously growing demand for aromatic plants and their derivatives in Venezuela justifies the increasing interest in cultivation and manufacture of native aromatic plants and their products which may partly be superior to even the imported goods and would not only be much cheaper, but additionally would motivate industrialization in Venezuela.

Some aromatic plants which abundantly occur in Tachira, are listed in the following table (Table 6).

Melaleuca leucadendron, for example, is used in popular medicine as an antibiotic, astrigent, expectorant, haemostatic, digestive, sedative and eupeptic. The leaves and the essence extracted from the leaves are used.

Ocotea barcellensis has antirheumatic, rubefacient and insecticidal properties. The essential oil of the wood, which is the useful substance, is extracted from the stem by perforation; in this way, about 250 ml can be obtained in a drip-catcher within 12 hours. With its insecticidal properties the aromatic oil protects the wood, even against termites.

Table 3. Medicinal plants which are sold at the herbolarios (little stalls) of the popular markets in the State of Trujillo.

Family, scientific name	Local name	Used parts	Traditional use	Way of preparation	Administration	Origin
Acanthaceae						
Justicia secunda Valh	Sanguinaria	Leaves	Blood purification, for weight loss	Decoction	Orally	Wild plant
Annonaceae						
Xylopia aromatica (Lam.) mart.	Manga larga	Fruits	Diabetes	Decoction	Orally	Wild plant
Boraginaceae						
Heliotropum indicum L.	Rabo de alacrán	Shoot, leaves and flowers	To strengthen the legs of children	Decoction	baths	Wild plant
Burseraceae						
Bursera graveolens (H. B. K.) Tr. & Planch.	Sasafrás	Shoot and leaves	Flu	Decoction	Orally	Wild plant
Bursera simaruba (L.) Sarg.	Indio desnudo	Bark of the stem	Arthritis, rheumatism, diabetes	Tincture in brandy, decoction	Orally	Wild plant
Cactaceae						
Opuntia ficus-indica (L.) Mill	Tuna real	Crystals of the shoot	Varices, to enhance blood circulation	Dissolved crystals	Orally	Wild plant
Caprifoliaceae						
Sambucus mexicana Presl.	Saúco	Flowers	Cough	Decoction	Orally	Cultivated
Celastraceae						
Maytenus laevis Reiss	Chuchuguasa	Bark of the stem	Arthritis, rheumatism	Decoction, macerated in wine	Orally	Wild plant
Commelinaceae						
Commelina diffusa Burm.	Suelda con suelda	Shoot and leaves	Diabetes	Decoction	Orally	Wild plant
Compositae						
Achyroline satureioides (Lam.) D. C.	Vira vira	Shoot and leaves	Skin infections, to control menstruation	Decoction	Baths, orally	Wild plant
Ambrosia cumanensis H. B. K.	Altamisa	Shoot, leaves and flowers	Pain in the legs, bruises and injuries. refreshing,	Decoction	Baths	Cultivated
Artemisia absinthium L.	Ajenjo	Shoot and leaves	Stomachache, diabetes, skin infections	Decoction	Orally	Wild plant
Bacharis tricuneata (L. f.) Pers.	Sanalotodo	Shoot and leaves	Diabetes	Crude sap of the leaves, decoction	Orally, cataplasm	Wild plant
Espeletia sp	Frailejón	Leaves	Cough	Decoction in syrup	Orally	Wild plant
Hypochoeris sessiliflora H. B. K. vel. aff.	Chicoria	Root, shoot and leaves	Fever	Decoction in syrup	Orally	Wild plant
Matricaria chamomilla L.	Manzanilla dulce	Flowers	Colic, stomachache	Decoction	Orally	Cultivated
Pluchea odorata (L.) Cass.	Salvia real	Shoot and leaves	For weight loss, laxative	Decoction	Orally	Wild plant
Taraxacum officinale Web.	Diente de león	Leaves and flowers	Diuretic, blood purification	Decoction	Orally	Wild plant
Vernonia brasiliana (L.) Dr.	Estoraque	Shoot and leaves, flowers	Malaria	Decoction	Baths	Wild plant
Crassulaceae						
Kalanchöe pinnata (Lam.) Pers.	Colombiana, Libertad	Shoot and leaves	To evacuate the colon	Decoction	Orally	Wild plant
Cruciferae						
Lepidium virginicum L.	Mastuerzo niño	Shoot and leaves, seeds	Fractures, aches of the bones	Decoction	Orally	Wild plant
Cucurbitaceae						
Momordica charantia L.	Cundeamor	Shoot and leaves	Diabetes, fever	Decoction	Orally	Wild plant
Chenopodiaceae						
Chenopodium ambrosioides L.	Yerba sagrada, Pasote	Shoot and leaves, seeds	Intestinal worms	Decoction	Orally	Wild plant
Equisetaceae						
Equisetum sp.	Cola de caballo	Shoot and leaves	Kidney stones, diuretic	Decoction	Orally	Wild plant
Euphorbiaceae						
Jatropha gossypifolia L.	Túa túa	Leaves	Diabetes	Decoction	Orally	Wild plant
Phyllanthus niruri L.	Flor escondida	Shoot and leaves, flowers	Fever	Decoction	Orally	Wild plant
Geraniaceae						
Pelargonium odoratissimum L.	Aroma	Leaves	Tranquilizer	Decoction	Baths	Cultivated
Pelargonium zonale L.	Geranio, novios	Shoot and leaves	Tranquilizer	Decoction	Baths	Cultivated
Gramineae						
Cymbopogon citratus (D. C.) Stapf.	Malojillo	Leaves	Tranquilizer	Infusion	Orally	Cultivated

Table 3. (Continued) Medicinal plants which are sold at the herbolarios (little stalls) of the popular markets in the State of Trujillo.

Family, scientific name	Local name	Used parts	Traditional use	Way of preparation	Administration	Origin
Labiatae						
Plecatranthus amboinicus Lour.	Orégano orejón	Shoot and leaves	Diuretic	Decoction	Orally	Cultivated
Marrubium vulgare L.	Marrubio	Shoot and leaves, flowers	Diabetes, for weight loss	Decoction	Orally	Wild plant
Mentha piperita var. citrata (Ehrh) Briq.	Yerbabuena	Shoot and leaves	Stomachache, meteorism	Infusion	Orally	Cultivated
Melissa officinalis L.	Toronjil	Shoot and leaves	Insomnia, tranquilizer	Decoction	Orally	Cultivated
Ocimum micranthum Willd.	Albahaca de monte	Shoot and leaves, flowers	Labour pain, stomachache	Decoction	Orally	Wild plant
Origanum majorana L.	Mejorana	Shoot and leaves	Refreshing	Decoction	Orally, baths	Cultivated
Rosmarinus officinalis L.	Romero	Shoot and leaves	Alopecia	Decoction	Lotion	Cultivated
Micromeria brownei (Sw.) Briq.	Poleo	Shoot and leaves, flowers	Cough	Decoction	Orally	Wild plant
Leguminosae						
Bauhinia sp	Pata de vaca	Shoot and leaves	Diabetes	Decoction	Orally	Wild plant
Cassia occidentalis L.	Brusca	Shoot and leaves	Fever	Decoction	Orally	Wild plant
Cercidium praecox (R&P) Harms	Retamo	Shoot and leaves	Scabies	Decoction	Baths	Wild plant
Hymenaea courbaril L.	Algarrobo	Fruits	Reconvalescence	Macerated pulp with milk	Orally	Wild plant
Pterocarpus acapulcensis Rose	Sangre drago	Bark of stem	Tonsillitis	Decoction	Gurgle	Wild plant
Liliaceae						
Aloe vera L.	Abila	Leaves	Cough, asthma	Crystals mixed with honey or sugar	Orally	Cultivated
Smilax sp	Zarzaparrilla	Roots	Blood purification	Decoction	Orally	Wild plant
Malvaceae						
Malva sylvestris L.	Malva	Shoot and leaves, flowers	Skin eruptions, refreshing	Decoction	Baths	Wild plant
Meliaceae						
Melia azederach L.	Aleli	Shoot and leaves	Antibiotic	Decoction	Orally	Cultivated
Moraceae						
Cecropia peltata L.	Yagrumo	Leaves	Swellings, parotitis	Leaves (lukewarm)	Cataplasm	Wild plant
Moringaceae						
Moringa oleifera Lam	Ben	Leaves	Urinary infections, laxative	Decoction	Orally	Cultivated
Musaceae						
Heliconia sp	Bijao	Leaves	Swellings, parotitis	Leaves (lukewarm)	Cataplasm	Wild plant
Myrtaceae						
Eucalyptus globulus Labill.	Eucalipto	Leaves	Respiratiory problems	Decoction	Orally	Cultivated
Punica granatum L.	Granada	Rind of fruit	Diarrhoea, amoebiasis	Decoction	Orally	Cultivated
Papaveraceae						
Argemone mexicana L.	Cardosanto	Leaves	Flu, fever, expectorant	Decoction	Orally	Wild plant
Umbelliferae						
Anethum graveolens L.	Eneldo	Fruits	Meteorism, colic	Decoction	Orally	Cultivated
Foeniculum vulgare Mill.	Hinojo	Shoot and leaves	Shingles	Macerated leaves	Cataplasm	Cultivated
Verbenaceae						
Aloysia triphylla (L. Her.) Britt.	Cidrón	Shoot and leaves	Tranquilizer	Decoction	Orally	Cultivated
Lippia alba (Mill.) N. E. Br. Ex Britton & Wilson	Toronjil, Toronjil mulato	Shoot and leaves	Tranquilizer, sedative	Decoction	Orally	Cultivated
Verbena litoralis H. B. K.	Verbena	Shoot and leaves	Fever	Decoction	Orally	Wild plant
Zingiberaceae						
Alpinia purpurata (Vieill.) K. Schum.	Paraiso	Flowers	Cough	Decoction	Orally	Cultivated
Zingiber officinalis Rosc.	Jengibre	Rhizome	Sore throat, hoarsenes, cough	Decoction mixed with honey	Orally	Cultivated

Table 4. Medicinal plants of the Táchira State

Family	Species
Asteraceae (Compositae)	1. *Ageratum conizoides* L. 2. *Ambrosia artemisifolia* L. 3. *Ambrosia cumanensis* H. B. K. 4. *Artemisia absinthium* L. 5. *Bidens pilosa* L. 6. *Chaptalia nutans* (L) Polak 7. *Elephantopus mollis* H. B. K. 8. *Emilia sonchifolia* (L) D. C. 9. *Galinsoga caracasana* (D. C.) Sch. 10. *Galinsoga ciliata* (Ref) Blake. 11. *Galinsoga parviflora* Cav. 12. *Pseudoelephantopus spicatus* (Aubl) Gleas 13. *Pluchea odorata* (L) Cass. 14. *Sonchus oleraceus* L. 15. *Spilanthes americana* (Mut) Hiern. 16. *Spilanthes acmella* L. 17. *Spilanthes ciliata* H. B. K. 18. *Spilantes ocimifolia* (Lam) Moore 19. *Siegesbecka jorulensis* H. B. K. 20. *Sinedrella nodiflora* (L) Gaertn. 21. *Taraxacum officinale* Welv.
Leguminosae	22. *Mimosa pudica* H. B. k. 23. *Indigofera fruticosa* Mill 24. *Zornia curvata* Mohl 25. *Desmodium adscendens* (Siv) D. C. 26. *Desmodium intortum* (Mill) Urban 27. *Stylosanthes guyanensis* (Aubl) Siv 28. *Crotalaria incana* L.
Euphorbiaceas	29. *Euphorbia hypericifolia* L. 30. *Euphorbia hirta* L. 31. *Euphorbia prostrata* Ait. 32. *Euphorbia thymifolia* L. 33. *Croton malambo* Karts 34. *Phyllanthus niruri* L. 35. *Phyllanthus corcovadensis* Muell
Poaceae (Gramineae)	36. *Eragrostis pilosa* Beauv 37. *Eleusine indica* (L) Gaertn 38. *Cynodon dactylon* (L) pers 39. *Pennisetum clandestinum* Hochst 40. *Andropogon bicornis* L. 41. *Melinis minutiflora* Beauv
Amaranthaceae	42. *Amarnathus spinosus* L. 43. *Amaranthus dubius* Mart 44. *Amaranthus gracilis* Desf. 45. *Alternanthera polygonoides* Gris. 46. *Iresine celocia* L.
Lamiaceae (Labiatae)	47. *Hyptis atrorubens* Port. 48. *Hyptis capitata* Jacq. 49. *Marsypianthes chamaedrys* (Vahl) Kuntze 50. *Salvia occidentalis* Sw. 51. *Stachys michelena* Briquet
Pteridophyta	52. *Lycopodium clavatum* L. 53. *Equisetum gianteum* L. 54. *Adiantum capillus veneris* L. 55. *Pteridium aquilinum* (L) Kuhn.
Malvaceae	56. *Sida rhombifolia* L. 57. *Sida acuta* L. 58. *Urena lobata* L. 59. *Urena sinuata* L.
Verbenaceae	60. *Verbena litoralis* H. B. K. 61. *Stachycarpeta cayannensis* (Rich) Vahl. 62. *Lantana camara* L. 63. *Lantana trifolia* L.

Family	Species
Urticaceae	64. *Urtica urens* L. 65. *Urera caracasana* (Jacq) Gand 66. *Parietaria debilis* Forst.
Brassicaceae	67. *Brassica alba* Boiss. 68. *Nasturtium officinale* L. 69. *Cardamine bonariensis* Pers.
Rubiaceae	70. *Borreria suaveolens* G. F. Mey. 71. *Hamelia patens* Jacq. 72. *Rebulbium hypocarpium* (L) Hams.
Solanaceae	73. *Physalis angulata* L. 74. *Solanum americanum* Mill. 75. *Solanum hirtum* Vahl.
Cyperaceae	76. *Dichromena ciliata* Vahl. 77. *Cyperus ferox* (L) Rich. 78. *Cyperus rotundus* L.
Commelinaceae	79. *Commelina diffusa* Burn 80. *Commelina virginica* L. 81. *Tripogandra cumanensis* (Kuhnt) Woods.
Loranthaceae	82. *Loranthus leptostachys* H. B. K. 83. *Phtirusa pyrifolia* (HBK) Eichl.
Polygonaceae	84. *Polygonum glabrum* Wild. 85. *Rumex crispus* L.
Portulacaceae	86. *Portulaca oleracea* L. 87. *Talinum paniculatum* (L) Gaerth
Oxalidaceae	88. *Oxalis corniculata* L. 89. *Oxalis latifolia* H. B. K.
Lythraceae	90. *Cuphea micrantha* H. B. K. 91. *Cuphea racemosa* (L) Spreng.
Acanthaceae	92. *Ruellia tuberosa* L. 93. *Thunbergia alata* Boj. 94. *Justicia pectoralis* Jacq.
Agavaceae	95. *Furcraea humboldtiana* Trel.
Hypoxidaceae	96. *Hypoxis decumbens* L.
Iridaceae	97. *Sisyrinchium bogotense* H. B. K.
Phytolacaceae	98. *Phytolaca thyrsiflora* Fhizl.
Nyctaginaceae	99. *Mirabilis jalapa* L.
Chenopodiaceae	100. *Chenopodium antihelminticum* L.
Piperaceae	101. *Piper aduncum* L.
Aristolochiaceae	102. *Aristolochia spp.*
Papaveraceae	103. *Argemone mexicana* L.
Zygophyllaceae	104. *Tribulus cistioides* L.
Polygalaceae	105. *Polygala paniculata* L.
Vitaceae	106. *Cissus sicyoides* L.
Begoniaceae	107. *Begonia spp.*
Cucurbitaceae	108. *Momordica charantia* L.
Apiaceae	109. *Hydrocotyle umbellata* L.
Asclepiadaceae	110. *Asclepias curassavica* L.
Scrophulariaceae	111. *Scopara dulcis* L.
Bignoniaceae	112. *Arrabidaea chica* H. B. K.
Gesneriaceae	113. *Kohleria spicata* (HBK)
Plantaginaceae	114. *Plantago major* L.

Table 5. Diseases, cured by the plants mentioned in Table 4

Diseases, medicinal properties of plants	No. of the species mentioned in Table 4
Antispasmodic	18, 46, 52, 81, 107
Depression	1, 84, 85
Asthma	3, 101
Diarrhoea	12, 16, 24, 27, 36, 75, 82, 94, 98, 105
Antiseptic	2, 33, 34, 36, 48, 49, 79, 80
Cancer	19
Conditions of the liver	8, 9, 13, 20, 21, 28, 39, 64, 65, 70, 76, 84
Snake bites	38
Pulmonary conditions	2, 19, 30, 47, 62, 63, 65, 72, 85, 89, 107
Acne	5, 47, 65, 114
Arthritis	13
Anaemia	32, 33
Kidney conditions	7, 13, 14, 15, 25, 34, 37, 38, 66, 74, 92
Astringent	2, 20, 24, 31, 32, 68, 69, 80, 82, 93
Bactericide/fungicide	2, 24, 96
Goitre	41, 42, 107
Bronchitis	15, 17, 19, 30, 77
High cholesterol levels	22, 58, 59
Baldness	44
Headache	68, 69, 73, 82
Detoxifying	19
Diabetes	2, 10, 11, 15, 20, 21, 58, 59, 98
Diuretic	2, 5, 13, 14, 15, 16, 20, 21, 26, 28, 31, 33, 34, 37, 39, 42, 45, 51, 58, 59, 60, 70, 76, 86, 89, 92, 114
Hypomenorrhoea	30, 34, 76, 78, 86
Emollient	4, 43, 62, 87
Fever	31, 62, 63, 77, 81
Gastritis	64, 65
Homeostatic	71, 80, 92
Hepatitis	13, 50, 51
Haemorrhage	6, 52, 59, 60, 72
Genital infections	22, 23, 24
Insecticide	24, 95, 96
Lithiasis	8, 9, 10, 20, 21, 58, 59, 100, 108, 109, 110, 111
With magic effects	78
Pneumonia	7, 89

Table 6. Aromatic plants found in Táchira

Family	Species
Burseraceae	*Bursera simarouba, Protium heptaphyllum*
Cactaceae	*Hylocereus venezuelensis*
Capparidaceae	*Capparis odoratissima*
Cucurbitaceae	*Cyclanthera pedata, Sicana odorifera*
Compositae	*Pluchea odorata, Tagetes pusilla, Porophyllum porophyllum, Pectis ssp.*
Dilleniaceae	*Curatella americana*
Gramineae	*Melinis minutiflora*
Guttiferae	*Mammea americana*
Labiatae	*Hyptis suaveolens, Ocimum micranthum, Satureja brownei, Minthostachys mollis*
Lauraceae	*Ocotea barcellensis*
Leguminosae	*Enterolobium cyclocarpum, Parkinsonia aculeata, Dipteryx odorata*
Loganiaceae	*Buddleia americana*
Myrtaceae	*Melaleuca leucadendron*
Orchidaceae	*Cattleya perciveliana, Odontoglossum spectatissimun*
Rutaceae	*Citrus aurantifolia*
Solanaceae	*Brunfelsia americana, Cestrum diurnum*
Verbenaceae	*Lippia micromera, Lippia origanoides, Aloysia triphylla*
Zingiberaceae	*Costus comosus, Renealmia aromatica*

Paria peninsula

SILVA (1999) studied a semi-evergreen forest on the Paria peninsula in the north-east of the Venezuelan Cordillera de la Costa, in the State of Sucre. The forest is characterized as a submontane forest of tropophytes extending between 400 and 600 meters above sea leavel.

SILVA studied 339 species of 271 genera belonging to 93 plant families. By her own observations and through consultation of the native people of the rural community of Maturincito, SILVA was able to discover the use of 127 species of 109 genera belonging to 55 plant families. Asteraceae, Euphorbiaceae, Fabaceae and Arecaceae supply 5–10 species each (Table 7).

Useful plants are mostly applied for construction, medicine, food, religion and witchcraft, householdarticles and as ornamentals.

35 species, some of which are even lianas or shrubs, are useful in construction. Roofing is carried out with the leaves of several palm species (*Sabal mauritiaeformis and Attalea butyracea*).

Table 7. Plant species used on the peninsula Paria/Venezuela.

Plant species (family)	Vernacular name	Description	Use	Collected in	Planted in
Acalypha macrostachya (Euphorbiaceae)	Curasarna	Small tree, used for skin diseases such as scabies, infections etc.	2	M2	
Aiphanes aculeata (Arecaceae)	Coquitos de cupiro	Palm tree with thorns. The fruits are assembled in panicles and edible ('red coquitos').	3, 7	B-1	
Allosidastrum pyramidatum (Malvaceae)	Malvapica	Herb, used to treat stomachache and haemocatharsis.	2	M2	b
Aloe vera (Liliaceae)	Zabila	The fleshy leaves arranged in rosettes are used for wound healing and against stomachache.	2,7		b
Alternanthera congesta (Amaranthaceae)	Té de jardin	Herb with a woody stem; cultivated to cure stomachache and the flu.	2, 5		b
Ambrosia cumanensis (Asteraceae)	Artemisa	Herb with a woody stem. The boiled leaves are used against stomach and abdominal pain.	2, 4, 6	M2	b
Amphilophium paniculatum (Bignoniaceae)	Bejuco mara	Climber, much used in construction and in arts and crafts.	1,5,7	B-1, B-2	
Ananas comosus (Bromeliaceae)	Piña	Ground covering plant, cultivated for the fruit, used for hepatocatharsis, against parasites and inflammations.	2, 3, 7		a
Annona sp. (Annonaceae)	Catuche	Small tree, cultivated for its fruits. The leaves are used to treat contusion and mumps, a decoction is used against diarrhoea.	2, 3, 7	B-2	b
Anthurium andreanum (Araceae)	Cala rosada	Ground covering plant, with attractive inflorescenses that vary in size and colour; used as decoration on grave yards and altars.	4, 6, 7	B-2	b
Anthurium sp. 1 (Araceae)	Malanga	Climbing plant, large herb, used together with other plants to treat snake bites. It is frequently planted in gardens and patios as an ornamental plant.	2, 6	B-1, B-2	
Attalea butyracea (Arecaceae)	Curumiche, yagua	Medium-sized to large palm tree, which produces very large panicles with oily fruits. The tree is relatively resistant to fire. The leaves are used on religious holidays to decorate churches and altars.	1, 4	B-1, B-2	
Ayapana amygdalinum (Asteraceae)	Hierba santa	Herb-like plant, used to treat infected wounds, mostly insect bites and stings.	2	M2	
Bactris cf. Macanilla (Arecaceae)	Macanilla, corozo	Medium-sized palm tree with thorns. The hard wood is used for construction and ship-building. The fruit is edible when boiled in water; oil can be extracted from it.	1, 2, 3	B-1	
Bauhinia colombiensis (Caesalpiniaceae)	Bejuco de cadena	Woody climber, used for house construction (wood lining of the walls). The leaves are used for haemocatharsis.	1, 5	B-1, B-2	
Begonia humilis (Begoniaceae)	Clavo pozo	Herb with soft stalks, used as a tranquillizer.	2	B-1, B-2	
Blechnum occidentale (Pteridophyta)	Helecho común	Herbal fern, the spores are used for skin infections.	4, 6, 7	M2, B-2	
Bourreria cumanensis (Boraginaceae)	Guatacare	Medium-sized tree, much used for its hard and resistant wood (construction of houses, ships and piers).	1, 2	B-1, B-2	
Bravaisia integerrima (Acanthaceae)	Mangle	Medium-sized tree, The wood is used for buildings and furniture.	1	B-2	
Brownea coccinea (Caesalpiniaceae)	Rosa de montaña	Bush or small tree, with beautiful rose-coloured or red inflorescences, found in patios as ornamental plant.	6	B-1	
Bulnesia arborea (Zygophyllaceae)	Vera	Medium-sized tree, its hard and dense wood is used to build ships and piers and in arts and crafts.	1	B-1	

Table 7. (Continued) Plant species used on the peninsula Paria/Venezuela.

Plant species (family)	Vernacular name	Description	Use	Collected in	Planted in
Calathea allouia (Marantaceae)	Laire	Perennial herb with edible tubers that taste like potatoes.	3, 7	B-1, B-2	
Carica papaya (Caricaceae)	Lechoza	Medium-sized tree, the fruit is used to treat stomach diseases.	2, 3, 7		a, b
Casearia aculeata (Flacourtiaceae)	Cuspa quina	Richly branched shrub, used to treat headache.	2	B-1, B-2	
Casearia silvestris (Flacourtiaceae)	Cuspa café	Medium-sized tree, used to for wall foundation (bajareque), for brooms and to repell insects.	1, 5	M2	
Casearia decandra (Falcourtiaceae)	Cuspa	Medium-sized tree, used for wall foundation (bajareque) and as fire wood.	1, 5, 7	M2, B-2	
Cecropia palmatisecta (Cecropiaceae)	Guarumo	Small to medium-sized tree, used for wall construction in houses, a decoction of the leaves is used against cough.	1, 2, 5	M2, B-2	
Cedrela odorata (Meliaceae)	Cedro	Large tree providing valuable timber which is used for furniture-making and for frames of doors and windows.	1, 7	B-1	
Cestrum latifolium (Solanaceae)	Puta de noche	Shrub with conspicuous lenticels often found in patios and gardens.	6	M2	
Chaptalia nutans (Asteraceae)	Sanalo	Herb with in rosettes arranged leaves. Used to treat flu and skin diseases.	2	M2, Hs	
Cissus erosa (Vitaceae)	Lava ojo	Climber, used to treat infections, irritations of the eye and muscular pain.	2, 5	B-1	
Citrus reticulatus (Rutaceae)	*Mandarina*	Small tree, cultivated for its fruit. The leaves have a tranquillizing effect and a tea of them is used for gastritic problems.	2, 3, 7		a, b
Coffea arabica (Rubiaceae)	*Café*	Cultivated shrub or small tree. The aroma of roasted beans may be used to treat paralysis of the face, the husks are used to accelerate wound healing.	2, 7	B-2	a
Colocasia esculenta (Araceae)	Ocumo chino	Herb, 0.5–1m high, cultivated for its edible, subterraneous bulbs.	3, 7		a, b
Cordia curassavica (Boraginaceae)	Pava canelilla	Medium-sized shrub, used for brooms and as firewood.	5	M2, B-2	
Cordia collococca (Boraginaceae)	Alatrique	Medium-sized tree, frequently used for house building and furniture-making.	1, 7	B-1	
Cordia alliodora (Boraginaceae)	Pardillo	Medium-sized tree, used for house and ship building and furniture-making.	1, 5, 7	B-1, B-2	
Costus scaber (Zingiberaceae)	Caña la india	Large herb with attractive flowers used as ornamental plant.	6	B-1	
Costus villosissimus (Zingiberaceae)	Caña la india	Large herb with attractive flowers used as ornamental plant.	6	B-1, B-2	
Croton lobatus (Euphorbiaceae)	Sangre de gallo	Annual herb, used for haemocatharsis and to regulate menstruation.	2	M2	
Croton gossypifolius (Euphorbiaceae)	Sangredrago	Medium-sized tree. Latex is used for wound healing and to treat infections of the throat.	2	B-1, B-2	
Cucurbita sp. (Cucurbitaceae)	Auyama	Climbing or creeping plant with tendrils, cultivated for its edible fruit.	3, 7	M2	
Cupania americana (Sapindaceae)	Maraquillo	Medium-sized tree, used for timber and as fire wood.	1, 7	B-2	
Cyperus luzulae (Cyperaceae)	Junco	large herb, used in rural house construction.	1	B-2	
Cyrtocymura scorpioides (Asteraceae)	Rabo de alacran	Woody plant with sweeping branches, used in wound healing and to treat infected insect bites.	2, 5	M2	
Dioscorea alata (Dioscoreaceae)	Ñame	Herb, cultivated for its edible tubers.	3, 7	B-1	b
Dioscorea trifida (Dioscoreaceae)	Mapuey	Climber, cultivated for its edible subterraneous tubers	3,7		a, b

Table 7. (Continued) Plant species used on the peninsula Paria/Venezuela.

Plant species (family)	Vernacular name	Description	Use	Collected in	Planted in
Dorstenia contrajerva (Moraceae)	Flor de plato	Herb, used to treat snake bites, insect bites and diarrhoea.	2, 7	B-2	
Enterolobium cyclocarpum (Mimosaceae)	Cara, caro	Medium-sized robust tree, the fruits are used to make soap, the seeds in arts and crafts.	5	B-1, B-2	
Epiphyllum phyllanthus (Cactaceae)	Pitahaya	Epiphyte, frequently found as decoration in gardens and patios.	6	B-2	
Eryngium foetidum (Umbelliferae)	Culantro	Herb, about 20 to 80 cm high. The root is used to treat cough and as a laxative.	2, 3	M2, B-2	b
Erythrina poeppigiana (Fabaceae)	Bucare	Large tree, used in coffee plantations to give shade. The seeds are applied against the 'evil eye' and to make necklaces.	4	B-1, B-2	
Erythrina glauca (Fabaceae)	Anauco	Large tree, used in coffee plantations to give shade, the seeds are used against the 'evil eye' and to drive away bad spirits.	4	B-1, B-2	
Eupatorium pycnocephalum (Asteraceae)	Curia	Woody, perennial herb, used to treat flu and headache.	2	M2	
Euterpe precatoria (Arecaceae)	Palma real	Medium-sized palm tree, the leaves are used to decorate churches and altars on religious holidays.	4	B-1	
Ficus obtusifolia (Moraceae)	Chara	Medium-sized to large tree. Latex is applied to remove corns.	2	B-1, B-2	
Ficus clusiaefolia (Moraceae)	Matapalo rojo	Medium-sized tree, used to cure asthma and contusions.	2	B-1, B-2	
Funastrum clausum (Asclepiadaceae)	Bejuco de ajo	Climber, latex is used in wall construction and to make ropes.	1, 5	B-1	
Gloxinia pallidiflora (Gesneriaceae)	Hierba buena	Climber, used to treat stomach diseases.	2, 4	B-1	b
Guaiacum officinale (Zygophyllaceae)	Guayacán	Small tree, cultivated as ornamental plant. The hard wood is used for ship and pier construction.	1	B-1, B-2	
Guzmania monostachya (Bromeliaceae)	Canareque	Epiphyte, found in gardens and patios as an ornamental plant.	6	B-1, B-2	
Gynerium sagittatum (Poaceae)	Caña brava	Herb, used for roofing in rural areas and in arts and crafts.	1, 7	B-2	
Hebeclinum macrophyllum (Asteraceae)	Concertadera	Small shrub, used to treat stomach diseases and flu.	2	M2	b
Heliconia hirsuta (Heliconiaceae)	Vijao	Large herb with a soft stem. The leaves are soaked in water and used for the wall foundation in houses ('bajareque'). The inflorescences are sold for decoration.	1, 6, 7	B-2	
Heliconia caribaea (Heliconiaceae)	Vijao	Large herb, leaves are soaked to make the 'bajareque', the inflorescences are sold for decoration.	1, 6, 7	B-1, B-2, M2	
Heliocarpus popayanensis (Tiliaceae)	Majagua	Medium-sized tree, used as construction wood and to build light boats, the bark is used for ropes.	1	B-2	
Hibiscus rosa-siniensis (Malvaceae)	Cachuplina	Cultivated shrub, used as an ornamental plant and for soap making, to treat nervous ailments (tea) and as anexpectorant, used in gardens and patios to repell snakes	5, 6		b
Hura crepitans (Euphorbiaceae)	Jabilla	Large tree, the timber is used for ship building and furniture-making.	1, 7	B-1, B-2	
Hyptis mutabilis (Labiatae)	Mastranto	Tall grass, used for brooms. Infusion is used against headache.	2, 5	M2, Hs	
Jatropha curcas (Euphorbiaceae)	Piñon	Latex-rich shrub, cultivated and used against parasites and for religious purposes. The latex is used for infected insect bites and for wound healing.	2, 4, 5		b

Table 7. (Continued) Plant species used on the peninsula Paria/Venezuela.

Plant species (family)	Vernacular name	Description	Use	Collected in	Planted in
Landenbergia moritziana (Rubiaceae)	Quina	Medium-sized tree, used as a remedy against hair loss and as an antipyretic.	2	B-1	
Lantana armata (Verbenaceae)	Ramoncillo	Much branched, small shrub, used to treat fever, the 'evil eye' and for house-cleaning and laundry.	2, 5	M2	
Lantana canescens (Verbenaceae)	Canelilla	Much branched, small shrub, used to treat asthma, abdominal pain and to wash clothes.	2, 5	M2	
Lantana caracasana (Verbenaceae)	Cariaquito blanco	Much branched, small shrub, used to decorate altars on religious holidays and to drive away evil spirits.	4, 5, 7	M2, B-2	
Lippia alba (Verbenaceae)	Oregano	Herb with a woody stem, remedy for flu, for the nerves and against stomach disorders.	2	B-2	b
Lonchocarpus hedyosmus (Fabaceae)	Aco blanco	Medium-sized to large tree, used for construction.	1, 7	B-1, B-2	
Malpighia glabra (Malpighiaceae)	Semeruco	Small tree with edible fruits, cultivated to sell the fruit.	3, 7		b
Mammea americana (Guttiferae)	Mamey	Small tree, not cultivated, esteemed for its edible fruit which is sold on markets.	3, 7	B-2	
Mandevilla moritziana (Apocynaceae)	Algodoncillo	Climber with white latex, used for house-cleaning.	5	M2, B-2	
Mangifera indica (Anacardiaceae)	Mango	Medium-sized fruit tree, cultivated for its fruit. The leaves are used to cure contusions and skin diseases, the resin of the fruit is applied for wound-healing.	2, 3, 7		b
Manihot esculenta (Euphorbiaceae)	Yuca	Shrub, cultivated fot its edible tubers, which are sold on local markets.	3, 7		a, b
Melanthera aspera (Asteraceae)	Engorda macho	Tall herb, very bushy, treatment for menstrual problems and against anaemia.	2	M2	
Momordica charantia (Cucurbitaceae)	Cundeamor	Climber, the herb is used to clean and desinfect the house and to treat scabies and other skin diseases.	5	M2	
Mucuna altissima (Fabaceae)	Pepa de zamuro	Woody climber, seeds are used in religious ceremonies, the stems to make ropes.	4, 5	B-1	
Musa sp. (Musaceae)	Cambur	The plants (many races and varieties) are cultivated for the fruits and used against parasites and anaemia.	2, 3, 7	B-2	b
Myrcia splendens (Myrtaceae)	Pajuicillo, Menudito	Small to medium-sized tree, its wood is used for building and in furniture making.	1, 5, 7	B-2	
Ochroma lagopus (Bombacaceae)	Tacarigua	Small tree, the soft wood is used in arts and crafts and to make kitchen utensils.	5	M2, B-2	
Olyra longifolia (Poaceae)	Bambu	Tall herb, the stems are used in house construction for the 'bajareque', a mixture of earth and plants to line the walls.	1	B-2	
Olyra latifolia (Poaceae)	Corocillo	Tall herb, the stem is used in house construction for the 'bajareque', a mixture of earth and plants to line the walls.	1	B-2	
Passiflora foetida (Passifloraceae)	Parchita del diablo	Climbing or creeping plant, the edible fruits are used to treat inflammations.	3	M2, B-2	
Passiflora quadriglandulosa (Passifloraceae)	Guaicoruco	Climber with edible fruits.	3	B-1, B-2	
Pavonia fruticosa (Malvaceae)	Cadillo de perro	Herb, remedy against flu and to treat liver and kidney disorders (diuretic).	2	M2	b
Persea americana (Lauraceae)	Aguacate	Medium-sized fruit tree with edible fruits that are very rich in oil and are used to treat skin and skalp diseases.	2, 3, 7		a, b

Table 7. (Continued) Plant species used on the peninsula Paria/Venezuela.

Plant species (family)	Vernacular name	Description	Use	Collected in	Planted in
Petiveria alliacea (Phytolaccaceae)	Mapurite	Herb with a strong and unpleasant odour, remedy for fever and flu, for manufacture of brushes to expel instects.	2, 5	M2	
Phyllanthus acuminatus (Euphorbiaceae)	Lagrima de San Pedro	Shrub, used against the 'evil eye' during prayers and to drive away evil spirits.	4	B-2	
Pimenta racemosa (Myrtaceae)	Bayrum	Small tree, used as a treatment against pulmonary and respiratory disorders and to cure contusions.	2	B-2	b
Pithecellobium samanea saman (Mimosaceae)	Saman	Medium-sized tree, the timber is used in house and ship building and in furniture-making.	1,7	B-1, B-2	
Plantago hirtella (Plantaginaceae)	Llantén	Herb, used for disorders of stomach and liver and used to cure diarrhoea.	2, 4, 7	M2	
Platymiscium diadelphum (Fabaceae)	Roble	Medium-sized tree, the timber is used in house and ship building and in furniture-making.	1,7	B-1, B-2	
Pluchea odorata (Asteraceae)	Salvia	Woody plant, used to regulate blood pressure and to cure stomachache and convulsions.	2	M2	b
Polypodium caceresii (Pteridophyta)	Helecho sano	Small clustered fern, the root is used together with other plants as an antidote to snake venom.	2, 6	B-1, B-2	
Pothomorphe peltata (Piperaceae)	Guayuyo corazon	Tall herb, used to cure infected wounds, stomachache, sunstroke and for house-cleaning.	2, 4	M2, B-2	
Psidium guajava (Myrtaceae)	Guayaba	Small tree, cultivated as a fruit tree, used to treat anaemia and menstruation problems.	2	B-2	b
Renealmia alpina (Zingiberaceae)	Conopio	Tall herb, used to treat backache and infected insect bites.	2	B-1, B-2	
Rhynchosia phaseoloides (Fabaceae)	Peonio cadeno	Climber, ornamental plant to decorate altars.	4	B-2	
Ricinus communis (Euphorbiaceae)	Higuereta, Tártago	Shrub, cultivated and used against parasites and skin diseases.	2, 4	M2, B-2	b
Ruta graveolens (Rutaceae)	Ruda	Tall herb, fragrant, cutlivated and used to cure fractures of the bones and bruises.	2		b
Sabal mauritiaeformis (Arecaceae)	Carata	Medium-sized to tall palm tree, relatively resistant to fire, used for house (roof) construction and in arts and crafts.	1, 4	B-1, B-2, Hs	
Senna (Cassia) *alata* (Caesalpiniaceae)	Tarantan	Medium-sized tree with attractive flowers found in gardens as ornamental plant.	6		b
Sida sp. 1 (Malvaceae)	Escoba	Herb, used to make brushes, also used as a remedy for eye diseases and an expectorant.	2, 5	M2	
Solanum nigrum (Solanaceae)	Yerba mora	Herb, used as a treatment for skin diseases, scabies, tropical lichen and infections.	2, 5, 7	M2, B-2	
Solanum sp. (Solanaceae)	Pechos de la reina	Much branched shrub, used to make insect poison.	2	M2, B-2	
Syzygium jambos (Myrtaceae)	Puma rosa	Small to medium-sized tree, cultivated for its edible fruits and as an ornamental plant.	3, 6	B-2	b
Tabebuia pentaphylla (Bignoniaceae)	Apamate	Medium-sized to tall tree, esteemed for its valuable timber for house and ship building and furniture-making.	1, 7	B-1	
Taraxacum officinale (Asteraceae)	Hierba de león	Shrub, latex is used for haemocatharsis, against anaemia and to stimulate the appetite.	2	M2	
Theobroma cacao (Sterculiaceae)	Cacao	Small tree, cultivated for its seeds (preparation of cocoa). The remains of the fruit are used as a menure and to make fuel. The fat obtained from the seeds is used to cure burns.	2, 5, 7	B-2	a

Table 7. (Continued) Plant species used on the peninsula Paria/Venezuela.

Plant species (family)	Vernacular name	Description	Use	Collected in	Planted in
Tillandsia recurvata (Bromeliaceae)	Barba de palo	Epiphyte, used for house-cleaning and to drive away evil spirits.	5, 6	B-1, B-2	
Trichilia cuneata (Meliaceae)	Cedrillo	Medium-sized tree, the timber is esteemed for house and furniture building.	1, 7	B-1	
Trichilia brachystachya (Meliaceae)	Caobano	Small to medium-sized tree, esteemed for construction and furniture building.	1, 7	B-1, B-2	
Trixis divaricata (Asteraceae)	Juan de la calle	Half-shrub, used to treat infected wounds and fever.	2	M2	
Vanilla pompona (Orchidaceae)	Vainilla	Climber and epiphyte, used for house-cleaning.	5	B-2	
Vismia cayennensis (Clusiaceae)	Lacre blanco	Medium-sized tree with latex, used to make wall-foundation ('bajareque') in houses.	1, 7	B-1, B-2	
Vismia baccifera (Clusiaceae)	Lacre	Small to medium-sized tree with latex, used to make wall-foundation ('bajareque') in houses.	1, 7	M2, B-2	
Vitex orinocensis (Verbenaceae)	Totumillo, Nazareno	Medium-sized to tall tree, the timber is used for house building and construction.	1	B-1	
Xanthosoma sagittifolium (Araceae)	Ocumo blanco	Perennial herb, cultivated for its edible tubers.	3, 7		a, b
Zea mays (Gramineae	Maiz	Herb, 3–4 m high (grass), cultivated for food, boiled spikes are used for kidney aches.	2, 3, 7		a, b
Zingiber officinale (Zingiberaceae)	Gengibre	Tall herb, cultivated for its aromatic rhizomes, used to treat sore throat and as a diuretic.	2, 7		b

Explanations:

Construction	1	Primary forest	B-1	Small tree	5–10 m
Medicine	2	Secondary forest	B-2	Medium size tree	11–19 m
Food	3	Secondary bush	M2	Large tree	20–30 m
Witchcraft	4	Secondary grassland	Hs		
Household	5	Plantations	a		
Ornamental	6	Housegardens	b		
Commercial	7				

58 species play an important part in popular medicine, and 24 of them are applied exclusiveley to cure diseases. All plants serving for medicinal purposes are still important for the rural population, because of the inadequate official medical provision in these regions.

23 species supply food, but only 2 species are exclusively used as such (*Passiflora quadriglandulosa* and *P. foetida*), while the other 21 species have multiple applications.

17 species belong to the category religion and witchcraft and 5 of these are exclusively applied for this purpose. The remaining 12 species also have other applications.

29 species fulfill the category household , 6 of which belong exclusively to this category.

17 species are used as ornamentals. Some are cultivated and sold.

All useful plants play a much more important part in the daily life of the population than they do economically. 47 of the 127 useful species are brought to the market.

Many of the useful plants come from the secondary forest. The use of tree species for timber, e.g. that of *Cordia alliodora, Cedrela odorata, Pithecellobium saman and Tabebuia pentaphylla* is very common.

Of the 127 useful species, 37 are cultivated. Cultivation of *Coffea arabica, Theobroma cacao* and *Ananas comosus* in the form of plantations is most frequent.

Ethnobotanical studies of the Paria peninsula were recently begin. The knowledge of useful plants in this region is probably derived from two ethnic groups who lived there before the Spanish conquest, namely the Acios (who possibly belong to the Arahuaca) and the Pariagotos, who are part of the Caribbean population.

Collection, cultivation and application of useful plants by women is an important factor on the Paria peninsula. Many species are cultivated in private gardens by women, but women also play their part in agriculture. They are acquainted with the biodiversity of nature (Diagram 3).

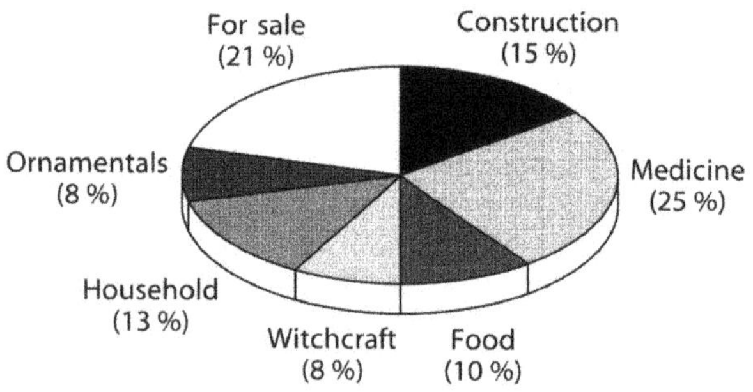

Diagram 3 Way of application of the 127 plant species used by the population of Maturincito (in percentages), according to A. SILVA (1999).

Plants used by Indian tribes

Yanomami Indians

FUENTES (1980) studied the Indian Yanomami tribe in the State Amazonas/Venezuela. This tribe migrates between the Sierra Parima, a mountainous region with humid forests, and the great plain of the meseta with savannas, emphasizing their relationship to plants. He describes how the tribes identify and classify the plants, how they denominate them in their language and he also gives examples and explanations of the meaning of the words used for plants and plant parts. FUENTES furthermore describes the use of the plants in the Yanomami culture, e.g. as a food, in technology (for construction, weapons, vessels, ropes, tissues, bridges and boats), for dyes, ornaments, toys, as well as for poisons, hallucinogenes, magics and medicines.

TILLET and FUENTES added a list of the plants known and used by the Yanomami and their names in the indigenous language.

Warao Indians

The Warao Indians of the Orinoco Delta in eastern Venezuela elaborated a pharmacopoeia of 100 plant species which pertain to 45 families and from which they obtain 259 remedies to treat 52 different diseases. Statistics indicate that gastrointestinal diseases are a principal cause of morbidity and mortality especially among infant and pre-adolescent age groups. The phytotherapeutical practice predominantly rests in the hands of women who – in a complementary function to the largely psychiatric shamanic healing methods dominated by men – treat an array of endemic diseases with vegetable drugs.

The Warao have incorporated into their pharmacopoeia products amounting to 49–74 % of all plant species within their particular surroundings. This pharmacopoeia is the product of 3000–4000 years of experimentation, as WILBERT & HAIEK (1991) emphasize. The authors emphasize that it would not have reached such a long tradition, if the plants had not had symptomatic and/or curative values.

Special contents

Alkaloids

Only a few studies were dedicated to certain substances in Venezuelan plants.

ARISTEGUIETA (1976) who studied the presence of alkaloids in Venezuelan plants, observed that certain plant families are particularly rich in alkaloids: the Apocynaceae, Aquifoliaceae, Asclepiadaceae, Bignoniaceae, Capparidaceae, Erythroxylaceae, Euphorbiaceae, Flacourtiaceae, Lauraceae, Malpighiaceae, Menispermaceae, Leguminosae, Myricaceae, Piperaceae, Rubiaceae, Rutaceae, Sapindaceae, and Solanaceae.

The alkaloid content of plants is however variable and depends mainly on the state of development of the plant, the site where the plant grows, the time of harvest (spring, autumn etc.), the part of the plant collected (leaf, axis, root, flower, fruit, seed), and on possible varieties or local forms. All these factors impede unambiguous statements and standardization.

At first, the author tried to detect alkaloids by a bitter taste, e.g. of leaves etc. But this test is fallacious so that the reagents of Dragendorf (orange colours) and of Meyer (precipitation) had to be applied; the methanolic extract of dry plant material was used for the tests. ARISTEGUIETA inclines to think that fruits contain alkaloids most frequently and in the largest quantities, even when the leaves show negative results; in *Homalium, Palicourea,* and *Solanum*, fruits gave positive, but leaves negative results. Unfortunately the species of the plants studied could not always be indentified either because they did not flower or because they belonged to a very complicated group of species.

In the following, a short summary of the results obtained by ARISTEGUIETA is given.

(+) indcates a positive result, (-) a negative result.

Apocynaceae: *Aspidosperma fendleri* and *A. cuspa* +. Most of the species studied are positive. *Rauvolfia heterophylla*: fruits and roots + (yohimban derivatives).

Aquifoliaceae: *Ilex paraguariensis* which supplies the Maté tea and other *Ilex* species are rich in purin derivatives.

Asclepiadaceae: Most of the species are poisonous and contain cardenolides. *Matelea maritima* with + fruits and *Macrocepis longiflora* with + leaves.

Bignoniaceae: Seeds of *Jacaranda filicifolia* +.

Buxaceae: *Buxus (Tricera) citrifolia* with strongly positive leaves (pseudoalkaloids?).

Erythroxylaceae: Cocaine is extracted from *Erythroxylon coca. E. orinocense* with a weak positive reaction in the leaves.

Euphorbiaceae: Latex with caoutchouc. Rich in alkaloids. Many species of *Croton* in Venezuela show positive reaction and often only the fruits.

Flacourtiaceae: *Homalium racemosum* with + fruits. *Laetia procera* +, *Banara nitida* +, *Casearia guayanensis* +, *Ryania* speciosa +.

Guttiferae: Almost all species studied proved negative. (*Vismia*?).

Lauraceae: Many species have useful wood, essential oils or are ornamentals. *Cassytha* sp. is only slightly positive.

Leguminosae: *Pthecellobium tortum* +, *Pithecellobium carabobense* with positive bark, *Pithecellobium pedicellare* +, *Pentaclethra macroloba* +, *Pentaclethra macrophylla* +, *Inga ruiziana* +, *Piptadenia psilostachya* +, *Hymenaea courbaril* +, *Dipteryx odorata* +, *Andira inermis* +, *Lonchocarpus sericeus* +, *Lupinus* (several species) +, *Sophora tomentosa* +.

Loganiaceae: *Strychnos* supplies curare, rich in alkaloids (indolalkaloids: strychnine). The roots are usually rich in alkaloids.

Moraceae: Rich in latex. *Cecropia* +, several species of *Ficus* +. Toxicity of certain milksap types is due to cardenolides.

Myrsinaceae: *Rapanea ferruginea* with positive leaves.

Myrtaceae: Rich in volatile oils. *Myrcia* sp. with positive leaves.

Ochnaceae: *Ouratea guildingii* with positive leaves.

Onagraceae: *Ludwigia* (*Jussieua*) with slightly positive leaves.

Proteaceae: *Roupala complicata* with slightly positive leaves.

Rubiaceae: China alkaloids in the china bark in species of *Cinchona*. *Chimarrhis microcarpa* +, *Cinchona pubescens* +, *Randia* sp.+, *Alseis labatioides* +, *Genipa caruto* +, *Palicourea fendleri* +.

Rutaceae: Volatile oils in *Citrus* species. Alkaloids are frequent in the family. All species studied of *Zanthoxylum* (*Fagara*) show a high content of alkaloids. *Pilocarpus alvaradoi* is slightly + in the leaves.

Solanaceae: Almost all fruits of the genus *Solanum* studied are positive, the leaves are however negative. Negative reactions in the leaves of *Cestrum latifolium, Cyphomandra* sp., and species of *Datura* were observed. The Solanaceae are generally rich in alkaloids.

Verbenaceae: *Citharexylum macrophyllum* +, Vitex staheli +.

Vochysiaceae: *Vochysia venezuelana* +.

Aromatic plants and perfumes

Aromatic plants and perfumes are the subject of a book written by ROTH & KORMANN (1996) in which a number of South American plants are mentioned, such as *Myroxylon pereirae* (Peru balsam), *Aniba rosaeodora* (Bois de rose femelle) from french Guiana, Surinam and Brazil, *Liquidambar styraciflua* (American sweet gum) from Central America, *Vanilla planifolia* (Vanilla) from South America/Venezuela, *Mirabilis jalapa* (Four a'glock) from South America/Venezuela, *Nicotiana alata* from South America, *Tropaeolum majus* (common Nasturtium) from South America, *Gaultheria procumbens* (Winter green) also from South America/Venezuela, *Myristica fragrans* (Nutmeg) mainly cultivated in the West Indies, and *Syzygium aromaticum* (clove) likewise cultivated in the West Indies. All these species are used for manufacture of perfumes.

Other plant species supply balsams, resins and gumresins, such as *Guaiacum officinale* (lignum sanctum) and *Guaiacum sanctum* (lignum vitae), both from South America/Venezuela. Species of *Copaifera*, Venezuela, supply copaivae balsamum or maracaibo resin (from *Copaifera officinalis*). *Myroxylon balsamum*, Venezuela, is the source of balsam of tolu or balsamum tolutanum.

Volatile oils are extracted from *Copaifera* species (*C. officinalis*: Oil of Copaiba), Venezuela, *Myristica fragrans, Syzygium aromaticum, Myroxylon balsamum* var. *pereirae* from El Salvador.

Even enzymes can be used as fragrance producers, e.g. lipases, esterases and proteases obtained from microorganisms or animals.

A survey of angiospermous plants with gum and resin ducts and cavities has been elaborated by NAIR (1995), who also studied the development, anatomical structure and mode of secretion of these secretory structures. Several of the cited plant species are of South American origin.

MAREN EGGERS of the Chair for Pharmazeutical Biology of the University of Hamburg made a list of Venezuelan medicinal plants with essential oils (personal communication 1999) (Table 8), mainly using the book of H. PITTIER 'Manual de las plantas

Table 8. Plants with ethereal oils found in Venezuela (according to Eggers 1999)

Family	Genus	Species	Vernacular name	Occurrence of ethereal oil	Application	Bibliography
Asteraceae	Lactuca	*L. sativa*	Lechuga	In traces	Leaves used to treat insomnia.	1
	Mikania	*M. guaco*	Guaco	M. cordifolia with ethereal oil	Leaves as febrifuge, diuretic and sedative.	1,2
	Vernonia	*V. brasiliana*	Palotal		Leaves used in a bath to cure malaria, young shoots against constipation.	1
	Spilanthes	*S. ocymifolia*	Corimiento		Sap for toothache.	1
	Pluchea	*P. odorata*	Salvia real		Antidote.	2
	Tagetes	*T. pusilla*	Anis cimarron		Blood purification.	2
	Chrysanthemum, today: Tanacetum	*C. parthenium*	Flor de Santa Maria	0.025 – 0.07 % in the flowering herb	For dehydration.	2
Apiaceae	Apium	*A. graveolens*	Apio España	1.9 – 3 % in fruits 0.1 – 0.8 % in the herb	Sap of the root is a laxative, macerated leaves and stalk are used to regulate menstruation.	1
	Daucus	*D. carota*	Zanahoria		Decoction of root together with milk and honey against asthma and flu; grated root as a poultice for burns.	1
	Foeniculum	*F. vulgare*	Hinojo	0.8 – 8.5 % in fruits	Alkalic extract against anaemia; sap and tea as a sedative; tea against colic.	1
	Petroselinum	*P. crispum*	Perejil	2 – 6 % in fruits	Tea of the leaves and flowers as a diuretic; leaves and grated roots against fleas.	1
	Pimpinella	*P. anisum*	Anis	2 – 6 % in fruits	Tea of the roots against angina. Tea of the fruits against colics and as a digestive stimulant, also lactiferous.	1
Lamiaceae	Ocimum	*O.micranthum*	Albanaca		Tea of the leaves calmative and against stomachache. The entire plant is used in a bath against rheumatism. The sap of the leaves in olive oil as a liniment for sprains and contusions.	1

Table 8. (Continued) Plants with ethereal oils found in Venezuela (according to Eggers 1999).

Family	Genus	Species	Vernacular name	Occurrence of ethereal oil	Application	Bibliography
	Rosmarinus	*R. officinalis*	Romero		Tea of the leaves against nervousness. Alkalic extract as a liniment for muscular ache, circulatory disorders and haematoma.	1
	Salvia	*S. palifolia*	Pega Pega		Tea of the leaves as a gargle to cure catarrh of the respiratory tract and sinusitis. Internally against meteorism and disorders of the liver.	1, 3
	Origanum		Dictamo real	Ethereal oil in the leaves	Blood purification.	2
	Mentha	*M. viridis*	Yerba buena		Spasmolyticum.	2
Verbenaceae	Lantana	*L. camara*	Cariaquito morado	0.05 – 0.2% in the leaves	Against leper.	2
		L. trifolia	Cariaquito morado		Tea of the leaves for fever and asthma, antirheumaticum.	1, 2
	Lippia	*L. stoechardifolia*	Te		Antiasthmaticum.	2
		L. micromera	Oregano		Antiasthmaticum.	2
		L. origanoides	Oregano		Tea of the leaves is digestive and tranquillizing, externally to clean wounds.	1
	Vitex	*V. compressa*	Aceituno, Guarataro, Piyaya		Tea of the leaves against headache, externally for wound healing and cicatrization.	1
Caesalpiniaceae	Copaifera	*C. officinalis*	Aceite, Cumcay, Copalba		Externally as plaster (coil or balsam) for ulcer and for wounds. Decoction of the bark as a bath for rheumatism. Blood purifier.	2
Lauraceae	Ocotea	*O. glomerata*	Lauree		Oil of the shoot for cicatrization.	1
	Persea	*P. americana*	Aguacate, Curo	0.5% in bark and leaves	Seeds for diarrhoea, leaves to regulate menstruation, aphrodisiacum.	1, 2
	Nectandra	*N. pichurim*	Capuchino		Antidiarrhoeicum.	2
	Laurus	*L. camphora;* today: *Cinnamomum c.*	Alkanfor		Antirheumaticum.	2
	Aniba	*A. canelilla*	Canelito		Antirheumaticum.	2
Myrtaceae	Psidium	*P. guajava*	Guayaba		Decoction of the bark for diarrhoea. Tea of the flowers to regulate menstruation. Decoction for infections in the mouth.	1
	Syzygium	*S. jambos*, Syn.: *Eugenia jambos*	Pomagas	0.18% in leaves	Tea to gargle with.	1
Piperaceae	Piper	*P. marginatum*	Rabo de mato		Against toothache: small pieces are put directly on the teeth.	1
Rutaceae	Citrus	*C. medica*	Cordoncillo, Limón		Spasmolyticum.	2
		C. decumana	Cidra		Spasmolyticum.	2
		C. aurantium	Limón, Naranjo		Hot sap against flu. Fruit flesh or sap to desinfect wounds or against warts. Spasmolyticum.	1, 2
	Cusparia	*C. trifoliata*	Quina, cuspa, Cascarillo, Quina amarilla	0.1 – 0.4% in the leaves	Decoction of the bark for malaria.	1
	Ruta	*R. graveolens*			Entire plant as a poultrice for sprains, bruises and fractures.	1
	Amyris	*A. balsamifera*	Quigua		Blood purification.	2
Zygophyllaceae	Guajacum	*G. officinale*	Guayacan	1.5 – 3.5% in leaves and wood	Antirheumaticum.	2

Bibliograpy used:

1. Delascio Chitty, F: (1985) Algunas plantas usuales en la medicina empirica venezolana. Direccion de investigationes biologicas, Division de vegetation, Jardin Botanico, Imparques
2. Véles-Salaz, F.: (1959) Plantas medicinales de Venezuela, Editoral Las Novedades, Carácas
3. Vareschi, V.: (1970) Flora de los Páramos. Universidad de los Andes, Ediciones del Rectorado, Merida
4. List, P.H.; Hörhammer, L.: (1972) Hagers Handbuch der Pharmazeutischen Praxis, Band 4–6. Springer Verlag
5. Hänsel, R; Keller, K; Rimpler, H.; Schneider, G.: (1992) Hagers Handbuch der Pharmazeutischen Praxis, Band 4–6. Springer Verlag

usuales de Venezuela' (1926/1970), citing 38 different species of 34 genera belonging to 10 different plant families. However, a number of medicinal herbs is included in this list which are not indigenous to Venezuela, but are only cultivated there. A botanist, familiar with the tropical flora would however suspect that not only a much larger number of species with essential oils should be found in Venezuela, but that quite a few genera should comprise not only one species with essential oils, but several of them with the same characteristics because aggregate species often occur in the Neotropics.

Mucilage and gum

In his publication 'Mucilage or gum in seeds and fruits of Angiosperms', GRUBERT (1981) gave a survey on the occurrence of mucilage and gum in seeds and fruits, mentioning 100 families of Angiosperms with several thousands of species and citing 2826 bibliographic works. He emphasized that in spite of the increasing demand of mucilage and gum, only few members of the large number of species with these substances have been studied and used to date. Many of the species studied are myxospermous, producing slime externally which, when wet sticks the seeds or fruits to the substrate.

Mucilage and gum find many medicinal applications, e.g. as laxatives, as volume increasing appetite suppressors, as expectorants in respiratory diseases, as plasma substitutes, as antidotes against snake bites and selenium intoxication, in the treatment of gastrointestinal diseases and as additives in dietary compositions; they also help reduce serum cholesterol.

Mucilage and gum of fruits and seeds are furthermore used in the food industry, e.g. as gelatinizing and thickening agents, as stabilizers in ice cream, mayonnaise and dressings, in milk and cocoa products as well as in catsup and related tomato products, and finally as additives in calorie reducing foods. Mucilage and gum also affect the baking properties of flour (cereal gum)- and the quality of beer (barley gum).

Besides, mucilage polysaccharides find wide applications in the technical field, being used as thickeners for printing paste and for explosive suspensions, as sizing substances for textiles, as vegetable glue, in photographic processing solutions, as adhesives for cellulose products in the paper industry, in compositions for oil recovery, for the aggregation of soil particles, as protective colloids, in drilling fluids, in the tobacco industry, in sealing material, as additives in insecticides and herbicides, as suspending or emulsifying agents, in cosmetic preparations and, finally, for the entrapment of larvae by slimy plant seeds.

A great number of species described by GRUBERT are of tropical origin. Genera and species such as *Annona muricata, Bixa orellana, Ceiba pentandra, Gyranthera caribensis, Carica papaya, Momordica charantia, Jatropha curcas, Samanea saman, Artocarpus heterophyllus, Psidium guajava, Capsicum frutescens, Theobroma cacao* as well as species of the genera *Ruellia, Euphorbia, Sporobolus, Acacia, Cassia, Plantago* etc. are found in South America/Venezuela. But the vast majority of South American/Venezuelan wild plants with slimy or gummy fruits and seeds are still waiting to be discovered, investigated, applied and appreciated.

Natural dyes

In his 'Manual of Natural Dyes' SCHWEPPE (1992) mentions the most important vegetable dyes used in the precolumbian America: e.g. indigo (anil) obtained from *Indigofera suffruticosa* MILL., preferentially used in Mexico and South America, and lignum campechianum from *Haematoxylon campechianum* L., growing in tropical America. Accoding to the type of mordant which is used, the colour becomes red, violet, brown violet, dark blue, or even black. Haematoxylin extracted from the wood of this tree is today used in microscopical staining techniques and is a source of ink. A red dye can be extracted from *Haematoxylon brasiletto* KARST. (brazilwood).

After the conquest of Mexico by Cortez in 1522, the above mentioned plants became important export articles for the dyeing industry in Europe.

Besides *Indigofera suffruticosa, Haematoxylon campechianum* and *H. brasiletto*, the following plant species were used by Indian cultures in Central America (Maya, Aztec), in Peru (Inca), and in Chile (Arauko):

- A *black* dye was extracted from the fruits of *Caesalpinia coriaria* (JACQ.) WILLD. (divi-divi). The pods contain 25–30% tannin which is used for tanning and as a dye; in Mexico it is also used for ink. The mesquite gum of *Prosopis juliflora* DC becomes black with ferric alum. A black ink was manufactured of the fruits of *Acacia farnesiana* (L.) WILLD. which are rich in tannins; it was also used to give leather a black colour. The fruits of *Jatropha curcas* L. were used by the Aztecs to obtain a black dye.
- A *purple* colour was extracted from *Mora* sp.

- A *red* dye, in a large variety of colour-tones, was obtained by the Aztecs from the bracts of *Euphorbia pulcherrima* and from the red flowers of the wild *Dahlia* (*D. variabilis* CAV.). Petals of *Cosmos sulphureus* CAV. were used as red and yellow dyes. The seed coat of *Bixa orellana* was used by the Indians to paint their bodies when going to war; it is still the source of a yellow pigment used for colouring butter, cheese and ointments.
- A *yellow* dye was extracted from the bark of *Erythrina americana* and from *Bursera mexicana*. The wood of *Diphysa robinioides* (palo amarillo) is also the source of a yellow dye.
- A *blue* dye was supplied by *Jacobinia tinctoria*. The fruit of *Vitis discolor* (synonym: *Cissus sicyoides*) and the wood of *Fuchsia parviflora* were used to extract a blue dye. From the bark of *Pithecellobium pachypus* which is rich in tannins, a dark blue dye was extracted.
- *Byrsonima crassifolia* and *Alnus accuminata* were used to obtain a brownish dye.

Dyes used for textiles in the ancient Peru are likewise mentioned by SCHWEPPE. The most ancient sample of a fabric is dated back to 2500 B.C. It is possibly made of *Gossypium barbadense*, the American cotton, and has a colour between ivory and dark brown. Indigo together with yellow wood and violet of cochineal was much used as a dye in the ancient Peru. A black dye was extracted from *Caesalpinia tinctoria*. Several species of *Relbunium* were also used for dyeing. The Inkas used mostly indigo of *Indigofera anil* (synonym: *I. suffruticosa*), a purple dye obtained from *Concholepas peruviana* (a purple snail), cochineal and the roots of *Relbunium*. These colouring matters were more recently identified by spectral analysis. Even lichens were used for dyeing by the Indians.

A table adapted from MAYOLO 1989 indicates the plant species applied for dyeing in the ancient Peru, as well as the plant parts used, and the respective colours obtained.

SCHWEPPE 1992 furthermore listed all the natural dyes known up to date, some of which are mentioned in the following:

- carotenoid dyes (yellow, orange, red);
- diaryloylmethane dyes (curcumin yellow of *Curcuma*);
- benzoquinone dyes (extracted from fungi and lichens);
- naphthoquinone dyes (lapachol of *Tabebuia, Tecoma* and other Bignoniaceae; juglone of *Juglans*);
- alkanna red of *Alkanna tinctoria* (droserone of *Drosera*);
- anthraquinone dyes (from species of *Rhamnus* and other Polygonaceae; from *Rubia, Galium, Morinda* and other Rubiaceae);
- indigo (of *Indigofera, Polygonum, Marsdenia, Lonchocarpus, Isatis*);
- flavonoid dyes (of *Reseda*; acacetin [apigenin] of *Robinia pseudacacia*; vitexin of *Vitex*; quercetin of *Quercus*; morin of *Chlorophora tinctoria, Artocarpus heterophylla, Morus alba*; myricetin of *Myrica*);
- anthocyanin as a red, violet or blue dye in the cell sap of many plants, depending on the acid or alkaline reaction of the medium;
- neoflavonoid dyes (of brazilwood, *Haematoxylon campechianum, Caesalpinia brasiliensis*);
- insoluble red woods (various species of *Pterocarpus*);
- xanthone dyes (of *Gentiana, Mangingifera indica, Garcinia, Symphonia globulifera, Iris, Arabidaea [Bignonia] chica*);
- alkaline natural dyes (of *Berberis, Hydrastis, Phellodendron* a.s.o.);
- alkaloid dyes (of *Erythrina, Symplocos* etc.);
- benzophenone dyes (of *Rhamnus, Chlorophora tinctoria, Aniba*);
- gallotannins for black, brown or dark blue colours (tannins in galls, bark, wood, fruits, leaves [of *Rhus, Coriaria, Quercus, Terminalia, Caesalpinia, Punica granatum, Astronium, Byrsonima, Persea, Tecoma, Malpighia*]);
- condensed tannins (in *Coniferae, Anacardiaceae*, catechu of *Acacia, Eucalyptus*, in mangroves such as *Rhizophora, Avicennia* etc., kino of *Pterocarpus*);
- naphthalene derivatives (gossypol of *Gossypium*, diospyrol of *Diospsyros*);
- chlorophyll, present in all green plant parts and the most important natural colour.

With the same dye, however, different colours may result in different materials, such as wool, cotton, silk or leather. Not all natural dyes are adequate for all textile fabrics, but special dyes are used for certain textiles. Most of the natural dyes are suitable for wool, but others are used for cotton or silk, although most of the dyes for wool are also good for silk. *Carthamus tincotorius*, for example, gives beautiful results with silk, but is not suitable for wool. That one and the same dye gives different results on different materials; we can even observe in our catalogues of modern fashion: a blouse is offered in the colours red, blue, orange and green, but the fabric of the blouse may change according to the colour; in red, the blouse is made of pure silk, in blue it is of a mixture of polyester and silk, in orange it is a mixture of silk and cotton.... because certain fabrics do not react well with certain dyes; a well-known problem colour is lilac.

Adequate plants for silk dyeing are, for example, *Carthamus tinctorius, Rubia tinctorum, Chlorophora tinctoria, Indigofera* species, *Haematoxylon campechianum, Rocella tinctoria, Punica granatum* and others.

Suitable for leather dyeing are *Carthamus tincotorius, Chlorophora tinctoria, Cotinus coggygria, Haematoxylon campechianum, Rubia tinctorum* etc. Black tones were most frequently used in former times; alum leather was preferred. A well preserved dyed leather was found in an egyptian tomb, more than 4000 years old. The quality of leather dyeing mostly depends on the type of tanning. In Venezuela, the bark of *Piptadenia peregrina* was not only used for tanning but also for dyeing leather with a red colour.

For dyeing processes, the water quality is of much importance, e.g. the content of calcium and particularly the presence of calcium hydrogen carbonate. Some dyeing processes are very complicated, such as the dyeing with turkish red, which also extends over a long time, as it includes oiling, fixation, dyeing, steaming and brightening. Other special techniques are batik (with wax) much used in Java, plangi (cutout technique: South Asia and Africa), and the ikat technique. The most important mordents for dyeing are alum and ferric salts; even plants are used as mordants. Good quality dyeing must be brilliant, unaffected by water, heat and air, and washable. As a consequence, a high price is unavoidable.

In spite of the complicated processes and the high prices, natural dyes are still used relatively frequently in our days, for example in food (butter, cheese, milk, sweets, pudding, liquor, farinaceous pastes, lemonade, fruit juice, wine, condiments), cosmetics (soap, tooth paste, shampoo, pomade, tinctures, hair tint), in vitamins, antibiotics, insecticides, wax, furs, ink and in microscopic techniques (haematoxylin), furthermore for dyeing paper and leather, as indicators (litmus paper) and analytic reagents, in pharmaceutics, and as artistic pigments. Some dyes such as alkannin, haematoxylin and quercetin are even used as antioxidants (SCHWEPPE 1992).

Natural dyes are principally applied by the native people for body painting, tattooing, as textile dyes and for painting and decorating wooden vessels, wooden fruits used as cups, for dyeing of fibrous material and of woven baskets, hats and the like.

A brilliant red is obtained from the seed coat of *Bixa orellana*, a dark violet from fruits of *Genipa americana*, a black colour from *Licania heterophylla* and *L. heteromorpha*, all growing in South America. However, the names of plant species which supply dyestuff are endless. UPHOF (1968) cites not less than 241 angiospermous genera (!).

As a body paint, the tattoo indicates to the native people the social status of the person. In certain cultures, a black dye is applied to paint the faces or bodies of corpses.

Dyeing of textiles and decoration with ornaments makes cloth and other personal belongings more attractive. But it is not only vanity that makes people adorn themselves, ornaments may also have a religious significance.

Tanning substances and leather tanning

For thousands of years, tanniferous plants have been used for tanning. Today however very different methods of leather tanning are known. Principally, artificial chemicals are used to make the leather durable and water proof. These processes are very rapid in contrast to the more time concuming and complicated tanning with plants. But artificial tanning products contain many toxic substances, whereas natural tanning substances are not toxic.

To make the tannin extraction profitable, plants which contain a high percentage of tannins have to be used for tanning. Tanning plants are therefore generally tropical trees, such as *Schinopsis*, occurring in Argentina (ROTH & BOLZON 1997), and *Rhizophora*, occurring in Venezuela (ROTH 1992). The most important tanning plant is the quebracho tree, *Schinopsis lorentzii*, native of the Argentine Chaco, of Bolivia and Paraguay, whose heartwood supplies tannin. However, fast-growing *Acacia* species with tanniferous barks are competing with the Quebracho, for example, Mimosa bark and, Wattle bark. However the plants are mainly cultivated in Africa instead of South America where they originated.

In the meantime, the leather industry is aware of the significance of ecologically perfect products, and mistakes of the past are being corrected. Certain heavy metals and other toxic substances were eliminated from the European production (Ciba Geigy). Leather tanning with vegetable tannins has been improved and can now be processsed within a few days. But conservation ingredients must be active against fungi and bacteria, on the one hand, and be biologically inactive, on the other. This is a contradiction in itself – only the one or the other requirement can be fulfilled. Furthermore, for leather tanning with chromium only 400 000 tons of chromium sulfate are needed yearly throughout the world, on the contrary for vegetable tanning, 2 million tons of tannin extract would be necessary world-wide, every year a quantity which is not available.

Specific healing properties of plants

Effects of monocotyledonous medicinal plants on cardiac, respiratory and cancerous diseases were studied by YAJIA et al. (2000) in Argentina. Approximately 40% of the Argentine monocotyledons are found in the Province Misiones. *Bromelia balansae* MEZ., Bromeliaceae, *Herreria montevidensis* KLOTZSCH EX GRISEB., Liliaceae, Eichhornia crassipes (MART.) SOLMS-LAUBACH., Pontederiaceae, *Acrocomia totai* MART., Arecaceae, and *Syagrus romanzoffiana* (CHAM.) GLASS, Arecaceae, are used in popular medicine for cardiac, respiratory and malignant or neoplastic diseases. The authors describe the used plant parts, healing properties, method of administration, dose and presentation, and also give anatomical data for plant organs. *Herreria montevidensis* and *Syagrus romanzoffiana* are used for diseases of the heart, and in both plants, the roots are the useful organs. Of *Bromelia balansae* and *Acrocomia totai*, fruits and roots are used to treat respiratory affections, and only *Eichhornia crassipes* has antitumour properties in its aerial parts.

YANES et al. (2000) studied the toxicity of aqueous plant extracts of *Melia azedarach, Moringa oleifera* and *Ocimum basilicum* on Trypanosoma cruzi (mal de chagas) and found a significant trypanocide activity.

Studies of this kind can only be carried out when healing properties of plants are already known. The active chemical compounds of these plants then have to be investigated experimentally and chemically.

Material and methods

The material studied for this book comes principally from Venezuela. However, species mainly occuring in other parts of South America are also included. Most of the species analysed occur as wild plants in Venezuela. Nevertheless, some plants that are cultivated in Venezuela but native to other countries of the world, have also been included. A large part of the species in question is still unknown outside of Venezuela, but several medicinal plants are already well known.

The plants were collected in different regions of the country, at different sites and in different plant associations. Besides, some drugs were bought in the so-called herbolarios, little stalls, along the streets or at markets in the interior of Venezuela. A herbarium sheet for each species was prepared, showing shoots, twigs, leaves, and, where possible, flowers and fruits as well. Vaucher specimens of the collected plants are preserved in the National Herbarium of Caracas and can be consulted there. Collector's names, numbers of their vaucher specimens, place and date of collection are enumerated.

Part of the material was studied fresh, another part was fixed in 70% alcohol, FAA or formalin. The fresh material was treated with several reagents, with Sudan III to demonstrate cutin, suberin and oily or fatty substances; phloroglucine in HCl has been used to test for lignin. Methylene blue was used to test for slime, I in KI to test for starch and $FeCl_3$ for tannins.

Each species is described morphologically and anatomically. Either all important organs of a species are described or only its useful parts, when it seemed sufficient.

For the description of the inner plant structure, we made either hand sections of fresh material or we used the time-consuming embedding method in paraffin. In this case, fixed material was first dehydrated by ascending solutions of alcohol and water, then put in pure alcohol and afterwards into xylol. The dehydrated material was finally embedded in paraffin. The ready paraffin blocks were cut with a Jung sliding microtome about 15 μm thick. The handslides, however, were about 20–25 μm thick. The slides were dyed with different stains, mainly with toluidine blue, methylene blue, safranin/fast green or haematoxylin. Then to preserve them the slides were embedded in Euparal, Caedax or other resinous stubstances.

Plant families, genera and species are arranged alphabetically to facilitate consultation. A short characterization is given, related to the use of the most important species of each family. Close botanical relations were occasionally indicated to stress which may be the possible constituents and uses of the species in question and in which direction further investigations should be undertaken.

Whenever possible, the following data were procured:

- vernacular name of the species in Venezuela, a short taxonomical description (including only the most important information),
- origin of the species,
- historical background,

- occurrence,
- anatomical description (stem, leaf, root, fruit, seed),
- ethnobotanical and general use induding a) nutritional use for men and animals, b) economical utilization (possible industrial use), c) medical use if advisable (leaf, stem, bark, root, flower and fruit with seed are discussed separately),
- method of use,
- healing properties,
- chemical contents,
- varieties and related species,
- cultivation,
- observations (specific characteristics for identification).

With reference to the bibliography, we laid most emphasis on the local publications. For foreigners, however, these may be difficult to obtain. The books most used were: the series of SECAB, UPHOF, GUPTA, RODRIGUEZ, BRÜCHER, SCHULTES & RAFFAUF, HOYOS, and the books dealing with local medicinal plants written by several Venezuelan authors.

To mention all the useful and medicinal plant species of South America known to date would fill about 100 volumes. The SECAB series including certain species of Bolivia, Columbia, Peru, Ecuador, Venezuela and Chile already comprises 15 volumes. But it is far from complete: Many economically useful and medicinal plant species and even genera or entire families are missing completely.

Each species was studied in the following way: The vernacular name or names of each species used in Venezuela are mentioned. The taxonomic description is followed by the mention of the probable origin of the species, the historical background and the actual occurrence of the species in the country. In the anatomical description, we either mention all parts of the plant or only the useful organs. As far as possible, economical utilization of the species, the names of the drugs, the directions for use, the healing properties, the chemical contents (as far as studied), toxicity, varieties and related species, possible cultivation, special characteristics which facilitate identification and the corresponding bibliography were described. Naturally the above mentioned information was not available for all species studied and certain information is missing here and there.

We laid weight on the intention to study only little known plants of medicinal or economical use, but we also included some better known ones. We placed emphasis on the South American or Venezuelan origin of plants. But we thought that also some plants not indigenous to South America, but cultivated in this region, should be added.

Most specimens were recently collected and studied, some however, have been adopted from our earlier publications.

The recently studied plants which were collected at different sites, are all preserved in the form of a professional herbarium which will be incorporated in the National Herbarium of the University Caracas, Instituto Botanico, Venezuela.

As to the chemical contents of already (partly or completely) studied medicinal plants, we were not able to mention all the secondary compounds found in each species, so we only mentioned the most common ones. Readers who are interested to learn all the results in detail are referred to the bibliography cited and/or may consult the SECAB series and GUPTA 1995.

The recipes for the preparation of drugs are adopted from other books, but no guarantee is given that the mentioned drug has the desired effects. Neither can the possibility of undesired secondary effects be entirely excluded. Our contribution has to be understood as an attempt to show the numerous possibilities of using neotropical plants medicinally or economically. It is suggested that their chemical structure should be studied thoroughly first and that their healing effects, their possible toxicity and secondary effects should be studied experimentally. Certainly, Indians carried out experiments on themselves, possibly for hundreds of years, but they may have already been accustomed to certain poisonous substances or may have been or may be more resistant to drugs than non-native people.

Plant families which supply medicinal and economically useful species

As already mentioned, we arranged the plants we studied according to their taxonomical relationship. Plant families, genera and species are presented in alphabetical order. In this way, certain characteristic contents or structures become obvious in respective families or genera. As aggregate species which are alike in their stucture and contents are frequently found in the tropics, our species arrangement facilitates the discovery of common secondary compounds or of other valuable products in related species.

Kapok, for example, is obtained from *Ceiba pentandra*, Bombacaceae; however, other species of

Ceiba as well as related genera supply the same fibers, some of even better quality. The knowledge of aggregate species may therefore be valuable for the detection of new chemicals, new chemical compositions or new economically valuable products.

Certain families are well-known for their latex, e.g. the Euphorbiaceae and Sapotaceae. The latex of the distinct genera and species is, however, variable in its composition so that latex of distinct species may serve different purposes. The basic substance used in chewing-gum manufacture is found in quite a few Sapotaceae, and not only in *Achras* (*Manilkara*) *zapota*. Citation of similar products in related (aggregate) species could be continued at will. Investigations in this direction may be very fruitful to find substances adequate to combat certain diseases.

On the following pages, the reader finds a selection of useful and medicinal plants we have studied. The following families are mentioned: Acanthaceae, Amaranthaceae, Anacardiaceae, Annonaceae, Apocynaceae, Araceae, Araliaceae, Asclepiadaceae, Berberidaceae, Bignoniaceae, Bixaceae, Bombacaceae, Boraginaceae, Brunelliaceae, Burseraceae, Cactaceae, Caesalpiniaceae, Cannaceae, Capparaceae, Caricaceae, Caryocaraceae, Celastraceae, Chenopodiaceae, Chrysobalanaceae, Cochlospermaceae, Combretaceae, Commelinaceae, Compositae, Convolvulaceae, Cucurbitaceae, Cunoniaceae, Dichapetalaceae, Ebenaceae, Elaeocarpaceae, Ericaceae, Euphorbiaceae, Flacourtiaceae, Gentianaceae, Gesneriaceae, Gramineae, Guttiferae, Hernandiaceae, Humiriaceae, Iridaceae, Krameriaceae, Labiatae, Lauraceae, Lecythidaceae, Loranthaceae, Malpighiaceae, Malvaceae, Marantaceae, Melastomaceae, Meliaceae, Mimosaceae, Moraceae, Musaceae, Myricaceae, Myristicaceae, Myrsinaceae, Myrtaceae, Nyctaginaceae, Ochnaceae, Olacaceae, Opiliaceae, Palmae, Papilionaceae, Passifloraceae, Phytolaccaceae, Piperaceae, Plantaginaceae, Polygonaceae, Punicaceae, Quiinaceae, Rhamnaceae, Rhizophoraceae, Rosaceae, Rubiaceae, Rutaceae, Sabiaceae, Sapindaceae, Sapotaceae, Scrophulariaceae, Simaroubaceae, Solanaceae, Sterculiaceae, Tepuianthaceae, Tiliaceae, Umbelliferae, Urticaceae, Verbenaceae, Violaceae, Vochysiaceae, and Zygophyllaceae.

About 120 species were studied very thoroughly, while many others (more than 200) were studied merely in some aspects or are only mentioned, but have not been exhaustively investigated.

PART II Plant Families

Acanthaceae

The family mainly comprises herbs and shrubs, often climbing, while trees are rare. The leaves are usually decussate. The flowers are bilabiate, the ovary is bilocular. The fruit is a loculicidal beaked capsule opening with 2 valves. The funicle develops a hook-like process which aids in the expulsion of the seeds. The seed coat often has a slimy epidermis. Seeds are usually without endosperm, but with a large embryo.

2600 species, mainly growing in tropical and subtropical regions, have been identified.

Polyphenols, ethereal oils, non-volatile isoprenoids and alkaloids have been identified. *Justicia* is best studied chemically, otherwise the family is only slightly investigated.

Diuretic and antiasthamtic activities have been reported.

Justicia

The genus has been best studied chemically.

About 300 species of herbs and small shrubs occur in the tropics and subtropics.

Aromatic oils of some species are employed as moth repellents or insecticides, as well as fixatives in perfumery and in the soap industry. The leaves can be used as an incense.

Lignans are very important constitutents, aromatic amines, kaempferol, sterols and salicylic acid are also found.

SCHULTES & RAFFAUF mention 10 useful species, UPHOF two.

Justicia stipitata WASSHAUSEN & ARROYO. is a very frequently found small shrub of the undergrowth of the Venezuelan cloud forest proper and was described in 1976 by WASSHAUSEN & KALIN DE ARROYO. The leaf anatomy has been studied by ROTH (1990). It would perhaps be interesting to study the chemistry of this species.

Justicia pectoralis JACQ. is a further species occurring in Venezuela in humid areas of hot and temperate regions. Medicinally, the plant is used as a tranquillizer, pectoral, astringent, vulnerary, antihaemorrhagic, antibacterial, antirheumatic, for gout and rheumatism, dysmenorrhoea, skin rashes, and as a curative of the liver.

The extracted coumarin has antiinflammatory and cicatrizing properties; Umbelliferone and swertesine have sedative and relaxing properties on the smooth muscular system.

Ruellia

All species are perennial shrubs or herbs.

Flavonoid devivatives, sterols and putrescine have been isolated. *R. tuberosa* contains flavonoids, sterols, n-alkanes and C_{28}–C_{34} acid esters.

Quite a few species are ornamentals and preferred green house plants.

Ruellia macrophylla VAHL. is a small shrub of the undergrowth of the Venezuelan transitional cloud forest, reaching a height of 3 m.

The leaf anatomy has been studied by ROTH (1990). Unicellular short and long hairs are characteristic of the lower epidermis.

Ruellia tuberosa L. (yuquilla, raiz de barreto, oreja de ratón, escopetilla, peonilla)

Taxonomical description

Perennial herb, 13–80 (150) cm high, ramified, erect (Fig. 4). The fusiform roots reach 0.3–0.7 cm in diameter. The fleshy-fibrous roots may attain more than that. The erect quatrangular axis which shows a white pubescene on young twigs, is woody down to the base. The simple leaves are opposite and have a short petiole. The foliar blades are of papery consistency and are obovate to elliptic-ovate. They measure 4.5–12 cm in length and 1.5–5 cm in width; the apex is acute to slightly acuminate; the entire margins are undulated; the leaf base is cuneate; the white pubescence on both leaf surfaces is covered by simple trichomes of a length up to 0.1 cm. The venation is pinnate, brochidodromous with 9–11 pairs of secondary nerves; the midrib is prominent on the abaxial side and slightly sunk below the upper side. The 0.5–1 cm long petiole is surcate and has a white pubescence.

The axillary inflorescences are dichasial and 3–10 cm long. The zygomorphous flowers have a profoundly 5-partite calyx with filiform lobules, about 1.5–2.0 cm long and show a scarce pubescence. The tubular crown is of violet colour (and only seldomly white), 2.5–3.5 cm long, with suborbicular lolubels, 1–1.5 cm long. The crown reaches a diameter of 1.3–1.5 cm. There are 4 stamens. The ovary is bilocular with several to many ovules in each locule. The stigma has 2 unequal lobules.

The capsular fruit is cylindric, and 1.8–3 cm long. The 20 ore more suborbicular seeds in each locule are flattened and measure 0.2–0.3 cm in diameter. The capsule has a hygroscopic dehiscence mechanism and opens explosively in wet weather (ROTH 1977).

Origin

Neotropics; possibly West Indies.

Occurence

New world subtropics and tropics. In Venezuela the species frequently occurs in hot and sunny places. It grows mostly at low elevations, but can go up to 500 meters above sea level.

Anatomical description

Leaf. Anatomical details described in this section can be found in Fig. 4 (b, c, f) and Fig. 5. The upper epidermis is larger-celled and slightly papillose. Most conspicuous are cystoliths in both epidermal layers; they are elongated parallel to the leaf surface, but in different directions; as seen in a surface view, their walls are less bent than those of the regular epidermis cells. Stomata are present in both, upper and lower epidermis, but are much more numerous in the latter. They are slightly elevated above the surface. They have 2 perpendicularly oriented subsidiary cells (subtype c, which type was found frequently in the Rubiaceae by ROTH & CLAUSNITZER (1969) and which is called cross-celled or diacytic by METCALFE & CHALK (1950). Small glandular hairs with a unicellular foot, a unicellular stalk and a head with 8 cells, are common in the lower epidermis and less frequent in the upper epidermis. Simple multicellular hairs occur on both surfaces.

As seen in a surface view, the cystoliths, which also occur in the mesophyll, are often bent and some have one pointed end. The cells of the upper epidermis have undulated anticlinical walls, while those of the lower epidermis are bent or even straight. As a general rule, the opposite is more common, namely that the cell walls of the lower epidermis are more undulated.

Both epidermal layers are much alike, but differ in the size and shape of their cells and in the frequency of stomata and of simple and glandular hairs which are more frequent in the lower epidermis.

There is only one layer of palisade parenchyma, composed of short occasionally funnel-shaped cells. The spongy parenchyma comprises about 3 layers of loosely arranged cells with short arms.

The mesophyll is thus little developed and the leaf is thin.
The veins of secondary order have a single arc-shaped vascular strand which is surrounded by a parenchymatous sheath. The midrib has also an arc-shaped vascular bundle and is reinforced on upper and lower side by an angular collenchyma which replaces the palisade parenchyma on the upper side.

Axis. A young stem, about 3.5 mm in diameter, shows a more or less quadrangular transection. The epidermis is relatively large-celled and contains cystoliths as well as simple uniseriate hairs. Beneath follow several layers of collenchyma. The majority of the primary cortex is parenchymatous and transformed into an aerenchyma with larger intercellular spaces. Cystoliths are found dispersed in the periphery of the cortex. The vascular cylinder is conspicuously tetragonal. The endodermis shows Casparian stripes.

The pith is well-developed and very large-celled. Some cells of the cortex and the pith contain crystal needles. The 4 angles of the axis are due to radially arranged rows of parenchyma cells: each cell divides periclinally several times (Fig. 4 d and Fig. 6).

Hypocotyl. The transitional region between axis and root shows an epidermis with abundant root-hairs. About 1–3 cell layers beneath are collenchymatous. The parenchymatous primary cortex is very well developed; its cells are radially arranged and form an aerenchyma with intercellular spaces between the cells. The endodermis has Casparian stripes. The vascular system appears in the form of a ring which is composed of individual bundles. The pith is likewise large and parenchymatous. Dispersed in pith and cortex are cystoliths and idioblasts with crystal aggregations.

Ethnobotanical and general use

Nutritional use

The seeds are a good source of unsaturated fatty acids which can be used nutritionally.

Medical use

Name of the drug: *Ruellia tuberosa* L. planta, folia, caule, radix, flos.

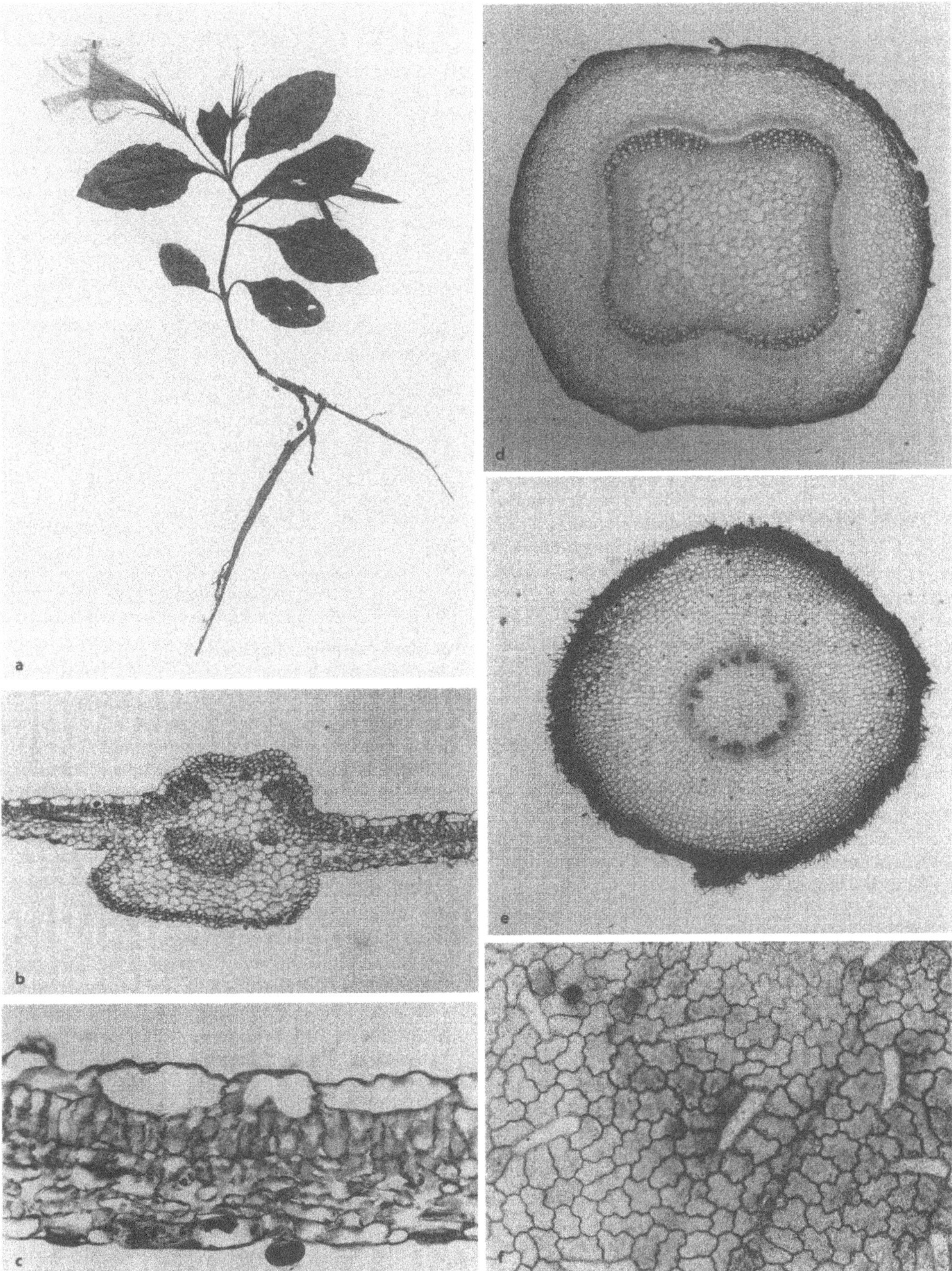

Fig. 4. *Ruella tubersa.* **a** habitus, **b** t.s. of midrib, **c** of blade, **d** of axis, **e** of root. **f** Lower leaf epidermis, surface view.

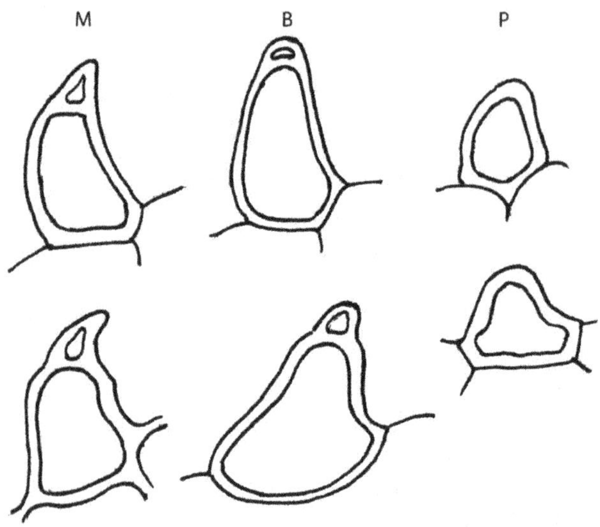

Fig. 5. *Ruellia tuberosa*. Marginal hairs (M), bicellular hairs (B), Papillas (P).

Entire plant. A decoction of the entire plant is used to cure wounds; it is also applied against high fever. A decoction of the plant with its root is used against diseases of the urinary tract.

Leaf. Locally applied the leaves have an anti-inflammatory effect and cure eruptions in the mouth. A decoction helps against diseases of the bronchi. Leaf and shoot (axis) in decoction are applied against inflammations of the mammary glands.

Axis. A decoction of leaves together with the axis is used against eruptions in the mouth and to cure inflammations of the mammary glands. The shoot is applied against inflammations and affections of the thorax.

Root. The root is applied against oliguria, diabetes, hypertension, heat, flu, headache, constipation, gonorrhoea, leprosy, as a purgative and vomitive, a disinfectant of the genitourinary tract, a febrifuge and purifier. The entire plants and the roots are usually applied in the form of a decoction; An infusion of the root is vomitive and said to cure syphillis and infections of the urinary tract.

The root is the organ of the plant which is mainly used ('raiz de barreto').

Flower. The sap of the flower heals wounds.

Method of use

The leaves may be applied fresh and locally. The sap of the flowers is used to heal wounds. Entire plant, leaves and root are usually applied in the form of a decoction, infusion, tisane, or a tea.

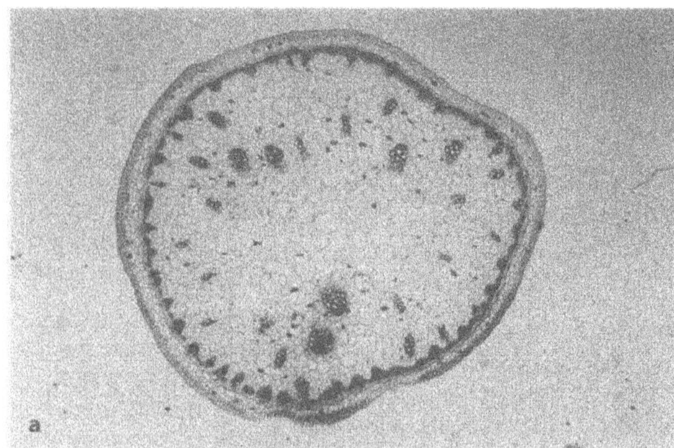

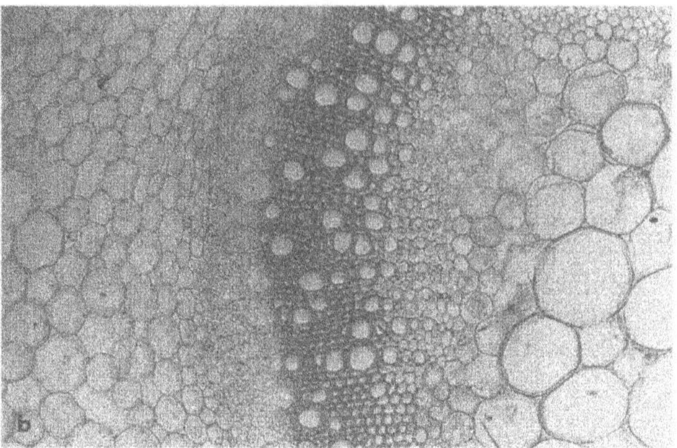

Fig. 6. a *Amaranthus viridis*, axis in t.s. 2.5 × 5.5. **b** *Ruellia tuberosa*, t.s. of axis. Note the large pith cells, xylem, phloem and part of the cortex.

A decoction of the plant is used to combat high fever. An infusion of the root helps against vomiting and cures syphilis.

Healing properties

Febrifuge, emetic, purgative, vomitive, disinfectant, purifying. These properties are mainly ascribed to the root.

Chemical contents

Among many other compounds the plant contains gallic acid, lauric acid, myristic acid, resins, phytosterols, lupeol, flavonoids, aspigenin glycosides, sitosterols.

Varieties and related species

The species is not very thoroughly studied yet. But all species of the genus *Ruella* seem to have similar medicinal properties. The roots of all species of the genus have purgative and emetic effects (SANTIAGO CORTES cit. by GARCIA-BARRIGA 1975).

Cultivation

The plant prefers hot regions up to 500 m a. s. l. and open sites. It is common in pastures and other non-arable grassy places, will tolerate most conditions and grows like a weed.

Observations

The most conspicuous peculiarity of the species are the cystoliths which occur in leaf, axis and hypocotyl. A further characteristic are the subsidiary cells of the stomata which belong to the 'cross-celled' type. The quadrangular shape of the axis with its cortical aerenchyma and the tetragonal shape of the vascular cylinder are further distinguishing characteristics.

Amaranthaceae

The Amaranthaceae are a small family closely related to the Chenopodiaceae. Anomalous secondary growth of stem and root is very frequent. The inflorescences are spikes, racemes, heads or panicles. Tepals and bracts of the flowers are often dry, strawy and brightly coloured. Several species are therefore ornamental plants. The leaves of *Amaranthus lividus* VAR. *ascendens* and VAR. *oleraceus* supply vegetables (Chinese spinach).

The genus *Amaranthus* L. includes domesticated species which produce grain, others are ornamentals or serve as colouring matter. Amaranthus also includes non-cultivated and non-domesticated species which are eaten as vegetables (leaves) by the natives. *A. hypochondriacus* L. is principally considered a producer of seeds, but was also selected as an ornamental, and in certain regions, the leaves are used as a vegetable. The species which are used as vegetable, generally represent weeds of the field or ruderal plants. At least 10 such species are used by the natives. The grain has an agreeable nutty taste.

The seed protein of *Amaranthus* is of high value and is suitable for use in bread, cookies, sweets, creme, cakes, muesli etc. *Amaranthus* was much used in precolombian times; it was cultivated by Mayas and Aztecs for about 8000 years, but after the conquest its cultivation was prohibited by the Spaniards. It is, however, very easy to cultivate.

Amaranthus viridis L. (pira, pira blanca, bledo blanco, cadillo de empedrados)

Taxonomical description

This species is a herbaceous annual plant, 30–60 (150) cm high, ramified and erect (Fig. 7). The stem is cylindric and fleshy. The alternating leaves are

Fig. 7. *Amaranthus viridis*, habitus.

simple, entire, 9–18 cm long (including the petiole) and glabrous. The foliar blades are slightly fleshy at the touch, of ovate outlines or elliptic to rhomboid, 5–10 cm long and 3–8.5 cm broad, having a rounded to retuse tip and a cuneate base. The venation is pinnate, broquidodromous, with 7–9 pairs of secondary nerves. The middle nerve is prominent on the abaxial side, but sunk below the upper surface. The petiole is 3.5–10 cm long, is ribbed and glabrous. The lateral stipules are linear-lanceolate, 0.2–0.3 cm long and caducuous. The spikes are axillary and terminal, 2.5–15 (29) cm long. The green bracts are ovate to apiculate and 1.5–2.5 cm long. The flowers are 02.–0.3 cm long. The brilliant black fruit has approximately circular outlines, but is flattened, and measures less than 0.9 mm.

Origin

The species is probably indigenous to Central and South America, e. g. Peru, Mexico, Eucador.

Historical background

Amaranthus species have been cultivated from time immemorial in both Central and South America, particularly by the Aztecs and Mayas. BRÜCHER (1989) reminds that cruel rituals were undertaken with amarant grains which were mixed with the blood of sacrified humans to form replicas of Aztec goddesses. As a consequence, the Christian 'Conquistadores' prohibited not only the ceremonies, but also the cultivation of amarant, so that only small plots survived which now represent an invaluable genepool (see also VÉLEZ & VÉLEZ 1990).

Occurence

According to PITTIER (1970), the species is found in central and east Venezuela where it is called 'pira blanca', while in Tachira it is known under the name 'bledo blanco'. It is a pantropical weed.

Anatomical description

Leaf. (Fig. 8 a, b, d and Fig. 9 a, b, c). The bifacial amphistomatic leaf has a 'Kranz' structure. The upper epidermis is single-layered and consists of large cells with somewhat undulated walls, as seen in surface view. The cuticle is delicate. Stomata occur infrequently and are found at epidermis level; they are without subsidiary cells (anomocytic). Capitate haris with an excentric ellipsoidal head and with 'yellow walls to the short basal cells of the stalk' (METCALFE & CHALK 1950, P. 1069, SOLEREDER 1908) occur sporadically.

The palisade parenchyma forms a single layer of more or less isodiametric cells, some of which are, however, longer than broad. The spongy parenchyma has about the same proportion as the palisade parenchyma and consists of large lobulate cells.

The middle nerve projects above the lower surface, but is sunk below the upper leaf surface. A horseshoe-shaped vascular bundle lies in the center which seems to be bicollateral. Above the vascular bundle (on the adaxial side) the palisade parenchyma is interrupted by large parenchymatous cells; even larger parenchymatous cells are observed on the abaxial side below the bundle. The secondary nerves are similar to the median nerve, but project less above the lower leaf surface. Nerves of higher order which are abundant in the leaf are surrounded by a parenchymatous sheath; the sheath cells are large, nearly globose, as seen in t.s., and contain large chloroplasts arranged along the inner wall side (close to the vascular bundle) in the form of a horseshoe. The chloroplasts are conspicously larger than those of the rest of the mesophyll The 'Kranz' structure of the leaf is thus very obvious. *Amaranthus viridis* thus seems to belong to the C4-type of photosynthesis plant. The lower epidermis which is also unistratified consists of large cells with strongly undulated walls, as seen in surface view. The cuticle is delicate. The stomata which are more frequent on the lower side are likewise 'anomocytic' and do not show companion cells. They are also larger than the stomata on the upper side. Idioblasts, each with a large druse of calcium oxalate, are frequent in the mesophyll.

Shoot. (Fig. 6 a) In a shoot of about 4–5 mm in diameter, 3–4 (5) layers of a small-celled angular collenchyma are situated below a small-celled layer of epidermis cells. Two layers of large parenchymatous cells follow towards the inside some of which are completely filled with crystal sand. A continous vascular cylinder in which the original vascular bundles (united by a continuous cambial zone) are still conspicous, encloses the ample pith. This peripheral bundle ring represents the first step of the beginning anomalous growth in thickness. The interfascicular cambium between the bundles mainly forms thin-walled parenchyma towards the outside and thicker-walled, later-on lignified, parenchyma towards the xylem side. The meristematic zone which proceeds from the pericycle thus produces

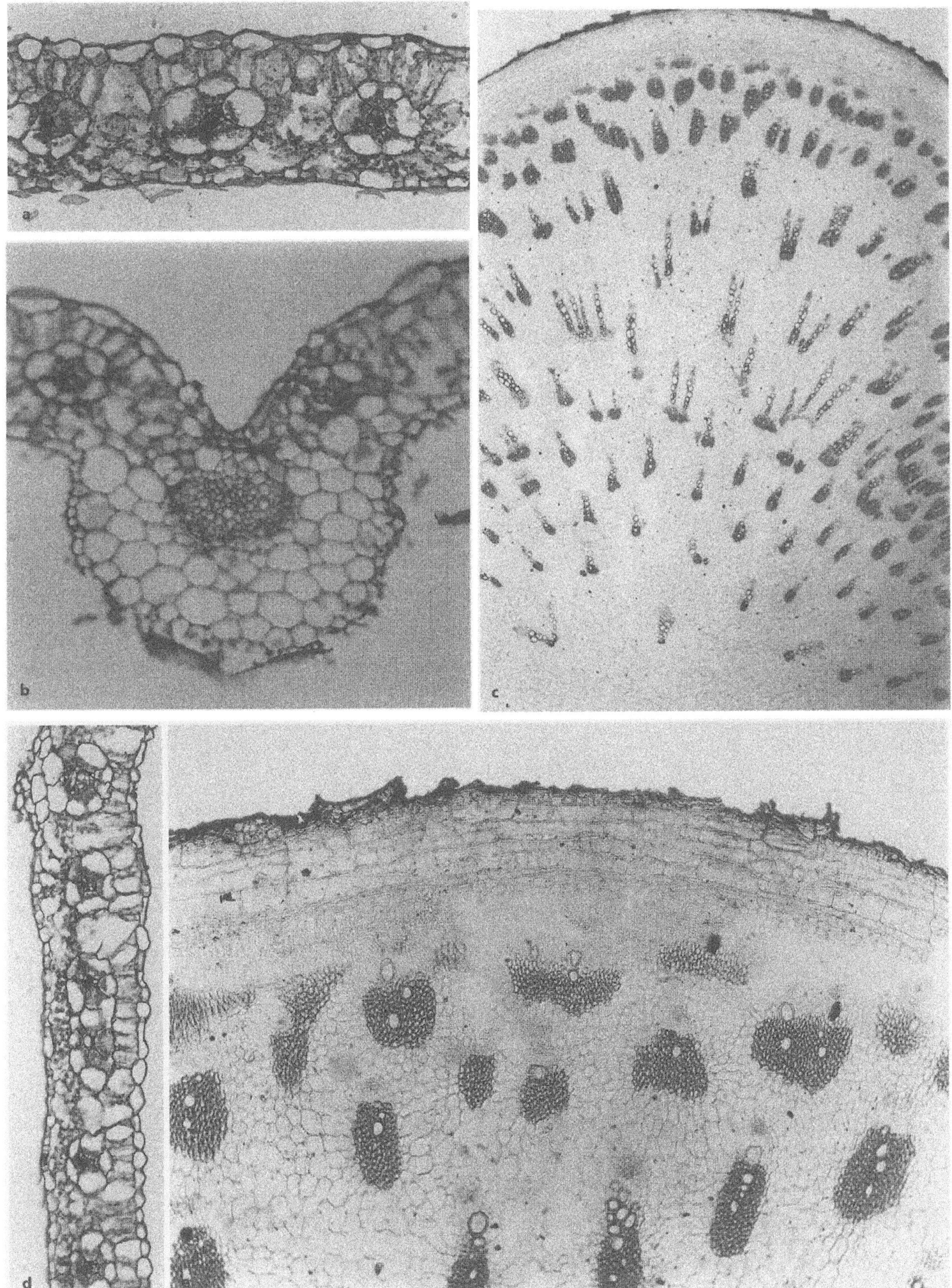

Fig. 8. *Amaranthus viridis.* **a** t.s. of lead blade, **b** of midrib, **c**, **e** t.s. of root, **d** leaf. **a**, **b** × 100, **d** × 88, **c** × 14, **e** × 44.

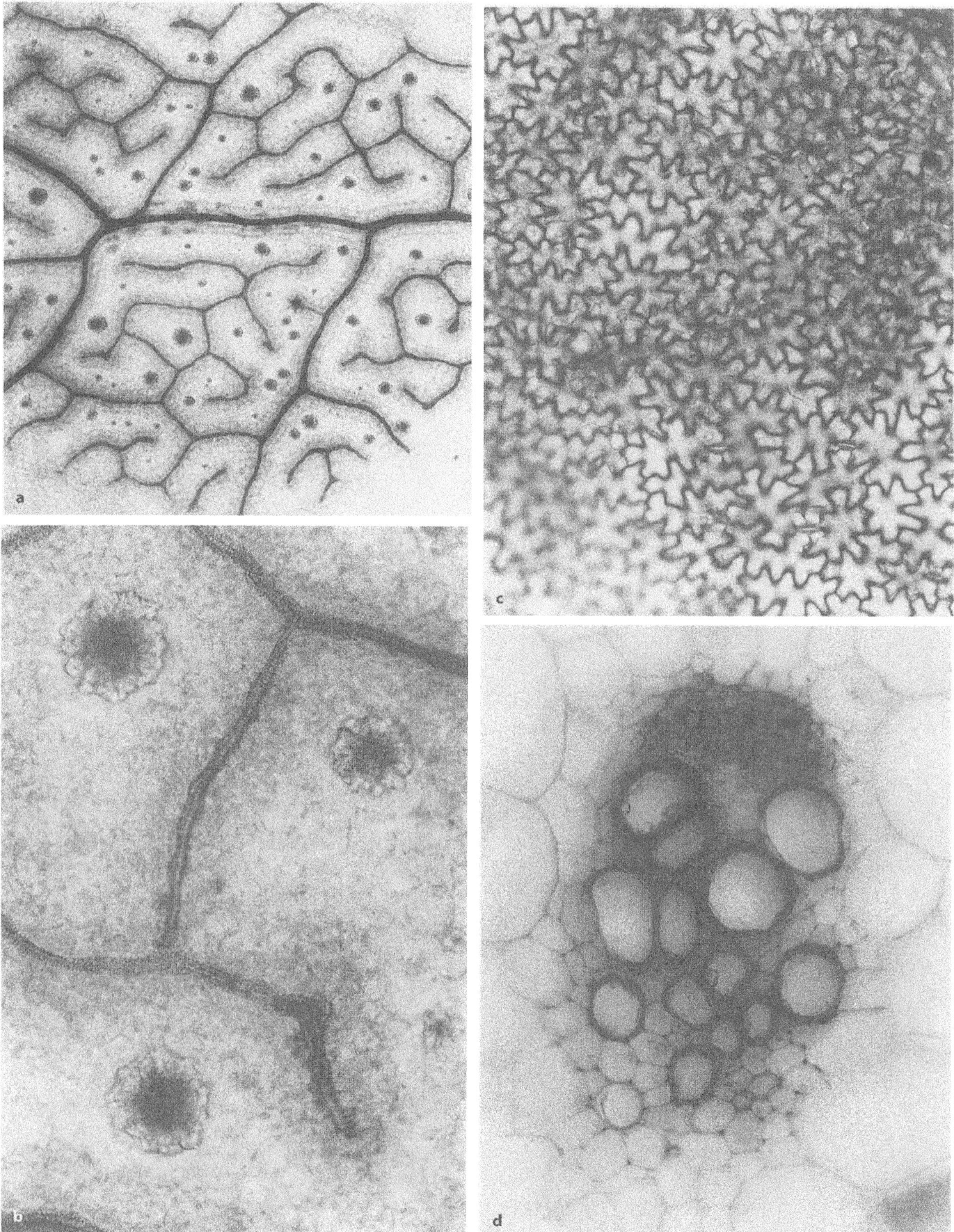

Fig. 9. *Amaranthus viridis.* **a** Venation pattern of leaf, **b** large druses of calcium oxalate in the leaf (× 250), **c** lower epidermis (× 160), **d** vascular bundle of the axis in t.s. (× 100).

collateral vascular bundles as well as conjunctive tissue towards the inside and consequently always remains in the periphery. However, the formation of bundles is not regular, being advanced in some spots and delayed in others. The pith in the center is very ample, being composed of large thin-walled parenchyma cells; the largest cells are found in the very center. Many of the pith cells are likewise filled with crystal sand. Embedded in the pith are irregularly dispersed collateral vascular bundles. The larger vessels seem to be partly filled with crystal sand (Fig. 6 a; 9 d).

Root. The structure of a principal root about 4–5 mm in diameter very much resembles that of the shoot (Fig. 8 c, e). The few-layers of cork show an irregular aspect and proceed from periclinical divisions of individual cells without formation of a phellogen. The primary cortex consists of large and thin-walled parenchyma cells which are tangentially extended and often contain crystal sand. An irregular and discontinuous vascular cylinder already broken up into arcs follows towards the inside. The cambium which proceeded from the pericycle forms entire collateral vascular bundles as well as parenchyma towards the inside, but it works asymmetrically so that only a half vascular ring is present on one half of the root, while in the other half up to 3 arcs are already visible. The conjunctive parenchyma formed by the cambium toward the inside is thin-walled; but in the outermost arc (the most recently formed), the parenchyma connecting the bundles tangentially is thicker-walled and lignified. The presence of crystal sand which lines the wall inside of many vessels is very strange.

The center of the root is occupied by the protoxylem rays (about 5). From there, lateral roots may ramify. The ample pith composed of thin-walled parenchyma accommodates irregularly dispersed collateral bundles. Many of the parenchymatous pith cells contain crystal sand.

The asymmetric growth of the cambium becomes even more conspicuous in a root of about 13 mm in diameter. While in one half of the root about 2–3 arcs have been formed, up to 7 arcs may be observed in the other half which then shows a favoured growth. The secondary growth is however very irregular and the xylem of the bundles on the favoured side mainly consists of long radial rows of vessels.

Both, shoot and root, thus show the same type of anomalous growth in thickness.

Ethnobotanical and general use

Nutritional use

There are quite a few species of *Amaranthus* which have nutritive value. The leaves are used as vegetables and the seeds as a substitute of cereals (Inca wheat) (Correa & Bernal 1989). The vegetable amarants are rich in vitamin A and C and in minerals, and have a fair content of protein, but oxalates and nitrates are antinutritional substances (Brücher 1989). The seeds contain 6 % fat, 10–17 % protein. The amarant starch is composed of very small granules, with amylopectin as its major component (Brücher 1989). Young plants of *Amaranthus viridis* are eaten in Venezuela like spinach. Meal can be obtained from the seeds which are also eaten raw with milk as muesli. The leaves of *Amaranthus viridis* contain 6.7 g% proteins, 0.5 g% fat, 28.8 g% carbohydrates and 10 g% fiber. The seeds show about the same proportions, but have more proteins (13 g%) and 65 g% carbohydrates, and furthermore contain phosphorus, calcium, iron, vitamin B & C, and linoleic acid (Albornoz M.A. 1993).

Medical use

According to Albornoz (1993), the plant is used for inflammations, fever, pectoral diseases, intestinal parasites, and to cure wounds. It is also applied for contusions, rheumatism, cough, and ulcer. Related species such as *Amaranthus bilitum* L., *A. caudatus* L. and *A. dubius* Martius have similar medicinal properties (Correa & Bernal 1989).

Varieties and related species

Related species which are used as vegetables are *Amaranthus hypochondriacus* L., *A. cruentus* L., *A. quitensis* H.B.K., *A. caudatus* L., *A. mantegazzianus* Pass. Brücher (1989) describes them as vegetable crops.

Cultivation

The above mentioned leaf-producing amarants are adapted to many different ecological environments. These plants use the C4 carbon fixation which enables them to support a hot and dry climate, and are thus easy to cultivate (Brücher 1989).

Observation

The species is well characterized by its anomalous growth in shoot and root, the Kranz type of the mesophyll with a conspicous sheath around the bundles, the small starch grains, the large druses in the leaf and the crystal sand, particularly in the vessels which is a very unique property.

Amaranthus caudatus L. (hierba caracas, pira, coime, amaranto).

The infusion of the inflorescence is used as an astringent to inhibit metrorrhagia. In Peru, the species is used for bronchitis and to treat tuberculosis and pulmonar problems. In Venezuela, the plant is considered useful to oxygenate the cerebrovascular system; 2 cups of a strong infusion are taken daily or the leaves are prepared as a salad. The plant is also applied to cure paranoia, hallucinations, autism and other neurotic diseases, as well as to ease mental disorders and nightmares, to help to recuperate the immune system after bacterial or viral

Fig. 10. *Anacardium occidentale*, habitus. **a** as a tree, **b** shrub like.

infections. The infusion stimulates the pituitary and the thymus gland.

Amaranthus spinosus. The entire plant in decoction is used for colics. Externally, it is applied on sores and ulcers and for cicatrization.

Anacardiaceae

The Anacardiaceae are mostly woody plants which occur mainly in tropical and subtropical regions of the New as well as of the Old World. The fruit is usually a drupe with a mesocarp rich in resinous substances (e.g. mango). The leaves may be simple or compound. The family is rich in tannins and always forms schizolysigeneous resinous canals with etheric oils or with a rubber-like balm (particularly in the bark). The family is best known for its phenols and phenolic acids (anacardol, anacardic acid) causing serious skin irritations. Anacardic acid has been reported to have anthelmintic activity. Terpenes, triterpenes and polyphenols are also common.

Many species are fruit trees (mango, cashew nut), others supply technical and medicinal raw material, some are appreciated as ornamental plants because of their red autumn tints. *Astronium* has a useful wood and the bark serves for tanning. ROTH studied the bark structure 1981.

Loxopterygium sagotii HOOK. has a useful wood. ROTH studied the bark structure 1981.

Anacardium occidentale L. (Merey)

Taxonomical description

Anacardium occidentale L. (Merey) (Figs. 10, 11) is a small tree, 3-6 (10-15 m) high. The simple leaves are arranged alternately, have a short petiole and are glabrous; their consistency is subcoriaceous to coriaceous. The lamina has obovate shape, an obtuse to emarginate tip, and a cuneate to obtuse base. The leaves have entire margins and reach a length of 7-20 cm and a width of 4-10 cm. The venation is pinnate, broquidodromous, with 10-12 pairs of lateral (secondary) nerves. The middle nerve is prominent on the lower surface and slightly vaulted above the upper leaf side. The petiole reaches 0.6-1.5 cm in length and is slightly hairy. The terminal panicles with a grayish pubescence are 5-25 cm long. The calyx is 5-lobed with elliptic-oblong lobules, 4-5 mm long. The yellowish to rose-coloured crown has pink stripes and is composed of 5 linear-lanceolate petals in imbricate position which reach a length of 1-1.4 cm. The ovary is unilocular and contains a single ovule. The ripe fruit is reniform, laterally compressed and is 2-3.5 cm long. The fleshy peduncle is pear-shaped, red, 6-10 cm long and 4-6 cm broad (Fig. 12 c). It is edible (see also SCHNEE 1960).

Origin

The species is probably native of the semi-arid coasts of the Caribean islands, Central America, and of Venezuela and Brazil, where many biotypes in the wild state can still be found (BRÜCHER 1989).

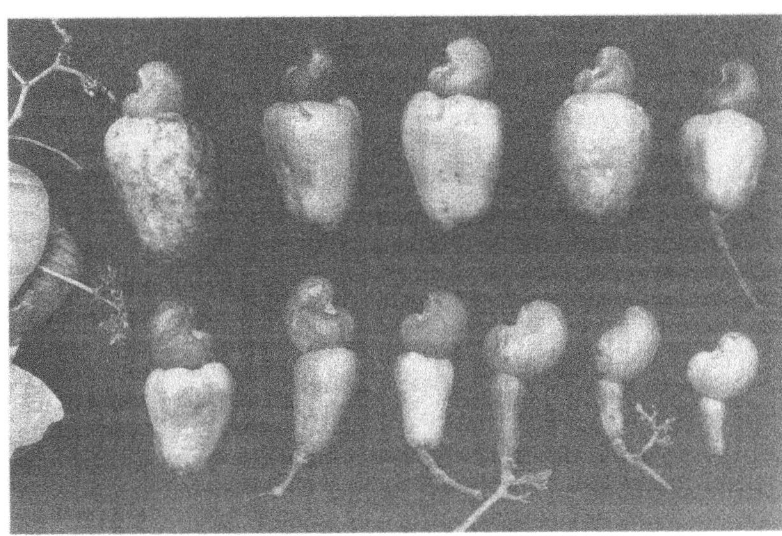

Fig. 11. *Anacardium occidentale.* Fruits with their stalks in different stages of development. Only when the real fruit (nut) is almost grown up, the stalk begins to transform into a false berry (Roth 1977).

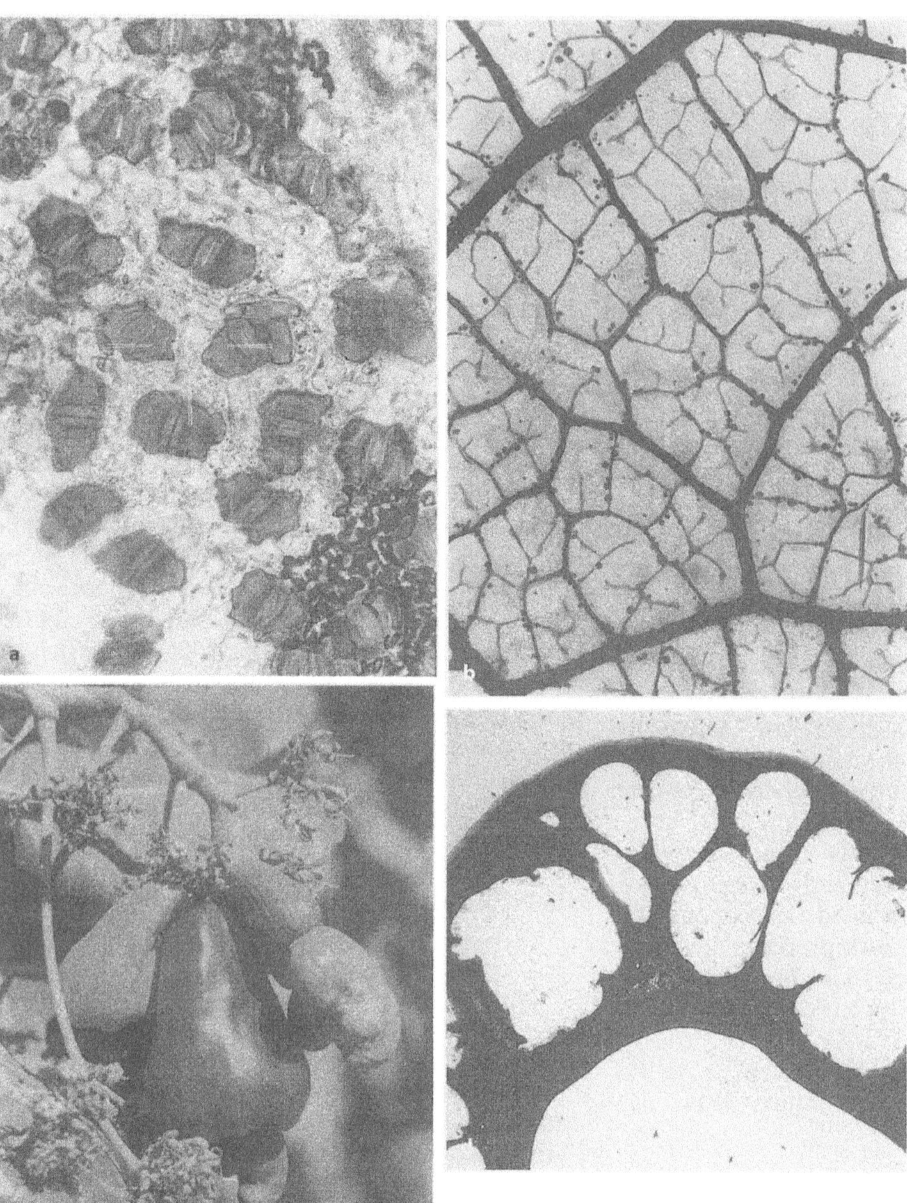

Fig. 12. *Anacardium occidentale.* **a** Lower epidermis with stomata. **c** Nuts and cashew apples in different stages of development. **b** Leaf venation. **d** t.s. of pericarp (× 12.5).

Historical background

The Tupi Indians from Brazil call the fruit 'acaju', a name which still exists in our modern languages as cashew nut. The first description of the plant was made by THEVET in 1558 who presented illustrations of the cashew tree (BRÜCHER 1989).

Occurrence

Tropical America. The species is common in the savannas of the 'Llanos' and in forests with a hot and humid climate of the north and south of Venezuela. The plant is now cultivated all over the tropical world.

Anatomical description

Leaf. (Figs. 12 a, b; 13 a, b; 14) The leaf is bifacial and hypostomatic. The single-layered upper epidermis has cells of regular size with somewhat tickened outer walls. As seen in a surface view, the anticlinal walls are slightly wavy. The palisade parenchyma is composed of 2 layers of cells which are longer than broad; the first layer is very compact, while the second layer is looser. The spongy paren-

Fig. 13. *Anacardium occidentale.* **a** T.s. of leaf blade, **b** midrib in t.s.

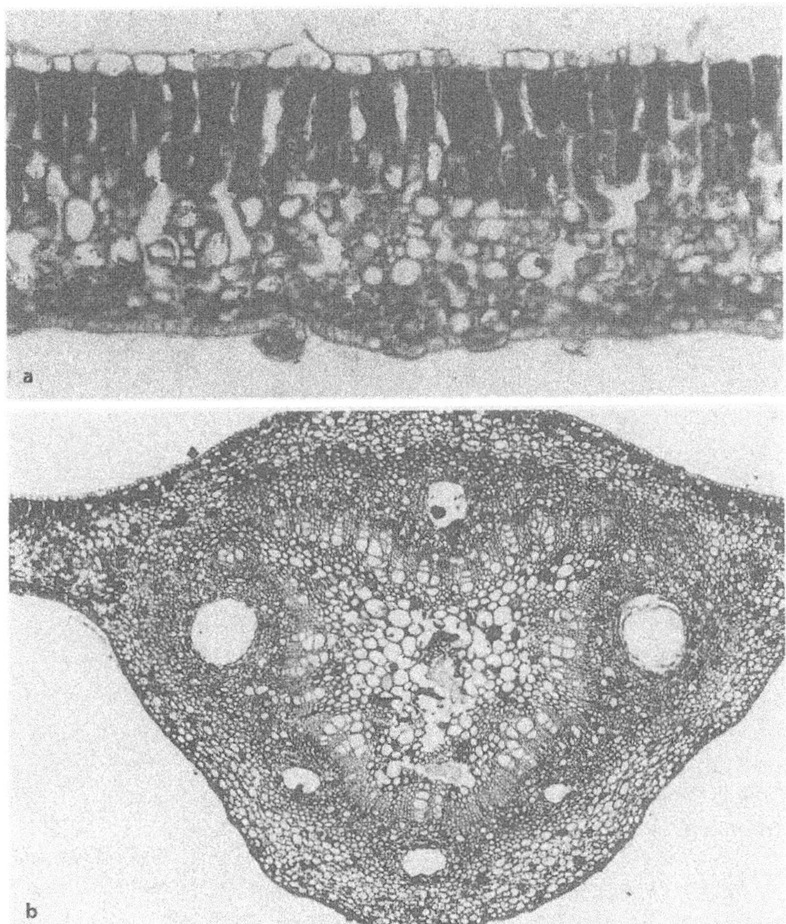

chyma which comprises about 3–4 layers is composed of small globular or slightly lobed cells. Palisade and spongy parenchyma are approximately equal in proportion. The cells of the lower epidermis are smaller than those of the upper side and their anticlinal walls are more strongly undulated, as seen in a surface view. The cell walls are also somewhat thicker in the lower epidermis and their pits are more conspicuous. The small stomata are abundant in the lower epidermis and occur at epidermis level. They have a subsidiary cell on each side (paracytic). The middle nerve projects conspicuously above the lower leaf side, but only slightly above the upper side, as seen in transverse section. The vascular tissue forms an irregular ring with a depression on tbe adaxial side. It is surrounded on the outside by a small sclerenchymatous ring. The secretory canals which lie in the phloem likewise form a ring. Very large canals are situated on both sides of the vascular ring and another very large canal lies on the adaxial side; it induces the adaxial depression of the vascular ring. The central part of the middle nerve is occupied by large parenchymatous cells. The palisade parenchyma is interrupted on the adaxial side of the middle nerve, giving room to a collenchyma; collenchyma is also present on the abaxial side. The secondary nerves and the nerves of a lower order consists of a collateral vascular bundle which is surrounded by a sclerenchymatous sheath, transcurrent to the upper as well as to the lower epidermis. In the larger nerves, there is a secretory canal in the phloem.

Small druses are abundant, particularly around the nerves. We must finally emphasize that we found stomata of the paracytic or rubiaceous type in the leaves of *Anacardium occidentale,* in contrast to the suggestion of METCALFE & CHALK, that the 'stomata are of the ranunculaceous type without clearly distinguished subsidiary cells'. As Fig. 12 a shows, the stomata are accompanied on either side by one or more subsidiary cells oriented parallel to the long axis of the pore and guard cells: the rubiaceous or paracytic (parallel-celled) type. As *Anacardium occidentale* has probably been cultivated for thousands of years, it is quite possible that varieties exist which are distinguished by certain anatomical peculiarities.

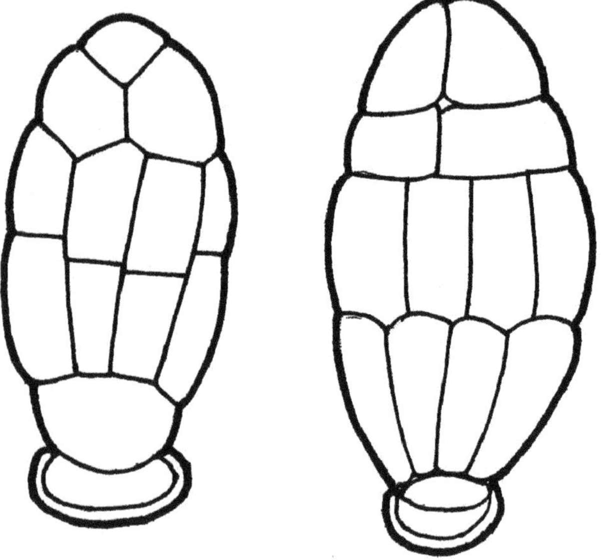

Fig. 14. *Anacardium occidentale.* Club-shaped glandular hairs of the lower leaf epidermis with basal cell, stalk cell and pluricellular head.

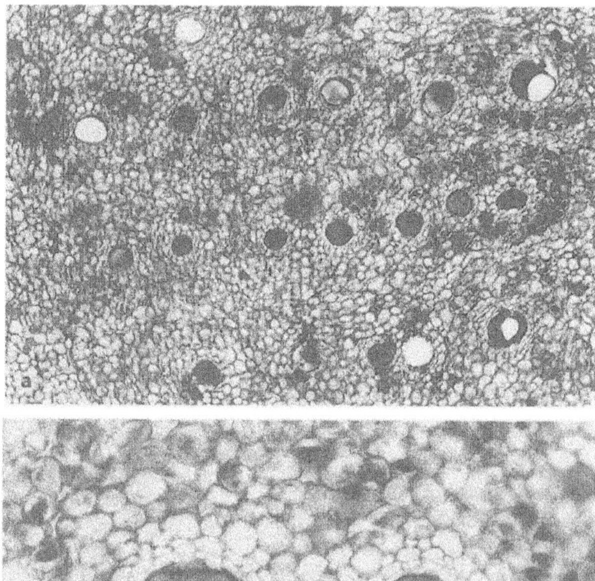

Fig. 15. *Anacardium occidentale.* Bark with oil canals surrounded by a special parenchyma. **a** × 40, **b** × 144.

Club-shaped glandular hairs with a basal cell, a stalk cell and a multicellular head are found on the lower leaf side (Fig. 14).

Bark. (Figs. 15; 18 a) In the sample studied, the rhytidome occupies a large part, reaching about 1.5 mm in thickness. About 7–9 periderms could be counted, in which the phelloderm was very small (1–2 layers thick); the cork was more or less conspicuously layered with layers of thin-walled cells alternating with layers of U-shaped wall thickenings. The periderms were strongly ramified, forming very small scales. The larger part of the bark consisted of thin-walled parenchyma. The main pattern imposed on the bark arises through the large secretory canals which are mainly arranged in the form of tangential rows. The individual canals are surrounded by a thin parenchymatous sheath composed of small cells. Additionally, stone cells, either solitary or in small groups, are dispersed within the parenchyma. In certain parts of the soft bark, a very conspicuous radial arrangement of the parenchymatous cells could be observed which apparently arises through periclinal cell divisions. Tannin sacs are also dispersed in the phloem. Crystals, mainly in the form of small druses, are abundant.

Fruit. (Figs. 12 c, d; 16; 17) The fruit anatomy of *Anacardium occidentale* L. has been described thoroughly by ROTH 1977. The entire fruit, considered a double fruit, is composed of 2 parts: the cashew nut which develops from the gynoecium, and the cashew apple, which originates from the floral peduncle. The latter becomes fleshy and in the mature state resembles a very attractive berry of red colour. In contrast to the false berry which may reach a considerable size of up to 10 cm in length and 5 cm in diameter, the nut remains much smaller with an approximate length of 3.5 cm and a width of 2.5 cm. In the mature state, the nut has a brownish colour and a hard consistence; however it develops a more rigid endocarp and is therefore regarded as a drupe by ENGLER (1896 AND 1964). Pseudoberry and true fruit, the kidney-shaped nut, have a very different structure and development. The true fruit reaches its final size first, before the peduncle begins to transform into a 'berry'. The development of both thus proceeds in accordance with phylogenetic evolution. The gynoecium is originally composed of 3 carpels, only one of which is fertile and shows dorsiventral symmetry from the beginning, being laterally compressed. By unilateral growth of the ovary, the style becomes directed towards the adaxial side; the scar of the style is still visible in the mature stage in the form of a false hilum; the nut thus resembles very much a legume seed of the *Phaseo-*

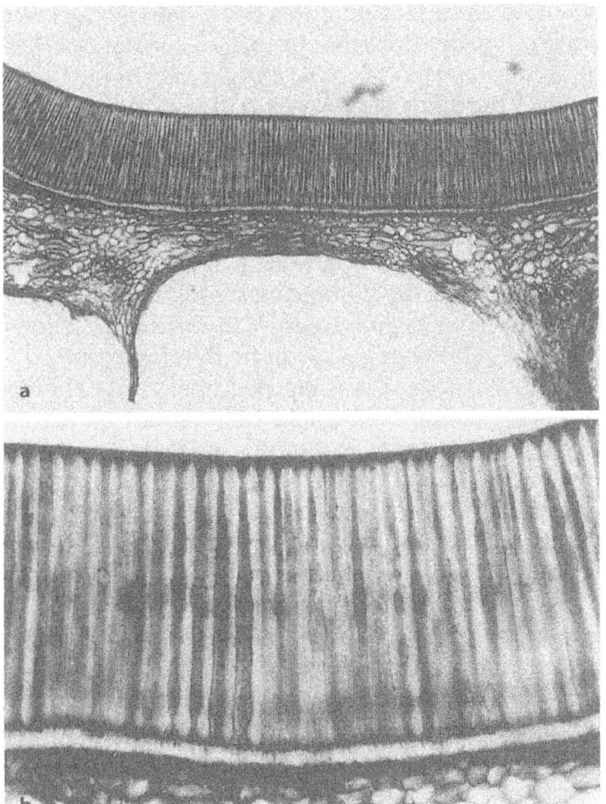

Fig. 16. *Anacardium occidentale*. Fruit endocarp. a ×40, b×160.

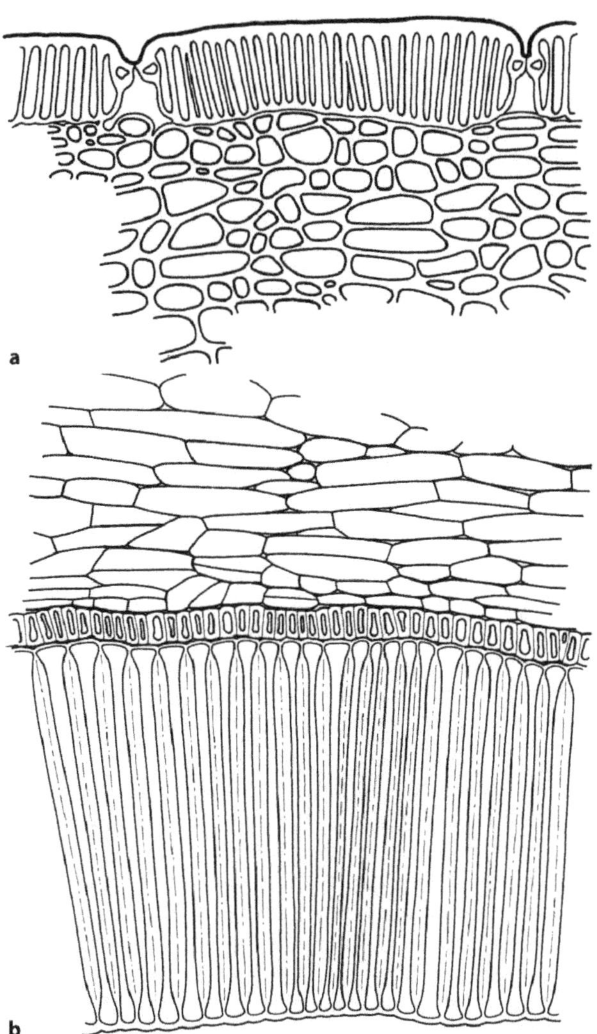

Fig. 17. *Anacardium occidentale*. T.s. of almost ripe fruit. a Outer epidermis with radially elongated cells and thick walls. b Endocarp in the form of palisade-like cells with thick radial walls (Roth 1974, 1977).

lus type. The vascular bundles of the fruit spread from the peduncle over the pericarp describing parabolic curves and uniting in the false hilum.

In a young fruit of about 7 mm in length, the vascular bundles are not differentiated yet, but the secretory cavities are already beginning to enlarge. Growth in thickness of the pericarp is principally due to the radial enlargement of the secretory cavities together with the surrounding parenchyma cells. The cells of the innermost layer, i.e. of the inner epidermis begin to enlarge now in a radial direction and finally transform into an inner palisade layer. This layer thus corresponds to the endocarp and is the hard layer which lead ENGLER to the suggestion that the fruit was a drupe. Now likewise the cells of the outer epidermis begin to enlarge radially to form an outer but considerably shorter palisade layer, the exocarp.

During development, the pericarp assumes a more or less hard consistence. The principal part of the mechanical tissue is however supplied by the endocarp. The radial walls of the inner epidermis which enlarged considerably in a radial direction also grew remarkably in thickness. The cells of the first subepidermal layer below the palisade cells of the inner epidermis, although still small, now also assume a mechanical function by forming thick walls. The parenchymatous cells of the pericarp finally also form thick walls which give them the aspect of a collenchyma. The outer epidermis cells too develop thick walls in their turn. At very late stages, the pericarp tissue takes on a brownish colour owing to certain cell wall impregnations, among others probably tannins.

The most curious feature of the fruit is the formation of inner hairs within the secretory cavities; this curious feature may also be observed in the peduncle during its transformation into a false berry. The secretory cavities of the true fruit are not continuous with the secretory canals of the peduncle, but are superposed, one above the other, forming longitudinal rows which are separated by small bridges of parenchyma. As seen in transverse section, the cavities surround the locule in the form of an ellipsis, originating and terminating at the same level; a pattern of regular sqaures becomes evidedent, when seen in tangetial section. The origin of the secretory canals and the secretory cavities is much the same, as both apparently develop schizogeneously (VENNING, 1948). Both secretory units are covered by a secretory epithelium, the cells of which are distinguished by their small size and their high content of protoplasm. These cells adapt to the continuous enlargement of the secretory cavities by anticlinal divisions. At more advanced stages, the convex inner tangential walls of the epithelium begin to protrude into the cavity so that they project above the surface. The tips of the cells elongate more and more and transverse walls may possibly develop. Multicellular simple and uniseriate hairs composed of up to 5–6 cells originate in this way. The tips of the hairs are rounded and sometimes bent, transforming them into hooks. A fine granular content possibly consisting of small oil drops is visible. However, not all of the epithelium cells transform into trichomes. The secretory cavities become filled with a yellowish juice, principally composed of an oily substance. With methylene blue, the tips of the hairs stain most intensely, suggesting that principally the apical cells secrete the oil into the cavity. PAULA & HAMBURGO (1973) found similar multicellular hairs in the secretory cavities of *Anacardium spruceanum*, while WEBER (1908) observed papilliform secretory cells in *Semecarpus anacardium* (synonym: *Anacardium latifolium*).

A very interesting formation which develops around the secretory cavities is a special sheath of parenchyma cells. At the beginning, all parenchyma cells surrounding the secretory cavity are alike; they are elongated parallel to the cavity surface and have thin walls. When the fruit begins to mature, several layers can be distinguished around the cavity; the first layer, in direct contact with the secretory epithelium, keeps its thin cell walls; the second layer, however, develops cell walls which become impregnated with lignin. This layer is therefore very clearly distinguished, when phloroglucine and hydrochloric acid are added. The cell walls and particularly the outer tangential walls then take on a red colour. This sheath must be of physiological importance and has been called a physiological sheat by ROTH (1974). The parenchymatous layers which follow further outwards have thicker cell walls but are not lignified. The sheath is comparable to the cutinized layers around nectaries or digestive glands, e.g. of *Nepenthes* and *Drosera* (ROTH 1954), called intermediate layers (see also ROTH 1974: *Passiflora*). In all these examples it represents a separating layer which impedes diffusion of certain substances. The lignified sheath of *Anacardium* is found only around the large secretory cavities of the nut, but does not surround the small secretory canals of the peduncle.

The gynoecium has almost reached its final length and breadth, when the peduncle begins its major enlargement period. It becomes evident, moreover, that the peduncle extends first in length and only later begins to grow in thickness, following the evolutionary steps in its development.

The mature nut is composed of an outer epidermis with radially elongated cells (about 60 μm in height); the stomata within this epidermis as in the peduncle remain open. The most conspicuous structures of the mesocarp are the secretory cavities which occupy most of the fruit wall. They are separated from one another by small divisory walls of parenchymatous cells which are radially elongated (with respect to the entire pericarp symmetry) and are partly transversely compressed. The inner surface of the cavities is partly lined with inner secretory hairs and partly with regular secretory cells. A lignified physiological sheath, which has originated from the second parenchymatous layer beneath the secretory epithelium, surrounds the cavities. Most of the vascular bundles are situated on the outside of the cavities; their xylem is composed of slightly lignified tracheids, most of which show helicoidal wall thickenings. Inside the cavities, only a few vascular bundles are found. Most of the parenchyma cells (except those between the cavities) are tangentially elongated. The inner epidermis, deprived of stomata, forms a conspicuous palisade layer with thickened radial walls about 250 μm in height. The first subepidermal layer, on the other hand, consists of small cells with thickened walls. Both layers together form the endocarp. In the fruit of *Anacardium spruceanum*, PAULA & HAMBURGO (1973) observed palisade-shaped inner epidermis cells, about 256 μm in length; the palisade cells of the endocarp of *Semecarpus anacardium* even reach a length of 300–400 μm according to WEBER (1907).

The true fruit is thus strikingly distinguished from the false berry which originates from the peduncle. As a nut, it develops mechanical tissue. Growth in thickness of the pericarp wall is mainly brought about by radial enlargement of the secretory cavaties, while growth in circumference is accomplished by anticlinal divisions of the parenchyma cells. Outer and inner epidermis remain meristematic for a long time. The inner epidermis becomes bistratified by periclinal divisons and forms the endocarp. This capacity of the inner epidermis to divide periclinally seems to be characteristic of the family of Anacardiaceae, as can also be observed in other fruits, e.g. of *Pistacia chinensis*, according to COPELAND (1955).

In the peduncle, on the other hand, growth in thickness is much more pronounced than in the nut. The symmetry of the peduncle is radial, while that of the nut is dorsiventral. Both organs also differ in their inner anatomy. Enlargement of the secretory canals does not play any part in the increase in size of the peduncle, as it does in the true fruit. But the division activity of the fundamental parenchyma is of greatest importance for growth in thickness of the peduncle. The arrangement of the vascular bundles in the false berry is of the dispersed type, while it occurs in the form of a ring in the nut. The structure of the vascular bundles also differs in nut and peduncle; in the latter it is of the concentric type (with xylem surrounded by phloem), but collateral in the true fruit. An inner epidermis is necessarily absent in the false berry, but plays an important part as endocarp formation in the nut. Although the outer epidermis of the peduncle is continuous with that of the fruit, it is distinguished by the number of stomata per surface unity in so far as stomata density is higher in the true fruit. The cashew nut and the cashew apple, which together form the double fruit, therefore differ considerably in their structure and development. The peduncle, however, which becomes transformed into a false fruit, completely looses its character as such and adopts as a whole the qualities of a fruit (ROTH 1974). The entire process of double-fruit formation is reminiscent of the phenomenon of the 'vegetative pear' of Mitschurin proceeding from the apical thickening of brachyblasts (short branches) together with the leaf bases (SCHWANITZ 1967).

Seed. The seed coat has an outer epidermis composed of more or less isodiametric cells. Beneath follow varying numbers of layers with thin-walled, colourless and compressed cells (outer region) which are partly elongated parallel to the seed coat surface. This tissue is relatively loose and contains elongated cells with a darkly staining content when methylene blue is added. These cells possibly correspond to tannin sacs. Towards the inside follow several layers of more or less isodiametric cells (middle region). Raphide bundles are found at the outside of this layer. The innermost region is composed of somewhat smaller isodiametric cells of a brownish colour.

The endosperm which is very much reduced as in most Anacardiaceae, contains little oil drops and aleurone grains. The embryo is the edible part of the seed and principally consists of parenchymatous cells which contain small oil drops, aleurone grains and starch grains. The aleurone grains reach a diameter of up to 6 μm and show crystalloids and globoids. The starch grains, up to 12 μm long, are elliptic, kidney-shaped or wedge-shaped and have an elongated hilum (VAUGHAN 1970). The oil cavities in the cotyledons are conspicuous and can reach a diameter of 50 μm and are surrounded by a ring of secretory cells. They are irregularly dispersed in the parenchyma. The epidermis of the cotyledons is single-layered and small-celled. Our description of the seed somewhat deviates from that of VAUGHAN (1970), this may possibly be due to the varying anatomy of certain varieties.

Ethnobotanical and general use

Anacardium occidentale is a widely known species which is cultivated in tropical regions all over the world. Best known of all useful parts of the plant is however the cashew nut which is the most important product of the tree.

Nutritional use

The cashew nut which reaches a length of about 25–40 mm, is rich in fat (40 %), protein (15 %), vitamin A and B2. However, the pericarp contains a caustic oil, cardol, which affects the skin. Cardol together with anacardic acid forms the content of the secretory cavities in the pericarp. The nuts are therefore toasted first so that the caustic oil evaporates. The pericarp and the testa of the seed are then removed by hand. Harvesting and shelling is usually done by cheap labourers who consequently suffer from skin diseases. The main exporting countries are India, Mozambique, Tanzania, Kenya, and Brazil.

The cashew apple is only known and eaten by the indigenous people. It is reddish or yellow (2 different races? HOYOS 1989), very aromatic, rich in vitamin C and sweet tasting. It has a fleshy consistency and reaches a length of 10–15 cm and 4–8 cm in diameter. It is eaten raw, as a juice or jam.

According to ECKEY (1954), the seed contains up to 47 % oil, which is a pale yellow liquid at ordinary temperatures. It should be excellent for edible purposes (WILLIAMS 1966), such as the production of salad oil or margarine, but because the kernels are used as dessert nuts, there is no commercial trade, although the oil has been used to some extent to harden chocolate (VAUGHAN 1970).

According to PURSEGLOVE (1968), the seeds contain 5 % water, 20 % proteins, 45 % fat, 26 % carbohydrates, 1.5 % fiber, 2.5 % minerals. The pericarp contains 50 % caustic oil, rich in anacardic acid

(90 %) and in cardol (10 %). The peduncle, in contrast, contains 88 % water, 0.2 % proteins, 0.1 % fat, 11.5 % carbohydrates and vitamin C.

Industrial utilization

The bark, but also the root and the leaves are rich in tannins and are therefore very astringent. The tannin is present in elongated sacs which occur abundantly in the phloem. The tannin can therefore be used for tanning of animal skin. From the tanniferous sap of the bark an indelible dye is prepared (CORREA & BERNAL 1989). Furthermore, a resinous exudate ('goma de acajú') which is brilliant and turns reddish when oxidizing, is obtained by incisions in the bark. It is the content of the resinous canals. The acajú gum is used for the fabrication of varnish. The exudate of the bark is also used as a substitute of arabic gum and as an insecticide.

The wood of *Anacardium occidentale* is strong, of great resistance, fine in structure and easy to work. It is used for the fabrication of boxes and in construction. It is of whitish, brownish or pinkish colour (HOYOS 1989).

The caustic oil of the fruit pericarp, cardol, is industrially utilized to protect cut wood, paper, book covers and other articles against insect attacks. In general, it can be used as an insecticide, it has however to be taken in mind that the oil is also toxic for animals and human beings. The Indians use it as a fish poison.

The oil obtained from different parts of the plant is used as a lubricant for brakes and for the production of resins, varnish and enamel. According to MORTON (1961), the different substances obtained from *Anacardium occidentale* can be applied in a variety of technical processes and for the fabrication of a large number of technical products, even as a raw material in plastics.

Medical use

All parts of the plant have hundreds of applications in popular medicine, the name of the drug is therefore *Anacardium occidentale* L., plantae. All descriptions and recipes have to be treated with great caution and in a very critical way, because great confusion exists between true fruit and pedicel, and even between bark and seed coat or pericarp. Furthermore, the plant is toxic and first of all, the caustic oil has to be applied with much care. Sensitive persons should probably not use it on their skin.

Leaf. The leaf is applied medicinally in different ways. When fresh leaves are frequently masticated, one's teeth can be conserved up to a great age. In a decoction and at a very low dose (because they intoxicate) the leaves are recommended for the treatment of scorbut (high vitamin C content) and to cure aphthae and ulcers in the mouth and a sore throat. The leaves are furthermore used as a diuretic, against gastric ulcers and for diarrhoea. A cup of tea, prepared with 2–3 leaves in a cup of water, and taken 3 times a day, is used against hypertension and diarrhoea.

Bark. An extract of the very astringent bark orally taken, lowers the blood pressure and has a hypoglycaemic effect. The Cuna Indians prepare a tea from the bark which they use against asthma, congestions and colds. The macerated bark, soaked in water for 24 hours until it turns yellow, supplies a drink which cures diabetes. An alcoholic extract of the bark helps against malaria and fever. Bark macerated in cold water cures diabetes, when a cup of it is taken 3 times a day. The same remedy is also used against dermatitis, inflammations, aphthae and furs of the throat. A decoction of the dry bark for 15 minutes and in a dose of 20 g per liter of water, stops diarrhoea. A double dose used as a bath reduces the swelling of the feet (JUSCAFRESCA 1975). The bark as well as the fruit are applied to combat cancer and as a remedy for coughs. Finally, a tea of the bark, which is rich in tannins, is also utilized as a remedy against swelling of the articulations caused by syphilis. Taken each month during menstruation, a bark decoction is believed to be contraceptive (SCHULTES & RAFFAUF 1990).

Flower. The flowers, which are much visited by bees, supply a good honey. In the form of an infusion, the flowers are used as an astringent and a tonic, due to their tannin content, and as an aphrodisiac and stimulant on the basis of anacardein (CORREA & BERNAL 1989; GUPTA 1995).

Fruit. The oil of the fruit is taken as an antidote against irritating toxicants; in the form of an emulsion it is applied as a demulcent. In tropical medicine, the oil applied externally serves as a rubefacient and vesicant in the treatment of leprosy, elefantiasis, psoriasis, acne, warts, callus and scorchings on the feet. It is furthermore a help to burn pimples and ulcers. It is also applied as a cosmetic to peel off the skin of the face so that a new one with a better texture can develop beneath. Women of the West-Indies use it for simple vanity. In West-Africa, the oil is utilized for decorative tattooing and to cover the holes of the teeth. In Cuba, the resin of the fruit is applied for the treatment of cold. In Costa Rica, the juice of the fruit is used to cure bleeding from the nose. Cardol, one of the active principles of the oil in the nut, is applied diluted and in small quantities as a vermifuge. The fruit pericarp contains a violet sap which turns black when oxidized and supplies an indelible dye. Curiously, it is also said that the sap may be used as a vesicant to extract carious teeth. A wine prepared from the fruit is one of the best antidysenteric remedies. According to GIRAULT(1987), the fresh fruit cut into 2 halves is used as a liniment against warts. The powder in decoction is applied internally to combat intestinal parasites (e.g. *Ascaris*). Fresh fruits cut into small pieces and macerated in water for several days serve against stomach aches. Fresh and cut into small pieces and macerated in aquavit, the fruits fortify the memory and the cerebral functions in general and are therefore applied as a stimulant in the treatment of impotence and debility. The fruits are also used to fight cancer and serve as a remedy against coughs. To cure inflammations of the throat, a fruit is eaten on an empty stomach. In the form of a syrup, the fruits serve as laxatives, expectorants, anticatarrhal, antidysenteric and to provoke abortion.

Peduncle. The sap of the green peduncle destroys warts. A wine is prepared from the sap of the ripe peduncle which has antidysenteric properties. Additionally, laxative properties are ascribed to the peduncle; it is also used as an expectorant and anticatarrhal.

Seed. The seed is mainly used for culinary purposes, but it also contains oil and proteins. It may therefore also have some healing properties.

Method of use

As indicated above, the bark, leaves and fruits are partly used in the fresh state, partly in the form of a tea or decoction or in pulverized form. Likewise, alcoholic extracts are utilized. In many cases, the exact doses and recipes are known (see also CORREA & BERNAL 1989; and GUPTA 1995). We have already mentioned that all these remedies have to be taken with care, as the caustic oil of the plant is toxic and can irritate the skin and the ectoderm in general.

Healing properties and chemical contents

The fruit pericarp is richest in the caustic oil which mainly consists of cardol and anacardic acid. This becomes quite clear from the anatomical data, as the enormous secretory cavities almost fill the entire space of the pericarp being only separtated by small parenchymatous bridges (Fig. 12 d). Resinous secretory canals are however also present in the fruit pedicel and in the cotyledons, as well as in the leaves and the bark. All these organs also have healing properties. Besides the oil, the plant contains large quantities of tannins located in so-called tanning sacs (METCALFE & CHALK) which correspond to elongated individual cells. Such cells are present in the bark, in the leaves, in the pericarp and the seed coat. The great confusion arises through the fact that the fruit pedicel is sometimes considered a true fruit. The fruit pericarp is possibly regarded as a seed coat, the entire nut is called a seed by some authors, so that it becomes quite obvious: the knowledge of plant anatomy is not an unnecessary requisite, as is generally accepted in our modern days. Furthermore, such an anatomomorphological mistake can have deadly consequences!

The resinous contents of the plant are suitable to prepare fire retardants and insecticides. Furthermore, varnish, resin and surface coating materials are manufactured with them. The use of these substances as plasticizers for polymerization is limited because of the colour resulting from quinone formation.

Several drug analogs could be prepared from the phenolic constituents of the pericarp liquid (GULATI & SUBBA 1964). In the distinct parts of the plant, the following curative properties could be tested: antifilarial (cardol), antihypertensive (bark), antiinflammatory, antimicrobial (essential oil), antitumoral, antimycotic, ichthyotoxic, molluscocidic (anacardic acid), as well as hypoglycaemic and analgesic. The hypoglycaemic effect of a decoction of the inner bark part (bast) was studied by COSTA & CAVALCANTI (1958). THUILLER & GIONO-BARBER (1971) observed an antihypertensive effect in a bark extract, obtained by maceration at −5 °C, kept in the dark.

An extract of the pericarp was shown to be effective as a molluscicide. Higher concentrations were toxic for certain fish. The antiinflammatory activity of (−) epicatechin, a bioflavonoid, in *Anacardium* was proved by SWARNALAKSHMI et al. (1981). The epicatechin significantly reduces oedema. The salt of anacardic acid, sodium anacardat, destroys in vitro the toxic substances produced by *Crotalus* and *Bothrops atrox*, as well as the tetanic and diphteric toxins (HNO. DANIEL 1984).

Varieties and related species

HOYOS (1989) reports that there exist 2 types or races of *Anacardium occidentale*: a scarlet-coloured and a yellow one. The red one has a better taste, is more juicy and contains more citric acid and tannic substances.

A related species is *Anacardium excelsum* (KUNTH) SKEELS.

However it lacks the same valuable properties as *A. occidentale*. The wood is used very much for the fabrication of panelling, of parquet and canoes. The fleshy peduncles are eaten by monkeys and bats, but are poisonous for men when eaten raw (ANGEHR et al. 1984). The bark is applied for fish poisoning and the stems exude a gum which also can be utilized. *A. giganteum* HANCOCK supplies edible 'fruits' originating from the red-coloured fleshy peduncle which are said to be superior to those of the common cashew.

A. excelsum and *A. giganteum* reach 20–30 m in height, but *A. rhinocarpus* DC is a really majestic tree which may become as high as 40 m with a circumference of 7.5 m. It grows in the Llanos and forests of Venezuela: its simple leaves are leathery. The fruits are similar to those of *A. occidentale*, but of a smaller size. The pedicels are also edible, but are used mainly as a pig fodder. *A. pumilum* and *A. humile* are smaller trees. The former is described as a low bushy tree native to the semiarid central parts of Brazil and its small 'fruits' and nuts are eaten by the native Indians. *A. humile* has a very short trunk, often twisted and sprawling on the ground. It can even grow on the salty soils of the torrid Brazilian interior (BRÜCHER 1989).

Cultivation

The cultivation of this valuable tree is easy, as it develops in poor soils, generally at the coastal frontier, where it also serves as a wind breaker. It supports sandy, rocky and lateritic soils and develops a strong radical system that even withstands heavy winds. In hot regions the tree is even used for afforestation, because it prevents drought.

Observations

Anacardium occidentale is already a species known world-wide, extensively cultivated in the tropics of the New and Old World. The world owes this gift to the Indians of Central and South America. However, morphology and anatomy were not completely understood up to the present so that mistakes are still made. The anatomical data may help to identify better the distinct parts of the tree which give rise to the different products supplied by this species. Identification of the species as such is not difficult in this case, but a profound knowledge of the anatomy, particularly of the bark, the true fruit, seed and peduncle can help to recognize adulterations and to localize correctly the different sources of the active substances and the raw material within the plant.

The pericarp is easy to recognize (Fig. 12 d) by its large secretory cavities separated by bridges of parenchyma. The innermost layer of the pericarp consists of thick-walled palisade cells (Fig. 17 b). The seed inside the fruit has a coat composed of several cell layers in which tannin in tanniferous cells is deposited. The endosperm has almost vanished, but the 2 cotyledons of the embryo are well developed, filling almost the entire seed cavity. The swollen peduncle beneath the nut, on the contrary, consists mainly of thinwalled parenchyma which is longitudinally traversed by small secretory canals. The bark is characterized by the large secretory canals which lie in tangential rows (Fig. 15). The midrib of the leaf is distinguished by the large secretory canals which approximately form a ring in the phloem circle; druses and club-shaped glandular hairs (Fig. 14) are additional peculiarities.

All in all, this species has a very exceptional morphology and anatomy.

Toxicity

Caution is recommended in the utilization of this species, because it is toxic (HEGNAUER 1964). The pericarp ist the most toxic part due to the high content of a caustic oil. Internally applied, this oil can cause gastroenteritis, loss of the muscular control and apnoea (MORTON 1961). The caustic oil contains irritating substances which cause dermatitis. The fruit is therefore vesicant on the skin when raw and untoasted.

Mangifera indica L. (mango)

The mango, although of Indian origin, is the most important fruit tree of the tropics which is much cultivated in Venezuela and therefore shall be briefly mentioned here.

The ovary is functionally one-carpellate and encloses a single seed, although this monocarpellary condition appears to have been derived from a three-carpellate one.

Fruit development and anatomy have been studied by a student of the senior author HECTOR LÓPEZ-NARANJO, in his thesis.

The exocarp or skin of the mango is composed of the outer epidermis together with several subepidermal layers of a thickwalled collenchymatous comparatively small-celled hypodermis. During development, the outer tangential walls of the epidermis cells become enormously thickened and cutinized. Stomata in the epidermis however, are rare.

The mesocarp mainly consists of parenchyma constituting the edible part of the fruit. Vascular bundles, resin ducts and tanniferous cells are dispersed in the parenchyma. The peripheral small-celled part of the pericarp is low in, or completely devoid of starch, whereas towards the inside of the mesocarp larger cells follow which contain the typical compound and simple starch grains (Fig. 19).

The resin ducts originate from a group of cells which differ from the surrounding parenchyma cells by their dense protoplasm and their conspicuous nuclei; these cells increase in number by cell division, separating from one another in the center so that an intercellular space arises which later transforms into the cavity. The processes of wall separation are schizo-lysigeneous. The mature duct which may reach a diameter of 150 µm or more is thus surrounded by a ring of numerous secretory cells. The ducts contain the resin characteristic of the mango which adds the flavour to the fruit, but has a taste of turpentine. The ducts are more concentrated towards the periphery. The vascular bundles which are composed of elements with helicoidal and reticulate wall thickenings, are surrounded by a sheath of small fibers with poorly thickened walls.

During endocarp formation, periclinal cell divisions are observed in the first subepidermal layer beneath the inner fruit epidermis. The mature endocarp consists mainly of fibers elongated in various directions, forming an intricate network of cells. Vascular bundles also occur in the endocarp. The inner epidermis transforms into sclereids with strongly thickened wavy walls. The endocarp lignifies and forms the hard kernel of the fruit.

Numerous varieties may be distinguished; the more primitive ones contain more fibers and more resin canals, whereas in the better quality varieties the proportion of fiber and resin is low.

Ethnobotanical and general use

Nutritional use

In the regions of production, the mango fruit is eaten fresh. The fruits contain a high percentage of sugar, proteins and large quantities of vitamin C, A, niacine, B_1 and B_2, as well as minerals such as iron, phosphorous and calcium. The fruits are used in beverages, and milk shakes. Mango is found in preparations such as mango pickle in oil, sweet mango chutney, mango preserves, mango jelly, mango compote, mango jam, mango squash and canned mango. The fruit is also a source of vinegar.

The thick cotyledons of the seed are recommended as a food for humans and animals.

The flowers yield a good honey.

Economical utilization

With its large crown, the plant makes a good shade tree. Wood is used for interior furniture, floor and ceiling boards, boat building, agricultural implements, planking, tea chests, rural construction, boxes and plywood.

A yellow dye is obtained from urine of cows that have been fed with leaves of the plant. It is called Indian Yellow and is being applied in aquarel and oil painting; it is resistant to light.

Medical use

The tree is the source of resin (oleoresin) which is present in all parts of the plant being called mango gum. Mixed with egg albumen and opium, it is considered antidysenteric. The skin of the fruit is regarded as helpful in haemorrhages of the uterus. Also the seed is applied for treating haemorrhages of the uterus, haemorrhoids and to expel ascaris.

The resin, smelling of turpentine, can also provoke allergies when fruits are consumed in large quantities and may produce stomachache.

Mangiferin, the C-glycoside of a xanthone, has antibiotic properties. Virucidal activity has been claimed for the consitutents of the leaves.

Spondias mombin L. (jobo, marapa, caimito, hobo, jobo amarillo, ciruelo jobo)

Taxonomical description

Large deciduous tree (Fig. 18b), about 6–25 m high, with a straight stem and a DBH of 60 cm. The bark is smooth and has vertical fissures. The alternate leaves are imparipinnate and 10–50 cm long. The leaflets are arranged in 3–15 pairs in opposite or subopposite position; they are oblong-lanceolate, 4–13 cm long and 2.5–5 cm broad, with an obtuse base and an obtuse to acuminate apex and a short petiole (5 mm). All leaflets are asymmetric except the terminal one which also has a longer petiole.

Inflorescences are in the form of pyramidal terminal panicles, about 30–40 cm long. Short hairs are covering the inflorescences, particularly the pedicels, bracts and bracteoles. The pedicels are about 4 mm long. The white flowers are fragrant. The yellow calyx has 5 broadly triangular acute segments, about 0.5 mm long, which are externally covered with short hairs. The 5 petals are white, show valvate aestivation, and are of elliptic subacute shape; they are 2.5–3 mm long and 1.5 mm broad, and are externally pubescent. The 8–10 stamens have about the same length as the petals and are inserted below the disc.

Fig. 18. **a** *Anacardium occidentale.* Bark × 16. **b** *Spondias mombin*, habitus.

The anthers are versatile. The ovary has 3–5 locules, each with an ovule; there are 3–5 styles. The yellow drupaceous fruit is ovoid to oblong and 2–4 cm long. The mesocarp is fleshy and edible. The seeds are large and up to 3 cm long.

In young plants, numerous spiny protuberances are observed on the stem.

Origin

The species is autochthonous of tropical America.

Occurrence

The tree grows wild from Mexico to Peru and Brazil, including the West Indies. It prefers hot regions, but can go up as high as 1200 m. In Venezuela it is very common.

Anatomical description

Leaf. (Fig. 20) The leaf is dorsiventral and hypostomatic. The upper epidermis cells are comparatively large and transparent and probably serve as a water reservoir. As seen in a surface view, the walls are straight and the cells polygonal. The palisade parenchyma is one-layered and consists of long and slender cells which have the nucleus in their center so that a line of nuclei can be seen in a transverse section through the leaf. Equally, as in the epidermis cells, the cells of the palisade and spongy parenchyma are thinwalled. The spongy parenchyma is loose and its cells form short arms; it

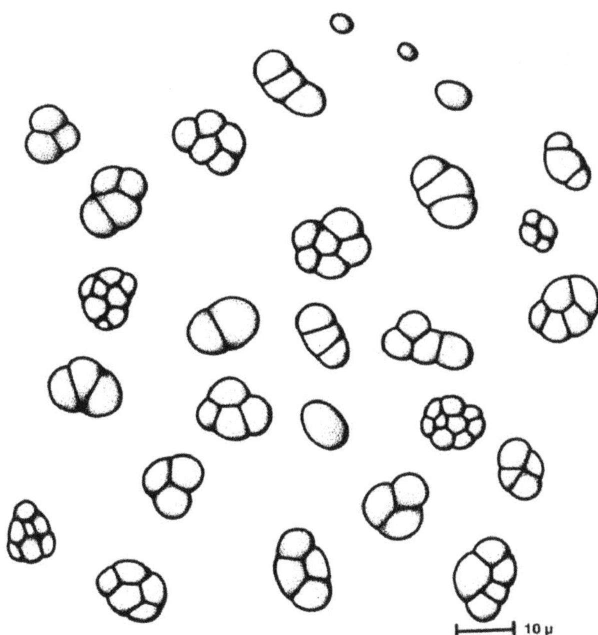

Fig. 19. *Mangifera indica.* Anacadiaceae, compound starch grains of ripe fruit flesh. The scale indicates 10 μ (ROTH 1977).

comprises about 6–7 cell layers. The stomata are found at epidermis level or are only very slightly elevated above the surface. As seen in a surface view, they are anomocytic and do not show a special arrangement of subsidiary cells. The lower epidermis cells which are smaller than those of the upper epidermis, are also polygonal and have straight or only very slightly bent walls. Small uniseriate hairs as well as unicellular hairs are observed here and there, but are most frequent on the midrib. Druses of calcium oxalate are present in the mesophyll as well as in the midrib.

The midrib structure is very characteristic and can be used for identification. There is a vascular semi-circle with endoscopic xylem on each, upper and lower side, however the lower circle is larger. Both circles are separated by parenchym in the middle of which a very large secretory (resin) canal lies. Besides the large secretory canal in the center, there is usually another large canal in the center of the phloem of the upper semi-circle and several canals are found in the phloem of the lower circle. With toluidine dark-blue staining larger cells sur-

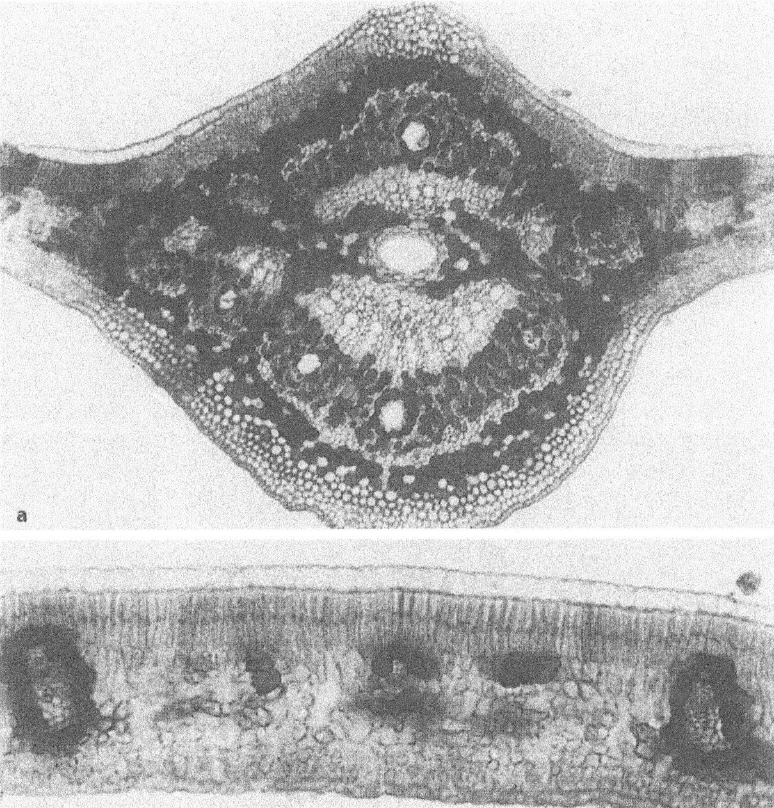

Fig. 20. *Spondia mombin.* **a** t.s. of midrib (× 10), **b** t.s. of blade (× 20).

round the two semi-circles and are also found in the center of the midrib; they may be interpreted as secretory (tannin?) cells. The palisade parenchyma ascends on both sides of the midrib, but is interrupted by collenchyma in the zenith of the midrib. Collenchyma also reinforces the midrib on the lower side. The midrib is thus prominent on both surfaces. Vascular bundles of secondary order may deviate from the lateral sides of the vascular semi-circles.

The collateral bundles of higher order are not reinforced, but are surrounded by a sheath of large secretory cells which stain deeply with toluidine blue. The vascular bundles which are frequent, are thus weakly developed. On the margins of the lamina, a very large secretory canal may be observed which possibly originally proceeded from a vascular bundle which was later on partly destroyed.

Bark. (Fig. 21, 22). The bark structure has been described by ROTH (1969 and 1981). The bark structure is very regular and stratified. The hard bast in the form of fibers develops very early in the inner conducting bark, becoming arranged in the form of very regular superposed plates which together with the neighbouring plates form continuous concentric rings. A certain layering may also be observed in the soft bast where the secretory canals form tangential rows or rings; each canal is surrounded by its own parenchymatous sheath which resembles the vasicentric type of vessel parenchyma. The parenchymatous rings which surround the canals, as seen in a transverse section, become more or less united in tangentially extending layers which alternate in radial directions with layers of collapsed sieve tubes and layers of fibers. The development of the sheath parenchyma is still perceptible in the adult state by the radial arrangement of the cells. The comparatively large pluriseriate 2–6 seriate rays are bent where they meet with a row of collapsed sieve tubes. The rays leave distances of about 4–15 (20) cells between one another, as seen in tangential direction. A very regular square-shaped pattern results through the layering of the hard and soft bast together with the crossing medullary rays. The layering of the hard and soft bast lends itself to the speculation of the existence of growth rings. ROTH(1981) estimated that about 4 layers of soft and hardbast are formed in a year, deducing this suggestion from the rhythmic formation of larger softbast layers.

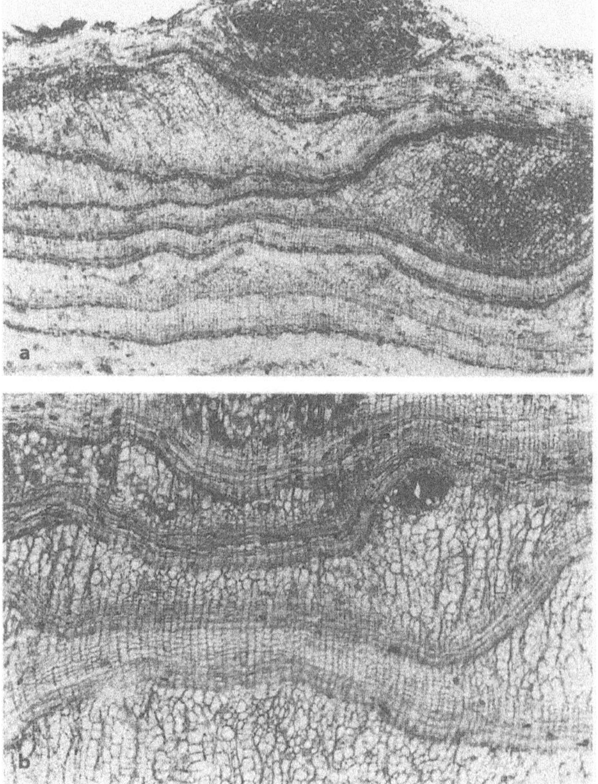

Fig. 21. *Spondias mombin.* **a, b** Bark with several peridem layers.

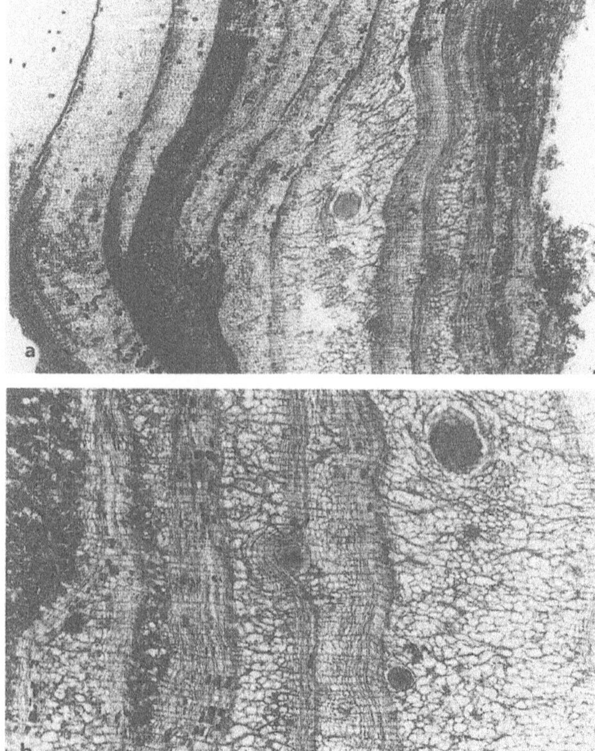

Fig. 22. *Spondias mombin.* **a, b** Bark (note the secretory canals).

Dilatation growth is irregular and mainly affects the axial parenchyma, while the rays do participate very little in it. Dilatation growth is concentrated in certain areas where abundant formation of stone cells also takes place, while in other places it is very reduced. Larger stone cell groups may unite with other ones in a tangetial direction so that also here a layering may be observed.

The rhytidome is very extensive. At first, relatively small periderms cut out small scales from the bark. The cork is delicate to regular in its size, similar to the phelloderm. The cork cells remain delicate or develop horseshoe-shaped wall thickenings (lower tangential walls mainly), while some of the phelloderm cells may transform into stone cells with uniformly thickened walls. The cork cells are radially shortened (brick shape), while the phelloderm cells are more or less isodiametric or become roundish. A large number of periderms could be observed, up to 15 or more.

The periderms frequently anastomose and ramify. In later stages, the cork may become somewhat stratified, forming layers of thin-walled cells which alternate with layers of cells with horseshoe-shaped wall thickenings. Furthermore, the phelloderms may attain a considerable thickness, so that the cut-out pieces of the secondary phloem may become small; they contain the secretory canals besides parenchyma and stone cells.

Very well developed barks may reach up to 22 mm in thickness.

The bark secretion is almost colourless, but turns reddish-brown when oxidized.

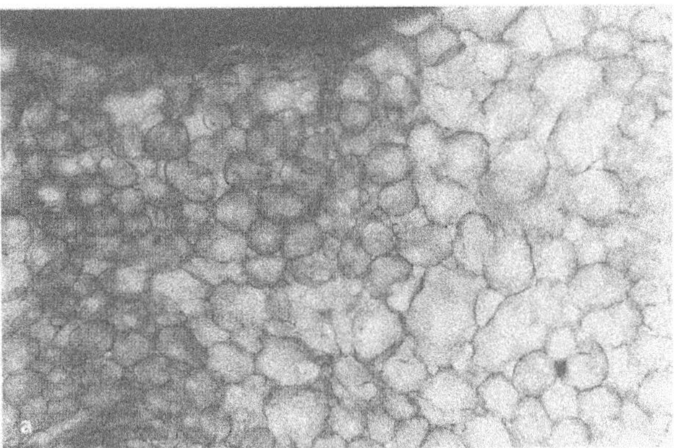

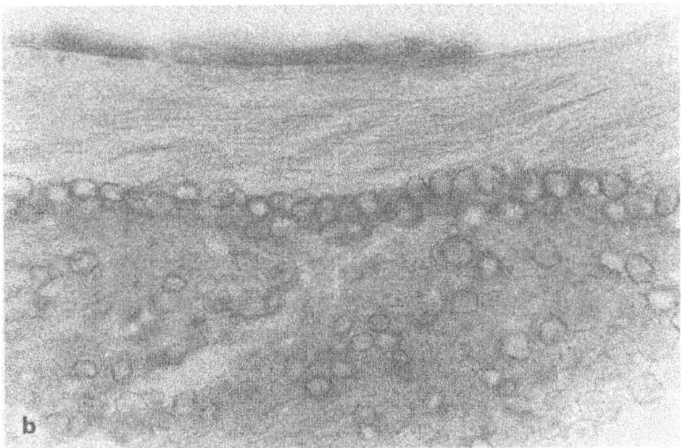

Fig. 23. *Spondias mombin.* **a** Fruit mesocarp in t.s., parenchyma cells with short arms. **b** Endocarp with inner fruit epidermis and crystals. × 20.

Fruit. (Fig. 23, 24) The outer fruit epidermis is composed of very small cells. Below follow several layers of tangentially extended and partly compressed parenchyma cells. These layers, together with the epidermis, form the exocarp (sensu lato) and adopt a yellow coloration. The mesocarp consists of thin-walled parenchyma cells which contain starch grains. The parenchyma cells are mostly radially extended.

The outer surface of the endocarp shows usually 5 furrows and 5 excrescences which penetrate into the mesocarp. The mesocarp parenchyma changes its aspect at the point where the spines of the endocarp end. The cells become more or less globular and develop short arms, so that intercellular spaces arise between them, rendering the mesocarp spongy in this way. Reduced vascular bundles surrounded by a thick fibrous sheath mainly lie where the endocarp spines end, but also traverse the mesocarp extending towards the endocarp. This structure renders the inner part of the mesocarp more spongy and fibrous. Dispersed in the mesocarp are secretory canals surrounded by a sheath (ring, as seen in a transverse section) of secretory cells.

The endocarp is very fibrous with fiber bundles which run in different directions crossing one another in different ways. The inner epidermis of the fruit, limiting the endocarp towards the seed coat, is small-celled and has a thin cuticle. Large solitary crystals are abundant mostly in the inner part of the endocarp.

Of the 3–5 original ovules, only one develops so that the ripe fruit becomes one-seeded.

Ethnobotanical and general use

Nutritional use

The sour-sweet pulp of the fruit is eaten raw or used for preserves, beverages and syrups. The fruits are also used as a fodder for cattle, but many other (wild) animals, such as monkeys, cats, deer, iguanas etc. like to eat them. The chemical content of the fruit has been analyzed (VELEZ & VELEZ 1990), but its content of vitamin C is comparatively low.

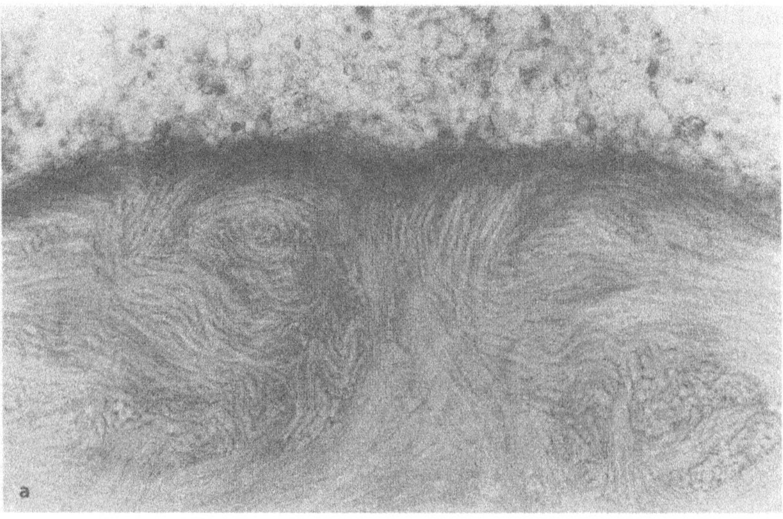

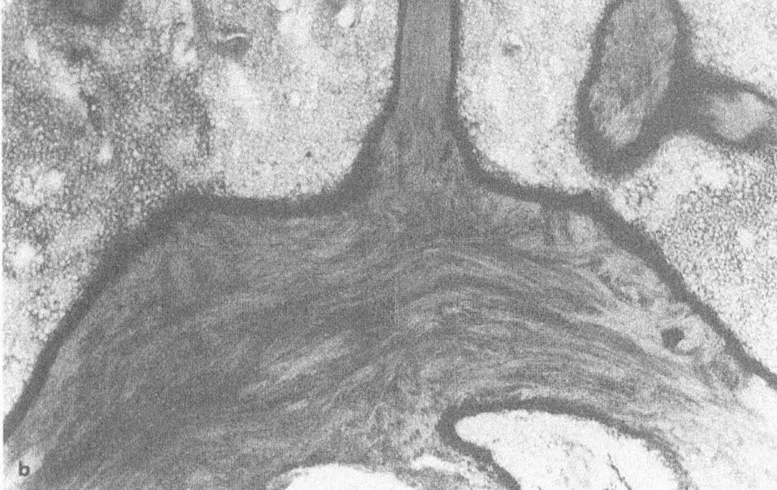

Fig. 24. *Spondias mombin*, endocarp, **a** Crossing fibers. **b** Endocarp: Spine penetrating the mesocarp.

The branches of the tree are used as a forage. The flower clusters are eaten as a vegetable. the roots are also edible and are processed like Yuca brava *(Manihot)*. The roots are rich in liquid and were used by the indigenous people to quench thirst; the Indians cut a piece out of the root, put one end in the mouth holding up the other one (OVIEDO in HOYOS 1989).

In former times (CODAZZI 1840) a fermented drink was prepared of the fruits.

Economical utilization

The wood is not of the best quality, but can be used for pulp and paper, for the fabrication of boxes, crates, match sticks, spattles, and even for posts. Locally, it is used for manifold purposes.

The smooth grayish bark is easy to polish and can be used for seals and stamps. Furthermore, the bark is rich in tannin and can be used for tanning leather.

Medical use

Drug. *Spondias mombin* L. cortex, fructus.

The part of the plant that is mainly used, is the edible fruit. However, the leaves, the bark and the root are also medically applied.

Leaf. The leaves when ground emanate an odour of turpentine or mango fruits; this smell is due to the resinous content in the secretory canals.

The aqueous extract of boiled leaves is used as a gargle for sore throat (infection with *Candida albicans*).

Decoctions are applied as a bath for sores and erysipelas. The decoction as a drink is used for colds, against diarrhoea, and fever. A decoction is also applied to clean wounds.

A bath even helps against gout.

Bark. The bark is used by the indigenous people of the Amazon against metrorrhagia and polymenorrhoea (excessive bleeding during menstruation); to cure these diseases, the bark is put into boiling water; women drink it when the preparation is cool; a cupful taken daily has contraceptive effects. A decoction is also taken for muscular pain, stomach pain, diarrhoea, nephritis and to cleanse wounds.

The bark is rich in tannins and secretes a whitish viscid liquid or 'gum'. This cures burns and favours wound healing.

Root. Roots and leaves are applied to cure fever, and colds, or to clean wounds.

Fruits. The fruits locally applied, cure inflammations of the knees. The macerated fruit taken in water is a vomitive. The fruit contains only a small amount of resin, as compared with the leaf and the bark.

Seeds are applied after child birth.

Method of use

As mentioned above, the fruits may be applied raw and locally. Of leaf and bark, a decoction is prepared which is either used internally as a drink, or externally as a bath.

Healing properties

Astringent, antiseptic, vulnerary.

Chemical contents

The active principles in the plant are tannins and resins. A detailed chemical analysis of the species has however not yet been presented.

Varieties and related species

BRÜCHER (1989) emphasizes that selections from the rich gene pool of wild growing biotypes in Brazil, Venezuela and the Caribbean Islands should be made, which could give a better crop, especially in view of the existing attractive Asiatic species *Spondias cytherea* SONNER, synonym: S. *dulcis* FORST. This species supplies the best quality fruits.

There are, however, also other species, such as *S. purpurea* L., *S. tuberosa* ARRUDA etc. which are used economically and medicinally.

Cultivation

The tree grows in semideciduous as well as in evergreen forests. It is apt for reforestation of poor soils.

It is also planted as an ornamental plant or as a shade dispenser.

Observations

The species is very well characterized by certain anatomical properties. There are the secretory canals in leaf, bark and fruit; the large tannin cells in the leaf, the characteristic structure of the midrib with 2 superposed semicircles of vascular bundles and endoscopic xylem, the abundant cork formation in the bark, the fibrous fruit endocarp with solitary crystals, and the spongy parenchyma in the mesocarp.

Tapiriria guianensis AUBL. The flowers are made into a tea which corrects painful urination. ROTH 1981, ROTH 1987.

Annonaceae

Cauliflory is not infrequent in this family. Flowers are spirocyclically arranged. The flower axis is occasionally elongated. Sepals mostly 3, petals 3–12 (usually 6), stamens many, connective not infrequently with appendix; carpels many to one. The usually aggregate fruits ROTH (1977) are composed of many individual berries or follicles. The seeds occasionally have a reduced aril.

Oil, slime and tannin in tanniferous cells are frequent. Mainly trees, shrubs or lianas which occur predominantly in rain forests, are known.

The family is well-known for its alkaloids, particularly of the benzylisoquinoline, aporphine and berberine types, as well as for its diterpenes, ethereal oils and polyphenols.

Insecticidal principles of unknown constitution are found in several species of *Annona*; cyanogenic compounds occur in a few other genera.

Quite a few species of *Annona* supply aromatic edible fruits of a high quality and are cultivated in Venezuela. They can be distinguished by disposition and outer aspect of the carpels of the ripe fruit. BENITEZ & RUIZ 1982 elaborated a key of distinction basing on these characteristics *for A. cherimola, A. muricata, A. reticulata, A. squamosa, A. purpurea, A. diversifolia.*

Besides some edible fruits and oils which are of economic importance, The Annonaceae in general show antitumoral, cytotoxic, antiparasitic, antimicrobial, antimalarial and plague-eliminating properties. Besides other substances, the family contains isochinolinic alkaloids, polyphenols, flavonoids, essential oils, terpenes and aromatic compounds. Of the 2300 species known, which belong to 130 genera, only 19 species of 6 genera have been studied in search of metabolites with cytotoxic activity. Recently, the biologically active acetogenins were isolated from Annonaceae; they are derivatives of fatty acids and show a strong cytotoxic activity.

Anaxagorea

The genus *Anaxagorea* mainly occurs in Central America, Trinidad and in the north of South America. The follicular fruits open along the inside (ventricidal dehiscence).

Anaxagorea acuminata (DUN.) ST. HIL. (majagua)

Shrub or small tree (up to 7 m); young branches ferruginous-tomentose. Leaves rigid, elliptic or lanceolate-oblong, largely acuminate, with an acute base, 8–17 cm long and 3–6 cm broad, young leaves ferruginous-pubescent on the lower side, older leaves glabrous. Petioles 3–7 mm long.

Flowers axillary; pedicels 5–10 mm long, provided with an amplexicaulous bract close to the calyx. Sepals united at the base, 3–5 mm long, more or less rounded and obtuse, ferruginous-tomentose on the outside. Petals thick, obtuse, ferruginous-tomentose on the outside; outer petals are aovate-lanceolate, up to 11–12 mm long and 6 mm broad; the inner petals smaller and only 8 mm long. Carpels free, when ripe about 2.5 cm long and 1–1.5 cm broad.

Fruit pedicels 0.6–0.8 cm long, tomentous. The follicles yellow tomentous; stipe about 1.5 cm long. Seeds $1.2 \times 0.7 \times 0.3$ cm with an acute and an oblique end.

Occurrence

Guiana, Venezuela and Trinidad. In Venezuela in the coastal Cordillera and on the hills of Guiana.

In the Guianas it is found in the rainforest, particularly along creeks.

Anatomical description

Leaf. (Fig. 25) The leaf is dorsiventral and hypostomatic. The upper epidermis is of a normal size and has a comparatively well-developed cuticle. There is only one layer of palisade parenchyma composed of relatively short and broad cells. Palisade cells may even divide periclinally, particularly in the region of the midrib. The spongy parenchyma occupies about 4–5 times the space of the palisade parenchyma and is aerenchymatous, including very large intercellular spaces (lacunas). The lower epidermis is about the same size as the upper one.

Stomata are frequent in the lower epidermis and are of the parallel-celled type (lateral-parallel, see ROTH & CLAUSNITZER (1969), although the regularity may become disturbed by the extreme sinuosity of the epidermis walls.

The lower epidermis cells have strongly wavy antichinal walls and show a conspicuous ribbing of the cuticle (Fig. 25). Isolated fibers below and parallel to the lower epidermis are also found in the parenchyma of the midrib. Small clustered crystals in the form of druses occur in the mesophyll. Scars of simple hairs are obvious in the lower epidermis.

The vascular system appears in the form of a ring in the midrib; it is surrounded by a sclerenchymatous ring. Besides the large sclereids (stone cells) which are conspicuous in the center of the midrib, a second xylem ring becomes evident; the midrib structure is very characteristic and may be used for identification. Stone cells are dispersed in the parenchyma surrounding the sclerenchymatous ring of the midrib. On the upper side, the space between the sclerenchymatous ring and the upper epidermis is occupied by photosynthetic tissue; cells of the palisade parenchyma are partly divided periclinally in this region, however a spongy tissue also seems to participate in the formation of the midrib.

The larger and the medium-sized lateral bundles are transcurrent to the upper and lower epidermis by sclerenchyma (fibers). However, the smaller bundles are not transcurrent.

Ethnobotanical and general use

Medical use
In several species of *Anaxagorea*, the bark of the trees was used as a kind of curare. Leaves and bark are used as a contraceptive.

Fig. 25. *Anaxagorea acuminata* Annonaceae. **a** Lower epidermis with cuticular striation and stomata (× 40). **b** Midrib in t.s. (× 10).

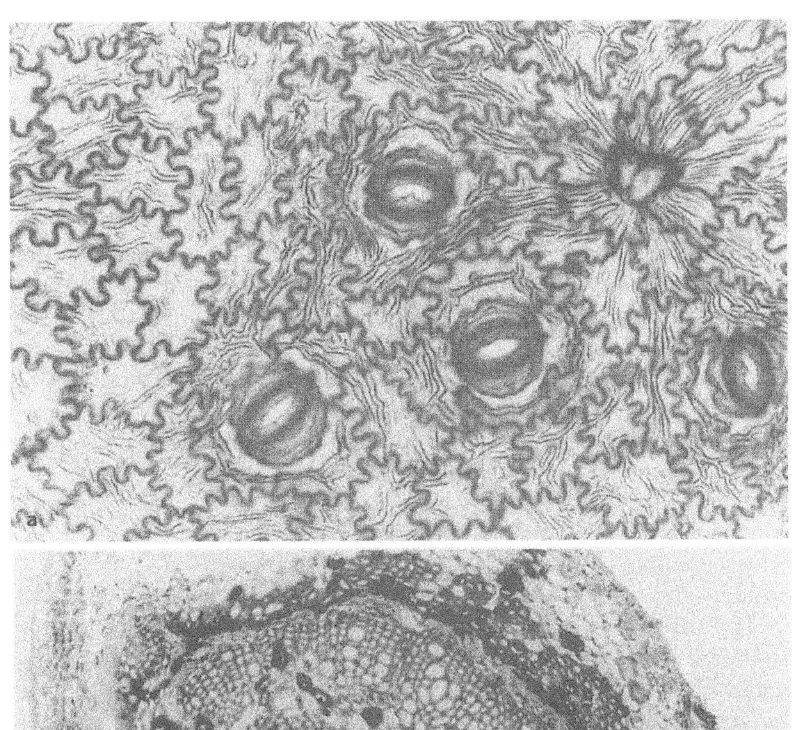

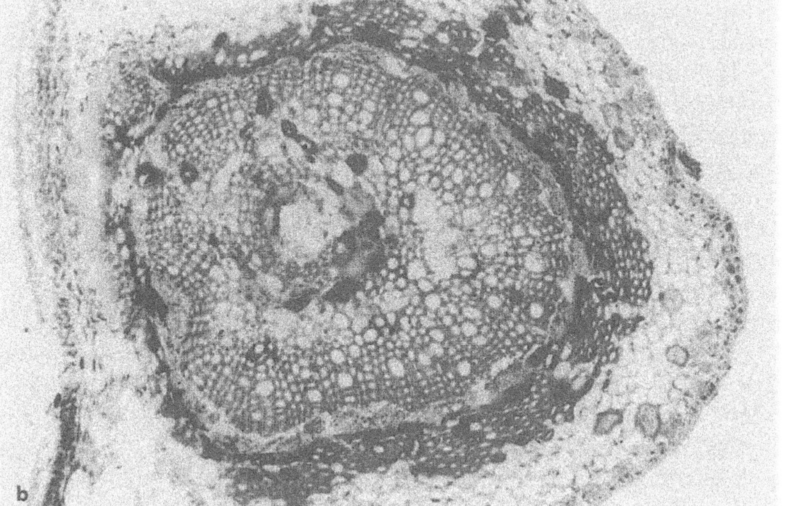

Varieties and related species

Related species are *A. dolichocarpa*, *A. petiolata*, and *A. phaeocarpa*. *Anaxagorea* sp., synonym: *Yarayara amarilla* (Guiana) see ROTH 1984, (pp. 23, 27, 342, 424, and Figs. 68 and 69), and ROTH 1987, (pp. 52, 53, 153, and Fig. 7). *Anaxagorea acuminata* was studied by LINDORF 1992, 1993.

Observations

The leaf of this species is very well characterized by its very well developed aerenchymatous spongy parenchyma, the wavy walls and the cuticular ribbing of the lower epidermis cells, the solitary fibers in the mesophyll, the vascular ring in the midrib, stone cells in the midrib parenchyma, and the transcurrent larger lateral vascular bundles.

Duguetia

D. has useful wood; leaves and stem are used for rheumatism, leaves and bark for ulcer; flowers for aromatic flavour.

ROTH studied the leaf structure 1984, fruit structure and dispersal 1987.

AMERO-FERNÁNDEZ et al. (1999) studied the cytotoxicity of some species of Annonaceae which are used in popular medicine in Venezuela. *Annona muricata* has diuretic, antidiarrheoic, sedative, soporific properties, regulates the blood pressure and is applied for kidney stones. The fruit sap helps against pellagra and scorbutus. The dry fruit is an antidiarrhoeic and the seeds have emetic effects.

The bark of *Annona purpurea* is applied to cure hydropsis and icterus.

The ground seeds of *Annona purpurea* serve as an insecticide and the decoction of the bark is purgative.

Annona squamosa is applied to cure rheumatism and is also used as insecticide.

The ground seeds of *Rollinia sp.* are used for enterocolitis and the fruit has antiscorbutic effects.

Xylopia is applied against malaria.

Cytotoxic tests were carried out with *Annona purpurea, A. jahnii, A. squamosa, A. muricata, A. montana, Xylopia aromatica* and *Rollinia sp.* The experiments showed that the extract of *Annona muricata* is most cytotoxic and consequently most suitable to combat tumoral cells.

The most extensive genus of the Annonaceae is *Guatteria* with 250 species, 35 of which are found in Venezuela. About 130 alkaloids were isolated from the genus. HASEGAWA et al. (1999) studied particularly *G. subsessilis, G. schomburgkiana, G. maypurensis* and *G. cardoniana* and isolated a variety of alkaloids from them. The antitumoral activity of oxostefanine and of protoberberinic alkaloids and their derivatives, were tested in vitro.

The protoberberinic alkaloids tested are of no therapeutical interest, as they showed more cytotoxicity against normal than neoplastic cells.

Oxoaporphine alkaloids, on the other hand, have antimicrobial, antibacterial, fungicide, antiprotozoan and antitumoral activities, but are only slightly toxic to normal cells.

ROTH (1977) described fruit morphology and anatomy of the Annonaceae. Starch is present in almost all parenchymatous cells of the fruit, either in the form of simple or compound grains of various sizes and shapes, but is transformed into sugar during ripening processes.

Guatteria

G. is a tropical American genus ranging from Mexico to southern Brazil. All species are small trees, shrubs or rarely lianas. Alkaloids and tannins have been found in several species.
The wood is valuable and may be used for tools. Quite a few species have a useful fiber in their bark. Others are medically used, mainly for their alkaloid content. Leaves are applied as an insect repellent and against rheumatism, as a contraceptive, and for indigestion.

Guatteria saffordiana PITTIER is a tree which reaches 10–12 (15) m in height. The species occurs in the Venezuelan cloud forest proper at an altitude between 1400 and 1500 m. The leaf structure has been studied by ROTH (1990).

Guatteria calva R.E. FRIES was studied by LINDORF (1992, 1993).

Rollinia

At least 7 useful species are known.

Fruits are edible, some resembling a Cherimoya, melting in the mouth like ice-cream, are recommendable for cultivation, BRÜCHER 1989.

Roth studied the bark structure 1987.

Unonopsis

At least 6 useful species are known.

Bark contains a resin which is used for rheumatism and as an antifertility agent. An arrow poison is also obtained.

ROTH studied the bark structure 1981, fruit structure and dispersal 1987.

Xylopia

Roots are used as a remedy for sterility, root bark for ulcers, fiber of bark for cordage and cables. Useful wood.

Leaves and stem for insomnia; leaves as a tranquillizer, as a diuretic and for oedema.

Flowers as a tonic, carminative, and source of essential oil.

Fruits as a condiment. Seeds are carminative and used for indigestion.

Fruits are also applied for colds.

X. frutescens AUBL. The bark is used for ropes, fruits are used as a condiment. Roth studied the bark structure 1981, fruit structure and dispersal 1987.

Apocynaceae

The family is mostly represented by trees and more rarely by large Herbs, shrubs or vines. The leaves are opposite and simple. The corolla is tubular; the fruit consists typically of 2 dry or fleshy carpels, which are dehiscent or indehiscent (Fig. 26). The seeds are occasionally winged or have an aril.

Intraxylary phloem and laticifers are characteristic.

The family supplies rubber-like products, fibers, drugs and many ornamentals. It is one of the richest in alkaloids. It is rich in alkaloids of several types, and also in stereoids, cardenolides, rubber, terpenes, polyphenols, coumarins, lignans and others.

The latex may contain caoutchouc. Many representatives have toxic contents. The medicinally used species contain indole alkaloids or cardenolides.

Aspidosperma

A. has at least 10 useful species. The wood is useful.

The roots are used as a repellent for termites and ants. Leaves and bark are febrifuge. The bark is carminative, stomachic and tanniferous. Latex contains alkaloids, is antifungal, febrifugal, is used as a protective coating and against leprosy.

A. excelsum BENTH. Wood is useful. Bark is carminative, stomachic.

ROTH studied the bark structure 1981, leaf structure 1984, fruit structure and dispersal 1987, leaf venation 1996.

A. cuspa H.B.K (BLAKE). LINDORF studied the wood 1994.

Fig. 26. *Allamanda sp.* Apocynaceae. Fruit with 2 follicular mericarps.

Himatanthus

Leaves as an antidote for poison, as a poultice for snake bites; febrifugal. Laticiferous bark for wounds. Latex is disinfectant and used to kill parasitic larvae.

ROTH studied the bark structure 1981.

Parahankornia amapa DUCKE. Latex of bark is used as a pectoral and to hasten healing of wounds, for general debility and as a poultice for contusions.

ROTH studied the bark structure 1981, fruit structure and dispersal 1987.

Tabernaemontana

With about 100 shrubs and trees T. is rich in alkaloids.

SCHULTES & RAFFAUF (1990) mention about 11 species of utility. Eight further species are mentioned by UPHOF (1968)

Tabernaemontana benthamiana MUELL. ARG. is a small shrub of the undergrowth in the transitional Venezuelan cloud forest. The plant contains latex. The leaf anatomy has been studied by ROTH (1990). The lower epidermis is furnished with glands which may function as hydathodes.

The genus is rich in alkaloids and has usually latex. Of the species studied, roots, fruits, leaves and bark may be useful.

From several species, a rubber-like substance is extracted from the latex. The latex is also used to destroy warts. A bark decoction is used to treat syphilis in the tertiary stage. Inflammation of the eyes, cough, snake bites, wounds, tumors and many other diseases are cured wiht the above mentioned parts of different species.

It is therefore possible that *T. benthamiana* also has certain healing properties.

Araceae

The monocotyledonous Araceae are herbs, often with tuberous rhizomes, lianas, egiphytes or are even shrubby or tree-like; they mainly occur in tropical and subtropical regions. The small monoecious flowers are numerous on the spadix, which is surrounded by a frequently showy spatha. The occasionally very large leaves have a basal sheath and show a pseudoreticulate venation pattern. The midvein is frequently composed of a bundle of parallel side nerves. Laticifers, tannin cells, spicular cells, mucilage cells, resin ducts, calcium oxalate crystals, as well as hydathodes and scent-glands are common.

Some members are important as ornamentals and are cultivated for interior decoration (*Aglaonema, Caladium, Anthurium, Monstera, Dieffenbachia, Philodendron*). In the tropics and subtropics of the Old World, tubers of *Colocasia* (taro) are consumed and used as a source of starchy food. In South America 'fruits' (infrutescences) of *Monstera deliciosa* are appreciated for their aromatic taste which resemble that of pineapple and banana.

LINDORF (1980) studied the foliar structure of *Anthurium bredemeyeri* SCHOTT, *Anthurium longissimum* PITTIER, *Dieffenbachia seguine* (JACQ.) SCHOTT and *Monstera henri-pittieri* BUNTING.

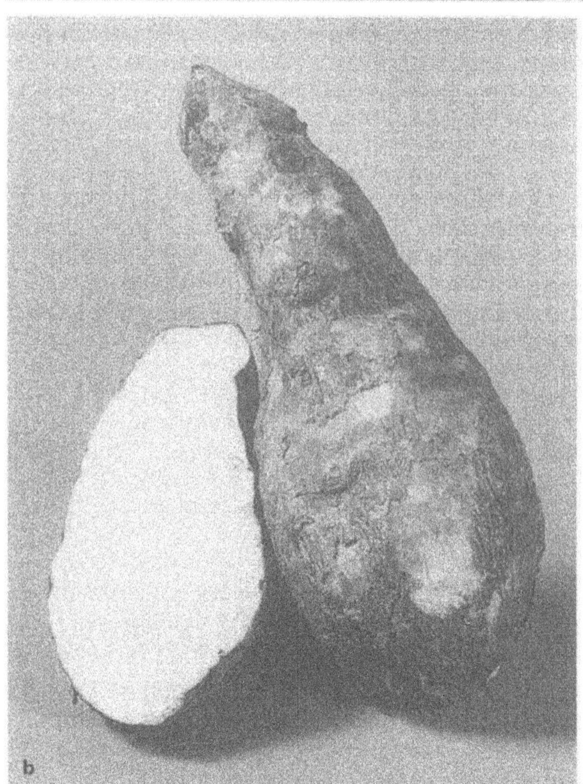

Fig. 27. *Xanthosoma sagittifolium*. **a** Young plant. **b** tubers, the left cut longitudinally.

Xanthosoma sagittifolium (L.) SCHOTT. (ocumo, yautia, tannia)

Taxonomical description

Large perennial herb with a tuberous edible rhizome, about 80–120 (200) cm high. Leaves with a basal sheath and a long petiole (Fig. 27). The leaf blade is sagittateovate, 40–90 cm long and 30–60 cm broad, shortly acuminate with a cordiform base, but is not peltate. Venation is pseudoreticulate with a prominent midvein. Spathe with an oblong-oviform tube, 6–7 cm long and 3.5–4 cm in diameter and a greenish-white limb, 15 cm long and 5–6 cm broad, with an acuminate tip. The fleshy spadix is much shorter than the spathe; the female part is almost as long as the male part.

The fruits are red berries.

From the fleshy principal rhizome, filamentous roots as well as tubers arise.

Origin

Xanthosoma is a neotropical genus, native to the American tropics.

Historical backround

Xanthosoma cultivars were selected from several American wild species many thousand years ago. There are, however, no archaeological remains, due to the quick decomposition of organic matter in the rainforests (BRÜCHER 1989).

Occurrence

West Indies and South America. Much cultivated in the tropics of the Old and New World.

Anatomical description

Leaf. (Fig. 28) The leaf is dorsiventral and amphistomatic. The upper epidermis is comparatively

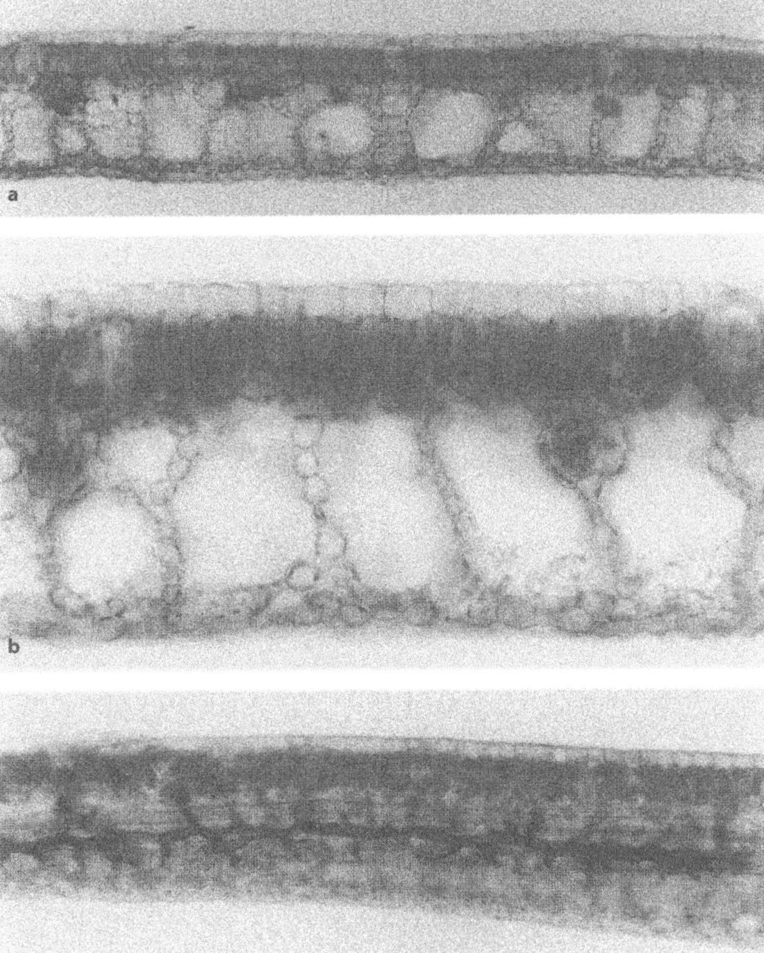

Fig. 28. *Xanthosoma sagittifolium.* Leaf in t.s. **a** × 10, **b** × 20, **c** × 10. Note the ramified laticifers (dark).

large-celled and thin-walled. As seen in a surface view, the cells are polygonal and have straight or only slightly curved walls. The palisade parenchyma occupies about one third of the entire mesophyll or slightly more. The palisade cells are arranged in the form of anticlinal rows comprising 2–3 or up to 4 cells. The spongy parenchyma is transformed into an aerenchyma with very large intercellular spaces. The spongy parenchyma is divided by these 'lacunae' into compartments, separated from one another by walls composed of a single layer of spongy parenchyma cells. Not infrequently, a single lacuna fills the space between the palisade parenchyma and a layer of spongy cells above the lower epidermis. (6) 8–10 layers of spongy parenchyma may be involved in the formation of the aerenchyma. The vascular bundles, lying below the palisade parenchyma, are small and are surrounded only by a parenchymatous sheath. They are accompanied by very well developed ramified and articulated laticifers. The lower epidermis is somewhat smaller-celled than the upper one and is papillose; a short papilla is found in the center of each cell. The small stomata are more numerous in the lower than in the upper epidermis; they occur at epidermis level and are paracytic, being accompanied by 2 lateral subsidiary cells. As seen in a surface view, the lower epidermis cells have the same aspect as the upper ones, but their anticlinical walls may be slightly more bent. The cuticle of the lower epidermis is slightly striated.

The midrib has a complex structure. The palisade parenchyma is continous on the upper side of the midrib. Stomata, although scarce, occur on both sides. The ground tissue of the midrib is an aerenchyma with large intercellular spaces separated from one another by walls which are one cell layer thick. Numerous vascular bundles are dispersed in the ground parenchyma. They have 1–2 very large vessels on the xylem side, but only the peripheral

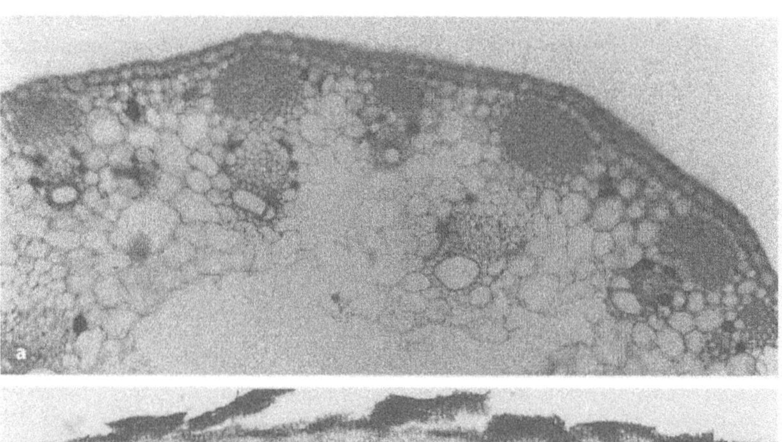

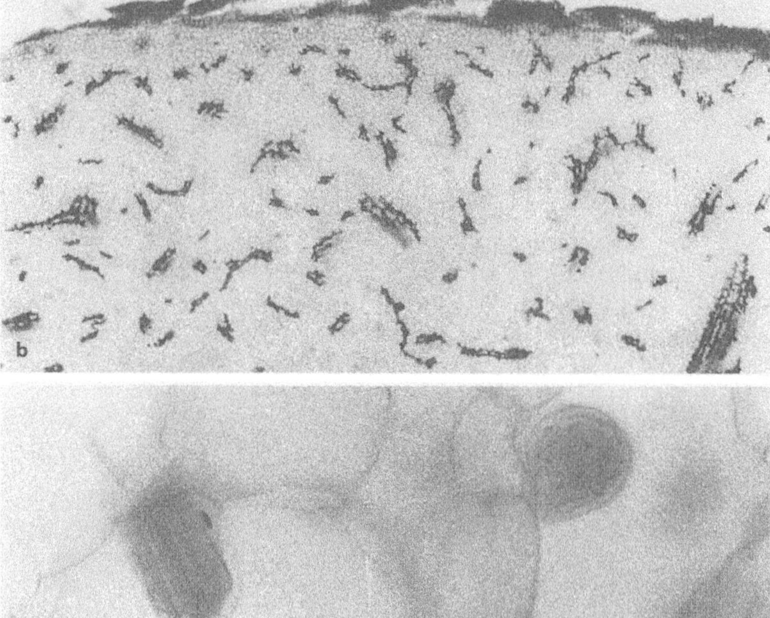

Fig. 29. *Xanthosoma sagittifolium.* **a** T.s. of Axis (× 20). **b** Rhizome with laticifers (in dark – × 2.5). **c** Raphide cells of the midrib protruding into the lacunas (× 40).

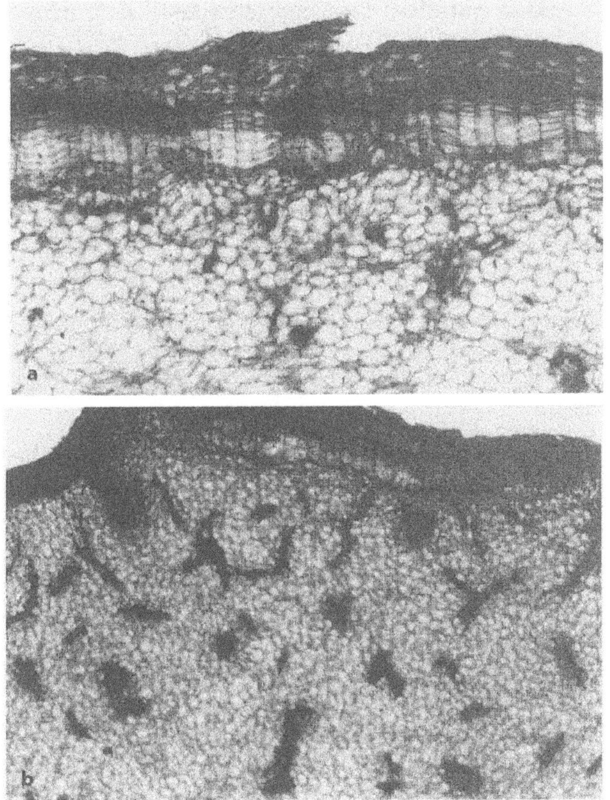

Fig. 30. *Xanthosoma sagittifolium.* **a** Cork of rhizome (× 10). **b** Rhizome with laticifers, stained darkly (× 5).

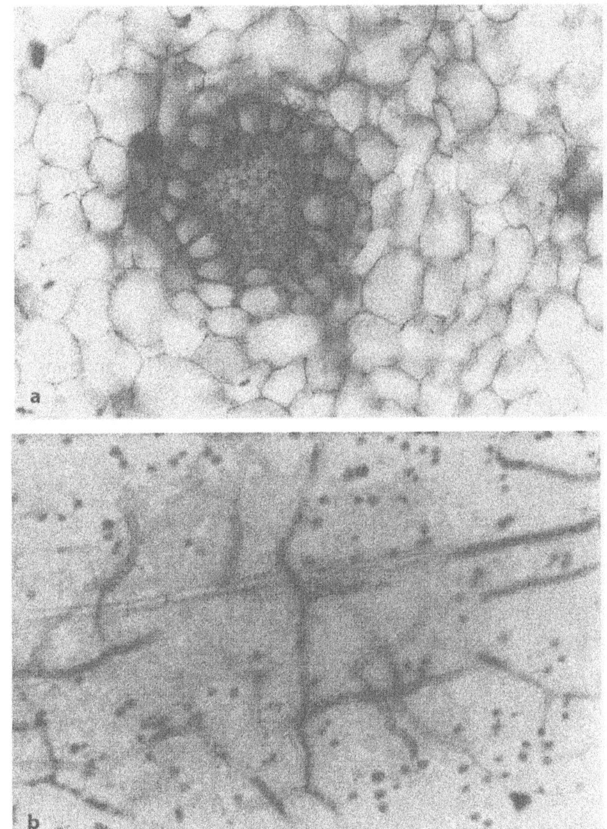

Fig. 31. **a** *Xanthosoma sagittifolium*, vascular bundle of rhizome (× 20). **b** *Asclepias curassavica*, laticifers in the axis (× 63).

bundles develop a fibrous cap on the phloem side. Crystalliferous cells of 2 types are found in the meshes of the aerenchyma: small cells with a single druse, on the one hand and large cells extending into the lacuna filled with raphides. Laticifers are also present in the midrib and partly accompany the vascular bundles.

Rhizome. (Fig. 29, 30, 31a) The rhizome is surrounded by a well developed cork with thin walls. The ground tissue is parenchymatous and consists of medium-sized spherical cells with thin walls. In the periphery, vascular bundles are small and comparatively rare and starch grains are less abundant in the cells. In the central part, on the other hand, vascular bundles are much more numerous and of concentric structure, i. e. of the amphivasal or leptocentric type (with the xylem surrounding the centrally situated phloem). Furthermore, the parenchyma cells are stuffed with starch in the central part. Peripheral and central parts are separated from one another by a small band of slightly smaller-celled tissue very rich in starch. Ramified laticifers are particularly frequent in the peripheral part. The starch grains are compound. Druses, raphides and prismatic crystals are also present in the rhizome.

Ethnobotanical and general use

Nutritional use

The edible parts are young leaves and the tubers. The young leaves are cooked and eaten like spinach.

Young shoots together with the stem are eaten as a vegetable called 'chou des caraibes' in the West Indies.

The tubers which have a high starch content (27 % to 39 % of the fresh weight) likewise serve as a food. But as they contain acrid substances and alkaloids, these have to be destroyed first by heat. For this reason, the tubers must not be eaten raw; the needle-shaped raphides of calcium oxalate irritate the tongue; but additionally these crystal cells contain acrid substances of unknown nature which provoke local irritations.

From the principal rhizome, filamentous roots and tubers arise which are eaten. The tubers are of high caloric value with 26–39 % carbohydrate content. Furthermore they contain up to 3.7 % protein, fat, fiber, calcium, phosphor, tiamine, riboflavine, niacin, Vitamins A and C, and up to 35.19 μg folic acid.

The tubers are cooked or roasted and used like potatoes, or are eaten in soups, or as pancakes.

Medical use
Leaf and rhizome are used.

Leaf. The sap of the leaf is applied for fungal infections of the skin.

Healing properties
Wound healing, emollient.

Chemical contents
Calcium oxalate in the crystals, mucilage in the raphide cells, tannins in the laticifers, acrid substances of unknown nature.

Varieties and related species

According to BRÜCHER, there are different varieties of *Xanthosma*: a) with yellow-orange corms, b) with white flesh and large corms, c) with small corms and purple flesh (*atrovirens, caracú,* and *violaceum*).

Cultivation

Xanthosoma needs strong continous rainfall and high temperatures and prospers best in heavy wet soils, even water-submerged (certain types), with much organic matter. It goes up to 1200 m above sea level.

Propagation is carried out by corm offshoots. Sections of the principal rhizome which already have buds are used. They should be planted at distances of 1–1.2 m. *Xanthosoma* is planted at the beginning of the rainy season and can be harvested after 6–10 months.

The plant is very resistant to diseases. Selections should be directed towards elimination of oxalate and mucilaginous substances (BRÜCHER 1989).

Observations

The leaf is well distinguished by the spongy aerenchyma with large lacunae dividing the spongy parenchyma into compartments; the midrib as seen in t. s. has a large number of irregularly dispersed vascular bundles, ramified laticifers, and druses and raphides.

The rhizome is characterized by the irregularly dispersed vascular bundles in the ground parenchyma, the amphivasal leptocentric bundles in the center; the ramified laticifers and the small compound starch grains.

Araliaceae

Mostly trees or shrubs, only seldom herbs or lianas. Leaves entire or divided, occasionally with a median stipule on top of the sheath. Flowers small, aggregated in heads, umbels or spikes.

The fruit is a berry or drupe with a fleshy meso and a hard endocarp, and a seed in each compartment, occasionally even separating into mericarps.

Stellate hairs and scaly hairs are frequent, as well as schizogeneous resin ducts.

Triterpene saponins are characteristic contents which are probably responsible for the antitussive and secretolytic effects. The asiatic Ginseng is said to be a tonicum, roborant, and geriatricum.

Dendropanax arborea DECNE. & PLANCH. Useful wood. Fruit structure and dispersal have been studied by ROTH 1987.

Didymopanax morototoni DECNE. & PLANCH. Chips of wood in oil are used for massage to cure vertebral pain. Bark structure has been studied by ROTH 1981, leaf structure 1984, fruit structure and dispersal 1987, leaf venation 1996.

Oreopanax

O. embraces numerous species of trees and shrubs, the latter sometimes epiphytic, and shows an extensive distribution in tropical America, particularly in the West Indies, in southern Mexixo, Central America and northwestern South America.

The white wood is easy to work with and can be used for small joinery work and, due to its texture and flexibility, is preferred for making candy boxes, wooden spoons, kitchen utensils, and guitars. The wood is of moderate density, but occasionally very light, soft and 'stringy'; the texture is rather fine, uniform, and the grain straight; working properties are good, durability is low.

Oreopanax capitatus (JACQ.) DCNE. is a shrub or tree up to 20 m high, which occasionally lives as an epiphyte. The tree grows in the transitional and real cloud forest between 900 and 1800 m. The leaves are 10–20 cm long and very variable in shape and size. The flowers are of creamy colour and very fragrant. The wood is very light and of white colour.
ROTH (1990) studied the leaf structure.

Asclepiadaceae

Mainly subshrubs or shrubs, leaves often reduced and succulent. The formation of a 'corona' is frequent. Androeceum and gynoeceum are united in a 'gynostegium'; 'translators' serve for the application of the pollinia. The 2 carpels are apocarpous and supply 2 follicles usually with many seeds; these are furnished with a tuft of hair.

Laticifers are very characteristic. They contain alkaloids, toxic glycosides and cardenolides. Also characteristic are bitter principles of the condurangine (from *Marsdenia condurango*) or vincetoxins (from *Vincetoxicum hirundinaria*); they are polyhydropregnanes which esterified with sugars form esterglycosides which possess properties of saponins.

Asclepias syriaca supplies 'vegetable silk' from seed hairs, as well as fibers from the bark.

Asclepias curassavica and other species of the genus are beautiful ornamental plants with interesting flowers of very complicated structure.

The genus *Calotropis* comes from the Old World, but has now a pantropical distribution. *C. procera* is native of West Africa, *C. gigantea* of South Asia.

The Asclepiadaceae are closely related to the Apocynaceae which are well known for the occurence of latex in latificers. The very peculiar flowers of the Asclepiadaceae are however much more specialized with their gynostegium, the pollinia and the nectariferous often ascidiform petals. While alkaloids (indole alkaloids) are frequent in the Apocynaceae, these are rather rare in the Asclepiadaceae. Glycosides acting as cardiac stimulants are however found instead. Toxic cardenolides frequently occurring in the Asclepiadaceae are of no therapeutical importance, according to FROHNE & JENSEN (1979).

Asclepias curassavica L. (yuquillo).

Taxonomical description

Taxonomical details described in this section can be found in Fig. 32. Errect perennial herb, 30–100 (200) cm high with a slightly ramified cylindric stem which is woody at the base and measures 0.5–0.7 cm in diameter. The glabrous simple leaves are opposed, lanceolate, oblong-lanceolate to linear-lanceolate, 5–15 cm long and 1–3 cm broad, with an acuminate tip, entire margins and an attenuate base. The petiole reaches 0.5–2 cm in length. The venation is pinnate broquidodromous with 11–15 pairs of lateral nerves. The middle nerve is prominent on the abaxial side and flat on the adaxial surface. The actinomorphic scarlet-coloured flowers are orange-red inside and measure 0.9–1.2 cm in diameter. The calyx is profoundly 5-partite and tomentose externally. The crown is lanceolate-lobate with 0.8 cm long recurvate (turned outwards) lobes. The gynostegium reaches about 0.6 cm in length. The axillary inflorescences are umbelliform and about 4–9 cm long. The fruit is fusiform, surcate, 5–10 cm long and 0.4–0.6 cm broad. The numerous seeds are oviform, of brownish colour, flattened with very reduced yellowish wings and with white bunches of hair (SCHNEE 1960, SPELLMANN 1975).

Origin

Probably Central and South America, from Florida, Panama, Colombia, Venezuela, Ecuador, Peru to Bolivia.

Occurence

The species is widely distributed and common in the northern part of Venezuela, occuring from sea level up to approximately 1600 m. It is found at the sea side as well as in the coastal Cordillera. It supports temperate as well as hot climates. It even grows on siliceous soils, and frequently occurs wild as weed.

Anatomical description

Leaf. (Fig. 32, 33, 34) The bifacial leaf is amphistomatic and shows mesomorphic and hygromorphic characteristics. The single-layered upper

Fig. 32. *Asclepias curassavica.* **a** Young plant, **b** t.s. of midrib, **c** t.s. of axis, **d** inflorescence.

epidermis has slightly papillose cells with a delicate cuticle, which is slightly ribbed. The anticlinal walls are straight or only slightly wavy, as seen in surface view. The single-layerd lower epidermis, on the contrary, shows only some papillose cells, but has conspicuously undulated anticlinal walls (Fig. 33). The anomocytic stomata (without subsidiary cells) are more numerous on the lower leaf side and occur at the same level as the other epidermis cells or are only slightly elevated above the surface. Simple hairs are infrequent and occur only here and there along the midrib. The palisade parenchyma composed of a single-layer of relatively short cells is not very compact, but occasionally may consist of longer palisade cells, which are more closely arranged; probably as a result of ecological adaption: the plant not only grows in shady but also in sunny places. The spongy parenchyma which occupies about 4–6 layers is loose and consists of irregularly shaped cells. Particular colourless cells with a druse are dispersed in the mesophyll. Laticifers accompany the vascular bundles and are likewise found in the middle nerve. A bicollateral horseshoe-shaped vascular bundle lies in the center of the middle nerve; the palisade parenchyma is interrupted above the vascular bundle. The nerves of higher order are likewise bicollateral and are surrounded by a parenchymatous sheath.

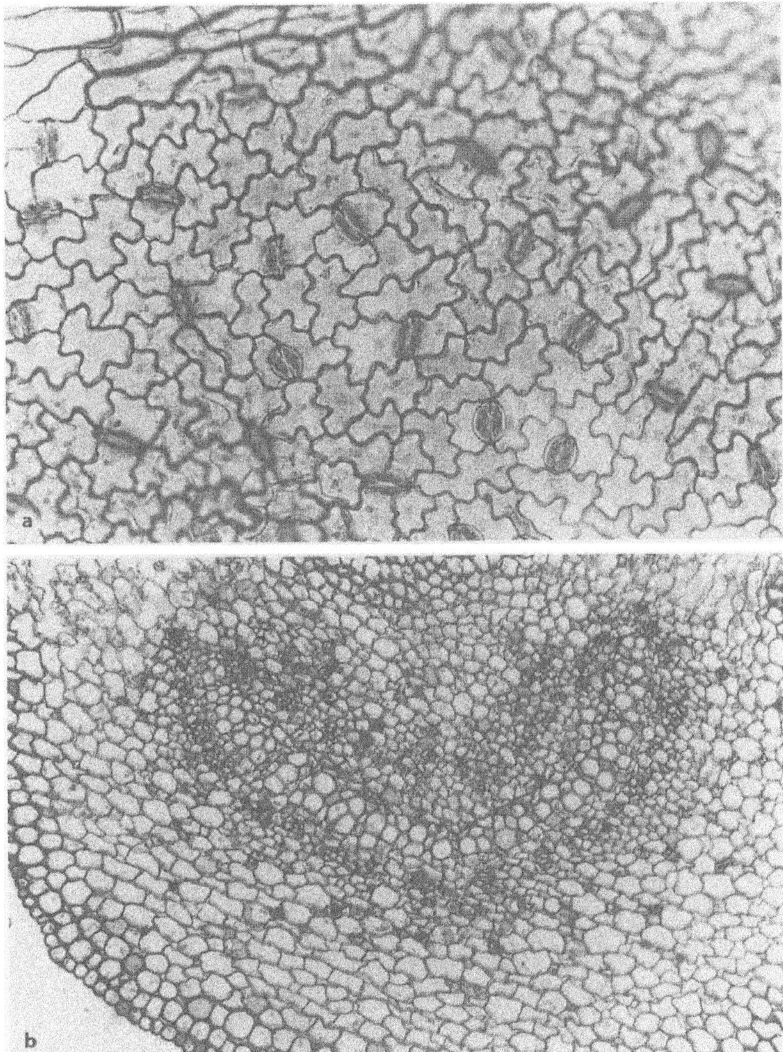

Fig. 33. *Asclepias curassavica.* **a** Lower leaf epidermis with stomata (× 25). **b** Midrib in t.s. (× 10).

As to the arrangement of the cells surrounding the stomata, there is no special order perceptible in the adult state, but it is possible that they belong to the ontogenetic parallel type (rubiaceous). METCALFE & CHALK 1950 also observe that the stomata are 'somewhat variable'.

Stem. (Fig. 31b, 32, 34) In a stem of 6 mm in diameter, the thick-walled epidermis is partially substituted by a cork of subepidermic origin; lenticels are already developed. The primary cortex comprises about 8–14 layers of parechymatous cells which are tangentially extended and show anticlinal walls. The pericycle lying in the middle of the cortex is already ruptured into larger fiber bundles; the fibers have a gelatinous aspect. The parenchymatous cells of the primary cortex become thicker-walled towards the inside and have conspicuous pits. The innermost cells of the primary cortex contain small starch grains. The cells of the secondary phloem are disposed in the form of distinct radial rows. The xylem is diffuse-porous with solitary vessels and vessels arranged in racemes or radial rows. The parenchyma is scanty paratracheal and apotracheal diffuse. The basic tissue is formed by fibers. The rays are uniseriate. Intraxylary phloem is present at the periphery of the pith in the form of a more or less continous ring. The pith is composed of more or less cylindric cells which leave small intercellular spaces between one another. The cells are filled with small starch grains. Laticifers occurring in the primary contex, the phloem and the pith, as well as cells containing a single druse are very conspicious (Fig. 31b). The laticifers have thickened walls. Rhombic crystals are also found here and there, and some seem to lie in the laticifers.

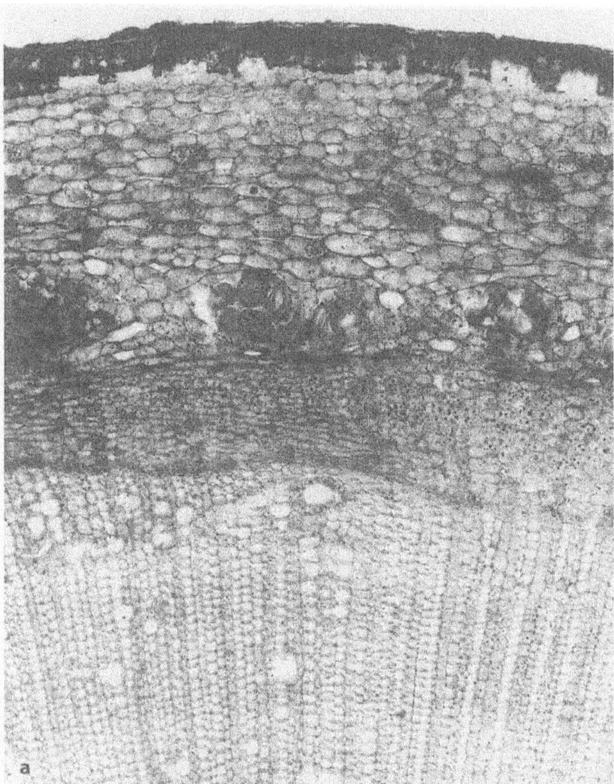

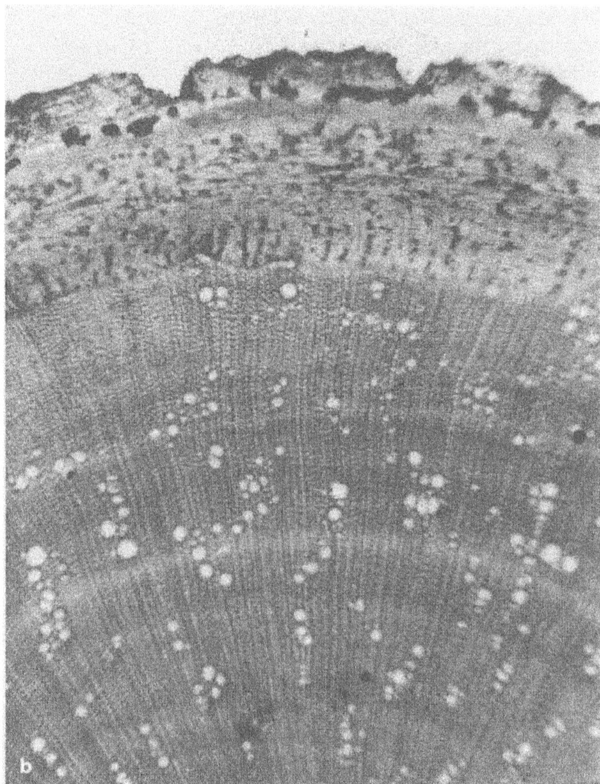

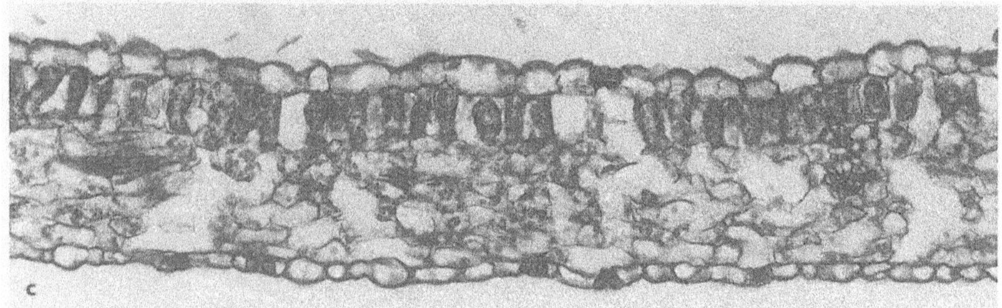

Fig. 34. *Asclepias curassavica.* **a** T.s. of axis, **b** t.s. of root, **c** t.s. of leaf blade.

Root. (Fig. 34) The structure of the main root is very similar to that of the stem, the cork is however better developed. The primary cortex consists of tangentially elongated parenchyma cells with slightly thickened walls and pits. The cells contain small starch grains. Thicker-walled laticifers and cells with druses are interspersed within the parenchyma. the pericycle is ruptured into small bundles of fibers. An endodermis cannot be distinguished. The phloem, arranged in radial cell rows, contains druses and starch grains. The xylem is better developed than in the stem at the expense of the pith. A certain formation of growth rings is better discernible than in the stem and the ring of the intraxylary phloem is smaller. Laticifers and cells with crystal druses are abundant in the pith, while the former are rare in the phloem. Rhombic solitary crystals occur here and there. Lateral roots ramify from the innermost xylem.

Ethnobotanical and general use

Nutritional use

The plant is much visited by bees and supplies good honey.

Economical utilization

There is no real industrial utilization, but the plant is cultivated as an ornamental. The seeds with their brilliant tuft of hair are occasionally applied for embroidery.

Medicinal use

The name of the drug is folia, caule, flos, radix.

Leaf. A decoction of the leaves together with the stem induces vomiting, a decoction of the root, on the other hand, is astringent and purgative (HNO. APOLINAR MARIA (1938). Leaves and tips of inflorescences in a dose of 40 g/l in a decoction of 5 minutes cure any stage of syphilis (JUSCAFRESA 1975). The leaves applied locally are used as an antiinflammatory and antipyretic remedy and are also employed to ease headache, to dry wounds and to soothe inflammation of the liver. Pulverized leaves cure abdominal cancer and act in a caustic way. Orally taken leaves ease haemorrhoids. A decoction of the leaves is used to cure wounds and ulcers. Fresh or dry leaves in a decoction serve as a vomitory in case of intoxication with spoiled food. Leaves are also anthelmintic.

Stem. A decoction of the stem and twigs produces vomiting. Fresh or dry branches in a decoction are used as a vomitory in the case of intoxication by spoiled food.

Root. The root acts astringent and purgative. The dry root pulverized has an emetic action. In popular medicine, the roots are applied in the form of a lavation for diseases of the skin and the mucous membranes. An extract of the root inhaled cures chronic catarrh. Fresh or dry roots cut into pieces and treated as a decoction, are a very strong vomitory against poisoning.

Flower. A decoction of flowers and leaves is used to cure wounds and ulcers. The terminal inflorescences are applied in popular medicine as a haemostatic and against gonorrhoea.

Seeds. Orally taken, the seeds are an antidote against snake bites. Fried seeds taken with milk cure green diarrhoea.

Entire plants. The entire plant orally taken cures rheumatism and excites the central nervous system. The use of the entire plant is however very dangerous, because the plant is toxic and can affect the heart by a sudden paralysation.

Latex. In certain cases, only the latex is applied. A wad of cotton soaked with latex is applied for the extraction of carious teeth. The latex put on a carious tooth immediately soothes the pains. The latex is also used orally as an anthelmintic; dried and pulverized it helps as a sternutatory; it is also said to be abortive. Orally taken, the latex acts as a cathartic, purgative and as an emmenagoge, aiding monthly menstruation. Locally applied, the latex cures abscesses of the ear; dissolved in milk and taken internally it is a vermifuge. Taken purely and applied locally the latex removes warts. The latex flowing from the wound of a plant may have important cardiotonic effects, but this property has not yet been studied carefully.

Toxicity

The toxic principle is calotropin.

The plant is toxic because of its latex content, not only for man but also for cattle, and therefore has to be handled with care. It can produce salivation, nausea, vomiting, diarrhoea, spasm and paralysation of the heart.

Method of use

Branches and leaves, either fresh or dry are used in decoction. Dry or fresh leaves are also used locally. A decoction of the root is taken orally or applied as a lavation or even as an inhalation. Furthermore the root is used dry and in a powder form mixed with liquids or taken directly. A decoction of the flowers is taken orally or applied as a bath. Fried seeds are taken with milk. The latex is locally applied, or mixed with water or milk, and taken orally; dried and in powder form, it is used for inhalation.

Healing properties

The distinct parts of the plant are emetic, cathartic, haemostatic, sudorific, anthelmintic, antisyphilitic, antileprous, abortive or are used against irritations of the skin, to extract teeth, to cure dysentery and haemorrhoids.

Chemical contents

Phytochemical analysis shows the presence of glucosides with purgative and emetic effects, such as asclepina; of cardiac poisons, ethereal oil, resins and peptic substances. The root has been used as a substitute or adulterant of radix ipecacuanhae (from *Cephaelis ipecacuanha*, Rubiaceae), although it does not contain the same alkaloids but instead those of asclepidina and vincetoxina.

Larvae of certain butterflies of the Danainae subfamily contain by ingestion of the plant the cardiac poisons of *Asclepias curassavica* and are protected in this way against predation by certain birds.

Alcoholic extracts of *Asclepias curassavica* showed *in vitro* a significant inhibitory effect on cells derived from human nasal-pharyngeal carcinoma (KUPCHAN et al. 1964).

Cultivation

The plant grows like a weed even on siliceous soils and is easy to cultivate, as it supports temperate as well as hot climates.

Observation

The plant is anatomically well characterized by the presence of laticifers in stem, leaves and root, the intraxylary phloem in stem and root, as well as by the abundance of druses.

Calotropis gigantea (L.) R. BR.

This is an erect shrub or small tree with milky latex. The leaves are opposite, sessile, ovate or obovate, with acute tips and a cordate base, about 11–14 cm long and 8.5–9.5 cm broad, and leathery. The pentamerous flowers occur in cymes. The corolla is pale lavender, the fruit large and fleshy. The seeds are provided with silky hairs.

Anatomical description

Leaf. The leaf is very thick and fleshy due to the extensive mesophyll. Upper and lower epidermis are relatively large-celled. The outer walls of the upper epidermis are considerably thickened. Beneath follow about 4 layers of elongated palisade cells which are densely packed. The cells have a lenght/width index of about 4.9–5.9. The mesophyll structure is isolateral, as palisade cells also occur beneath the lower epidermis, where 3–4 layers of somewhat looser palisade parenchyma are observed. The vascular bundles which occur very frequently are embedded in the spongy parenchyma which fills the space in the center of the leaf between the upper and lower palisade parenchyma. The cells of the spongy parenchyma have either a more or less globular shape or develop short arms so that larger intercellular spaces arise. Intermediate mesophyll layers with anticlinally elongated cells are conspicuous between the spongy parenchyma and the lower palisade parenchyma. The spongy parenchyma comprises about 7–8 layers. Stomata are present on both surfaces. Uniseriate hairs occur on both leaf sides, but are more numerous on the lower side. The succulent consistence of the blade is due to the extensive mesophyll composed of turgescent cells. For further details see ROTH 1992.

Ethnobotanical and general use

Economical utilization

The bark supplies the very strong yercum fiber. The silky hairs of the seeds 'floss' are too short for spinning textiles but are used for stuffing quilts and upholstery; they are also spun into fishing lines and nets. An extract of the petals in kerosen is used as an insecticide. The latex can be used as a fish poison.

Medical use

Used parts: folia, latex, radix, flos.

The plant is used against tooth ache, liver problems, muscular pain, rheumatism, dropsy, epilepsy, hemiplegia, obstruction of the intestine, convulsions, fever, neuralgia, pleurisy, pneumonia, for difficulties during birth, for wounds and bites from rabid dogs.

The cortex of the root is emetic. A powder of the root is used as an expectorant, tonic and diaphoretic.

The latex is sudorific and purgative. It is used locally for the treatment of leprosy, syphilis and other severe cutaneous diseases.

The latex contains histamine and poisonous principles. Three glycosides, calotropin, uscharin and calotoxin as well as proteolytic enzymes are present. It also contains gigantin, a cardiac and fish poison. A bitter toxic compound has been found in the floss of the seeds. The plant is said to have been used in India as a suicidal and infanticidal poison. Calotropin, calotoxin, uscharin and the proteolytic enzymes exert a powerful digitalis-like action.

The healing properties of the plant are antiinflammatory, antineuralgic, emetic, vulnerary, and vesicant.

Toxicity

As the plant is highly toxic, ALBORNOZ (1993) advises against the internal use of the plant, and external use is considered only when there are no other remedies.

Observations

The leaf can be recognized by its anatomical structure.

Berberidaceae

The Berberidaceae are usually spiny shrubs. The leaves of the Venezuelan species are rigid and leathery. The flowers are yellow. The fruit is fleshy. The wood is mostly of yellow colour.

Berberis vitellina HIERON. is the only Berberidacea found at the Avila of the Cordillera de la Costa in Venezuela; it is a shrub, 1–2 m high, sometimes with 2 spines at the leaf base. The glabrous leaves have an ovate shape and their margins are devoid of teeth. The leaf tip is more or less apiculate.

The flowers are fragrant, the fruits are edible. The species occurs in the subpáramos at about 2200 m (STEYERMARK & HUBER 1978).

The leaf is leathery and xeromorphic.

The leaf anatomy has been described by ROTH 1995, PP. 191–193..

UPHOF (1968) mentions 11 species of *Berberis* which are useful. In several species, the roots, the bark or the wood are source of a yellow dye. The wood is used for mosaic-work, turnery and toothpicks. Fruits of several species are edible and made into preserves. Root and bark may have medicinal properties being used as a tonic, febrifuge, carminative, as a gentle aperient, as an alterative and against haemorrhoids (tannins.).

Bignoniaceae/Crescentieae

The family is mainly represented by trees, shrubs and lianas.

The leaves are opposed and in lianas may be partly transformed into tendrils. The showy flowers occur in inflorescences. The bicarpellary gynoecium forms a capsule frequently with winged seeds. The heart-shaped embryo has 2 deeply indented cotyledons. Glands are frequent. Anomalous secondary growth is frequently observed in lianas. The family is tropical and subtropical and richly represented in South America.

The family is of horticultural importance, as many trees and climbers are ornamental. At least 7 neotropical countries have chosen a Bignoniacea, *Jacaranda mimosifolia* D. DON, as their national tree or flower, respectively. *Spathodea campanulata* is cultivated everywhere in the Neotropics. There are also several vines, such as *Macfadyena* which are ornamentals in gardens and glass houses, or *Pyrostegia venusta*. The climbers additionally have a very interesting secondary growth in the stem.

Other utilitarian attributes are edible fruits and seeds: e. g. the pulp of *Crescentia* and its seeds which are made into a 'refresco'. The shells of *Crescentia* are used for 'maracas' rattles.

Capsule valves of *Jacaranda copaya* serve as tools to shape pottery, the timber is used for many purposes and even dyes are extracted.

Many species supply medicines, hallucinogens and aphrodisiacs. There are medicines for better immunity, against AIDS and cancer, for conjunctivitis in ophtalmia, for diabetes, syphilis, malaria, hepatitis, rabies, leishmaniasis and allergy (GENTRY 1992).

The Crescentieae have a unilocular indehiscent fruit and seeds without wings. No tendrils are developed on the leaves. The genus *Crescentia* with 6 species has flowers pollinated by bats and is of neotropical origin.

ALLAIN ARBE (1994) studied the bark structure of *Jacaranda, Tabebuia* and *Tecoma*. The identification of the species was carried out with the aid of an identification key of barks elaborated by ROTH (1981). and most emphasis was laid on the distribution of the hard bast (as seen in t. s.), on the type and course of rays, on storied structure, composition and course of periderms, as well as on the presence and location of crystals. ALLAIN ARBE elaborated a special key for the distinction of the 3 above mentioned genera considering their bark structure (Figs. 40, 41).

Apart from a taxonomic description, ALLAIN ARBE characterized each drug according to the plant anatomy and indicated its medical use. In a separate table, the microscopically identified and microchemically tested cell components of each species were listed.

However, identification of species still remains difficult on account of the very homogenous bark structure of the Bignoniaceae with their layered hard and soft bast, their layered periderms and their interlaced network of fibers, characteristics which are found in almost every species studied (Fig. 40, 41).

As ROTH (1969 and 1981) has already observed, a regular alternation of hard and soft bast in the form of layers is frequently found in Bignoniaceae, e. g. in the genera *Tabebuia* and *Tecoma* (Fig. 40c and d), whereas , lenticular hard bast groups seem to be characteristic in the genus *Jacaranda* (Fig. 40b). The hardbast is usually composed of fibers (Fig. 41 a).

A storied structure, for example of rays, is not infrequently found (Fig. 40a). Through collapse of sieve tubes in the soft bast, an undulated or wavy course of the rays may result (Fig. 41b). Crystals in the form of rhomboids frequently occur in septate crystal strands, e. g. of *Tabebuia* (Fig. 41d, e), whereas druses are found less frequently (Fig. 41c *Jacaranda acutifolia*). By its bark structure, *Jacaranda* is thus easily distinguished from *Tabebuia* and *Tecoma*.

Crescentia cujete L. (totumo, taparo, cucharo, guire).

Taxonomical description

It is a small tree, up to 10 m high, with a diameter of 30 cm (Fig. 35). The larger branches are usu-

Fig. 35. *Crescentia cujete*. **a** ripe fruits, **b** tree.

ally torsive, while the smaller branches on which the brachyblasts arise, are of cylindrical shape. The sessile or subsessile leaves - arranged in fascicles of about 4-7 - originate from brachyblasts which form on knotty enlargements of the supporting axis. The leaves of each brachyblast (which differ in size) are simple, obovate, cuneiform-oblanceolate or spathulate, with an obtuse to acute apex and an attenuate base, 3.4-26 cm long and 1-7.6 cm broad, glabrous on the upper side and pubescent above the veins of the lower side and of a papery to cardboard-like consistency. The venation is pinnate, brochydodromous, with 13-17 pairs of secondary nerves. The middle nerve is prominent on the abaxial side, and only slightly enlarged on the adaxial side. The plane-convex petiole is 0.2-0.5 cm long.

The flowers are solitary or occur in inflorescences of 2-3 which are 1.5 cm long and cauliflorous, originating from the stem or from strong branches. The calyx, 1.5–2.2 cm long, is profoundly lobed. The tubular bell-shaped corolla is white to whitish-yellow with a somewhat purple venation, and 4–7.4 cm long. The stamens are inserted below the middle of the tube. The disc is large and hemispheric. The ovary is oviform-elliptic, unilocular and has numerous ovules on 4 parietal placentas. The large fruit is a pumpkin (pepo), spherical to ovoid, 13–20 (40) cm in diameter with a hard and smooth indehiscent shell. The small seeds are embedded in a pulp.

Origin

Central and South America

Historical background

According to Brücher, *Crescentia* was domesticated by Indians at distant places and independently, due to the easy propagation. Attractive mutations in size and shape have been selected and survived with the indigenous population „even in face of the present competition of cheap plastic containers‘ BRÜCHER (1989).

Occurence

Tropical America. In Venezuela the plant has an ample distribution and is also cultivated in hot regions.

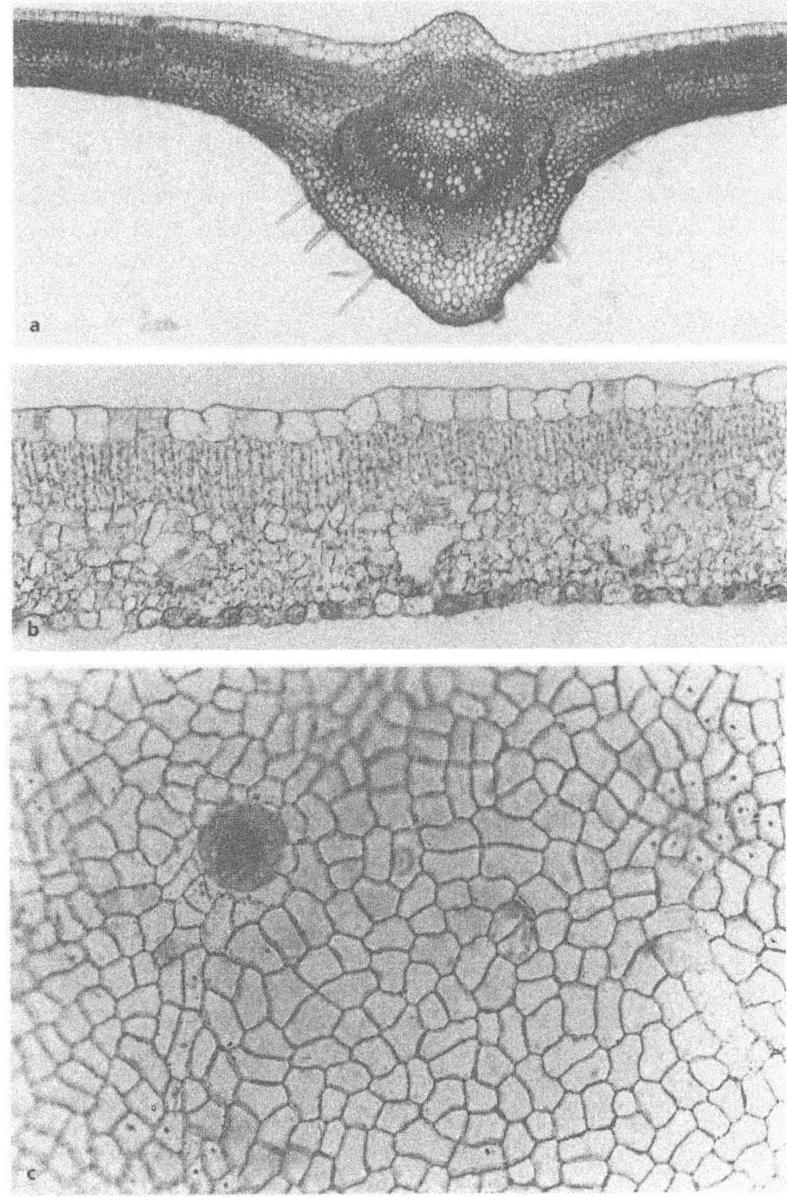

Fig. 36. *Crescentia cujete.* **a** t.s. of midrib, **b** t.s. of leaf blade (× 16). **c** upper epidermis in surface view.

Anatomical description

Leaf. (Fig. 36, 37) The leaf is bifacial and hypostomatic. The single-layered upper epidermis consists of large transparent water-storing cells; the outer walls are slightly thickened and cutinized. As seen in a surface view, the cells are polygonal with slightly curved anticlinal walls. Peltate glandular hairs occur on the upper as well as on the lower side, but are less frequent on the upper leaf side. As seen in transverse section, they consist of a bicellular foot, a strongly cutinized middle cell and a discoidal head composed of 16 cells. The palisade parenchyma comprises 2 layers although it is obvious that the second layer originated by periclinal cell divisions. The spongy parenchyma is more or less compact, comprising 3–4 layers of isodiametric cells. The proportion of the spongy parenchyma is somewhat larger than that of the palisade parenchyma.

The lower single-layered epidermis has smaller cells; as seen in a surface view, the cells are polygonal with slightly stronger curved anticlinal walls. The glandular hairs are more abundant in the lower epidermis. The stomata are without subsidiary cells (anomocytic). The guard cells are at epidermis level

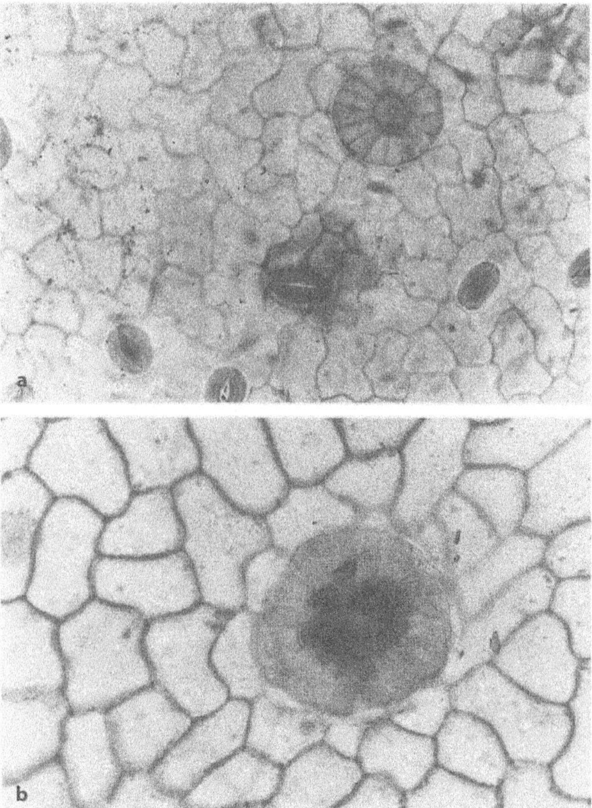

Fig. 37. *Crescentia cujete*, **a** upper epidermis in surface view. **b** glandular hair in upper epidermis.

with the anterior cuticular 'horns' slightly elevated above the surface.

The middle nerve projects above both sides, but more so on the lower leaf side. The vascular tissue is arranged in the form of 2 opposed arcs, the larger one on the lower side. The phloem is externally surrounded by a sclerenchyma. The palisade parenchyma is laterally substituted by a chlorenchyma of isodiametric cells and by collenchyma in the central part. Collenchyma is also found on the lower side. The rest of the tissue is a parenchymatous filling tissue. In the very center of the middle nerve, a large rhexigenous lacuna may be found.

The secondary nerves consist of a single vascular bundel surrounded by sclerenchyma and by a parenchymatous sheath. The veins of higher order are surrounded only by a parenchymatous sheath.

Simple unicellular hairs are found on the upper and lower side of the middle nerve.

Bark. (Fig. 38) The bark is well developed and shows alternating layers of hard (fibers) and soft bast. The hard bast is arranged in the form of continous concentric rings. As seen in a transverse section, the hard bast is accompanied on its outer and inner side by crystalliferous strands which contain raphides; this is a very extraordinary characteristic which may be well used for identification. The plates of hard bast comprise 2–4 cell layers in width, while the layers of soft bast alternating with the hard bast are about 4–5 times as wide. The rays are uni- or biseriate and are somewhat curved, due to the obliteration of sieve tubes.

The rhytidome is very well developed. The radial distances between the successive periderms are small so that up to 15 periderms could be observed. The cork is partially stratified into layers of thin-walled cells alternating with layers of thicker-walled cells with pits. The periderms run parallel to each other and to the hard bast layers. Dilatation growth of the axial parenchyma could only be observed sporadically and very regionally. No secondary formation of stone cells was observed. The structure of the bark is very regular and coincides with that of the bark of other Bignoniaceae studied (ROTH 1981).

Fruit. (Fig. 39) The pericarp consists of a very small-celled outer epidermis with thick outer walls. Below follow several layers of stone cells, which are roundish, as seen in transverse section. The „woody' layer of the pericarp consists of a large zone of fibers, arranged in fascicles crossing each other in different directions, as it is known of many endocarp structures (see ROTH 1977, 1979 e. g. in *Theobroma cacao*). The fibers are thus cut partly transversely, partly longitudinally or obliquely as

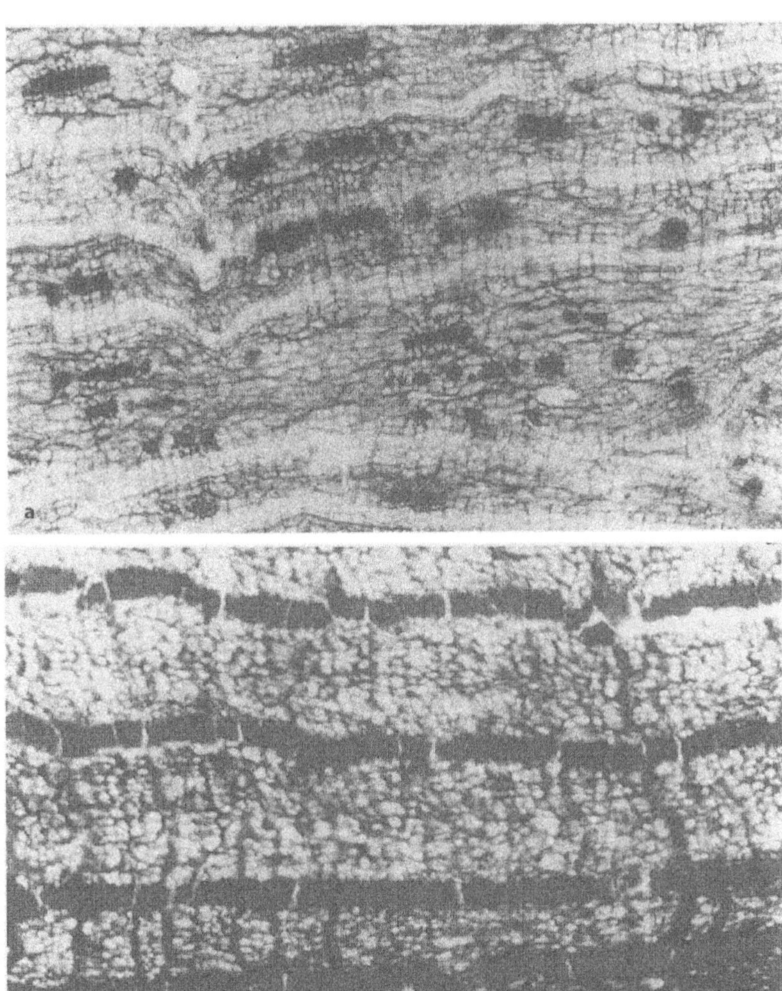

Fig. 38. *Crescentia cujete.* **a** Rhytidome (× 10), **b** bark, closer to the cambium (× 10). Note the fibrous hardbast layers (dark).

seen in transverse section; a large number of the fibers are however oriented in a longitudinal direction. Stone cells also penetrate this fibrous zone and bundles of them alternate with fibrous fascicles. This crisscross structure adds much strength and stability to this woody zone.

Towards the inside follow several layers of parenchymatous cells of roundish shape and thick walls. Innermost, distorted parenchyma cells with thin walls corresponding to the spongy part of the fruit were observed. Vascular bundles traverse all these regions in different directions.

Seed. (Fig. 39) The outer epidermis of the seed consists of more or less roundish cells with thin walls. Below follows a layer of anticlinally elongated cells with thick walls and large pits. Towards the inside follows an aerenchymatous-parenchymatous tissue with lobed cells which leave small intercellular spaces between one another.

Ethnobotanical and general use

Nutritional use

Young seeds which are rich in oil, are edible and taste good when cooked and fried or toasted. The seeds contain sugars (2.6 %), proteins (8 %), and oil (37 %: oleic acid, linolic acid and saturated oleic acids).

Economical utilization

The firm and tough timber is used for ribs and keels of small boats, for hubs and felloes of wheels, for handles of instruments, and for firewood and charcoal.

The fruit (calabash) was the basic household vessel for the Indians. They are still used as practical containers for food and liquids, smaller ones serve as cups, spoons and scoops, dishes and other household utensils, some of them are ornately carved or painted. Even musical instruments are prepared from the fruit shells.

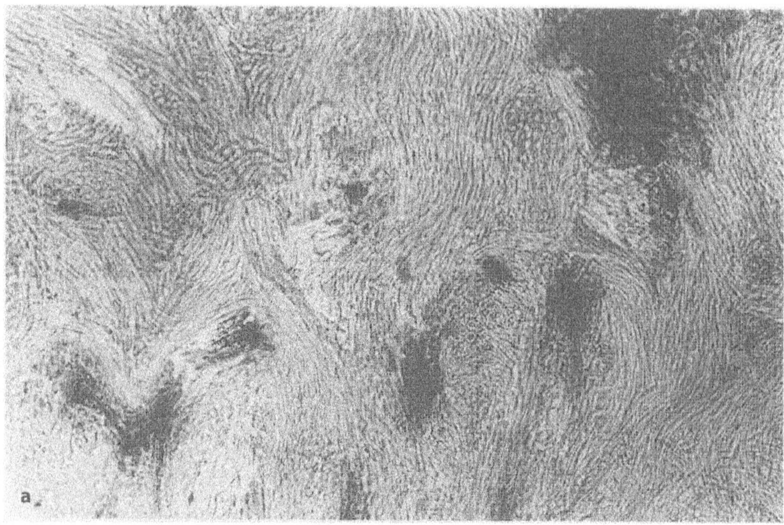

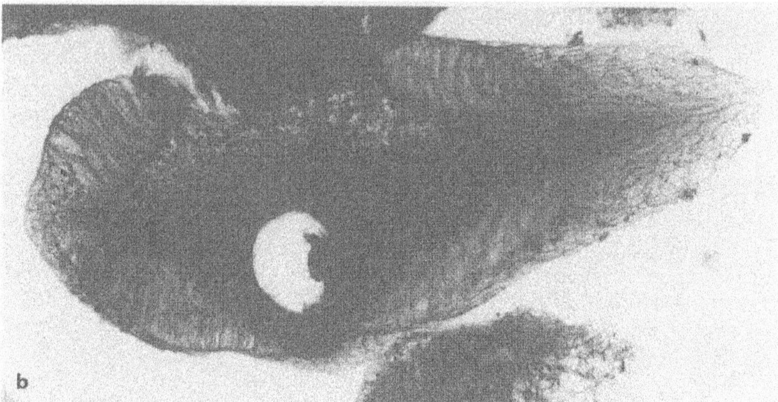

Fig. 39. *Crescentia cujete.* **a** part of the fruit wall with interwoven fibers. **b** Seed × 6.3.

The plant is also used as an ornamental and as a supporting tree for vanilla cultivations as well as a source of shade.

Medical use

Name of the drug: *Crescentia cujete* L. semen, folia, fructus, cortex.

The fruit is most frequently used in popular medicine, the leaves and the cortex are also applied, but the flowers are more rarely used.

Leaf. Haemorrhage, cicatrization and ulcers are cured by a sap of the leaves or by macerated leaves. In infusion, the leaves are used against palpitations, hypertension, flu, pneumonia, coughs, catarrh, and diarrhoea. Diarrhoea, diabetes, indigestion, palpitations and nervous irritations are cured with a decoction of the leaves. A less concentrated decoction is taken orally for women's diseases. Besides the leaves, also flowers and entire twigs serve for a vaginal bath.

Leaves applied locally on the abdomen facilitate child birth. The sap sqeezed from the leaves removes pimples and freckels from the skin. Locally applied, the leaves also serve as a hair tonic. The heated sap of the leaves is locally used against earache. The leaves are chewed to treat toothache.

Cortex. Fresh cortex is used to ease child birth. In decoction and taken orally it accelerates child birth. Cortex is furthermore applied against haemorrhoids, prurigo of the genitals, vaginal colic, and whoopig cough.

The cortex is said to have calmative effects. It has been proved that the fresh cortex has antibacterial properties (50 mg/ml: GUPTA 1995).

The juice squeezed from the inner bark is used to treat bruises of the eyes.

Fruit. A decoction of the fruit cures respiratory irritations and phthisis; it has also purgative and abortive effects, particularly in cattle. The sap of the fruit toasted and mixed with honey stimulates mestruation. Fruit juice with honey favours induction of birh and ejection of the placenta or dead foetus. Pulp taken as a syrup or as an infusion calms asthma. The pulp as a syrup taken orally is likewise used against respiratory (bronchitis, asthma) and gastrointestinal (colic, constipation, hepatitis) irritations, inflammations and urethritis. Diluted in water, the pulp is taken against irritations of the liver and to cure palpitations, hypertension, flu, and pneumonia. The cooked pulp (plain or in sweet soup) is taken as a febrifuge, purgative, expectorant, vermifuge, against diarrhoea and to soothe head ache. The raw pulp is effective against red ascaris.

Clear distinction has to be made between the use of the raw and the cooked pulp: The fruit eaten raw is antidiarrhoeic and antiinflammatory, whereas the cooked pulp has purgative effects. A certain conformity with the use of the apple - raw or cooked - becomes obvious in this case!

Crude pulp used externally and locally as an ointment cures psoriasis and sunstroke; it also kills mites. Fresh fruit and twigs locally and externally applied cure irritations of the skin and help as a haemostatic. The pulp is locally applied against dermatitis, leucorrhoea, haemorrhoids, tumors, and to eject the placenta. The slightly heated pulp put on the breast as a cataplasm cures angina pectoris and pulmonar diseases.

In the following recipe, the fruit is drug and vessel at once: the dried fruit is filled with cognac, the content is left to macerate for one week and then 2 cups are taken of it every 3 days, to cure tuberculosis and astma.

Seeds. The seeds are not only edible, when toasted, they are also used medically.

The sap squeezed out of the seeds mixed with honey is applied against menstrual pain and to regulate menstruation. It is furthermore used as an abortive for cattle. The seeds have, however, no antibiotic effects.

Method of use

The plant is used externally and internally. Locally, it is used as a cataplasm. Crude pulp, leaves, seeds, twigs are useful. A decoction or infusion (fruit, leaves), possibly mixed with honey or as a syrup, is applied, as well as the entire fruit macerated with cognac. The cortex is used fresh or in decoction, either externally or internally.

Healing properties

Leaves: antiinflammatory, antibacterial, abortive, emollient.

Cortex: Has mainly calmative and antimicrobial effects.

Fruit: The fruit shows most of the healing properties being purgative, laxative, emetic, emollient, pectoral, antidiarrhoeic, expectorant, astringent, vermifuge, febrifuge, contraceptive, aboritve, antidiarrhoeic, antiinflammatory, antibacterial, antiseptic, calmative, tonic, antidiabetic, reconstituent, sudorific, vulnerary, analgesic, emmenagogic, haemostatic, aperitive, and child birth accelerant.

Seeds: Calmative, abortive.

Chemical contents

The leaves contain flavonoids, triterpenes, phenolic compounds, sterols, and caffeic acid.

The wood contains naphtho quinones.

The fruit pulp contains crescentic, tartaric, citric, tannic and cyanhydric acid. Alkaloids and polyphenols were also found in the fruit.

The seeds contain oleic acid, sugars (2.6 %), proteins (8 %), oil (37 %).

The antiinflammatory effects are ascribed to the flavonoids apigenine and quercetin. The antiinflammatory and antiallergic properties of flavonoids are well known. The effect of the antiinflammatory property depends on the dose used; 1200 mg/kg are effective for 24 hours.

Toxicity

The plant is toxic for birds and small mammals.

The pulp can produce severe diarrhoea and is suspected to be carcinogenic.

Related species

According to BRÜCHER (1989), there are altogether 6 wild species of *Crescentia* known in South America. *Crescentia alata* HBK of Central America is a small tree with edible fruits. Dried fruits are likewise used as containers (bowls and vessels)

Cultivation

According to BRÜCHER (1989) *Crescentia* was domesticated independently and at distant places due to the ease of its propagation. Mutations, attractive in fruit size and fruit shape, have been selected by the Indians and survived with the indigenous population. The diversity of calabsh forms is astonishing. The natives grow them between their cocoa groves.

The plant is propagated by seeds or cuttings; it grows rapidly, has a profound radical system and is very resistant to drought. It can be used as an ornamental in gardens, parks and at places in cities.

Observations

The plant is well known and frequently cultivated in hot regions of Venezuela. Its bark has the typical structure of Bignoniaceae (ROTH 1981).

The peltate structure of the glandular hairs and the structure of the midrib of the leaf are very characteristic. Likewise the structure of the fruit shell is peculiar to the species. The hardbast fibers in the bark are accompanied by crystalliferous strands which contain raphides (see bark anatomy).

Jacaranda

At least 8 useful species are known.

Wood is useful. Antibiotic, antiseptic, and antitumor properties are ascribed to several species. Tannins are present.

The bark is used for colds and pneumonia, for skin infections, as a diuretic, emetic, cathartic, sudorific and antisyphilitic.

The leaves are alterative, cicatrizing, antiseptic, diaphoretic, diuretic, emetic; they have a sedative effect on the nervous system.

The roots are diaphoretic.

J. copaia D. DON (Guayana) (Fig. 42) Wood is used for pulp. Bark serves for colds and pneumonia; bark and leaves are used for skin infections, leaves for wounds to hasten healing. The bark is emetic and cathartic and also used for leishmaniasis. ROTH studied the bark structure 1981, fruit structure and dispersal 1987.

J. obtusifolia H & B (Venezuela, Guayana). The timber is used for furniture, the bark to heal wounds. The tree is ornamental.

Roth studied the leaf structure 1984, fruit structure and dispersal 1987, Venation patterns of leaves 1996. CASTILLO (1995) studied the plant taxonomically.

J. acutifolia HUMB. ET BONPL., Peru (1000–2300 m), Brazil, cultivated in Venezuela, is adstringent, diuretic, used for wounds and dermatitis (RUTTER 1990). The bark is mostly used (ALLAIN ARBE 1994).

J. glabra (DC.) BUREAU ET K. SCHUM. (Colombia, Bolivia, western Brazil). ALLAIN ARBE 1994 (Fig. 40.b).

Tabebuia

At least 7 useful species are known (Figs. 40 and 41).

Wood is useful.

The leaves are a bleaching agent. Leaves are used for flatulence.

Bark is applied for stomach ulcers, for malaria and chronic anaemia. Bark is antirheumatic and has anticancer fame (*T. obscura*).

Roots are used to treat snake bites.

Flowers serve for regular menstruation.

Antiinflammatory, antimicrobial and antineoplastic activities have been observed.

T. alba (CHAM.) SANDWITH. Brazil to Argentina (ALLAIN ARBE 1994).

T. aurea (MANSO) BENTH. & HOOKER occurs in the Brasilean Cerrado. The plant is called paratodo which signifies 'for everything'. A decoction of the bark is used as an antipyretic, as an expectorant as well as for influenza and diarrhoea; it is also put on ulcers. The bark powder together with quinine is applied for the treatment of malaria, rheumatitis, lumbago, gout, sciatica, as well as for inflammations of the stomach and intestine.

T. barbata (E. MEY.) SANDWITH occurs in inundated forests along the Rio Negro and the upper Orinoco, the Anazonas and its affluents. The tree is described by CASTILLO (1995) for the humid forest along the Cataniapo river, State Anazonas of Venezuela. The plant has a hard and fine structured timber of great economic value. The Kuripakos use the plant against flatulence after heaving eaten tapir meat (GENTRY 1992 B).

T. fluviatilis (AUBL.) DC. occurs along rivers in the brackish water, in the Amacuro delta of the Orinoco (Venezuela) up to Marantiao, Brazil (ALLAIN ARBE 1994).

T. heterophylla (DC.) BRITTON is very frequent, often found on the Antilles from 1–1000 m.

It is used as a plaster to remove callous layers of the skin. A decoction of the bark and the leaves helps against fainting (GENTRY 1992B).

T. impetiginosa (MART. ET DC.) STANDL. occurs in the tropical rain forests between Mexico and Argentina. Watery extracts of the bark were already used by the Incas to cure a variety of diseases. The bark is still used to treat leukaemia. The plant is said to have antibacterial, antitumoral, antiviral,

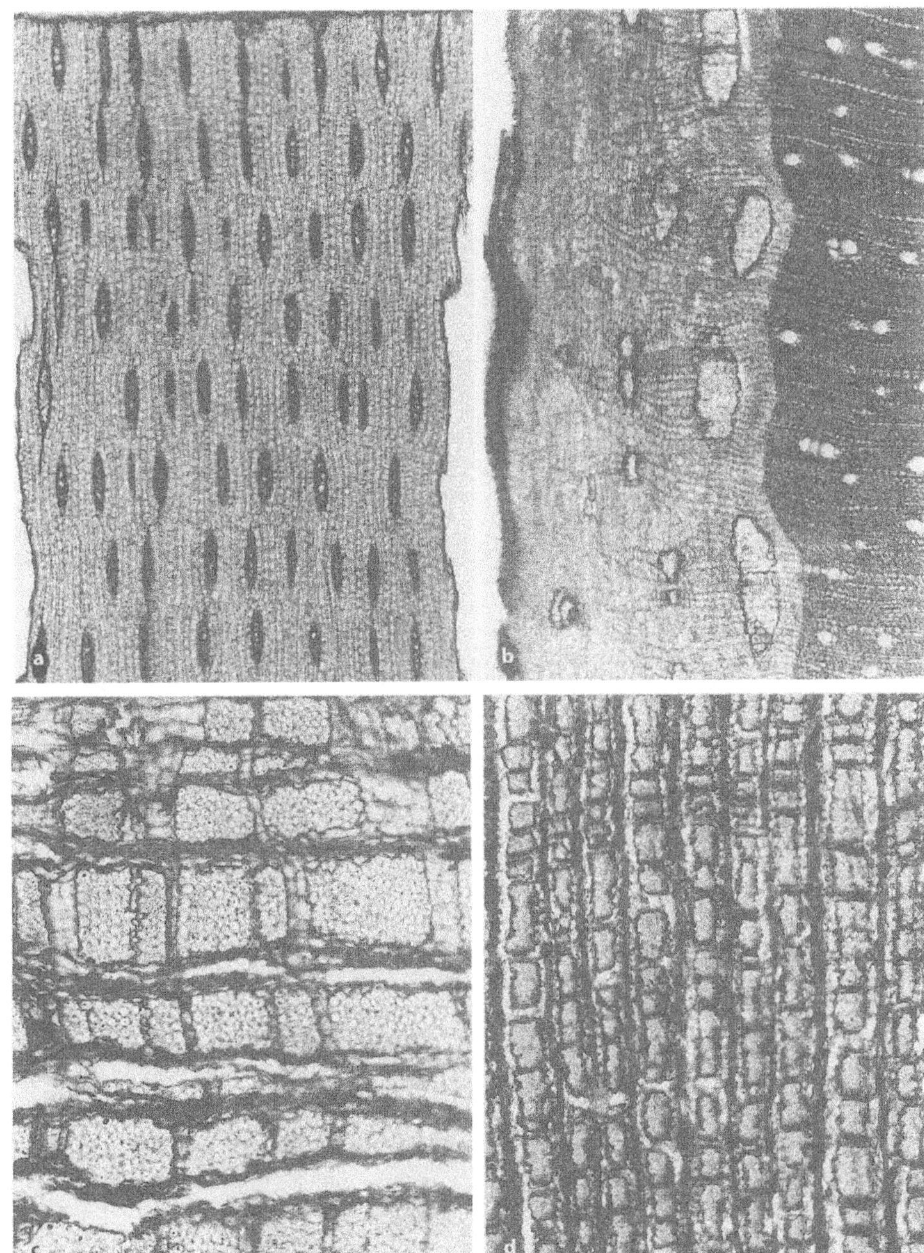

Fig. 40. Bignoniaceae: Barks. **a** *Tabebuia avellanedae*, longitudinal-tangential section with storied structure of rays. 6.3 × 6.3. **b** *Jacaranda glabra*, t.s. of a young twig with lenticular hard bast groups. 6.3 × 6.3. **c** *Tecoma guche*, t.s. with broad fiber layers. 6.3 × 9. **d** *Tabebuia insignis*, t.s. with layers of fibers radially interupted by rays. 6.3 × 9.

antifungal, antiinflammatory and antipyretic properties; it was formerly also used as an antisyphilitc. The Campas in Peru apply it against cancer (GENTRY 1992B, KREHER 1988/9, REYNEL et al. 1999).

T. insignis (MIQ.) SANDWITH occurs from the south of Colombia and Venezuela to Guiana and in Brasil and Bolivia.

The bark is used to treat conjunctivitis (SCHULTES & RAFFAUF 1999).

T. rosea (BERTOL.) DC. occurs from the south of Mexico to Venezuela and to the coast of Ecuador. The plant has an excellent timber of great economic value. LINDORF (1998) studied and compared stem wood with root wood anatomically. The bark of the tree is applied by the Mayas of Mexico for the treatment of cancer. GENTRY (1992B) mentions that the plant is also used against rabies and malaria.

T. serratifolia NICHOLS. Bark is medicinal. ROTH studied the bark structure 1981, leaf structure 1984, fruit structure and dispersal 1987, leaf venation 1996.

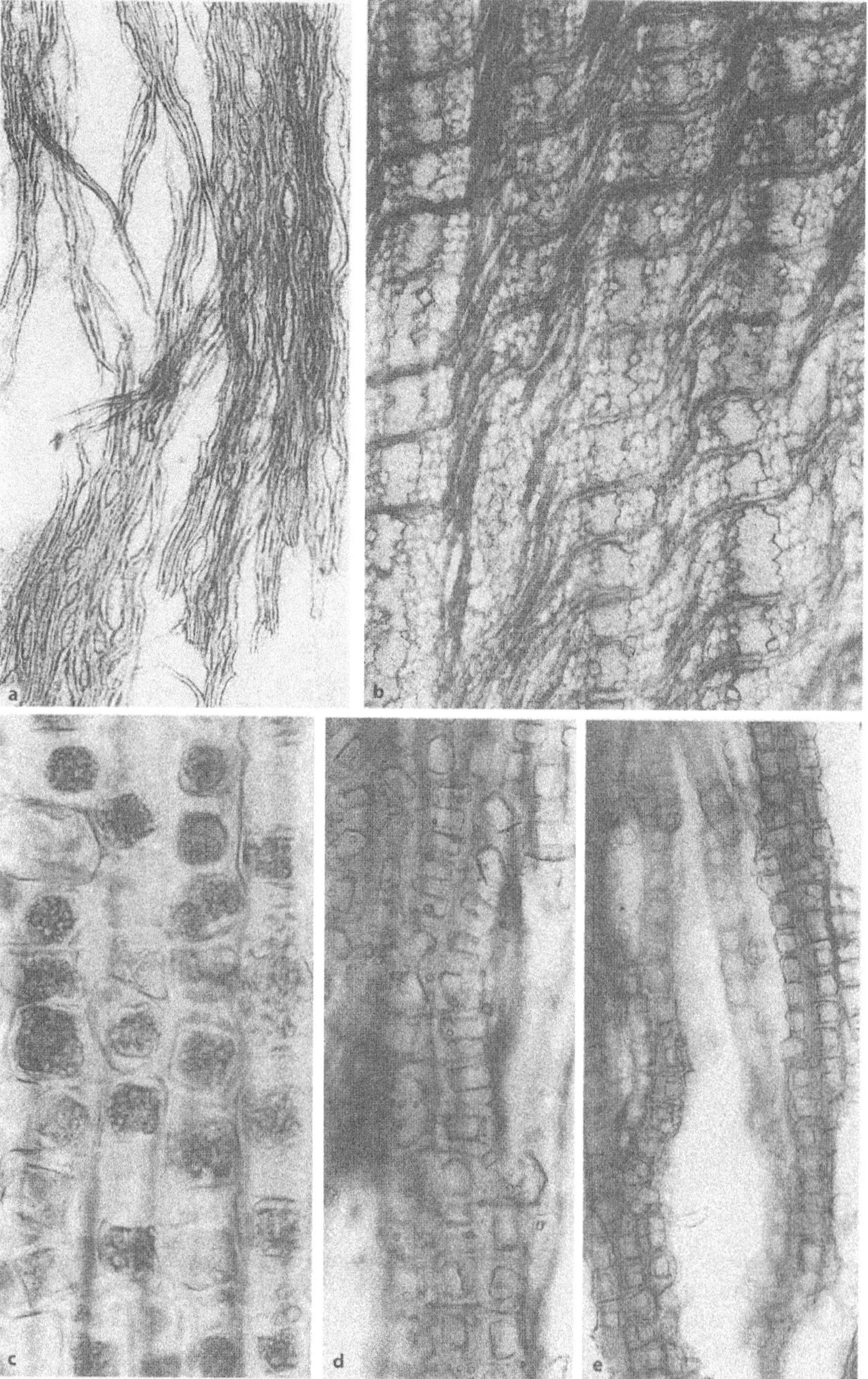

Fig. 41. Bignoniaceae: Barks. **a** *Tabebuia alba*, macerated preparation with fibers. 6.3 × 7. **b** *Tabebuia rosea*, t.s. showing undulated course of rays. 16 × 5, **c** *Jacaranda acutifolia*, longitudinal section with crystals in the form of druses. 25 × 12.5 × 16. **d** *Tabebuia avellanedae*, longitudinal section with septate crystal strands. 16 × 12.5. **e** *Tabebuia* caraiba, longitudinal section with rhomboid crystals in the septate crystals strands 16 × 9.

Bixaceae

The Bixaceae are a very small family which possibly has affinities to the Cochlospermaceae and Tiliaceae (METCALFE & CHALK 1950, CRONQUIST 1988, BERNAL & CORREA 1989). This view coincides with certain anatomical characteristica such as the presence of a stratified phloem with hard and softbast and rays dilated in the funnel form as well as the frequent occurrence of secretory cells and canals.

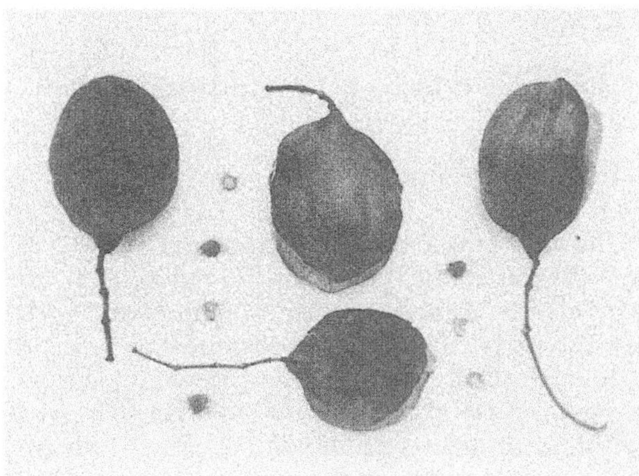

Fig. 42. Fruits and seeds of *Jacaranda copaia*, Bignoniaceae. The bicarpellary fruits are already dehisced to release winged seeds (ROTH 1987).

Large resinous cells and lysigenous slime pockets are also characteristic of the Bixaceae (ENGLER 1964). But there is no concordance at the chemical level between Bixa and Cochlospermum.

Bixa orellana L.
(onoto, achiote, bijal, anatto)

Taxonomical description

Taxonomical details described in this section can be found in Fig. 43 and Fig. 44. Bixa orellana is a shrub or small tree, 3–5 up to 7 (10) m in height. The stem is small and contains a latex of yellow-reddish colour. The alternating simple leaves are

Fig. 43. *Bixa orellana*. a habitus, b open capsules with numerous seeds.

Fig. 44. *Bixa orellana.* **a** flowers. **b** twig.

broad-ovate, 10–23 cm long and have an acuminate to cuspidate tip; the leaf base is truncate or subcordate and the lateral stipules are small. The petiole measures 4–6 cm. The flowers are of rose colour or white and are arranged in terminal panicles. The pedicels are 6 mm long. The 5 sepals are free, concave, rounded and 10–13 mm long. The 5 free petals are obovate and 2.5 cm long. The numerous stamens are inserted in a ring-shaped disc. The ovary is unilocular, spiny and accommodates numerous ovules. The only style has a bilobulate stigma. In Venezuela, the flowers appear between June and November. The fruit is ovoid to globose, up to 5 cm long and beset with prostrate flexible spines or bristles, 7–9 mm long and of dark-red colour, when ripe (SCHNEE 1960). The fruit is a capsule which opens with 2 valves. The numerous orange-red seeds are surrounded by an orange-red aril; they reach 4.5–5.3 mm in length.

Origin

The species is native to tropical South America.

Historical remarks

The colouring matter obtained from the aril of the seeds was formerly used by the Indians, particularly of the Caribbean region, to paint their bodies; this was partly done for adornment and partly for relief from insect nuisance; the ointment also gave protection against strong sunlight. The red skin of the Indians painted with achiote or anatto led to the erroneous conclusion that a 'red race' existed (SECAB II 1989, P. 270).

Occurence

The species occurs partly wild and partly cultivated over most of tropical America and prefers high temperatures. It is, however, not pretentious concerning soil and climate, occuring from sea level up to 1200 m above sea level. It is widely planted in tropical regions of the whole world.

Anatomical decription

Leaf: (Fig. 45, 46) The leaf is bifacial and hypostomatic and may be considered a 'sun leaf' (for the definition see ROTH 1984). Upper and lower epidermis are unistratified with outer walls of regular thickness; as seen in surface view, the anticlinal walls are straight. The stomata (without companion cells) are at epidermis level or are only slightly elevated above the surface. Peltate hairs with 4 basal cells, possibly 4 stalk cells and a pluricellular head appear in depressions of the upper and lower epidermis. As seen in surface view, 4 central cells are surrounded by a single ring of about 12 cells or by 2 rings of approximately 24 cells altogether. The outer and inner rings are probably formed by periclinal divisions of the original ring cells. As seen in a longitudinal section through the hair, the pluricellular head shows several stories probably also developed through periclinal cell divisions.

The palisade parenchyma consists of about 2–4 layers of relatively short cells, whereby the 3rd and the 4th cell layer proceed from periclinal cell divisions. The loose spongy parenchyma comprises about 3–4 layers of cells with short arms which form a network surrounding larger intercellular spaces. Druses of calcium oxalate are abundant in the mesophyll, particularly below the upper and lower epidermis.

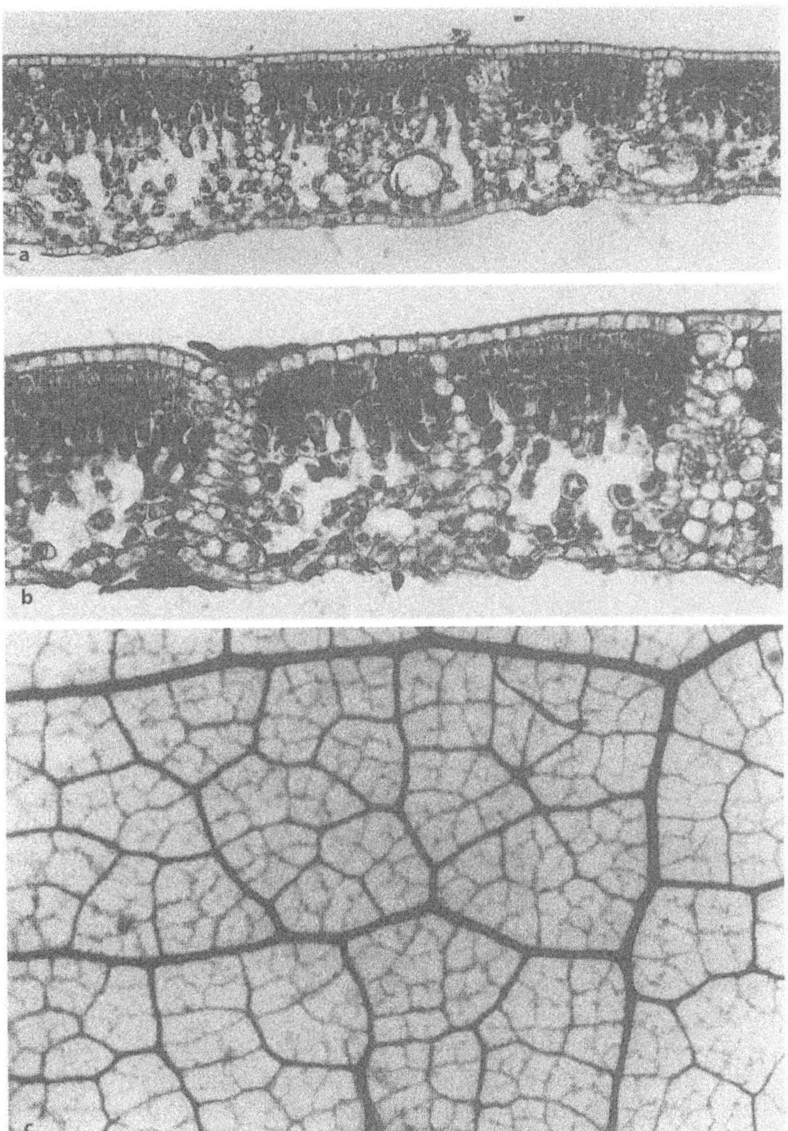

Fig. 45. *Bixa orellana*. **a, b** t.s. of leaf blade. **c** leaf venation

Small secretory cells can be observed dispersedly in the palisade parenchyma. Large idioblasts and 'giant cells' frequently occur in the mesophyll, and particularly in the spongy parenchyma; they are secretory cells filled with a dark refractive content which resembles resin or slime.

The large lateral nerves of first order become slightly vaulted above the surface on both leaf sides; the palisade parenchyma is interrupted in these nerves. A large horseshoe-shaped vascular bundle occupies the center of the lateral nerves. The fibers in the phloem are conspicous. Secretory cells are dispersed in the parenchyma. Several large secretory cells or idioblasts may occur on the abaxial side, below the phloem. However, on the adaxial side, one or two giant idioblasts, sometimes of enormous size, characterize the lateral nerves. At certain levels, the deviation and separation of nerves of higher order disturbes the characteristic picture of the first order nerves. The smaller vascular bundles of higher order are transcurrent to the upper as well as to the lower epidermis by parenchyma. Some stronger nerves of higher order may however be surrounded by a transcurrent sclerenchymatous sheath.

Stem. (Fig. 47) A twig of 4–5 cm in diameter shows a periderm only a few layers thick with lenticels. The primary cortex is proportionally large and comprises the following elements: several layers (5–9) of smaller cells frequently with a druse of calcium oxalate in which secretory cells and canals are embedded; a zone in the form of a continuous ring (as seen in t. s.) in which tangentially extended and

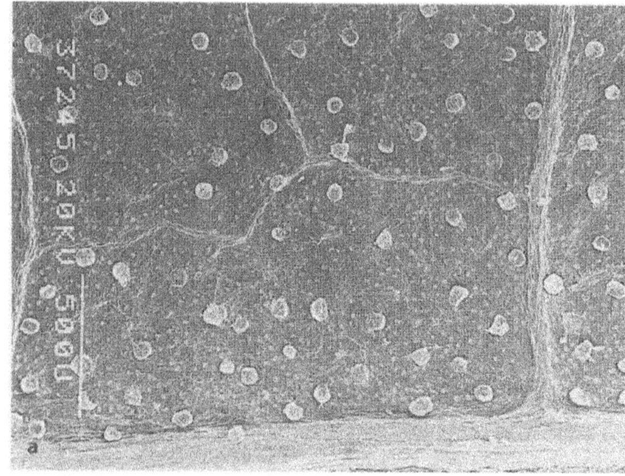

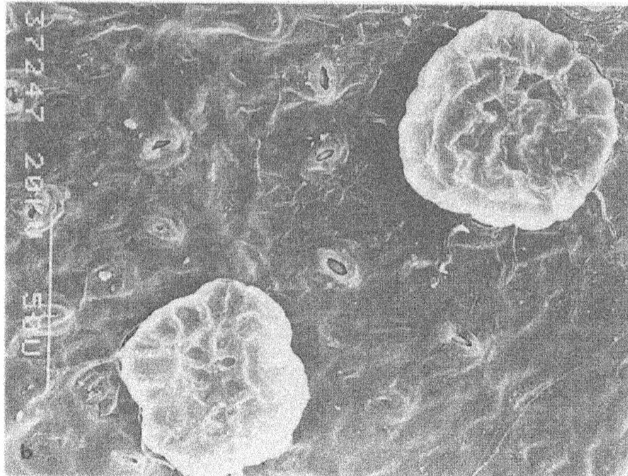

Fig. 46. *Bixa orellana.* **a** Leaf surface with glands and stomata, **b** glands and stomata. Scanning microscope.

anticlinally divided parenchyma cells as well as obliterated cells are found. Most conspicous in this zone are the large secretory canals surrounded by a ring of secretory cells; this zone is followed by 3–4 layers of regular parenchymatous cells. The following pericycle ring is broken up already into small portions of fibers due to the action of tangentialy expanding parenchyma cells. The regular dilatation growth in this zone is very striking; it is brought about by the formation of abundant anticlinal cell divisions.

The phloem shows the typical stratification with smaller hardbast bands (about 2–6 cells thick) and broader softbast layers. Very characteristic are certain medullary rays which dilate very regularly towards the outside in the funnel form; these rays leave larger distances between one another; the intervening rays do not dilate or only little, so that wedge-shaped phloem sectors traversed by small rays are separated from one another by the funnel-shaped rays. The rays are originally 1–2, seldom 3-seriate and are filled with starch grains. The phloem is well developed compared with the xylem. The xylem is diffuse-porous; the comparatively small vessels are arranged in radial multiples of 2–5 or more cells. Very conspicuous is the radial arrangement of all vessels, as seen in a general view. Fibers are the basic element of the xylem. The axial parenchyma, on the other hand, is very scarce, apotracheal and partly paratracheal, occuring in the form of scattered cells.

Most conspicous is the pith with a single very large secretory canal in the center and a ring of large secretory canals (about 10) at the periphery. This pattern may, however, be variable. The content of the secretory canals stains intensely dark-blue with toluidine blue. Druses of calcium oxalate are frequent, not only in the pith cells, but also in the medullary rays.

Our anatomical description differs from that of METCALFE & CHALK (1950) in some minor details. The same authors also emphasize 'the somewhat conflicting statements in the literature' concerning the anatomy of *Bixa orellana.* These differences may however be ascribed to the existence of varieties of this species.

Seed. (Figs. 49, 50a) As outermost layers (outer epidermis), the very large balloon-shaped cells (papillas) of the seed coat can be observed. They have delicate walls and contain the colouring matter bixin (carotenoids in the plastids) responsible for the red-orange colour of the seeds. Towards the inside, a thin cuticle is observed. A layer of macrosclereids which are likewise covered by a thin cuticle, follows. This layer probably corresponds to the outer epidermis of the inner integumet. The macrosclereids are palisade-like, as seen in transverse section, and have thickened walls with a cell lumen enlarging towards the inside. The next layer consists of compressed cells, followed by distorted cells with a larger cell lumen and very thick anticlinal and inner walls. As seen in transverse section, the cells have irregular outlines and irregular wall thickenings. As seen in surface view, the walls of these cells are strongly undulated and have pits. This layer which corresponds to the inner epidermis of the seed coat is in direct contact with the endosperm and seems to be slightly cutinized on the inside. The endosperm consists of large cells with thin walls which are completely filled with starch grains; these show a cleft of dryness in the center (hilum). The layering of the starch grains is concentric. Around the raphe, the cells are radially arranged. The limit between the raphe and the endosperm consists of cells with irregularly thickened and undulating walls.

The embryo is straight and has flat cotyledons. In the very mature state, the papillas of the seed

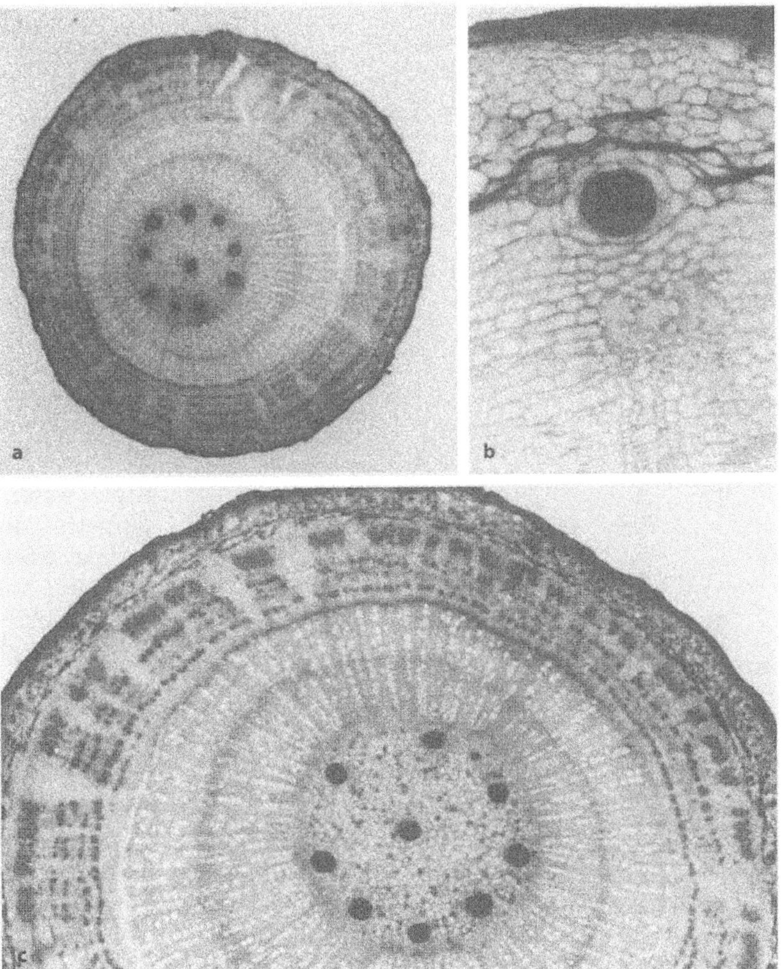

Fig. 47. *Bixa orellana.* **a, c** Axis in t.s., **b** secretory canal.

coat unite in a pulpy mush, resembling an aril. However, the aril itself is very small, being restricted to the funicle.

Fruit. (Fig. 48) Fruit morphology and dispersal are described by ROTH 1987.

Ethnobotanical and general use

Nutritional use

The arillike seed coat is used as a particularly tasty condiment and to give colour to food stuff such as soups and dishes. It is also used as a substitute for saffron. It facilitates digestion when added to dishes. The seed is rich in calories, carbohydrates, fiber, minerals (salts), provitamin A, and 2 pigments, bixina and orellina, which are responsible for the colour. Vitamin A is generated from bixina by oxidation. Chickens are fed with the testa to give the egg yolk a more intense colour.

Industrial utilization

Industrially, the seed coat is applied as a dye for cheese, butter, margarine, other fats and oils, rice, candy, chocolate and other foods, as well as for dying ointments and cosmetics, as a natural dye for lipsticks, soap, pomade, shampoos, laquers and varnish. It is furthermore employed in drinks with carbonic acid, such as orange juices, to give them a more brilliant colour. The seed coat also gives a beautiful colour to textiles such as wool, cotton and silk. It could very well be used industrially as an insect repellent and as a protection for the skin against excessive solar radiation.

In the same way as the testa and the leaves, entire twigs are also used as dyes for fabrics such as silk and cotton which then adopt a pleasant yellow, orange or red colour. The seed coat is also applied as a natural dye in fine handicraft arts, and in ceramics.

To obtain the dye, the seeds are macerated in hot oil or water and the strained liquid is used as a

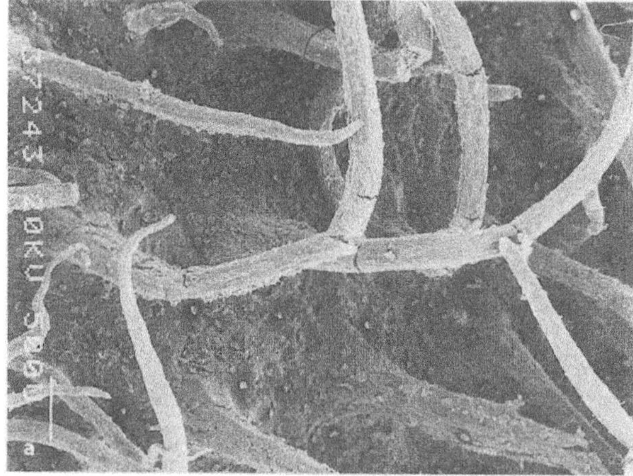

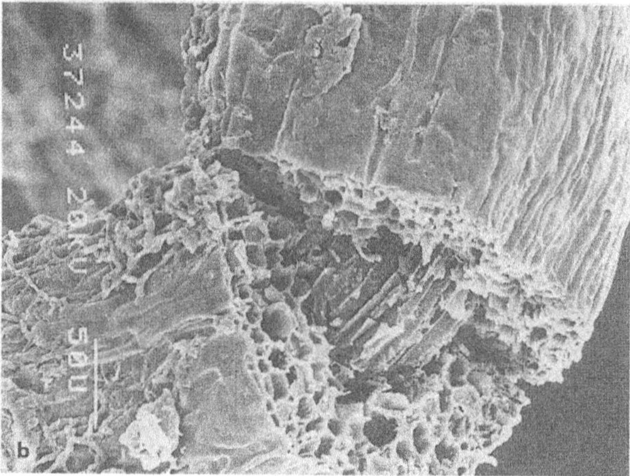

Fig. 48. *Bixa orellana*. **a** spines of fruit. **b** part of a spine. Scanning microscope.

colouring matter. There are however other methods. After maceration and washing the waterexcess is removed by evaporation and the residue is moulded into cakes or rolls for the market.

Peru, India, Panama, Ecuador, Jamaica and Brazil are the main exporting countries. The exported stuff is mainly employed as a dye for fabrics and food.

The fiber of the bark is useful for making ropes. By incising the bark of the living tree, it exudes a gum that is similar in its qualities to 'arabic gum' and which is applied in the same way.

Medicinal use

The following parts of the plant used as a medicine are stem, leaves, root, flowers, fruits and seeds and the corresponding names of the drugs are: cortex, folia, radix, fructus, and semen.

In the 'herbolarios' (herb shops) of Venezuela mostly the twigs with fruits are sold.

Bark shredded and macerated in water is applied against jaundice.

Leaves have digestive effects and are used as antipyretics. A decoction of the leaves (30 g per liter) for 2 minutes is used as a gargle and has also haemostatic effects on wounds; it is furtermore considered a remedy for diseases of the liver. A decoction of the leaves is also used against diarrhoea and against parasites of the intestine. Fresh leaves placed on the forehead relieve headache and migraine. Fresh leaves applied locally are also said to cure erysipelas. Leaves and twigs shredded and macerated in water supply a slime similar to that of arabic gum and with the same stimulant properties. A decoction of leaves and root is applied against dysentery.

Leaves shredded and macerated in a small quantity of water, supply a gummy substance which has purgative, diuretic, antiinflammatory, and antigonorrheal effects. Further healing effects ascribed to the leaves are healing of wounds, curing of diabetes, of heamorrhage and anuria. A decoction of the leaves is applied for loss of hair (alopecia).

An infusion of the root has diuretic effects and cures icterus. A tea of the root alleviates cough. A powder of the root has an antispasmodic effect and raises blood pressure. Healing properties ascribed to the root are those of curing hepatitis and jaundice, of reducing debility and sleepiness; and having digestive effects.

The flowers which are rich in honey are used in an infusion that has laxative effects, and also to remove the mucus from new-born children.

The inner pulp of the fruit is used to cure tonsillitis. A decoction of the fruit is used to cure tonsillitis. A decoction of the fruit is diuretic, when locally applied it cures haemorrhoids. Fried it is used against erythema. A decoction of fruit and leaves has depurative effects.

A spoonful of seeds in a cup of boiling water is an expectorant and helps against bronchitis. Seeds are applied against diseases of the skin and for burns. A mixture of seeds and fruit pulp is used for herpes zoster and eczema. An infusion of the seeds is used as a diuretic, against diabetes, against venereal diseases and haemorrhage. Orally the seeds are applied against measles, smallpox and pleurisy. The oil of the seeds taken orally is good for the stomach, has refreshing effects and is purgative; it is also used against leprosy. Seeds taken orally promote the monthly menstruation (emmenagogum). The part of the plant mostly used is the seed coat which is rich in carotenes. It is used as a condiment, a digestive and to cure the stomach. As an antidote it is furthermore used against poisoning with Yuca

Fig. 49. *Bixa orellana*. **a, b** Seed surface. Scanning microscope.

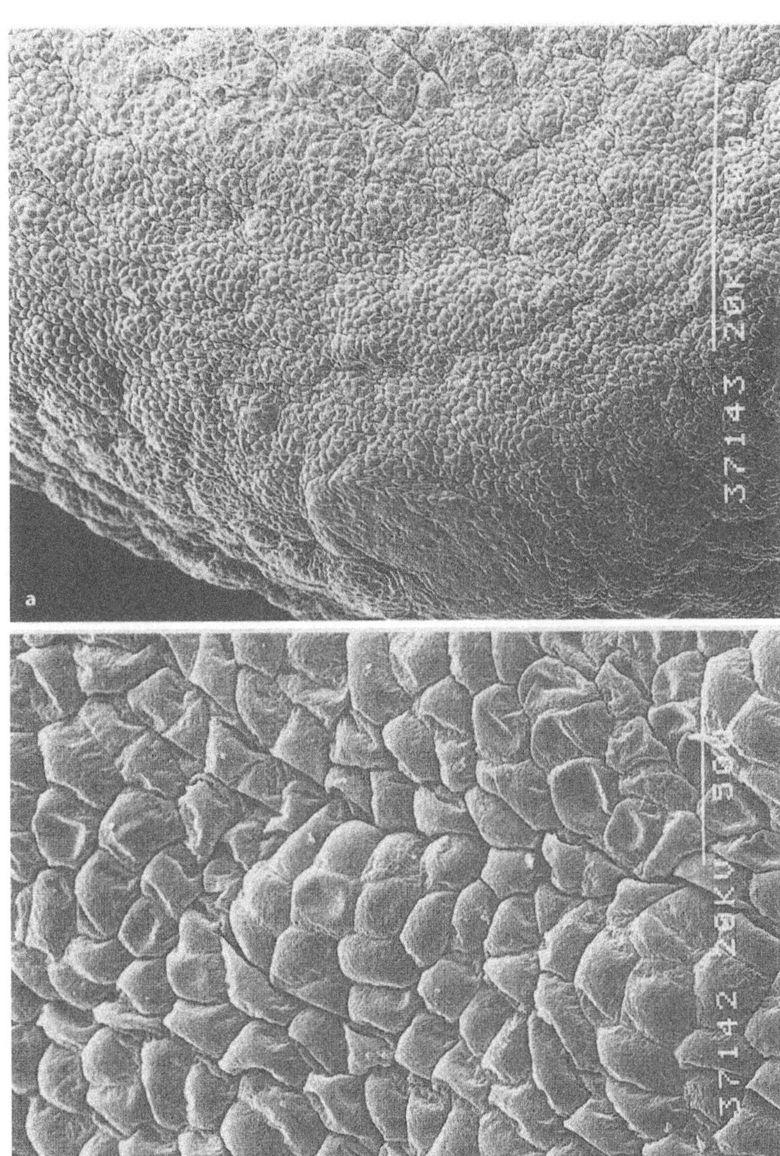

brava. Put on the skin it causes better cicatrization. A powder of the testa mixed with oil is used as an ointment and cures burns and ulcers.

Four grams of the dried testa taken a day are said to have an aphrodisiac effect.

The healing substances of the distinct plant parts are mainly located in the secretory cells and canals which contain resin and slime.

Method of use

Decoctions or infusions are prepared from the root, either in pure water or in aquavit. Entire twigs are used for bath preparations. The grind bark dissolved in water supplies a drink. The boiled leaves are taken as a remedy either orally or as a gargle; fresh leaves are applied locally. Flowers in an infusion, fruits as a decoction and seeds as an infusion are used in the same way as the leaves. Pulverized and mixed with oil all these parts also serve as an ointment. See also the methods of application described above.

Healing properties

The plant is used as an aphrodisiac (seeds), as a laxative and diuretic, against diseases of the liver, against angina, dysentery, tonsillitis and asthma, as an expectorant, against hedache, to cure burns, ulcers and eczema, against gonorrhoea, smallpox, leprosy, and diabetes; as an antidote against intoxication caused by yuca brava or túa-túa, it is also

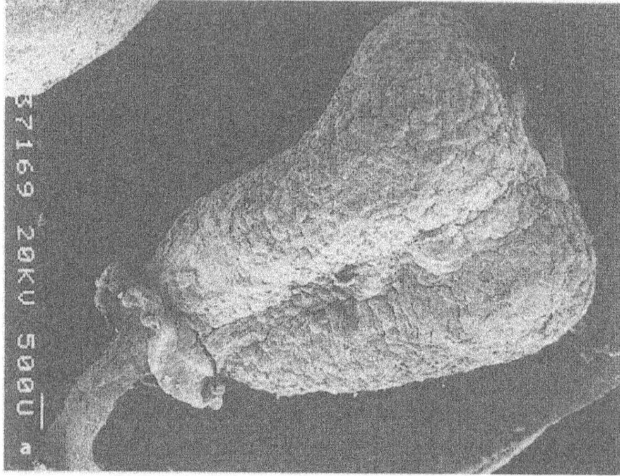

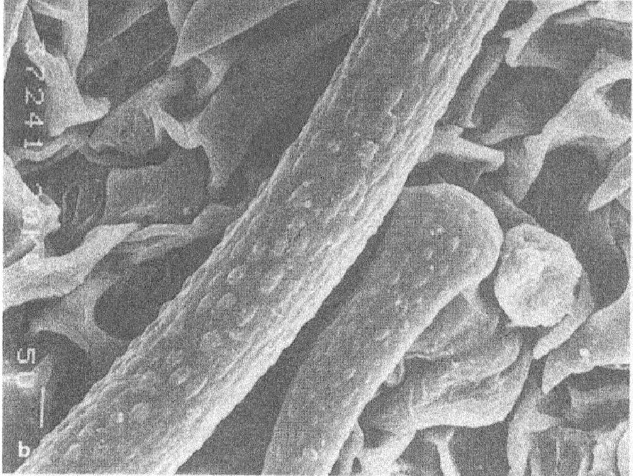

Fig. 50. **a** *Bixa orellana*, entire seed. **b** *Heliotropium indicum* with cystolith hairs on lower leaf surface.

applied as a haemostat for slight wounds, against haemorrhoids and even against insect bites and as an insect repellent.

Toxicity has not been reported. The plant is not carcinogenic and is therefore preferred as a colouring for food.

Chemical contents
Although the plant is well known in the tropics of the whole world, relatively little is known of its contents. The species is rich in carotene, in a red pigment (bixin) and a yellow pigment (orellina); it also contains slime, tannins, 7 different flavones, mono- and sesquiterpenes, tomentosic acid, oil, gummy substances and saponins (SECAB II, 1989).

Varieties and related species

A black, a variegated and a yellow variety are known in Venezuela. Related species are *B. excelsa* GLEASON & KRUKOFF, *B. platycarpa* RUIZ & PAVON, and *B. urucurana* WILLD. from Ecuador with similar characteristics as *B. orellana*; this species is still collected by the Indians to paint their bodies (BRÜCHER 1989, MOLAU 1983). METCALFE & CHALK 1950 suggest affinities of *Bixa* with Cochlospermaceae and Tiliaceae for anatomical reasons; CRONQUIST 1988 suggests to unite the genera *Cochlospermum* and *Bixa*, but there is no concordance at the chemical level.

Cultivation

The plant is easily propagated by seeds and develops rapidly.

In only 2 years it already produces fruit, so that about 135–270 kg fruit may be harvested per tree. Propagation by cuttings is also possible. However, the distinct varieties differ in production. No disease or plague is known which affects this plant. The plant is also cultivated as an ornamental tree for its beautiful white or rose-coloured flowers.

Observations

The species is easily recognized by its very characteristic stem transection. Likewise the fruits with their red spines and the seeds with the balloon-like cells of the aril are very typical and unmistakable.

Bombacaceae

The Bombacaceae are a tropical family which developed particularly richly in America. The representatives are trees, often with a thick stem (bottle trees) in which water is stored. The leaves are digiate or undivided. The flowers are frequently very big and showy. The fruits are capsules; the nude seeds are often embedded in a woolly mass, the hairs of the pericarp, which supply kapok (*Ceiba pentandra*). The endosperm is weakly developed or completely absent.

Mucilaginous ducts are often present. Stellate hairs and scales are also frequent.

Important genera of South America are *Ceiba*, *Ochroma* and *Chorisia*. The trees are usually very large with enormous buttresses and very thick trunks.

Very little is known of the chemistry of this family.

Ceiba pentandra (L.) GAERTN. (ceiba, ceiba yuca, ceibo jabillo, habillo, cumaca, jabillo, paraná)

Taxonomical description

Ceiba is a very large tree which reaches 30–50 m and forms extensive buttresses towards the ground, attaining an enormous trunk diameter; the crown is likewise large. Young trees and twigs are spiny. The palmate leaves have 6–9 leaflets; they are shed during the dry season, giving the tree a barren aspect. At this time, the flowers appear with delicate white or pink petals. The fruit is a large capsule, about 12–18 (24) cm long and has a coriaceous pericarp which dehisces loculicidally to expose numerous seeds. The small black seeds are embedded in a huge dense white, fibrous and silky mass of hairs which proceed from the fruit endocarp. The fruit outside is brown and shows 5 longitudinal ribs. The calyx is persistent on the fruit. The endocarp hairs are commercially used as kapok. The capsules contain about 17% fibrous hairs which are 12–15 mm long and are covered with a thin layer of wax, inhibiting spinning.

The tree lives many hundreds of years; as it attains such a large size, it grows very isolated and many other trees have to be destroyed, when it is cut.

The seeds are dispersed by wind together with the hairs, which considerably reduce their specific weight. A single tree may cover a very large area with its seeds and woolly wad.

The species is pantropical and grows perfectly in secondary forests as well as in cultivation.

Bark anatomy has been studied by ROTH (1981), fruit morphology and dispersal by ROTH 1987.

Ethnobotanical and general use

Nutritional use

The seeds have a high (20–25%) content of oil used for culinary purposes; also as a lubricant, for illumination and the manufacture of soap.

Seeds are eaten fried or fresh and ground. They are also a food for cattle. Even the flowers are eaten by cows.

Economical utilization

Seed. As indicated already above, the oil of the seeds is applied as a lubricant, for manufacture of soap and for illumination.

Wood. The wood is white, light and soft and easily worked, but not very durable; it is similar to balsa in its structure and applications. It is not only used for packing-boxes, matches, plywood, toys, light furniture, cabinet work, turner's work and canoes (by the natives), but also for drums and trays and finally as a combustible or pulp for paper manufacture. The ash of the wood could be used as a substitute of salt.

Trunk. The stems can be used for fences and the entire tree as a shade dispenser due to its extense crown.

Bark. The bark contains a reddish fiber which could be used for paper pulp.

Kapok. The fiber is resilient, buoyant, water resisting and highly moisture proof. It is also appropriate for stuffing.

Kapok makes a good thermic and acoustic insulating material, particularly suitable for airplane equipment, as an equipment for automobile trucks and for refrigeration, as well as a stuffing material for pillows, mattresses, life belts (due to its water repellent properties), and jackets. It is even appropriate for needle work and weaving. It is also very suitable for shipping equipment, as real wool becomes fusty due to humidity.

The trash of kapok gathering (mainly the shells) are added to fertilizer combinations.

Medical use

Caule, cortex, folia.

Twigs and stem. They are applied locally, are used as an antiinflammatory for tumors and abscesses.

Bark. Bark is applied to treat wounds. Bark is said to have emetic, diuretic and antispasmodic properties when applied in infusion.

The resin. Resin of the bark is used to cure intestinal problems.

A decoction of the bark strengthens the hair as a wash and taken as a drink it helps to expel the placenta after birth in humans as well as in cows.

A bath with the decoction cures oedema. Bark crushed and boiled helps to cure rheumatism when applied in a bath.

Leaves. With leaves in decoction, pickles and insect stings are treated, while skin conditions are cured with a cataplasm or bath.

Root. The root is used to cure leprosy.

Cultivation

The plant can be propagated by seeds or cuttings; it grows fast, reaching 5 m in 5 years. Fructification takes place after 5 or 6 years, and production can continue until the plant reaches 50 or 60 years or even longer. From 100 fruits apporximately 3.5 kg fiber are harvested. It is calculated that from a hectare with 280 trees, which are 8 years old, 450 kg fiber can be obtained.

The silky fibers contain 64 % cellulose and 13 % lignin.

Gyranthera caribensis PITTIER (cucharón, candelo, niño) (Fig. 51, 52, 53)

This is the highest tree in the transitional cloud forest, 30–50 m high, but some individuals even reach 64 m. The leaves are palmately compound with 7 leaflets in the adult stage. The flowers are 20–30 cm long and of creamy colour. The staminal tube is cylindric and has numerous anthers. The ovary is 5-locular with numerous ovules. The fruit is an elliptic dark-brown capsule, 25–30 cm long and contains 8–12 large winged seeds of brown colour.

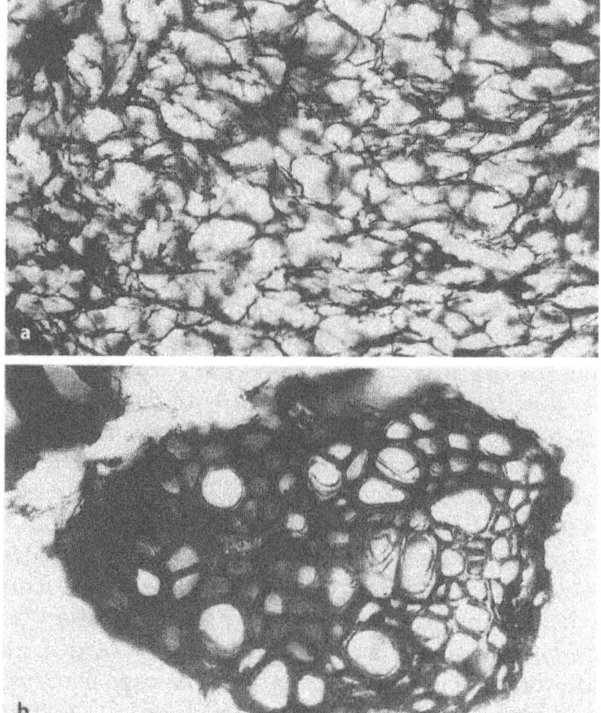

Fig. 52. *Gyranthera caribensis.* Wing of the seed in t.s. **a** Dead tissue which contains air. **b** Vascular bundle (ROTH 1977).

Fig. 51. *Gyranthera caribensis.* **a** Seedlings. The left one with the fleshy macrocotyledon. **b** Seeds with large wings (ROTH 1990).

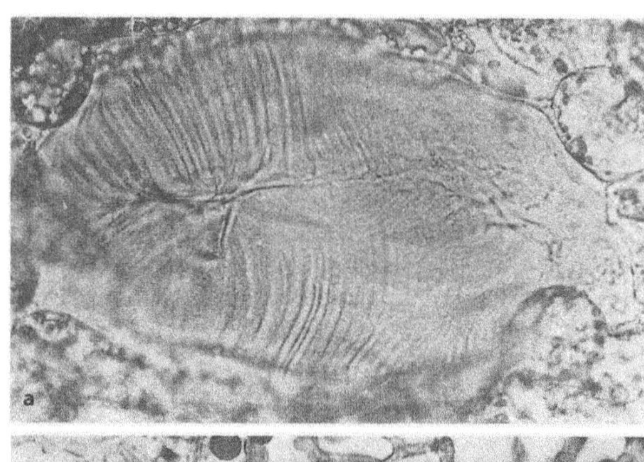

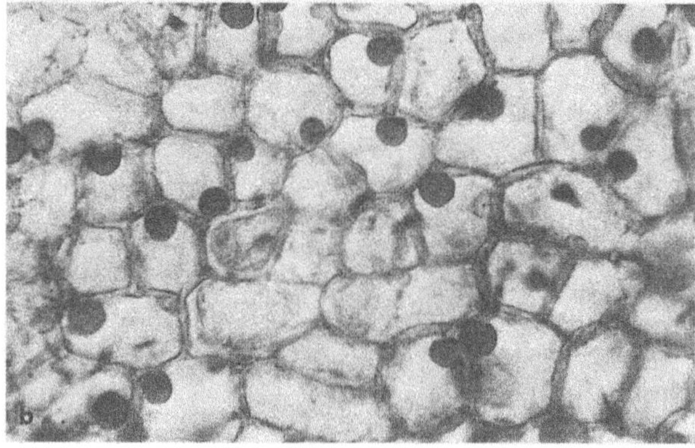

Fig. 53. *Gyranthera caribensis.* **a** Slime cavity of the macrocotyledon. **b** Surfce view of the upper epidermis of the macrocotyledon with large oil drops (ROTH 1977).

The leaf anatomy has been described by ROTH (1992).

The structure of the seed coat including the large wing has been studied by ROTH (1977). The wing consists of dead air-containing parenchyma, reducing the specific weight of the seeds. The seedling shows anisocotyly with a reduced microcotyledon and a very large macrocotyledon, which contains large slime cavities as well as oil drops and starch; the cell walls are impregnated with tannins, equally as those of the seed coat. The seedling, rich in nutritive substances, can survive for several months in the shade of the mother plant. The tannins prevent it from rotting (ROTH 1977).

The tree is endemic in the Cordillera de la Costa of Venezuela occurring at an altitude between 900 and 1100 m.

The wood, of light colour and very soft, could be used for general carpentry, boxes and crates or as pulp for paper manufacture; it is tough and strong, light in weight and easy to work but not resistant to decay.

The bark is rich in tannin and fibers. Even the slime of the macrocotyledon could be used in pharmacy.

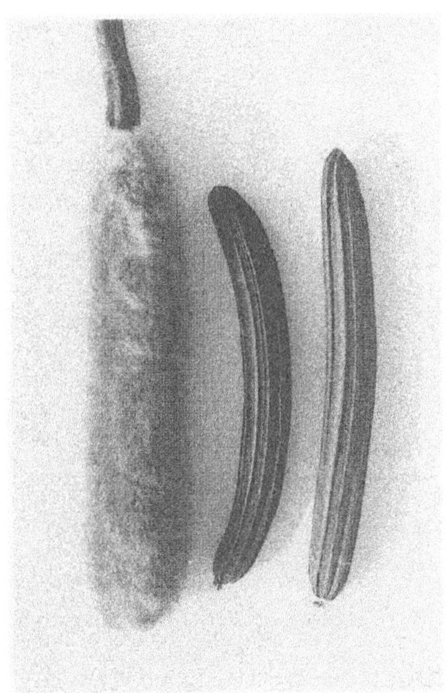

Fig. 54. *Ochroma* sp. Closed and dehisced fruits with Kapok-like locular hairs (ROTH 1987).

The tree is native to the Caribbean and tropical America.

Ochroma sp., Ochroma pyramidale (CAV.) URB. (identical or related to *Ochroma lagopus* SW.) (Fig. 54)

The small to medium-sized tree is known by its exceedingly fast growth; it is therefore regionally planted by the natives to serve as a fodder plant for cattle, being cut as long as it is still green.

The vernacular name 'balsa' was given to the wood because it was used by the Indians, to construct little canoes (ferries) to cross small river beds; the Spanish word balsa signifies 'ferry'.

The very light and soft wood with a specific weight of 0.12–0.30 (air dry) is used by the natives for canoes, floating rafts for fishing, for bowls, for toys and carving. In modern industry it could be applied or is already used as an insulating material and for airplane construction. It is also useful for novelties, cement forms and temporary boarding.

The cucumber-like, approximately 30 cm long capsule with a papery pericarp contains a great number of seeds embedded in a cotton-like 'wad' which proceeds from the fibrous hairs of the endocarp (Fig. 54). This wool could be used in the same way as 'kapok' of *Ceiba pentandra* for filling material, for stuffing pillows, matresses, life belts etc. or as an insulating material for cold, heat and sound.

As the plant grows very fast, it could easily be used for large plantations.

Pachira

There are only 2 species of *Pachira* in Venezuela and few in the enitre tropical region of the New World. Both Venezuelan species are ornamental trees cultivated for their large and showy flowers.

The leaf structure of the 2 species has been studied by ROTH & LINDORF (1991).

Pachira aquatica AUBLET and *Pachira insignis* SAVIGN. (Figs. 55, 56 and 57) are both indigenous to Venezuela and have a very similar leaf structure. *P. aquatica* grows in inundated regions (e. g. of the Alto Orinoco), while *P. insignis* is more resistant supporting drought and strong sunlight. Both species keep their leaves throughout the year. But the leaf of *p. insignis* is slightly more xeromorphic and has a multilayered palisade parenchyma, typical of sun leaves. The vascular network is comparatively dense; bundle extensions supply the upper and lower hypodermis with water. Lenticels are more abundant in *P. aquatica* where they serve for aeration. Crystals of calcium oxalate are more frequent in *P. insignis*. Likewise, secretory cavities filled with slime are more frequent in *P. insignis*. The cuticular crests on the lower leaf epidermis which produce particular patterns are very characteristic.

A detailed description of th 2 species, their leaf anatomy and their distribution in Venezuela (with a map) is found in ROTH & LINDORF 1991.

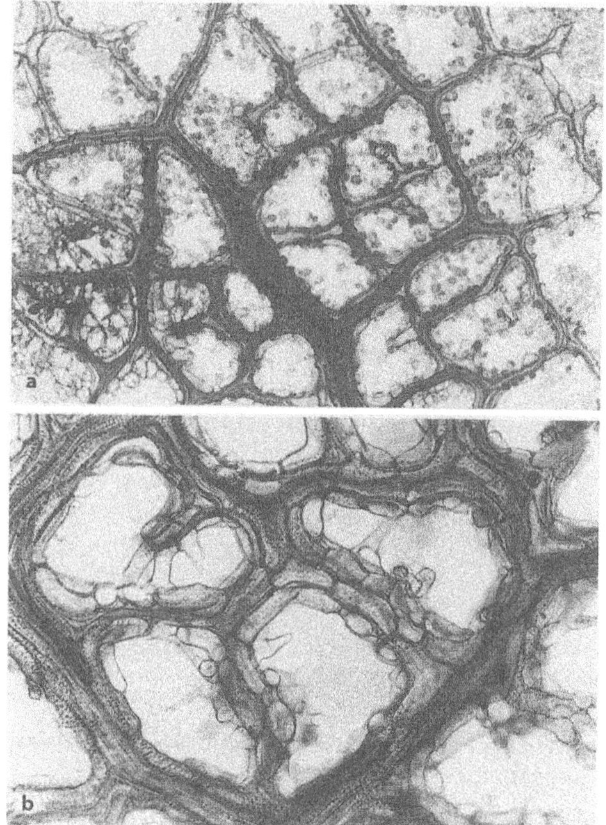

Fig. 55. a, b Venation pattern in the leaf of *Pachira aquatica*, Bombacaceae. The meshes are very small and the free endings few (ROTH & LINDORF 1991).

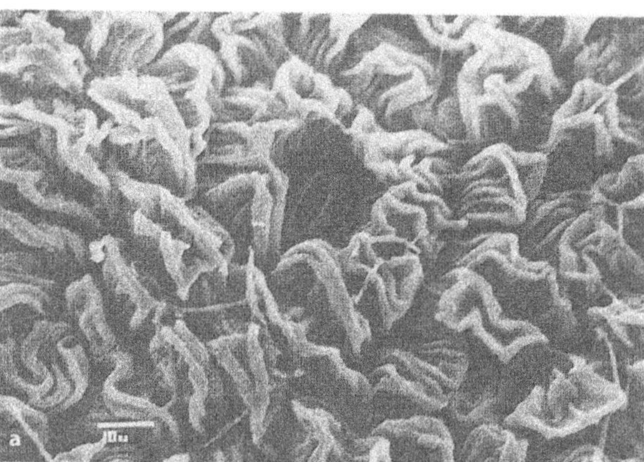

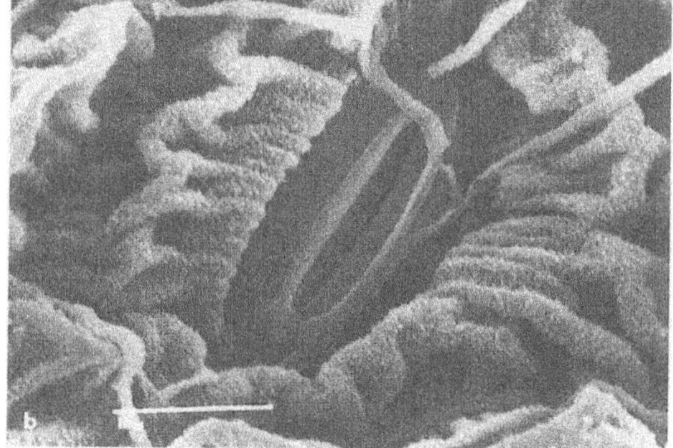

Fig. 56. *Pachira aquatica.* **a, b** Lower leaf epidermis with strongly sculptured surface; note the cuticular crests and the stoma (ROTH & LINDORF 1991).

Ethnobotanical and general use

Pachira aquatica. The seed oil contains cyclopropene, fatty acids and β-sitosterol.

The seeds are called 'chestnut of America'; they are very starchy and are not only edible, but taste like real chestnuts.

Fruits of *Pachira aquatica* are applied for respiratory diseases, cough and whooping cough, extracting the liquid from the locules.

The leaves serve for fever with headache; a handful leaves are boiled in a liter of water for half an hour.

Pachira insignis produces a similar seed oil. The seeds are also eaten and likewise called 'chestnut of America'. The defatted seed (removing sterculic acid) is rich in amino acids, including lysine, and could be a nutritional supplement for the world, suggest SCHULTES & RAFFAUF (1990).

The seeds have to be cooked or fried first (see above!) before being eaten.

Fruits and leaves also have a medicinal use. The leaves contain about 10 % tannins, also slime, carbohydrates and minerals.

They have antidiarrhoeic and expectorant properties and have the effect of a bronchodilator. They are applied against diarrhoea, and cough.

To prepare the tincture, 200 g leaves and fruit are put into 1 liter of alcohol (60 %) for 7 days. The liquid is filtered and taken.

Mainly *P. insignis* is planted as an ornamental along highways and in gardens and parks. The flowers are very large, showy and fragrant, they are about 30 cm long; the 5 petals are red, the filaments yellow (up to 1000 were counted! (VELEZ & VELEZ 1990); the fruit is a capsule up to 30 cm in diameter, and contains many brown seeds, 2 cm in diameter.

The plant is reproduced by seeds; it has a slow growth and a profound radical system, but has a long life and is one of the most resistant trees of the country.

Pseudobombax maximum ROBYNS (sibucara)

P. maximum (Figs. 58, 59, 60) is a medium-sized tree with a bottle-shaped trunk (bottle tree). The

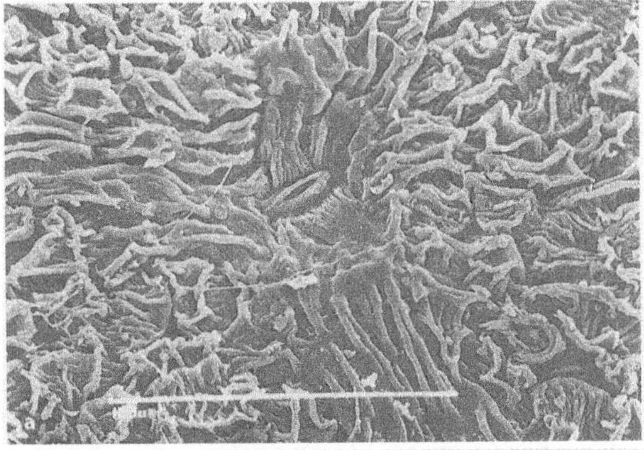

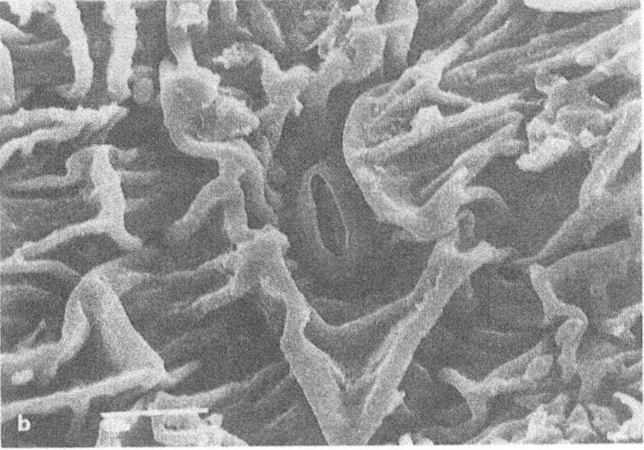

Fig. 57. *Pachira insignis.* **a, b** Lower leaf epidermis with cuticular folds and stoma (ROTH & LINDORF 1991).

swollen stem represents a water reservoir; the wood has an elevated water content storing the water in the thin-walled xylem parenchyma. Stem and branches remain partially green throughout the life time of the tree. The white flowers are large and very showy. The stamens are very numerous. The fruit is an oblong greenish capsule, about 12–18 cm long containing numerous seeds. These are wrapped in a grayish silky wool with which they are dispersed. The wool consists of long curled hairs which arise from the ovary inside and could probably be used in a similar way as 'kapok' of *Ceiba pentandra.*

The plant can be propagated by seeds. It is very resistant to extreme heat and drought and supports very poor soils.

The tree is deciduous and bare of leaves most of the year, but uses the large trunk and the branches for photosynthesis.

The species is characteristic of Central America and the north of South America. In Venezuela it grows in arid regions and is a representative of dry and thorny woodlands of Falcon, Zulia and Lara.

Ethnobotanical and general use

The tree can be used as an ornamental in gardens and along highways.

The silky wool could be used for stuffing pillows and mattresses, or as an insulating material.

The wood could be used for manufacture of boxes and as a firewood, if the tree were cultivated.

Even the slime could be useful in pharmacy etc.

Boraginaceae

The representatives of the Boraginaceae are trees, shrubs or herbs. The undivided leaves are seldom opposite. The flowers occur in boragoid or scorpioid cymes. The hairs are characteristic: bristle hairs, stellate hairs, hook-shaped hairs and cystoliths are frequent and generate the rough surface. The ovary is 2–4 celled and transforms into a drupe, nut or into mericarps (nutlets, synonym: 'Klausen').

The walls of the bristle hairs as well as the bases of the cystoliths may be impregnated with SiO_2 and/or $CaCO_3$. Further chemical contents are the boragoid alkaloids (cynoglossin), allantoin and alkanna-pigments.

Allantoin is a preferred form for the storage and transportation of nitrogen. It improves granulation and is consequently valued as a wound-healing remedy in Boraginaceae.

Red dyes corresponding to esters of phenolic compounds are frequent in Boraginaceae with a thick taproot; they probably develop from colourless precursors when the plant is dried. *Alkanna tinctoria* TAUSCH. yields a red dye, alkannin, in its roots (alkanet root); it is a naphthoquinone derivative, which is permitted as a food dye.

Certain species (Lithospermeae) have oestrogenic and contraceptive effects; phenolic compounds are probably the active principles.

Cordia

At least 14 useful species are known.

For this genus a large variety of applications is reported.

The wood is useful.

The bark contains tannin and a gum. Medically it is used for coughs. Bark fiber is also used.

The leaves are stomachic and tonic. Rough leaves of some species are applied as sand paper. Leaves are likewise helpful to kill fly larvae.

Fig. 58. *Pseudobombax maximum*, Bombacaceae. **a** Habitus with bottle-shaped stem. **b** Green bark with long cream-coloured stripes formed by the medullary rays. Note the longitudinal extention of the rays (ROTH 1981).

Leaves and shoots are astringent and haemostatic.

Flowers are used for leis, and to induce perspiration.

Fruits are edible. Fruits and leaves help with coughs and colds. Some mucilaginous fruits supply a glue. Medically, fruits are emollient and help against fever.

An ointment obtained from the seeds is applied for cutaneous diseases. Herbal parts are used for intestinal and stomach complaints and bronchial conditions.

Bark structure has been studied by ROTH 1981, as has structure and dispersal 1987, and leaf venation 1996.

Cordia alliodora CHAM. The wood is useful.

Leaves are stomachic and tonic, they help catarrh.

An ointment of the seeds is applied for cutaneous diseases.

The fruit is edible.

Bark structure has been studied by ROTH 1981, as has fruit structure and dispersal 1987, and leaf venation 1996.

Corida nodosa LAMARCK. Leaves are used to kill larvae of bot flies embedded in the flesh. Leaf venation has been studied by ROTH 1996.

Heliotropium angiospermum MURRAY (alacrán, hierba alacrán)

Taxonomical description

Annual or perennial herb or suffrutex, erect, 15–80 cm high, somewhat pubescent. Leaves alternating or some opposite, blade elliptic ovate to lanceolate, with acuminate or obtuse tip 1.5–9 cm long and 0.7–4 cm broad, somewhat hairy, particularly along the nerves, and with cystoliths. The relatively broad petiole 4–15 mm long. Leaf margins decurrent on the axis.

Terminal inflorescences in scorpioid cymes; flowers solitary or in pairs. 20–90 sessile flowers per cyme. Lobules of the calyx lanceolate 0.9–2 mm long, slightly unequal or the largest 2 times the size of the smallest, 0.5–1 mm broad. Corolla white, infundibuliform, inner surface with crests and papil-

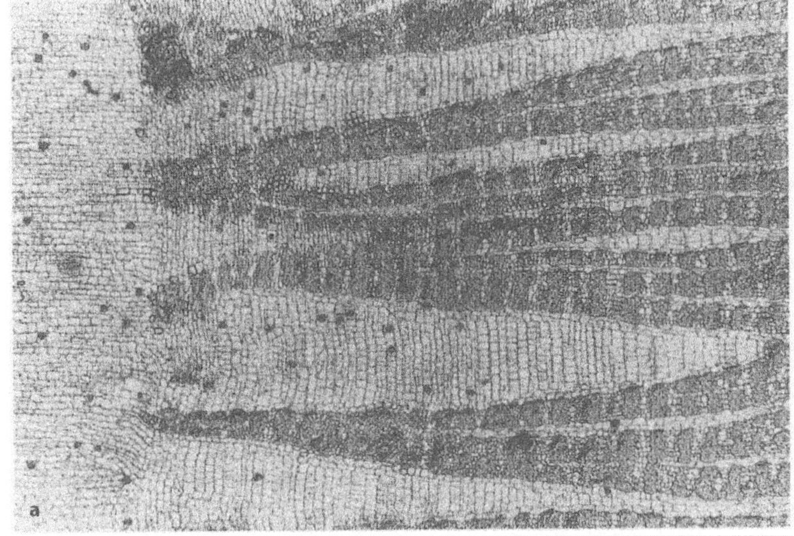
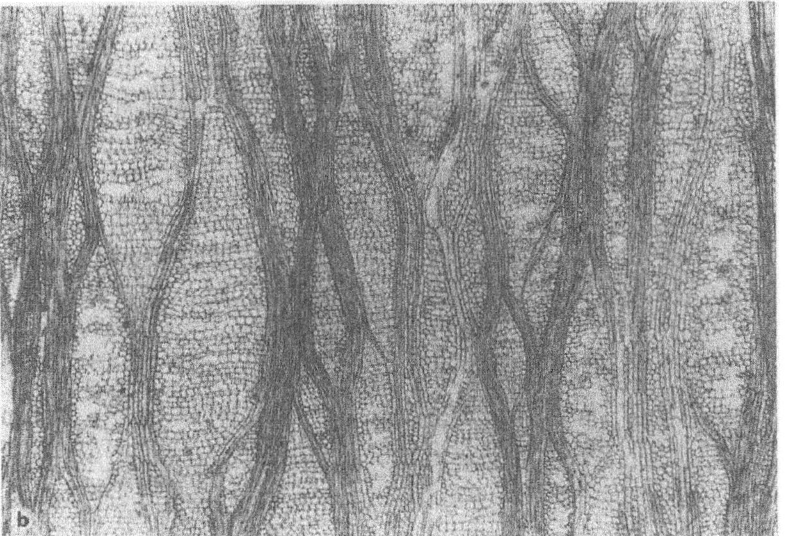

Fig. 59. *Pseudobombax maximum*. **a** T.s. of bark with enlarged rays and hardbast in the form of short plates. **b** Longitunidal-tangential section of bark with enlarged rays and fibers of the hard bast forming an intrinsicate network (ROTH 1981).

lose trichomes, tube inflated, 1.2–2 mm long inside yellowish. Diameter of crown 1.8–3.3 mm, rounded lobules 0.3–0.6 mm long. Stamens with short filaments. Style less than 0.2 mm long, disc 0.7–1 mm in diameter, apical appendix sterile.

Fruit laterally compressed, bilobed, 2–3 mm in diameter, rugose, and densely covered with small vesicles. The fruit produces 2 mericarps, each with 2 seeds.

Occurence

From Texas and Florida over Mexico, the West Indies, Venezuela, Colombia, Ecuador, Peru, Bolivia and Chile, to Brazil. It is found between 0 and 1320 m a. s. l. in the Andean region.

In Venezuela it occurs in the coastal Cordillera from 900 to 1200 m, particularly in the deciduous forest and the savanna. It also occurs in alterated and cultivated regions.

ROTH & MÉRIDA found the plant in Venezuela along the sandy beach of the former 'Playa Grande' (great beach) close to Maiquetia, which has been transformed into an international airport, some years ago.

Anatomical description

Leaf. The leaf structure of this species is very simple. The leaf comprises only a few cell layers, as seen in transverse section and is, therefore, thin. The upper epidermis cells are relatively large, but some of them are enlarged more than others and

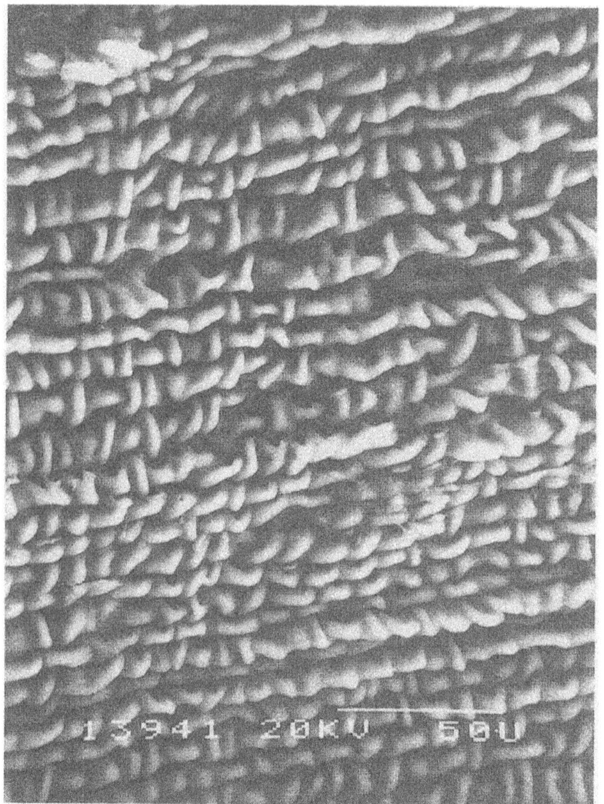

Fig. 60. *Pseudobombax maximum*, stem epidermis showing cutinized undulated cell walls. Scanning microscope (ROTH 1981).

contain cystoliths. The palisade parenchyma is only one-layered and consists in elongated, somewhat broadened cells with a lenght/width index of about 4–5. The spongy parenchyma is composed of approximately 2–3 layers of more or less globular cells which leave small intercellular spaces between one another. The lower epidermis is smaller-celled and has glandular hairs with a unicellular head. Stomata which are small are confined to the lower epidermis. Unicellular hairs with a bulbous base and with cystoliths – typical of the Boraginaceae – give a rough feeling to the leaf surface.

Ethnobotanical and general use

Medical use

Plantae, folia, radix, flos.

The plant is said to cure flu.

The entire plant, crushed and mixed with petrol is used to wash the hair, to combat headache and to cure wounds.

The leaves in wine are a remedy against fainting. The leaves applied in the form of a plaster heal wounds. A decoction of the leaves is used to wash the face, internally taken it cures meteorism.

A decoction of the root applied locally reduces swellings.

The infusion of leaves and flowers (a mouthfull) is used against dysentery and as a stimulant.

The buds are applied to combat nose-bleeding and the cutaneous eruption of 'San Antonio'.

Method of use

Externally, as a plaster, a wash or orally. Leaves and entire plants are macerated and mixed with petrol. Infusions and decoctions are used externally as well as internally. Leaves are also used with wine.

Healing properties

As MANFRED (1977, 1982) has emphasized, the healing properties are much the same in all Heliotrope species: Stimulant, tonic, against headache, to cure swellings, wounds, eruptions of the skin, dysentery, haemorrhage and meteorism.

Chemical contents

Necines, aminoalcohols of pyrrolizidine alkaloids, putrescine, spermidine, spermine and others.

Heliotropium indicum L. (borrajón, rabo de alacrán, heliotrope).

Taxonomical description

Taxonomical details described in this section can be found in Fig. 61 a. 10–150 cm high, puberscent pereannial herb. Leaves alternate, the larger ones of ovate to deltoid shape and the smaller ones ovate to broad-lanceolate. Blade 5–15 cm long and 2–10 cm broad, pubescent, with creante (or entire) margins, an acute tip and an obliqely acute to subcordate or sinous asymmetric base. The venation is pinnate with 5–6 pairs of very prominent secondary nerves and a midrib which is prominent on both sides. The petiole is 3–10 cm long and slightly winged below the blade. The flowers are arranged in terminal scorpioid spikes with a long peduncle beset with tiny bracts altogether 5–30 cm long, flowers developing in acropetal direction and thus following the sequence of maturation. About 50 or more sessile flowers occur per cyme, either solitary or in pairs.

Calyx with subulate or cuneiform lobules, 2–2.5 (4) mm long, enlarging together with the fruit ripening. Corolla blue, or violet, more rarely white, occasionally with a yellow center, disc-shaped, with a diameter of 2–4 mm and a tube, 2.5–4.5 mm long,

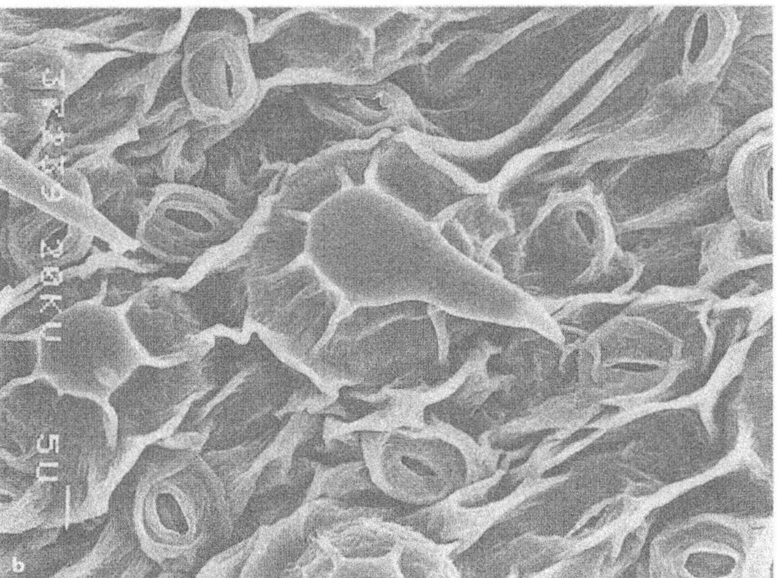

Fig. 61. *Heliotropium indicum.* **a** Habitus, **b** lower leaf epidermis with stomata and cystolith hairs.

stigma disc-shaped, 4-lobed with sterile apical appendix. Stamens 5, with short filaments.

Fruit glabrous, laterally compressed, 0.2–0.35 cm long and 0.15–2 cm in diameter, strongly ribbed, profoundly bilobed and dividing into 4 nutlets at maturity, about 2–3 mm long, each with a single seed.

Pantropical weed, possibly native of America.

Origin

The species is typical of the tropical South American coast line. It is supposed that it is of South American origin.

Occurrence

Pantropical weed.

In Venezuela, the species is frequent along the Caribbean beaches. It is however also found in the savanna, in the deciduous forest and in non-cultivated places. STEYERMARK & HUBER 1978 found it on the Mt. Avila at altitudes between 900 and 1100 m. It often grows on dry and sandy soils, but also in inundated regions. It occurs from Texas, Florida, Mexico, West Indies down to the north of Argentina, as well as in tropical Africa, Asia and Australia.

Anatomical description

Leaf. (Figs. 61b, 62, 63,64). The leaf is dorsiventral and amphistomatic. Upper and lower epidermis are single-layered and have undulated anticlinal walls, as seen in a surface view. The stomata are without subsidiary cells (anomocytic) and occur at epidermis level or are only slightly elevated above it, on the lower leaf side. Cystolith hairs with a cystolith in the swollen basal part are frequent on both sides; they are either short and conic or long with a pointed end and consist of a single cell. Glandular hairs composed of a foot cell, a stalk cell and a unicellular head occur in the neighbourhood of the veins or above them, but are rare. Stomata and cystolith hairs are more frequent on the lower than on the upper side, and the anticlinal cell walls are slightly more undulated on the lower side.

The palisade parenchyma consists of 1–2 layers of long and slender cells and is very compact. Likewise the spongy parenchyma is comparatively compact, consisting of 2–3 layers of cells with short arms.

The midrib as well as the lateral veins of first order project considerably above the surface. The midrib is reinforced with collenchyma beneath the epidermis on both sides and the palisade parenchyma is interrupted in this region. The vascular system is principally arranged in the form of an open arc, but may become almost closed to a ring by a small arc on the upper side. This arc however changes in its outlines or almost disappears when

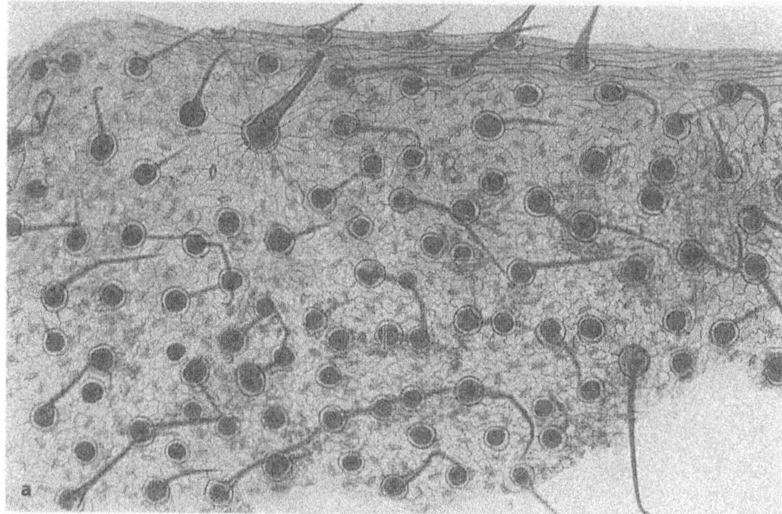

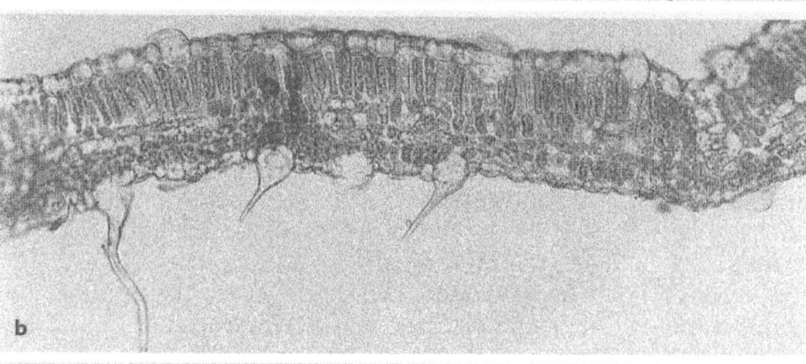

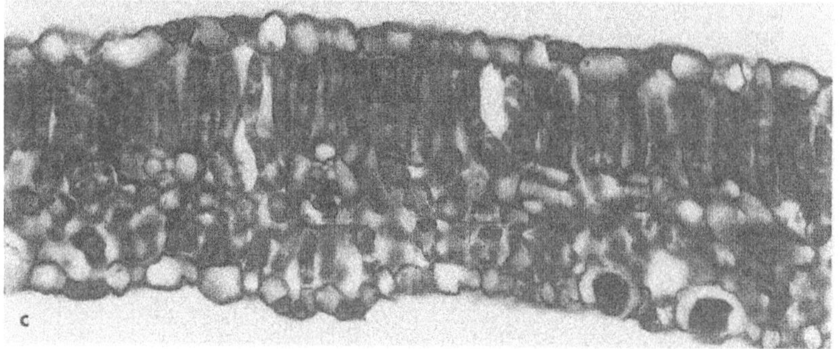

Fig. 62. *Heliotropium indicum.* **a** lower leaf epidermis in surface view. **b, c** t.s. of leaf blade. Note the cystolith hairs with the swollen basal part.

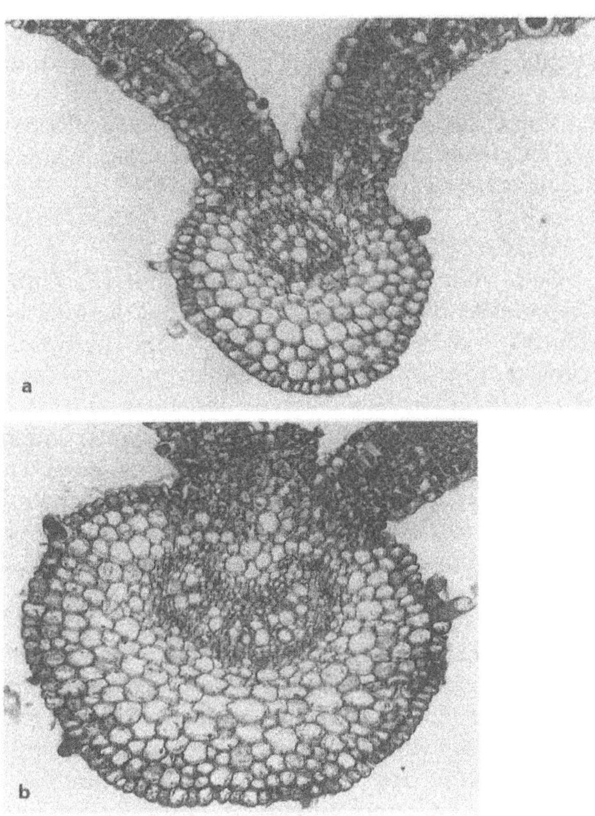

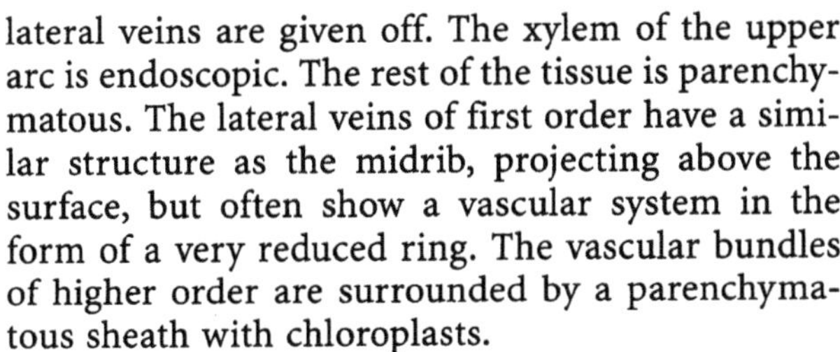

Fig. 63. *Heliotropium indicum.* **a, b** Lateral ribs of secondary order in t.s.

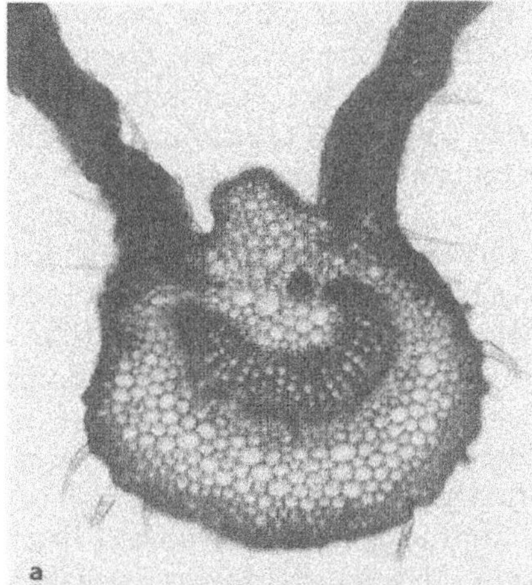

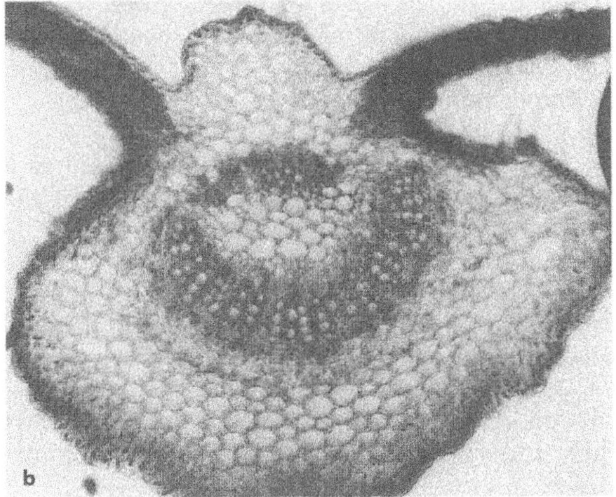

Fig. 64. *Heliotropium indicum.* **a, b** Midribs in t.s. Note the variation in midrib and side rib development (Fig. 63).

lateral veins are given off. The xylem of the upper arc is endoscopic. The rest of the tissue is parenchymatous. The lateral veins of first order have a similar structure as the midrib, projecting above the surface, but often show a vascular system in the form of a very reduced ring. The vascular bundles of higher order are surrounded by a parenchymatous sheath with chloroplasts.

Axis. The epidermis of an axis about 6 mm in diameter, consists of cells of regular aspect. some simple uniseriate hairs and some glandular hairs can be observed in the epidermis. Below the epidermis, there are 2–3 layers of photosynthetic parenchyma. Subsequently, an angular collenchyma follows which comprises about 8 layers. Between the collenchyma and the phloem, an interrupted ring of fibers may be observed. The xylem forms a continuous ring, as seen in transverse section, traversed by uniseriate rays. The pith consists of parenchyma cells which continuously enlarge towards the center. A large rhexigenous hole is found in the very center.

Ethnobotanical and general use

Medical use

Plantae, folia, radix, flos.

Leaf. The sap of the leaves is used to promote the dentition of children and to cure scabies; it is also used as a febrifuge. The sap of the leaves squeezed into the eyes cures ophtalmia. Equally, the decoction of the leaves is applied as an eye bath.

The leaves are also used to cure pruritus. The sap of the leaves is said to act as a febrifuge, aperitive and helps to cure sunstroke. The leaves are locally applied as analgesic. They have also antiinflammatory effects (especially on feet) and are used against coughs. A plaster of the leaves is applied locally to cure infections of the gum, erysipelas and bites of noxious insects. An infusion of the leaves taken orally is used against a cold.

Root. A foot-bath prepared with the infusion of the root cures frostbites. The roots are the medically most effective organs of the plant. A decoction of the root is used as a remedy against flu. The infusion of the root regulates menstruation.

Flower. The flowers in infusion are taken as a sudorific and for gargles. Flowers in decoction used internally cure whooping cough. Flowers in decoction are also applied in massages to soothe pains and swellings.

Flowers are often used in the same way as leaves: Flowers in decoction taken orally cure pains and tumescences. A bath in the decoction of flowers heals pimples of the skin. Flowers and leaves are taken against cough. The sap of the flowers mixed with oil is applied as an ointment against haemorrhoids. The decoction, left alone for 2–3 days and taken on an empty stomach is abortive.

Twigs. A decoction of the branches used as a sitz bath cures haemorrhoids. The vapour of the plant applied to the feet cures flu.

Entire plant. Eruptions of the skin, allergic eruptions, coughs and fever are cured with the decoction of the entire plant. The entire plant acts as an astringent, diuretic and solvent. The sap or decoction of the leaves and the entire plant is febrifuge, aperitive and cures sunstroke and haemorrhoids.

The sap of the plant is applied to combat aphthae, angina, pharyngitis, ulcer and burns.

Method of use

The sap of the crushed and squeezed leaves is taken externally or internally. Leaves are applied locally in the form of a plaster. The sap squeezed out of the leaves is used as an ophtalmic. Leaves in decoction are applied as an eye bath. Infusions and decoctions of the leaves are used externally as a bath or internally. For the decoction, 2 handful of leaves are boiled in 1 liter of water for 10 minutes. This decoction is used for gargles.

An infusion of the roots is taken internally. Infusions of the root are also used for a foot-bath. Decoctions and infusions of the root are used externally as well as orally.

The entire plant is used in infusion or decoctionor in the form of a squeezed out sap to be applied externally or internally in the same way as the distinct organs (leaves, roots, flowers).

Flowers in infusion are used for gargles or orally, e. g. as a sudorific. Flowers in decoction are applied internally or for massages and a bath. The sap of the flowers mixed with oil is used as an ointment. Decoctions are also applied for a bath; a decoction orally taken after 3 days is said to be abortive.

For a cholagogue, 30 g leaves with 20 gr flowers and 20 g root are boiled in 1 liter of water; a small cup is taken every 4 hours (BAUDI 1987).

Healing properties

Sudorific, astringent, aperitive, diuretic, abortive, analgesic, febrifugal, scabicide, ophthalmic, resolutive, choleretic, dissolving, antispasmodic (root), antiinflammatory, as a cholagogue, regulator of menstruation, against scorpion bites (spanisch name of the plant: 'rabo de alacrán'), against cough and whooping cough, against erysipelas, acne, allergy and eruptions of the skin, aphthae, pruritus, frost-bite, sun-stroke, swellings and tumescences, ulcer and burns, haemorrhoids, pharyngitis, cold, flu, angina, and against pain.

Chemical contents

Indicine-N-oxide extracted from *Heliotropium indicum* apparently exercises anticancerous effects. Other anticancerous substances found in *Heliotropium indicum* are taxol, philantoside, homoharringtonine. Indicine-N-oxide is clinically studied in chemotherapy of human leucaemia (NASH & MORENO 1981).

Heliotrine, another alkaloid isolated from the seeds of *Heliotropium indicum*, has ganglion-blocking activities.

Other alkaloids found in *Heliotropium indicum* are echinatine, supinine, heleurine, heliotrine, lasiocarpine. Furthermore, precursors of pheromones, β-sitosterol and many other compounds were found in *Heliotropium indicum* (MEHTA et al. 1981, and the bibliography of BERNAL & CORREA 1989.

Varieties and related species

Heliotropium develops a very rich aroma in our gardens and for this reason it is cultivated all over the world, observes MANFRED (1983). There exist varieties of heliotrope, but all the heliotropes show more or less the same characteristics and are used medically, although they are distinguished by differences in their fragrance. All parts of the plants can be used in decoction giving good medical results.

MANFRED (1983) mentions particularly *Heliotropium peruvianum*; a handful of the plant prepared as a decoction in one liter of water heals wounds, cures and disinfects varicose ulcers and ulcers in general especially those with a tendency towards malignancy. A tea prepared of 10–15 g of the plant with a cup of water taken with honey on an empty stomach cures chronic cystitis.

In popular medicine, the flowers prepared in the form of a tea are mostly used; sweetened with honey the tea is indicated in the case of palpitations, migraine, headache and depression. Two cups taken with honey are recommended a day.

Cultivation

The species grows wild on dry and sandy soils of hot regions.

Observations

The species is easily recognized by its habit, the cystoliths and glandular hairs, and particularly by the structure of the midrib and the lateral nerves.

Heliotropium polyphyllum LEHM. is an annual or perennial herbaceous plant with lanceolate to ovate sessile leaves with an acute tip. The very small leaves have a length of about 2–2.5 cm and a breadth of 0.6–0.7 cm. The leaf surface is covered with bristle hairs.

The leaf anatomy was studied by ROTH & MÉRIDA (unpublished) and ROTH 1992.

The upper epidermis is very large-celled. The outer walls are strongly thickened. There is only one layer of palisade cells which are short and very broad. Their length/width index varies between 1.8 and 4. The rest of the mesophyll is looser and the cells are partly of globular shape and partly palisade-like, particularly those which surround the vascular bundles. The vascular bundles are extremely frequent and leave about 2 mesophyll cells between one another. There are usually 1–2 layers of spongy parenchyma between the vascular bundles and the lower epidermis. The largest intercellular spaces are in contact with the lower epidermis. The vascular bundles are well developed. They are surrounded by a parenchymatous vascular sheath in which the chloroplasts are arranged in a horsehoe-shaped form on the inner and radial walls. Elongated palisade cells partly radiate from the vascular bundles in all directions. But this 'Kranz' structure is not always very conspicuous. The lower epidermis is smaller-celled and has thinner walls. Stomata are present on both leaf sides, but those of the lower surface are much more numerous. Furthermore, the stomata of the upper side are at epidermis level, while those of the lower side are elevated above the epidermis level. Unicellular hairs (bristels) are abundantly present on both surfaces. They are very thick-walled and have many incrustations in the form of little knobs on the wall outside. The upper epidermis has probably a water-storing function. This leaf seems to belong to the photosynthetic C_4-type.

Heliotropium polyphyllum was found by ROTH, MÉRIDA & LINDORF 1986 in the same region as *Heliotropium angiospermum* (Playa Grande near Maiquetia). This species has a very interesting leaf structure of the 'Kranz' type with a parenchymatous sheath surrounding the vascular bundles in which the chloroplasts usually adopt a horseshoe-shaped position at the inner walls of the cells. It is consequently very likely that the species belongs to the C_4 photosynthetic type. It can also be very easily recognized by its particular leaf structure.

As has been emphasized by the authors, all species of *Heliotropium* have similar chemical contents and can be used medically, it is to be expected that also *Heliotropium polyphyllum* has similar chemical properties. It is therefore suggested that the plant should be thoroughly studied chemically.

Tournefortia scandens MILL.

This is a climbing plant, at times reaching up to 3.5 m in length. The elliptic to lanceolate leaves have an acute tip and are 2–7 cm long. The inflorescence is composed of various cymes. The corolla is whitish-green and 2–3 mm long. The fruits are 2–3 mm in diameter, being composed of 2–4 roundish nutlets. In Venezuela, the species is found along the Caribbean littoral.

Anatomical description

Leaf. The leaf is thin. The upper epidermis is relatively large-celled and thick-walled. There is only one layer of palisade cells with an approximate lenght/width index of 3.2–4.5. The palisade cells are long and broad. The spongy parenchyma underneath consists of spherical cells which leave intercellular spaces between one another. Stomata are confined to the lower epidermis. Very conspicuous are large unicellular hairs which are frequent in the upper as well as in the lower epidermis.

These hairs with a very broad base and an elongated bent tip are very characteristic and of enormous size compared with the entire leaf section.

Ethnobotanical and general use

There are quite a few species of *Tournefortia* which are known as medicinal plants or applied for other purposes. Pyrrolizidine alkaloids have been reported for several species. The infusion of *T. angustifolia* is used as a purgative. A tea is prepared from the leaves of *T. cuspidata*. The leaves of *T. fuliginosa* are regarded as an excellent haemostatic; a decoction is taken for internal bleeding. The leaves of *T. argentea* are occasionally smoked instead of tobacco.

T. gnaphaloides is applied internally as a decoction of the branches for the kidneys. Against rheumatism, the plant is used in a bath, or branches in alcohol are applied externally. Some kind of a balsam is extracted from the plant (RODRIGUEZ 1983: FOR VENEZUELA).

A cooling drink is made of *T. hirsutissima* for the bladder. 150 gr of the plant are needed for every liter of water to make the infusion. A stimulating bath is also made of the same species together with some other herbs (SEAFORTH, ADAMS & SYLVESTER 1983).

It may thus be worthwhile to see whether *T. scandens* has similar medicinal properties.

Brunelliaceae

B. is a monotypic family with a single genus Brunellia and some 50 species of shrubs, small and medium-sized trees in the Andes of Bolivia up to Venezuela, Central America and the West Indies.

Flowers small, fruits follicular with 1–2 seeds.

Brunellia comocladifolia CUATR. SSP. FUNCKIANA is a tree up to 20 m in height occurring in the cloud forest at an altitude between 1500 and 2000 m. Young twigs, petiole and rhachis are ferrugineous-pubescent. The compound leaves are imparipinnate and 15–20 cm long. The leaflets (11–23 on each leaf) are oblong-lanceolate or oblong-elliptic, 4.5–15 cm long and 2.2–6 cm broad. The flowers are inconspicous, of greenish colour, and without a corolla. The folicles are 3–3.5 cm long. The leaf anatomy has been described by ROTH (1992). The wood is of little use. The chemistry of the family is not known.

B. sibundoya Cuatr. The leaves are used together with the mucilaginous fruits of *Saurauria brachybotrys* for a tea to treat pulmonary congestion in cases of pneumonia or influenza.

Burseraceae

The Burseraceae are characterized by compound leaves and small flowers arranged in panicles. The fruit is usually a 5–2 locular drupe. In some species, the exocarp opens with valves. The schizogenous secretory canals in the bark are very typical. All species are valuable useful plants because of their content of resin and balsam. The pantropical family comprises trees and shrubs. It has a very homogenous structure.

The Burseraceae supply resinous extracts such as incenses (*Boswellia sacra* and others) and oil of myrrh (species of Commiphora). The balsams and resins of the family are rich in terpenes.

Bursera

The genus *Bursera* is the source of perfume materials. Antimicrobial activity is associated with essential oils, antitumoral activity is ascribed to *B. klugii* and *B. morelensis*.

Protium

The genus Protium supplies balsamic resins. Resin to clear the nasal passages from colds. Balsam to heal wounds. Resin for incense. Elmira resin for varnishes (elemi occidental). Leaves are used for intestinal ailments. Wood is useful.

ROTH studied the bark structure 1981, leaf structure 1984, fruit structure and dispersal 1987.

Tetragastris

The genus Tetragastris supplies a bitter resin used for fever, and chronic ailments of the mucous membrane.

Bursera simarouba (L.) SARG. ,
synonym: B. gummifera L.
(indio desnudo).

The 'naked indian' (indio desnudo) is a small tree of 8–20 m in height with imparipinnate leaves having 3–13 leaflets. The oblong-ovate leaflets reach a length of 6–10 cm. The small whitish flowers are arranged in panicles. Male and hermaphrodite flowers occur. The fruit is a drupe with a red exocarp which opens with 3 valves. Very characteristic is the smooth red-brown bark which resembles the skin of a naked Indian.

Origin and occurence

The tree is found in Florida, Mexico, Central America, the West-Indies and in tropical continental America.

In Venezuela it is common in xerophilous forests of the so-called 'tierra caliente' (thorny woodland); however its range of occurence is very ample.

Anatomical description

Bark. (Fig. 65, 66) The vernacular name 'naked indian' refers to the red-brown very smooth bark which constantly exfoliates in a way comparable to that of birch. When it has recently peeled off, the underlying green cortical layers appear. The tree sheds the leaves during the dry season.

The entire bark width amounts about 6–10 mm. As seen in transverse section, the bark of an old stem has a very regular structure becoming apparent in a regular stratification of alternating layers of soft and hard bast (Fig. 65). Outside the cambial zone, the layering may start with a zone of soft bast where sieve tubes prevail; then follows a stratum of parenchyma which was called by ROTH (1969A) a sheath parenchyma, because the secretory canals are embedded in it. The sheath parenchyma is distinguished from the normal parenchyma by smaller globular cells which leave very small intercellular spaces between one another. The secretory canals embedded in this layer reach a diamter of 0.08–0.1 mm and are surrounded by a variable number of secretory cells (10–15); they are filled with a transparent yellowish and sticky resin, which smells like copal and has a slightly sweet taste. The canals are very frequent so that 3–5 of them may closely lie side by side. Average distance between 2 neighbouring canals is about 0.5 mm. The sheath parenchyma cells are easily recognized by their arrangement in radial rows. Exterior to the sheath parenchyma another stratum of sieve tubes follows. The sieve tube strata are comparatively large and about 0.3 mm in width. The sieve tubes are relatively large and very frequent. Towards the outside, a small layer of fibers is developed, comprising about 3–5 cells in a radial row; chambered crystal strands accompany the fibers. Then the same pattern is repeated towards the outside starting with a layer of sieve tubes. As seen in longitudinal-tangential section, the fiber layers form a continuous network. The secretory canals form a continuous system, also in the form of a network. Most of the canals take a longitudinal or tangential course, but some even pass in a radial direction through the rays. These are 1–4 seriate and occur at distances of 3–15 cells. As seen in longitudinal-tangential section, the secretory canals form a dense network of horizontally and longitudinally running tubes (Fig. 65). Rays are 10–20 cells in height.

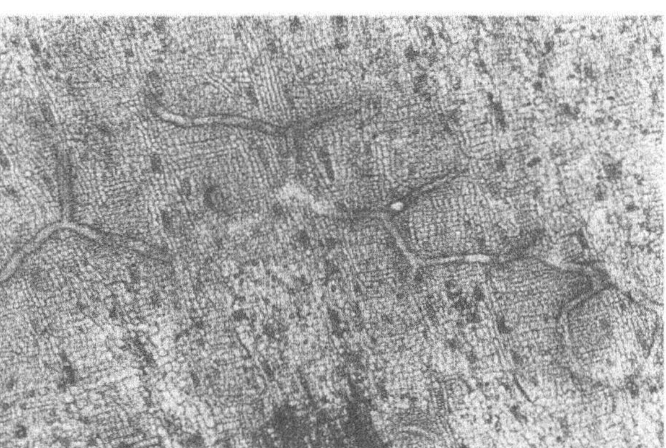

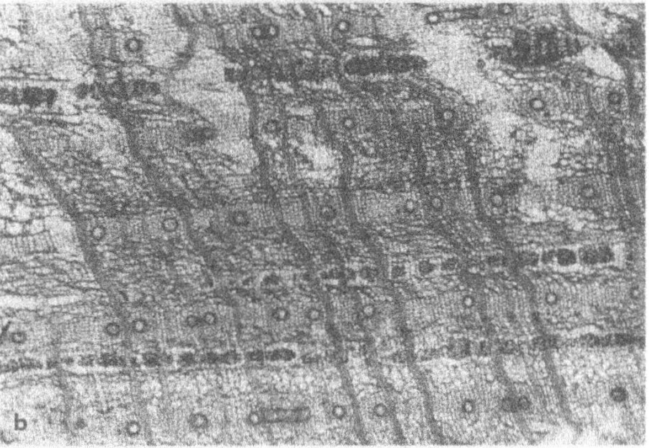

Fig. 65. *Bursera simarouba*, Burseraceae. **a** longitudinal tangential section through bark showing a continuous network of secretory canals. **b** T.s. through bark. Regular layering of hardbast, soft bast and sheath parenchyma in which the secretory canals are embedded. The rays dilate towards the outside (ROTH 1981).

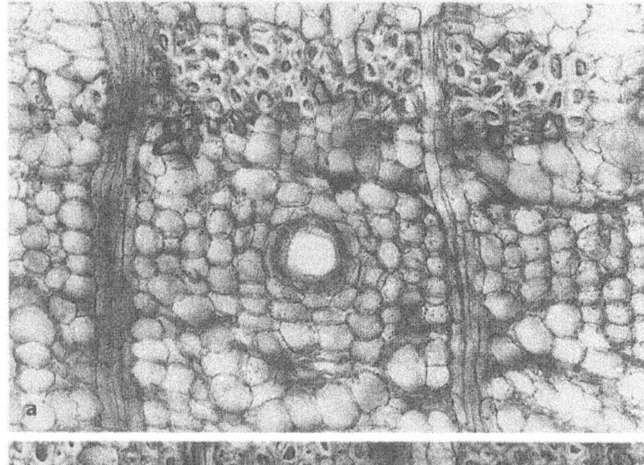

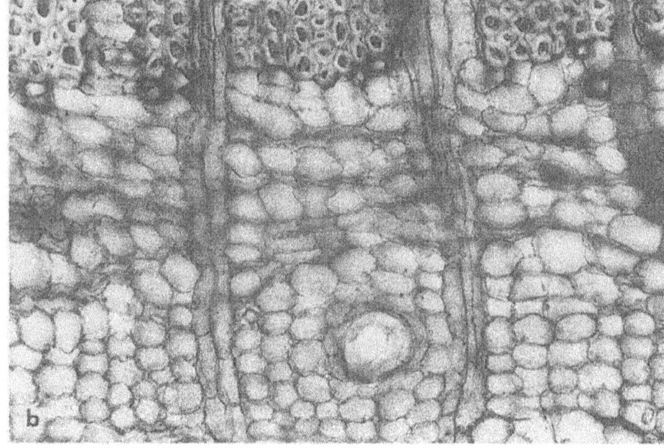

Fig. 66. *Bursera simarouba.* **a, b** T.s. of bark. Secretory canals embedded in the sheath parenchyma (radial cell rows), fibrous hardbast and partly compressed softbast (ROTH 1981).

During dilatation growth inflated parenchyma cells sclerify transforming into sclereids in the outer bark region. The cork layer adhering to the bark outside is small comprising about 8 cells in a radial row, but the phelloderm is much better developed including crystal cells with solitary crystals and irregular layers of stone cells. During growth in thickness of the trunk, several cork layers arise. They easily separate from one another, particularly the slightly thicker-walled outer cell layers from the thinner-walled inner layers. The outermost layers are more strongly suberized.

The exfoliating cork layers comprise 2–4 cells in a radial row. The papery consistence of the exfoliating scales is thus due to the structure of the cork which may be so thin that the scales are almost transparent. True bark scales usually form in response to a lesion of the bark, but very small scales also develop spontaneously.

Very old trees develop scales with abundant lenticels in their basel part of the trunk.

Ethnobotanical and general use

Nutritional use

The leaves are used as a tea.

Economical utilization

The wood is used as firewood. It is also used for fabrication of matches and for cabinet work. Cuttings are used for fences.

The resin is not only used as an incense, but also as an insecticide. It has a semll of turpentine. The gumresin even serves to glue together broken china, ceramic and glass; boats are impregnated with resin to avoid destruction by termites.

The main source of resin is in the bark which peels off easily in the form of delicate streaks, although the entire plant is supplied with resin. The resin called elemi, elqueme or tacamahaca, is used also in varnishes and as a substitute of Arabic gum.

Medical use

Name of the drug: Cortex, folia, caule, fructus, semen. But also flowers and roots have medical applications.

The twigs in decoction are good for the liver and the thyroid gland, but also aid to weight loss; 1–2 cups a day are sufficient.

Wood is likewise resinous and an infusion is sudorific, antirheumatic and heals ulcers.

Flowers in infusion cure diarrhoea.

The sap of the plant is applied on infected wounds; it is also taken by persons who suffer from haematuria and intestinal heamorrhage. The sap is antiinflammatory and antidiarrhoeic; it furthermore helps to treat spider bites.

The fruit is applied for diarrhoea.

The seeds cure snake bites.

The leaves are diuretic, purgative, antiveneral and antirheumatic. A tea of the leaves cures amygdalitis, asthma, inflammations of the gums, and of the knee; a cataplasm helps against gangrene and obesity, and accelerates delivery in birth; a tea of the leaves may also be applied.

Leaves, bark and root together are used for diarrhoea, flu, hypertension and for the kidneys. They are equally used as a stomach tonic, a diuretic, a purgative, an antispasmodic, and a sudorific.

The bark. also has many applications. It is used to cure wounds, stomach disorders, gastrointestinal pain, fever, nosebleed, measles, and as a depilatory. Particularly fresh bark is applied on burns.

A decoction of the bark applied to the body 3 times a day causes peeling of the skin. The same decoction is taken internally for cancer of the stomach.

A decoction taken orally is applied for gastric ulcer; half a cup is taken twice a day for 9 days; it is taken before breakfast and in the afternoon. It can also be used for indigestion, as a fungicide and for insect bites.

A bath taken in the decoction is good for the skin. Wounds are likewise cleaned with it to accelerate cicatrization.

For diarrhoea, the bark is crushed and soaked in water over night; the water is taken during the day.

A decoction is also taken orally for intermittent fever.

The resin cures infected wounds, abscesses, inflamed ganglia, ulcers, venereal diseases, diarrhoea, intermittent fever, rheumatism, gastric haemorrhages, ulcer, diarrhoea; it is even used in the extraction of spines and thorns. In the form of cataplasm it is used for gangrene. It is put on the navel of new born children.

The resin ash put on wounds inhibits inflammations.

According to UPHOF (1968), the resin is furthermore applied for oedema, dysentery and yellow fever.

The resin is aromatic and smells of turpentine.

Healing properties

Extracts of the aerial parts of the plant are aromatic, diaphoretic, spasmolytic, vasodilatory, fungicidal, molluscidial, and act as a stimulant on the smooth muscular system (duodenum). Particularly the axial plant organs have fungicidal effects. The dry bark shows molluscicidal activities.

Cytotoxic activity is ascribed to dry fruits.

Leaves and axial organs show relaxing effects on the smooth muscular system as well as spasmolytic and vasodilatory activities.

A triterpene has antitumoral activities in certain types of cancer.

Protium heptaphyllum (AUBL.) MARCH. (tacamahaco, aceitico, anime, caraña, currucay)

Taxonomical description

The Protieae (Fig. 67) develop a drupe with 5–2 stones. *Protium* is pantropical, but occurs predominantly in tropical America.

The 5–20 m high tree has pubescent young branches of a rusty colour. The imparipinnate compound leaves with 2–3 pairs of leaflets (frequently 7 leaflets altogether: heptaphyllum!), are 15–20 cm long; the petiole is 4–5 cm long. Leaflets oblong-lanceolate to oblong-elliptic, attentuate towards the apex to form a 'drip tip', 8–10 cm long, 3–4 cm broad and of semicoriaceous consistency. Inflorescences are axillary in the form of small paniculate glomeruli (1–4 cm long); pedicels 1.5–2.5 mm long, slightly hairy or glabrous. Flowers tetramerous, only seldom pentamerous, 2.5–3 mm long, of yellowish-green to reddish-green colour. Calyx cup-shaped with triangular lobules. Petals oblong, generally glabrous, but at times slightly hairy. Disc in the form of a ring, 8-lobed. Pistill glabrous, in the male flowers about 1 mm long, in the hermaphrodite flowers 2–2.5 mm long; ovary of globose shape and with 4 locules; style with 4 furrows and a 4-lobed stigma.

Drupe obliquely oviform and monospermous or globular with 2–3 seeds; the pericarp is red.

The species is characteristic of the north of South America. In Venezuela it is frequently found in the hot regions.

Origins

Tropical South America

Occurence

Protium heptaphyllum is the most widely distributed South American species of the genus (RECORD & HESS 1943) and is frequently found in Venezuela (SCHNEE 1960).

Anatomical description

The genus *Protium* is anatomically homogeneous, according to METCALFE & CHALK (1950).

Fig. 67. *Protium heptaphyllum.* **a** Habitus. **b** Leaves and fruits.

Bark. (Fig. 68) As seen in transverse section, the rhytidome is scaly, being composed of several periderms. The scales are small. The phelloderm, composed of thin-walled cells, comprises 1 to several cell layers. The cork is about 8–15 cell layers thick. The cells have U-shaped wall thickenings with thickened inner tangential walls which are traversed by large pits so that the inner wall surface may adopt a dentate structure (Fig. 68b).

The course of the rays is very irregular, due to the sieve tube collapse. The rays are small, mostly 1–2 seriate, and dilate very irregularly towards the outside, mainly by tangential extention of cells.

The hardbast occurs in the form of fibers which are accompanied by septate crystal strands. As seen in t. s. the hardbast fibers form irregularly superposed plates which vary extensively in shape and size. The plates are about 2–10 cells high and 5–25

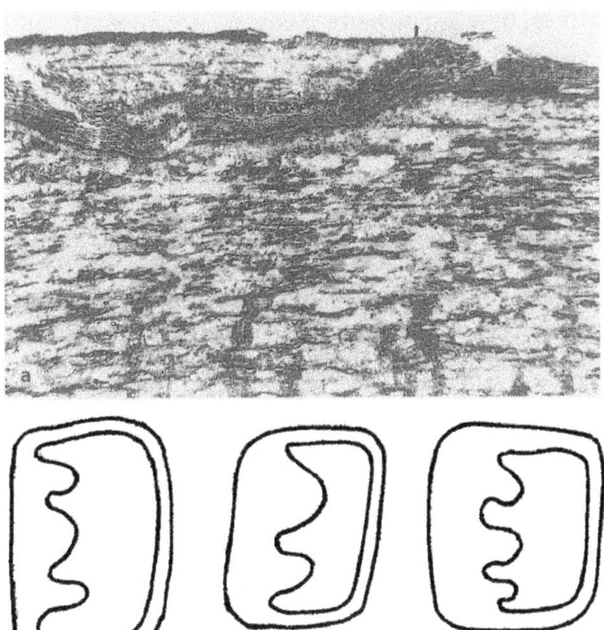

Fig. 68. *Protium heptaphyllum*. a T.s. of Bark (rhytidome development). b Cork cells with teeth in the walls.

cells wide so that a very irregular pattern arises. The septate crystal strands accompany the hardbast plates on their outside.

Collapsed sieve tubes with their companion cells together with axial parenchyma occur in the form of small tangential bands.

Secondary formation of stone cells occurs in the dilating zones of the outer bark. Very conspicous are the abundant large secretory canals which are surrounded by a secretory epithelium. They are embedded in the tangential strands of axial parenchyma and dominate the pattern of the bark.

The outermost part of the bark is very irregular, being dominated by the dilatation growth. Parenchyma cells here are tangentially extended and partly form tangential cell rows. Stone cells occur in the form of very large solitary cells or in small irregular groups.

Ethnobotanical and general use

Nutritional use

In Colombia and Panama, the fleshy pulp of the fruit is eaten. In the argentine Chaco, a refreshing drink is prepared from the fruits.

Economical utilization

The tree emanates a pleasant scent due to the incense-like resin that drops from any wound in the bark and collects abundantly on the ground around the tree. In the northern parts of South America, the Indians use the resin to scent the oil with which they anoint themselves. In Para, Brazil, the resin is collected and enters the world market under the name 'elemi', a term applied also to other burseraceous gums (RECORD & HESS 1943). The bark particularly has an aromatic fragrance reminiscent of turpentine. It exudates a transparent semiviscous resin. The resin is used for varnishes and as an incense. Indians add clay to the resin for making pots.

Medical use

Name of the drug: cortex, fructus.

The resin extracted from the plant is known as almaciga, anime, tacamahaca, tey and more commonly as elemi of Brazil. It is applied in many ways in pharmacy, e. g. as an antiinflammatory.

Leaf. An infusion of the leaves is helpful against catarrh.

Bark. The bark of the stems and twigs, as well as the fruit, contain an abundance of a perfumed oleoresin which is soluble in alcohol (PITTIER 1926). This is localized in the large secretory canals. A decoction of the bark is helpful as a purifyer and an astringent.

The exudate mixed with tallow of cows and applied in the form of a cataplasm, cures catarrhal affections.

In Brazil and Colombia, the decoction of the gum-resin of the bark is used to cure tumors.

The essential oil obtained from *Protium heptaphyllum* shows a cercaricidal activity (FRISCHKORN ET AL. 1978). A similar essential oil can be extracted form *P. marginatum* which kills 90–96 % of the cercarias within the first 15 minutes. This natural substance provides an attractive alternative to prevent schistosomiasis and to keep the natural waters healthy.

The essential oil of *Protium heptaphyllum* has some molluscicidal activity against *Biomphalaria glabrata*.

Fruit. The fruit produces a yellow, liquid and oily resin which is called oil of sassafras and is applied in popular medicine to cure syphilis, ulcers, acne, swellings and headache.

All species of the genus *Protium* could be useful in industry, not only for their perfumed durable wood, but also for their edible fruits and their oily seeds which could be used like olives (CORREA & BERNAL 1990).

Method of use

The leaves are used in an infusion.

From the bark, a decoction is prepared. The resin is extracted from the bark and is used pure or mixed with talcum and applied as a cataplasm.

Healing properties

Purifying and astringent. Used against headache, catarrh, swellings, tumor, acne, ulcers and syphilis.

Related species

Related species are *P. aracouchini, P. carana, P. copal, P. icicariba* which contain resin; this may be used as an incense for its pleasant scent or its wound-healing properties. *P. guianense* (AUBL.) MARCH likewise contains a resin in the bark which is used against contusions and aches. The resin of the bark of *P. unifoliatum* is used for headache.

Observations

The genus *Protium* is anatomically homogenous (METCALFE & CHALK 1950). The distinct species are therefore not easy to distinguish. The dentate wall thickenings of the cork cells may be helpful in identification; furthermore, the irregular structure of the bark and the secretory canals embedded in irregular strands of axial parenchyma may facilitate identification.

Tetragastris panamensis O. KUNTZE. supplies resin. ROTH studied the bark structure 1981, leaf structure 1984, fruit structure and dispersal 1987, leaf venation 1996.

Cactaceae

The more than 2000 species of Cactaceae are almost exclusively restricted to America (from Canada to Patagonia). They grow predominantly in dry regions and occur in the Andes up to a height of 4700 m a.s.l. Their representatives are mostly stem succulents, ramified or unramified, and of a cylindrical, disc-like, columnar, globular or flattened and leaf-like shape. The axis takes over photosynthesis, while the leaf blade becomes reduced and the leaf base transformed into a kind of cushion. Not infrequently, the leaves are modified into spines, bristles, or hairs. Reduced brachyblasts covered with spines, bristles or hairs, the areoles, are frequently found on stem and fruit.

The flowers (usually solitary) are often large and attractive, being composed of a great number of petals. Certain species open their flowers only during the night (Queen of the night).

The evolutionary tendency of this family is the reduction of the vegetative phase.

In contrast to the Euphorbiaceae where the succulent types with a similar habit almost always develop laticifers, the Cactaceae rarely have latex, an attribute which very much facilitates distinction of the 2 families.

Nitrogenous anthocyanins connect the Cactales with the Centrospermae. The Cactaceae are rich in polysaccharide or slime, found in slime cells and in lysigenous slime cavities. Dried shoots of the mexican cactus *Lophophora williamsii* supply the halucinogenic drug peyotl which contains the protoalkaloid mescaline. Extracts of *Selenicereus grandiflorus* have curative properties for the heart and are used in homoepathic drugs.

Some species have fleshy and juicy fruits which are edible, many others are ornamental.

Cereus hexagonus (L.) MILL. (cardón, dato, pitahaya, pitajón, reina de noche)

Taxonomical description

The species is of arboreus-candelabra aspect, having a woody base and a columnar stem which reaches 10–15 m in heigt and 30–40 cm in diameter (Fig. 69). The stem is usually ramified close to the base from 3 m upwards. The branches are generally 6-angular (see the species name!) in the adult stage, but have only 4–5 ribs in the juvenile stage, and rarely 7 ribs. The delicate ribs are 3–5 cm deep and have undulated margins. The areolae are small and occur at distances of about 2 cm. Young branches are spineless or have only few short spines, about 2–3 cm long. The older branches, on the contrary, have more and longer (5–6 cm long) spines; these occur clustered in fascicles of 8–10 and are of very different lengths.

The flowers are 20–25 cm long and open at night (common name: Queen of the night); the tube reaches 10 cm in length, the outer rose-coloured slightly fleshy lanceolate segments are 6–7 cm; the inner oblong-lanceolate segments are 7–8 cm long and white. The stamens are very numerous, the style is green-coloured. The egg-shaped fruit is 3–6 cm in diameter and 5–15 cm long, slightly

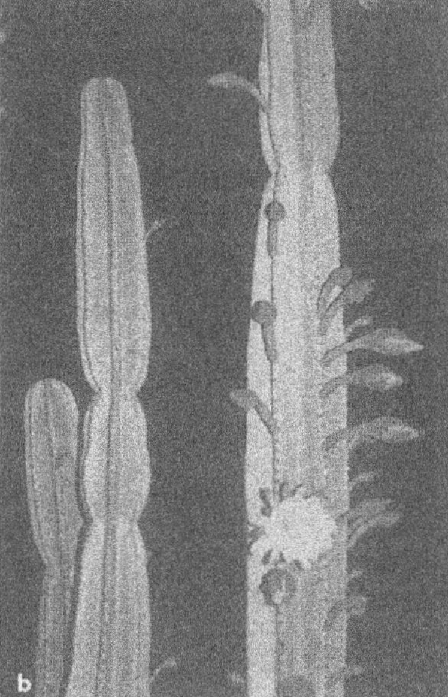

Fig. 69. *Cereus hexagonus*: **a–c** Habitus (flowers and flower buds).

oblique, with a truncate or slightly depressed apex, of purple, rose or withish colour and is edible. Its surface is smooth and the areolae are small.

Origin

The species is native to the north of South America and possibly of a part of the West Indies (Trinidad and Tobago, Margarita), the lesser Antilles.

Historical background

OVIEDO & VALDEZ described the species in 1548. CASTELLANOS found and described it in 1589 for Coro and the island of Margarita. LINNÉ described it as *Cereus hexagonus* in 1768 (VELEZ & VELEZ 1990).

Occurrence

The species is found wild in the hot regions and in the xerophytic formations along the coastal area of Venezuela. It is sometimes cultivated for its fruits or as an ornamental plant.

The species is also found in Colombia and Ecuador.

Anatomical description

Stem. (Figs. 70, 71) A 4 cm deep and 11–12 mm thick rib has been studied. The comparatively small epidermis cells have thickened outer walls and a thick cuticle which is ribbed on the surface. The stomata are slightly sunk below the surface and have an enormous substomatal chamber which is channel-shaped and lines the entire underlying hypodermis. The hypodermis comprises about 4 layers of very thick-walled cells with pits; they stain intensely blue with toluidine blue. This tissue is water storing. The substomatal chamber ends at the level where the hypodermis is followed by parenchyma. The underlying parenchyma is arranged in the form of anticlinal rows of cells; each row consists of about 16 cells which are slightly anticlinally elongated and have short arms so that small intercellular spaces arise. This is the region of the cortex. Beneath it is the pith, starting with a region where the cells are partly tangentially elongated. Towards the inside, the cells take on a more or less globular shape and have very large pits. The vascular bundles are irregularly dispersed in the pith; they are reduced and do not show sclerenchyma. Solitary and clustered crystals are observed mainly in the pith.

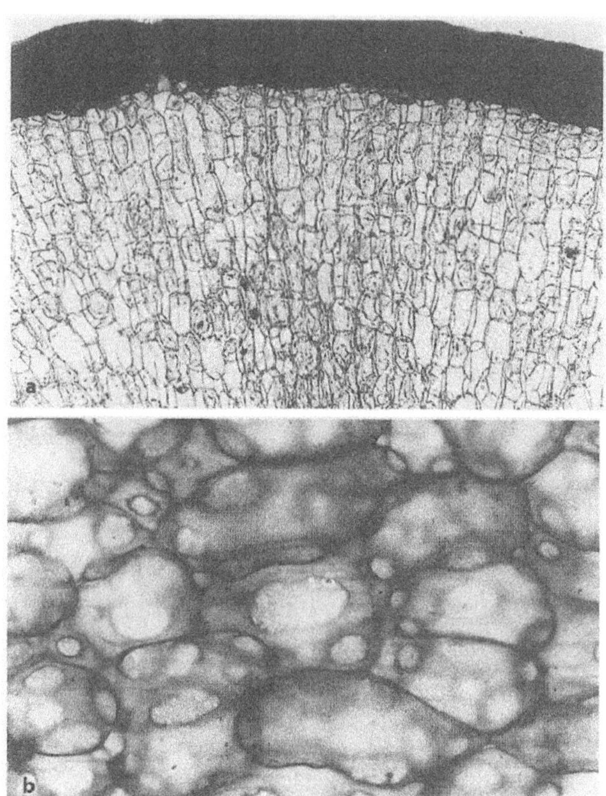

Fig. 70. *Cereus hexagonus*. **a** T.s. of axis with subepidermal meristem. **b** Pitted parenchyma cells of the axis.

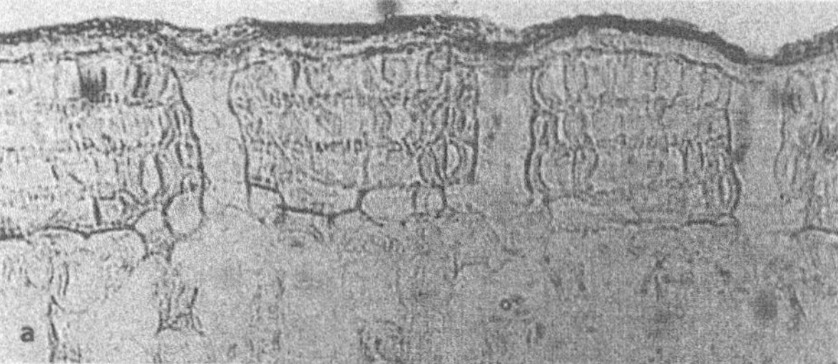

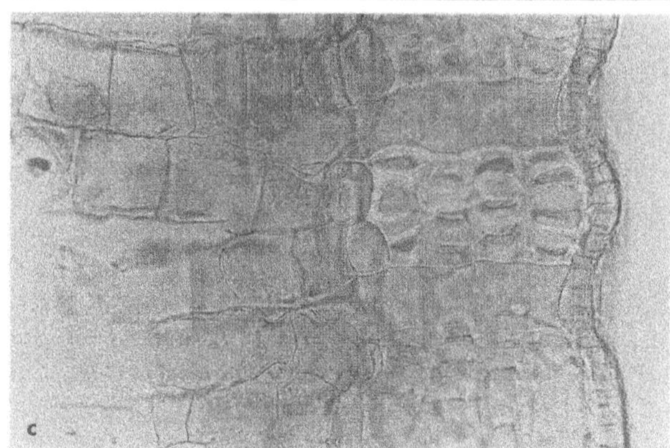

Fig. 71. *Cereus hexagonus*. **a–c** t.s. of axis. Note the channel shaped substomatal chamber.

Fruit. The pulp of the fruit is white and consists mainly of parenchyma. The fruit has a pleasant taste. Numerous black seeds are dispersed in the pulp. The surface of the fruit is smooth and without spines.

Ethnobotanical and general use

Nutritional use

The fruits are usually eaten raw, or a drink or jam is prepared from them

Economical utilization

As an ornamental tree and as a living fence.

Medical use

The branches and the flowers are used in popular medicine.

Method of use

An infusion of the branches is applied to control inner haemorrhage.

Healing properties

The plant is used against toothache, for the eyes, for heart diseases and as a protection against strong sunlight. The healing properties are antiseptic, anti-inflammatory, diuretic, emollient.

Chemical contents

Oleanolic acid, a lactone, proteins, fat, fiber, carotene, thiamin, riboflavin, niacin, ascorbic acid, iron, phosphor, calcium, and 89–95 % water (VELEZ & CHAVEZ 1980, see CORREA & BERNAL 1990, T. III).

Related species

There are also other Cactaceae in Venezuela which are called Pitahaya, e.g. *Cereus undatus* HAW., *Cereus margaritensis* J. R. JOHNSTON, or *Acanthocereus tetragonus* (L.) HUMMELINK which have edible fruits.

Cultivation

Although the plant can be reproduced by seeds, it is faster and more efficient to propagate it by short slips of branches. After cutting, it is advisable to let the slip dry out for some days before it is planted to avoid rotting.

The growth is slow. As all Cactaceae, the plant needs strong sunlight and light soils with a good drainage. But it easily resists long periods of drought.

Observations

The species is well defined by its outer candelabra appearance, its large flowers which open for one night, its large fruits and particularly by its stem anatomy.

Pereskia grandifolia HAW. (guamacho morado, guamacho rosadeo)

This is a shrub or small tree, 3–5 m high, with a fleshy spiny stem. The oblong fleshy leaves are (4) 6–16 cm long and have an obtuse or acute tip and an attenuate base. The flowers are white or rose coloured, measuring 4–6 cm in diameter, and are arranged in small globular clusters. The stamens are numerous. The pear-shaped fruits are 4–6 cm long and contain numerous seeds; they are spineless.

The species is common in hot regions of Venezuela.

Anatomical description

Leaf. Upper and lower epidermis are of normal structure. The stomata at epidermis level are confined to the lower epidermis. Simple hairs and peltate glands are additionally present. The photosynthetic tissue is not differentiated in palisade and spongy parenchyma, but consists of thin-walled turgescent cells which are partly somewhat elongated perpendicularly to the leaf surface; it comprises about 7 layers. The succulent cells are water-storing and add the fleshy consistence to the leaf. Clustered crystals of calcium oxalate in the form of druses abundantly occur in the mesophyll.

The leaf thus has isolateral structure. The glands possibly correspond originally to hydathodes, but it is doubtful whether they function as extrafloral nectaries, as reported for other genera (*Cereus*, *Opuntia*). It would be more plausible if they would serve for water absorption.

Ethnobotanical and general use

In Venezuela, the plant is often found as an ornamental in gardens (PITTIER 1970). The soursweet fruits are edible.

Varieties and related species

Three other species of *Pereskia* additionally occur in Venezuela.

P. guamacho F. A. C. WEBER IN BOIS with edible fruits of orange colour, and *P. bleo* (HBK) DC. the leaves of which are used as a vegetable in certain regions.

P. aculeata MILL. is a woody vine the fruits of which are eaten raw or preserved. The leaves are used as a pot-herb. The fruit is soursweet and slightly sticky. As the plant is spiny and can easily be propagated by cuttings, it is often used a a living fence. The fruits are a good source of vitamin A.

P. guamacho is also used medically. A decoction of bark and leaves is applied to clean wounds and ulcers, and to cure rheumatism. Its properties are diuretic, antirheumatic and vulnerary. Leaves and shoots macerated are said to cure asthma. A decoction of the shoots is applied externally as a bath for inflammations and rheumatism.

Observations

The leaf may be recognized by its isolateral structure, the druses of calcium oxalate and the presence of simple hairs as well as of peltate glands.

Caesalpiniaceae

This family of the Legumiosae mainly consists of trees and shrubs.

The leaves are mostly simple pinnate. The flowers are often large and attractively coloured. They are of zygomorphic symmetry with one flag-like petal. Reductions in the floral region by the lack of members are frequent. The aestvation is ascending. Stamens 10 or less.

The family principally occurs in the tropics and subtropics.

Laxative anthraglycosides are found in certain species of *Cassia* (*C. sennae*, *C. angustifolia*). Leaves and fruits are medicinally used as or in evacuants. The dianthronglycosides (sennosides) prevail in quantity. Adulterants of leaves of other species which contain less anthraglycosides also occur on the market. A soft laxative is the fruit pulp of *Cassia fistula*.

Some representatives supply balsam, e.g. species of *Copaifera* the Copaiva balm.

The fruits of *Tamarindus indica* are used as a purgative due to their high content of fruit acids. The fruits of *Ceratonia siliqua*, which are rich in slime and tannins, are utilized as a substitute for coffee.

Some trees are ornamental, e.g. *Delonix regia*, the 'Flamboyant of Madagascar', but frequently planted in Venezuela.

Bauhinia

Shrubs, lianas and small trees (Figs. 72, 73). The simple palmately veined leaves which are more or less deeply notched at the end suggesting the hoof of a deer or goat, are very characteristic. Lianas have often winged or flattened wavy stems and anomalous growth in thickness. The flowers are generally large and showy.

At least 18 species are useful.

The wood is useful. Bark fiber is used for cordage, cloth and ropes. Bark is used for tanning. Bark is astringent, febrifuge and stimulating. Stem is applied for kidney ailments.

Roots supply a dye. Roots are used for colds, cough, fever and intermittent fever.

Leaves are used as cigarette covers. Leaves are applied for eye disease, for ulcers and for flavouring food. Flowers are eaten as a vegetable. Seeds and pods are edible and supply a blue dye. Seeds are diuretic. Ashes of the plant are used for soap manufacture (ROTH & ASCENSIO 1977, ROTH 1987, ROTH 1990).

Fig. 72. *Bauhinia guianensis* ('monkey swing'), Caesalpiniaceae. Note the undulation of the axis.

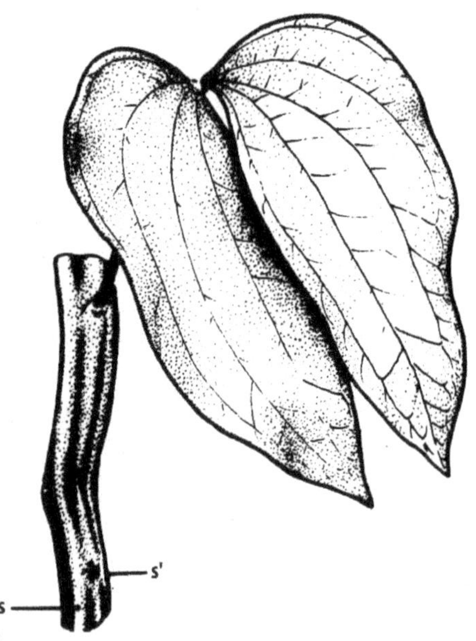

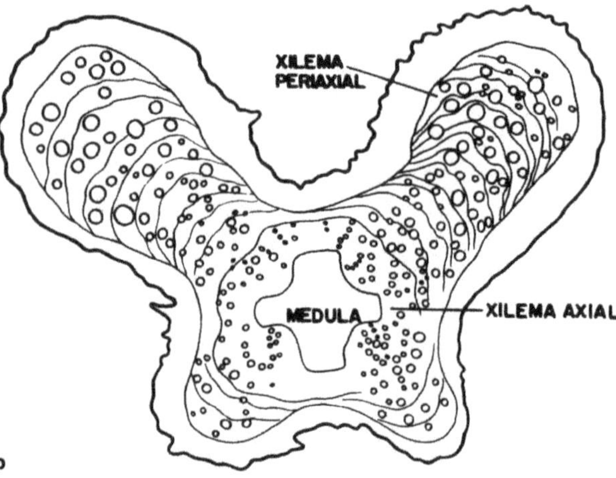

Fig. 73. *Bauhinia guianensis*. **a** Leaf. **b** T.s. of axis. Note the 'butterfly shape'. Medula=pith, xylema periaxial=periaxial xylem. Xylema axial=axial xylem.

Bauhinia cumanensis H. B. K. and/or *B. guianensis* AUBL. (escalera de mono, bejuco de cadena, monkey swing)

The identification of some *Bauhinia* species is difficult. The commonly known bejuco de cadena with a band-shaped stem, which is undulated and has an anomalous growth in thickness, may correspond to *B. cumanensis* or *guianensis* or to both. The undulation of the stem and the anomalous growth in thickness are related to the position of the leaves. ROTH & ASCENSIO (1977) profoundly studied the stem anatomy of several representatives of the genus (Fig. 73). A short review of this study is found in ROTH 1990. pp. 149–152. The leaf anatomy has been described by ROTH 1990. The stomata are hidden in holes of the lower leaf side.

Ethnobotanical and general use

Leaves and bark are used as a vulnerary and against dandruff. A 5% decoction in water is prepared.

Pieces of the plant in a decoction, taken 3 times a day, have a purifying effect. To cure straining at stool and dysentery 3 spoonful of honey of roses are added to the decoction. For haemorrhage, leaves of *Plantago* are added to the decoction.

The plant is used as an astringent, antipyretic and for venereal diseases, against diarrhoea, diabetes, blenorrhagia, headache, snake bites, rhinitis, dermal ulcer and kidney ailments. The seeds are said to be diuretic.

There are several other species which show similar effects, e.g. *B. manca* STANDL, used in cases of rheumatism, debility and for parasites.

Chemical contents

The hypoglycaemic effects are ascribed to the presence of lectins. Tannins and saponins are also found besides catechols, flavonoids, glycosides and alkaloids.

Brownea

Brownea is an ornamental with beautiful headed inflorescences. Flowers are used as a vomitive and to cure bleeding and excessice menstruation. Twigs and bark are contraceptive.

Brownea grandiceps JACQ. (rosa de montaña, rosa de Venezuela) is a small tree or shrub 3–8 m high, with simply pinnate leaves, 5–45 cm long, having 5–15 pairs of pinnae. The fruits are large legumes, about 25 cm long and coriaceous. The inflorescences are head-shaped and large, about 10–15 cm in diameter. The flowers are of a showy red. The leaf buds show the very interesting phenomenon of discharge of foliage; (ROTH & CLAUSNITZER 1969B). The leaf buds, enveloped in scales, turn geotropically positive down, then enlarge enormously and finally the scales are ruptured and a whole bunch of leaves comes out, at first flabbily limping. The phenomenon is almost exclusively observed in the Caesalpiniaceae. The opened limping buds can reach a length of more than one meter. The phenomenon was studied by ROTH & CLAUSNITZER 1969; the hanging position is not due to a lack of turgescence, but to a lack of mechanical tissue which is developed during the following weeks or months. Then the leaves rise into horizontal position by a negative geotropism. ROTH & MÉROLA studied the consistency of the cataphylls of *Brownea grandiceps* 1969.

The plant is very beautiful and of high ornamental value. SCHULTES & RAFFAUF (1990) reported that the plant is used as a contraceptive by the Indians.

Reproduction occurs by seeds. The plant has a slow growth. Its radical system is profound and its life span long. But it needs shade below other trees (HOYOS 1978).

Cassia

At least 40 useful species are known. Wood is useful. Dyewood. Bark is used for tanning, twigs for brooms. Bark for earache and headache. Roots as a substitute for soap. Roots for rheuma, oedema and gonorhoea. The root is diuretic. Leaves and pods supply Alexandrian Senna of commerce, a purgative (anthraquinone glycosides). Leaves as insect repellent and for lice. Leaves for skin diseases (chrysophanic acid), as antidote for snake bites and for insect stings. Leaves laxative and wound healing. Leaves, bark and seeds cathartic. Leaves as a vegetable.

Entire plant supplies green manure, producing large amounts of humus. Flowers are a source of honey. Fruits are purgative, laxative.

Seeds as mordant for dyeing. Seeds as a substitute for coffee. Seeds tonic, stomachic, diuretic, febrifuge, cathartic, emetic, for oedema, rheumatism, eczema, ringworm and for ophthalmia. Seeds vermifugal and an emmenagogue.

Cassia grandis L. The bitter fruit pulp is laxative and febrigufal.

ROTH studied the bark structure 1981, fruit structure and dispersal 1987.

Cassia moschata H. B. K. has edible fruits which are also used as forage for livestock. Medically, the plant is applied for affections of the liver and to combat intestinal parasites. The timber is resistant to fungi and termites.

Cercidium praecox (R. & P.) HARMS. (yabo, yabita, jano, cuíca, brea, retama, palo verde)

Taxonomical description

The plant is a small treelet with an umbelliform crown, 2–4 (6) m high, with spines (2–10 mm long) and a green cortex. The bipinnate leaves have 1–2 (occasionally 3) pairs of leaflets of first order. The leaflets of 2nd order are arranged in (3) 6–8 pairs. The oblong, obtuse or slightly acute leaflets are 3–4 mm long and 1–1.5 mm broad. the floral racemes, 1–2 cm long, develop 2–6 yellow flowers. The pedicels reach about 4 mm in length and are slightly hairy or glabrous. The calyx tube is bell-shaped and is divided into 5 yellow segments which are 6 mm long, have an acute tip and are slightly hairy. The 5 petals, oblong to orbicular, and more or less unguicular, are unequal, about 1 cm long and of yellow colour; the upper petals additionally have red spots. The 10 stamens are free and somewhat hairy at the base.

The flattened legume is of oblong or oblanceolate shape (3–5 (6) cm long, 1 cm broad, and has a short stalk.

The flowering period occurs between April and May. The yellow flowers appear before the leaves which come out in spring when the rain begins. the tree is bare of leaves during the driest months.

Occurrence

The species is characteristic of tropical America.

In Venezuela, it frequently occurs in arid regions, particularly of the west (in the States Zulia, Lara, Falcón and Trujillo), as well as in the east (the States Sucre and Nueva Esparta). Its ecological niche is the spiny tropical woodland.

Anatomical description

Axis. (Fig. 74) The species is leafless most of the year. Leaves develop in March/April after the first heavy rain falls and are shed shortly afterwards, so that the tree is dependent on the photosynthetic activity of the twigs and the stem. First, the leaflets are shed and then the rachis. The tree keeps the leaves for only about 6–10 weeks.

Cortex. The cells of the primary cortex which contain chloroplasts take over the function of photosynthesis. Formation of a cork or a bark is impossible. However, the plant is able to form cork, e.g. after an injury. The epidermis cells maintain the ability to divide and to extend permanently. In fact, the epidermis becomes multilayered. The epidermis cells also become very thick-walled.

Additionally, a thick-walled hypodermis forms beneath the epidermis which functions as a water reservoir. Multilayered epidermis and hypodermis together form a very effective protection against excessive transpiration and desiccation. The tree therefore supports strong and long desiccation periods; even cut branches keep fresh for many weeks.

Cercidium praecox is deciduous, although it may be stated in bibliography that it is evergreen. It is

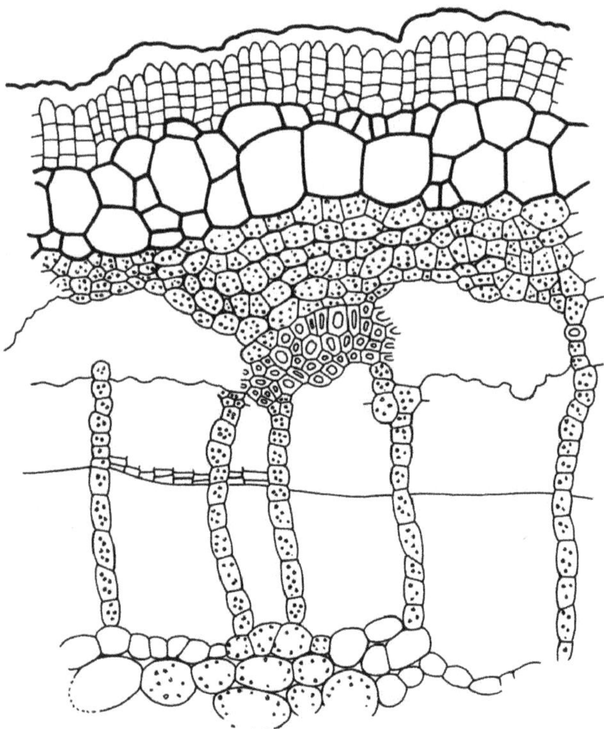

Fig. 74. *Cercidium preaecox*, Caesalpiniaceae. T.s. of young stem with multilayered epi and hypodermis, photosynthetic cortical tissue, uniseriate rays and globular pith cells in the center.

only evergreen as far as the green cortex of stem and branches is concerned. The stem anatomy of *C. praecox* and of *C. torreyanum*, which grows in the deserts of southern California, is very similar (ROTH 1963). In an advanced stage, the epidermis may comprise as many as 11 layers. The outermost walls of the epidermis reach an enormous thickness and are heavily cutinized and covered by a waxy layer. The hypodermis cells are of a giant size and become divided periclinally to form several layers. Beneath follow several layers of photosynthetic tissue with chloroplasts. Towards the inside lies a ring of stone cells which gives strength to the twig or stem. This is again followed by photosynthetic parenchyma. Enlarging rays and phloem are in contact with it. Even the rays contain chloroplasts as well as the pith cells in the center of the twigs.

For more details see ROTH 1992, 1963.

Ethnobotanical and general use

Economical utilization

During certain months of the year, stem and branches are covered with a semi-transparent astringent greenish gum or lac which is soluble in alkali and is used locally for making soap.

Natives use the cortex together with oil of marine turtles and salt as a lotion to cure contusions and dislocations.

The gum could be used industrially for lacquer production.

Medical use

The cortex is used in popular medicine as an antidysenteric, principally for live-stock and other domestic animals.

A drink is prepared of a decoction of the cortex in water to assist the expulsion of the placenta.

Chemical contents

The cortex contains large amounts of arabinose and xylose.

Related species

Cercidium torreyanum SARG., also called 'Palo verde', is similar in its structure and grows in the deserts of south California (ROTH 1963). The ground seeds are made into cakes and consumed by the Indians of Arizona and California. A beverage is also prepared of the seeds.

Observations

The multilayered epidermis and the large-celled hypodermis are very good characteristics for identification.

Copaifera

At least 13 useful species are known. Wood is useful. Copaiba balsam for varnish, photographic paper, removal of old varnish; medically it is used for skin diseases, eczema, gonorrhoea, for pomades and in perfumery. It is diuretic, stimulant, expectorant, genito-urinary disinfectant, and is applied also for diarrhoea and chronic dystentery.

Source of copal resin.

C. pubiflora. ROTH studied the fruit structure and dispersal 1987.

Crudia

A tea of the bark is prepared to induce emesis following food poisoning. ROTH studied the bark structure 1981.

Dialium

At least 6 useful species are known. Wood is useful. Bark is used as a masticatory. Fruits are edible.

D. guianensis SANDWITH. Wood is useful.

ROTH studied the bark structure 1981, leaf structure 1984, fruit structure and dispersal 1987, leaf venation 1996.

Dimorphandra

Wood is useful. Seeds are used as a dye.

D. gongrijpii SANDW. has useful wood.

ROTH studied fruit structure and dispersal 1987.

Eperua

E. supplies resin, balsam for varnish. Bark emetic. Oil of bark for painful joints. Flowers for hair care and better hair growth.

E. falcata AUBL. Wood is useful. ROTH studied the bark structure 1981, fruit structure and dispersal 1987.

Hymenaea

Resin for fungal infections of the feet. Source of Copal used in varnishes. Terpenes and phenolics are responsible for inhibition of leaf fungus growth on plants.

H. courbaril L. has a yellowish or orange resinous gum, Copal, in the bark, which is used as incense, for varnishes and leather. Pulp around the seeds is edible. Wood is useful.

A decoction of the bark is used for a bath and to alleviate contusions. Internally it is taken as a sedative and vermifuge. The burning pericarp is an insect repellent. Timber is resistant to fungi and termites (CASTILLO 1995).

ROTH studied the bark structure 1981, leaf structure 1984, fruit structure and dispersal 1987, leaf venation 1996.

Macrolobium acaciaefolium BENTH. has a useful wood. The leaves cure ulcers. ROTH studied the fruit structure and dispersal 1987.

Mora excelsa BENTH. has a useful wood. ROTH studied the bark structure 1981, fruit structure and dispersal 1987, leaf venation 1996.

Parkinsonia aculeata L. (retama, mata linda, chiguare, espinillo, espinillo de Espana, espinito, cují extraniero, yabo zuliano, pinopino)

Taxonomical description

The 3–8 m high plant is a spiny shrub or small tree, only occasionally reaching 10 m. The crown is globular being topped by slender somewhat pendulous spiny branches. The trunk is short and smooth-barked, sometimes reaching 38 cm in diameter. Stem and branches are of green colour and important as photosynthetic organs, particularly during the time when the leaflets are shed. The leaves are bipinnate with several (2–3) pairs of pinnae of first order clustered at the rudmentary main axis (Fig. 75). The flat persistent rhachis of the pinnae of first order is 20–30 (40) cm long and 0.5–0.8 cm broad and develops numerous rather widely spaced pairs of minute oblong or linear leaflets of second order which have a dense indumentum in a young stage, but are caducous. The main rhachis of the compound leaf terminates in a short spine.

Fig. 75. *Parkinsonia aculeata*, Caesalpiniaceae. Leaf. Rhachis enlarged and leaflets reduced.

The fragrant and showy yellow flowers, 1–2 cm in diameter, are borne in axillary racemes. The few-seeded legume is 5–15 cm long and 8 mm broad; it is constricted between the seeds.

The radical system is profound.

Occurrence

The plant is common in sandy littoral areas of Venezuela. It is common in tropical America.

Anatomical description

Leaf. (Fig. 77a) Leaflet of second order: Upper and lower epidermis are small-celled. The outer epidermis walls are thickened. Beneath the epidermis follow 2 laysers of palisade cells. The cells of the spongy parenchyma are spherical and frequently

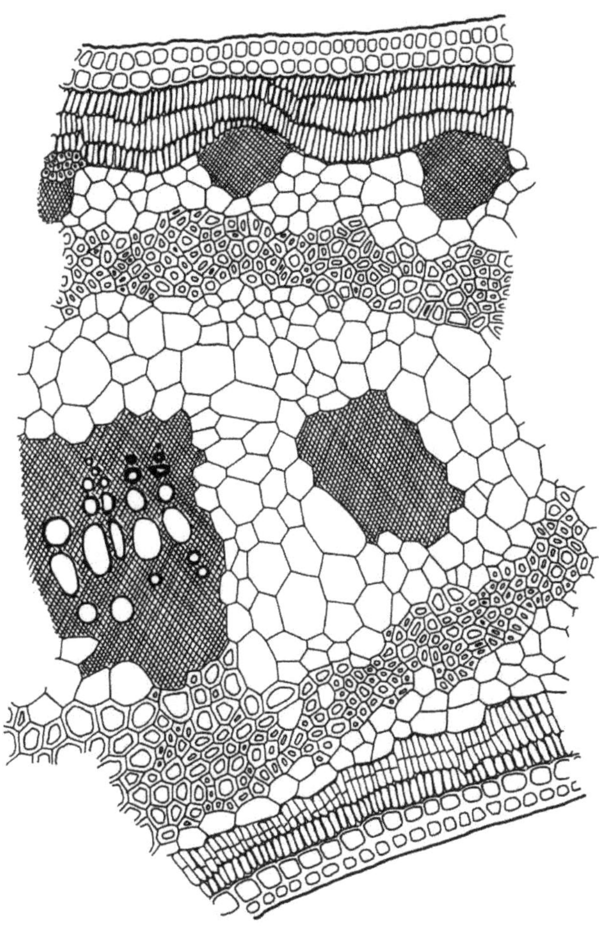

Fig. 76. *Parkinsonia aculeata.* T.s. of rhachis with 2-layered epidermis. 3 layers of palisade parenchyma, beneath vascular bundles (hatched), several layers of fibers, ground parenchyma and main vascular bundle, fiber layers and photosynthetic tissue also on the lower side.

contain druses of calcium oxalate. The vascular bundles (accompanied by fibers) are embedded in the spongy parenchyma. An acuiferous hypodermis consisting of enlarged thin-walled cells is confined to the lower leaf side. A palisade parenchyma, also consisting of 2 layers, is found above the hypodermis. The leaflet has thus equifacial or isolateral structure. Stomata occur on both leaf surfaces.

Rhachis of the pinnae of first order. As seen in a transverse section (Figs. 76, 77b), the phyllode-like rhachis appears as a flattened, but very thick organ of a strange structure. The main vascular bundle in the center is surrounded by an upper and lower sclerenchymatous arc. Small vascular bundles are additionally present forming an incomplete ellipsis beneath the palisade parenchyma which is interrupted on the lower side by the lower sclerenchymatous arc. The phyllode has thus a petiole-like structure. The smaller vascular bundles are capped by sclerenchyma on the xylem side. The ellipsis of vascular bundles is surrounded by the photosynthetic tissue which is better developed on the upper leaf side, but becomes smaller on the lower leaf side. It is composed of about 3 layers of palisade cells. The connecting or filling tissue in the leaf center consists of thin-walled polygonal cells which are tightly packed together and hardly leave intercellular spaces between one another. This tissue forms the main bulk of the phyllode. Stomata are present on both leaf surfaces, but are more numerous on the lower side. The structure of the rhachis is likewise isolateral. Its leathery consistence is due to the abundant sclerenchyma as well as to the large-celled hypodermis and to the dense filling tissue.

Ethnobotanical and general use

The plant is used as an ornamental in gardens, parks and on city places. It can also be used for afforestation, as it is an undemanding plant with a deep root system and rapid growth. It can be propagated by seeds.

The wood is said to be of good quality. But it is only used for fuel, because of its rare occurrence.

The plant is also medicinally applied. An infusion of the leaves has febrifugal effects, is stomachic and may even be applied against rheumatism.

Related species

The seeds of *P. microphylla* TORR. are eaten fresh or made into flour by the Indians of S. W. USA and Mexico.

Observations

The leaf can easily be recognized by its morphology and inner anatomy, particularly by its isolaterality, and by the arrangement of vascular bundles and sclerenchyma in the rhachis.

Peltogyne

The wood is useful. Pulverized roots are used as a repellent of dogs. Bark is astringent. ROTH studied the bark structure 1981, leaf structure 1984, fruit structure and dispersal 1987, leaf venation 1996.

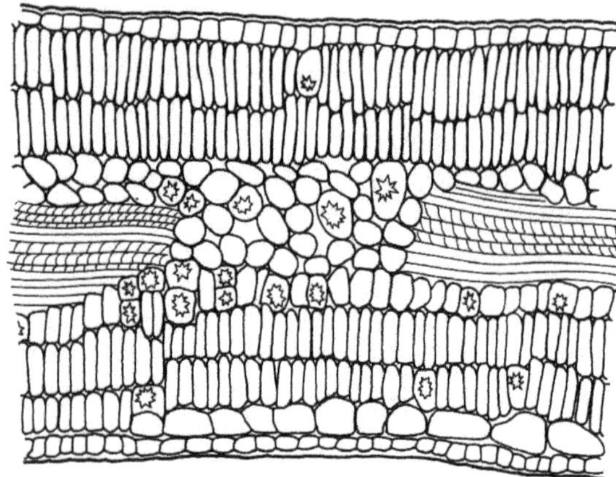

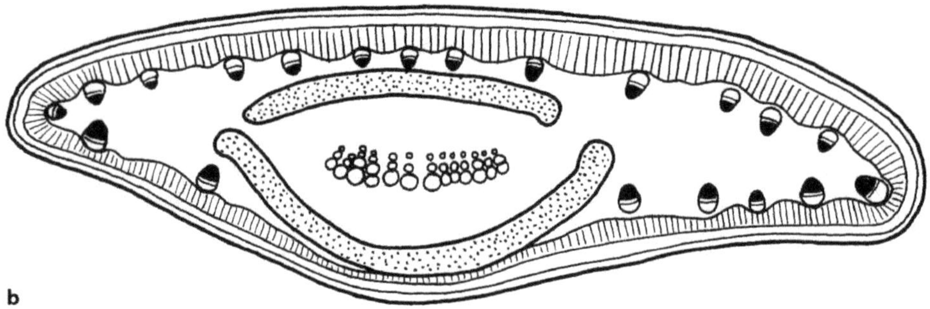

Fig. 77. *Parkinsonia aculeata.* **a** Leaflet in t.s. Note the photosynthetic tissue on both sides. **b** T.s. of rhachis, general view. Xylem of vascular bundles black, fibrous sclerenchyma dotted (ROTH 1992).

Swartzia

The wood is useful. Bark as a fish poison. Ashes of bark for pottery. Stem for diarrhoea. Leaves vermifugal. Leaves for an antiamenorrhoeal bath. Fruits for diarrhoea, to cure physical debilitation. Pods as a vermifuge. Seeds for intestinal parasites.

S. schomburgkii BENTH. Pods are used as a vermifuge and for diarhoea with bloody stools. ROTH studied the fruit structurre and dispersal 1987.

Tachigalia

At least 6 useful species are known. The bark is antiseptic and used to cleanse serious cuts and wounds and is also applied for chancres of the mouth. Leaves are used for gastric upset due to poisoning and to bloody stools. Leaves serve to alleviate chest pains. A tea of leaves is antiinflammatory and helps ease arthritis. Leaves are used as an insect repellent and for pain through stings of the fierce fire ants. Ashes of leaves are used for pottery. Flowers are applied for tuberculosis. Unripe pods are an aphrodisiac.

T. cavipes is called 'no-children-medicine' by the Indians, because it contains an antifertility agent.

T. paniculata. AUBL. Leaves in infusion are applied for pains of limbs and chest pains. A tea of the leaves is stimulant. An emetic tea is made of the seeds. Unripe pods are an aphrodisiac. ROTH studied the bark structure 1981, leaf structure 1984, fruit structure and dispersal 1987.

Cannaceae

The monocotyledonous Cannaceae are perennial herbs with subterraneous rhizomes and leaves provided with a large sheath. The flowers are usually large and highly derived concerning their corolla which partly originates from the petaloid staminodia and the petaloid style. The fruit is a trilocular capsule. The only genus Canna occurs mainly in tropical and subtropical America.

Canna indica L.,
synonym: **Canna edulis** KER.-GAWLER
(capacho)

Taxonomical description

The plant is a perennial herb (Fig. 78), about 50–150 (300) cm high with a subterraneous rhizome. The alternating simple leaves are oblong to ovate-oblong, 26–50 (15–65) cm long and (10) 15–30 cm broad, with an acuminate tip, an entire margin and an attenuate asymmetric base, abruptly followed by the sheath. The blade has a slighty leathery consistency and shows a false pinnate venation – the plant is monocotyledonous! – with about 24–30 (or more) pairs of secondary nerves; the middle nerve is prominent on the abaxial side and flat on the adaxial side. The terminal inflorescence is 7–15 cm long. The zygomorphic somewhat asymmetric flowers are about 5 cm long, of red and orange colour partially with red spots and are arranged in spikes or panicles. The 3 red sepals are slightly unequal; they are followed by 3 red and yellowish petals, one of which is smaller than the others; the androecium is composed of 3 petaloid staminodia of reddish-yellow or orange colour with red spots; the reddish labellum corresponds to the innermost staminode; the fertile stamen has 2 pollen sacs on the one half, while the other half, developed from the broadened connective, is also staminodial and petaloid. The components of the androecium thus contribute to the formation of the corolla; even the style is petaloid and of orange or reddish colour. The trilocular capsular fruit is 1–3 (2–7) cm long and 1.5–2.5 cm in diameter; it becomes dark-brown when ripe and is densely covered with warts (ENGLER 1964).

Fig. 78. *Canna indica*. **a** rhizome with sprouts. **b** Flowers and fruits (with seeds).

Origin

According to BRÜCHER (1989), the species is of South American origin. Brücher assumes that domestication occurred in the subtropical eastern valleys of the Andean Cordillera, as he observed wild-growing samples in the north of Argentina and in Bolivia. Very old prehistoric finds were made along the coastline of Peru. The species was certainly known to the Incas and has its origin in Peru and Colombia (BERNAL & CORREA 1990). It has been in use for at least 2500 years.

Historical background

For prehistoric studies see UGENT & POZORSKI (1984) and GADE (1966) cited by BRÜCHER (1989), VELEZ & VELEZ (1990), UGENT, POZORSKI & POZORSKI (1986), BERNAL & CORRE (1990) and the bibliogrphy cited there.

Occurrence

According to STEYERMARK & HUBER (1978), the species is found in Venezuela in gallery and cloud forrests, usually in humid gorges and at a height between 950 and 1600 m.

Anatomical description

Rhizome. In the rhizome (Fig. 79), a peripheric part consisting of slightly smaller cells is distinguished from a central part with somewhat larger cells. At the border line of both, the endodermis, composed of tangentially elongated cells with Cas-

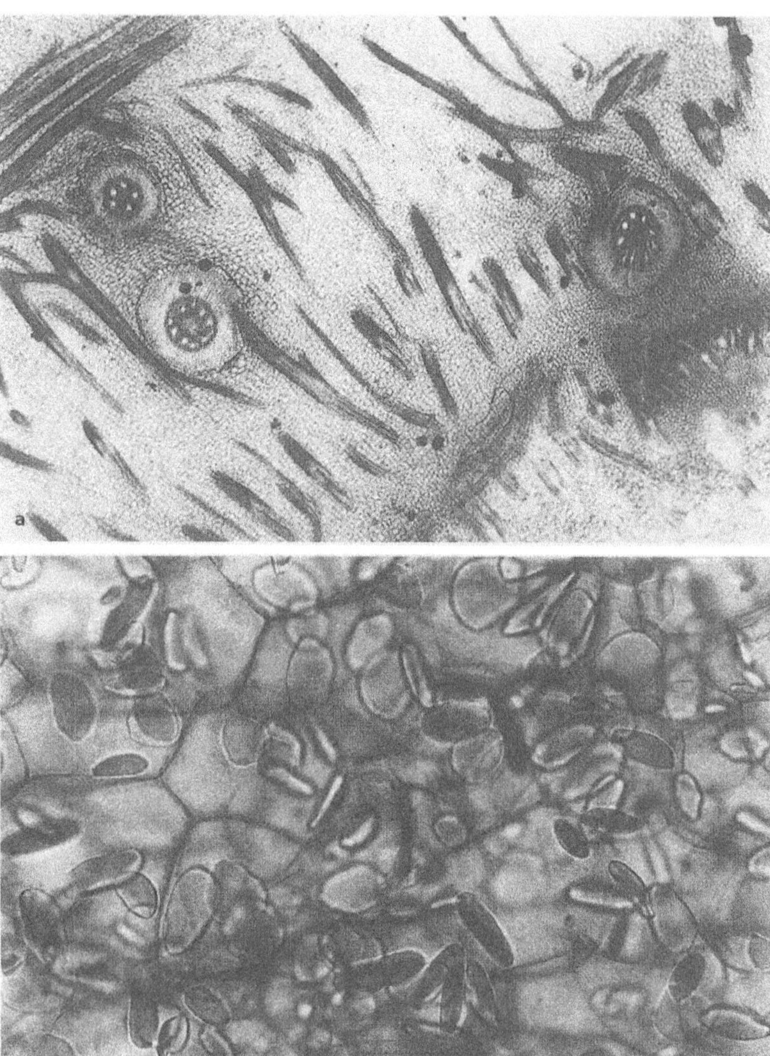

Fig. 79. *Canna indica.* **a** part of the rhizome in t.s. with vascular bundles. **b** starch grains of the rhizone (× 20).

parian strips in the radial walls, is found. Adventitious roots originate in the peripheric part, and may partly be observed in transverse, oblique or longitudinal section. Peripheral and central parts are composed mainly of parenchymatous cells, but the cells of the central part are slightly larger, particularly towards the very center. Irregularly dispersed over the section are collateral vascular bundles, which are found in the peripheral as well as in the central part. A concentration of bundles in the form of a ring, as seen in transverse section, is found beneath the endodermis where the bundles may even cross one another, running mainly in tangential-transversal and longitudinal direction. The parenchymatous cells of the peripheral zone contain small starch grains, while those of the central part have very large grains. These vary in shape and size, however *Canna* starch grains pertain to the largest known in the plant kingdom. They are eccentric broad-ovate, roundish or elongated and reach a size of 60–95 μm, exceptionally even as much as 130–150 μm. They are more rounded on one end which frequently contains the center of condensation; this always lies in a very eccentric position. The flat grains are lenticular to club-shaped, as seen in transverse section. They are well distinguished from potato starch by their larger size, the denser layering and the more eccentric shape.

Additionally, there are secretory canals surrounded by an epithelium of secretory cells in the central part of the rhizome, as well as cells with crystals.

Ethnobotanical and general use

Nutritional use

Canna is mainly known as a starch plant. The dried rhizome (Queensland-Arrowroot) contains about 75–80% starch, 6–14% dried sugar, and 1–3% protein and a high concentration of potassium, but little phosphorus and calcium (BARNAL & CORREA 1990). *Canna* is planted in Australia (therefore the name of the rhizome), Guadalupe, and India as a promising starch crop with yields 25–50 t/ha. Only recently, the fecula has been appreciated as a food for babies and sick people; because it is easy to digest.

Canna has a higher content of amylose than maize and potato. The rhizomes are used as a substitute for potatoes and are offered for sale at the markets of La Paz. In Peru, baked rhizomes are prepared for the 'fiesta' of Corpus Christi. With the flour extracted from the rhizome, delicate cookies are made. At certain Peruvian and Argentine markets the starch of *Canna* is used as an adulterant of cocoa powder.

The leaves serve as an envelope for certain types of food, such as cheese, butter or maize pies, and to keep food warm.

In Hawaii, rhizomes and leaves, raw or cooked, are used as a fodder for cattle.

Economical utilization

The raw material used in industry is the rhizome for extraction of starch. During processing, the rhizome is cut into strips, ground, mixed with water and the pulp is separated from the starch by decanting and sifting through a fine strainer of linen; finally, the purified starch is dried in the sun.

The seeds which are of very regular shape and almost equal in weight, are still used by gold diggers and native goldsmiths as a weight unit (1/16 ounce) for small gold quantities.

Although *Canna indica* is native to South America and has been used there by the Indians for thousands of years (at least since 2500 AC), in its native countries the species is only used on a small scale. It could however be cultivated in the Andean regions for exportation. Instead it is planted in the Caribbean area, in Indonesia, Taiwan, the Philippine Islands, Australia, Madagascar, Sri Lanka, Burma, and Hawaii.

In its native countries the species is mainly cultivated as an ornamental for its beautiful flowers and decorative leaves.

Medical use

The name of the drug is *Canna indica* L. radix, caule, folia.

Leaf. The leaves in decoction are used in Brazil to cleanse ulcers and in the form of vapour they are used against rheumatism. Fresh or slightly toasted (dried in the sun) leaves put on the breasts of women who have had their first baby prevent the exhaustion of the milk flux. Leaves in decoction are put on wounds as an antiseptic which at the same time promotes cicatrization. A decoction of the leaves is diuretic and antiabortive.

Stipe. The natives of Colombia, Cuna, prepare with 4 pieces of the stipe an infusion which is applied as a complete bath; this helps to recover from sicknesses and to regain high spirits.

Rhizome. The rhizome is applied as an emollient cataplasm and as a diuretic when given in decoction. The dried and cut rhizome boiled for 15 minutes and applied in a dose of 30 g/l of water relieves pain at the onset of menstruation; it also helps as a demulcent and prevents infections of the kidney and the bladder. The starch is said to have a sedative effect; it is furthermore easy to digest. The inhabitants of Samoa treat inflammations with *Canna*. The rhizomes which have a slightly aromatic flavour, originating from the secretion in the secretory canals, combat fever and oedema. The decoction of leaves and rhizomes has diuretic and antigonorrhoeic properties.

Method of use

Leaves, stipes and rhizomes are used fresh, in decoction, in powder form or as a cataplasm. To prepare the decoction for the emollient cataplasm, a spoonful of the rhizome is diluted in 2 cups of water forming a viscous fluid. When taken internally, the dried and cut rhizome is boiled for 15 minutes and applied in a dose of 30 g/l. The decoction for diuretic purposes is prepared with a handful of the rhizome and 1 l of water; this mixture is boiled and 3 cups of the drink are taken a day. The same decoction also helps against painful menstruation and strengthens its flow. In the last cases, 3 cups of the liquid are taken dayly a week before menstruation begins.

Healing properties

Canna indica has diuretic and emmenagogic properties as well as an antiseptic and calmative effect.

Chemical contents

Of all the chemical contents of *Canna*, it is the starch that has been mainly studied. Among other components, amylose and amylopectin, flavonoids, carotenes, xanthophylls, maltose, sucrose, fructose and glucose, as well as L-arabinose, have been found. Although *Canna* is a very well known plant from time immemorial, the content of the secretory canals has not yet been studied separately; this content is probably responsible for the special medicinal properties of the plant.

Varieties and related species

Canna edulis KER-GAWL., *Canna lutea* MILLER and *Canna indica* L. are considered as pertaining to the same species, as the variations in the floral regions, and particularly in the androecium (labellum, staminodes) are not sufficiently constant to distinguish 3 species. Another synonym is *Canna coccinea* MILLER (see also the list of synonyms in BERNAL & CORREA (1990). In the Andes, a green variety is distinguished from a violet one (see below). The rhizome of *Canna glauca*, likewise from the Andean region, has also nutritional and medicinal properties (BERNAL & CORREA 1990).

Cultivation

The plant is propagated by cuttings of the rhizome. These are planted at a depth of 15 cm and at a distance of 80 cm between the slips. After 2 weeks the sprouts appear and adult leaves are formed after 2 months. Planted simultaneously with coffee, 2 crops of *Canna* are obtained, before the coffee sets fruit. The plant grows even as a ruderal in coffee, banana, mango or sugar-cane plantations.

In its natural environment, it flowers all the year round and grows from sea level to 1730 (2900) meters, a.s.l. in the high evergreen forest, and in the low deciduous forest, together with *Scheelea liebmannii*, and is abundant as a ruderal plant. The species is easy to cultivate, because it is neutral to the day length, and resistant to rainfalls between 250 and 4000 mm, it tolerates temperatures from 32 °C to 0 °C, supports the majority of the soils including those having an acidity between pH 4.5 and 8.0 and even grows in completely exhausted soils. The plant may be affected by heat or drought, but tolerates inundations, excessive humidity, snow and frost.

The cultivated plants however have larger leaves and flowers. Triploid plants are sterile which is an advantage, when rhizomes are cultivated to harvest the starch. Fertilization, on the other hand, increases production. The plant is easy to cultivate and to propagate, it grows rapidly and the rhizomes may already be harvested in the 6th month, when they are most succulent; the main harvest however is gathered after 8–10 weeks, when the rhizomes have attained their maximal size. The most productive and most domesticated types do not produce seeds.

In the Andes, 2 types of cultivated varieties are distinguished: the green one with a white rhizome and larger starch grains, and the one with violet rhizomes. There exist plenty of variations as to leaf colour, height of the plant, size of the rhizome, time of flowering and fruiting. In Ecuador the plant reaches a height of up to 2 m or more and is therefore cultivated as a windbreaker for other planted crops.

In South America, the leaves of *Canna* are eaten up by different types of 'defoliators'. In Peru, the plant is often infested by fungal diseases, such as *Puccinia cannae, Fusarium* and *Rhizoctonia.*

The cultivation of *Canna* in South America is recommended, as the plant could become of great importance from the economic point of view by its ornamental, nutritional and industrial value. The plant is easy to propagate and does not need special treatment. Furthermore, leaves and stipe can be used as a fodder for cattle.

Observations

The rhizome of *Canna* is very easily recognized by its large starch grains, and by its division into a peripheral part in which the adventicious roots are developed, and into a central part where the secretory canals are found.

Capparaceae

The gynophore of the flower is very characteristic of this family, it is usually elongated in the form of a stipe. The Capparaceae are closely related to the Cruciferae: They frequently have 4 sepals, 4 petals and an ovary composed of 2 carpels; the fruit may resemble the cruciferous siliqua and develop a replum. The leaves are simple or compound and often have stipules. Myrosin cells may be observed in the cortex, bark and pith.

Many Capparaceae are xerophytes and live in very dry regions and deserts.

Very typical contents of the Capparaceae and the closely related cruciferae are the sulphur-bearing glucosinolates, the matrix of the pungent mustard oils. In the myrosin cells, thioglucosidase is deposited; it can become active only after the destruction of the cells. Due to the pungent taste, many representatives of the Cruciferae and Capparaceae are used as spices: e.g. mustard, horseradish, caper. Allyl mustard oil is a rubefacient. Some mustard oils have antibiotic effects.

The seeds are rich in faty oils.

Worldwide, the Capparaceae plant family comprises 36 genera and about 600 species, mainly distributed in tropical and subtropical regions. In Venezuela, the family is represented by about 8 genera and 51 species occurring from sea level up to an altitude of 1400 m.

Some of the species are medicinal, others ornamental, certain species have edible fruits, others are shade plants, supply timber or are toxic.

The Capparaceae are appreciated for their evergreen foliage, their resistance to drought, and to pathogens.

RUIZ ZAPATA (1999) mentions 19 useful species, indigenous to Venezuela.

Belencita nemorosa (JACQ.) DUGAND. The fruit pulp is edible.

Capparis amplissima LAM. The fruit pulp is edible.

Capparis flexuosa (L.) L. The leaves in decoction are used for baths against muscular pain. The stems are applied in construction of rustic houses and for fences or as firewood. The bark of the root is diuretic and an emmenagogue; young soft twigs are useful as tooth-picks.

Capparis hastata f. coccolobifolia (MART. EX EICHL.) H.H. ILTIS. The stems serve for rustic constructions. The timber is used for posts, and occasionally, the plant is cultivated as an ornamental.

Capparis indica L. *Druce.* The plant is occasionally cultivated as an ornamental. The hard timber which is of fine texture and of light yellow colour is used for cylinders and firewood. The bark of the root has a stimulating effect. Of the bark of the stem and of the leaves a decoction is prepared which cures hysteria and hypochondria. The infusion of the leaves is applied to cure nervous diseases.

Capparis linearis JACQ. A decoction of the leaves is used for a sitz bath and for diarrhoea. The bark mixed with edible oil is applied as sinapism. Ground leaves and twigs are used for tooth ache and ear ache, putting the powder directly in the outer ear. The species also supplies timber.

Capparis muco H. H. ILTIS, L. CUMANA, R. DELGADO ET G. AYMARD. The pulp of the fruit is eaten. The leaves are applied to bake Venezuelan sweets.

Capparis odoratissima JACQ. The plant is cultivated as an ornamental. It also serves as a shade tree for livestock. During the holy week the plant is used as a subsitute for the olive tree to give the picture of Jesus in the garden of olives an escort. The infusion of the leaves is used for gargles as well as for baths to cure skin problems and to be lucky. The twigs in decoction combat pruritus and stomachache. With the wood, shores are manufactured.

Capparis osmantha DIELS. The pulp of the fruit is eaten.

Capparis pachaca KUNTH. The fruit pulp is eaten. The plant serves as an ornamental and shade tree. From the wood, paddles are fabricated.

Capparis pulcherrima JACQ. The fruits are toxic and can cause the death of children.

Capparis stenosepala URB. The plant is used as an ornamental and shade tree. The fruits mixed with food kill rodents and skunks.

Capparis tenuisiliqua JACQ. The stems are used to fabricate crosses in the Holy week. The stems are also applied in witchcraft.

Cleome gynandra L. In Venezuela it is considered a weed. In other parts of the world, however, it has many applications. Leaves, twigs and young fruits are eaten boiled. The plant is also used in popular medicine. The seeds in decoction are prepared for a drink which expels worms. Externally applied, the leaves cure irritations. The sap of the leaves is applied in the form of drops to cure ear ache.

Cleome hasslerana CHODAT. The plant is cultivated for its showy white or lila flowers; of all Capparaceae it is most used as an ornamental in gardens, parks and public places of the temperate regions.

Cleoma speciosa RAF. The plant is used for ornamental purposes.

Cleome spinosa JACQ. It is considered as a weed. However it is used as a barbasco to kill fish. An infusion of the leaves is used to calm aches of the ear in the form of a washing. It is also applied as a barbasco to kill cockroaches, bedbugs and similar insects. Furthermore, it has rubefacient properties.

Crataeva tapia L. The fruits are eaten. The plant is ornamental and a shade tree. The fruit is besides febrifuge, tonic, stomachic and antidysenteric. The leaves in infusion are an antithelmintic. The plant is said to be irritant. The sweet, refreshing fruits are appropriate for the confection of refreshing drinks and particularly for an antifebrile syrup.

Morisonia americana L. The fruit is edible. The wood suits for rustic constructions. The flowers have medicinal properties.

Capparis

Capparis comprises a number of useful species.

The root is used as an uterine tonic, for oedema and as an emmenagogue. The leaves serve for cutaneous diseases, for swellings, boils and piles. The bark is diuretic. The root bark is stomachic, sedative, antihydrotic and helps for excessive perspiration and gastric irritation.

Flower buds are well-known as capers. Fruits are medically applied for epilepsy, hysteria, chorea, and as a sedative. The fruits are edible. The seeds are used for coughs. The leaves serve as a fodder for animals.

ROTH studied the bark structure 1981, leaf structure 1984, fruit structure and dispersal 1987. LINDORF (1994) studied the wood structure of *C. flexuosa*, *C. hastata*, *C. odoratissima* and *C. tennisiliqua* LEÓN, AGOSTINI and RODRIGUEZ studied the leaf structure of *C. flexuosa* (1988). An infusion of the roots, which taste of horseradish, is used for oedema and as an emmenagogue. A decoction of the leaves is used for cutaneous diseases. Fruits are sedative and antiperiodic. Bark is diuretic and an emmenagogue.

Capparis odoratissima JACQ. (olivo, olivo macho, olivillo, olivo hembra, kapüchirii, guajira).

Taxonomical description

The species is a shrub or small tree, 3–5 m high. The branches are densely covered with very small scales. The coriaceous leaves, ovate, oblong or elliptic, are (1.5) 2–11 cm long and 1–4 (6) cm broad and have a rounded tip. The upper leaf side is glossy, the lower side is densely covered with scales. The fragrant flowers, 2 cm in diameter, occur in corymbs. The 4 petals, first of a creamy colour, but later turning to purple, are about 5–8 mm long. The stamens are numerous.

The fruit is cylindric, 4–12 cm long and 5 cm in diameter, covered with brown or silvery scales, and of a warty aspect. The fruit pulp is scarlet. The fruit dehisces along one or both sutures and contains 1–3 seeds.

The radical system is profound.

Occurrence

The species is found in Central America (Panama), Colombia, Peru and Venezuela.

It is very resistant to drought and lasts long. It is frequent in dry xerophytic forests and deciduous thorny woodland of Venezuela, along the northern coast line. It keeps the leathery leaves even during the driest periods.

Anatomical description

The leaf structure is characteristic of xeromorphic species of the genus *Capparis* (Fig. 80, 81). The upper epidermis is larger-celled than the lower one and has very thick outer cell walls which are strongly cutinized. Below it, there are 2 laysers of palisade parenchyma. The first layer consists of long and very slender cells with a length/width index between 7 and 9. The cells of the underlying second layer, on the other hand, are shorter. An intermediate layer between spongy and palisade parenchyma may possibly be present. The spongy parenchyma is composed of more or less roundish cells which leave small intercellular spaces between one another. Sclereids passing through the mesophyll from one epidermis to the other are very characteristic. They have very thick walls and a narrow

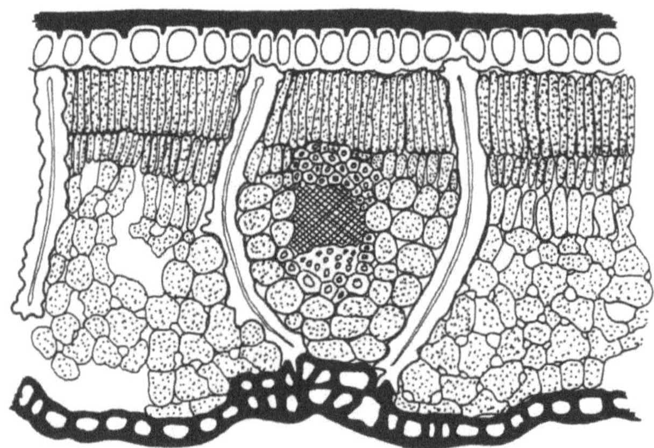

Fig. 80. *Capparis odoratissima*, Capparaceae. Note the sclereids in the mesophyll which function as pillars. Photosynthetic tissue dotted (ROTH 1992).

Fig. 81. *Capparis odoratissima.* Lower leaf surface with scales which cover one another on the margins (ROTH 1992).

cell lumen. Their outlines are irregular, as they adapt to the outlines of the surrounding cells by protrusions. They serve as pillars, strengthening the mesophyll to prevent it from collapsing when the tissue suffers from water stress during dry periods. As seen in paradermic section, the sclereids appear as light dots below the upper epidermis and may have the effect of little windows. The vascular bundles are capped on both sides by sclerenchyma.

The stomata which are confined to the lower leaf surface are small, but not very numerous. They not only hide in depressions, but additonally are protected by peltate hairs in the form of large scales. These cover one another on the margins. They certainly reduce transpiration drastically. In some places, the thick-walled lower epidermis becomes 2-layered.

The leaf has a strongly xeromorphic structure and is of a coriaceous consistency.

Ethnobotanical and general use

Economic utilization

The plant is very beautiful with its permanent lustrous foliage and is therefore used as an ornamental in gardens and parks.

It is furthermore applied in the form of living fences and as a shade dispenser for cattle.

Medical use

Name of the drug: *Capparis odoratissima* JACQUIN, branches.

A decoction of the branches helps to combat pruritus and stomach aches. The decoction is used in the form of a bath and is supposed to bring good luck.

There are at least 10 species of *Capparis* which are medically applied. The fruits of quite a few species are also edible.

Cultivation

Should be very easy.

Observations

The species is recognized by its leaf structure and the large unbranched sclereids (see also ROTH 1992).

Cleome spinosa JACQ. (garcita, aleli, clavellina blanca, desdicha, uña del diablo, barba del chivo, barba de tigre)

Taxonomical description

Cleome spinosa (Fig. 82) is an annual herb, 60–150 cm high, somewhat spiny and glandular-pubescent. The compound leaves are palmatisect with 5–7, sometimes only 3 leaflets, which reach a length of 3–9 cm. The leaflets are ovate, elliptic or oblanceolate and pinnately veined. The lower regular leaves (nomophylls) have long petioles (about 5–8 cm long), while the upper leaves (hypsophylls) are transformed into simple sessile bracts. At the base of the petiole, two stipular spines may be found.

The slightly zygomorphous flowers which occur in terminal racemes 10–40 cm in length, are large and showy. They are composed of 4 narrow lanceolate, 4–6 cm long sepals and of 4 obovate 2.5 cm long petals of white or pale rose colour; one of the petals has a claw-like structure (unguis) which reaches about the length of the blade of the leaflet. The 6 stamens are rose-coloured and reach 2 times the length of the petals. The pistil is found on top of the 2–6 cm long gynophore.

The more or less spiny, linear cylindric dry silique (supplied with a replum) is 5–12 cm long. The seeds are kidney-shaped and 2 mm long.

The species is very variable in its morphology.

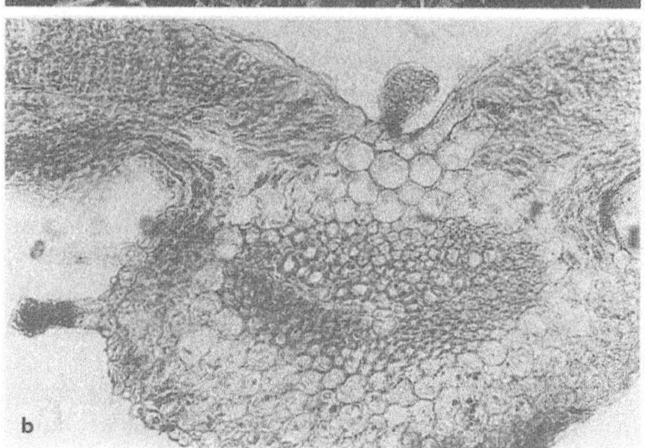

Fig. 82. *Cleome spinosa*. **a** Habitus. **b** Midrib in t.s. with gland on top (×20).

Origin

Tropical America, Andean subregion.

Occurrence

The species is common in Venezuela and is mainly found in hot and arid regions. It also occurs in Colombia, Panama, Peru and Bolivia.

Anatomical description

Leaf. (Fig. 82) The dorsiventral, amphistomatic leaf is very thin. The midrib is vaulted above the lower leaf side, but shows a depression on the upper side. The upper epidermis cells are of regular size and slightly papillose with strongly vaulted outer walls. The very small stomata lie at the upper epidermis level. They do not show subsidiary cells. The anticlinal walls of the epidermis cells are slightly wavy, as seen in surface view. The palisade parenchyma is well developed. It consists of anticlinal cell rows, each cell comprising about 4–7 (8) cells. The spongy parenchyma occupies about half the space of the palisade parenchyma and consists of more or less globular cells which leave small intercellular spaces between one another.

An arc of vascular bundles lies in the center of the midrib which is surrounded by parenchyma. The palisade parenchyma is interrupted in the midrib. The lower small-celled epidermis is relatively thick-walled in the midrib.

The veins of first order resemble the midrib in their structure, but are smaller and have only a single vascular bundle.

The vascular bundles of higher order which occur frequently, are weakly developed, but are surrounded by a large-celled parenchymatous sheath.

The lower epidermis is slightly smaller-celled than the upper epidermis and its cells are more papillose. The stomata are at epidermis level or are slightly elevated above the surface. However, the lower epidermis very much resembles the upper epidermis in its structure. Very characteristic are the glandular hairs which are more frequently found on the lower leaf side. they consists of a foot cell, a pluricellular often comparatively long and thick stalk and a pluricellular head of several stories (Fig. 82). A large glandular hair regularly occurs in the depression formed by the midrib on the upper leaf side (Fig. 82b) The gland possibly contains etheric oils.

Ethnobotanical and general use

The infusion of the leaves is applied as an insecticide for cockroaches, bugs and other insects and as a fish poison.

Medical use

Name of the drug: *Cleome spinosa* JACQUIN, folia. The leaves, stipes, roots and seeds are used in popular medicine. In Cojedes, Venezuela, an infusion of the leaves applied as a lotion cures ear-ache. The sap extracted from the leaves and mixed with olive oil is introduced into the acoustic duct by the Kallawaya in Bolivia to cure otitis. The Kallawaya also apply the fresh or dry leaves pepared as an infusion and in small doses against stomache ache and flatulence. Ground and boiled in water, the leaves are used as a very hot cataplasm which applied to the temples cures migraine. Fresh or dried leaves and stipes, ground and macerated in 40 % alcohol are applied as massage to cure rheumatism of the limbs. The sap extracted from the stipes when inhaled helps against sinusitis and very strong head ache. Boiled in water and provided with some fresh urine, the decoction of the leaves serves as a bath for rickety children.

The species has also rubefacient properties in the leaves, the cortex of the root and the spines, due to a volatile principle allyl mustard oil, present in the myrosine cells. The plant is also used to cure diseases of the liver.

Toxicity
The seeds, if ingested, are said to be toxic.

Cultivation

The species grows wild gregariously and is often found at uncultivated places with a fertile soil, especially in residential areas.

Observations

The species is easily recognized by its characteristic flowers and leaves.

Anatomically it may be identified first of all by its conspicuous large glandular hairs, the long anticlinal rows of palisade cells, the structure of the veins and the papillose epidermis cells.

Crataeva

The root is tonic and stomachic. Leaves serve for rheumatism. Fruit or bark help for a skin disease called lobosisso. The bark is appetite stimulating and laxative. Pickled flowers are digestive.

C. tapia L. The bark is tonic, stomachic, antidysenteric, febrifuge. The leaves are stomachic; their sap is used externally for rheumatism. ROTH studied the bark structure 1981, leaf structure 1984, fruit structure and dispersal 1987.

Caricaceae

A very small family with only 4 genera.

Trees, shrubs or giant herbs. Flowers unisexual by reduction. Ovary mostly unilocular. Numerous seeds in parietal placentation. Fruit a berry with a juicy pulp.

Articulated laticifers with a white milk sap are very characteristic.

Carica papaya L. (lechosa, paw paw)

Taxonomical description

The plant is a giant herb reaching a height of 3–8 m and a stem diameter of about 20 cm. The hollow trunk covered by the very conspicuous scars of the abscised leaf petioles is unramified in the vegetative region. The leaf blade reaches a length and width of 30–80 cm and is deeply palmately lobed, while the petiole is about 0.5–1 m long. The plants tend to be dioecious with female and male flowers on individual plants, but hermaphrodite flowers also occur; reduction of sexes in the unisexual flowers may be observed in different degrees. The large yellowish or withish female flowers arise solitary in the leaf axils or form pauciflorous inflorescences. The male flowers, which are a creamy white or greenish colour, are smaller and occur in pendulous paniculate and long-stalked inflorescences which reach a length of 0.5–1 m. The petals show contorted aestivation in the flower bud stage.

The ovoid fruit reaches a length of 0.6–1 m in some cultivars and a weight of up to 10 kg, but on average, the fruit has a length of 15–30 cm. Its shape is ovoid to pear-shaped or elongated and in some varieties cucumber-like. At maturity, the fruit adopts an orange colour or becomes green-yellowish or even red. The soft and sweet pulp is orange or red in some varieties contrasting with the numerous 5–7 mm long black seeds in parietal position. The sarcotesta is mucilaginous, while the sclerotesta shows many protuberances on its surface.

Origin and occurrence

Carica papaya does not exist in the wild. However, a large number of hybrids, varieties and types are known.

The species may have originated from Central America and the south of Mexico, but comes more probably from the mountainous regions (Andes) of Peru, Colombia and Ecuador, according to BRÜCHER (1989).

Anatomical description

Fruit. (Fig. 83, 84). The paw-paw is sometimes considered a pepo-like fruit, but the ovary is superior with marginal-parietal placentation. As compared with the fruit size, the pericarp remains relatively thin and often measures not more than 2.5–3 cm in the ripe fruit. A big hollow or cavity therefore extends in the fruit center. As seen in transverse section, the fruit has a pentagonal shape with a large vascular bundle in each edge corresponding to the median dorsal bundle of the respective carpel. Alternating with these bundles 5 other vascular bundles appear in the placentae, each of which represents a product of fusion of 2 marginal bundles pertaining to two adjacent carpels. Anatomy and development of fruit and seed have been described by ROTH & CLAUSNITZER (1972).

The ripe pericarp of the paw-paw consists mainly of parenchyma which is more or less differentiated into three regions. The peripheral layers are composed of much smaller cells than the rest of the parenchyma and are distinguished by their chloroplast content. Together with the outer fruit epidermis they give some support to the pericarp and do not show any larger intercellular spaces. The outer epidermis cells remain small and their outer walls are thickened.

The median zone of the pericarp is different from the peripheral layers, as it is composed of

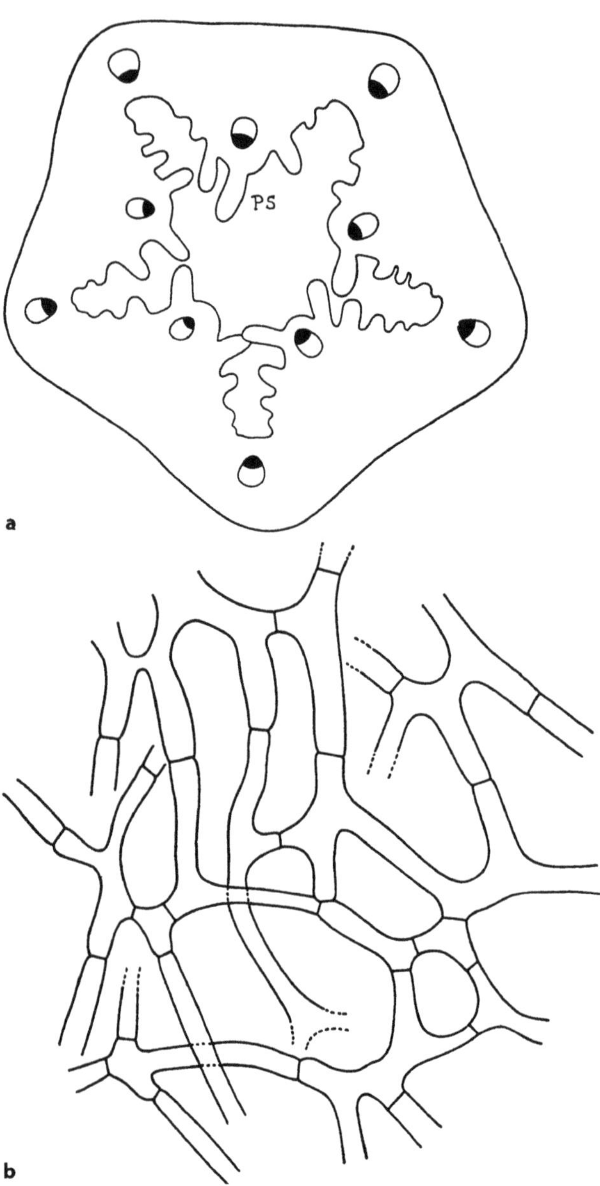

Fig. 83. *Carica papaya*, Caricaceae. T.s. of young fruit. **a** general view with dorsal (outer ring) and ventral (inner ring) of carpel bundles, seed primordia (PS) and a hollow in the center. **b** Spongy tissue of inner pericarp part of a ripe fruit (ROTH 1977).

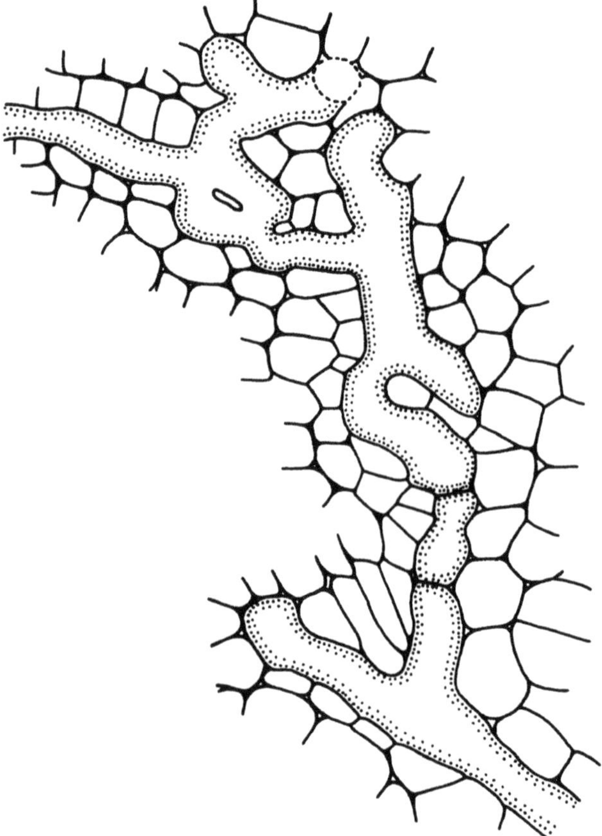

Fig. 84. *Carica papaya.* Cross section through pericarp of young fruit with laticifers (ROTH 1977.

much larger cells of roundish shape, leaving intercellular spaces between one another. Many orange-coloured elongate or needle-shaped chromoplasts in the cells add the characteristic colour to the fruit flesh.

In the innermost region of the pericarp however, the cells are elongated tangentially and develop large arms meeting with those of adjacent cells. A spongy parenchyma with large intercellular spaces arises in this way which almost resembles an aerenchyma. The laticifers which represent a characteristic feature of the flesh develop early in the young fruit and are of the articulated-anastomosing type. their latex contains the well-known proteolytic enzyme papain; its action is comparable to that of pepsin in the animal stomach. The latex, i.e. the papain, is therefore used for a variety of purposes.

In the ripe fruit, the seeds are embedded in a mucilaginous mass derived from the pluristratified outer epidermius of the outer integuments of the seeds.

The laticifers often develop in the vicinity of the vascular bundles and seem to be related to them. They are at first septate by transverse walls which are absorbed later during development. In the mature stage, the laticifers correspond therefore to series of superposed fused cells.

We refer to ROTH AND CLAUSNITZER (1972) and ROTH (1977) for information on the fruit development.

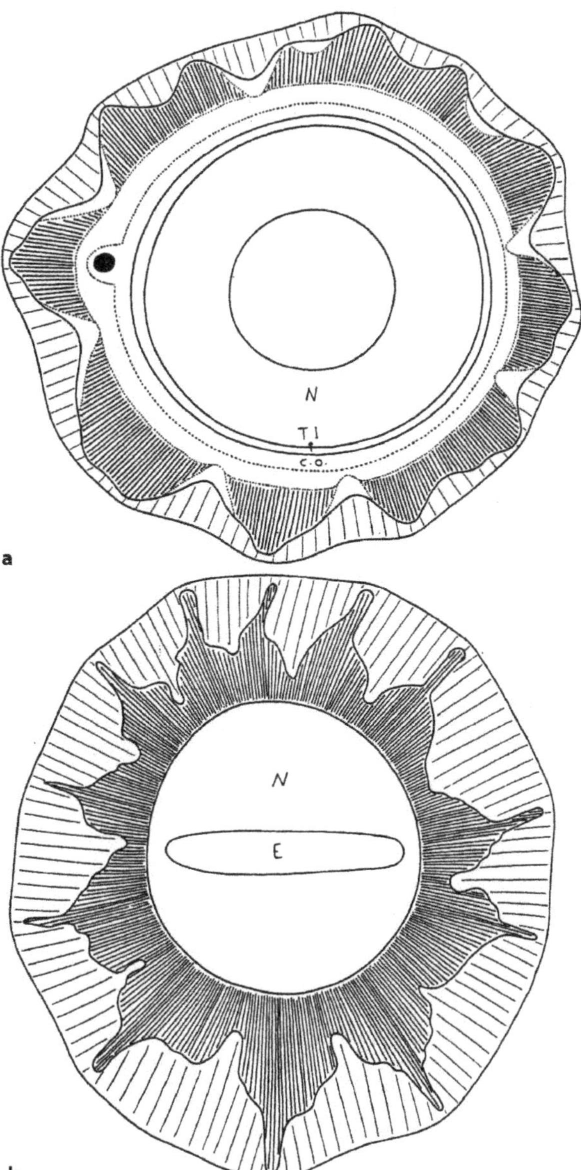

Fig. 85. *Carica papaya*, seed. **a** Transverse section of a seed measuring 2.5 mm in diameter. **b** Seed, 6 mm in diameter. Sarcotesta = hatching at large distances. Endotesta = hatching at small distances. N = Nucellus. E = Embryo. C.O. = Dark layer which retains much air bubbles. TI = Inner integuement.

Seeds. (Figs. 85–88c). The seed primordia develop in centrifugal direction from the vicinity of the marginal bundle towards the dorsal bundle of the carpel (Fig. 83). In their major part, they originate from the epidermal and 2 subepidermal layers of the fruit inside. The outer integument develops before the inner one. At first, both integuments comprise about 5–6 layers including both epidermal layers. The inner integument hardly augments the number of laysers and comprises 6–8 (10) layers in the mature stage. The additional layers originate by periclinal cell divisions.

The outer integument, on the other hand, very much increases in thickness developing a great number of layers. First, the cells of the outer epidermis increase in size, enlarging anticlinally to the surface. However, not all cells enlarge in the same way, but some enlarge more than others. To compensate for this unequal growth, a subepidermal meristem develops below the outer epidermis, which is more active where the epidermis cells enlarge less anticlinally. In this way, depressions are formed in the subepidermal tissue in which the epidermis cells enlarge more so that alternating with the depressions, subepidermal protuberances arise, on top of which the epidermis cells remain shorter (Fig. 86).

The outer epidermis cells do not only enlarge, but also repeatedly divide periclinally forming 2–4 periclinal walls causing anticlinal series of cells to develop. Above the subepidermal protuberances however, the epidermis remains uniseriate. The tissue of the depressions could therefore be called an epidermal meristem.

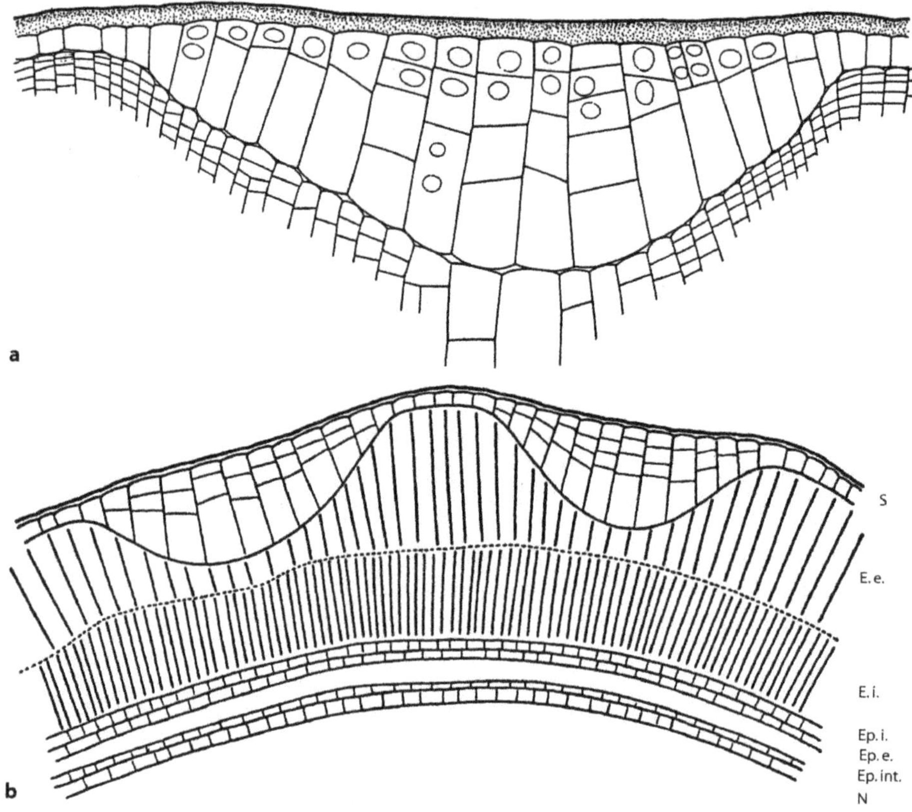

Fig. 86. *Carica papaya.* **a** Sarotesta of a small seed showing the enlargement of the epidermal cells and the formation of periclinal walls. Below the epidermis, there appears the abaxial meristem. **b** General view of the same seed (2 mm in diameter). S = Sarcotesta. E. e. = Outer part of the endotesta. E. i. = Inner part of the endotesta. Ep. i. = Inner epidermis of the outer integument. Ep. e. = Outer epidermis of the inner integument. Ep. int. = Inner epidermis of the inner integument. N = Nucellus.

To compensate this meristematic activity of the epidermis, a subepidermal meristem is developed above the future protuberances. It is the first subepidermal layer below the outer epidermis which starts to divide periclinally, so that anticlinal cell rows arise. As the epidermal cells enlarge considerably in the depressions, long anticlinal rows of subepidermal cells may be observed in the protuberances. In the depressions, on the other hand, only little meristematic activity can be seen in the first subepidermal layers; here, the cells increase in size and only divide periclinally a few times.

Besides the epidermal and the subepidermal meristem a third type of meristem starts its activities at the same time. It arises in the first subepidermal layer below the inner epidermis. The cells which remain small and are rich in protoplasm maintain their meristematic state from the beginning on. In the inner part of the outer integument, anticlinal or radial rows of cells are therefore also formed, but the cells are distinguished by their smaller size. The inner epidermis of the outer integument remains uniseriate and is only distinguished by the presence of solitary crystals of calcium oxalate in each cell.

The inner integument, on the contrary, suffers only few transformations. Outer and inner epidermis remain uniseriate, while the cells of the „mesophyll‘ enlarge tangentially. The inner epidermis develops a well delimited cuticle which forms the border line between inner integument and nucellus.

During the process of differentiation, the outer epidermis cells of the outer integument still enlarge considerably in the depressions. Their walls however, remain thin, only the outer most tangential walls become thicker and covered by a thin cuticle. The epidermis cells of the depressions are very turgescent being filled with cell sap, and represent the sarcotesta together with a few large cells below the depressions which arose by the activity of the subepidermal meristem.

The protuberances of the outer integument maintain the anticlinal cell rows for much time; differentiating in a centrifugal way. They enlarge in size, round off and leave small intercellullar spaces between one another. Later their walls become thickened and large simple pits develop.

Likewise the cells of the inner part of the outer integument become differentiated, but only slightly increase in size; particularly the innermost cells round off their walls, become thickened and pits develop in them. They can be distinguished from the outer part of the integument by their smaller size, even in the mature seed.

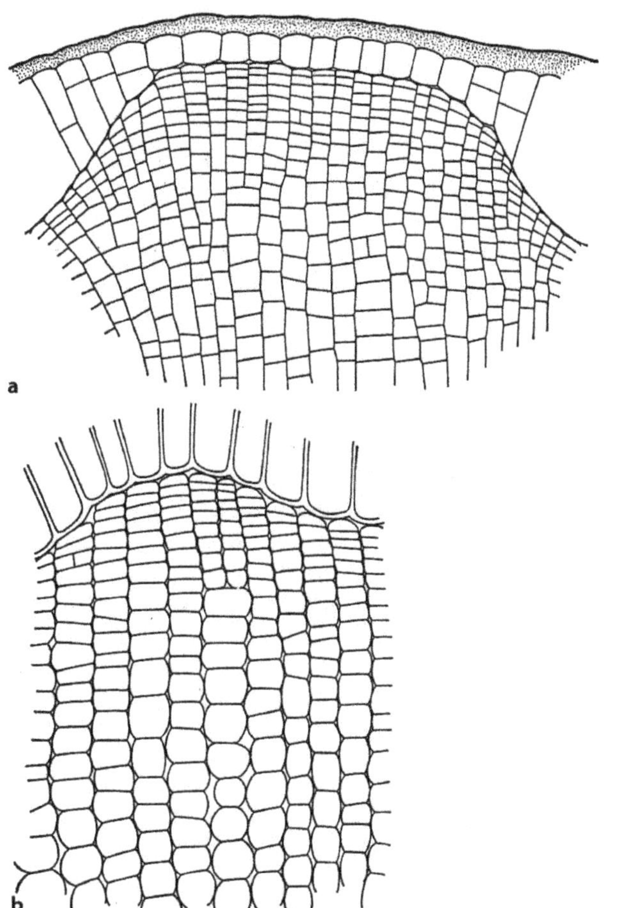

Fig. 87. Carica papaya. Subepidermal abaxial meristem of the outer integument. **a** Younger stage of a seed measuring 2 mm in diameter. **b** Slightly older stage of a seed measuring 2.5 mm in diameter; in this stage, the cells begin to differentiate, to round off so that intercellular spaces are left between them.

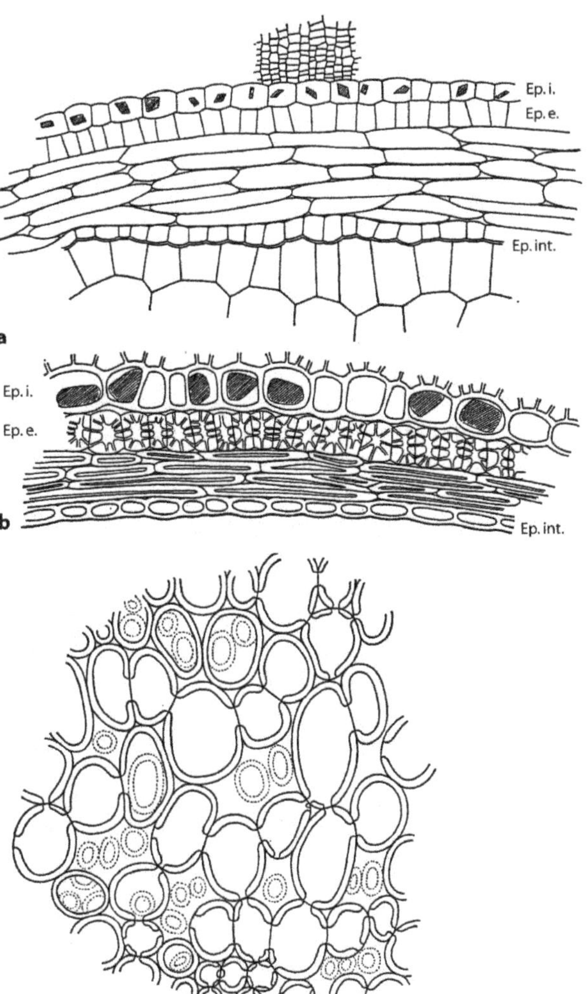

Fig. 88. Carica papaya, seed coat. **a** Inner part of the testa of a young seed, 2 mm in diameter, with subepidermal abaxial meristem and inner epidermis of the outer integument (Ep. i.), outer epidermis of the inner integument (Ep. e.), tangentially enlarged cells of the mesophyll, inner epidermis of the inner integument (Ep. int.), and nucellus. In the inner epidermis of the outer integument there are rhomboid crystals. **b** The same part of the testa but of a mature seed. In the inner epidermis of the outer integument, the crystals are still obvious. The outer epidermis of the inner integument is transformed into a layer of stone cells, and the tangentially enlarged cells adopted thickened walls. **c** Outer part of the endotesta of the mature seed composed of globular cells with thickened walls and large simple pits, as seen in a transversal section (dotted: in surface view).

The protuberances are first truncately rounded, but the tip of the protuberances becomes longer and longer, as some central cell rows continue to divide periclinally, while the others cease to divide so that spines arise, as seen in transverse section through the seed. The cells of the spine enlarge in a radial direction in their turn, so that the spines are finally pointed. The compact tissue of the protuberances together with the inner part of the outer integument form the so-called endotesta or sclerotesta. The inner epidermis of the outer integument can still be recognized in the mature seed by the solitary crystals which lie in each cell.

The inner integument is distinguished by its outer epidermis the cells of which transform into stone cells with few large pits. The walls of the tangentially enlarged cells of the mesophyll become only slightly thickened. The inner epidermis shows a regular aspect and forms a well-distinguished cuticle.

In the mature stage, the sarcotesta is composed of large radially extended cells; the cell size diminishes towards the periphery. The radial (anticlinical) cell rows are maintained.

The existence of the 3 different types of meristems in the outer integument was first described by ROTH & CLAUSNITZER (1972).

Ethnobotanical and general use

Nutritional use

The sweet ripe paw-paw fruit is predominantly eaten fresh, but is also prepared in the form of a juice, milk shake, ice cream or is candied. It makes an excellent additive in mueslis and fruit salads. It is also used in salads, sherbets, jellies, preserves, pickles, and pies. The green fruit is cooked as a vegetable.

The fruit however has a short durability and cannot be stored for a long time, but may be canned. It is rich in vitamin A and C, carbohydrates (sugar content: 10 %) and certain minerals.

The seeds are sometimes utilized as a condiment.

Even the shoot is edible as a vegetable.

Economical utilization

Besides a large variety of other uses, the plant is also grown for ornamental purposes.

The leaves were used by the natives as a substitute for soap.

The papain present in the white latex is useful for many purposes, e.g. for leather tanning, to soften fibers of wool and silk, to prepare cheese, chicle, soaps, hairshampoos and other cosmetics, to purify beer. In Africa and Asia, *Carica papaya* is used as a mosquitoe repellent. Papain is also commercially used as a meet tenderizer. The papain of the latex has even been recommended for dry cleaning and dyeing. Cheese is prepared with papain as a coagulating agent for milk.

The papaya cigarettes prepared from the leaves soothe the respiratory tract of humans instead of irritating it, as happens with tobacco. The leaves are processed by the same conventional methods, and it is maintained that the papaya cigarettes have the same taste, aroma, appearance and qualities as tobacco, but do not contain the noxious nicotine.

Medical use

Name of the drug: *Carica papaya* L. plantae, folia, radix, fructus, semen.

The plant principally has digestive, stomachic, vermifuge and vulnerary effects.

In general medicine, the plant is used for amoebiasis, asthma, certain types of tumor and cancer, elephantiasis, arthritis, headache, psoriasis, splenitis, ulcers. toothache, colitis, gastrointestinal problems, callus, warts, pectoral trouble, and scorpion stings. It is furthermore applied as a colagogum, emmenagogum, vermifuge, ecbolic (abortive) anthelmintic, as an insecticidal and ascaridicidal, and against ring worms. Generally, it could be shown that anthelmintics may also be antifertility agents. Extracts of various plant parts show effects on malaria and some viral infections. The growth of certain tumors is inhibited. the plant also has certain calmative effects.

Noxious effects were however also observed: the plant may produce allergic rhinitis, bronchial asthma and hypersensitive pneumonitis.

Leaf. Although the leaves are principally used as a tenderizer of meat – meat is wrapped up in leaves for a night – they are also of medical interest. They have anthelmintic properties, cure purulent wounds and asthma.

A decoction of leaves is applied as a purgative for horses and as an antimalarial remedy for humans. In former times, leaves were also used as a substitute for soap. Their tannin content is low. However, extracts of leaves have antimicrobial properties and act on gram-positive bacteria and mycobateria (e.g. *Staphylococcus aureus, Escherichia coli* etc.).

The papaya cigarettes are recommended as a healthy alternative to tobacco cigarettes. They do not affect the pulmonary tract since they do not contain nicotine, but they are equal in taste, aroma, appearance and quality to the cigarettes made of tobacco.

Stem (shoot). The shoot is locally applied against haemorrhoids and for diseases of the eyes.

Root. The root has a weak antibiotic effect, is rubefacient and vesicant.

A decoction of the root is used for haemorrhoids, an infusion for syphilis.

Extracts of the root act on gram-positive bacteria and mycobacteria.

Decoctions and infusions of the root of male individuals (!) are applied for oliguria, veneric diseases in general, flu, heat, and constipation.

Locally applied (e.g. in the form of an ointment) it cures eruptions on the skin and acne.

The root emanates an offensive smell which repels cats.

Flowers. An infusion of the flowers is applied as an emmenagogum, febrifuge and pectoral.

Fruit. The most important organ of the plant is the fruit, and the substance particularly wanted is the latex in the fruit. Besides, the fruit contains resins, vitamin A and C.

To obtain the latex, immature green fruits are tapped by longitudinal incisions; the exuding latex is left on the fruit to dry out and then the dried product is scraped off the fruit. All properties ascribed to the fruit are also found in the latex. The most important substance found in the latex is papain, a proteolytic enzyme.

The fruit has first of all digestive properties. Besides gastric disorders, urinary problems are also cured with papaya fruits. The fruit is furthermore considered to be antidiarrhoeic, laxative, anthelmintic, acting also as an emmenagogum. The juice of green fruits is furthermore utilized for the preparation of cosmetics, to remove pimples etc.

However, the fruit may provoke colics in newborn children and abortus in pregnant women.

Seed. The toxicity of the seed is low.

The seeds are said to be vermifuge and anthelmintic, green seeds are abortifacient and are applied as an emmenagogum.

A decoction is prepared of the dried and pulverized seeds to combat intestinal worms (*Ascaris, Taenia..*).

The seeds also have carminative properties and are used for irritations.

Extracts of seeds exercise antimicrobial activities on *Streptococcus aureus, Bacillus cereus, Escherichia coli* and other bacteria.

The aglucone of glucotropeoline in the seeds has bactericidal and fungicidal properties. This antibiotic activity may be used to cure certain intestinal and urinary infections.

Latex. All parts of the plant contain latex, but latex is most abundant in the green fruit and in the trunk. The white latex occurs in the laticifers.

The latex is caustic and therefore applied to remove callus. The purgative effect of the latex is used to expel intestinal parasites. Cleaning the skin with latex causes a burning feeling; likewise, the contact of open wounds with latex hurts. Latex is applied to disinfect and cure wounds. Infected wounds and furuncles are treated with latex as well as pimples and warts. The latex is furthermore employed against cancerous and lymphatic tumors. Latex is also helpful to cure infections such as flu or to lower hypertension.

A very interesting attribute of the latex is that it combats pollen allergies, due to the proteolytic action of the papain.

Besides proteolytic enzymes, there are also several alkaloids present in the latex, one of which is carpaine. The latex also contains an anticoagulant of blood, inhibiting the action of thrombin on the fibrinogen.

Latex in decoction soothes asthma and indigestion.

It is estimated that about 1 pound of dry latex is obtained from 8000 pounds of fruit. About 100 g dry latex are gathered from a fruit annually. One kg latex produces about 200 g crude papain.

Method of use

Fruits are used fresh or the latex is extracted by incisions. Papain is obtained from the latex and industrially purified. Digestive drugs exist on the market in the form of dragees.

Digestive effects of the fruits are best when these are eaten fresh.

Seeds are dried and pulverized and a decoction is prepared of them. A nose spray with papain is available against pollen allergies.

Healing properties

Papain is mainly a digestive, but it may also be applied wherever proteolytic enzymes are helpful.

Carpaine reduces blood pressure and has properties similar to those of *Digitalis.*

The latex is caustic and helps to remove callus, pimples and warts.

Chemical contents

Papain (synonym: papayotin) is a protease or proteolytic enzyme. The crude papain is purified by dissolution in water and precipitation with alcohol.

Papain alleviates gastrointestinal diseases, colitis, dyspepsia, and dissolves crusts from burns, and chronic purulent otitis. It eliminates exudates of tracheobronchitis, aids with diphteria, inflammations, infected wounds, acne, skin ulcers, psoriasis, cicatrization of wounds and furuncles.

A nasal spray with papain alleviates certain pollen allergies due to its protein denaturing capacity. Papain cures peritonial adhesions after abdominal surgery. Medical interest is being increasingly concentrated on proteolytic enzymes which are used to eliminate proteic material of haemorrhage (haematoma and bodily fluid). In certain types of arteriosclerosis abnormal deposits in the arteries are eliminated by papain digestion.

Papain is also the active principle in purifying beer, in tanning skins, treatment of woollen and silky fibers, in cheese production, manufacture of soap, cosmetics and chicles, in dry cleaning, dyeing and tenderizing of meat.

However, papain, chymopapain and other proteolytic enzymes of the plant do not exhibit antimicrobial activities.

A very important property of chymopapain is that of curing discal hernia.

Carpaine. Carpaine is an alkaloid present in all green parts of the papaya plant mainly in leaves and in seeds. It also occurs in other members of the family and in certain Apocynaceae.

Its activity is cardiocinetic and diuretic. Carpaine has pharmaceutical properties similar to those of *Digitalis.* It also has hypotensory effects and reduces the pulse frequency; it depresses the central nervious system. Although similar to Digitalis in its action carpaine has no noxious effects chemically not being related to digitalin or emetine, it is a heart stimulator and a diuretic. But it not only reduces the contractile force of the heart and lowers the blood pressure; but also exercises a relaxing effect on the uterus.

Furthermore, in low dilution, it inhibits the growth of *Mycobacterium tuberculosis.* Besides, it is amoebicid and comabts dysentery.

Tropeoline. The aglucone of glucotropeoline inhibits the growth of many gram-positive and gram-negative microorganisms, such as of *Escherichia coli, Staphylococcus aureus,* and of *Penicillium notatum* etc.

Adulterations

Papain is occassionally adulterated with starch, coco, gum arabic, or with resin from the brazil nut.

Varieties and related species

According to BRÜCHER (1989), there are several wild species which may have played a role in the phylogeny of the cultivar papaya, e.g. *Carica cauliflora* JACQ., *C. microcarpa* JACQ., *C. monoica* DESF., *C. parviflora* (A. DC) SOLMS, *C. pubescens* LENNE & KOCH, *C. stipulata* BADILLO, *C. quercifolia* SOLMS.

Cultivation

The plant is mainly propagated by seeds. As male and female individuals exist, only flower formation reveals the sex of the plant. This takes about 5–6 months. Whether crossed or self-fertilized, the hermaphrodite flowers produce 2/3 hermaphrodite individuals and 1/3 female individuals. When, on the other hand, a hermaphrodite flower is crossed with a female flower, 50 % of the resulting individuals will be hermaphrodite and the other half female.

Germination of seeds occurs 20–25 days after dissemination. 8–12 months after transplantation, the plants start to produce fruits and continue to do so throughout the year. 15–20 fruits are formed by an individual during a year. Cultivation is very easy when the plants receive enough water and have a well-drainined soil.

Caryocaraceae

There exist at least 6 useful species of the genus *Caryocar.*

The edible seeds are a source of oil. Likewise the fruit pulp is edible. The bark containing saponins is used to wash clothes and hair.

Saponins are responsible for the ichthyotoxicity and atticidal activity. Indians use the plants as an insecticide and repellent. *Caryocar nuciferum* L. has edible nuts which produce an oil. The wood is also useful.

Bark anatomy has been studied by ROTH 1977 and 1981, fruit morphology and dispersal also by ROTH 1987.

Celastraceae

Goupia glabra AUBL. Source of goupi wood. Leaves to dye skin and hair; for the eyes to treat cataracts.

ROTH studied the bark structure 1981, leaf structure 1984, fruit structure and dispersal 1987.

Maytenus

Useful wood. Foliage with bitter principle and tannin, used for intestinal ailments, as a stomachic, cicatrizant, aperitive, analgesic, tonic, astringent. Leaves are febrifuge. Buds for blenorrhagia. Roots for wounds and as an aphrodisiac. Bark is stimulant, for rheumatic pains and arthritis. Aril of the seed contains caffeine and is used as a diuretic.

ROTH studied the bark structure 1981, leaf structure 1984, fruit structure and dispersal 1987.

Chenopodiaceae

The Chenopodiaceae, which are considered a derived family, are mostly herbs and more seldom shrubs or small trees. The leaves are often succulent or even reduced; the inconspicuous flowers are united in cymose inflorescences. Many representatives are ruderal plants, grow in steppes and deserts or along beaches, being adapted to dry and salty habitats (halophytes).

The genus *Chenopodium* comprises more than 250 species mostly of temperate regions.

Chenopodium

The genus Chenopodium comprises more than 250 species, mostly considered weeds; some of them, however, are called pseudo-cereals because of their special use as a cereal substitute, particularly in the Andean regions and in Mexico, where they are also cultivated.

Chenopodium ambrosioides L. (pasote, pazote, paíco, apazote, yerba santa, yerba sagrada, hormiguera)

Taxonomical description

The species (Fig. 89), an annual or perennial herb with a taproot, is about 50–150 cm high, simple or ramified, with erect or decumbent branches. The plant is distinguished by a very strong garlic-like or camphor-like smell. The branches are cylindric and of reddish colour. The alternating entire leaves are sessile or subsessile and glabrous. The blades are fleshy, lanceolate or oblong-lanceolate, 3–8 (10) cm long and 0.8–2 (5) cm broad, with an acute apex and a decurrent base. Leaf shape and leaf margins are somewhat variable and either sinuate-dentate, crenate or entire. The lower leaves (nomophylls) have mostly sinuate margins, while the uppermost leaves (hypsophylls) are much smaller and have entire margins. The lower leaf side is very glandular. The venation is pinnate, mingled craspedodromous, with 8–10 secondary nerves; the middle nerve is prominent on the lower side, but flat on the upper side. The petiole is 0.1–0.4 cm long and plane-convex.

The inconspicuous inflorescences are axillary or terminal, and 1.5–6 cm long. The very small flowers which are of greenish colour are assembled in panicles; they are hermaphrodite or unisexual; the perigone is 1 mm long and has 5 members. The hermaphrodite flowers have 5 stamens and an ovary with 2–3 short stigmas; the female flowers, on the contrary, have 3 long stigmas; the ovary is supplied

Fig. 89. *Chenopodium ambrosioides.* **a, b** Habitus.

with glands. The male flowers which occur more rarely, have 5 stamens. The spheric fruit reaches less than 0,1 cm in diameter. The only seed is black-brilliant and has a diameter of about 0.7 cm.

Origin

South America (Andean region) and Mexico.

Historical background

In the 'Historia del Nuevo Mundo' Cobo (1654) refers to the use of *Chenopodium ambrosioides* (apasote) as a very important medicinal plant the leaves of which were used as a plaster against any type of tumor. However, the species had many other medicinal applications and also served as a food to South and North American Indians.

Occurrence

The plant is now almost a cosmopolitan weed. In Venezuela it is frequently found in uncultivated regions or in disturbed areas near housing.

The plant occurs from 0 to 2760 m a.s.l. STEYERMARK & HUBER (1978) found it at the Avila between 900 and 1000 m.

Anatomical description

Leaf. (Fig. 90a, b, c; 91c) The leaf is dorsiventral and amphistomatic. The upper epidermis consists of thin-walled cells which have strongly convex outer walls. As seen in surface view, the anticlinal walls are slightly curved. The palisade parenchyma comprises anticlinal rows of 2–4 cells. The spongy parenchyma, which is about the same size as the palisade parenchym, is looser and shows intercellular spaces. The lower epidermis is somewhat smaller-celled and slightly papillose. the stomata are elevated above the surface in both, the upper and the lower epidermal layer. Clustered crystals (druses) are found in large spheric cells of the mesophyll; additionally, cells with crystal sand are present.

The midrib shows a ring or flattened ellipsis of vascular bundles with a larger bundle on the upper side and an arc of 3–4 bundles on the lower side (with endoscopic xylem). In the central part of the midrib, the palisade parenchyma is interrupted and substituted by a weak collenchyma. A slightly stronger collenchyma is found on the lower side. The veins of secondary order only contain one vascular bundle which is surrounded by a parenchymatous sheath. The palisade parenchyma is also interrupted here. The veins of higher order are reduced, but have a parenchymatous sheath and occur very frequently.

The stomata have no subsidiary cells and occur more frequently on the lower than on the upper side. The anticlinal walls of the lower epidermis cells are more undulated than those of the upper side.

Hairs (Fig. 90c; 92) are very characteristic of the species and represent a valuable diagnostic feature. Three different types of hair may be distinguished: hairs with very large balloon-like heads (vesicular heads); hairs with an asymmetric head so that one arm is very elongated and possibly rolled inwards; and hairs with a head composed of several storeys. The hairs occur on the upper as well as on the lower leaf side, but are more numerous on the lower side.

Axis. (Fig. 91a, b) The ribbed axis (with about 10 ribs) is herbaceous and has a small-celled epidermis. The many angles are collenchymatous, while the small photosynthetic tissued lies in the 'valleys'. The stomata are elevated above the surface and the hairs are of the same type as in the leaf. As seen in transverse section, the collenchyma is angular. The cell size of the cortex increases towards the inside. Crystal sand may be observed in the largest innermost cortex cells. Below follows an interrupted ring of fibers (with a large lumen) only one cell layer thick. The phloem cylinder is very small and is only 2–4 cell laysers in thickness. The xylem forms a continous cylinder in which few islands of phloem are included; the intraxylary phloem is associated with large vessels. Medullary bundles are in direct contact with the inside of the xylem cylinder. The pith is very extensive and its cell size increases towards the center of the pith. Crystal sand is also found in the parenchymatous pith cells. This type of axial structure is very characteristic of the Chenopodiaceae. Medullary rays are absent. The stem studied had a diameter of about 5 mm.

Ethnobotanical and general use

Nutritional use

The infrutescences are edible. The plant is already mentioned in prehispanic times in the texts of náhuatl as 'epázotl', an expression derived from the word épatl which means skunk; the speculative association between plant and animal is the bad smell which both emit.

The infrutescences are eaten raw or cooked. The 'epazote del zorillo' is still used as a condiment in Mexican cuisine: The young and tender leaves of

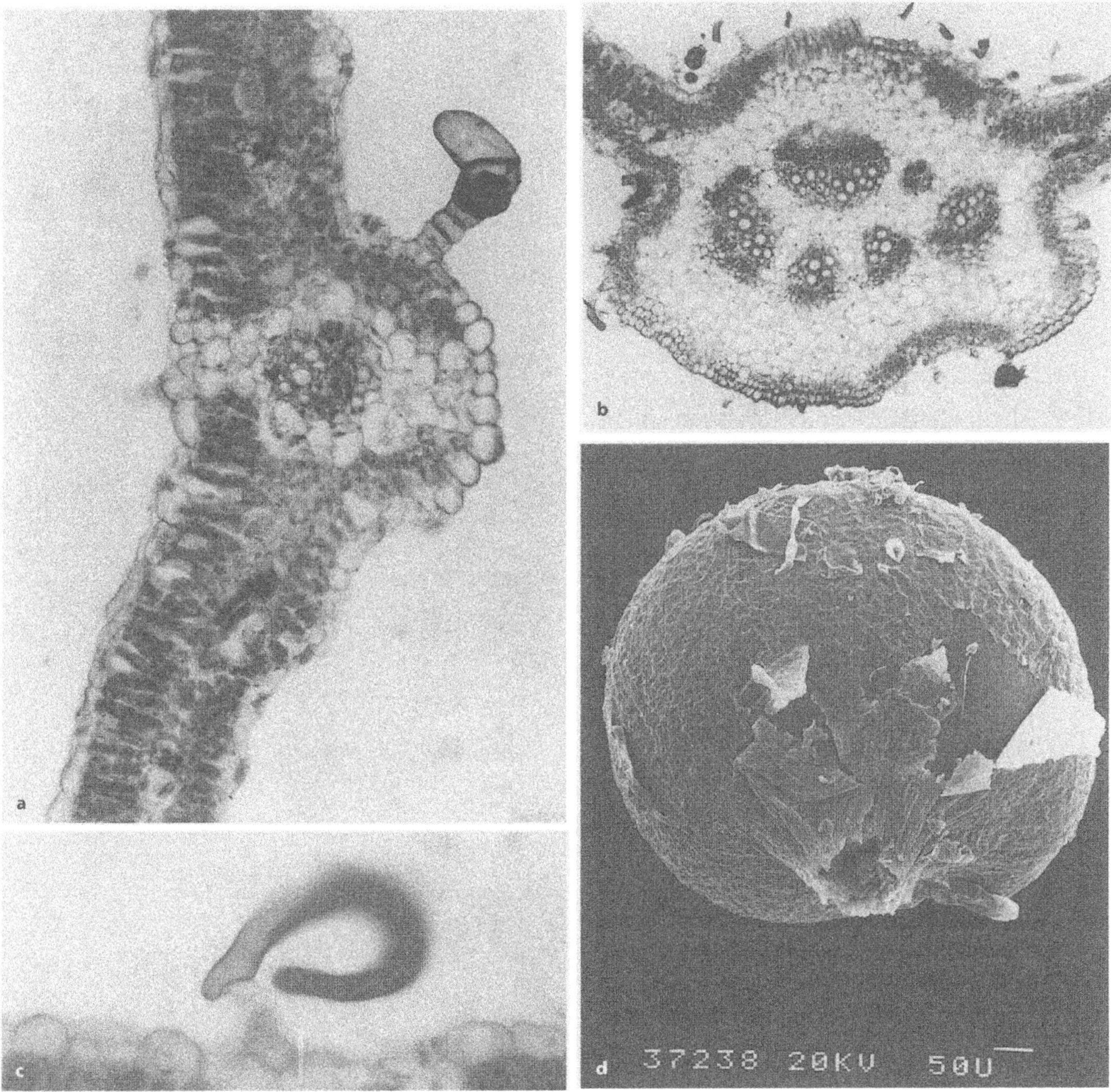

Fig. 90. *Chenopodium ambrosioides.* **a** T.s. of leaf with siderib (× 16). **b** Midrib. × 6.3. **c** hair. **d** Seed.

the plant are added to the meal at the end of the preparation. The species is esteemed for its nutritional and curative properties. In mediaeval times, the species was therefore widely planted and cultivated in Mexico (FRANCISCO HERNÁNDEZ 1571).

Medical use

All parts of the plant are used: Leaves, stem, root, flowers, fruits, seeds and the entire plant.

Leaf. The leaves in infusion are used as a tonic and stomach remedy or even against chorea. In latinoamerican pharmacopoeia, limonene was long ago obtained by distillation of leaves and sold as an oil rich in ascaridole (HERRERA 1921). Leaves applied as a plaster heal any kind of tumor.

Root. According to FRANCISCO HERNÁNDEZ (1571) the cooked roots were utilized by the Indians to cure dysentery or inflammations and were applied as a vermifuge.

Flowers. In Brazil, the flowers are used as a vermifuge.

Fruits. In Mexico, there exists an anthelmintic variety which is superior to the traditional species; in this case, the fruits are used as a vermifuge.

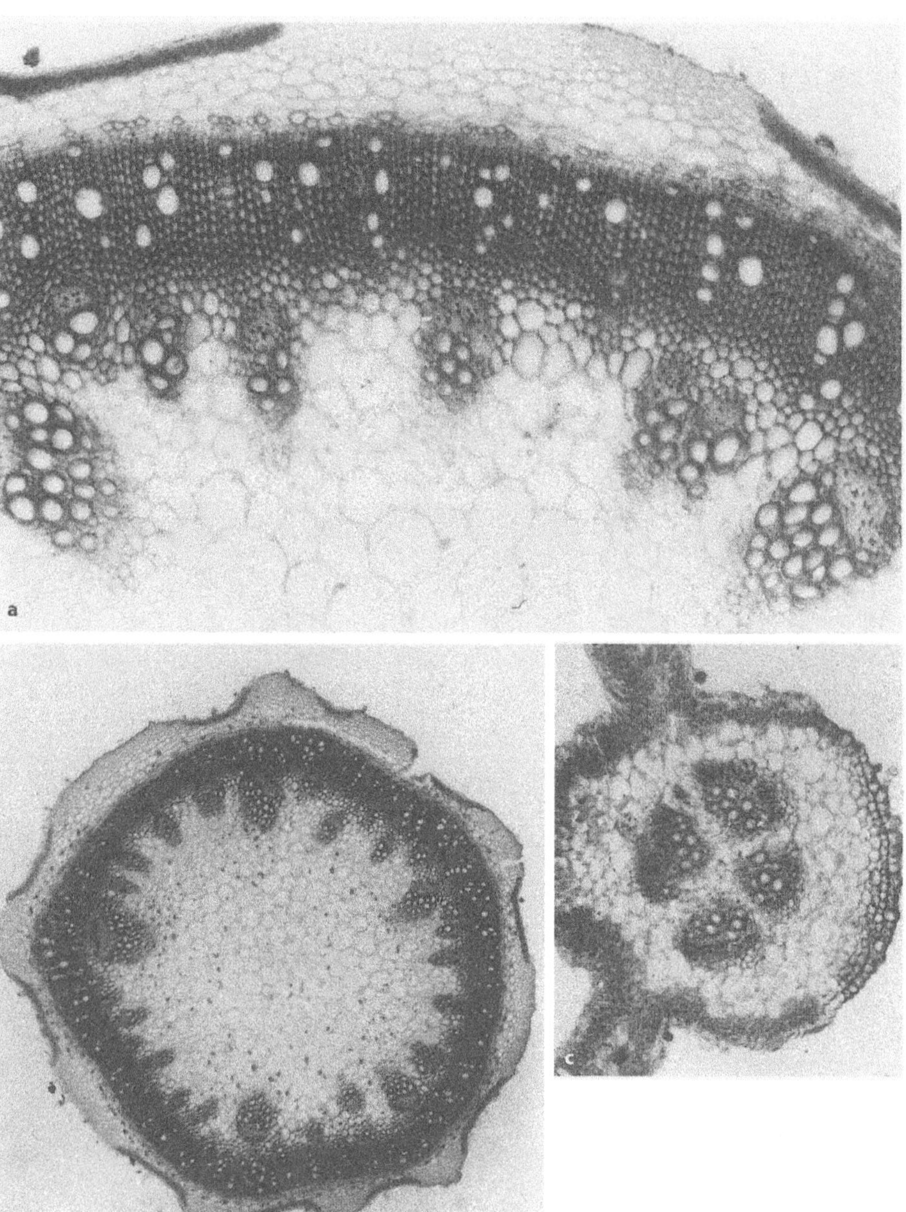

Fig. 91. *Chenopodium ambrosioides.* **a, b** axis in t.s. **c** t.s. of midrib (6.3).

Entire plant. In many cases, no distinction is made between the different plant parts; the entire plant is used as a tonic, a stomachic, a vermifuge, against indigestion, fatigue, palpitations, dyspnoea, dysentery, cough, asthma, as well as a postpartum depurant, and to cure sores.

A decoction of the plant with much salt produces detumescence of podagra. Further applications are against headache and toothache. The plant is also said to be sudorific, diuretic, febrifuge, and emmenagogic. It is used against irritations of the skin, against kidney problems, colics, stomachache, haemorrhoids, high blood pressure, inflammations, stings of poisonous insects and spiders, diarrhoea, and meteorism. It is applied as an antirheumatic, anthelmintic, amoebicide, antispasmodic, antiasth matic, and as cough-easing. It is said to be eupeptic, diaphoretic, carminative, antipalduic, aperitive, and abortive.

Method of use

Usually the entire plant is either applied orally or externally. A decoction of the entire fresh plant is taken orally.

In the case of haemorrhoids and stings of poisonous insects, the irritated area is washed with the decoction of the branches. A decoction of the roots, branches and leaves orally taken was a very effec-

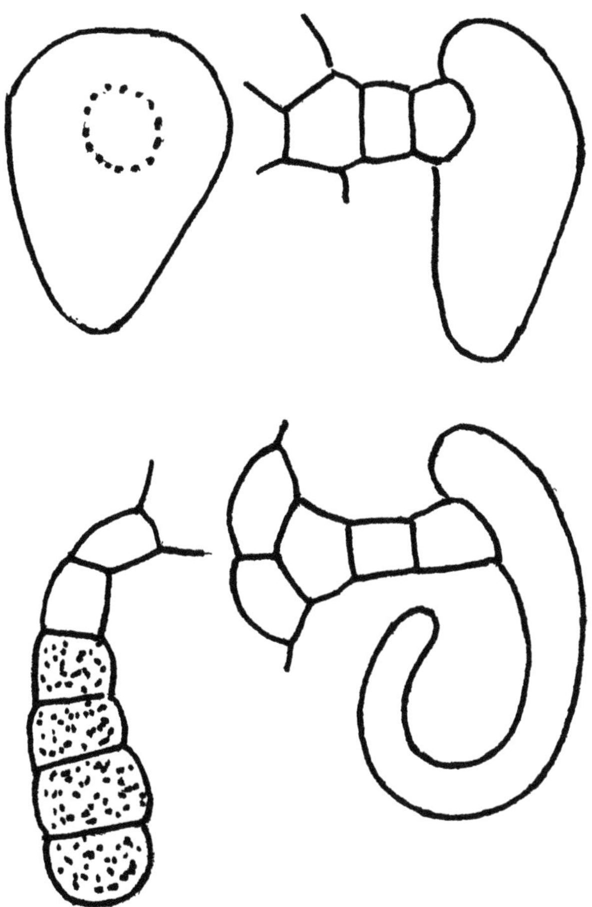
Fig. 92. *Chenopodium ambrosioides*. Different types of hairs.

tive remedy against dysentery, asthma, intestinal inconveniences, toothache and a vermifuge during colonial times. A decoction of the plant taken together with 'atole', a dish prepared with corn meal, sugar and water, controls headache, flu and rheumatism. In Uruguay, the infusion of branches, leaves, fruits and seeds is applied as an eupeptic, diaphoretic, carminative, emmenagogum, anthelmintic, antipaludic, febrifuge and against bronchitis.

For anthelmintic effects, a decoction of the entire plant or of the sap of the leaves is prepared to which milk or water is added; the remedy is preferentially taken on an empty stomach. As antiparasitic, the oil extracted from the seeds is taken, followed afterwards by ingestion of oil of *Ricinus*. In Trinidad and Tobago, a teacup of an infusion of 50 g plant in water is used as a remedy for the treatment of intestinal worms in children. This must be followed by a purgative. Something sweet is often taken before, to stimulate the worms prior to taking the decoction mixed with garlic (*Allium sativum*) and parsley (*Portulacca oleracea*).

Externally, the plant is used in the form of a sitz bath to cure haemorrhoids, rheumatism and as a resolutive.

To wash out wounds, a decoction of the leaves together with leaves of *Carica papaya* is applied.

Healing properties

The antiparasitic property of ascaridole could be widely demonstrated. The drug is very toxic to *Ascaris* and *Ancylostoma*. Pharmacological studies were carried out to prove the curative properties of *Chenopodium ambrosioides* related to ulcer, malaria, muscular relaxation, as a respiratory stimulant, and against cardiac debilitation, as well as its antifungal and antibacterial (*Staphylococcus aureus* and *Pseudomonas aeruginosa*) effects. In all experiments, the studied activities could be confirmed. The lethal dose of ascaridole in rats is 0.075 mg/kg.

The oil is however also toxic to humans and can produce nausea, vomiting, head ache, nervous depressions, injuries of the liver and the kidneys, visual disturbanmces, deafness, cardiac and respiratory problems, such as tachycardia and complete respiratory break down. Very high doses can even cause death. A carcinogeneous action has been observed in rats.

The lack of standardization of the drug still makes dosage difficult, due to the proportional variations of the active principle in the plant and to the existence of numerous varieties of the species.

Chemical contents

The plant is rich in essential oils. One of the components is ascaridol which exercises the antiparasitic effects on man and animals (dogs). The highest content of this oil is found in the seeds. Further contents are limonene, camphor, ambroside, Chenopodium saponin A, chenopodioside, the rhamnoside kaempferol, santonin and others (GUPTA 1995).

The plant has a very penetrant smell of camphor and a bitter taste. 1 % of the essential oil is found in the fruit and seed, and only 0.4 % in the leaf.

Varieties and related species

As already mentioned above, there exists a large array of varieties of the species in which the proportion of the active principle varies.

Related pseudo-cereals are *Chenopodium quinoa* WILLD., *Chenopodium pallidicaule* AELLEN and *Chenopodium nuttaliae* STAFFORD. They are alimentary and medicinal plants. Many varieties of these species likewise exist.

Cultivation

The plants are easily propagated by seeds. The plant is very resistant to drought and even grows at high altitudes.

Observations

The species is easily recognized by its anatomical peculiarities e.g. the absence of medullary rays in the xylem and the characteristic hairs.

Chrysobalanaceae (Rosales)

The Chrysobalanaceae are usually trees or shrubs with small inconspicuous flowers aggregated in inflorescences and with stipules on the leaves. They are of pantropical origin, but occur mostly in South America and in the Amazon region, where they comprise one of the dominant plant families. Characteristically, they have siliceous deposits and silicified cell walls. Tannins are also frequently found.

Chrysobalanus icaco L.
(icaco, hicaco, gicaco, jucaquillo, jicaquillo, jicacillo, and coco-plum)

Taxonomical description

Chrysobalanus icaco is a shrub or treelet, 5–8 m high, with a short stem and a delicate reddish-brown bark with white spots. The simple glabrous leaves which are elliptic to obovate, 3–12 cm long and 2.5–7 cm broad have an obtuse to emarginate tip and an obtuse to attenuated base. The small white flowers are arranged in cymes (of 3–10 flowers). The campanulate calyx is 3.5 mm long and silky pubescent; its 5 lobules are short, broad, obtuse and as long as the tube. The 5 obovate spathulate petals are 5 mm long. There are 15 or more stamens. The glabrous fruit is a rose-coloured or purple drupe of spheric or ovoid shape, slightly ribbed on the outside and 2–4 cm long. The thick white pulp tastes slightly sweet and astringent and surrounds a single seed, which has pentagonal outlines; it is covered by a fibrous coat.

Origin

The species is indigenous to tropical America, the Caribbean region and the South of Florida.

Historical background

The species was first mentioned in 1760 by CAULÍN.

Occurrence

The plant occurs in tropical America and in Venezuela, it is sometimes cultivated for its edible pulp. It prefers a hot climate and also occurs in the vicinity of the sea. It is normally cultivated along the northern coast line of Venezuela, but also appears spontaneously.

Anatomical description

Axis. (Fig. 93) A twig, about 5 mm in diameter, shows a well-developed superficial cork, several laysers in thickness, with lenticels. The parenchymatous primary cortex is of regular size; its cells enlarge towards the inside and partly extend in a tangential direction. Cells with silicified walls and a siliceous content are irregularly dispersed over the section and cells which contain clustered crystals are also found here and there. Some cells, probably tanniferous, stain intensely with artificial dyes.

The pericycel is composed of fibers and stone cells with U-shaped wall thickenings, a characteristic of the Chrysobalanoideae. U-shaped stone cells also occur in the phloem and in the uniseriate rays. The fibers of the hardbast are scarce and irregularly dispersed. The phloem is poorly developed in comparison with the xylem.

The comparatively small vessels are solitary and irregularly distributed over the section. The rays are uniseriate. The axial paerenchyma occurs in the form of small tangential bands, usually one, 2–3 sometimes, cell layers in thickness and is apotracheal. However, the tangential bands are very numerous.

The ample pith is composed of cells with thickened walls many of which contain large clustered crystals, probably of calcium oxalate of lime. Tannin cells which stain darkly are dispersed in the pith in a similar way as in the cortex.

Fruit. (Fig. 94, 95) The epidermis is composed of cells of regular appearance. Beneath follow 2–5

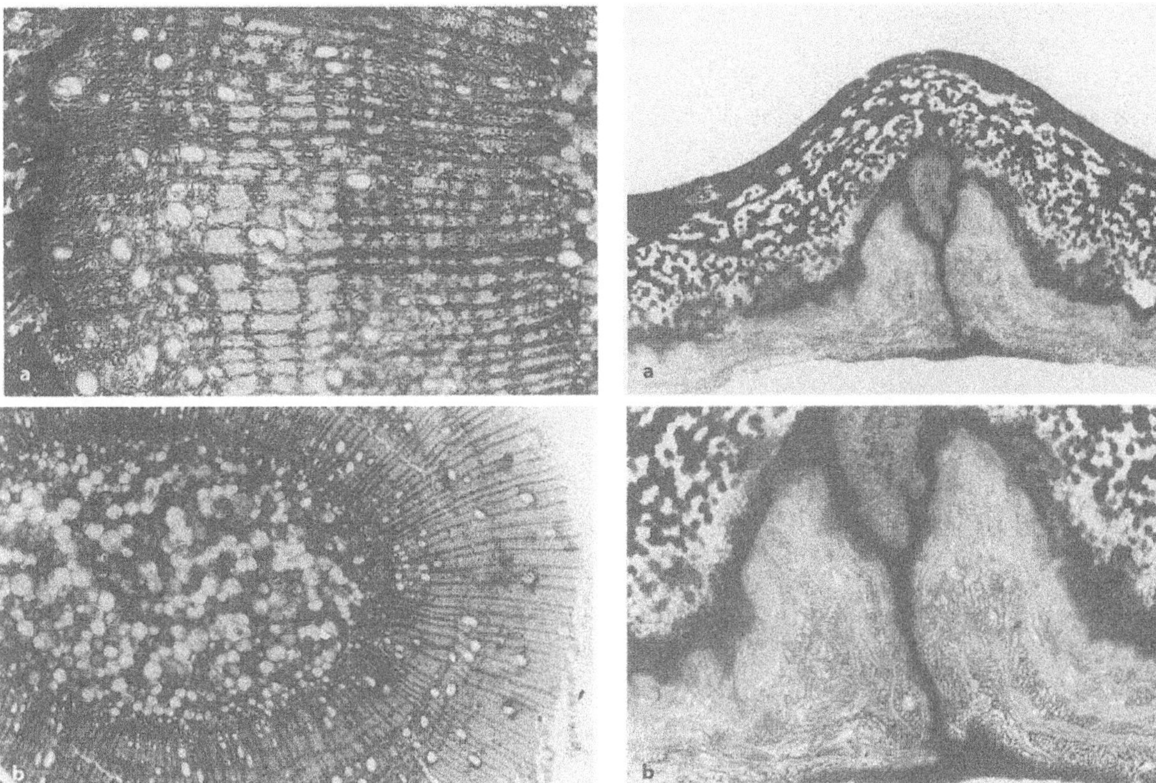

Fig. 93. *Chrysobalanus icaco*. **a** Xylem of the axis in t.s. **b** Pith of the axis in t.s. (both × 6.3).

Fig. 94. *Chrysobalanus icaco*. **a, b** Fruit in t.s. Note the fibrous endocarp and the intensely staining cells (tannin cells?) of the mesocarp.

layers of smaller cells, which stain intensely with artificial dyes; they probably correspond to tannin cells. Together with the epidermis they form a compact tissue that corresponds to the exocarp in a wider sense (see also ROTH 1977). Clustered crystals are conspicuous in aggregations of epidermal and subepidermal cells which are slightly elevated above the surface.

Towards the inside, the cells enlarge more and more. The cells are rounded off, leaving small intercellular spaces between one another. The innermost parenchyma cells of the mesocarp (about 3–5 layers) are small and of globular shape. All parenchyma cells are thin-walled. The vascular bundles surrounded by a strong fibrous sheath lie near the endocarp. Intensely staining cells of the mesocarp which are dispersed over the section may correspond to tanniferous cells.

The endocarp consists principally of fibers arranged in the form of fascicles which cross one another in different directions becoming interwoven as in textile fabric. Moving the micrometric screw of the microscope up and down, the picture changes and seems to move as on a movie card. The border line between meso and endocarp is irregular in such a way that the mesocarp is protruding into the endocarp at places, and vice versa. At certain distances, the parenchyma of the mesocarp interrupts the endocarp in a radial direction uniting with an inner layer of small-celled parenchyma which lines the inner side of the endocarp. The fibers of the endocarp are comparatively short and have a large cell lumen (compare the endocarp of coffee: p. 391, 392 in ROTH 1977).

The several ribs (5–6 or more) on the outside of the fruit are due to the protrusion of the endocarp into the mesocarp.

Ethnobotanical and general use

Nutritional use

The fruit is edible. However it is seldom eaten raw, but cooked as a preseve. The seed which is rich in fatty acids is likewise edible. The flavour of the preserve is increased, when the seed coat is opened or removed before cooking so that the taste of the 'nut' (cotyledons) can spread over the compote. When being cooked, the fruit and the syrup adopt a reddish-purple colour. In Venezuela, the 'preserve of Maracaibo' is famous. When the fruit is well cooked, the seed softens and becomes edible like an

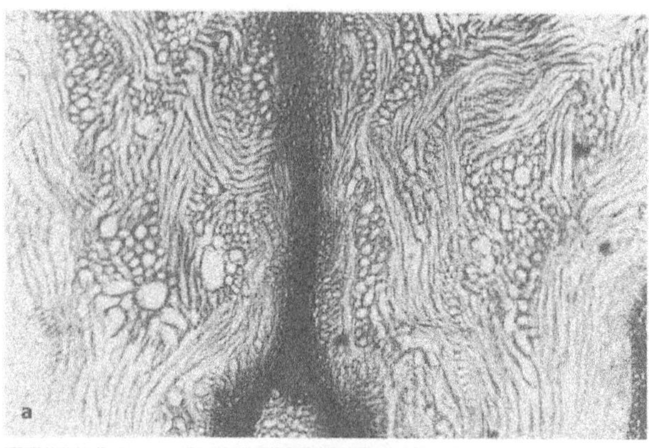

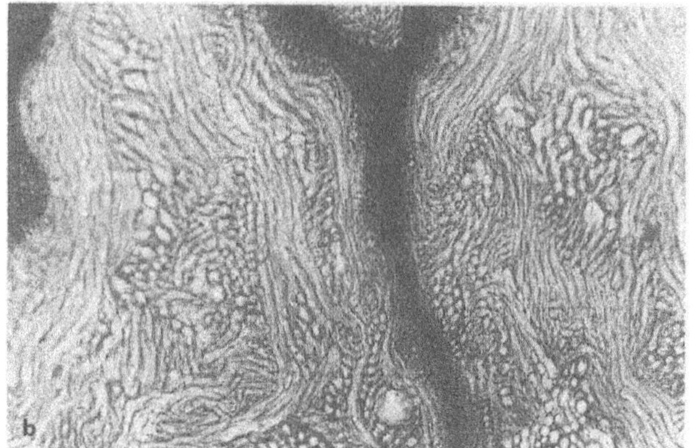

Fig. 95. *Chrysobalanus icaco.* **a, b** Fruit endocarp. Note the interwoven fibers (20 × 6.3).

'almond'. The fruit is usually cooked in syrup, and a jelly can also be prepared of it.

Economical utilization

The plant is cultivated as an ornamental tree.

The wood is very hard and heavy and is only seldom used, as the stem is very short.

Medicinal use

The fruit contains vitamins of the B and C group. It is rich in tannins leading to an astringent taste and therefore used in popular medicine as an antidiarrhoeic and antidysenteric. Fruits, bark and root are astringent, hypoglycaemic, and are applied against the acquired immune deficiency syndrome (AIDS).

Varieties

A variety has been obtained by cultivation which is smaller, has purple to blackish fruits, smaller leaves and is known as *Chrysobalanus icaco* var. *pellocarpus* (G.F.W. MEY) DC.

Cultivation

The plant is propagated by seeds. However, it shows a slow growth. It prefers sandy soils, a hot climate along the coast line or along river beds or where some humidity is available. This observation of HOYOS (1989) is astonishing, as the tree has a somewhat xerophytic appearance and possesses coriaceous leaves. However, it is able to keep the leaves, the flowers and the fruits for most of the year.

Observations

The species is easy to recognize by its general habit. Anatomical characteristics which facilitate identification are the intensely staining tannin cells, the peculiar clustered crystals, the silicified walls of silicia cells, the U-shaped wall thickenings of stone cells in the axis, the small tangential bands of axial parenchyma in the wood and the uniseriate rays.

Couepia

At least 7 useful species are known.

In this genus edible fruits and seeds occur. Seed oil is manufactured into inks and varnishes. Astringent fruits are used to cure sore gums. Bark is used as a hardener for pottery. Astringent bark is used medically for cough. Seeds are applied for dysentery.

C. glandulosa MIQUEL has edible fruits with an astringent agent which are used to cure sore gums. A fermented drink is made of the seeds. Seeds are also used for indigestion.

ROTH studied the bark structure 1981, leaf structure 1984, fruit structure and dispersal 1987, leaf venation 1996.

Hirtella

At least 9–10 useful species are known.

Leaves are used for sore throats and sores of the head.

Astringent bark for sores of the mouth and for gargles.

Ashes of branches for pottery. Tea of roots and stem is abortifacient. Roasted seeds for digestive problems. Seed oil for respiratory problems.

H. racemosa LAMARCK. Oil of the flowers for earache. Tea of flowers for the throat.

H. elongata MARTIUS & ZUCC. Bark is used for diarrhoea (CASTILLO 1995).

Licania

At least 20 useful species are known.

Useful wood. Bark for toothache, astringent bark for bleeding gums. Leaves for infections on the feet. Tea of leaves for digestion after an attack of malaria. Bark ashes for pottery.

Edible fruits. Fruits as a stomach tonic and after malarial attacks. Seed oil for inks and varnishes and for fungal infections of the ear. Seed oil for care of the hair and as protective coating. Seed oil for manufacture of candles, soap and grease.

L. apetala FRITSCH. Ashes of the bark are used for pottery, seed oil for sores on the skin. ROTH studied the leaf structure 1984, bark structure 1981, fruit structure and dispersal 1987.

L. hypoleuca BENTH. Seed oil for festering wounds. ROTH studied the bark structure 1981, leaf structure 1984, fruit structure and dispersal 1987, leaf venation 1996.

Parinari

At least 12 useful species are known.

Wood is useful. Bark and leaves as a mouth wash and for inflamed eyes. Bark for tanning and preparation of skin and dyeing (darkbrown).

Roots for cataracts.

Edible fruits with pleasant flavour. Fruit rind as perfume. Seeds as an additive to vegetables. Kernels are a source of fat with parinaric acid used for varnish. Flowers for cosmetics.

P. excelsa SAB. Useful wood. Wood and bark used for preparation of skins. ROTH studied the bark structure 1981, leaf structure 1984, fruit structure and dispersal 1987, leaf venation 1996.

Cochlospermaceae

At least 6 useful species of *Cochlospermum* are known.

Some species are ornamentals. Seeds are the source of a red dye, the tubers a source of a yellow dye. The gum is used as a substitute for Tragacanth. Fibers of the bark serve for ropes. Subterranean parts are tonic. Plants used as an emmenagogue.

C. orinocense SPRUCE supplies fire wood. The bark is applied as a febrifuge. Fruit morphology and dispersal have been studied by ROTH 1987.

Combretaceae

The fruit is always one-seeded, leathery, often winged, rarely a drupe; endosperm is absent.

The representatives are trees, shrubs or lianas with hairs of different kinds (glandular, scaly or simple). Pollination occurs by water or wind (formation of floating tissue or wings).

Terminalia

Terminalia is a frequently occurring pantropical tree. *T. catappa*, found along the coast lines, is distinguished by its growth and ramification in the form of storeys; edible seeds and a bark rich in tannins are gathered from it.

Lumnitzera and Laguncularia

Lumnitzera and Laguncularia are mangroves.

Buchenavia capitata. EICHL. has a very attractive wood.

ROTH studied the bark structure 1981, leaf structure 1984, fruit structure and dispersal 1987.

Conocarpus erecta L. (mangle botoncillo, botoncillo, mangle botón, mangle lloroso, mangle blanco, mangle)

Tree or shrub, 3–8 m high. Alternate leaves elliptic or angustielliptic, 4–10 cm long, and 1.5–3 cm broad, acute, glabrescent or glabrous. Hermaphrodite and male flowers mixed in dense subsessile or pedunculate capituli. Calyx in the form of a cup with 5 triangular lobes. Petals absent. Between 10 and 5 stamens. Disc in the form of 5 fleshy glands surrounding the base of the ovary. Ovary velvety, generally with 2 ovules.

Fruit reddish brown, in the form of a scale, curved, 3–7 mm long, with 2 wings and a single seed. The fruits are united in a globose or conical infrutescence, 1.5 cm long and 1.3 cm in diameter.

The species occurs in tropical America and in the western part of Africa. In Venezuela it is frequent along the Caribbean beaches.

Anatomical description

Leaf. The leaves are of coriaceous-succulent texture and usually pubescent when young, but become glabrous with time, although varieties exist which have a persistent indumentum. The short petiole bears 2 prominent glands on its base.

The upper epidermis is relatively large-celled, but the cuticle covering the outer cell walls is comparatively thin. As the leaf is amphistomatic, stomata are also present in the upper epidermis, but in a reduced number. Glands are observed in depressions of the upper as well as of the lower epidermis. No hypoderm is developed on either side. The leaf is isolateral or equifacial and palisade parenchyma is found on the upper as well as on the lower leaf side. On both sides it consists of 2 layers of elongate cells which stain blue with Giemsa. However, the palisade cells of the upper side are more elongated. The spongy parenchyma has an extraordinary structure. The cells are elongated perpendicularly to the leaf surface, just like the palisade cells, with only small intercellular spaces. The spongy parenchyma comprises about 5-6 layers of cells which contain chloroplasts and starch, as well as oil drops.

The venation of the leaf is dense and the vascular bundles lie directly below the upper palisade parenchyma. Their blind terminations within the meshes consist of tracheids with thickened walls and many pits. The lower epidermis is slightly smaller-celled than the upper one. Druses of calcium oxalate are present in the entire mesophyll. The palisade parenchyma on the lower side is notably smaller than that of the upper side.

The head of the glands is of spherical or bulbous shape and composed of 6 secretory cells. Below follows an intermediate cell and then several collecting cells. The head is covered with a thick cuticle. Between the intermediate cell and the collecting cells there is a zone of transfusion. The entire glandular complex is not larger than an epidermis cell (for more details see ROTH 1992).

The foliar domatia on the lower leaf side which develop in the angle between the middle nerve and the lateral nerves are very conspicuous (Fig. 97.b). They arise from the activity of a dorsal meristem. Open cavities arise from the activity of this meristem and become filled with hairs and glands. It is possible that these domatia are used by ants in the same way as other domatia (ROTH 1976).

Besides the domatia, a pair of glands occurs on the base of the leaf stalk which are interpreted as extrafloral nectaries. They are furnished with an extensive secretory tissue and seem to be in contact with the midvein (for further information see ROTH 1992, 1976).

Ethnobotanical and general use

The timber is used to a minor extent for durable construction, but its chief value is for fuel and charcoal.

The bark is bitter and astringent and used for tanning leather.

Leaves and bark have tonic properties and are used by weak, anemic or reconvalescent persons. A decoction gives excellent results against Basedow's disease.

A handful of leaves and bark is crushed and boild for 10 minutes in a bottle of water. The filtered decoction is taken by the cup during the day. Sugar can be added for flavouring.

Observations

The leaf can be well recognized by its anatomical structure and the particular glands.

Laguncularia racemosa (L.) GAERTN. (mangle blanco, mangle amariilo)

Taxonomical description

The plant is a small shrubby tree (occasionally up to 20 m high) (Figs. 96, 97) with coriaceous opposed leaves which are oblong, obovate or elliptic, 3–10 cm long and 2–4 cm broad and glabrous. The petioles are up to 2 cm long.

The pentamerous greenish flowers are polygamous and arranged in large axillary spikes. The urceolate calyx has 5 persisting lobules. The 5 obovate

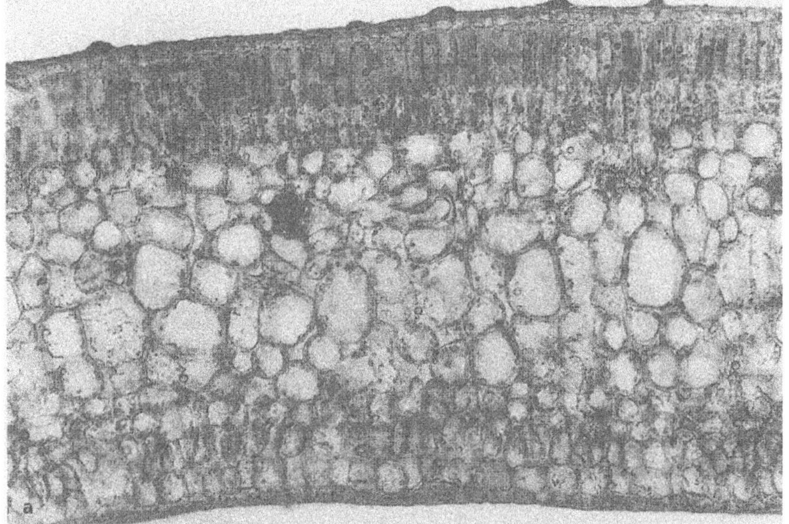

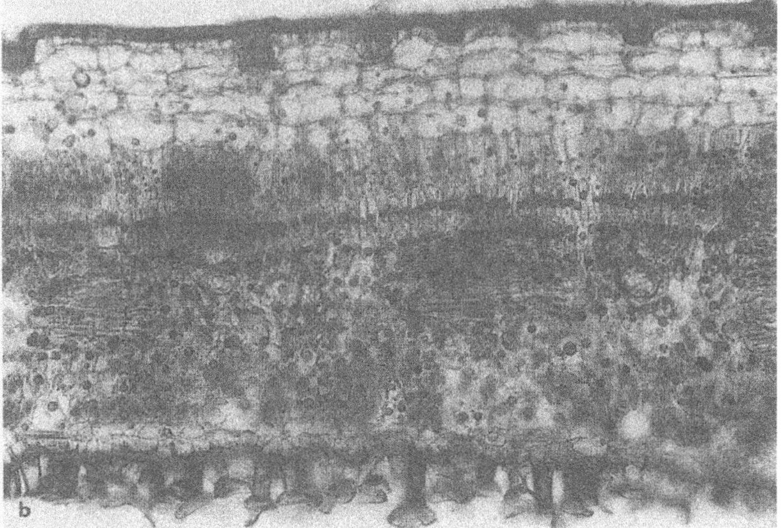

Fig. 96. Leaves of mangroves. **a** *Laguncularia racemosa*, Combretaceae. Isolateral leaf with a well developed central mesophyll. **b** Leaf of *Avicennia germinans* in t.s. Note the hammer hairs on the lower leaf side (ROTH 1992).

petals are 1.5 mm long. The 10 stamens occur in 2 series. The stigma is bilobed.

The fruit is angusti-oboviform, up to 15 mm long and 7–8 mm in diameter, first silky, but later more or less glabrescent. The calyx is persistent on the fruit; the fruit outside is longitudinally ribbed, the base is cuneate, the apex truncate. The fruit contains only one seed, which germinates while on the tree (viviapry).

In Venezuela, the species is frequently found along the Caribbean beaches.

Anatomical description

Leaf. (Fig. 96 a) The petiole develops 2 glands in its upper part. The leaves are more or less vertically oriented on the branches, a position that has an influence on the blade structure (equifaciality).

The upper epidermis is composed of small cells with strongly cutinized outer walls. Stomata occur on both leaf sides, but are more numeorus in the upper epidermis. Glandular hairs – hidden in cavities – occur in the upper as well as in the lower epidermis; their density is however lower on the upper side. The palisade parenchyma comprises 2 layers; the palisade cells contain large oil drops. A 2-layered palisade parenchyma is also found on the lower leaf side, so that the leaf becomes isolateral or equifacial. The centrally situated mesophyll (between the upper and the lower palisade layers) is small-celled beneath the upper palisade parenchyma, but cells increase in size towards the middle of the blade. Very large and thin-walled cells, which contain only few chloroplasts and oil drops, are characteristic of this middle zone. This tissue has the function of water storage. The frequently occurring vascular bundles either lie directly in the cen-

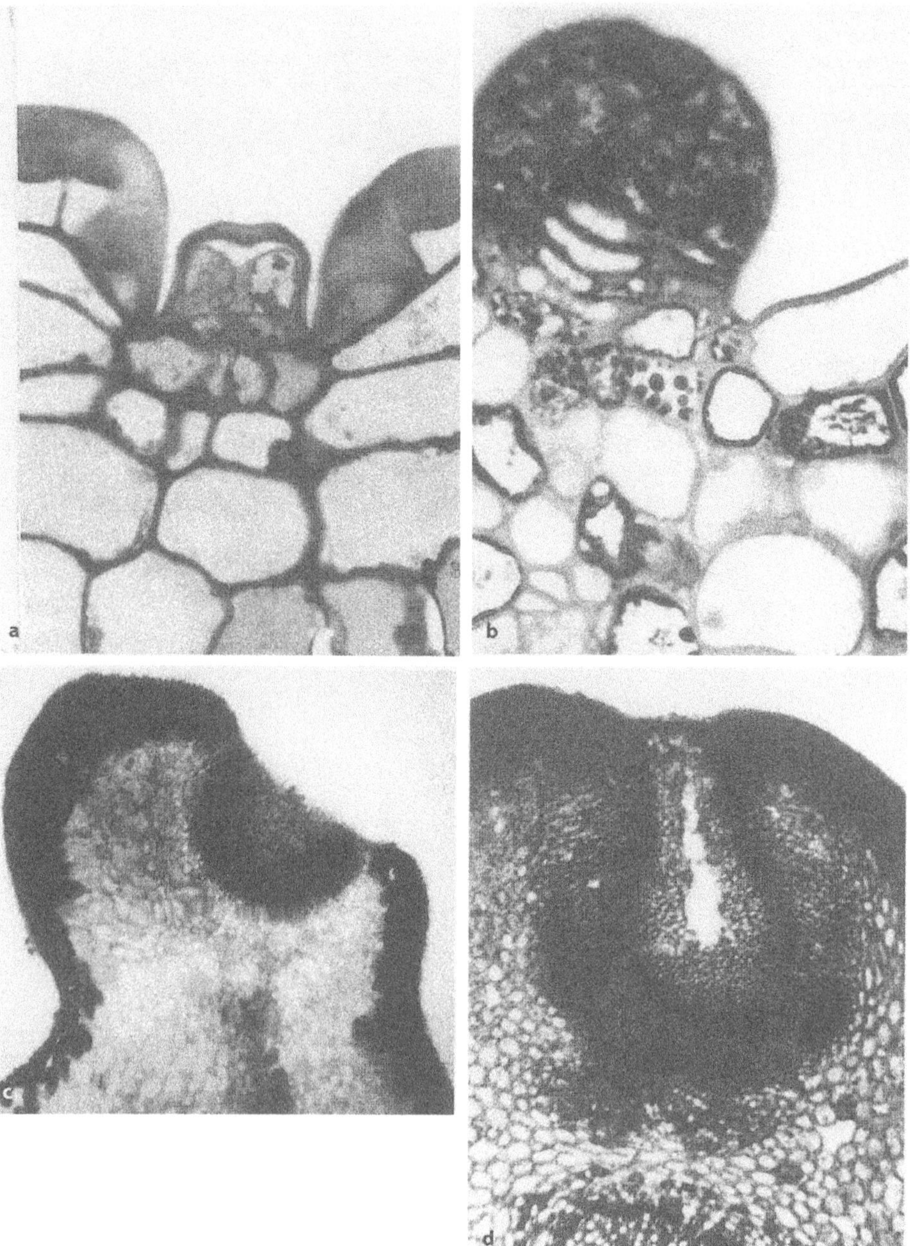

Fig. 97. Different types of glands of mangroves. **a** t.s. leaf. *Avicennia germinans*, **b** *Laguncularia racemosa*. **c** *Conocarpus*, extrafloral nectary of the leaf stalk. **d** *Laguncularia*, extrafloral nectary of the petiole (ROTH 1992).

ter of the leaf or, more frequently, move closer to the upper surface, coming in better contact with the palisade parenchyma. Between palisade parenchyma and the large-celled water-storing parenchyma, a few layers of smaller cells replacing a normal spongy parenchyma are found. In this region the vascular bundles may also be present. Large quantities of druses of calcium oxalate occur in the mesophyll, particularly in the palisade parenchyma of the upper side. The cells integrating the lower palisade parenchyma are often shorter than those of the upper one. There may also be a lower hypodermis below the lower palisade parenchyma, consisting of a single layer with few chloroplasts. However, this layser is not always present. The cells of the lower epidermis may be somewhat larger than those of the upper side. Stomata occur in lower numbers in the lower epidermis than in the upper one; secretory glands are however more numerous in the lower epidermis.

The glands are separated from the vascular bundles by only 1–2 mesophyll cells. The spherical head of the gland is composed of numerous secretory cells. The cavities at the bottom of which the glands are embedded are comparatively large, but communicate with the exterior by only a very narrow channel.

A second type of gland is found in the vicinity of the leaf margins which appear as dark-brown dots to the unaided eye. They are extrafloral nectaries with a secretory epithelium. A pair of extrafloral nectaries is likewise found on the leaf petiole.

The leaf of *Laguncularia* is thus thick and fleshy, hence the coriaceous consistency. It is able to store a large amount of water in its mesophyll cells. (For further information and details see ROTH 1992).

Ethnobotanical and general use

Bark, leaves and galls contain 10–17% tannin (of dry weight) and are used for tanning and medicinal purposes.

Bark and leaves are astringent and tonic and used as such in popular medicine.

The wood is very strong, hard and atrractive; it is even resistant to dry-wood termites and durable in contact with the soil. It is used for carts, gates, fences, sometimes for construction and repair of farm buildings. It is suitable for heavy-duty flooring, work benches, machinery platforms, heavy durable exterior construction work, railroad ties, piling in nonteredo areas, house posts, bridge lumber, firewood and fuel.

In Cuba, withes of the tree are seasoned in salt water and twisted into cables (cujes) from which tobacco is suspended for curing.

Observations

The leaves are easily recognized by their isolateral structure with a water-storing tissue in the centeral part of the mesophyll and by the presence of very particular glands in holes.

Terminalia

At least 20 useful species are known.

Fine useful wood. Edible fruits and seeds. Oil from seeds. Leaves are food for a silk worm. Leaves, roots, bark and fruits for tanning. Fruits and bark source of a dye. Bark for perfume.

Bark antiinflammatory and antiarthritic, used as cardiac and stimulant. Astringent bark for bilious fevers and dysentery. Bark for diarrhoea and for hepatitis. Source of astringent gum used in cosmetics.

Fruits for sore eyes. Fruits tonic, astringent, aperient, used for trush and diarrhoea.

ROTH studied the bark structure 1981, leaf structure 1984, fruit structure and dispersal 1987.

Commelinaceae

The monocotyledonous Commelinaceae are usually more or less succulent herbs. Their stalks are erect or procumbent. The leaves are often distichously disposed and have a sheath, the lamina is frequently fleshy with an epidermal or subepidermal water-storing tissue. The stomata are surrounded by 2 (4) lateral and 2 polar subsidiary cells. The flowers are usually 5-cyclic and have mostly verticils with 3 members; they are arranged in cincinnal cymes. The pantropical family is frequently found in tropical America and Africa.

The genus Commelina embraces about 150 pantropical and subtropical species.

Commelina diffusa BURM, synonyms: *Commelina cayennensis* RICH., *Commelina nudiflora* L. (canutillo, azulillo, suelda con suelda)

Taxonomical description

Commelina diffusa (Fig. 98) is a decumbent perennial herb with glabrous branching stipes. The fleshy leaves are lanceolate, 2–7 cm long and 1–2 cm broad, with an acute apex and a rounded base; they are sessile or almost sessile. The sheath is membranaceous, cylindric, 13–16 mm long, ciliate at the upper rim and slightly hairy along the ventral suture.

The inflorescences occur in the axils of the upper leaves. The flowers have 3 sepals, 3 mm long, all of the same size; 3 petals of blue or violet colour, one of them pale or withish and smaller or completely absent. There are 3 fertile stamens and 3 sterile ones. The tri-locular fruit is ellipsoid, 5 mm long, apiculate; one locule has a single seed, while the other 2 locules have 2 seeds each.

The species occurs in almost all tropical and subtropical regions. In Venezuela it is frequent in hot areas.

Origin

In almost all tropical and subtropical regions.

Occurrence

SCHNEE (1960) found the plant mainly at hot places in Venezuela. STEYERMARK & HUBER 1978

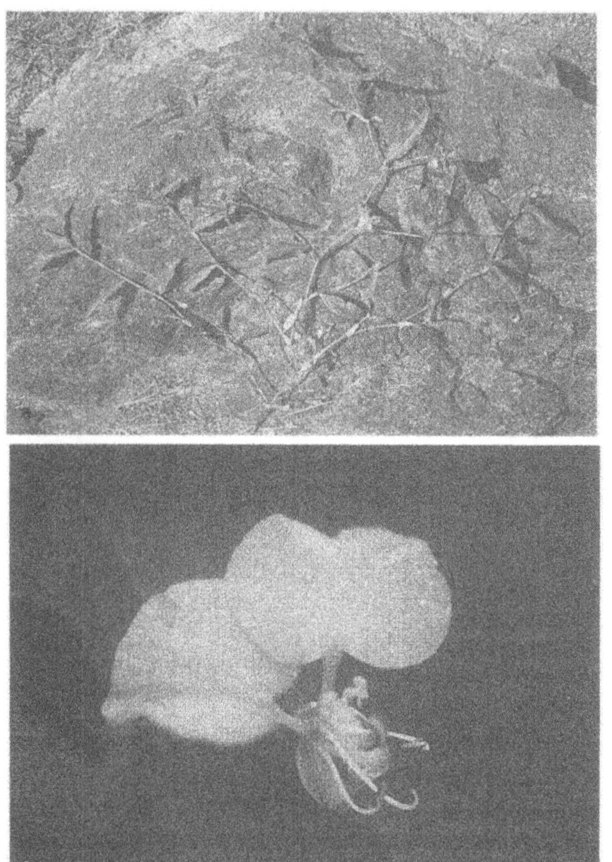

Fig. 98. *Commelina diffusa*. **a** Entire creeping plant, **b** flower.

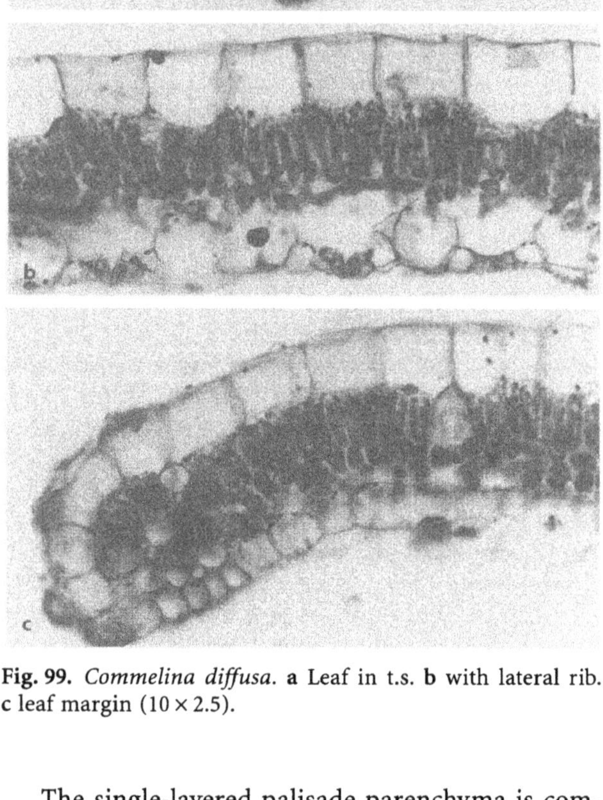

Fig. 99. *Commelina diffusa*. **a** Leaf in t.s. **b** with lateral rib. **c** leaf margin (10 × 2.5).

mention that the plant prefers humid sites and areas disturbed by man and that it grows in the deciduous and transitional forests. In the region of the Avila, the species occurs at a height between 900 and 1100 m above sea level.

Anatomical description

Leaf. (Fig. 99, 100). The delicate leaf is bifacial and amphistomatic. The upper epidermis consists of very large transparent cells which occupy about the same proportion as the entire mesophyll. Their walls are thin and their function is water storage. As seen in surface view, the cells are arranged in longitudinal rows; they are isodimatric, rectangular or even square-shaped. The stomata are scarce and mainly occur above the veins. They have 2–4 lateral and 2 polar subsidiary cells. Additionally, some short papilla-like unicellular hairs with more or less pointed ends are sporadically found in the upper epidermis.

The single-layered palisade parenchyma is composed of long and slender sometimes slightly conical cells. The spongy parenchyma comprises about 2–3 layers of lobulate cells which leave intercellular spaces between one another. Spongy and palisade parenchyma are about equal in proportion. The lower epidermis is similar to the upper epidermis in its structure. It consists mainly of large water-storing transparent cells which are only slightly smaller than those of the upper epidermis. The stomata are, however, much more numerous on the lower epidermis. They are surrounded by 4 lateral and 2 polar subsidiary cells. While the stomata of the upper epidermis are occasionally reduced, having only 4 or 5 subsidiary cells altogether, the stomata of the lower epidermis are usually surrounded by 6 subsidiary cells. The stomata occur at the upper epidermis level or are even slightly elevated above the surface. Short unicellular papillary hairs are rarely found on the lower epidermis, but occur along the leaf margins. The pointed ends may occasionally be separated by a wall so that the hair becomes bicellular.

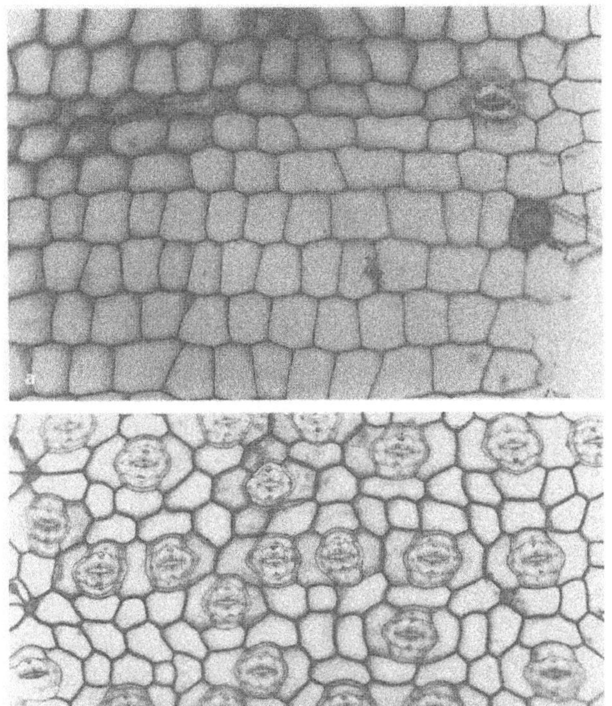

Fig. 100. *Commelina diffusa*. **a** Upper epidermis in surface view, **b** lower epidermis with stomata.

The middle nerve very slightly projects above the lower leaf side, but shows a depression on the upper side. It has only a single collateral bundle without sclerenchyma. The bundle is surrounded by parenchyma. The palisade parenchyma is substituted by isodiametric chlorenchyma cells in the region of the middle nerve.

Above this tissue, a small-celled hypodermis is found; a similar hypodermis is also present on the lower side. Upper and lower epidermis are very small-celled in the middle nerve.

The secondary veins are similar to the middle nerve with a single vascular bundle surrounded by parenchyma. The veins of higher order show parenchymatous extensions towards the upper epidermis.

Hyaline slime cells which are elongated parallel to the long axis of the leaf, but appear roundish, as seen in transverse section, are conspicuous between the mesophyll and the epidermal layers; they are arranged in longitudinal rows and contain raphides. These laticifer – like ceullular chains are called by TOMLINSON (1969) 'articulate raphide canals'.

Ethnobotanical and general use

Economical utilization

The plant may be used as a ground cover in gardens and parks to avoid desiccation and erosion of the soil.

It can also be grown in pots as a hanging ornamental.

Medical use

Name of the drug: *Commelina nudiflora* L. FLOS, plantae, folia. In popular medicine, the leaves, the flowers, the rhizome and the entire plants are used.

Leaves. The leaves in an infusion soothe colic. Leaves are also used as a cataplasm on fractured or dislocated bones.

Flowers. The flowers in an infusion are emollient and help as a pectoral.

Rhizome. The rhizome in decoction is effective against diarrhoea and haemoptysis, and cures burns.

Floral bracts. The floral bracts always hold water in their axils and thus offer a very favourable environment for symbionts and epiphytes. The water stored in the axils of the bracts is used in popular medicine as a eyewash ('colirio').

Entire plant. The entire plant in decoction is applied to soothe headache. A decoction of the entire plant is also applied to cure gonorrhoea and diseases of the kidney and it is taken as a refreshing drink. It is furthermore used for retention of the urine and to cure urinary tract infections. The juice of the plant is dispensed for gastric ulcer and in the form of a cataplasm it acts as an antiseptic and vulnerary. It is likewise used to stop haemorrhage of wounds.

Method of use

Leaves in an infusion or as a cataplasm. Flowers in infusion. Entire plant as a decoction, as a cataplasm or sap.

Healing properties

Antialgid, antiseptic, emollient, vulnerary.

Chemical contents

The chemical contents have been little studied. There is slime in the raphide cells. The leaves contain a derivative of the glucoside delphinidine. Coumaric acid and acetic acid were also indentified.

Related species

The common name suelda con suelda (SCHNEE 1960) is a collective name for certain species of *Commelina* and related species of the Commelinaceae family, e.g. *Commelina elegans* HBK (synonym: *Commelina virginica* L.), an Andean plant, is used in popular medicine in the form of a raw sap or as cataplasm; leaves and stipes are applied on fractures, luxations and dislocations of bones as well as an emollient on swellings due to dislocations.

Furthermore, to cure irritations of the eyes, and particularly conjunctivitis, some drops of the sap are put directly into the eyes. Equally well, fresh leaves may be put directly on the irritated eyes. The fresh sap squeezed out of the plant soothes the irritations produced by urticaria, herpes or exanthemas in general. Three cups a day are taken of the 2% decoction of the plant to cure bloody sputum and vaginal discharge.

It is also applied in fomentation and in a bath against nervous diseases of the skin (BERNAL & CORREA 1990).

SCHULTES & RAFFAUF (1990) report that the Ketwas (quechuas) of Ecuador take the sap of *Commelina erecta* L. orally as a contraceptive.

Cultivation

The plant is easily propagated by cuttings which rapidly form roots and grow fast, even in plain sun. The plant does not require good soil.

Observations

Following TOMLINSON (1969), the Commelinaceae in general have a very uniform anatomical structure. The genus *Commelina* is distinguished by its leaf structure and by certain particularities of its flowers: Fleshy leaves with an upper and lower epidermis specialized as a water storage tissue, 4–6 subsidiary cells around the stomata developing from cells adjacent to the stomatal mother cells, and raphide cells superposed in longitudinal rows, may be mentioned here.

For anatomical distinction of species, the variability of the hairs in *Commelina diffusa* may be used, as well as the structure and occurrence of the stomatal apparatus. The uni or bicellular hairs which pertain to the type 'macrohairs', following the classification of TOMLINSON (1969) are very variable and occur along the margins as well as on the surface of the blade, mainly of the upper side (Fig. 100 a). They sometimes have a rounded apex, or even have a bristle-shaped tip. The different hairs occur side by side. The occurrence of stomata on both leaf sides is rather rare and the fact that the stomatal apparatus on the upper leaf side is more reduced, being composed of only 4–5 subsidiary cells instead of 6 (as on the lower side), may also be used as a distinguishing feature. Furthermore, epidermis cells are arranged in longitudinal rows on the upper as well as on the lower side, however the rows on the lower side are somewhat disturbed by the presence of the stomata.

But we agree with TOMLINSON (1969) that divergence in the Commelinaceae has not proceeded very much at the anatomical level and that much more detailed investigation is necessary.

Phaeospherion

There are about 6 species of these herbaceous plants in tropical South America. The leaf anatomy of *P.* sp. growing in the transitional cloud forest of Venezuela has been studied by ROTH (1990). The lower epidermis is furnished with 2 types of hairs: simple pluricellular hairs and unicellular hairs with thick walls.

No chemical studies of this genus are available.

Phaeospherion persicariaefolium (DC.) CLARKE, synonym: *Commelinopsis persicariaefolia* (DC.) PICHON. is a medium-sized herb. The species grows in the Venezuelan transitional cloud forest at altitudes of 600–900 m. The leaf anatomy has been described by ROTH (1990). The plant is also mentioned by SCHULTES & RAFFAUF (1990), but no chemical studies are available.

Compositae (Asteraceae)

Sepals are often modified into pappus hairs. The crown is tubular, the ovary is inferior. The fruit, called achene, in which fruit wall and seed coat are intimately united, is dry, indeshiscent and nut-like. the seed is devoid of an endosperm. The flowers are usually united in the form of heads, surrounded by an involucrum of bracts. Marginal flowers are often sterile and differ from the other flowers in their outer appearance so that pseudanthia develop. The family is represented mainly by herbs, perennial herbs and shrubs, but seldom by trees. Members of this family are therefore comparatively rarely found in the tropics except for the mountainous regions (e.g. *Espeletia* in the Páramos of the Andes).

Few members of the Compositae supply foodstuff (vegetables, salad, oil), some are medicinal plants, many are ornamental.

Articulated laticifers and schizogeneous secretory cavities and canals with oily or resinous contents are common. Inulin is characteristic of the family.

Ambrosia cumanensis H. B. K. (artemisia, altamisa, artesima)

Taxonomical description

The plant is a fruticose perennial herb, 0.4–2 m high, erect or subdecumbent and frequently with a woody base (Fig. 101). The twigs are more or less circular, as seen in t.s., and are pubescent with long fine hairs. The alternating leaves are bipinnatifid, 4–8 cm long and 2–4 cm broad, being divided into obtuse-dentate segments. The upper and lower surfaces are hairy. The statement that minute glands appear as small dots on the leaf is erroneous (CORREA & BERNAL 1990; STUESSY, 'Flora of Panama', 1975). Probably the very large hair bases were taken for glands or another species was studied. The hairy petiole is 1–4 cm long. The inflorescences are unisexual. The male heads, arranged in terminal spiciform racemes, are 2.5–3 mm long and have a short pedicel. The sessile female heads appear in the form of glomeruli in the axil of reduced leaves (bracts), situated below the male inflorescences and reach a length of about 3 mm. The involucre of the female heads is tuberculate and spiny when mature. The subglobose achene is about 2 mm long and becomes completely surrounded by a bract.

Fig. 101. *Ambrosia cumanensis.* **a** Habitus, **b** leaves.

Origin

The species occurs from Mexico to Peru. It is possibly indigenous to Central America.

Occurrence

The plant develops better in hot regions, but also grows in cooler zones. In Venezuela it occurs in the Andes, the coastal Cordillera, the Llanos and in Guiana.

Anatomical description

Leaf. (Fig. 102; 103 b, c) Bifacial, amphistomatic. The upper epidermis has cells of regular size with walls of regular thickness. As seen in surface view, the cells are more or less polygonal, but have slightly wavy anticlinal walls. Long simple and uniseriate hairs with a large basal cell and an acute tip are frequent. Stomata are without subsidiary cells. The palisade parenchyma is composed of 2 layers, but the outer layer has longer cells. The spongy parenchyma comprises about 2–3 layers, but is slightly smaller than the palisade parenchyma. The composing cells are small and have short arms, leaving small intercellular spaces. The lower epider-

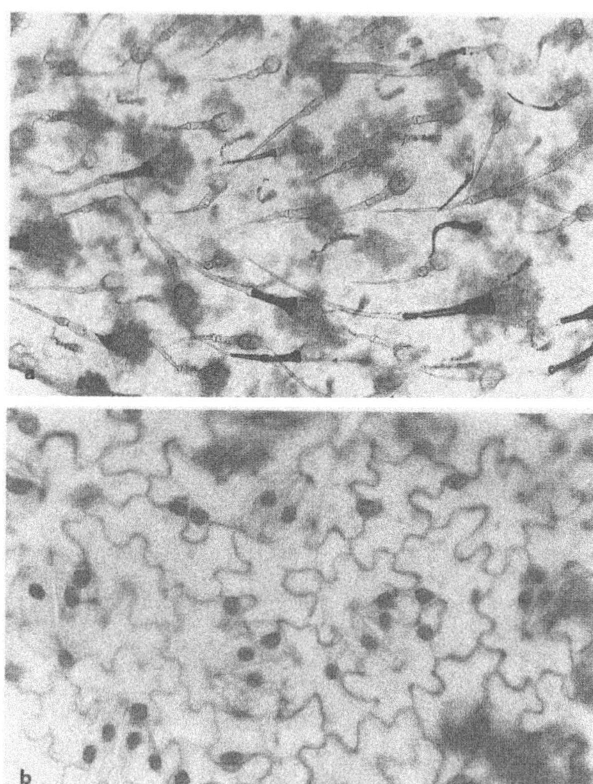

Fig. 102. *Ambrosia cumanensis*. **a** Upper epidermis with hairs. **b** Lower epidermis with stomata.

mis has cells of more irregular shape with conspicuously undulated walls. The stomata - without subsidiary cells - are more numerous on the lower than on the upper side and are somewhat elevated above the surface. Likewise the long delicate hairs are more frequent on the lower side.

The vascular bundles are surrounded by a parenchymatous sheath which is transcurrent to the upper epidermis. The midrib is slightly sunk below the upper surface, but is elevated above the lower leaf side. The palisade parenchyma is interrupted at the top of the middle nerve. The principal vascular pattern of the midrib consists of a horseshoe-shaped arrangement of individual bundles which is open towards the upper leaf side; the median bundle is the strongest. This pattern may however become reduced to a large median bundle with two smaller lateral bundles, one on each side of the median bundle, where the midrib becomes weaker; this reduction may even lead to the formation of only a single vascular bundle in the midrib. Collenchyma is present below both epidermal layers of the midrib, but is better developed towards the lower side.

Stem. (Figs. 103 a, 104, 105). A stem or twig of about 1 cm in diameter shows secondary growth. The cork is of subepidermal origin and discontinuous. Its cell walls are thin. The primary cortex is composed of parenchyma cells which leave larger intercellular spaces between one another so that some kind of an aerenchym arises. Secretory canals surrounded by a ring of secretory cells are conspicuous in the primary cortex. The cambium ring shows depressions where the phloem more strongly develops. A cap of fibers is found above the larger phloem portions. A discontinuous endodermis layer with Casparian stripes can still be recognized above the phloem. The radial arrangement of the cells in the phloem is very clear. The phloem only consists of softbast. Small secretory canals surrounded by an epithelium are frequent in the phloem. The original number of vascular bundles can still be recognized in the vascular cylinder by wedges of xylem which penetrate the pith. The pith cells between the xylem wedges are tangentially extended. Opposite the xylem wedges, the phloem is crowned by the above mentioned fiber caps. A ring of 1-2 layers of sclereids forms to complete the original pericycle ring between the fiber caps.

The vessels are principally arranged in radial rows (radial multiples) or in small racemiform groups. The medullary rays are pluriseriate. The pith is ample and has a hole in the center which is surrounded by distorted and torn cells. Small secretory canals are also dispersed within the pith.

Root. A root, about 8 mm in diameter, is in the state of secondary growth. As seen in t.s., the structure of the root differs from that of the stem in some characteristics. There is no pith in the center and there is no regular cork surrounding the root, but cell destruction and cell division is visible. The primary cortex consists in tangentially extended and anticlinally divided parenchyma cells. Some tangential cell rows separate from one another (as in the stem) to form larger intercellular spaces. The radial arrangement of the phloem cells is less regular than in the stem, and hardbast cells (sclereids) are dispersed here and there as groups or solitarily. Secretory canals are frequent in the primary cortex, in the phloem and in the phloem rays. The arrangement of the vessels is similar to that of the stem. The rays are pluriseriate 2-3 (1-4) seriate.

Ethnobotanical and general use

The plant is considered a malodorous weed which is used as an insecticide, a disinfectant, and to repel caterpillars. Brooms manufactured of the plant are used to combat fleas in the house.

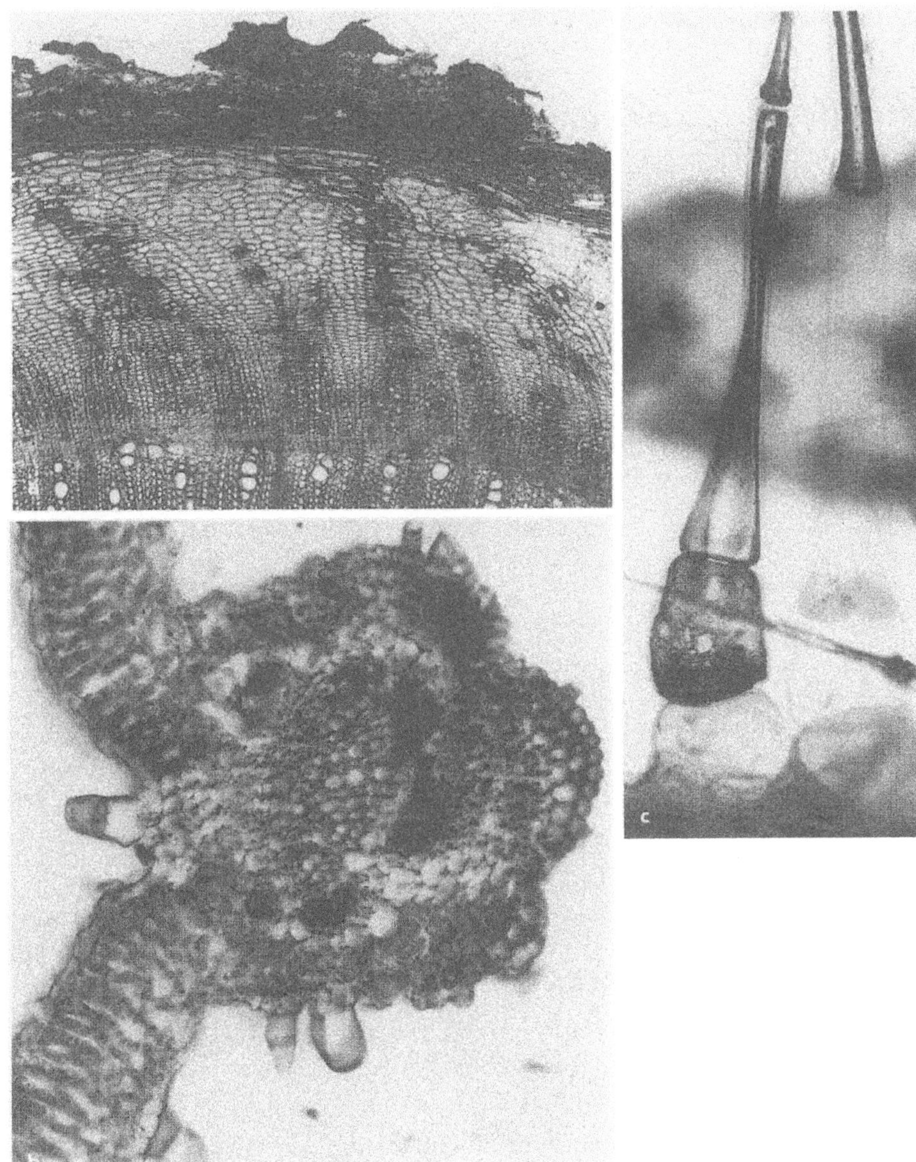

Fig. 103. *Ambrosia cumanensis*. **a** Phloem of the axis. **b** Midrib of the leaf. **c** hairs of the leaf (× 16).

Medicinal use

The drugs used are: Plantae, folia, caule, radix.

Leaf. Crushed leaves are used locally or in a decoction together with *Pluchea symphytifolia* that is taken 3 times a day to cure stomachache. Against stomach pain and to remove intestinal parasites a cool drink is prepared of the crushed aerial plant parts, and 2 spoons full of it have to be taken 3 times a day. To cure intestinal colics, 3 fresh buds are pounded, strained and mixed with honey. Fresh ground leaves are also used before and after a birth. A cool drink of ground leaves helps against spasms. To cure gastritis, a cup of the decoction of the leaf in a liter of water is taken 5 times a day; a daily cup of this decoction taken over 5 days helps against head ache. To cure any pain, a wettened leaf is fried and put on the painful part. A bath in a decoction of leaves together with bitter orange and lemon, 2 times a day and for 3 consecutive days, is recommended to combat fever. Pain in the shoulder is healed by a decoction of the leaf with salt, used locally as a friction. A bath in a decoction of leaves 3 times a day and for 8 consecutive days is used against allergies. To fight intestinal parasites, a cup of fresh ground leaves is taken in the morning on the empty stomach. To combat muscular pain, rheumatism and swollen tissue, the entire plant is toasted and applied directly on the affected part or is fried in oil and used as a friction. The plant is also used against arthritis, epilepsy, yellow fever, constipation, haemorrhagia, as an emmenagogum

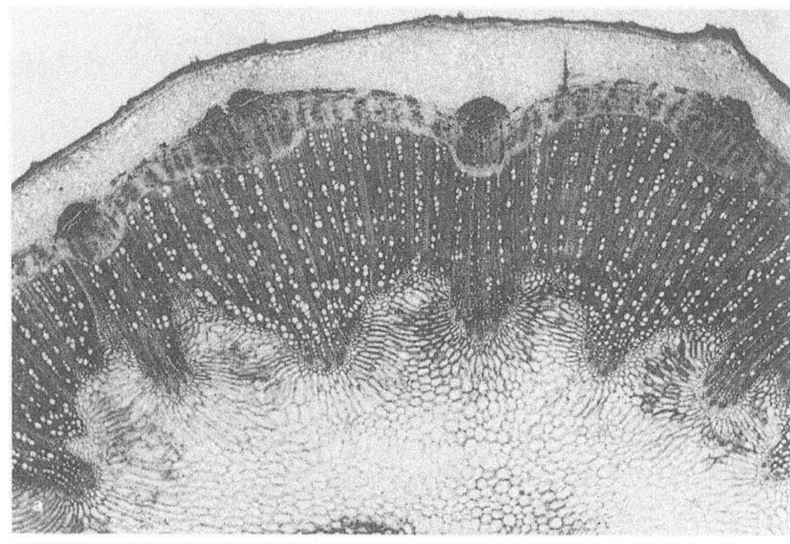

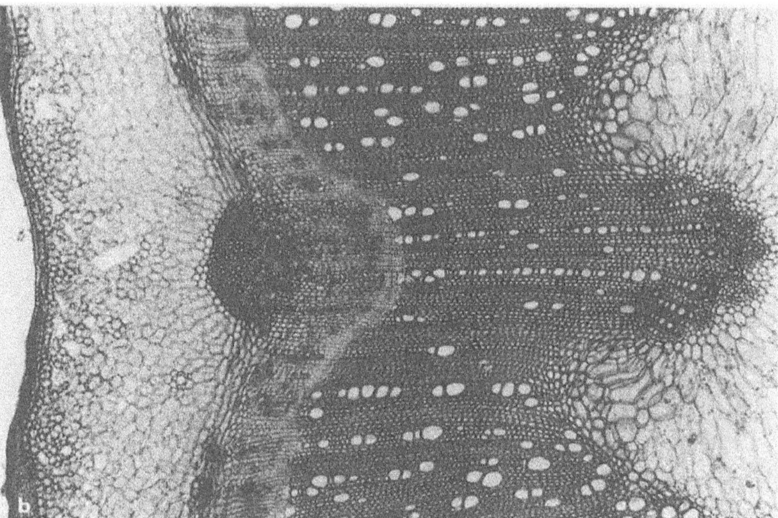

Fig. 104. *Ambrosia cumanensis.*
a, b Axis in t.s.

and for purification (as a remedy for monthly periods and for blood purification). It is furthermore used against white and yellow vaginal discharge and earache, to cure heart pain, to heal nervous attacks, hypertension, fainting and to induce sleep.

Method of use

The parts of the plant mainly used are the leaves, the buds, the entire plant and less commonly, the root.

The plant parts are either used fresh locally, unprepared, or macerated (in water or aquavit), pounded or as a decoction (with salt, honey, bitter orange and lemon or together with *Pluchea symphytifolia*) or as an infusion, or even pulverized. Leaves are also used as cataplasms. In each case, there is an exact indication, as to how the plant parts have to be prepared and in which quantities they have to be taken (GUPTA 1995, CORREA & BERNAL 1990, RODRIGUEZ 1983).

Healing properties

The plant is useful against stomachache, gastritis, intestinal pain, against parasites (antihelmintic) and spasms, for flatulence, pain in general, rheumatism, arthritis, headache, nervous attacks, muscular pains, fever, vaginal diseases.

Chemical contents

Among others, the plant contains coumarin, ambrosin and altamisin, essential oils, quercetin and alkaloids. In plant extracts, antibacterial and antiviral properties were found; the plant develops activities against *Staphylococcus aureus* and has insecticide properties. Toxicity was not observed.

Related species

Ambrosia peruviana WILLDENOW has similar properties to *Ambrosia cumanensis* and is used in a

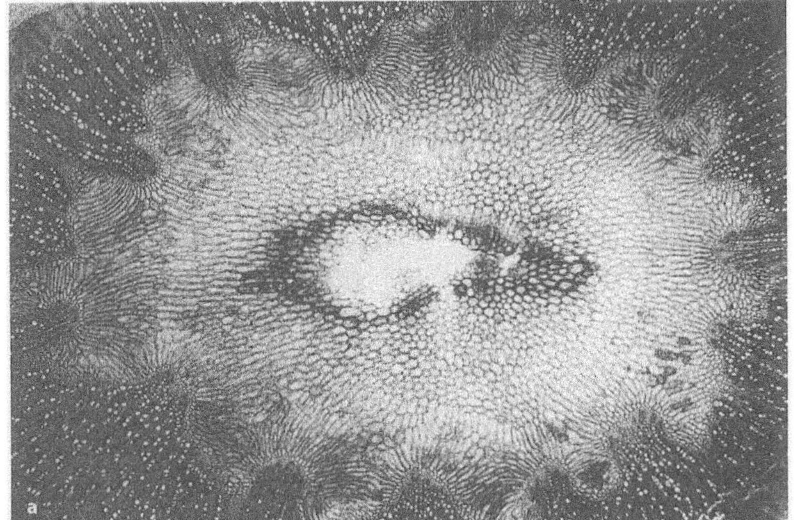

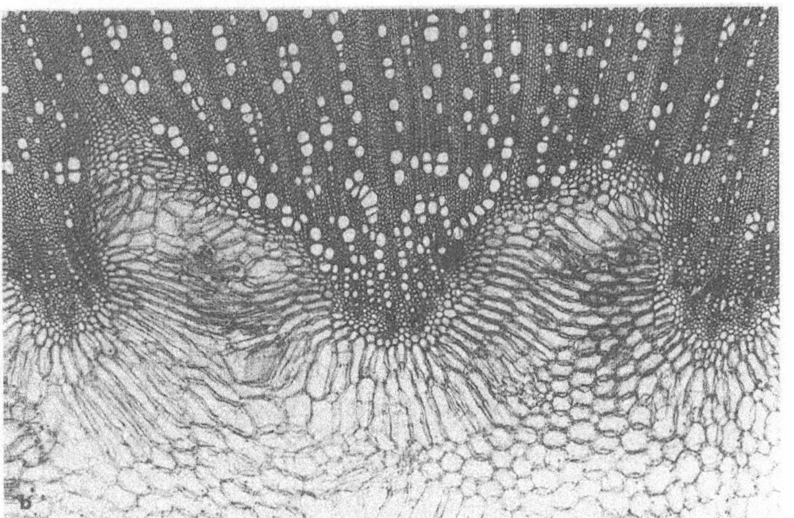

Fig. 105. *Ambrosia cumanensis.* **a** Pith of the axis in t.s. **b** Part of xylem and pith in t.s.

similar way. Leaves, stems, flowers and roots are used as medicinal drugs. The healing properties are similar.

Observations

The plant is easily recognized by the strongly undulated walls of the epidermis cells of the leaf, the characteristic hairs with a large basal cell, and the stem transection with xylem wedges protruding into the pith and tangentially extended pith cells. Secretory canals are likewise characteristic of the species.

Espeletia (frailejón, little monk)

Espeletia, a typical páramo plant, is endemic in the Andes of Colombia, Venezuela and Ecuador and actually represents one of the most important genera of the páramos due to the number of species and to the abundance of individuals. According to ARISTEGUIETA (1964), 54 species are known in Venezuela. More recently, CUATRECASAS has separated the genus *Carramboa* from *Espeletia*. *Espeletia* may be considered a dominant or leading genus in the páramos which partially gives shape to the landscape (VARESCHI 1970). The Espeletia belt reaches in the Andes from about 3000 m to 4700 m. As the genus is rich in species, these replace one another in the distinct altitudinal zones. *E. atropurpurea* grows even at 2000 m. *E. schultzii*, reaching from 2900 m to 4300 m is the most common species in the Venezuelan páramos. That *Espeletias* also occur as weeds in arable land proves their vigour. The only species found outside the Andes is *E. neriifolia*, growing in the Cordillera de la Costa.

Most species form large rosettes with long and small pubescent leaves. The rosettes are either at-

Fig. 106. *Espeletia* in the Andes: **a** *E. timotensis*, arboreous form growing at 4000 m height. **b** *E. schultzii*, one of the most frequently occurring species (ROTH 1995).

tached to the ground or are borne on erect trunks. Some caulescent species (*E. timotensis*) reach a height of 2 m or more.

The floral heads have usually a yellow colour.

Espeletia atropurpurea A. C. SM. (frailejón oscuro)

The species is described by VARESCHI (1970) as an undershrub or 'suffrutex'. It grows in the Andes at the lower height of 2000–3000 m.

The leaves are about 40 cm long and 12 cm wide.

As compared with *E. timotensis*, the leaf of *E. atropurpurea* has a less xeromorphic aspect. The green upper leaf side is almost smooth because hairs are extremely rare here. The upper epidermis is protected by thicker outer cell walls. On its lower side however, the leaf is covered and protected by hairs. The general inner leaf structure of *E. atropurpurea* resembles that of other species of *Espeletia*. Crypts are also present on the lower side. The midrib somewhat differs from that of *E. timotensis*, as the lower side is favoured in its development. The vascular bundles are arranged in the form of a more strongly bent horseshoe and secretory canals surrounding the bundles are scarce and inconspicuous. Secretory canals are absent from the upper and lower subepidermal regions, but an angular collenchyma is formed here instead. Differences in the structure of the midrib thus supply a good characteristic for identification (ROTH 1973, 1995).

The majority of the chlorenchyma is found on the lower leaf side, just as in *E. timotensis*, and consists of 2–3 layers, but can increase to 5 layers above the crypts. Slightly curled hairs and stomata elevated above the epidermis level are likewise found in the crypts. Secretory canals do not however develop at the tip of the rib projection, but a large canal is found on the base of the rib. The central part of the projection is occupied by parenchyma, the size of the cells increases towards the tip of the projection. A vascular bundle embedded in the parenchyma occurs in the very center of the projection. It is surrounded by a parenchmatous sheath which is more conspicuous in *E. atropurpurea* than in *E. timotensis*. The large hairs found on the tip of the projection are not of the complicated and very regular type as in *E. timotensis*, but are simple pluricellular somewhat irregularly curled hairs resembling those of the crypts.

Beneath the upper epidermis, a hypodermis composed of larger cells with thickened walls and simple pits is observed. It comprises 2–3 subepidermal layers and has a water-storing function (ROTH 1973, 1995).

Espeletia neriifolia (H. B. L.) SCH. BIP. (incienso, frailejón de arbolito, frailejón de palito)

The species represents a shrub or small tree between 4 and 10 m in height with a stem diameter of

25–30 cm. The twigs are very hairy. The eliptic-oblong leaves which are coriaceous are 15–25 cm long and 3–5 cm broad, have an obtuse, acute or mucronate tip and an attenuate base with a petiole; they are glabrous on the upper side, and have a white indumentum on the lower side. The corymbose paniculate inflorescences have a diameter of 15–25 cm. The heads are about 1 cm in diameter. The involucre consists of about 7 ovate-lanceolate bracts which are 4 mm long and 2 mm broad. The white 'ligules' are linear-oblong and measure 8 mm in length and 2 mm in width.

The species occurs in the Cordillera de la Costa between 1600 and 3000 m. It is also found in the Venezuelan and Colombian Andes.

The wood is rich in resin and burns easily, even when it is still green. The common name 'incienso' of the species refers to this property (incense).

Anatomical description

Espeletia neriifolia has the typical leaf structure of the genus. The upper leaf surface is glossy. The upper epidermis is single-layered and composed of isodiametric cells with straight anticlinal walls, as seen in a surface view. Glandular hairs with a 4-cellular head are conspicuous in the upper epidermis.

Below follows a 2-layered hypodermis consisting of cells with somewhat thickened walls which serve as a water-storing tissue. As seen in paradermal view, the hypodermis cells have undulated anticlinal walls, a relatively rare feature in the plant kingdom.

The palisade parenchyma comprises 1–2 cell layers. The cells of the first layer are long and slim with a length/width ratio of about 4:1, whereas the cells of the second layer are shorter.

The spongy parenchyma occupies less room and its cells are of a globular shape and sometimes form short arms towards one another.

As in other species of the genus, the ribs are very conspicuous and project strongly above the lower leaf surface, producing small chambers on the lower side, as seen in surface view. As seen in transverse section, the vascular bundles are transcurrent from the upper hypodermis to the lower epidermis by a colourless parenchyma; the cells which are similar to those of the hypodermis have slightly thickened walls with large pits. They also function as a water-storing tissue which is in contact with the water-storing hypodermis. Due to the projection of the ribs, crypts arise in the lower leaf side to which the stomata are restricted. These are of the hygromorphic type (e.g. of *Nerium oleander*), being elevated above the surface. Simple pluricellular hairs are mainly found in the crypts.

The basic structure of this leaf is thus very similar to that of other species of the genus studied. Without doubt, the leaf aspect is very xeromorphic. The coriaceous consistency of the leaf is mainly due to the well-developed acuiferous tissue (ROTH 1995).

Espeletia schultzii WEDDELL. (frailejón de octubre)

E.s. is a rosette subshrub covered by a white or grayish indumentum (Fig. 106 b). The basal leaves are oblong lanceolate and measure 20–40 cm in leght, 3–6 cm in width. The veins are very conspicuous on the lower side. The inflorescence axis reaches up to 1 m in length, having erect paniculate corymbs. The radiate heads are 4–5.5 cm in diameter, the radial flowers are numerous and of an intense yellow colour; they are 1–1.5 cm long. The flowers of the disc are also numerous and 6–7 mm long.

It is the most common species in all Andean páramos of Venezuela and grows from 3000–4000 m where it is frequently dominant. It also goes down to 2000 m behaving like an invasive plant in cultivated land and pastures.

The common name 'frailejón de octubre' indicates october as the main flowering period.

Ethnobotanical and general use

Small pieces of the leaf are put into the ear to soothe ear ache. An infusion of the leaves is applied to alleviate rheumatism and renal diseases. A decoction or a cataplasm of the leaves is used against asthma and pulmonary conditions. Resin diluted in water taken as a syrup is also used for asthma and pulmonary diseases.

Butter and cheese are wrapped with the leaves of *Espeletia* to give them a special aroma and to preserve them due to the resinous content (ROTH 1995).

Espeletia timotensis

This is an arboreous form reaching a maximum height of 3–4 m (Fig. 106 a, 107, 108). The leaf anatomy was studied by ROTH 1973 (1995). The long and relatively narrow leaves are of the lorate type and have a pronounced midrib. The leaf of *E. timotensis* is distinguished from the general type of *Espeletia*

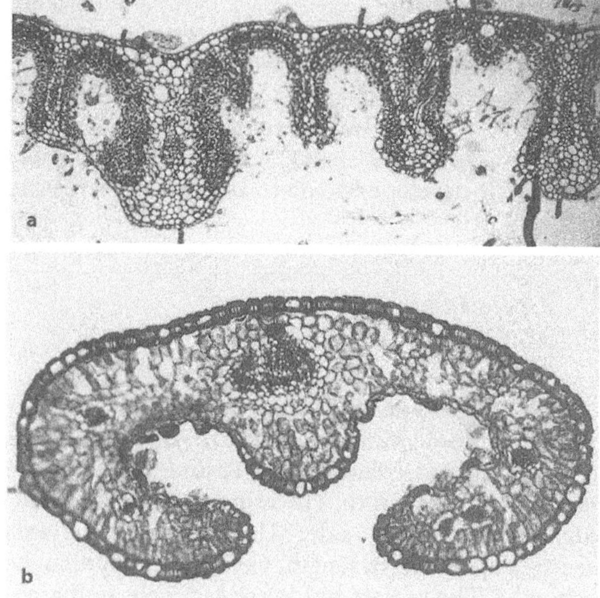

Fig. 107. a Leaf of *Espeletia timotensis* in t.s. showing the ribs on the lower side. b Leaf of *Hinterhubera imbricata*, likewise a characteristic plant of the Páramos (ROTH 1974).

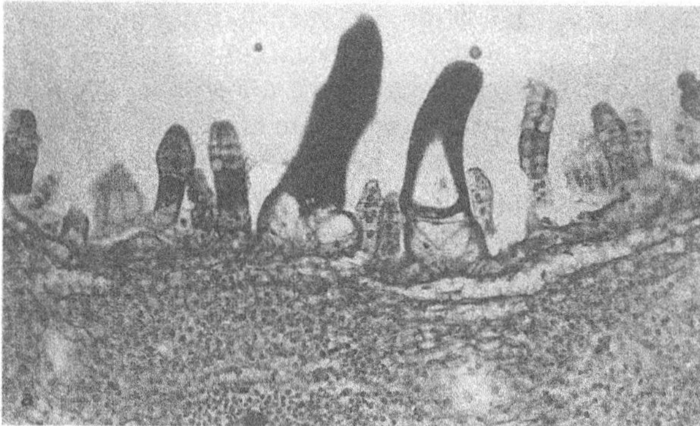

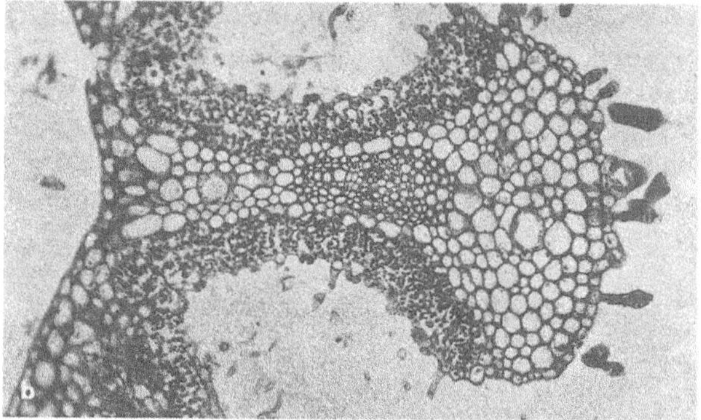

Fig. 108. *Espeletia timotensis*, t.s. of leaves: a Hairs and glands of the rib. b rib with secretory canal, vascular bundle and photosynthetic tissue (ROTH 1995).

by a large sheath with parallel venation. The upper side of the sheath is smooth and hairless, while the lower side has a thick indumentum. *E. timotensis* may be easily distinguished from other species by this characteristic feature. The epidermis cells of the sheath inside are arranged in longitudinal rows, parallel to the veins.

E. timotensis shows the most pronounced xeromorphic aspect of all species studied in the Venezuelan páramos. Its most striking feature is however the thick indumentum. In a young leaf, the indumentum may be 7–10 times thicker than the compact part of the wings of the blade. In this stage, the hairy layer may reach more than 3 mm in thickness. The hairs are curled like human hair. The simple pluricellular hairs are also slightly twisted. The wall thickness of the hairs influences the type of curling (ROTH 1973, 1995).

Altogether three types of hairs are found on the leaves. Besides the simple pluricellular hairs of the indumentum, the hairs of the crypts and the glandular hairs are distinguished. The glandular hairs and the curled hairs of the indumentum are found side by side on the upper and lower leaf surface, while the hairs of the crypts are restricted to the cavities between the ribs of the lower leaf side and are distinguished from the indumentum hairs by their smaller size and their simple structure.

The curly hairs of the indumentum may reach the considerable size of one to several centimeters, depending on their position on the leaf; the largest hairs are found on the outer side of the sheath. The hair base is surrounded by a cup originating from the neighbouring epidermis cells (Fig. 108 b). The walls of the hairs reach a considerable thickness and show stratification; their micellae seem to be arranged helicoidally; this fact may also contribute to their curling. The transverse walls of the hairs are perforated by large simple pits so that they resemble a sieve plate. The basal cell of the hair has a thinner wall being in contact with the surrounding epidermis cells through simple pits. In the adult stage, the cell walls of the hairs are lignified.

The glandular hairs are of the twin type, which is often found in Compositae. They originate from a single epidermis cell which at first divides anticlinally to the surface, while the following divisions in the 2 daughter cells are periclinal so that 2 cell rows arise, which only become visible in a front view.

The third type of hair restricted to the crypts is simple, pluricellular and slightly curled, but in a more irregular way. The curling is certainly also due to the limited space in the crypts. The hairs are much shorter than those of the indumentum and thinner-walled at the same time.

The formation of crypts on the lower leaf side is the second xeromorphic characteristic of the leaves of *E. timotensis*. The crypts are hidden beneath the indumentum and can only be seen, when the indumentum is removed. Then a fine venation network becomes obvious. The salient nerves produce and partly surround the crypts which fill the intercostal fields (Fig. 107 a). The crypts are lined by a small-celled lower epidermis with stomata elevated above the surface and with pluricellular curled hairs that fill up the entire cavity. The subsidiary cells of the stomata are elongated and curved upwards, lifting the stomata above the surface (Fig. 108 b). The salient nerves are responsible for the development of the crypts, but the way they project above the surface is extraordinary.

The growth in surface of the leaf is reduced so that the ribs remain close together. The intercostal fields therefore remain very small. The dense vascular network may be considered a further xeromorphic criterion. The nerves enlarge extraordinarily in their distal ends, particularly those of 2nd and 3rd order, as seen in t.s.; in this way, crypts of first and second order develop. Larger crypts are thus subdivided by nerves which emerge less. The nerves broaden considerably towards the outside of the crypt in order to narrow the entrance of the crypt.

The crypts represent calm or windless spaces where the gaseous exchange is rendered difficult not only by the hairs filling the cavities, but also by the felt of long undulated hairs of the indumentum which forms a dense cover on the lower leaf side. The waves of the undulated hairs act as additional border lines to protect the leaf against excessive transpiration.

The salient ribs are composed principally of parenchymatous cells which considerably increase in size towards the outside. Very conspicuous in the ribs are 2 secretory canals one of which is situated near the distal end or tip of the projection, while the other one lies near the base of the rib. The conducting tissue is composed of relatively small cells and seems to be conspicuously reduced in comparison with the total size of the projection. The xylem is principally composed of tracheids with helicoidal or scalariform thickenings. On the phloem side, the vascular bundles are accompanied by fiber-sclereids with thickened walls.

The secretory canals which are found in the wings of the blade, as well as in the midrib, are of schizogeneous origin. They possibly arise from a single mother cell. It is quite obvious that they originate at definite sites of the leaf. In the ribs, a vascular bundle is situated between 2 secretory canals; it thus seems that the secretory canals have a certain relation to the bundle. In the midrib, the vascular bundles are distributed in the form of an arc which is completely surrounded by a zone of secretory ducts. However, other secretory ducts are formed without a visible relation to vascular bundles, lying in the entire subepidermal zone of the midrib.

The distribution of the chlorenchyma in the leaf of *Espeletia* is most extraordinary. In normal bifacial leaves, the palisade parenchyma is oriented towards the upper leaf side, whereas the spongy parenchyma lines the lower leaf side. In *Espeletia* however the crypts considerably alter the gross anatomy of the leaf. The projections of the leaf are enormous, as compared with the entire blade thickness. The chlorenchyma corresponding to a palisade parenchyma develops along the surface of the crypts taking advantage of the length of the ribs; the chlorenchyma thus lines the crypts and the photosynthetic surface is considerably augmented in this way. The larger part of the chlorenchyma thus pertains to the lower leaf side, so that almost an inverted structure results. But between the ribs, in the intercostal fields, the chlorenchyma completely fills the space between the upper hypodermis and the lower epidermis. Part of the chlorenchyma thus theoretically pertains to the lower leaf side, while the upper part (above the crypts) belongs to the upper as well as to the lower leaf side.

The large-celled hypodermis beneath the upper epidermis which is connected with the vascular bundles by parenchymatous bridges is interpreted as a water-storing tissue and represents a further xeromorphic characteristic. All these xeromorphic properties enable a leaf to survive separated from the mother plant in the open air without drying up, so that the cells remain alive even weeks after! (Roth 1973, 1995).

Comparison of *E. timotensis* and *E. atropurpurea*

Relatively few differences in the leaf structure of both species can be noted. Differences are found in the midrib and in the arrangement of the vascular bundles in the form of an open arc (*E. timotensis*) and in the horseshoe form (*E. atropurpurea*). Furthermore secretory canals are much more abundant in the midrib of *E. timotensis*. Likewise the ribs differ slightly in their structure: while a secretory canal is found in the tip of the rib projection of *E. timotensis*, it is absent in *E. atropurpurea*.

The water-storing tissue on the upper leaf side is better developed in *E. atropurpurea*, consisting of up to 3 layers of large cells. The upper leaf side of *E. atropurpurea* is smooth and almost devoid of hairs, while a thick indumentum is present in *E. timotensis*. The better developed upper hypodermis seems to take over the protective function of the indumentum in *E. atropurpurea*. However, there are other species of *Espeletia* in which the waterstoring hypodermis is still better developed than in *E. atropurpurea*. Besides in the hypodermis, water storing tissue is also found in the rib projections. The parenchyma of the projections likewise serves as a waterstoring tissue. The parenchyma cells lack chloroplasts, their walls are thickened and provided with large simple pits.

The most outstanding difference between the 2 species is however the almost complete absence of hairs on the upper leaf side in *E. atropurpurea* and the complete absence of glandular hairs. A reduction of the number of hairs, on the one hand, and of certain structural peculiarities, on the other, is thus obvious in *E. atropurpurea* which grows at lower altitudes! (ROTH 1973).

Carramboa

More recently, CUATRECASAS (see ACOSTA DE OBANDO 1979), the taxonomist and specialist of the *Espeletias*, has separated the genus *Carramboa* Cuatrecasas from *Espeletia*. This is the reason why the alphabetical order is interrupted here: *Carramboa* is mentioned after *Espeletia*, since *Espeletia* is the original genus. A former student of the senior author and Professor of the Department of Pharmacy of the Universidad de Los Andes in Mérida, has studied 5 species of *Carramboa* which grow in the Venezuelan Andes (ACOSTA DE OBANDO 1979). A short description of each species is given in the following.

Carramboa badilloi (CUATR.) CUATR. As in the other species, the upper leaf epidermis is single-layered. The cells are thin-walled and have a thin cuticle. The anticlinal walls are slightly curved, as seen in a surface view. A hypodermis is absent. The one-layered palisade parenchyma consists of very long palisade cells. The spongy parenchyma is somewhat loose. The stomata which are confined to the lower epidermis are slightly elevated above the surface. Simple pluricellular hairs as well as glandular twin hairs are present. A secretory canal is found above the xylem of the lateral vascular bundles.

In the midrib, an almost circular arrangement of the vascular bundles may be noted; additionally, 2 groups of vascular bundles are observed close to the wings; from there lateral veins detach towards the wings. Some irregularly grouped vascular bundles are also conspicuous on the ventral leaf side. The xylem mainly points towards the inside. The distribution of the bundles is very dependent on the ramification of the lateral nerves, as can easily be seen. Consequently, the distribution of the bundles in the midrib at a certain height (or section) depends on the ramifying lateral bundles and therefore varies.

The species was collected at an altitude of 2600 m (ACOSTA DE OBANDO 1979).

Carramboa littlei (ARISTEG.) CUATR. The leaf has an upper strongly cutinized single-layered epidermis composed of polygonal cells with straight anticlinal walls, as seen in surface view (Fig. 109 a). Beneath follow 1–2 layers of palisade parenchyma; the outer cells exceed the inner ones in length by about one third. The palisade cells of the second layer are lobed on their upper side through indentations of the cell wall and may therefore be considered as arm palisades. The incisions increase the inner surface of the cells so that more chloroplasts may be accomodated on the upper leaf side. A more effective photosynthesis is thus warranted by the formation of 'arms'. Undoubtedly, these cells may also be regarded as a transitional form between palisade cells and cells of the spongy parenchyma. The latter develops arms in various directions which meet with the arms of the neighbour-

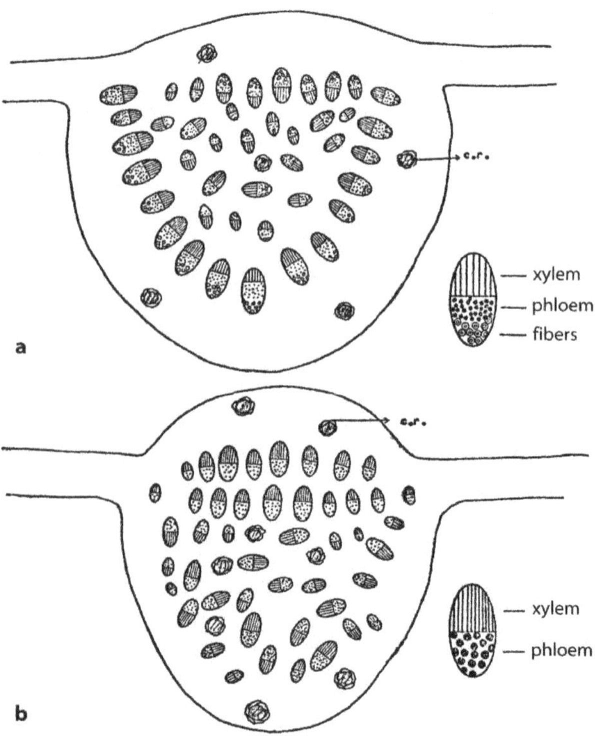

Fig. 109. T.s. of leaves of *Carramboa*. **a** *C. littlei*, **b** *C. pittieri*. Distribution of the vascular bundles in the midrib (ACOSTA DE OBANDO 1979).

ing cells so that a looser tissue with intercellular spaces results. The vascular bundles are capped by sclerenchyma (fibers) on the phloem side. The larger bundles have a resinous duct on the xylem side. The ribs enlarge towards the lower leaf side, but crypts are not developed. This is a very important distinction from *Espeletia*. The cells of the lower epidermis are smaller than those of the upper epidermis and have thin walls; their anticlinal walls are wavy, as seen in surface view. Stomata are only found on the lower leaf side. Hairs mainly occur in the lower epidermis, but are not very frequent.

Most important is the structure of the midrib in which an open arc of collateral vascular bundles is present; additional bundles are irregularly dispersed within the arc. Each bundle has a sclerenchymatous cap on the phloem side. The xylem is mainly directed towards the center of the midrib. Collenchyma is present on the upper as well as on the lower leaf side. Glandular hairs are likewise found on the midrib.

This species is mainly distinguished from the species of *Espeletia* studied by its lack of crypts, the arrangement of the vascular bundles in the midrib and the lobed walls of the palisade cells, as well as by the absence of the subepidermal upper water-storing hypodermis. The species was collected in the páramos near Mérida at an altitude of 2250 m (ACOSTA DE OBANDO 1979).

Carramboa pittieri (CUATR.) CUATR (Fig. 109 b). The upper epidermis of the leaf is single-layered and strongly cutinized on the outside. 1–2 layers of a water-storing hypodermis are found beneath the epidermis. The palisade parenchyma is 1–2 layered; where 2 layers of palisade parenchyma occur, the hypodermis is reduced to one layser or disappears completely (and vice versa). The spongy parenchyma consists of roundish cells which leave intercellular spaces between one another. The lower epidermis cells are smaller than those of the upper epidermis and have thinner walls. The stomata are confined to the lower leaf side. The leaf is very hairy through long pluricellular simple hairs. Glandular hairs of the twin type with a head are found on the midrib.

The midrib has a more complicated structure than that of *C. littlei*. Two horizontal rows of vascular bundles may be dinstinguished on the adaxial side, while additional bundles are irregularly dispersed over the abaxial side. Secretory ducts surround the vascular bundles, just as in *C. littlei*.

The vascular bundles of the blade (wings) are likewise collateral and have a sclerenchymatous cap on the phloem side, while a secretory duct occurs above the xylem.

C. pittieri was collected at an altitude of 2850 m (ACOSTA DE OBANDO 1979).

Carramboa rodriguezii (CUATR.) CUATR. A bilayered acuiferous hypodermis occurs beneath the single-layered upper epidermis. The outermost cells of the three-layered palisade parenchyma are the longest. The spongy parenchyma is relatively loose. The vascular bundles have a cap of sclerenchyma on both sides, phloem and xylem, but the fiber cap on the phloem side is better developed. A resin duct is found on the xylem side. The ribs formed by the lateral nerves project above the lower surface of the blade in a way similar to that of *Espeletia*, so that crypts arise. The stomata which are slightly elevated above the surface are confined to the crypts; these are filled with hairs. The lower epidermis cells are small and thin-walled.

In the midrib, an adaxial as well as an abaxial arc of vascular bundles is present, however, the xylem of the adaxial arc is not endoscopic, but directed towards the upper leaf side. In the center of the midrib, between the 2 arcs, the vascular bundles are irregularly dispersed over the section with the xylem pointing in different directions. Secretory canals surround the vascular bundles, but are also dispersed irregularly over the section. As the secretory canals surrounding the vascular bundles are very regularly arranged, an almost continuous circle around the vascular bundles results. Collenchyma is present on both upper and lower side of the midrib. Glandular hairs also occur in the epidermis of the midrib.

The species was collected at an altitude of 2600 m (ACOSTA DE OBANDO 1979).

Carramboa trujillensis (CUATR.) CUATR. The upper epidermis cells of the leaf are quite large, compared to the entire blade thickness, and their outer walls seem to be slimy. The cuticle is thin. The palisade parenchyma is one-layered and the palisade cells are short and broad. The spongy parenchyma is loose. The lower epidermis is small-celled and the stomata, which are confined to the lower leaf side, occur at epidermis level. The ribs of lateral veins of first order show a very rare arrangement of the vascular bundles, similar to that of the midrib. Several vascular bundles (3 or 4) are arranged in the form of an arc with the xylem ponting towards the center. On top of this arc (on the ventral leaf side) a single inverse vascular bundle is found with the xylem pointing towards the abaxial side. The ribs of lower order veins which contain only a single vascular bundle may show a secretory canal on the xylem side.

In the midrib, many vascular bundles are irregularly dispersed over the section; their xylem points mainly towards the inside.

The species was collected in the Andean páramos of Venezuela, at an altitude of 2850 m (ACOSTA DE OBANDO 1979).

Distinction of species of Carramboa and differences between Carramboa and Espeletia

The various species of Carramboa are mainly distinguished by the different arrangement of the vascular bundles in the midrib. As indicated already by ROTH (1996), the midrib supplies – besides the leaf petiole – very good criteria for the identification of species. The ribs of the lateral veins of first order of *C. trujillensis* show an arc of vascular bundles with an inverse bundle on top – instead of a single vascular bundle, as observed in most species of dicotyledons.

A water-storing hypodermis is only present in two species of *Carramboa* (*C. pittieri* and *C. rodriguezii*). The stomata are elevated above the surface in 3 species (*C. pittieri*, very conspicuous in *C. badilloi* and only slighly in *C. rodriguezii*). The palisade parenchyma is one-layered in *C. trujillensis*; 1–2 layered in *C. littlei*, *C. pittieri* and *C. badilloi*; and 2–3 layered in *C. rodriguezii*. The palisade cells are very short and broad in *C. trujillensis*, very long in *C. badilloi* and have incisions in *C. littlei*. Strong projection of ribs above the surface and formation of crypts is only observed in *C. rodriguezii*.

Secretory ducts with a resiniferous content are observed in all species of *Espeletia* and *Carramboa* studied. As compared with the species of *Espeletia* studied, the leaves of *Carramboa* possess a midrib with a different arrangement of the vascular bundles. Pluricellular hairs are present in *Carramboa* as well as in *Espeletia*, but in *Carramboa* no thick felt-like indumentum is developed.

Ethnobotanical and general use of Espeletia and Carramboa

Economical utilization

Stems, leaves and shoots are used by the native people. Leaves of Espeletia are occasionally used as a substitute for an overcoat by Indians, on chilly nights (HUMBOLDT 1849). The woolly leaf type is used in this case. The nights may be very chilly in the páramos and the low atmospheric pressure as well as the low oxygen content in the air can easily lead to circulation problems in humans. The leaves and their hairs however are not only fashionable as an alternative garment they are also stuffed into mattresses. The wood of the arboreous types is used as a building material for walls and roofs of their ranchos (huts). Leaves of Espeletia are also used for roofing of the huts.

The strong smelling resin, present in the resin canals, is locally applied for the manufacture of soap, of incense and of gun powder or fire works – products which do not fit very well with a moral point of view! The resin has a smell of turpentine. Shoots and leaves which contain a large quantity of resin canals are used as candles during religious processions and feasts.

Besides, the plants with their showy yellow flowers are also used as ornamentals and bouquets and other decorations are made of them.

Medical use

The etheric oils in the glands and the resinous balms in the secretory canals are the principal substances applied in popular medicine. It is asserted that these substances cure asthma, a sickness not infrequently occurring in the high mountains of the Andes. They also have a healing influence on rheumatism and bronchitis and even on 'hysterical paralysis' (CORTES 1917).

Leaf. A decoction of the leaves is sudorific and cures pulmonary diseases as well as asthma. Leaves in decoction are also used as a cataplasm for the mentioned diseases.

To heal venous diseases (phlebitis), the natives use fresh leaves.

Put into the ear the indumentum of the leaves relieves earches.

Fresh leaves applied on the skin cure telangiectasis, small dark red spots on the skin formed by swollen capillaries.

Inflorescences. Inflorescences in decoction taken for 2–3 days with water heal injuries of the matrix.

At the German Homoeopathic Union, *Espeletia* is listed as an officinal plant of European standard called *Espeletia* D3, D4 dil. and is applied for stenocardia of various origin, against nicotine addiction (D3), claudicatio intermittens and smoker's leg.

Method of use

Fresh leaves, leaves or inflorescences in decoction, applied externally or internally. Indumentum put into the ear.

Healing properties

Sudorific, against asthma and bronchitis, phlebitis, telangiectasy, rheumatic pains.

Chemical contents

Etheric oil (cajeputol), balsamic resin, new hydroxyacetophenone derivatives, acetoxykaurenoic acid, kauradienoic acid, kaurinic acids, grandifloric acid, kauranoid diterpenes.

Related species

About 54 species of Espeletia in the Andes. Related genus: Carramboa.

Observations

Espeletia has been known in popular medicine for hundreds of years and formerly was much used by the Andean Indians. Pittier (1926) already knew and described 14 different species and their popular use. *E. schultzii* WEDD. and *E. grandiflora* HUMBOLDT & BONPLAND are very well known. Most of the Espeletias are rosette plants and only a few are arboreous. The species are best distinguished by their flowers and leaves; particularly the leaf anatomy supplies good criteria of distinction: e.g. the structure of the midrib, formation of crypts, indumentum, glands and distribution of secretory canals.

In Mérida, the capital of the Andean region, little stalls called herbolarios have small assortments of Andean medicinal plants, including *Espeletia*. However, the sellers only offer the useful parts (organs) of the plants so that it becomes difficult, even for a botanist, to identify the material.

The natives use all species of Espeletia in the same way; also the genus Carramboa which formerly belonged to Espeletia.

Eupatorium Steetzii ROB. (ayapana, diapana, guaco)

Taxonomical description

Eupatorium Steetzii ROB. (Fig. 110) is a shrub 1.5–2 m high which occurs at open sites of the subpáramo and of the cloud forest at elevations of 2000–2200 m (1900–2600 m) a.s.l. It has more or less heart-shaped leaves with acute tips and slightly serrate margins. Both leaf sides are pubescent. The leaves reach 6–7 cm in length and have a 2 cm long petiole.

E. Steetzii var. *macrocephalum* ARIST. is similar to the former species, but has larger floral heads.

Anatomical description

Leaf. The leaf is very thin and only consists of a few cell layers. The epidermis cells are large and transparent. The cuticle is relatively thin. Glandular hairs of an enormous size are very conspicuous on the upper side. They have a thick pluricellular stalk and a very large unicellular head which may contain volatile oils.

Two layers of palisade cells follow beneath. The cells of the outer layer are longer, while those of the second layer are shorter and possibly have a collecting function.

The spongy parenchyma occupies less space than the palisade parenchyma, being composed of more or less roundish cells. The lower smaller-celled epidermis accommodates the stomata which lie very close together almost side by side. They are of the hygromorphic type being elevated above the surface. The substomatal chamber is very large. Short pluricellular hairs as well as glandular hairs of the type *Chrysanthemum cinerariaefolium* (METCALFE & CHALK 1950) with a short stalk and a 2-cellular head are found in the lower epidermis.

The leaf is covered by a thick indumentum on both sides which is an excellent protection against excessive insolation and drought.

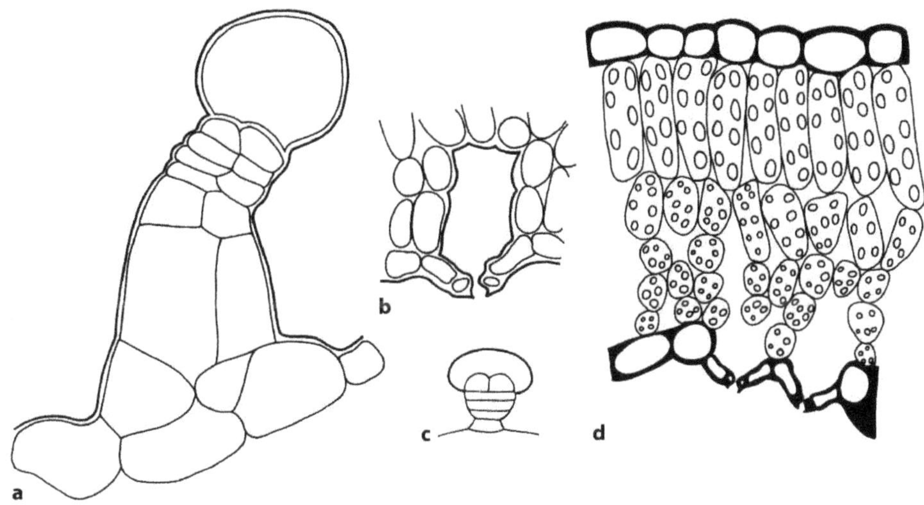

Fig. 110. *Eupatorium Stetzii*, Compositae. **a** Glandular hair. **b** stomatal apparaturs. **c** Glandular hair with 2-celled head. **d** T.s. of leaf (ROTH 1995).

Ethnobotanical and general use

More than 70 species of *Eupatorium* are known and used in popular medicine. About 90 % of all the species found in Colombia are used in popular medicine (CORREA & BERNAL 1990), because many are very much alike. Most of the species therefore have the same applications.

Volatile oils in the glands are possibly the most important substances. Besides the rootstocks, it is the leaves that are mainly applied either fresh, macerated, dried, as cataplasms, in infusions, decoctions and as teas.

The medicinal properties are manifold: aromatic, astringent, sudorific, antispasmodic, diuretic, digestive, carminative, digestive, stimulant, tonic, haemostatic, ophthalmic, disinfectant, antiseptic, febrifuge, vulnerary, cicatrizing, antirheumatic, diaphoretic, cathartic, emetic, antiinflammatory.

The plants are applied for oedema, diabetes, diarrhoea, cholera, ulcer, colds, kidney ailments, eczema, pruritus, erysipelas, sarna, colics, cough, inflammation of the throat, as an expectorant, against asthma, gangrene, malaria, epilepsy, as an antitumoral, against cancer, as an emmenagogum, and even against syphilis.

The species of *Eupatorium* recently aroused public interest, due to the antitumoral properties found in *E. formosanum, E. rotundifolium, E. semiserratum, E. cuneifolium* and *E. rhomboideum.*

The juice of the plant is also used to cure insect stings and against bites of poisonous snakes.

Fresh or dried entire plants are put into closets or fastened on doors and used as an insect repellent.

The bitter-aromatic leaves of *E. collinum* DC are used instead of hops in beer manufacture.

The coumarin-scented leaves of *E. dalea* L. are used as a substitute of vanilla.

The leaves of *E. indigofera* Parodi and *E. laeve* supply a blue dye. A green dye is extracted from *E. albicaule.* Fresh leaves of *E. ayapana* macerated with human urine supply a blue dye.

The plants are even used in veterinary medicine against colics.

MANFRED (1982) ascirbes to almost all species found in Venezuela the same medicinal properties. He recommends them against congestions and to cure bruises; the application is made externally. Excellent results were obtained with tumours and scrofula. It is applied for colics and stomach pain as well. In these cases, 3 to 4 soupspoonful of the fresh sap are taken on an empty stomach. Women apply it for a regular menstruation without pain: three glasses of the sap are taken with sugar or syrup a day, 3 or 4 days before menstruation is expected.

A strong decoction of the plants gives excellent results with gout and ulcer, particularly of a syphilitic origin. Cicatrization is very rapid in these cases.

A tea is prepared with one or two handfuls of the herb in half a liter of boiling water which is taken with sugar one cup after another.

MANFRED (1982) also emphasizes the curative properties of *Eupatorium* on snake bites; this action was discovered by MUTIS in Bogota already in 1788. In Venezuela, *Eupatorium* has even the reputation of being an antidote to the poison of snake bites. In certain regions, where snakes occur very frequently, the natives apply a preventive treatment against bites of poisonous snakes: First, they take 3 spoonfuls of the recently extracted sap of *Eupatorium* during 3 consecutive days. Then they make some form of a vaccination with the sap on different parts of the body. After this inoculation, they take 2–3 spoonful of the sap a day during 5 days every months.

The chemical compounds recently obtained from *Eupatorium* are numerous: essential oils, bitter glycosides, resins, di and triterpenes, sesquiterpenes, flavonoids, β-sitosterol, kaempferol, saponins (beer manufacture), coumarin, rutin, euserotonin, quercetin, euparin, eupatolin, eupatolitin, eupatoretin, eupatoriopicrin, eupatoriochromene, epifriedelinol, laevigatin, feroxone, kaurenic acid, pyrrolizidine alkaloids (lindelofine and supinine) etc. etc.

A new germacardenolid with cytotoxic activity has been isolated from *E. rhomboideum: euparhombin.* In other species, cytotoxic flavones which are tumour inhibitors were detected. Eupatorin acetate, a novel guaianolide tumour inhibitor was found in *E. rotundifolium.* Eupachlorin acetate is a new chlorosesquiterpenoid lactone and tumour inhibitor from *E. rotundifolium.* Eupacunin is a new antileukaemic sesquiterpene lactone from *E. cuneifolium.* Eupasserin and deacetyleupaserrin are new antileukaemic sesquiterpene lactones from *E. semiserratum.* Eupatolide is the major cytotoxic principle and an antitumour agent of *E. formosanum.* These cytotoxic compounds have been detected relatively recently, when the antitumoural activity of several species of *Eupatorium* became known from popular medicine (see also CORREA & BERNAL and the cited bibliography: KUPCHAN et al. 1969–1973, and other authors).

An antimicrobial activity of several species of *Eupatorium* is also known. An antibiotic principle was found in *E. capillifolium.* Immunologically active polysaccharides could be isolated from *E. cannabinum.*

Eupatorium can also be applied as an insecticide. Euponin is a new epoxy sesquiterpene lactone inhibiting insect development, isolated from *E. japonicum.* Likewise coumarin acts as an insect larval growth inhibitor.

Besides its manifold qualities as a medicinal agent, an insecticide, an adulterant of hops and vanilla, and a dye, the genus also serves for the preparation of biogas.

Most of the species have the same applications.

Eupatorium steetzii has very large glandular hairs which occur in abundance on the leaves. They contain volatile oils in large quantities and these may probably be used in the same way as those of other species. The species is therefore a really promising plant.

Gnaphalium antennarioides DC. (viravira paramera)

Taxonomical description

Gnaphalium antennarioides DC. (Fig. 111 a) is a perennial herb which propagates itself by stolons. The aerial axis reaches about 30–60 cm in length. The caulinar leaves which are alternately arranged, are of linear shape and 5–8 cm long by a width of 2–4 mm. Axis and leaves are covered with a fine white indumentum.

In the Venezuelan páramos, the species occurs between 3300 and 3600 m (3700 m), according to VARESCHI 1970.

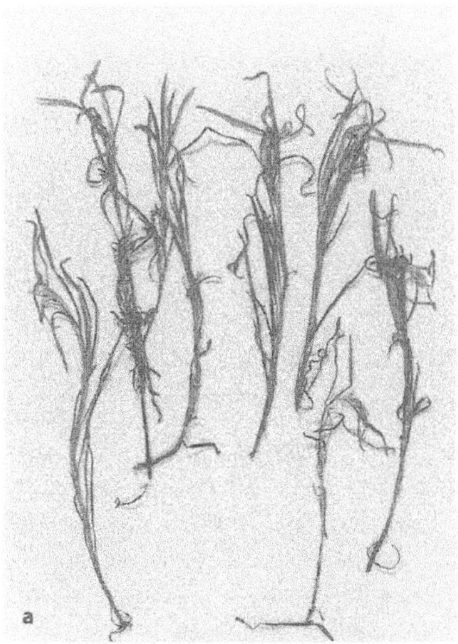

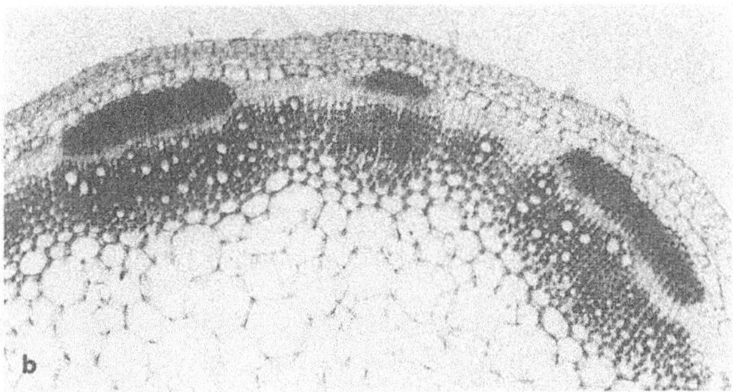

Fig. 111. *Gnaphalium antennarioides.* **a** habitus. **b** T.s. of axis.

Anatomical description

Leaf. A cover of uniseriate thick-walled very long and curled hairs surrounds the leaf. This dense indumentum reaches the thickness of the entire transverse section of the leaf. The hairs are pluricellular and their base is enlarged. The upper epidermis is very largecelled and sometimes slightly papillose, has somewhat tickened cell walls and serves as a water-storing tissue. It may occupy as much as one third of the entire transection of the leaf. The palisade parenchyma is partly single-layered and the palisade cells are relatively broad reaching a length/width index of about 3.5–4.5 which is low. The spongy parenchyma consists of cells with an irregular shape and is comparatively compact. Between the ribs, a transitional layer between palisade and spongy parenchyma may be found, which can be interpreted as a second palisade layer, since the cells are anticlinally elongated. The spongy parenchyma is small, giving room to an abaxial layer of palisade parenchyma on the lower leaf side. The spongy parenchyma thus becomes reduced to 2–3 layers of cells which occupy less space than the palisade parenchyma.

The midrib is very well developed, projecting over the lower leaf side; two laterals of second order likewise become pronounced on this side. The leaf margins are slightly curved backwards, so that together with the projecting ribs 4 slight furrows arise on the lower side which are arched over by 4 blade sections. A transitional case of crypt formation thus occurs. Each of the ribs contains a collateral vascular bundle, while the margins possess a bundle of reduced size which may be considered submarginal. Further small bundles of a third order are found in the fields between the pronounced ribs, about 3–5 in each field, as seen in transverse section. Very conspicuous is the parenchymatous sheath which surrounds each bundle of a third order. The bundles are transcurrent to the upper as well as to the lower leaf side by transparent cells with thick walls that may serve for water conduction and storage. The principal nerve and the 2 pronounced laterals are strengthened by caps of sclerenchyma on both sides. Below the bundles, the lower epidermis is also larger-celled and thicker-walled resembling the water-storing upper epidermis. But between the median bundle and the 2 proncounced laterals, the lower epidermis becomes small-celled and the stomata – confined to these sites – are somewhat elevated above the surface. In addition to the long curled hairs, which are found on both leaf sides, and dispersed between them, are glandular hairs with a pluricellular stalk and likewise a pluricellular head composed of about 2–4 cells.

The leaf is thus very xeromorphic and of the sun type.

Axis. (Fig. 111 b, 112) An axis of about 2 mm in diameter shows an extremely well developed parenchymatous pith and a ring of collateral vascular bundles which is already closed by the action of the interfascicular cambium.

The phloem side of the bundles is reinforced by caps or arcs of sclerenchyma. The epidermis is small-celled and the long pluricellular hairs as well as the short glandular hairs with a pluricellular head are present in the epidermis; both hair types are intermingled with one another.

The primary cortex is very small consisting of about 3–4 layers of small parenchyma cells, which have occasionally thickened walls and large pits. Very conspicuous is the endodermis below, which is extraordinarily large-celled and has well defined Casparian strips in the radial walls. The pericycle is small-celled and inconspicuous. The size of the pith cells increases towards the center.

Ethnobotanical and general use

Medical use

The various species of *Gnaphalium* are often used in the same way. A decoction of the entire plant has sudorific and febrifuge effects and serves also as a purifier and against syphilis. It is applied in diseases of the breast, bronchitis or pulmonary catarrh, and influenza. The decoction is a specific expectorant.

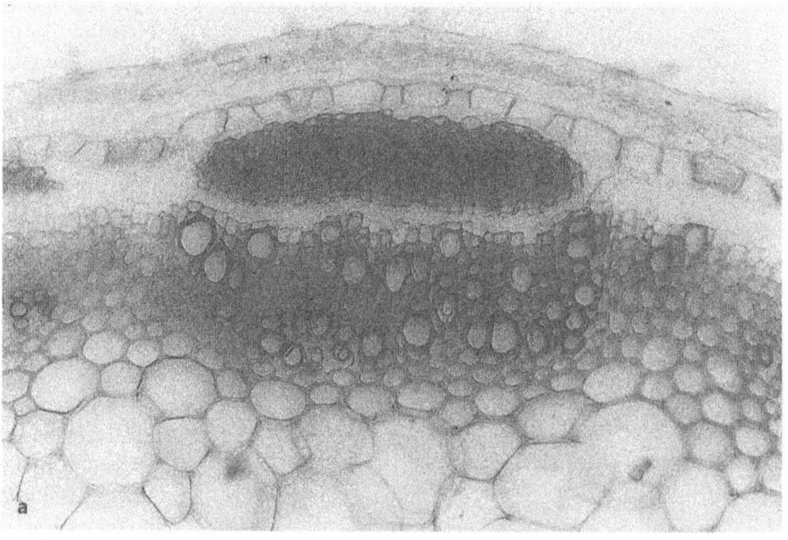

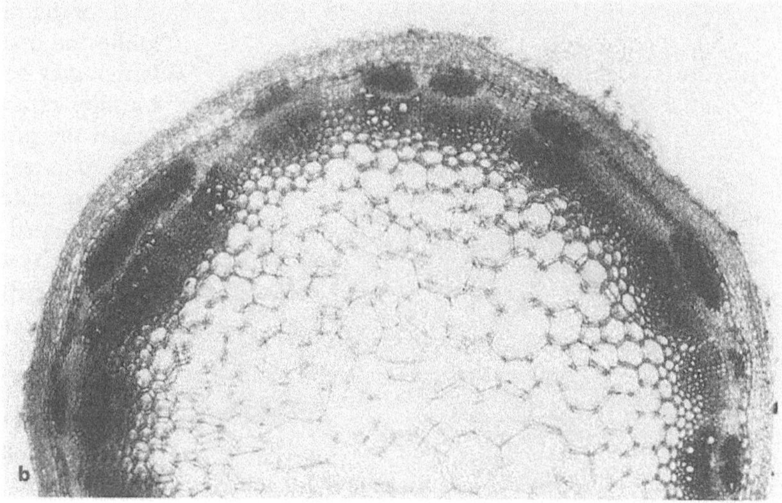

Fig. 112. *Gnaphalium antennarioides.* T.s. of axis. **a** Details with endodermis above the phloem. **b** General view.

Fig. 113 a, b. *Monticalia rex*, twigs. Partly defoliated axis to the right. Note the coriaceous leaves with the woolly inside.

A decoction of equal parts (1 g) of *Gnaphalium*, flowers of *Sambucus*, of *Viola* and *Malva* in 2 cups of water or a decoction of *Gnaphalium*, *Origanum* and cacao butter equally serves as an expectorant.

Gnaphalium elegans HBK (Yerba de la vida, Mirabira, Viravira), growing in the savanna and the subpáramos between 900 and 2200 m, gives good results against cancer and inflammation of the prostate. In infusion, it is used against dyspepsia and gastralgia. As a gargle, it is applied against mouth ulcers. As a 20% decoction it facilitates menstruation. In the form of a bath or cataplasm, it is applied against skin conditions, for wounds, swellings, and as a vulnerary.

In infusion, it is used against influenza, and as a vaginal bath. A decoction is used to cure cancer and inflammations of the prostate.

Observations

Very characteristic is the endodermis of the axis, the water-storing epidermis of the leaf with 2 kinds of hairs, the occurrence of palisade parenchym on both leaf sides, the transcurrent vascular bundles, and the pronounced ribs on the lower side, almost forming crypts.

Hinterhubera imbricata CUATR. & ARISTEGUIETA. A further species found in the Venezuelan páramos of the Andes is *Hinterhubera imbricata* (Fig. 107 b), a dwarf shrub, with ericoid 'rolled' leaves densely arranged on the axis. The plant is found at altitudes of 4000 m together with *Espeletia timotensis*, an arboreous species.

The leaf anatomy has been thoroughly described by ROTH (1974). A review in English is given in ROTH (1995). The leaf is furnished with large biseriate glandular hairs on its lower epidermis. They probably contain volatile oils which may possibly be of medicinal use. The chemistry is not yet studied.

Monticalia rex (SANDW.) JEFFREY (pelo lindo, beautiful hair, reinosa)

Taxonomical description

The plant (Fig. 113) is a dwarf shrub with very characteristic leaves which are about 3.5–4 cm long and 0.8–1.2 cm broad. The leaves are extremely coriaceous and have strongly backwards rolled margins, a peculiarity frequently found in xerophytes (ericoid with revolute vernation). The pinnate venation is conspicuous on the upper glossy side. Sixteen to twenty lateral nerves of secondary order ramify from the principal nerve in alternating positions. Between the laterals there is a dense network of meshes, characteristic of xeromorphic leaves. The lower leaf side is completely covered with a dense woolly indumentum having the aspect of 'angel hair', similar to that of *Espeletia*. The web-like hair covering may reach a thickness of up to 2 mm. The shoot is completely covered by the leaves and has additionally the same indumentum so that it is sometimes difficult to trace back the origin of a hair. The upper leaf side, on the contrary, is glabrous and devoid of hairs. The leaf base is more or less cordate and the apex obtuse. The leaves are sessile and alternately arranged on the axis, covering one another.

The axis is woody and erect.

The species is typical of the Andean páramos, occurring between 2500 m and 4400 m.

This species was formerly included in the genus *Senecio* and has only recently been separated from it. It is however closely related to *Senecio*.

ARISTEGUIETA (1964) mentions 45 species of *Senecio* in the páramos, 11 of which are described by VATESCHI (1970).

Anatomical description

Leaf (Fig. 114, 116). The leaf is about 0.8–0.9 mm thick.

Fig. 114. *Monticalia rex.* **a** T.s. of leaf (× 16). **b** Leaves with their woolly lower side.

The upper epidermis cells have heavily thickened and cutinized outer periclinal walls which may occupy about the same or even the double width of the cell lumen, as can be seen in a transverse section. The entire outer walls stain orange-red with Sudan III, a reagent of Cutin. The palisade parenchyma is extremely well developed and generally comprises 3, sometimes 4, layers of very long palisade cells with extraordinarily thick walls, interrupted by more or less ovalate pits so that the cells remain in good contact with one another. The length/width index of the palisade cells of the outermost layer is 8–11. The outermost cells are the longest. The spongy parenchyma is small. The cells of the spongy parenchyma are tangentially elongated and thick-walled. The spongy parenchyma is however relatively loose. Secretory canals are dispersed in the mesophyll. The lower epidermis is small-celled and has numerous stomata which are conspicuously elevated above the surface. Very long curled hairs occur abundantly on the lower leaf surface; they are pluricellular and thick-walled. Vascular bundles are very frequent, the larger ones are accompanied by secretory canals on the phloem side. Additional sclerenchyma is not developed. Towards the margins, the upper epidermis becomes multilayered (2–3 layered).

Although the leaf of *Monticalia rex* is very coriaceous, it has no sclerenchyma and is therefore not a sclerophyll in the strict sense. The coriaceous nature of the leaf is due to the extremely thick outer walls of the upper epidermis cells and particularly to the extremely thick-walled palisade cells; thick walls in the palisade parenchyma represent a very outstanding feature, but in this way they prevent the palisade cells from collapsing. The spongy parenchyma cells are also thick-walled. Finally, the revolute margins of the leaf are strengthened by the thick-walled multilayered epidermis. Thus the leaf attains a very hard and leathery texture.

Stem (Fig. 115). The stem anatomy of the arboreous species of *Monticalia, Senecio* and *Espeletia* resembles rather that of very large (giant) herbs with a megaphytic habit. Parenchyma is abundantly present, e.g. in the form of a large pith or wide rays. The ground tissue of the wood consists of fibers. The vessels are small and have a relatively low density. The plenitude of parenchyma allows storage for water and reserve substances.

The very large pith of a twig of *Monticalia rex* measuring 0.6–0.7 cm in diameter is thick-walled and lignified, and has many pits. This tissue seems to serve mainly as a water reservoir. The many originally separated vascular bundles are still recognizable, but are already united in a continuous woody ring, as seen in a transsection. The medullary rays are mainly pluri-seriate (2–5) or more; their pitted cells are mostly thick-walled and lignified. The phloem is relatively well developed, and almost every original bundle has a fiber cap on the phloem side. Resin ducts are found above the larger well-developed fiber caps. The primary cortex is parenchymatous consisting of roundish cells which are mostly arranged in tangential rows; the rows partly separate from one another at their tangential walls so that large intercellular spaces arise between them. The epidermis with the curled hairs is still preserved. The cork is of a subepidermal origin, has thickened cell walls and is lignified. It serves as a supporting tissue. Vessels can not be well distinguished in a transection because of their very small size, but fibers are abundant.

On the phloem side, the radial rows of the pluriseriate rays which have thin-walled cells partly separate from one another on the radial walls, so that additional intercellular spaces arise. In this way, an aerenchyma develops which consists partly of phloem rays and partly of primary cortex. It is very necessary, as the stem is densely covered by the leathery leaves and the thick indumentum.

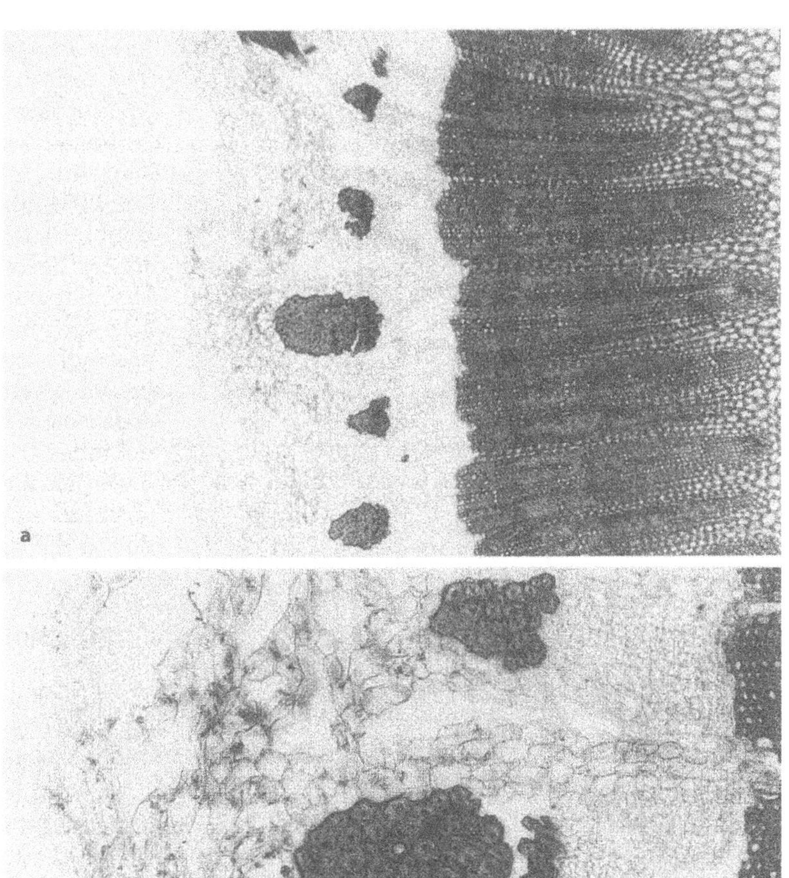

Fig. 115. *Monticalia rex.* **a, b** Axis in t.s. Note the arenchyma in the cortex **b.**

Ethnobotanical and general use

Medical use

The active ingredient of the plant is the resinous substance in the secretory canals of leaf and axis. The use is similar to that of species of *Senecio* and of *Espeletia*.

It is used for rheumatic pains, bronchitis and pulmonary conditions. The thick indumentum is put into the ears for ear ache.

Observations

The plant is easily recognized by its habitus: Axis densely beset with coriaceous leaves which have a glossy upper and a woolly lower side. Microscopically characteristic are the thick cutinized outer epidermis walls and the palisade cells with extraordinarily thick walls and large pits. Furthermore, the occurrence of large secretory canals surrounded by a wreath of secretory cells is notable.

Convolvulaceae

This family mainly occurs in the tropics and subtropics. It is represented by herbs and climbers or shrubs, rarely by trees. The showy crown is gamopetalous and often tubiform. The vernation is induplicative-valvat. The disc is annular. Carpels 2. The fruit is frequently a loculicidal capsule. Bicollateral vascular bundles are not infrequent. Articulated laticifers are characteristic.

Glycoretins which are characteristic of the family and have a strong laxative effect occur in the

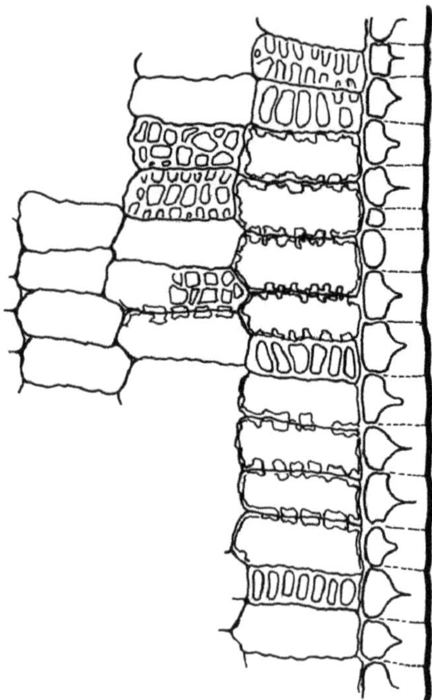

Fig. 116. *Monticalia rex*, t.s. of leaf. Note the thick outer walls of the cuticle and the thick-walled palisade parenchyma with large pits (ROTH 1995).

laticifers. Derivatives of lysergic acid are found in some species of *Ipomoea* (seeds). This explains the effects of the 'magic drug ololiuqui' prepared in Mexico. LSD is a derivative of lysergic acid.

Ipomoea pes-caprae SWEET. (bejuco de cadena)

This is a creeping herb with erect shoots and long stolons which form adventitious roots on the nodes with underground tubers. The leaves are succulent, 3–10 cm long, and of very variable shape, often bilobed (pes caprae), emarginate, truncate or heart-shaped. The corolla, 4–5 cm long, is white but has a purple center. The globular capsule, 1–1.5 cm in diameter contains 4 woolly seeds.

The pantropical species almost always occurs on sea shores.

In Venezuela it is common along the Caribbean beaches.

Anatomical description

Leaf. The leaf is thick and succulent and comprises about 12–14 mesophyll layers. Upper and lower epidermis are large-celled and water storing. The cells are thick-walled and cutinized and even the substomatal chamber is lined with thickened cutinized walls. The mesophyll is composed of short and broad palisade cells with an approximate length/width index of 1.8–2.6. Druses of calcium oxalate occur in the mesophyll cells. Intercellular spaces become larger towards the lower leaf side. Vascular bundles are very frequent and capped by sclerenchyma. Stomata occur on both surfaces, but are more numerous on the upper side. They are placed at epidermis level. Glandular hairs with a stalk and a pluricellular head are interpreted as hydathodes. They are larger on the upper than on the lower leaf side. The leaf is of equifacial or isolateral structure. The succulent consistence of the leaf is due to the turgescent mesophyll and epidermis.

Ethnobotanical and general use

Nutritional use

Young shoots are used as a vegetable and green fodder for animals.

Medical use

Root. A strong decoction of the root with 3 spoonful of honey is given in case of dysentery. The root is also purgative.

It is said that this species is more effective against venereal diseases and more active than *Smilax sp.* The shoots cut into pieces are put into cold water and the infusion is taken 3 times a day; it is warranted that any symptom of the sickness disappears within 9 to 14 days (PITTIER 1970).

Vapors of a decoction are used against rheumatism.

Leaf. The leaves are used as ingredients for a ritual bath by the Caribs.

Seed. The seeds may be used as a cathartic.

Varieties and related species

At least 20 species of *Impomoea* are used in one way or another. The most important species is however *I. batatas* (L.) POIR., the sweet potato, which has very starchy edible tuberous roots, prepared in different ways (more or less like potatoes). The tubers are sometimes even canned. They are also a source of starch, dextrine and alcohol.

There are several archaeologic proofs that *Ipomoea batatas* comes from the coastal Peru and was used there as far back as 4000 years B. C. Numerous varieties exist. The high chromosome number of the plant suggests a polyploid origin.

There are also other species, of which the roots, leaves or shoots are eaten. Other species are of medical use: emetic, purgative, cathartic, or are used for treating bites of rattlesnakes, for nervousness, paralysis, as a hydragogue, a cathartic, a tonic, alterative or aphrodisiac.

A few species are ornamental because of their large bell-shaped flowers.

Observations

The species is easily recognized by its outer habit, as well as by the thick succulent leaf structure.

Ipomoea phyllomega (VELL.) HOUSE. is a climbing herb. The leaves are simple and heart-shaped. The leaf anatomy has been described by ROTH (1992). The plant probably has similar properties to other related species. The glandular hairs of the leaf may contain volatile oils. Their chemistry is not yet known.

Ipomoea carnea JACQ. is used for rheumatism in the form of vapor. The root is purgative.

UPHOF (1968) mentions 21 species of *Ipomoea* which are useful. Several are cathartic and purgative, others aphrodisiac, curative for nervousness, for paralysis and to cure snake bites.

Cucurbitaceae

The Curcurbitaceae are frequently annual prostrate or scandent herbs with tendrils, distinguished by a rapid vegetative growth. The gynoecium, is 3-carpellary and unilocular, the fruit is usually a juicy berry, often called 'pepo'.

Bicollateral vascular bundles and large sieve tubes outside the vascular bundles, as well as cystoliths, are anatomical peculiarities.

Most members of the family are pantropical or subtropical.

The gourd family contains triterpene bitter principles, which are often localized in special idioblasts. Cucurbitacins are triterpenoid tetracyclic bitter principles which are responsible for the toxicity of many species. They also develop a laxative activity and were formerly used in laxantia. They have a necrotic effect on tumors.

Bitter tasting pentacyclic triterpene saponins are also found.

The cucumber, *Cucumis sativus*, the pumpkin, *Cucurbita pepo*, the sweet melon, *Cucumis melo*, and the water melon, *Citrullus lanatus*, with a red flesh belong to this family.

Several of the cucurbits were cultivated in pre-Columbian times. Cucurbits belong to the first cultivars of mankind and a great many varieties have since developed. Particularly the fruits show great variability in size, form, colour, flesh consistency and sugar content.

The seeds of *Cucurbita* have a very high fat content of 40–50 % and a protein content of about 30 %; they are very nutritious (for further information see BRÜCHER 1989).

Cucumis anguria L. (pepino de monte, pepino de sabana)

Taxonomical description

Cucumis anguria is an annual procumbent or climbing herb with simple tendrils (Fig. 117). The branches are ribbed and rough with hairs. The leaves are profoundly 3–5-lobular, cordate at the base and obtuse at the apex, with conspicuously dentate margins, 5–10 cm long and 4–8 cm broad, and of membranaceous to slightly leathery consistency. Venation on the lower side prominent and hairy. Petiole about 4–5 cm long and hairy.

The small flowers are yellow. The male flowers are agglomerated in axillary fascicles of 2–10, sessile or with a short peduncle of 1–2 cm, or, more seldom, solitary. Calyx 5–7 mm long and 5-lobed; corolla bell-shaped, stamens 4–5 mm long. Female flowers solitary with a peduncle 5–10 cm long (calyx and corolla as in the male flowers) and 3 staminodia.

Fruit globular or ovoid of the size of a hen's egg, yellowish, 2.5–5.2 cm long and 2–3.5 cm in diameter, covered with flexible spines up to 5 mm long. Seeds numerous elliptic, oblong, 4–5 mm long.

Origin

Tropical America. Of the 40 species known of *Cucumis* only *C. anguria* is considered indigenous to tropical America.

Occurrence

In America, it is found from Texas to Brazil and from sea level up to 1500 m. In Venezuela, the spe-

Fig. 117. a *Cucumis anguria*, fruits. b *Momordica charantia*, leaves, flower and fruit.

cies frequently occurs in hot regions of Nueva Esparta, Sucre, Monagas and Anzoátegui.

Anatomical description

Fruit (Fig. 118). The epidermis cells have anticlinal walls thickened in the form of pegs which are slightly elevated above the surface. Below follows a small-celled tissue the cells of which may transform into a tabular collenchyma. Small groups of stone cells arranged in the form of an interrupted ring lie below. The rest of the pericarp is parenchymatous; the tissue is more compact in the periphery, while it becomes more loose towards the center of the fruit. At the same time, the cell walls become more delicate. Bicollateral vascular bundles cross the pericarp in various directions. Glandular hairs with a spherical pluricellular head and stomata, elevated above the surface, are found in the fruit epidermis. Additionally, cystoliths raised above the surface by epidermis cells occur here and there.

Most conspicuous, however, are long emergences (Fig. 118) which originate from epidermal and subepidermal tissue and which may reach a length of up to 5 mm or even more. The epidermis cells are elongated parallel to the long axis of the protuberance. Stomata are found dispersed in the epidermis, but are more frequently found towards the base of the emergence. The rounded apex of the emergence terminates in a uniseriate simple hair with an acute tip and a slight sculpturing on its outside. Cystoliths in the center of wart-like projections are mainly found on the surface near the basal part of the protuberance. With these characteristics, the species is anatomically well defined as a Cucurbitacea.

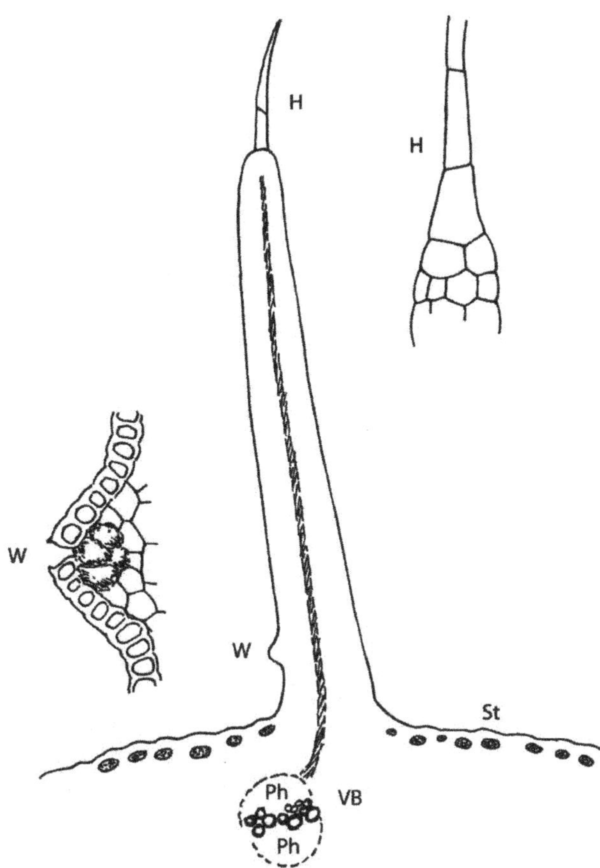

Fig. 118. *Cucumis anguria*, outer part of fruit in t.s. Note the small groups of stone cells (st), bicollateral vascular bundles (vb) with phloem on both sides (ph), large emergences (H), stomata on wart-like protuberances (w). The emergences are vascularized and terminte in a fine hair.

Ethnobotanical and general use

Fruit. The fruit is edible and mainly used in the local cuisine. The fruits have an agreable sour-sweet taste. They can be eaten raw, but are mostly used in the preparation of sweets with syrup and to add flavour to cooked meals with legumes or meat and to salads.

Chemical contents

BROWN et al. (1969) used paper chromatography of flavonoids to identify species of *Cucumis* and came to the conclusion that the 5 species studied (*C. anguria, C. ficifolius, C. metuliferus, C. sativus* and *C. zeyheri*) belonged to the same group, although they are morphological distinguished.

Only the chemical contents of the cotyledons of *Cucumis anguria* were identified (STAUB et al. 1987) (see also BERNAL & CORREA 1991).

Varieties and related species

The variety *Cucumis anguria* var. *anguria* is known (STAUB et al. 1987).

Related species are *Cucumis sativus* L., the cucumber, *Cucumis melo* L., the melon, both with edible fruits. Other Cucurbitaceae with edible fruits are *Citrullus vulgaris* SCHRAD., the water-melon, and *Cucurbita pepo* L., the vegetable marrow. The fruits of *Luffa cylindrica* ROEM. supply the vegetable sponge (loofah) which consists of the resistant vascular system, the 'skeleton of the fruits' (SINNOTT & BLOCH 1943).

White bryony or english mandrake, formerly used to allay coughing in pleurisy, is the root of *Bryonia dioica* JACQ.

Lagenaria vulgaris, the bottle gourd, with many different fruit forms supplies another type of calabash which is often mistaken for the fruit of *Crescentia cujete*.

Cultivation

The species is propagated by seeds; it shows a rapid growth and needs well drained soils, heat and sunlight.

Observations

It is not absolutely certain that the material collected corresponds to *Cucumis anguria* – it could possibly also be *C. dipsaceus*. More anatomical studies would be necessary to clarify this point.

Gurania spinulosa (POEPP. & ENDL.) CIGN. is a climbing herb of the transitional cloud forest. The leaves are 3–5 lobed, each segment being of elliptic shape. The leaf anatomy has been described by ROTH (1990).

A tea of the roots is employed to correct faulty menstruation. Boiled leaves are applied to scrapes and wounds.

Luffa cylindrica (Fig. 119) not only supplies the sponge obtained from the fibers of the fruit, but also a homoeopathic remedy Luffa D 3 nose drops used for inflammations of the mucosa, rhinitis with formation of crusts and a dry, easily bleeding mucosa, for dry and slimy nose and pharynx catarrh, ozeana, and as an assistance in the treatment of pyogenic inflammations of the nose and paranasal sinus.

Luffa operculata. The fruits are used against hay fever and other allergies (WOLTERS 1992).

Fig. 119. *Luffa cylindrica*, Cucurbitaceae, fruit. Right: Peeled fruit exposing the vascular system which supplies the 'sponge'. Left: intact fruit (Roth 1987).

Momordica charantia L. (cundeamor, balsam pear, maravilla)

Taxonomical description

Momordica charantia is an annual climbing herb with tendrils (Fig. 120 a). The pentagonal stalk reaches about 2–4 mm or more in diameter, and is slightly pubescent or glabrous. The alternate palmatifid leaves are 4–8 (19) cm long and broad, having 5–7 segments. The membranaceous segments are obovate, and glabrous, except the veins, and have serrate margins. The tendrils occur opposite to the leaves. The male flowers are solitary; the peduncle is 5 cm long; the broad bell-shaped calyx is 3 mm long, while the 5 ovate lobules reach 4 mm in width. The female flowers are likewise axillary and solitary, but have shorter peduncles. Calyx and ovary

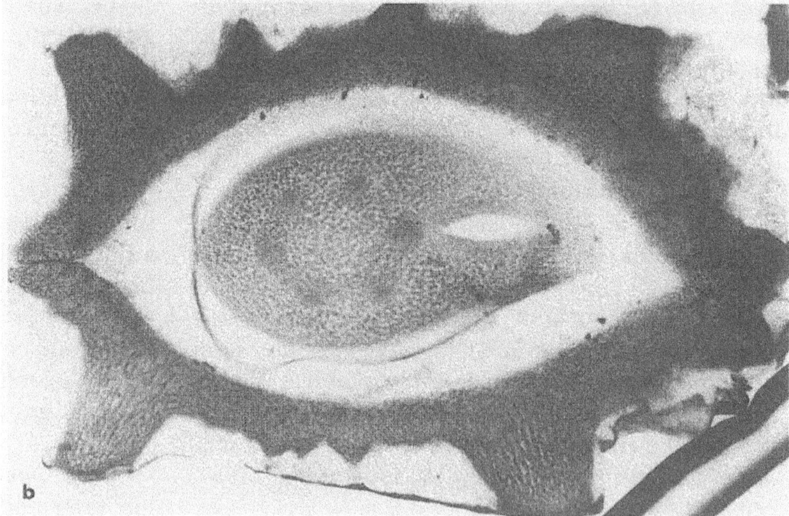

Fig. 120. *Momordica charantia*. **a** Flower and opened fruit. **b** Seed cut transversely.

are 1 cm long, the oblong lobules of the calyx reach 2–3 mm in length. The 5 petals are of yellow colour.

The fleshy orange-coloured fruit is 4–6 (15) cm long, constricted towards the top and the base, and has 8 ribs with prominent warts. It opens with 3 irregular valves to reveal the brilliant gray or chocolate-coloured pendulous seeds which are 8–10 mm long. They are surrounded by a scarlet-coloured sarcotesta (aril) or pulp.

Origin

It is of tropical/subtropical origin and probably introduced into the New World.

Occurrence

In Venezuela, the plant is common in hot as well as in temperate regions. It is found in forests and disturbed places at low altitudes.

Anatomical description

Leaf (Fig. 121). The leaf is very thin, dorsiventral and hypostomatic. The upper epidermis is papillose. There is only a single layer of palisade cells with a length/width index of about 3/1. The somewhat loose spongy parenchyma comprises about 2–3 layers. The cells develop arms towards one another so that small intercellular spaces arise. The cells of the lower epidermis are of the same size as those of the upper one or only slightly smaller. The stomata are slightly elevated above the surface. The leaf is more or less of the shade type.

As seen in a surface view, the anticlinal walls of both epidermal layers are wavy, but those of the lower epidermis more strongly so. The stomata which are confined to the lower epidermis do not show particular companion cells; they are anomocytic. Very large-celled simple uniseriate hairs with an acute tip, occasionally of a giant size, infrequently occur on the upper as well as on the lower leaf side. Club-shaped glandular hairs with a pluricellular head are likewise rare on both surfaces (Fig. 121c).

Very conspicuous are cystoliths which number 2, 3 or 4, in the lower epidermis and in the spongy parenchyma. They have a layered body (Fig. 124).

The midrib of the segments is very weakly developed. It is prominent on the lower surface, but sunk below the upper surface. It contains a single collateral vascular bundle with one (2) large vessel. Palisade paremchyma and spongy parenchyma are interrupted in the midrib and replaced by regular parenchyma. Towards the lower leaf side, a very weak collenchyma may be developed. The veins of higher order have a similar structure.

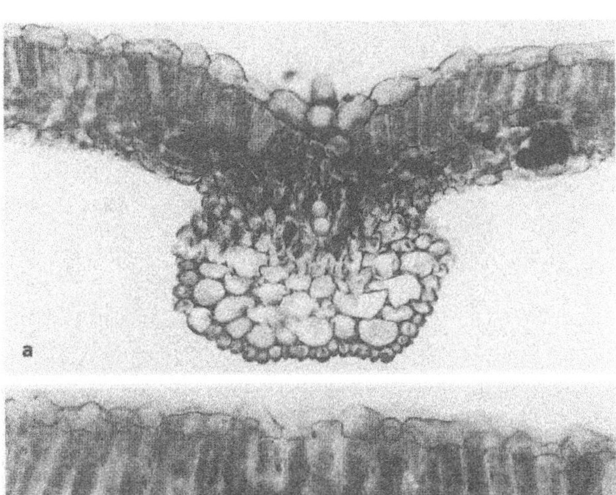

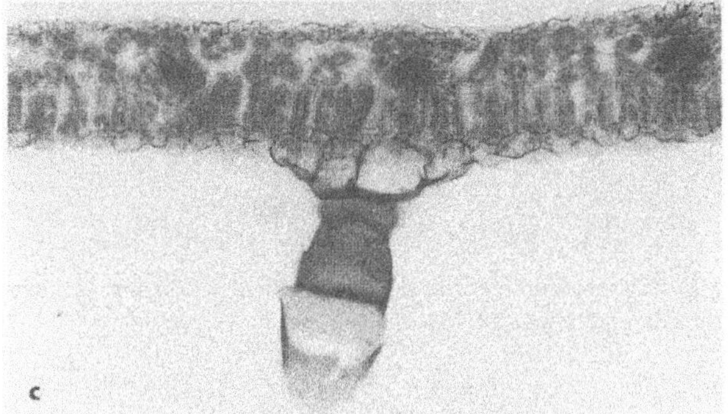

Fig. 121. *Momordica charantia.* **a** t.s. of midrib. **b** + **c** t.s. of leaf blade. Note the large glandular hair below (× 25).

Stalk (Fig. 122, 123, 124). As usual in the Cucurbitaceae, the stalk is pentagonal. There is a large bicollateral vascular bundle in each rib. Other 5 bundles which alternate with the outer ones, lie further inwards. However not all of them are fully developed: In the present case, 3 were bicollateral, while in the 2 others strong reductions took place; in one bundle, the outer phloem was almost completely missing, while the other bundle only consisted in a few phloem cells.

The ribs of the stalk are strengthened by collenchyma. The outer 5 bundles are surrounded by a continuous sclerenchymatous ring following the pentagonal outlines of the stalk. The filling tissue inside is parenchymatous, so that the largest cells lie in the central pith. Simple uniseriate hairs occur in the epidermis, which is papillose.

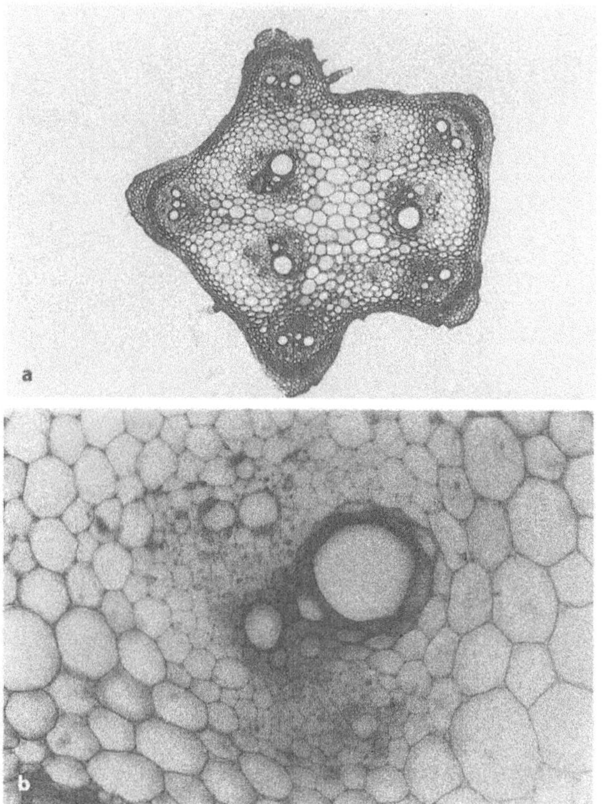

Fig. 122. *Momordica charantia*. **a** T.s. of axis. **b** inner vascular bundle of the axis (× 16).

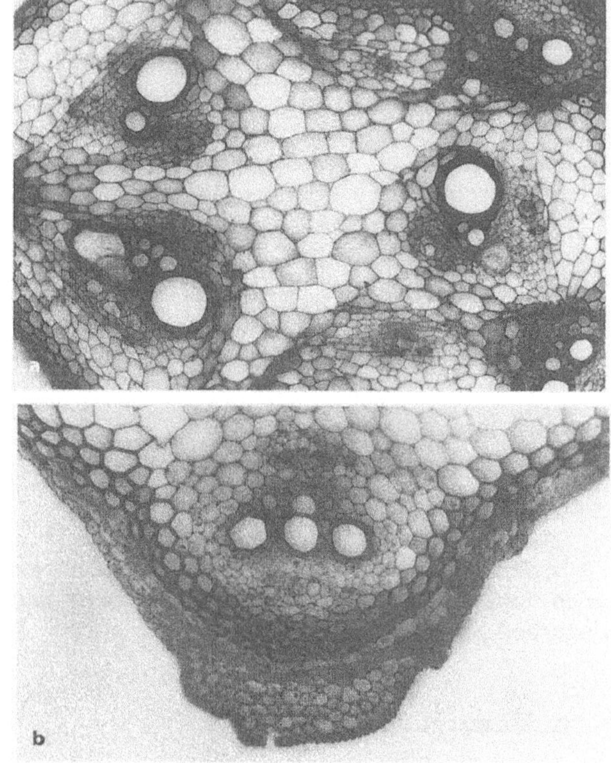

Fig. 123. *Momordica charantia*. T.s. of axis. **a** Inner part. **b** Edge (× 16).

Seed (Fig. 120 b). The seed coat shows the common typical pattern; it is largely derived from the exotesta. The outer epidermis is however regionally transformed into a sarcotesta consisting of radially much elongated thin-walled palisade cells with a juicy vacuole. Ourside, the cells are covered by a cuticle. The underlying tissue is slightly sclerenchymatous and small-celled. The main sclerenchymatous region consists of one to several layers of large cells with irregular outlines and very thick walls with many pits. In contact with this layer is the aerenchyma, composed of thin-walled cells which form lobes and short arms towards one another so that intercellular spaces arise. The chlorenchyma beneath consists of several layers of radially shoretend cells, while the inner epidermis is small-celled and inconspicuous. Vascular bundles lie inside the aerenchyma.

As seen in transverse section, the outer small-celled sclerenchymatous layer protrudes towards the outside by longer and shorter ribs: above the shorter ribs, the sarcotesta is reduced in radial length, while above the long ribs, the sarcotesta completely disappears to transform into a more or less regular epidermis. The sclerenchymatous cells of the ribs become larger and their walls may become sinuous and thicker-walled and pits are abundantly formed.

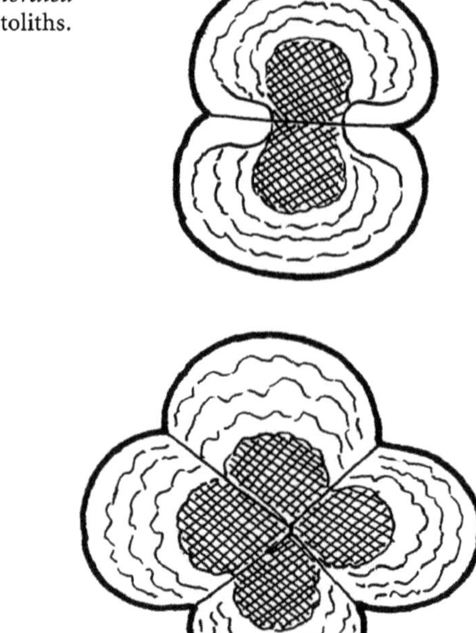

Fig. 124. *Momordica charantia*, cystoliths.

As seen in a longitudinal section, the sarcotesta is very well developed in the micropylar region.

Seeds with a sarcotesta are usually dispersed by zoochory. The sarcotesta therefore adopts a very attractive scarlet-red colour produced by anthocyanins in the vacuole. When the pericarp opens with 3 irregular valves, the sarcotesta is exposed. The attracting nutrient in the sarcotesta is probably a sugary solution. The sclerenchymatous layers beneath the sarcotesta protect the embryo from destruction, when the seed is consumed by animals.

Ethnobotanical and general use

Nutritional use

Young leaves and fresh shoots as well as fruits are consumed as vegetables. Green fruits and tender shoots contain bitter substances, a taste which is much appreciated by certain natives. The seeds are removed from the fruit before boiling them in hot water; they are also fried, while the leaves are par-boiled or used for salad. The sarcotesta of the seed has a sweet taste.

Economical utilization

In certain tropical countries, the plant is used as an insecticide.

In Haiti the plant is of great importance as an insecticide; it could be industrially used in the whole world. Processed fruits are applied as a substitute for soap. In the Philippines, the plant is used to prepare a poison for arrows.

The plant is cultivated, mainly for its medicinal properties.

Medical use

Leaves, stalks and shoots, roots, fruits and seeds, as well as the entire plants are used medically. Name of the drug: *Momordica charantia* L. plantae.

Leaf. The leaves are pungently aromatic due to the presence of glands. The sap is used against fever, colics and as a vermifuge. The leaves are emetic, purgative, and heal diseases of the gall bladder and burns on the sole of the foot. In decoction with lemon juice they are used as a purgative. As a decoction they are also applied against infections of the skin, psoriasis, dark spots on the face, as an emollient and astringent of the cutis. An infusion is used as a hypoglycaemic. A decoction is antipyretic and used against malaria and as a bath for haemorrhoids. Pulverized leaves are applied for cicatrization. To cure diabetes, 3–5 g/liter water are taken. The leaves are furthermore appetite stimulant, stomachic, have vermifuge effects and serve as an emmenagogue. An infusion is antipyretic, choleretic, antihypertensive, and helps to purify blood, against malaria, and exanthem. The juice is emetic and purgative. To soothe pain during menstruation, leaves are boiled for 2 minutes, a piece of *Euphorbia hirta* and a pinch of salt and a strip of orange peel are added; half a cup of this mixture is taken. Leaf tea is also applied against high blood pressure. A leaf decoction is against diabetes, hypertension, dysentery, malaria and as a vermifuge. Bath and poultice are used for rheumatism and haemorrhoids.

Branches. In decoction they expel stones from the kidney and reduce high blood pressure.

Root. The root is astringent; it is used against haemorrhoids. An infusion of crushed roots helps against malaria. The root in decoction is abortive. It also expels stones of the bladder. The root develops certain antibiotic activities.

Fruit. The juice of the fruit is employed against fever, colics, as a vermifuge and as a stomachic. It is locally applied against snake bites. It is furthermore used against tumours and malignant ulcers. In the form of jelly or syrup, it is given in cases of fever and malaria. In decoction it is utilized as an emetic and is an antivenereal. An infusion is laxative and anthelmintic. It helps against leucorrhoea. A cataplasm of the ripe fruit acts haemostatically. The ripe fruit mixed with oil is used to clean wounds and to heal bruises. The fruit juice is also used against diabetes through its hypoglycaemic effects.

Seeds. The seeds are slightly poisonous. They are emetic, purgative, vermifuge, have aphrodisiac effects and help against jaundice and excess of the bile production.

Entire plant. A decoction of the entire plant is used as a bath to cure haemorrhoids. The powdered plant helps against leprosy and malignant ulcers. The plant in infusion is taken as an contraceptive. The aerial parts of the plant in decoction taken with lemon juice are purgative. A decoction of the plant is also used as an antidiabetic: During the first days 3 cups a day are taken, then one cup a day only; urine should be controlled at the same time.

Method of use

As already indicated above, the plant parts and the entire plant are used in decoction, in infusion, as a cataplasm or a bath. The pure sap or juice is also applied directly. The decoctions and infusions have external as well as internal applications (see the respective organs).

Healing properties

A large number of effects is ascribed to the plant which is pantropical, being cultivated in several countries. It has therefore been studied scientifically.

The most important medical property of the plant is probably its hypoglycaemic action. It has an antifertility effect which could be interesting for the future. The antileukaemic effect also seems to be of much importance. It is anthelmintic, antimitotic, antirheumatic, antihypertensive, coleretic, antipyretic; it acts as an emmenagogum, an abortive, an aphrodisiac, is used against flatulence, ulcers, malignant tumours, fever, malaria, flu and many other diseases.

The plant has antimitotic effects and this may explain its antitumoural and antispermatogenic action; furthermore, inhibition of protein synthesis observed in *Momordica charantia* may have the same effects. An antiviral action against vesicular stomatitis virus was observed besides an antimutagenic activity and androgenic effects. Besides an antihyperglycaemic action, the plant is also efficient as an antihypercholesterolic. A gradual decrease of the sugar level in the blood of 42 % of patients was noted four hours after administration. The anthelmintic activity (against *ascaris*) could also be shown. The cytostatic activity of the plant could likewise be used medically in the struggle against cancer. A marked diminution of sperm formation was observed in dogs when the plant was taken for 20 days. The fruit extract has an apparent effect on the leucaemic lymphocytes of humans.

Chemical contents

An insulin-like compound was obtained by BALDWA & COL. 1977. WELIHINDA & COL. observed extrapancreatic effects of *Momordica charantia* (1986). Antimutagens (GUEVARA & COL. 1990), antitumour activities (MOTO 1983), antifertility activities (SAKSENA 1971) antilipolytic activities (WONG et al. 1985), have been proved and an inhibitor of HIV-I infection and replication as well (LEE-HUANG et al. 1990). The hypoglycaemic principle is called charantin.

Among others the following substances have been found: *Momordica* charantia lectin, *Momordica* agglutinin, cystostatic factor of *Momordica*, inhibitor of *Momordica* elastase, inhibitor of *Momordica* trypsin, neroldiol, V-insulin, P-insulin, β-sitosterol, derivatives of stigmasterol, 5-hydroxytriptamine, verbascoside, vicin, and an zeatin alkaloid.

The presence of saponins is responsible for the utilization of the plant as a soap substitute. The plant contains furthermore slime and volatile oil. The bitter substance is removed, when the leaves are shortly boiled and washed, but most of the ascorbic acid is lost during this process. The fruits are rich in ascorbic acid reaching a level of 188 mg/100 g fruit. The bitter alkaloid is momordicine; there are also bitter saponins in the plant. The seeds contain 2 alkaloids, one of which is momordicine, as well as a resin and a saponin, besides other compounds. The reddish-brown oil is the purgative factor; up to 32 % oily substances were found in the seeds. The unripe fruit is particularly rich in ascorbic acid and may be maintained in the refrigerator for 4 weeks at a temperature of 32–35 °F and a relative humidity of 85–90 %.

Toxicity

The fruit is toxic, has a drastic purgative effect which also causes vomiting, and its consumption has to be handled with care. It also induces abortion. The juice in India given to children causes vomiting, diarrhoea and death (PERKINS & PAYNE 1978). This remark of the cited authors throws an interesting light on the Indian mentality concerning women and children and the consequences of overpopulation!

Observations

The inner structure of the plant is very characteristic, particularly the cystoliths in the leaf, the angular outlines of the stalk, the bicollateral vascular bundles and the red sarcotesta of the seed.

Cunoniaceae

Shrubs or trees with simple or compound leaves and stipules. Flowers small, united in inflorescences; often without petals and hypogynous. Stamens twice as many as sepals, carpels 2. The fruit is a capsule often with winged seeds and a fleshy endosperm.

Tannins are present, occasionally cyanogenic compounds, saponins and alkaloids.

Weinmannia with about 190 species that are principally of South American origin, and mainly from the Andes.

Weinmannia lansbergiana ENGLER synonym: *W. venezuelensis* KILLIP & SMITH (sai)

The plant is a shrub of 1–2 m height. It occurs in the subpáramo of the Cordillera de la Costa between 2400 and 2600 m. The leaves are compound-imparipinnate. Each leaf has about 5–13 leaflets. These are glabrous and of ovalate shape, their margins are slightly serrate and the apex is obtuse. The rhachis is winged.

Sai is a collective name for *Weinmannia* in Venezuela. SCHNEE (1960) mentions 3 species which grow in the páramos or subpáramos. STEYERMARK & HUBER 1978 mention 4 further species for the Cordillera de la Costa (Avila), including *W. lansbergiana*.

Anatomical description

Leaf. The upper epidermis cells are very large; their outer walls are somewhat thickened. The inner walls are slimy; the layering of the inner walls is very conspicuous. This tissue serves as a water reservoir. Mucilaginous inner walls of the epi and hypodermis are characteristic of the Cunoniaceae and are found in all species studied from this family.

The well developed palisade parenchyma is 2-layered. The palisade cells of both layers are of about the same length. The spongy parenchyma occupies the same proportion of the leaf as the palisade parenchyma and consists of irregularly shaped cells. Druses of calcium oxalate occur in the entire mesophyll. The lower epidermis is small-celled. The stomata which are restricted to the lower leaf side, are at epidermis level. A very interesting feature is the occurrence of lenticels on the lower leaf surface.

The leaf is of the sun type and slightly xeromorphic (slime epidermis) (see also ROTH 1995).

Ethnobotanical and general use

About 14 useful species of *Weinmannia* are known (UPHOF 1968, SCHULTES & RAFFAUF 1990).

Several species have a fine hardwood with a pinkish hue which is used in turnery, for machinery bearings, in coachbulding, for tools, carving, for fuel and construction.

The bark is considered to be astringent and the secreted gum is used in Cuba to adulterate quinine (*W. pinnata*).

In other species, the bark is a source for tanning material or is used to give leather a red colour. The astringent bark is also applied on wounds.

The resinous leaves of quite a few species are applied against rheumatic pains. They are also an excellent remedy for painful joints; the leaves are rubbed on the afflicted area to reduce inflammation. The leaves are also boiled in water to clean the teeth and to harden the gums. A tea of the leaves with sugar serves as a stimulant and stops diarrhoea. Crushed fresh leaves are treated with warm water to prepare a poultice for open wounds. These crushed leaves are massaged on swollen and painful joints.

A decoction prepared by boilong flowers in water serves as an astringent wash for open wounds and helps to hasten healing.

The bark of *Weinmannia glabra* L. F. is rich in tannins and used for leather tanning; the vernacular name is therefore 'curtidero' (tan-bark). This Venezuelan species grows between 1600 and 2100 m. Its wood is used on a small scale (PITTIER 1970).

Dichapetalaceae

Tapura

The wood is useful. The fruits are edible. The leaves are toxic. Bark structure has been studied by ROTH 1981.

Ebenaceae

Diospyros

At least 39 useful species are known.

Ebony wood. The plant is used for dying and tanning. A gum is obtained from the fruit. Edible fruits. Fruits are partly used for insomnia and as an emmenagogum. Leaves are vermifuge and vermicid. Root and bark as a fish poison. Fruit for arrow poison. Seedlings are used as stock.

Fruit structure and dispersal have been described by ROTH 1987, leaf structure 1984.

Elaeocarpaceae

Sloanea (wood), Elaeocarpus, Aristotelia and Muntinga are useful genera.

Six species of Sloanea are cited by ROTH 1987, where fruit structure and dispersal are also described (Fig. 125). Bark structure has been studied by ROTH 1981.

Fig. 125. *Sloanea grandiflora*, Elaeocarpaceae, fruit. Opened fruit showing thickness of pericarp wall and length of the bristles (ROTH 1987).

Ericaceae

Subshrubs or shrubs with evergreen leaves and often with mycorrhiza. The flowers are obdiplostemonous. The calyx is persistent. The thecae of the stamens open with apical pores. The pollen appears in tetrads. The leaves are often xeromorphic.

The family belongs to the Bicornes with 2 hornlike protuberances on the anthers. It mainly occurs in temperate regions, and in the tropics it is mostly found in the mountains.

Gaultheria (procumbens) (laurel, albricias)

G. is a small shrub, popularily called laurel, because the leaf resembles a laurel leaf of coriaceous texture (Fig. 128 b). The ericaceous leaves are usually of hard and leathery texture. They can therefore be persistent and evergreen. The plants are mainly woody, often in the form of shrubs, and in the tropics they usually occur in mountainous regions. However, the species in question is possibly not correctly identified. *G. procumbens* is characteristic of atlantic North America and furnishes the fragrant 'oil of winter green'. It therefore seems doubtful to us that the species studied is really procumbens. VARESCHI (1970) does not mention this species as frequently occurring in the páramos. However, he comments on the difficulties identifying the species of this genus; STEYERMARK AND HUBER (1978) likewise point to the difficulties in identifying species of *Gaultheria*. VARESCHI (1970) refers to *G. cordifolia* HBK, synonym: *G. odorata* WILLD., a collective species in which also *G. coccinea* HBK and *G. brachybothrys* WEDD. are included as microspecies, distinguished by very minute differences so that he recommends a revision of the genus. Perhaps, the species in question also belongs to the collective group *Cordifolia*. This should be checked by taxonomists.

Taxonomical description

The species in question is a perennial shrub. Two samples were at our disposal: one with more or less smaller obovate leaves of a surface area between 1.5 and 4 cm^2, and the other with larger almost orbicular leaves with a surface area between 5 and 14 (15) cm^2. One sample is thus nanophyllous, the other microphyllous. The smaller leaf has an apiculate tip, the larger one an obtuse to almost retuse apex. The base is obtuse and the margins are slightly dentate in both cases. The leaves of both samples are glabrous and their dense venation is very conspicuous on both surfaces. Besides the principal nerve, there are about 4–5 strong laterals on each blade halve which are connected by a dense network of meshes. The leaf stalk is short. The phyllotaxis is alternate. The twigs are woody.

The species comes from the Andean region and was bought at a herbolario in Mérida, the capital of the Andean State.

Anatomical description

Leaf (Fig. 126). As seen in transverse section, the leaf has a strong midrib with a single collateral bundle, surrounded by sclerenchyma, and more or less transcurrent to the upper and lower epidermis by sclerenchyma on the upper side, and by collenchyma on the lower side. The stronger lateral bundles are transcurrent too. However, some thick-walled parenchyma cells may be in direct contact with the epidermis in both midrib and strong laterals. The upper epidermis is small-celled, but has very thick cutinized outer tangential walls, so that in some samples the wall thickness exceeds the cell lumen in width. As seen in surface view, the cuticle is ribbed or striated and the anticlinal walls are straight. A single-layered hypodermis follows beneath; the composing cells have thick walls with pits and are much larger than the epidermis cells: their function is water-storage. Locally, the hypodermis may become 2-layered. The underlaying palisade parenchyma is mainly 3-layered, but the

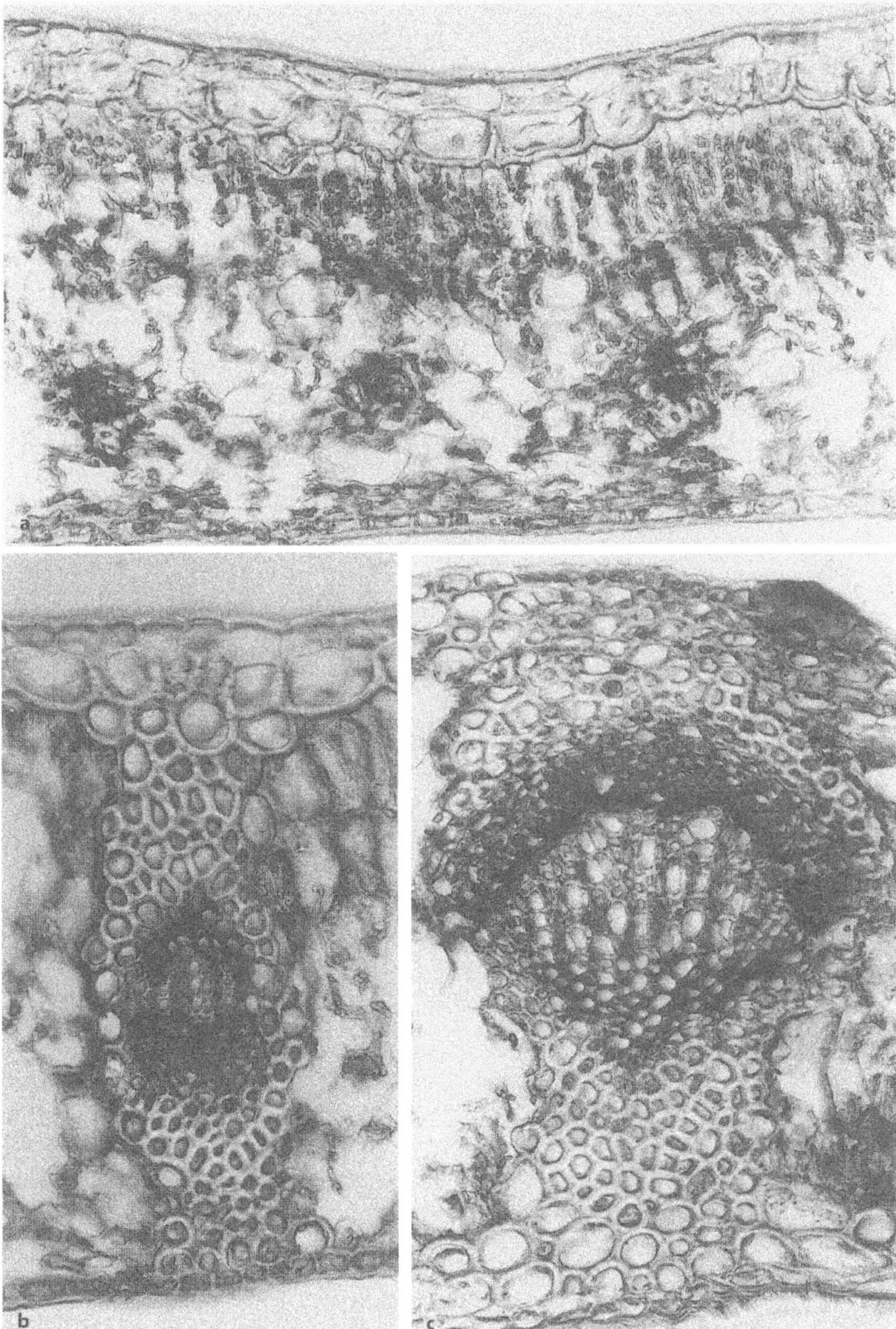

Fig. 126. *Gaultheria procumbens.* T.s. of Leaf. **a** with vascular bundles of higher order. **b** Stronger vascular bundle. **c** Midrib.

innermost layer may be considered a transitional one between palisade and spongy parenchyma; the cells of this layer are the shortest, while those of the uppermost layer are the longest. The length/width index of the palisade cells lies between 4.5 and 6. The spongy parenchyma is smaller than the palisade parenchyma and consists of irregularly shaped cells which extend short arms towards one another so that intercellular spaces develop. The consistency of the spongy parenchyma is thus loose. The lower epidermis is smaller-celled than the upper one and has thickened walls, particularly the outer tangential walls are thickened. The stomata confined to the lower epidermis – lie at the lower epidermis level being very frequent. Small druses of calcium oxalate are found here and there in the mesophyll, particularly in the spongy parenchyma. The leaf, which is relatively thick, receives its strength

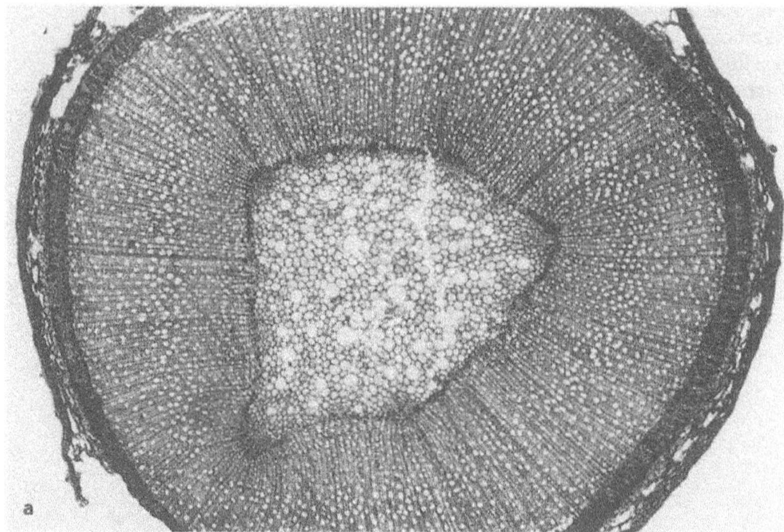

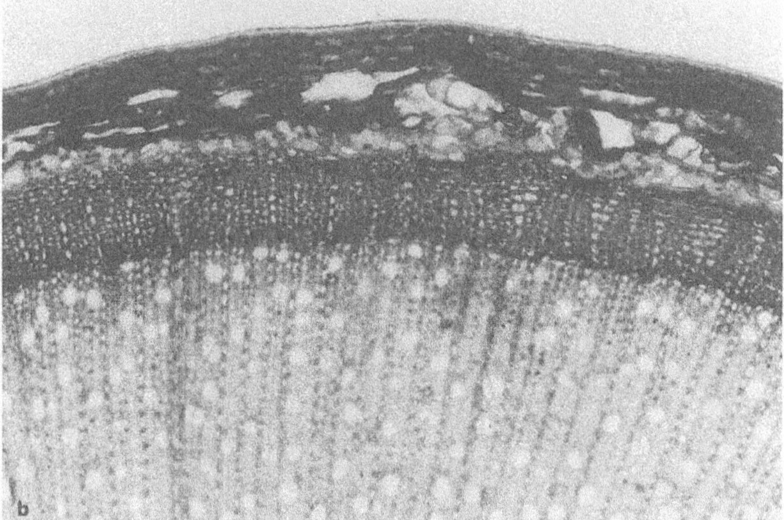

Fig. 127. *Gaultheria procumbens*. T.s. of axis. **a** Note the triangular pith. **b** Note the lacunar primary cortex, the small phloem (darker stained) and the extensive xylem.

mainly from the sclerified transcurrent midrib, the laterals, the thick-walled partly 2-layered hypodermis and from the thickened upper and lower epidermis cells.

Anatomically, the 2 morphological distinct samples do not differ from one another, so that it is probable that they belong to the same species.

A characteristic of importance found in both samples are the lenticels which are conspicuous on the lower leaf side, appearing as little brown dots; they may be useful as a feature of identification. They may arise from the scars of the glandular hairs.

Glandular hairs with a pluricellular bottle-shaped head occur at greater distances on the lower leaf side. Unicellular hairs and papillae are rarely found on the ribs.

Axis (Fig. 127). In an axis, 0.5 cm in diameter, the small-celled epidermis with thick outer tangential walls is still preserved. Beneath follow about 7 layers of the primary cortex; the outer 3 layers consist of thick-walled parenchymatous cells, while the inner 3–4 layers are composed of tangential rows of tangentially elongated cells which partly separated from one another along their tangential walls so that larger intercellular spaces arise. The fibrous pericycle is disrupted by the dilatation growth, but the original sclerenchymatous ring (as seen in transection) is already closed by the replacement of fibers through large stone cells.

The phloem is reduced, as compared with the xylem, and some of the uni to triseriate rays dilate towards the outside in the funnel form. The xylem is very well developed and the comparatively small vessels, solitary or in twins, are numerous.

Fibers are abundant: The large pith is more or less of triangular shape (a form induced by the phyllotaxis) and consists of slightly thick-walled

Fig. 128. **a** *Gentiana nevadensis*, entire plant with root. **b** *Gaultheria procumbens* with 2 different types of leaves (ROTH 1995).

parenchymatous cells with small pits. The cells are filled with solitary globular or compound starch grains, frequently in the form of twins or triplets.

The perianth of *Gaultheria* is fleshy (see ROTH 1977, Fig. 128.

Ethnobotanical and general use

According to LOPEZ PALACIOS (1987), the plant is sold at herbolarios in Mérida under the vernacular name laurel or albricias, in the form of branches. The plant has a strong aromatic smell of methyl salicylate.

It is used in the form of a bath against rheumatic pain, backache and sciatic pain.

Against rheumatism, equal parts of *Gaultheria, Pimenta racemosa* and *Ambrosia cumanensis* are put into 5 liters of hot water; a bath is taken in the morning and at night.

Observations

The plant has very characteristic leathery leaves. The leaf structure is distinguished by lenticels visible as minute brown spots on the lower surface, as well as by the transcurrent vascular bundles, as seen in a transverse section of the leaf. Glandular hairs are a further characteristic observed on the leaf.

Vaccinium meridionale SW., synonym: *V. caracasanum* H. B. K. is a small shrub of 0.4– m height or a small tree up to 8 m high with small simple leathery and glabrous leaves of oval shape; the leaf margins are delicately dentate and the tip is acute. The tubular flowers are white with pink or colourless stripes. The fleshy globular fruits are reminiscent of the European blueberry; they are of dark purple colour and edible, but have an acid taste which may be compensated for by adding milk and sugar. The species occurs in the cloud forest and in the súbparamos of the Cordillera de la Costa between 2000 and 2670 m (STYERMARK & HUBER 1978).

The leaf anatomy of this species was described by ROTH (1995, PP. 198–199). The leaf has a xeromorphic structure.

The plant is called billberry in Jamaica. The edible berries are made into a jelly or are used for tarts and pies.

UPHOF (1968) mentions 17 species of *Vaccinium* with edible berries, most of which however grow in more temperate regions.

Euphorbiaceae

The vegetative appearance of this family is very variable. The leaves are mostly simple and have stipules. Succulents are similar to cacti, but are distinguished by the presence of latex in laticifers. Latex is often poisonous. Flowers are small and inconspicuous, unisexual or dioecious, united in inflorescences. Dichasial ramification is frequently found. Partial inflorescences are often reduced to a pseudanthium (cyathium).

Euphorbia: involucral bracts occasionally showy (hypsophylls of *Poinsettia*) or with glandular appendages. Ovules with obturator and frequently with caruncle which in some species of *Euphorbia* is the basis of an elaiosome, dispersed by ants (myrmecochory). The fruits usually separate into 3 mericarps (cocci: Euphorbiales, synonym: Tricoccae) each dehiscing with 2 valves.

Tropical and subtropical, main distribution in tropical America and Africa.

The family is morphologically and chemically very heterogeneous. A double perianth and vestiges of the other sex are still found in *Jatropha curcas* and in many species of *Croton*.

Latex and tannins are very common in Euphorbiaceae. The most important compound of the milksap is caoutchouc (species of *Euphorbia, Hevea brasiliensis*). In certain species the latex contains specific ossiform starch grains. Alkaloids and cyanogenic compounds, such as linamarin, are frequently found, e.g. in the tuberous roots of *Manihot esculenta.*

The seeds are usually free of starch, but rich in fatty oils (oleum ricini). Ricinoleic acid (12-hydroxyderivative of oleic acid) exercises the purgative effect. The dangerous purgative effect of *Croton tiglium*, on the other hand, is brought about by a resinous substance. Seeds of Euphorbiaceae may contain very toxic polypeptides, such as curcin of *Jatropha curcas* and hurin of *Hura creptians* or ricin of *Ricinus communis.* Only a few seeds of *Ricinus* eaten by children can be fatal.

Acalypha alopecuroidea JACQ. (meona)

Taxonomical description

Annual herb, 20–40 (60) cm high, with quadrangular to roundish shoots, young twigs pubescent (Fig. 129). Simple leaves in alternating position, ovate (to triangular) with a cloth of hair on the adaxial side and above the nerves of the lower side (in older stages almost glabrous); 2.5–6.5 cm long and 2–4.5 cm broad; apex acuminate, base rounded off or truncate, margins conspicuously serrate. Venation pinnate, camptodromus, brochidodromus, with 7–8 pairs of secondary nerves; middle nerve prominent on the abaxial side. Petiole 2–3.5 (4) cm long, hairy, and flattened on the adaxial side. Lateral stipules of linear-lanceolate shape, approximately 0.2 cm long. Inflorescences spiciform; female spikes terminal, sessile 3–12 (20) cm long and

Fig. 129. *Acalypha alopecuroides.* **a** with inflorescences. **b** entire plant.

0.5–0.8 cm in diameter. Male spikes axillary, borne on a pedicel, 0.5–1.5 cm long, pedicels 0.3–0.8 cm long. Male flowers with 4 sepals and 8–16 stamens. Female flowers with 3–4 sepals and a trilocular ovary with 3 free ramified styles. Ovules solitary in the locules. Floral bracts triangular-ovate with an elongated setiform apex, divided into 3–5 segments, 0.7–1 cm long. The fruit is a tricoccous capsule, about 0.2 cm long.

Occurrence

The species occurs in the West-Indies and in Central America; in Venezuela it is found in the Tierra Caliente of the north.

Anatomical description

Leaf. The leaf (Fig. 130) is very thin, bifacial and amphistomatic, however the stomata are more numerous on the lower side. The upper epidermis is composed of relatively large cells with undulated anticlinal walls (as seen in a surface view). The cuticle is delicate. Stomata are relatively abundant and occur at epidermis level or are only slightly elevated above the surface. The guard cells are usually surrounded by 2 (1) subsidiary cells on either side (parallel type of development according to ROTH & CLAUSNITZER 1969 or paracytic according to METCALFE & CHALK 1950). As seen in a surface view, the lower epidermis cells are larger than those of the upper epidermis and their walls are more strongly undulated. Thick-walled long and slender unicellular hairs with acute tips mainly occur above the nerves of the lower leaf side, but some of them develop a few delicate transverse walls, becoming pluriseriate in this way.

Glandular hairs with a long pluricellular stalk and a pluricellular head occur on both leaf sides.

A single layer of palisade cells is found below the upper epidermis. The cells are about 2–3 times longer than broad, corresponding to a low length/

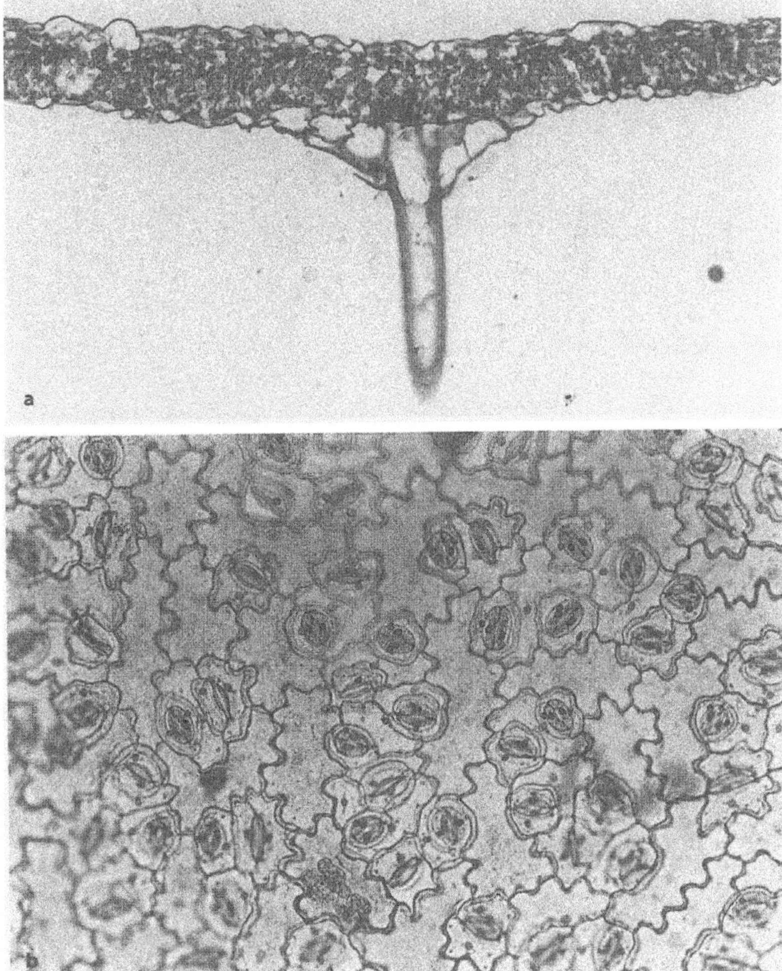

Fig. 130. *Acalypha alopecuroides.* **a** T.s. of leaf blade with enormous hair (× 25). **b** Surface view of lower epidermis.

width index. The spongy parenchyma composed of small cells occupies about 3 cell layers, reaching the width of the palisade parenchyma or slightly less. Very conspicuous are very large cells in the mesophyll (palisade parenchyma) which contain a single star-shaped druse of calcium oxalate.

The veins rise above the surface on both leaf sides, but more strongly so on the lower side. The middle nerve is composed of one or two very close collateral bundles. The palisade parenchyma is not interrupted above the bundles, but is continuous. Neither mechanical tissue nor a vascular sheath could be observed. The veins of higher order are possibly surrounded by a poorly defined parenchymatous sheath.

In its anatomical structure, the leaf belongs to the shade-type.

Shoot (Figs. 131, 132.1). An axis of about 0.6 cm in diameter is surrounded by a small periderm of subepidermic origin in which numerous lenticels can be observed. The primary cortex consisting of tangentially extended parenchyma cells in which druses of calcium oxalate frequently occur, is small. Small groups of fibers are intercalated in the parenchyma. The phloem is likewise small; darkly stained cells are conspicuous in it. The rays are mostly 1–2 seriate. The vessels are medium-sized and often occur in longer or shorter radial rows or in irregular groups. The main bulk of the xylem consists of fibers. The small pith in the center is composed of large parenchymatous cells and forms a hollow in the center.

Root (Figs. 131 b, 132.1). The root is very similar to the stem in its anatomical structure. A small periderm surrounds the root on its outside. The primary cortex is small, being composed of tangentially enlarged parenchymatous cells; druses of calcium oxalate are seldom found. As in the axis, the small phloem contains darkly stained cells. Most conspicuous is a ring of primary xylem surround-

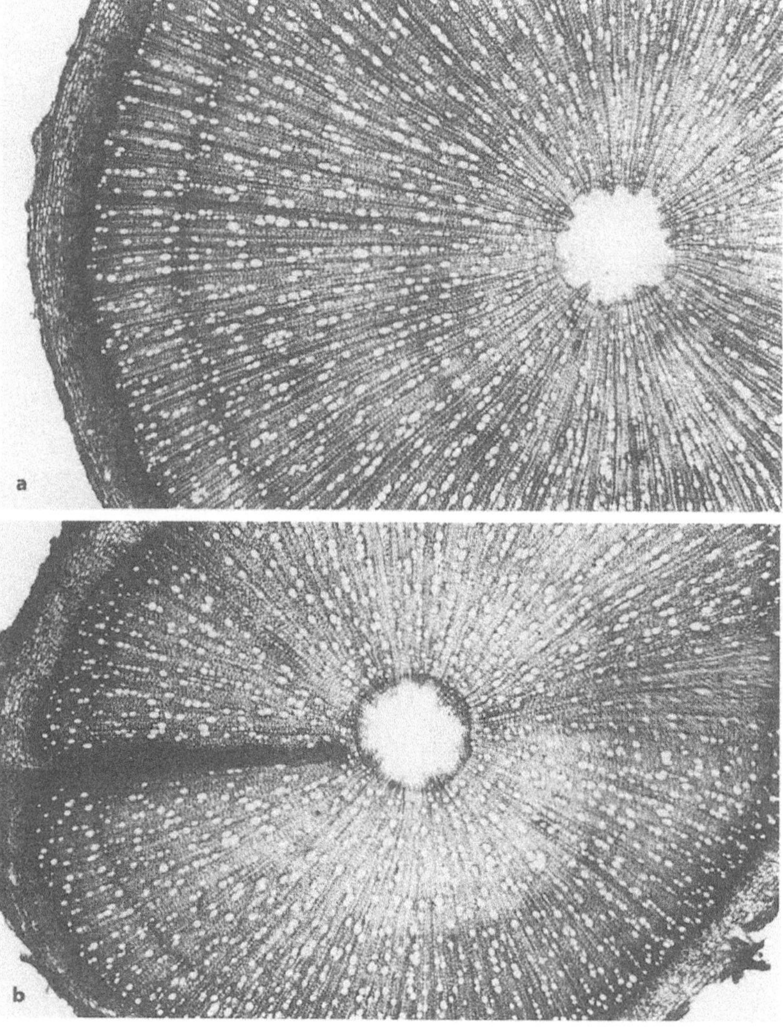

Fig. 131. *Acalypha alopecuroides*. **a** T.s. of Stem. **b** T.s. of root (both × 2.5).

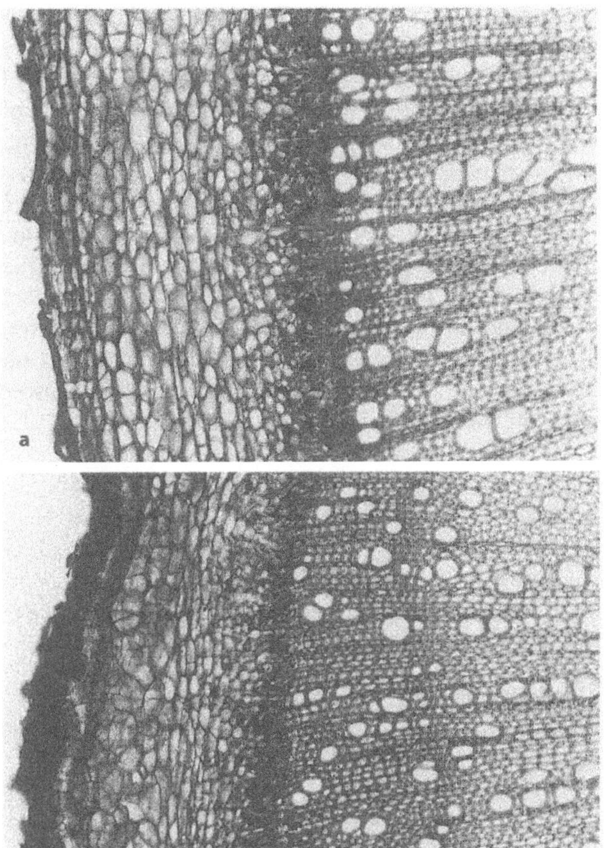

Fig. 132.1. *Acalypha alopecuroides.* a T.s. of axis in detail b of root.

ing the pith in which the cells have a meristematic aspect being arranged in very regular radial rows. A small pith with a central hollow is present in the center. Lateral roots are ramifying from the innermost primary xylem ring, indicating in this way the true nature of the root.

Ethnobotanical use

The medicinal drug of the leaves is called 'hierba meona' (SCHNEE 1960) and is sold in so-called herbolarios. The leaves have diuretic, carminative, and vomitive application in Venezuela and Cuba (PITTIER 1926, 1970; RODRIGUEZ 1983). In the Dominican Republic, a decoction of the leaves combined with salt is used against flatulence. According to GUPTA (1995) the root is also used as a diuretic in Cuba and Venezuela (MORTON 1981), as well as against renal calculus and venereal diseases in the Caribean room and in Latin America (POMPA 1974). According to ALBORNOZ (1993), the leaf has also anti-inflammatory effects.

Chemical contents

GUPTA (1995) reports flavonoids, saponins, polyphenols and tannins (WENIGER et al. 1984), as well as cyanogenic derivatives (HEGNAUER 1964).

Related species

Acalypha arvensis POEPP. & ENDL.; the leaves or twigs of this species are also sold in Venezuelan herbolarios as a medicinal drug (RODRIGUEZ 1983). *Acalypha guatemalensis* PAX & HOFFM. is a medicinal plant used in Central America. Best known is however *Acalypha indica* L. indigenous to tropical Asia and Africa; fresh or dried plants are used medicinally as a gastro-intestinal irritant, large doses are emetic; the contents are acalyphin, an essential oil (which is possibly the effective substance), resin, tannin and an alkaloid (UPHOF 1968). *A. indica* is also known as a poisonous plant (DUKE 1988). The flowers and leaves of *Acalypha evrardii* GAGNEP. are used in Vietnamese medicine for the preparation of a diuretic beverage (UPHOF 1968). Other species are said to have antibacterial (*Staphylococcus, Salmonella*), antimycotic or other antibiotic effects or are even used against cancer (GUPTA 1995).

Observations

The long acute hairs as well as the glandular hairs with a long stalk and a pluricellular head together with the abundant druses of calcium oxalate in the palisade parenchyma of the leaf facilitate identification. The presence of a very small pith in the center of stem and root is a further valuable diagnostic characteristic.

Conceveiba guianensis AUBLET. An infusion of the leaf buds increases the fertility of women. Fruits are edible (CASTILLO 1995). ROTH studied the leaf structure 1984, Fruit structure and dispersal 1987, leaf venation patterns 1996.

Croton

At least 26 useful species are known. Lacs and resins are obtained. Morphine type of alkaloids in the latex. Bark to make the skin young. Bark may be tonic bitter and aromatic with a volatile oil. Bark anthelmintic, for intermittent fever, ailments of the eyes; and for colics. A perfume is obtained from the bark.

Roots are used for constipation, as a purgative and against fever. Roots for skin diseases and urinary problems.

Oil of seeds is used as a purgative and an insecticide. Croton oil is a drastic purgative.

Leaves are used for infected sores and cuts and for eczema. Leaves are put on bleeding gums, as a tea, help for influenza. Leaf tea as anthelmintic (latex). Leaves are also used as a fish poison; a tea is used for upset stomach and for an eye wash.

The resin is used for varnishes, the gum of the trunk to clean the teeth and for toothache.

Latex is applied for ulcers and boils to reduce pain. The caustic latex is used to cure skin diseases. A bath helps for hysteria and rheumatism.

ROTH studied the bark structure 1981, leaf structure 1984, fruit structure and dispersal 1987, leaf venation 1996.

Croton rhamnifolius H. B. K. (carcanapire, carcanapire macho, salvia, barredero, amagoso, salvia muñeca, punta de lanza)

Taxonomical description

The species is a shrub, 1–2 m high. The oblong to ovalate leaves, 4–8 cm long and 2–6 cm broad, frequently have a rounded obtuse or slightly acute tip; the margins are entire or slightly dentate. The leaf base is somewhat heart-shaped and covered with small stellate hairs. Stellate hairs also occur on both leaf surfaces, but are more numerous on the lower side. The petiole reaches 2 cm in length.

The numerous small flowers, arranged in racemes, are unisexual.

Occurrence

The species occurs in the West-Indies and the north of South America. In Venezuela, it is frequent in the hot regions of the west and in the thornbush of the littoral and on the island of Margarita.

Anatomical description

Leaf. The upper epidermis cells are thin-walled. The palisade parenchyma consists of a single layer of very elongated cells, some of which may be divided by anticlinal walls; it occupies about 1/3 of the entire leaf volume. The length/width index of the palisade cells varies between 7 and 10.8. Very large water-storing cells with enormous druses of calcium oxalate abundantly occur in the palisade parenchyma. The spongy parenchyma is loosely arranged and consists of spheric cells. Intercellular spaces are partly large. The cells in the center of the mesophyll are the largest, while cell size diminished towards the lower epidermis. The very large cells of the spongy parenchyma are possibly water-storing. Small druses are also found in the spongy parenchyma. Vascular bundles occur infrequently. The lower epidermis is similar to the upper one. Stomata of the upper epidermis occur at epidermis level, while those of the lower epidermis project somewhat above the epidermis level. Stellate hairs, a characteristic of the genus Croton, are present on both surfaces, but are more numerous on the lower than on the upper side.

Ethnobotanical and general use

Economical utilization

Besoms made of the branches are applied to expel fleas.

Medical use

To cure fainting, the aromatic smell of the leaves is inhaled.

Against spasms, the fresh leaves are directly placed on the affected region.

Against ear aches, the leaves are put on the ear.

Small pieces of the plant in alcohol are rubbed on the skin to cure varices.

Parts of the plant ingested are laxative.

Chemical contents

Essential oil in the leaves (see GUPTA 1995).

Toxicity

The plant is toxic to cattle.

Related species

There are about 30 species of *Croton* which are well-known medicinal plants used as purgatives and tonics and for many other purposes. Others produce valuable lacs and resins utilized in the manufacture of varnishes. Best known is the Croton oil of *Croton tiglium* L., a drastic purgative. The seeds of this species are used to stupefy fish.

Observations

The large crystal cells with druses and the stellate hairs are good characteristics for identification. A further peculiarity are the slight differences of the stomata on both leaf surfaces.

Euphorbia

All species of Euphorbia contain a white latex in laticifers. At least 40 species of this genus (with about 2000 species altogether) are well known for their useful properties. Quite a few are succulent and cactus-like, but are usually distinguished from Cactaceae by their stipules modified to thorns and by the presence of the latex. Dichasial ramification is characteristic of the floral region. The flower-like pseudanthia called cyathia correspond to complicate inflorescences in which reduction of sexes takes place. Even zygomorphic pseudanthia are known within the Euphorbieae (e.g. *Pedilanthus*).

The most famous source of natural rubber is the Euphorbiacea *Hevea brasiliensis*. Rubber is also extracted from *Euphorbia intisy* (intisy rubber), *E. fiha, E. rhipsaloides* (almeidina rubber), and from some other species, although it is of poor quality. *E. marginata* is a source of chewing gum. The resin of *E. resinifera* (and of other species) is used in paints; the sticky latex of this species supplies a waterproof paint for ship's hulls. A cement is prepared from the latex of *E. cattimandoo.*

E.antisyphilitica ZUCC. (synonym: *E. cerifera*) is a source of wax.

E. kamerunica is used for tattooing by the natives of Guinea.

Toxicity and caustic properties of the latex are used in arrow poisons, rat poisons and to stupefy fish.

Young shoots of some species are boiled and eaten.

Other species have very attractive and intensely coloured hypsophylls and are cultivated as ornamentals.

The medicinal properties ascribed to many species of *Euphorbia* are: purgative, emetic, abortifacient, cathartic, diuretic, expectorant, diaphoretic, antispasmodic, antiasthmatic, sedative, haemostatic, soporific, allaying dyspnoea; some species are applied for dysentery, diarrhoea and as an emmenagogum.

The latex is also used in cosmetics to remove warts and pimples by rubbing the juice on the skin.

However, all species of *Euphorbia* have to be handled with care because of the toxicity and caustic effects of the latex.

Euphorbia buxifolia LAM. (Fig. 132.2) This species has been described by ROTH (1992). A short taxonomical description is found there besides the description of the leaf anatomy. The leaf structure is very typical and of the 'Kranz' type, belonging to the photosynthetic C_4 type. The weakly developed vascular bundles are surrounded by a large-celled parenchymatous sheath. Additionally, a 2-layered large-celled water-storing hypodermis is found on the lower leaf side. Stomata occur only on the upper leaf side and are sunk in deep depressions.

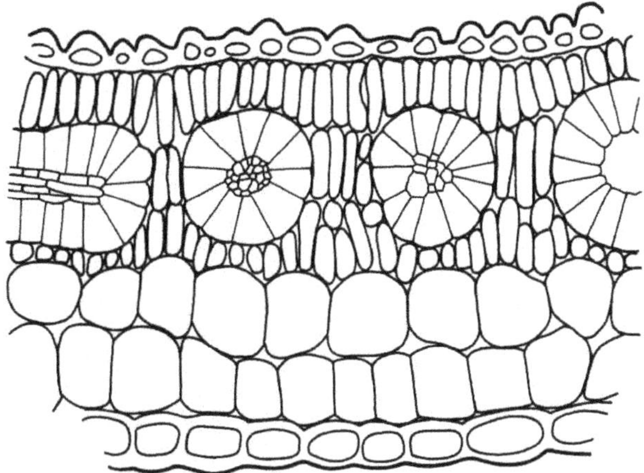

Fig. 132.2. *Euphorbia buxifolia*. T.s. of leaf. Equifacial structure, photosynthetic C4 type and 'Kranz' structure. Note the large cells of the lower water-storing hypodermis (ROTH 1992).

Perhaps this is another plant with promising properties.

Hieronyma

Wood is useful. Bark and leaves are used for skin eruptions and sarna. Bark is used for pruritus. Seed oil is anthelmintic. Roots are depurative.

ROTH studied the bark structure 1981, leaf structure 1984, fruit structure and dispersal 1987, leaf venation 1996.

Hura creptians L. (habillo, jabillo, ceiba blanca)

Taxonomical description

The plant reaches 10–40 m in height (Fig. 132.3). Stem and branches are spiny. The leaves are more or less orbicular or broadly ovate, 5–10 cm long, with a cuspidate-acuminate tip and a heart-shaped or rounded base; the margins are entire or dentate; the lower leaf side is often hairy along the nerves. The stipules are linear-lanceolate, pubescent and caducous.

The unisexual flowers are apetalous. The male flowers are arranged in cylindric-oviform spikes of

Fig. 132.3. *Hura creptians.* **a, b** Stem with spines (ROTH 1992).

red-brown colour, 6 cm long and 2 cm in diameter, the calyx is irregularly dentate; filaments and connectives are united in a thick column; the anthers are arranged in 2–5 verticils around the column. The female flowers occur solitarily in the axils of the upper leaves or at the base of the male spikes; the calyx has entire margins; the ovary shows 5–20 locules. The styles are united in a large and fleshy column; the stigmas appear in a star-shaped form and are of red-violet colour.

The capsules which are 3–4 cm long and 6–8 cm broad are profoundly surcate indicating the position of the numerous cocci inside. Their shape is depressed-globose. They contain about 11 one-seeded cocci.

The capsule opens in an epxlosive manner through a throwing mechanism, when the fruit dries out. Each mericarp separates from the adjacent mericarps as well as from the central column, dehiscing abruptly along the dorsal suture so that the mericarps are thrown far away, even as far as 14 m (ROTH 1977).

Occurrence

Tropical America. In Venezuela the tree is very common in hot regions as well as in the lower belt of the temperate regions.

Anatomical description

This tree is very easy to identify by its spiny trunks and twigs. Its bark consists mainly of parenchyma specialized as a storage tissue rich in starch, while sclerenchyma in the form of hardbast fibers and stone cells is very scarce. Solitary rhomboid crystals are frequently found as a cell content. A system of laticifers penetrates the parenchymatous tissue.

The bark of a large tree with a diameter of 30–35 cm measures 10–12 mm in width and is composed mainly of phloem parenchyma rich in starch. The starch grains differ greatly in size, some being very large, others small; solitary rhomboid crystals are abundant, while druses are rare. Only few solitary fibers and small stone cell groups occur here and there. In the outer bark, obliterated softbast strips form an intercommunicating network, enclosing small fiber groups within their meshes. An ample system of laticifers – some taking a radial course, others an axial one – penetrate the parenchymatous tissue. The milky sap is so abundant that it flows in streams when the bark is cut. It is caustic and the bark has therefore to be handled with care.

Dilatation growth is very irregular and becomes apparent through tangential expansion of parenchymatous cells which later divide anticlinally, but also ray cells may participate in the dilatation. The primary cortex with 2 interrupted sclerenchyma rings, as seen in a transverse section, is still present in older barks. The cork is small comprising about 10 cells in a radial row. The phelloderm with about 20 cells in a radial row is better developed and also contains some sclerified cells. The bark usually does not desquamate, so that the prickles are permanently maintained on trunk and twigs. There is no formation of a rhytidome, but all bark cells remain alive, except lignified and suberized cells; this is apparently necessary to maintain the prickles.

The prickles (Fig. 132.3; 132.4) have a length of 10–12 mm, but reach even 25–35 mm in old trees. They correspond to pure cork formations. To form the prickle, the cork cambium bulges outwards through increasing phelloderm production concentrated in a certain place, a processes comparable to the formation of lenticels. At the same time, concentrated cork formation takes place in the middle of the calotte-shaped phelloderm outgrowth owing to elongation of the cork cells parallel to the future long axis of the prickle. In the mature prickle, the phelloderm below consists of radial rows of thin-walled cells with occasionally intercalated lignified sclereids. The cell walls have a great number of pits.

The thin-walled cells of the prickle contain small oil drops and even small starch grains. The cork which forms the prickle is very thick-walled and lignified. Starch is plentiful in the cells. The cells are elongated parallel to the long axis of the prickle. Towards the tip of the prickle, cells diminish in size and elongated cells divide by transverse walls. These cells undergo a second wall thickening so that the cell lumen almost disappears. Likewise the periphery of the prickle has very thick-walled cells so that true sclereids strengthen the base as well as the periphery of the prickle. These hard prickles supply the tree with a very effective defense mechanism; moreover they grow very close together, leaving about 0.5 cm distance between one

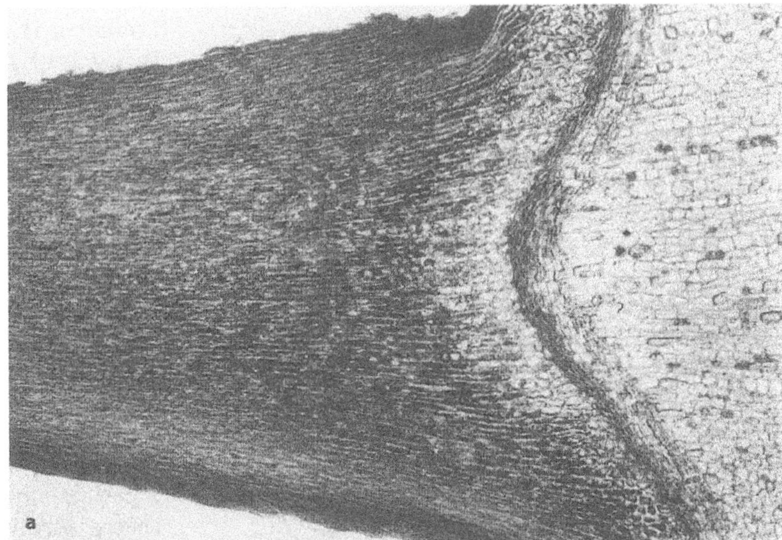

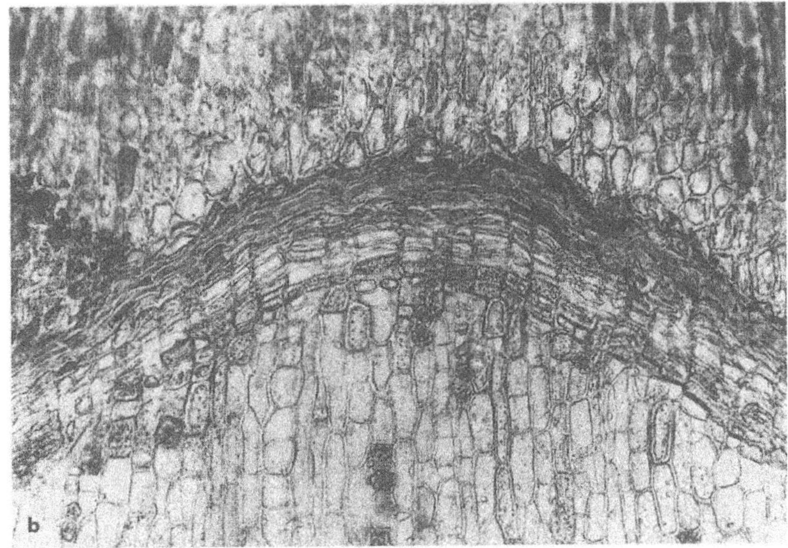

Fig. 132.4. *Hura creptians.* **a, b** Basal part of spine, longitudinal section (ROTH 1992).

another. Only on the base of very old stems some scales may occasionally be shed, together with the prickles, and new prickles arise afterwards from a new periderm (ROTH 1981 and 1992).

The fruit of *Hura crepitans* has a very interesting dispersal mechanism. Many Euphorbiaceae develop capsules with a throwing mechanism. In *Hura crepitans*, each mericarp separates from the adjacent mericarps as well as from the central column, when the fruit dries out, dehiscing abruptly along the dorsal suture, so that the seeds are thrown far away, up to a distance of 14 meters. Crossing elements of the inner hard layer generate dehiscence, which results suddenly, because the separation zone does not consist of thin-walled elements, but of fibers orientied parallel to the dehiscence line, these strenously resist the opening (GILLES 1905, v. GUTTENBERG 1971, ROTH 1977).

Ethnobotanical and general use

Economical utilization

The tree is sometimes cultivated as an ornamental plant.

The plant is also used as a fish and arrow poison. The latex is fermented before use by some South American tribes.

The fruits are occasionally used as sand boxes to dry ink-writing. The English name of sandboxtree is derived from the early practice of hollowing out the immature capsules and using them as containers of blotting sand.

The timber is used locally for interior construction, carpentry, boxes and crates.

Medical use

A decoction of the leaves heals colic.

A decoction of the leaves mixed with castor oil is employed to burst pustular tumours.

The seeds are a violent purgative (2–3 seeds are sufficient).

Toxicity

The seeds of the plant contain the very toxic polypeptide hurin. The latex is extremely caustic; it produces irritations on the skin and can even cause blindness. It contains hurin and crepitin.

The seeds have been employed to poison coyotes and other animals; however, they are the favored food of red and blue macaws in Costa Rica.

The latex, the seeds and a decoction of the bark have emetocathartic properties; in large doses, they are violently poisonous.

Abortion may occur when pregnant cows feed on this plant.

Besides the common storage proteins, the seeds of Euphorbiaceae may contain very toxic polypeptides, such as curcin in *Jatropha curcas*, hurin in *Hura crepitans*, and ricin in *Ricinus communis*.

Huratoxin of *Hura crepitans* is an arrow poison.

Jatropha

The genus occurs mainly in tropical America and Africa. Of the 175 species more than 10 are of economic use. The constituents found in this genus are tannins, sapogenins, alkaloids, ethereal oils, cyanogenic compounds and toxalbumins. Antitumour triterpenes and a substitution pyrrolinone with antitumour properties have been isolated from *J. podagrica*.

Young leaves of some species are consumed as a vegetable. When not toxic the oil of the seeds of other species is used for cooking. The decoction of *J. cinerea* is a source of a mordant in dyeing. The rhizome of *J. zeyheri* contains 22 % tannin and is used for tanning.

Jatropha curcas L. (piñón, piñol, physic nut, purging nut)

Taxonomical description

The laticiferous deciduous shrub is about 1–5 (6) m high (Fig. 133). The stem is (DAB 14–18 cm) cylindric, glabrous, with lenticels; and has white latex in the bark. Leaves are simple, alternating, with a petiole, about as long as the blade. Blades coriaceous, glabrous, pubescent above the nerves according to SCHNEE; of ovate shape, 3–5 lobed, 5–10 (15) cm long and 3–9 (15) cm broad with an acuminate tip, entire margins and a cordiforme base. Venation palmate actinodromous, the 5 principal veins very prominent on the lower surfaces; petiole 3–8 cm long, ribbed.

The small unisexual flowers are united in cymose inflorescences, the bracts are linear or lanceolate. Calyx 5-lobed, lobules 4–5 mm long; 5 petals, oblong-obovate, 8–9 mm long, and hairy on the inside; glands of the disc free. 8 stamens, outer filaments free, the inner filaments united. Ovary glabrous, styles short. Hermaphrodite flowers exist besides the unisexual ones and the number of stamens varies between 6–10, as does the number of petals.

The globular fruit is a capsule, 2.5–4 cm in diameter with 3–2 cocci. The black seeds are 1 – 2 cm long, reniform, and used in popular medicine.

Fig. 133. *Jatropha curcas*. a Habitus. b Flowers and fruits.

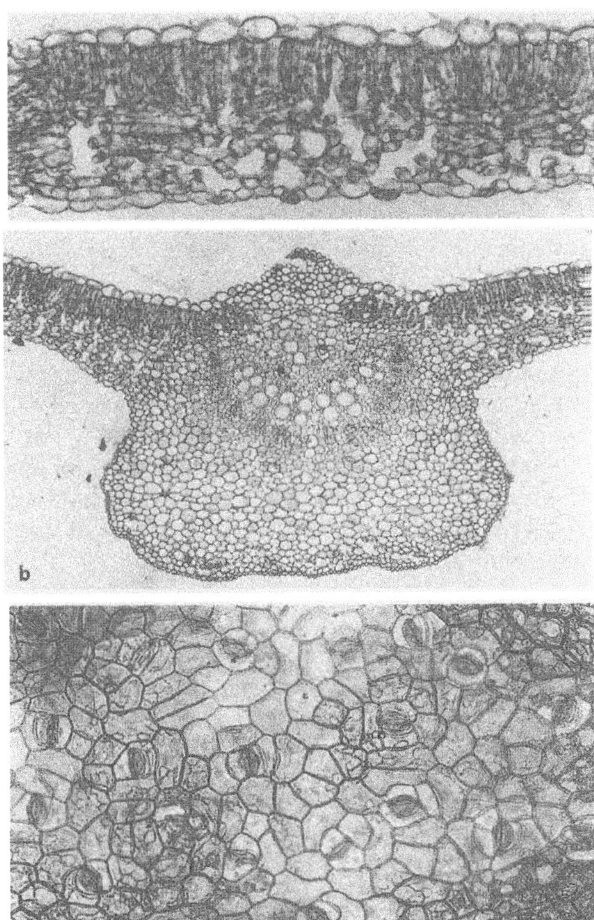

Fig. 134. *Jatropha curcas*. a T.s. of leaf (× 16). b Midrib in t.s. (× 6.3). c Surface view of lower epidermis (× 16).

Origin

The genus Jatropha is of neotropical origin, with more than 100 representatives in America.

Occurrence

The species is common in semi-arid regions of South and Central America; in Venezuela it occurs in the hot regions. Now it is cultivated all over the tropics, from 0–1700 m a.s.l. It supports poor rocky soils, and endures low precipitation, yielding fruits after 2–3 years. It reaches an age of 30–50 years.

Anatomical description

Leaf (Fig. 134). The leaf is dorsiventral and amphistomatic. The upper epidermis cells are comparatively large and somewhat papillose, with watchglass-shaped convex outer walls. As seen in a surface view, the anticlinal walls are straight. Stomata are rare in the upper epidermis and occur at the lower level of the epidermis.

There is a single layer of very long and slender parenchyma cells. Their length/width index reaches about 8–10/1. The palisade parenchyma is compact, while the spongy parenchyma is much looser with intercellular spaces of medium size. The spongy parenchyma is also larger in proportion than the palisade parenchyma, reaching 1 ½-twice the width of the palisade parenchyma. Very conspicuous are large to very large colourless cells which contain a large druse of calcium oxalate and which probably also have a water-storing function. Druses are also found in many cells of the spongy parenchyma and of the veins. Laticifers (tubes) occur in the mesophyll and are associated with the vascular bundles. The cells of the lower epidermis are slightly smaller than those of the upper epidermis and the stomata are slightly elevated above the epidermis. As seen in surface view, the anticlinal walls are straight or only very slightly bent.

The midrib is only slightly elevated above the upper side, but projects considerably above the lower side. The epidermis is small-celled and only slightly papillose in the region of the midrib; palisade as well as spongy parenchyma are interrupted in this region. The vascular system consists of an open arc. The filling tissue is composed of parenchyma. Laticifers are dispersed in the parenchyma. 2–3 layers situated below both epidermal layers may transform into a weakly developed angular collenchyma.

The lateral veins of the first order have a similar structure, whereas the veins of the higher order consist of a simple vascular bundle being transcurrent by parenchyma to the upper as well as to the lower epidermis.

Stomata are accompanied on either side by one (or 2) subsidiary cells running parallel to the long axis of the pore (rubiaceous type, according to METCALFE & CHALK 1950).

Stem (Fig. 135, 137 a). A stem of 10 mm in diameter is woody. The thin-walled cork is of a superficial origin, and compirses only a few layers. The primary cortex which consists mainly of parenchyma is very ample. Small fiber bundles as well as large laticiferous tubes with slightly thickened walls are abundantly dispersed over the section. Many parenchyma cells contain druses of calcium oxalate. The secondary phloem is comparatively small. The xylem is well developed and consists mainly of parenchyma. The medullary rays are uni or biseriate. The vessels are solitary or occur in shorter or longer radial rows. Some vessels have a content which stains intensely with toluidine blue. The pith is also ample and parenchymatous and many cells contain druses. Laticifers are dispersed within the pith.

Seed (Fig. 136). The seed is about 18 mm long, of oval shape and has a caruncle. As seen in surface view, the outer epidermis of the outer integument consists of polygonal cells with slightly thickened walls. The cells are stuffed with a dark-brown content, while the anticlinal walls appear to be of a

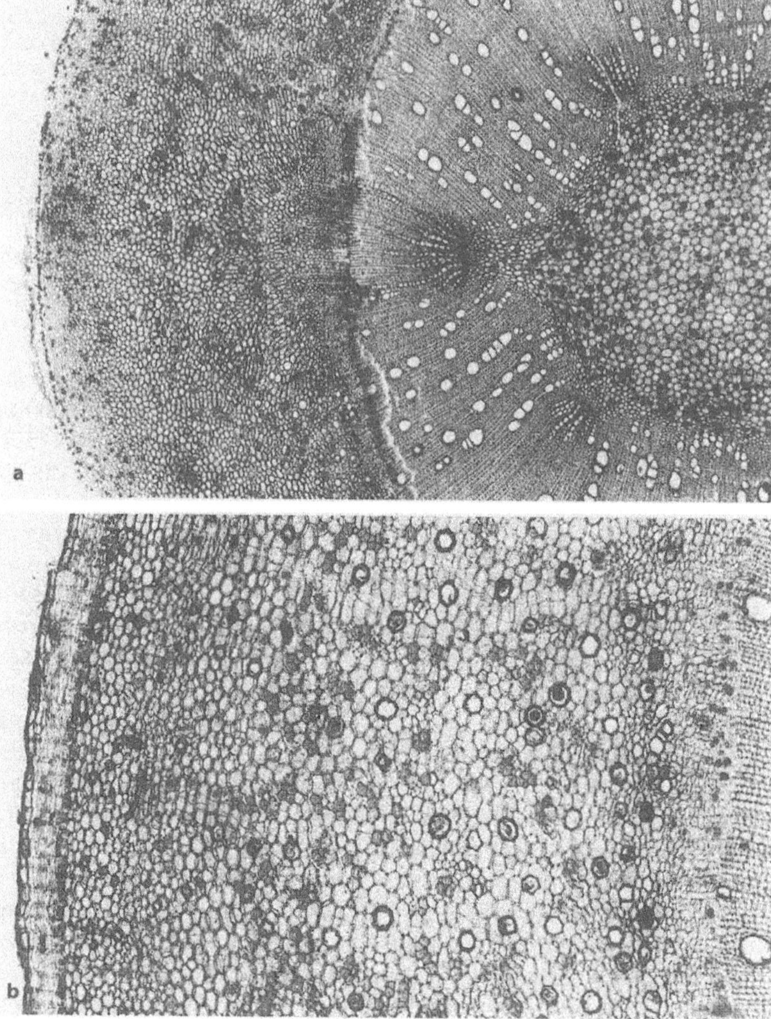

Fig. 135. *Jatropha curcas.* T.s. of axis. **a** Entire section. **b** Detail of primary cortex with small phloem band (× 6.3).

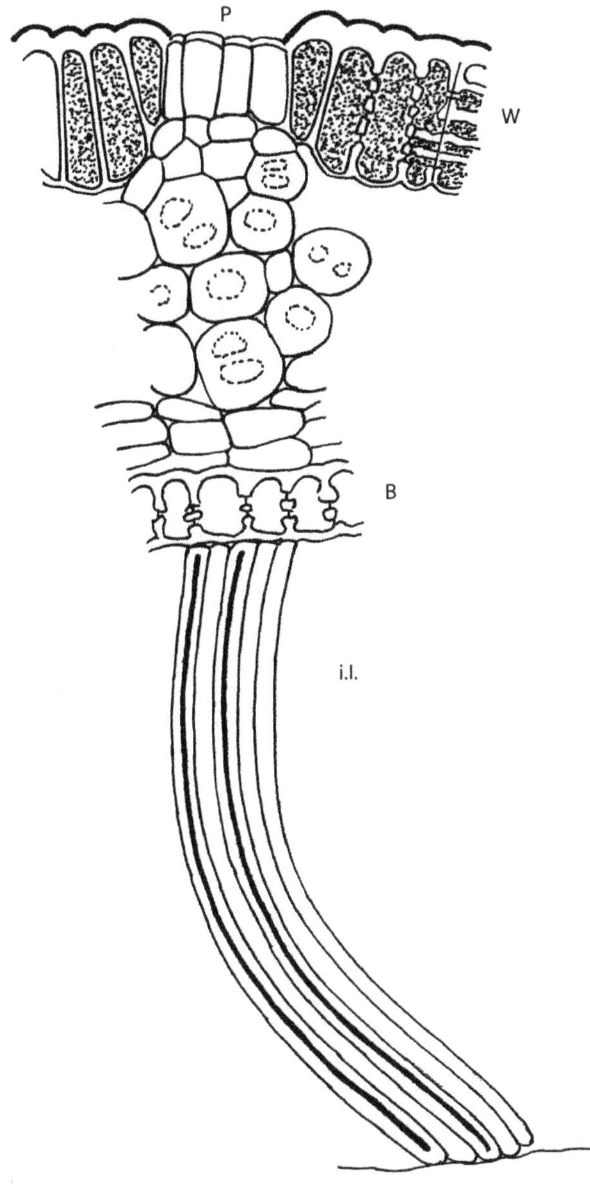

Fig. 136. *Jatropha curcas*. Detail of seed coat in t.s. Note the epidermis cells with their dark content, the 'cream pits' (p), the parenchyma cells with arms (dotted), the layer of 'buckled' cells (B) interpreted as the inner epidermis of the outer integument; the 'fiber layer' below (i.I.) which would be better interpreted as a palisade layer, corresponding to the outer epidermis of the inner integument (i.I.). W = wall thickenings in the outer epidermis of the outer integument.

creamy colour. This surface structure is interrupted by smaller or larger irregular spots of creamy colour where the cells do not show the dark-brown content; for this reason, they appear as cream pits (VAUGHAN 1970). The spots may be due to tensions and ruptures in the epidermis during development, being comparable to those observed by ROLLET, HÖGERMANN & ROTH 1990 in the leaf epidermis of certain tropical species. The normal cells around the patches may become arranged in the rosette form (VAUGHAN), also observed around stone cells, idioblasts etc. which are 'foreign bodies' in the epidermis (ROTH, earlier publications). The clear spots may function as windows for light entrance and gaseous exchange (no pigments, thinner cell walls). The outer tangential walls of the palisade-like cells of the outer epidermis are strongly thickened, while the radial walls are thinner, being interrupted by numerous pits. The subepidermal tissue is bulging out towards the outside where the cream patches arise, such that the cells which compose the patches are shorter in a radial direction than the other palisade cells; they also have thinner outer tangential walls. Perhaps the patches are comparable to lenticels.

The so-called subepidermal layers (VAUGHAN) comprise about 7–8 layers of roundish parenchyma cells with arms which become somewhat compressed towards the inner integument. Laticiferous tubes may be interspersed between the parenchyma. Further inwards follows a layer of cells with buckled walls (VAUGHAN) corresponding to cells with somewhat thickened, slightly wavy radial walls and pits in the walls through which the buckled appearance arises. This layer is interpreted as the inner epidermis of the outer integument (VAUGHAN). The following fiber layer belongs to the outer epidermis of the inner integument (SINGH 1954). The fiber layer consists of very long and slender cells with thickened walls in a curved arrangement, probably due to reasons of space. The layer may possibly be better designated as a palisade layer which is generally very common in seed coats (see above: the outer epidermis of the outer integument). Without doubt, the cells are very long reaching a height of 130 μm and a width of only 18 μm (VAUGHAN).

Ethnobotanical and general use

Nutritional use

The plant is toxic and therefore seldom used for food. Heat treatment would however remove the toxic substances, but this has not been done on a large commercial scale.

Seeds of the related *Ricinus communis* can be detoxified by fermentation and are then used as food. This would also probably be possible with seeds of *Jatropha*. Toasted seeds are not toxic and are eaten by the natives. There are also varieties with a low content of phorbol esters which can be eaten toasted.

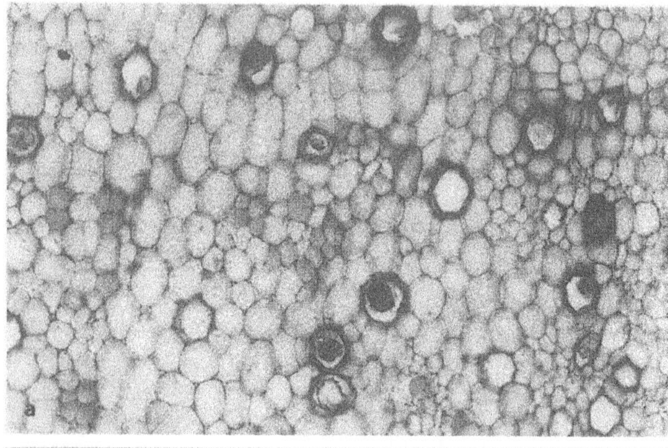

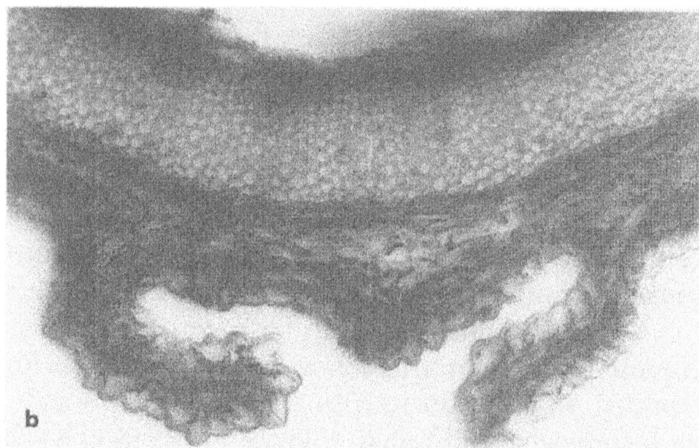

Fig. 137. **a** *Jatropha curcas.* T.s. of axis, cork formation in the primary cortex. **b** *Gentiana nevadensis.* T.s. of axis with a pair of leaves (× 20).

Economical utilization

Jatropha curcas is commonly used as an ornamental or hedge plant in tropical countries (living fences). It furthermore stabilizes shifting dunes.

Its seeds contain 50–60 % of a semi-drying oil which is used as an illuminant, for manufacture of candles, soap, lubrication, and in wool-industry.

The kernel contains 35 % protein, but the meal is toxic and is therefore used only as a fertilizer. The species could be grown as an oil-seed plant. The seed coat could serve as a combustible biodiesel.

Medical use

Leaves and seeds are mainly used in popular medicine. However, branches, roots and the latex (used separately) have their applications. Medicinal drugs: plantae, latex, semen.

Leaf. The use of the different plant organs is manifold. The leaves are used externally as cataplasms or as a bath, as well as internally. Many curative effects are ascribed to the plant.

A decoction of the leaves is locally used for a bath in case of fever and against catarrh. Taken orally, the decoction cures gastrointestinal irritations (such as constipation, diarrhoea, dysentery, stomach pain, intestinal parasites) and haemorrhoids, cataract, venereal diseases, toothache, eczema, erysipelas, gonorrhoea, gout, leprosy, neuralgia, malaria, sunburn and rheumatism.

The fresh leaves are applied in the treatment of erysipelas, wounds, swellings, burns, ulcers and jaundice. They are rubefacient and revulsive.

The leaves in decoction are applied as a cataplasm for external ulcers.

For infected wounds, ulcers and indigestion, some leaves of a twig end are boiled in water and the wounds are cleaned with this decoction. The same decoction prepared together with the leaves of *Hyptis suaveolens* and of orange is taken against pain of the stomach.

Yellow abscised leaves are boiled in water to cure kidney problems.

Extract and latex of leaves and twigs are externally applied to cure haemorrhoids, ulcers, herpes and other infections of the skin. The decoction is applied as a cataplasm and a galactogogum.

Fresh levaes in decoction are laxative, fresh and ground they are used as a cataplasm to mature abscesses and tumours.

To cure headache, the leaves of *Jatropha curcas* are ground together with leaves of *Petiveria alliacea* and the mixture is soaked in water for 2–3 hours. The water is used to wash the hair.

The sap of the petioles is rubbed on gums of children to cure mouth ulcers.

Stem (twig). The following healing properties are ascribed to the leaves and the bark: stomachic, antiinflammatory, stupefacient, narcotic, as a galactogogum, a vulnerary and an odontalgic. The decoction of the bark is used against colics and as an appetite stimulant.

Latex. The latex of the leaves and the stem which is slightly rubefacient and antiseptic, is locally used for the treatment of gingivitis, to heal wounds, fractures, haemorrhoids, haemorrhage, herpes, ulcer, insect bites, burns, and warts. It is also applied for gargles and mouthwash against infections of the mouth and inflammations of the tongue, as well as against toothache and leprosy. Wound healing, stomachic, haemostatic antihaemorrhoidal and verrucocidal properties are ascribed to the latex as well.

The latex of the stem is used in certain regions to cure infections of the mouth of children and as a keratolytic.

The pure latex of leaves and stems is applied on wounds and poisonous insect bites for cicatrization. In the form of drops it controls nasal haemorrhage, herpes, ulcers, haemorrhoids and burns.

Peasants put the pure latex on a wad of cotton and apply it locally in the mouth, although it is well known that the latex is caustic. Applied to the temples for several days the latex cures inflammations of the eyes.

Root. The root ground and boiled in hot water is used as a specific remedy against dysentery. The ethanolic extract of the root has antibacterial activities against *Staphylococcus aureus, Streptococcus pyogenes,* and *Streptococcus viridans.* The methanolic extract of the root has anticonvulsive properties.

Seed. The seed and particularly its oil, is the part of the plant mostly used. The oil of the seed is applied to treat fractures, gout and burns. However, the seeds have first of all purgative and emetic properties and are used in case of oedema. With good reason, the plant is called higuera infernal in popular medicine, which means infernal fig, referring to the drastic purgative effects. The oil is applied for oedema, chronic paralysis and against intestinal worms. It is a very drastic vermifuge and a laxative.

4 or 5 fresh or dry seeds in decoction are taken as an emetic and a purgative; after which 2–3 cups of tea or coca-cola have to be taken.

The fruit and seeds are reported to have contraceptive properties.

The oil of the seeds is extracted by pressure in a hot medium. Although it has a purgative effect, it can be used in the treatment of gout and toothache.

The phorbolester causes vormiting and diarrhoea.

Method of use

As already mentioned, the different plant organs can be applied fresh, dried, in decoction and infusion, as a cataplasm or a bath. They are applied externally and taken orally. For exact details see above (the mentioned organs and the latex). Latex and oil are used either partly diluted or concentrated. The toxic principle which is present in all plant organs can be eliminated by heating.

Healing properties

The latex is caustic and the seed oil is emetic and purgative. Antibacterial activity has been demonstrated in the root. Leaves and branches have effects on lymphocytic leukaemia. The root shows anticonvulsive properties.

In some species of Jatropha, an antitumour compound has been reported.

Abortive properties are attributed to the leaves.

For further healing properties see the description of the different organs and of the latex and oil.

Chemical contents

Bark, fruit, leaf, root and wood contain hydrocyanic acid. The latex eliminates warts very effectively, and its activity seems to be selective for verrucose tissue, acting as a chelating asgent.

The most toxic substance is the protein curcin. Other contents are jatrophine, curcasine, tannins (latex, bark), sapogenins and others.

Recently, slimy substances were extracted from the seeds. In *Jatropha curcas,* magnesium shows the highest concentration of all other elements and its degree of concentration in the different organs varies most conspicuously. Ripe feminine flowers have the highest magnesium concentration of all organs.

The oil extracted from the seeds has a brilliant yellow colour, is liquid at ambient temperature and has neither a peculiar smell nor taste; its melting-point is very low at −8 to −6 °C.

The irritant constituents of *Jatropha curcas* are polyunsaturated esters of 12-dioxi-16-hydroxiphorbol.

The calorific value of the oil of *Jatropha multifida* is 13.647 kcal/g. Oil of *Jatropha curcas* contains up to 72.2–84.0 % of oleic acids and linoleic acids. The oil can be used as fuel for agricultural machines.

The constituents of the genus include tannins, sapogenins, alkaloids, toxalbumins, cyanogenic compounds ethereal oils.

Toxicity

It is suggested that 4–5 seeds are lethal for humans.

The oil is not only toxic for humans, but also for cattle, fish, rats and molluscs. It is used as a fish poison by the native population. The toxic elements of the seed are the protein curcin, tiglinic acid and crotonic acid. They provoke acute abdominal pain, nausea, vomiting, diarrhoea, inflammation of the mucosa of the stomach and even coma after several hours. The minimum lethal dose of seed oil for rats is 1 ml per animal injected subcutaneously.

The latex is a skin irritant and rubefacient when externally applied and is toxic when internally taken. It has a caustic effect.

An abortive property is ascribed to the leaves.

Leaves and roots have pest-control properties and can be applied against *Aulacophora foveicollis, Lipaphis erysimi,* termites, mosquitos, *Musca domestica* and snails.

In some species of Jatropha an antitumor compound has been identified. New antileukemic jatrophone derivatives have been found in *Jatropha gossypifolia.*

Curcin, similar to ricin, is the most important toxic principle of *Jatropha curcas.*

A resin which has been found in the oil of the seeds of *Jatropha curcas* produces redness and pustular eruptions on the skin.

All species of *Jatropha* are probably toxic to cattle, but the animals seem to avoid eating them.

The common symptoms associated with intoxication by *Jatropha curcas* are burning of the throat, followed by distension of the stomach, giddiness, vomiting, diarrhoea, drowsiness, collapse and death (BLOHM 1962).

Interesting information on the variability of toxicity in plants and the toxic effects exercised on man are found in the bibliography. BLOHM (1962) states that the effect of the poison varies with the individual. Native people may already be accustomed to the toxicity, when they frequently ingest the poison in small doses. Immunization of animals against the action of ricin has been achieved through repeated injection of sublethal quantities of this substance; animals treated in this way may ingest up to 10 times the fatal dose without showing signs of intoxication; the serum of these animals is an effective antidote for persons poisoned by ricin. A similar procedure could also be developed for curcin.

In the poisoning by ricin the following treatment has been recommended by the 'Dispensatory of the United States': gastric lavage, administration of saline cathartics, maintenance of fluid and electrolyte equilibrium and symptomatic measures.

According to GARCIA-BARRIGA (1975) alcoholic drinks are antidotes against the toxic substances of the entire plant.

Furthermore, toxic alkaloids which disturb the nervous system are present in *Jatropa curcas*; saponins are also present, but in contrast to the alkaloids they are not absorbed, but exercise local effects on the digestive tract. LIENER (1969) found that the lethal dose of certain saponins is less than 100 mg/kg body weight, and that death is caused by inflammation of the digestive system.

HEIN (cited by Liener) demonstrated that distribution and content of saponins in the distinct parts of the plant depend on the temperature of the environment. GESTETNER et al. (1972) were able to show that the toxic action of the seed can be inhibited by cholesterol and β-sitosterol, which is present in *Jatropha curcas.*

It is interesting to note that the plant seems to be very toxic in certain geographic regions, while in others, wehre *Jatropha* is part of the human diet it is not poisonous. CANO & HERNANDES (1984) suppose that in regions where the seeds are eaten, the content of saponins is inactivated by the presence of β-sitosterol. A further suggestion is that different varieties of *Jatropha curcas* contain different amounts of toxic substances and it would be very important to compare seeds from regions where they are eaten with those where they are refused.

For the plant itself, the toxic compounds are a mechanism of self defense, as certain saponins are growth inhibitors of some microorganisms and of pathogenic protozoa (GESTETNER et al. 1972).

Varieties and related species

Details on varieties have not been studied thoroughly yet. But there are several species with similar chemical properties, such as *Jatropha gossypiifolia* L., *Jatropha multifida* L., *Jatropha urens* L., *J. glandulifera, J. dulcis, J. pandurifolia, J. costaricensis, J. zeyheri, J. pohliana, J. podagrica, J. macrorhiza.*

All of these species have medicinally interesting chemical contents. One of the most interesting compounds is certainly jatrophone, a novel macrocyclic diterpenoid tumour inhibitor from *Jatropha gossypifolia* (KUPCHAN et al. 1970).

Cultivation

The plant is easy to cultivate, it adapts easily to diverse climatic conditions and soils and is easy to handle. It grows from 10 to 1400 (1500) m a.s.l. and has 2 periods of flowering and fruiting. Propagation is realized either by seeds (germination takes about 30 days with a yield of 80 %) or by cuttings, one meter high and 5 cm in diameter; shooting starts after about 20 days.

The plant is commonly cultivated as a 'living fence' and in gardens. Commercial plantations at a small scale have been practiced up to date with populations of 400 individuals per hectare and a distance of 5 m between the shrubs.

Production starts after 3 years, when the shrubs reach a height of 4–6 m and the crown begins to extend. The estimated production per individual including 2 harvests a year is about 30 kg of fruit, corresponding to 12 kg seeds. The yield of fruit per hectare is 12,000 kg fruit or 4,800 kg seed. 1979 the value of the seed was 25–30 pesos/kg.

It is claimed that 100 t of seeds produce 32 t of crude oil.

Observations

The species is easily recognized by its outer morphology and its anatomical structure, particularly of the seed: The outer epidermis of the outer integument with palisade character and patches of creamy colour, the inner epidermis of the outer inteegument with wavy cell walls, and the very characteristic fiber layer consisting of very long and slender fibrous cells.

Jatropha gossypifolia L. (tuatúa, sibidigua, tuatúa morada, tuatúa blanca)

Jatropha g. L. is a shrub, 1–2 m high. The membranaceous leaves are 3–5 partite, 6–10 (15) cm long and 7–15 cm broad and have a heart-shaped base. The segments are oblong and 2–4 cm broad, with acute tips. The margins are denticulate or ciliate. The petiole reaches a length of 6–8 cm.

The flowers occur in paniculiform cymes. The bracts are 8–10 mm long, linear-oblong and have glandules. The flowers are unisexual. The 5 sepals are ovate, acute and with glands on the margins. The 5 petals are obovate, glabrous, slightly longer than the sepals and of purple colour. The disc consists of glandules. There are usually 8 stamens in a male flower. The ovary is pubescent.

The capsule measures 1 cm in diameter.

Occurrence

In Venezuela, the species is common in hot regions.

Anatomical description

Leaf. The upper epidermis is large-celled and thin-walled and serves as a water-storing tissue. The only layer of palisade parenchyma consists of very long and slender cells with a length/width index of 6.3–7.6. Interspersed in this layer are cells of enormous size with giant durses of calcium oxalate. The spongy parenchyma comprises 4–6 layers and is composed of roundish cells which leave intercellular spaces between one another. The lower epidermis is of regular structure. Stomata are present on both leaf sides, but are more numerous on the lower side; on this side, they are also slightly elevated above the surface.

Ethnobotanical and general use

Medical use

Leaf. A decoction taken orally is purgative, vomitive and stomachic; it cures disturbances of liver and bladder. The leaves also have diuretic effects and are applied for kidney trouble and for diabetes; additionally they have anticatarrhal effects.

Locally applied they cure arthritis.

A decoction or infusion of the leaves is employed for colics (CASTILLO et al. 1992).

Sap. A sap of the leaves is applied on wounds to control haemorrhage. To drain purulent wounds, a decoction of the leaves is mixed with castor oil.

Persistent diarrhoea is cured with a leaf decoction. A leaf decoction is also applied for colics and disorders of the spleen, for venereal diseases, stomach aches (emetic), biliary disturbances, diseases of the liver and fever.

A leaf decoction mixed with castor oil is also applied to the skin to make pustules burst. The sap of the leaves is haemostatic and cures mouth ulcers. A cataplasm of fresh crushed leaves is used for swollen breasts, fractures and bruises. The leaf is also used for anorexia and as a vulnerary. A decoction of the leaf buds orally taken (together with some other healing plants) is used to cure diarrhoea.

Bark. The bark in decoction is abortive and an emmenagogum and cures venereal diseases. A decoction of the bark is also antiblennorrhagic (CASTILLO et al. 1992).

Latex. The latex is used as an antipyretic and for buccal ulcers. It has haemostatic effects and is applied for haemorrhoids and burnings. A wad of cotton-wool soaked in the latex soothes tooth ache.

Branches. To cure diabetes, young branches are macerated in water.

A decoction of branches and leaves prepared as a bath is used for exanthema and eruptions of the skin.

Root. The root in infusion it taken as an emmenagogue and as an abortive.

A decoction of the root cures oedema and diabetes. It is also used in cases of indigestion.

Flower. Flowers are said to cure asthma. Flowering branches are also diuretic and cure the urinary tract ailments.

Fruit. The fruit has molluscicidal properties.

Seed. Seeds and leaves are purgative and emetic. The oil of the seeds is purgative, emetic, antihelmintic, antidiabetic, and is even used to cure cancer. Applied locally, the oil has an irritating effect.

Entire plant. The entire plant has stimulating effects on the uterine muscles.

Gargles are recommended for tonsillitis. Possibly, the plant can also be applied to cure leprosy.

Healing properties

Emetic, purgative, abortive, diuretic, antipyretic, vulnerary, antidiabetic, antibiotic, antitumoral, as an antidote against convulsions and as an emmenagogue.

Chemical contents and healing effects

The leaf contains jatrophone, flavonoids, mucilage, a macrocyclic diterpenoid tumor inhibitor and antileukemia jatrophone derivatives.

The bark contains lignan which can be toxic.

The root has antitumoural activities, cytotoxic activity, is molluscicidal and is effective against convulsions (jatrophone).

The seeds contain curcin, a toxic albumin.

The shoot has antimicrobial effects.

The entire plant is also molluscicidal.

Toxicity

The plant is toxic and may provoke an allergy on the skin (dermtitis). Intoxication may cause drastic diarrhoea, vomiting, abdominal pain, muscular contraction, palpitations, hypotension. More than 2 g of the plant per kilogram weight are toxic. Alcoholic drinks are used as an antidote.

The infusion of the leaves of *Jatropha gossypifolia* is an antidote to urticaria induced by *Cnidoscolus kunthiana*.

Observations

The large cells with giant druses of calcium oxalate interspersed in the palisade parenchyma are very characteristic in the leaf.

Mabea

Wood is useful. Seed oil is used for hair care and to prevent loss of hair.

Mabea piriri. AUBL. is a source of rubber. ROTH studied the bark structure 1981, leaf structure 1984, fruit structure and dispersal 1987, leaf venation 1996.

Pedilanthus tithymaloides (L.) POIR (ponopinito, zapatico de la virgen, piñipiña, tuturutú, pinipini, ipecacuana, ponoponito)

Taxonomical description

The plant is a small, 1–2 cm high shrub with succulent branches of green or bluish colour. The alternate leaves are ovate or oblong, 4–10 cm long and 2–4 cm broad, with an acute or acuminate tip and a cuneate base; they are subsessile, of fleshy consistency and have entire or slightly undulated margins. The stipules are transformed to small caducous glands. The inflorescence corresponds to a zygomorphic cyathium and is of carminic colour; the involucre is bilobed and contains several male flowers surrounding a single female flower. The involucre corresponds to 4 united hypsophylls, and is 10–15 cm long. The capsule is 8 mm long and 7–9 mm in diameter. The oviform seeds are about 5 mm long.

Occurrence

Tropical America.

In Venezuela, the species occurs in hot and dry regions among xerophytes. It has been collected in the Distrito Federal, in Aragua, Falcón, and the island of Margarita.

Anatomical description

ROTH (1969) described the growth in circumference of the axis in its different developmental steps. A summary of this description is found in ROTH (1992 P. 85/86).

The leaf develops storage tracheids in the mesophyll with helicoidal wall thickenings or large pits; these have the function of water storage.

Ethnobotanical and general use

Economical utilization.

The plant is used as a natural green fence or hedge as well as an ornamental for its variegated leaves. Certain arrow poisons are made of the latex.

Medical use

A decoction of the entire plant had a temporal reputation as a remedy against syphilis. It is now applied as an emmenagogum. The plant is also used against toothache. A decoction of the entire plant is vomitive and purgative.

The very acrid and caustic latex may cause severe blistering and vomiting. The latex of *Pedilanthus* is largely known as a drastic purgative and emetic. In Caracas, the milky sap is used externally to remove callosities, and rough spots of the skin.

Ingestion of the seeds is followed by repeated and violent vomiting.

In Cuba, the root is used as an abortifacient.

The main properties of the plant are purgative, emetic, keratolytic, abortifacient, narcotic and menstruation promoting.

In northern Peru, a hallucinogenic drink called 'cimora' is prepared from 6 plants, one of which is *Pedilanthus tithymaloides* (SCHULTES 1990).

Toxicity

The latex is toxic and caustic and causes severe dermal injuries. The plant should be used with much precaution.

Phyllanthus

At least 12 useful species are known.

Wood is useful. Roots are source of a dye. Leaves are used as an insect repellent. Medically, roots are applied for stones in the bladder. Roots and leaves are used for intermittent fever. Leaves are diuretic and help against gonorrhoea. Leaves are also used for headache and backache.

Bark contains tannin. Fruits are edible.

The diverse species are tonic, stomachic, febrifuge, diuretic, antidysenteric, astringent, used as emmengagogum, for oedema, blenorrhague, diarrhoea, affections of the liver, against hair loss and as a fish poison.

Phyllanthus acuminatus VAHL. is a shrub or small tree, 3–7 m high. The brachyblasts imitate a pinnate leaf. The leaves are ovate to elliptic and have an acute tip. The species frequently occurs in hot regions of Venezuela, but was also found in the cloud forest of Rancho Grande.

The leaf structure has been studied by ROTH (1990). The leaf is a typical hygromorphic shade leaf with lenticular cells in the lower epidermis which resemble ocelli.

The glycosides isolated from this species have antineoplastic activity (SCHULTES & RAFFAUF 1990).

The crushed leaves are prepared for a wash of the head to treat headache. A decoction of the leaves and flowering parts is taken for backache (SCHULTES & RAFFAUF 1990).

Quite a few species of this genus have medical use, several are cultivated as fish poisons (e.g. *P. piscatorum*). The leaves of this species are also applied as an insect repellent. For further medically used species see: SCHULTES & RAFFAUF 1990, UPHOFF 1968, GUPTA 1995, RORIGUEZ 1983, MANFRED 1982, ALBORNOZ 1993, DELASCIO CHITTY 1995.

Sapium

At least 12 useful species are known.

Wood is useful. Rubber is obtained from latex. A fish and arrow poison is extracted from branches. A red and orange dye is obtained from bark and leaves. Fruits supply 2 different kinds of fat. Fat is also obtained from seeds and manufactured in candles and soap; it furthermore has antifungal activities.

ROTH studied the bark structure 1981, leaf structure 1984, fruit structure and dispersal 1987, leaf venation 1996.

Flacourtiaceae

Banara

Banara contains fatty acids; *B. guianensis* AUBL. possibly contains alkaloids ROTH 1981, 1984, 1987, 1996.

Casearia

At least 8 useful species are known. Useful wood. Root is used for wounds and leprosy. Bitter bark supplies a red dye. The fruit is a fish poison.

The seed aril is edible. Oil is obtained from seeds. ROTH studied the bark structure 1981, leaf structure 1984, fruit structure and dispersal 1987, leaf venation 1996.

Homalium

At least 7 useful species are known. Useful wood. Bark for calking boats, roots for blenorrhagia. *H. racemosum* JACQ. Useful wood, resistant to dry-wood termites. ROTH 1981, 1984, 1987, 1996.

Laetia

Leaves as a repellent of flies, bark for flu. *L. procera* EICHLER leaves for persistent cramps. ROTH 1981 (bark), 1984 (leaf structure, 1996 (leaf venation).

Xylosma

Zuelania

Gum as a vomitive. *Z. guidonia* BRITT. & MILLSP., leaves, bark and resin are aromatic and diuretic. ROTH studied the bark structure 1981, fruit structure and dispersal 1987.

Gentianaceae

The Gentianaceae mainly comprise herbs and hardy annuals while woody forms are rarer. The entire leaves are opposite and glabrous. The fruit is mostly a septicidal capsule. The endosperm is fleshy. Intraxylary phloem is present in most of the species.

Bitter principles such as gentiopicrin are found.

Gentiana nevadensis GILG. (dictamo riñón)

The species occurs in the Andes in the form of cespitose cushions (VARESCHI 1970). It is a short-lived annual herb which is found at altitudes of 3200 to 4300 m, growing in little snow valleys on shifting and temporally freezing soils. The plant remains very low; it attains about 3.5–5 cm in height when flowering and has a conspicuous main taproot up to 7 cm long. The root is thus longer than the aerial part of the plant. The leaves in opposite position are very small and of the leptophyllous size category. The blade is thin; the leaf is of lanceolate shape, but has a broad base. It has an acuminate to rounded tip, but no petiole. The flowers are whitish to lilac-pinkish, but in the dried material appear yellowish.

The plant is a therophyte which completes its development in one season.

Taxonomical description

Leaf (Fig. 137 b). The inner structure of the blade is surprising. Upper and lower epidermis are small-celled and have thin walls. The mesophyll is not differentiated in palisade and spongy parenchyma, but the entire mesophyll is spongy and very loose. The cells are of irregular shape and form arms in various directions. There is only a single layer beneath the upper epidermis which is more compact. Towards the lower side, the mesophyll becomes more loose and large substomatal chambers arise. The stomata are somewhat elevated above the lower surface. The vascular bundles are of regular structure and the principal nerve as well as the stronger laterals have a small cap of fibers on the phloem side. There are no special reinforcements nor are there water-storing tissues. Furthermore, the leaf is of the shade type and of a herbaceous texture.

Axis (Fig. 138). An axis of about 1–2 mm in diameter shows a big hollow in the center that is partly filled by a central column in which a reduced vascular bundle occurs. From the column, fine cell tiers depart towards the vascular cylinder. In later stages, the central column as well as the parenchymatous cell tiers may be torn completely so that a hollow with residues of the cells remains.

The outlines of a transverse section of the axis show 4 wings which are formations of the epidermis and correspond to the descending margins of the leaves. Through the position of the wings it

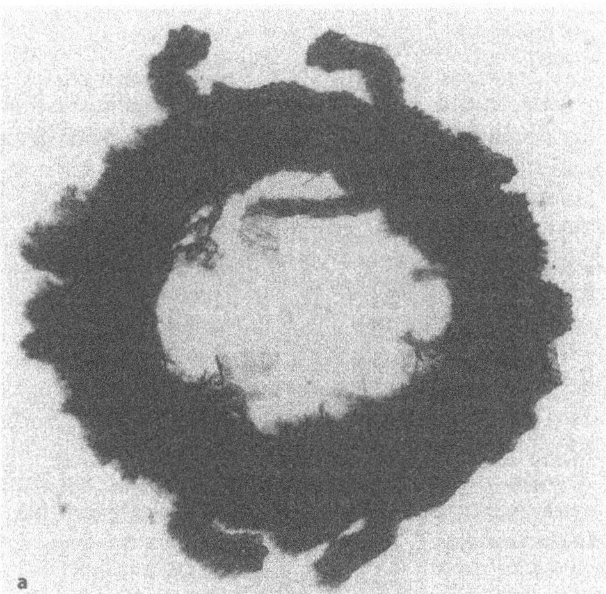

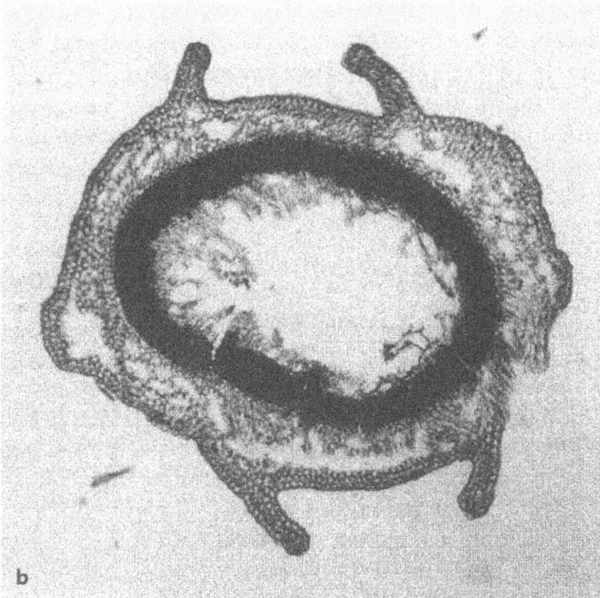

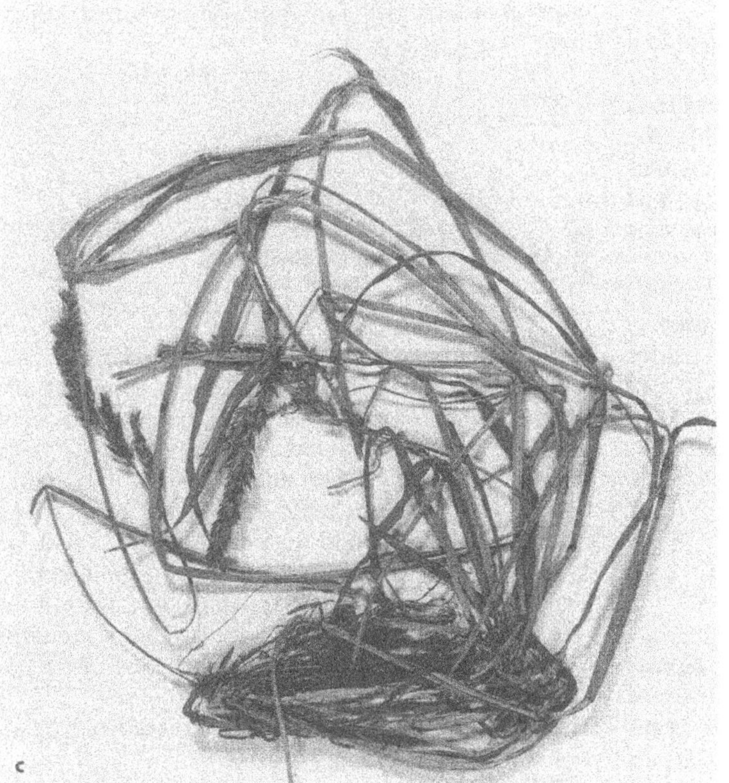

Fig. 138. a, b *Gentiana nevadensis*, t.s. of axis with leaves. **c** *Hierochloe mexicana*.

becomes obvious that the leaf position is opposite, as usually in Gentianaceae. The epidermis of the axis is still maintained and has extraordinarily thick outer tangntial walls. As seen in surface view, the epidermis surface is strongly ribbed; in transverse section the ribs appear like little teeth. The 1-3 parenchymatous cell layers beneath are also thick-walled and may be considered a collenchyma.

As seen in transection, the vascular system appears in the form of a completely closed ring, without traces of original individual bundles. The phloem is reduced to a few layers of parenchymatous cells, surrounding small groups of sieve tubes. The xylem is very homogeneous without distinct medullary rays and the vessels are inconspicuous, having the same diameter as the fibrous elements of the surrounding ground tissue. The intraxylary phloem occurs in the form of many isolated strands on the inside of the vascular ring (xylem side). The axis thus shows many peculiarities which are characteristic of the family. In more advanced stages, the primary cortex may transform into some kind of an aerenchyma through the formation of larger intercellular spaces.

Ethnobotanical and general use

Medicinal use

Many species of *Gentiana* contain a bitter principle, used medically as a tonic, a stomachic and as a stimulant of digestion. A bitter glucoside, gentiopicrin, is not only found in *G. lutea*, a plant which is very well known in home-medicine. In this species, the root is the most useful part. All medically used species of *Gentiana* are more or less applied for the same purposes.

CORREA & BERNAL (ED. SECAB 1993) presented a complete list of all species of *Gentiana* studied concerning their chemical contents. The list comprises 68 pages and contains more than 80 species (some of them are not identified), indicating the used analyzed parts and their chemical compounds. The best studied organ is the root, followed by leaf, axis and flower.

G. lutea is certainly the best studied species concerning its chemical contents. Gentianin, gentiopicrin, amarogentin, gentialutin were found besides countless other substances, such as bellidifolin and even chinic acid.

Gesneriaceae

This family is represented by small trees, shurbs or herbs; many are epiphytes. They are mostly of tropical or subtropical origin. The crown is zygomorphic, 5-lobed or bilabiate. Disc present. Carpels 2. Fruit a capsule or berry with many small seeds. Glandular hairs are frequent. The ornithogamous or entomogamous flowers are showy, the family therefore comprises many ornamentals, e.g. *Saintpaulia, Streptocarpus, Gloxinia* and others.

Flavonoids and cinnamic acid derivatives have been recorded. However, little is known about the chemistry.

Besleria

About 150 species belong to this genus which are distributed in tropical and subtropical America and the West Indies.

Alkaloids are found in *B. standleyi.*

Other species have purgative properties or are used as antidotes against bites of poisonous snakes or ant bites.

Besleria disgrega MORTON is a herb of the undergrowth of the Venezuelan cloud forest proper. The leaves are of herbaceous consistence and show a blue brilliance on the upper surface.

Anatomy of leaf, aerial stem, root and rhizome have been studied by ROTH (1990). The plant contains tannins.

Besleria affinis MORTON is a small shrub of the undergrowth of the transitional cloud forest, reaching a height of 3 m.

The leaf anatomy has been described by ROTH (1990).

Gramineae (Poaceae)

This monocotyledonous family comprises annual and perennial herbs usually with leaves in distichous arrangement. The leaves have a conspicuous sheath and a ligule at the border line between sheath and blade. The inflorescences are compound, the spikelets being aggregated in panicles, racemes or spikes. The flowers are small and reduced in their structure, due to the anemogamous fertilization. The fruit is typically a caryopsis in which seed coat and pericarp are united; it contains a single seed. The embryo is surrounded by an endosperm rich in starch.

The gramineae form the basis of the steppe, savanna and pasture-land vegetation.

The most useful species supply cereals, such as wheat, rye, rice, oats, barley or maize. Cereals are rich in carbohydrates and proteins. *Saccharum officinale* supplies sugar.

The woody bamboos are used in construction and in carpentery for tables and chairs etc.

Some species are of ornamental value.

The chemical contents of grasses are among others silicia bodies in short cells of the epidermis, proteins in the aleuron layer of the endosperm, starch in the seeds, fructosan, coumarin in *Hiero-*

chloe odorata (fragrant grass) and volatile oils in the Andropogoneae, which supply perfumes (palmarosa oil, lemongrass oil).

Further chemical constituents found in the grasses are: polyphenols, flavones, glucovanillin, methoxyarbutin, alkaloids, cyanogenic compounds, N-containing antimicrobial substances, saponins, sterols and triterpenes.

Hierochloe mexicana BENTH. (dictamo, paja maria)

The species is a perennial grass (Fig. 138 c) growing in the most elevated parts of the Sierra Nevada of Mérida between 3000 and 4000 m a.s.l. It is an erect herb, about 60–70 cm high, forming tufts. The spike is about 7 cm long. The total length of the leaves amounts to 17 cm, 7 cm of which occupies the sheath completely surrounding the stalk. The breadth of the blade is 3–5 mm. The ligule is about 2 mm long. The venation is longitudinal-parallel. The stalk is cylindric.

Anatomical description

Leaf (Fig. 139 a + b). The leaf has a structure typical of the Gramineae and the genus Hierochloe. The frequently occurring vascular bundles form ribs on the adaxial side which alternate with furrows. The upper as well as the lower epidermis are slightly large-celled. Short unicellular hairs occur principally above the ribs. Fan-shaped groups of 4–5 bulliform cells are found on the upper leaf side in the valleys between the ribs. The stomata which are confined to the upper side, are slightly sunk below the surface and often are found close to the bulliform cells; this seems to be reasonable, as the bulliform cells are subject to a turgor mechanism which causes the blade to roll up when the leaf looses too much water. The 3–5 layered mesophyll is more or less homogeneous without differentiation into palisade and spongy parenchyma. The cells are more or less isodiametric or tangentially extended. The vascular bundles are surrounded by an inner sheath with U-shaped wall thickenings and an outer parenchymatous sheath. The larger vascular bundles are transcurent to the upper as well as to the lower leaf side by sclerenchyma (fibers). The inner sheath is continuous around the bundles, except in the smallest vascular bundles, where it only occurs above the phloem. The outer sheath is continuous in the small and medium-sized bundles, but is interrupted at the poles of the largest bundles, where the inner scelerenchymatous sheath continues to the epidermis in the form of adaxial and abaxial girders.

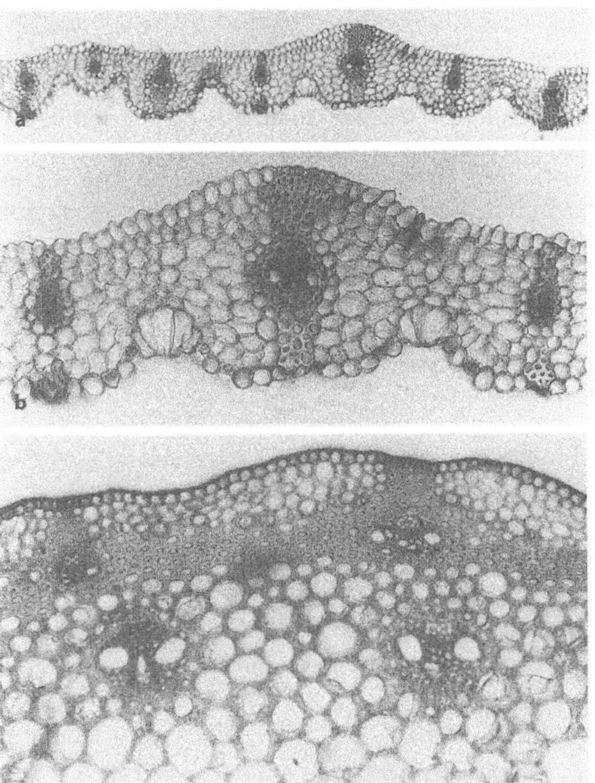

Fig. 139. *Hierochloe mexicana.* **a, b** Leaf in t.s. **c** Axis, t.s.

The plant has thus some xeromorphic features: The large-celled water-storing epidermis, stomata slightly sunk below the surface, the frequently occurring vascular bundles with their sclerenchymatous girders, and the ability to roll up the blade during water stress.

The species in question thus has the peculiarities typical of the genus *Hierochloe* (see also METCALFE 1960); stomata are sometimes absent from the abaxial side, the bundle sheath is double, the chlorenchyma is not radiate. The ribs on the upper surface are occassionally rounded and pronounced; the bulliform cells may form somewhat fan-shaped groups at the bases of the furrows. Furthermore, ribs of 2 sizes alternating with one another were observed in South American species. This is also the case in *H. mexicana.*

Axis (stalk or culm) (Fig. 139 c). The epidermis is of regular structure and has slightly thickened outer walls. Beneath follow about 3 layers of outer cortical parenchyma. The sclerenchymatous ring below is about 4–5 layers thick, as seen in t.s. Embedded in the ring are smaller vascular bundles, many of which are transcurrent to the upper epidermis by sclerenchymatous girders. A ring of larger solitary vascular bundles is embedded in the inner cortical parenchyma; this comprises about 5–10 layers of larger parenchyma cells. The vascular bundles are surrounded by a sclerenchymatous sheath. They have the typical structure of the monocotyledons with 2 large lateral vessels and a phloem with a chess board like pattern. The parenchyma cells are slightly thick-walled and have large pits. In the center of the stalk, there is a large hollow.

Ethnobotanical and general use

Medical use.
The root in infusion is used as an effective remedy for stomach trouble (PITTIER 1926/70).

Related species

Hierochloe odorata (L.) BEAUV. is called sweet grass. The dried foliage is burned as an incense during ceremonies by the Kiowa Indians. It is also used as a perfume for cloth and for the manufacture of baskets.

Observations

Most characteristic are the stomata sunk below the upper leaf side which are absent from the lower side, the vascular bundles with griders to the epidermal layers, the ribs of 2 sizes, bulliform cells in fanshaped groups, and non-radiating chlorenchyma.

Sporobolus indicus (L.) R. BR. (paja de gallina, tucupén, jeguey)

Taxonomical description

The plant is a perennial herb. The erect stalks occur individually or in clusters, reaching 30–100 cm in height. The ligule is very small and ciliate. The small blades are rolled up; they are about 1–3 mm broad and 8–10 cm long. The sheath is glabrous. The panicle is cylindric and over 30 cm long. The spikelets are 1.5–1.8 mm long. The caryopsis is reddish and slimy.

Occurrence

The plant is found in the whole of tropical America.

In Venezuela, it prefers hot and temperate regions; it is common in the savanna and at disturbed places.

Anatomical description

Sporobolus indicus has a typical leaf structure of the Gramineae (Fig. 140).

The upper epidermis is characterized by very large colourless bulliform cells. The arrangement of the bulliform cells and of the associated colourless cells in fan-shaped groups penetrating into the mesophyll is characteristic of the *Sporobolus* type. Additionally, papillae as well as some prickle hairs are present on the adaxial side. The long cells have sinous walls and are thick-walled over the veins. The mesophyll has the 'Kranz' type structure. The palisade cells radiate in a single layer in all directions from the vascular bundles. Their length/width index oscillates between 2.3 and 3.8. Shorter cells are observed on the abaxial side, while the longest palisade cells develop towards the upper side. The mesophyll cells between the vascular bundles are globular, leaving small intercellular spaces between one another and between the regions of contact with other tissue types.

The vascular bundles are very crowded and leave only a single cell layer between one another, as seen in transverse section. They have angular outlines and are surrounded by an inner small-celled and an outer large-celled parenchymatous sheath. The smaller bundles are surrounded by only a single triangular sheath which extends towards the adaxial side. Two layers of fiber strands are present on the ribs of both sides. The strands of the adaxial side are smaller, while those of the adaxial side are smaller, while those of the abaxial side have a larger tangential extension.

The lower epidermis is characterized by short and long cells and micro-hairs. Stomata are present on both leaf surfaces, but are more numerous on the lower side.

The leaf is of the photosynthetic C4 type.

Ethnobotanical and general use

Economical utilization.
The species furnishes a good cattle feed.

The plant is so strong that it is used locally as a rope to tie animals.

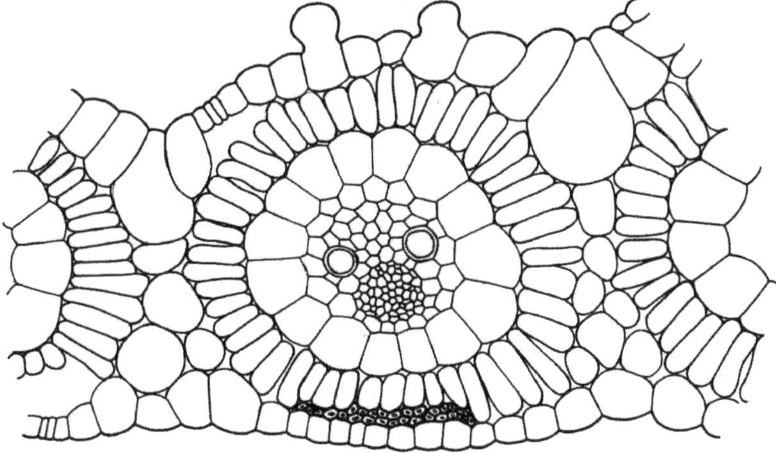

Fig. 140. *Sporobolus indicus*, t.s. of leaf ('grass type'). Vascular bundle surrounded by a prenchymatous sheath and a wreath of palisade cells (Roth 1992).

The golden-yellow glossy plant is used locally for the manufacture of hats.

Medical use

RODRIGUEZ (1983) indicates that branches of *Sporobolus* sp. are sold in herbolarios of Barquisimeto; used in infusion they are said to be good for the kidneys.

Toxicity

The poisonous fungus *Helminthosporium ravenelli* CURTIS which is said to cause trembling in animals has been reported in Venezuela as a pathogen of *Sporobolus indicus*.

Sporobolus virginicus (L.) KUNTH. (Fig. 141). The taxonomical and anatomical descriptions are found in ROTH 1992. The plant is a fodder of live-stock.

Sporobolus sp. (grama) is medically used. An infusion of the branches is taken for kidney problems.

Neurolepsis pittieri MCLURE is a grass of the undergrowth of the Venezuelan cloud forest. The leaf anatomy has been studied by ROTH (1990). The leaf is of a papery consistence and receives its rigidity from the vascular bundles. The leaf structure is very complicated and typical of this grass genus.

Pariana stenolemma TUTIN is a frequently found grass in the undergrowth of the transitional cloud forest of Venezuela. The leaves are leathery and slightly brittle. They are of lanceolate to elliptic shape and have an acute tip.

The leaf anatomy has been studied by ROTH (1990). Arm palisade cells are found in the mesophyll; the stomata are of the Pariana type of development (ROTH & CLAUSNITZER 1969). The leaf structure is slightly similar to that of *Neurolepsis pittieri*, but simplified. The plant is suitable for ground cover.

Guttiferae (Clusiaceae)

The family is represented mainly by trees and shrubs, and rarely by herbs. A few species are climbing epiphytes or stranglers. The leaves are simple, and frequently evergreen and leathery. The flowers are often large and attractive, spirocyclic or cyclic. Staminodia may be united in the form of cups or rings. The stigma is occasionally peltate. Placentation is usually central. The fruits are septicidal or septifragal capsules, berries or drupes. The integuments are thick. The seeds contain large embryos but no endosperm. They may be provided with arils or arilloids.

Schizogeneous oil or resin cavities or canals (name of the family: gutta) are universally present.

Tannins, anthraquinones, polyphenols, coumarins and xanthones are common in the family. Guttiferin is toxic. A highly toxic vermifugal constituent has been found in the seeds of a species of *Caraipa.*

Several species are sources of durable woods, drugs, dyes, gums and resins, fat, oils and of edible fruits.

Of the genus *Clusia* about 200 species exist in tropical and subtropical America, often as epiphytes and tree stranglers with aerial roots, or as lianas.

The family occurs in the Tropics and Subtropics, and only *Hypericum* is found in temperature regions.

Calophyllum

At least 16 useful species are known.

Wood is useful. Mainly bark and leaves supply gum, resin, balsam and oil. Gum is used for wound

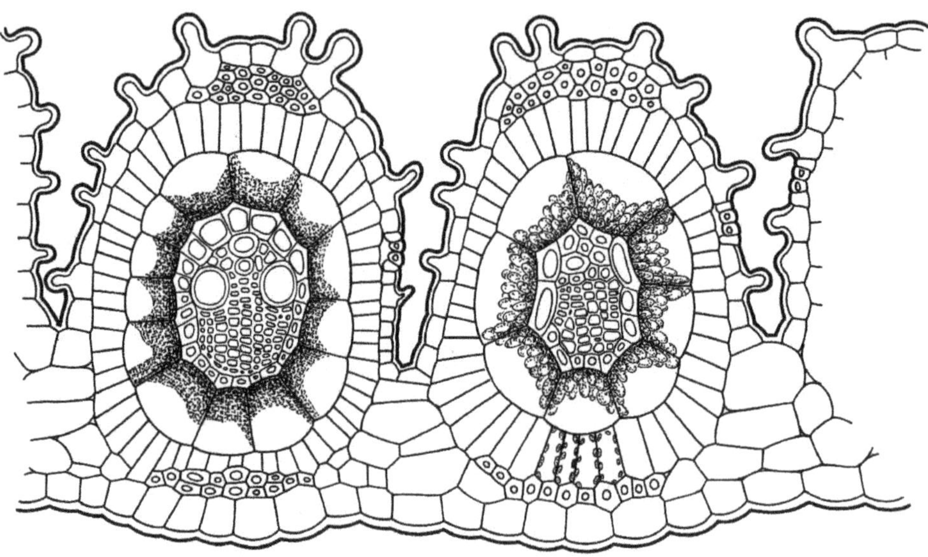

Fig. 141. *Sporobolus virginicus.* T.s. of leaf. The palisade parenchyma surrounds the vascular bundles in the form of a single layer; it contains small chloroplasts. On the inside of the palisade parenchyma there is a parenchymatous sheath consisting of large thin-walled cells in which the chloroplasts are arranged in the horseshoe shape; the chloroplasts are larger than those of the palisade parenchyma (vascular bundle to the right) (ROTH 1992).

healing, resin for incense and wound healing. Resin is applied for skin diseases. Resinous bark is used for aching joints and to hasten wound healing. A bark exudate is vermifuge. A seed oil is used for pruritus. Doma oil helps skin diseases and rheumatism. Oil relieves pain in leprosy patients. Leaves serve for fungal infections and tooth ache. Flowers are antidysenteric and, ingested, purgative. Fruits are diuretic. Tannins, saponins, cyanogenic compounds and coumarins with piscidial activity are present.

C. brasiliense CAMB. The wood is the source of a yellowish-green sandel oil. ROTH studied the bark structure 1981, leaf structure 1984, fruit structure and dispersal 1987, leaf venation 1996.

Caraipa

Caraipa with about 30 species of trees and shrubs, mostly occurs in the Amazon basin. The alternate leaves are pinnately nerved. The fragrant flowers are borne in terminal and axillary panicles, often large and conspicuous. The fruit is a woody capsule and when dehiscing the valves often separate from the central column (septifragal and columnicidal dehiscence, ROTH 1977) (RECORD & HESS 1943).

The largest and most widely distributed species is *C. densiflora* MART. which grows in Guiana, Surinam, and the states of Amazonas, sometimes attaining a height of 30 m. It is the source of Tamocoari balsam; the inner part of the bark contains a viscous yellow latex which is said to be highly caustic. The timber is of good quality, but apparently is too scarce to be of much importance (RECORD & HESS 1943).

ROOSMALEN (1985) described *C. densiflora, C. punctulata* and *C. richardiana* for Guiana. All 3 species have a yellow latex in the bark. Likewise bark of *C. fasciculata* growing in Brazil supplies a yellow balsam, put on wounds for healing. The resin or a decoction of the bark is used to treat skin ailments such as eczema, scabies and mange. In a similar way, the latex (sap) of *C. grandiflora* (herpes, mange, itches), *C. laxiflora* (fungal diseases of the skin), *C. paraensis* (herpes, mange, itches), *C. parvielliptica* (skin irritants, sores of the mucous membranes of the mouth) and *C.* sp. (hastens healing of sores of the skin) is used in traditional medicine (SCHULTES & RAFFAUF 1990).

The seeds of *C. psidifolia* are the source of an oil, used in Brazil for skin diseases and in ophthalmia. *C. lacerdiae, C. minor* and *C. palustris* have similar applications (UPHOF 1968). According to SCHULTES & RAFFAUF (1990), the gum resins found in the seeds of *Caraipa* species seem to be responsible for their use in treating skin diseases. A highly toxic vermifugal constituent has been found in the seeds of a species of *Caraipa.*

Xanthones, vanillin and triterpenes have been isolated from *C. grandiflora*, xanthones and triterpenes from *C. densiflora.*

At least 9 useful species are known.

Caraipa llanorum

Anatomical description

Leaf (Figs. 142, 143). The leaf is dorsiventral and hypostomatic. As seen in t.s., the upper epider-

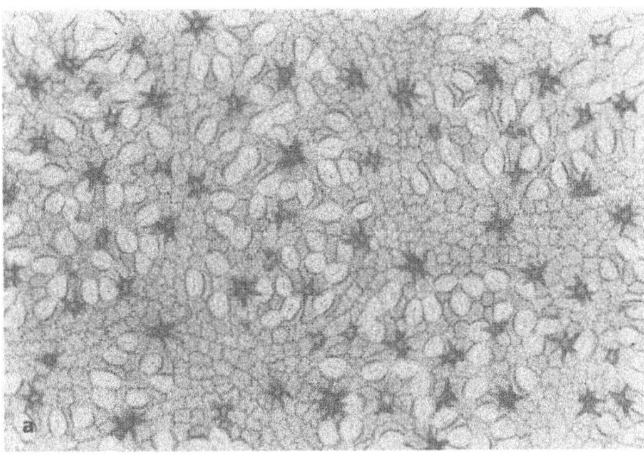

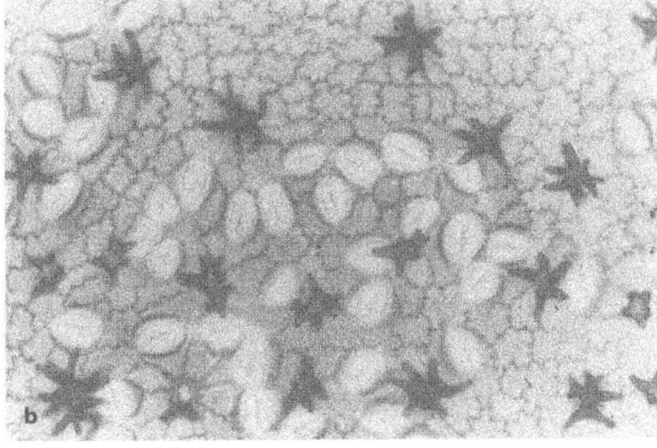

Fig. 142. *Caraipa llanorum.* **a, b** Lower epidermis with stomata and stellate hairs.

mis is composed of large cells with extremely thick outer tangential walls. The palisade parenchyma comprises 2 layers of elongated cells (and occasionally 3 layers); the outermost layer has the longest cells. The spongy parenchyma, partly composed of thicker-walled cells, is well developed and very loose in its structure with large intercellular spaces. The lower epidermis is small-celled and has thick walls. The mesophyll is divided into compartments by transcurrent vascular bundles of higher order which develop fibrous pillars towards the upper and lower epidermis. The pillars are small, as seen in a transverse section, and comprise 1-2 rows of fibers, enlarging around the vascular bundles and towards the upper and lower epidermis. Very conspicuous are sexretory cavities of enormous size which extend from the first palisade layer over 5-6 mesophyll layers and often have pear-shaped or egg-shaped outlines. Druses of calsium oxalate are frequent in the spongy parenchyma. Starch grains are small and elongated or ovoid.

The structure of the midrib is very characteristic. A large secretory duct occupies the center. The vascular system is composed of a ring of vascular bundles with endoscopic xylem, surrounded by a sclerenchymatous ring, and a row of vascular bundles in the center above the secretory canal. However, the entire center of the midrib inside the sclerenchymatous ring is sclerenchymatous and lignified. The palisade parenchyma is interrupted in the midrib and the parenchyma surrounding the sclerenchymatous ring is also partly lignified. Secretory canals are embedded in the parenchyma of the lower side. Upper and lower epidermis cells are likewise lignified in the midrib as well as in the lamina. A positive lignin reaction is also observed in the uppermost palisade layer.

Stomata, comparatively large and very frequent, as seen in surface view, and of the anomocytic type, are often without neighbouring subsidiary cells, but lie side by side. Anticlinal walls of the epidermis cells are wavy. Stellate hairs with about 4-8 (10) relatively short arms, some of which may occasionally be branched; are frequent on the lower leaf side.

Observations

The leaf structure is very characteristic and can be used for identification by the following peculiarities: The epidermis with the extremely thick outer walls, the formation of compartments by the fibrous sheath of the transcurrent vascular bundles, the secretory cavities of enormous size and the peculiar structure of the midrib with a peripheral ring of vascular bundles and a central row of bundles. A further peculiarity are the partly lignified walls of the mesophyll cells.

Also very characteristic are the stellate hairs with short arms and the numerous stomata, often side by side, without any epidermal subsidiary cells, on the lower leaf side.

The bark structure of *Caraipa richardiana* has been studied by ROTH 1981.

The plant contains a brownish-yellow resin and induces hypotension (CASTILLO 1995).

Clusia

At least 22 useful species are known.

The wood is useful. Resin, gum and latex are obtained from the different species. The resin is used to hasten healing of wounds and as an incense. Bark is used for leprosy and for aching joints. Leaves are applied for fungal infections, for toothache, and muscular sprains. Likewise, the latex helps for toothache. Latex is also used for skin problems and as a laxative. Bark and latex are a vermifuge. Flow-

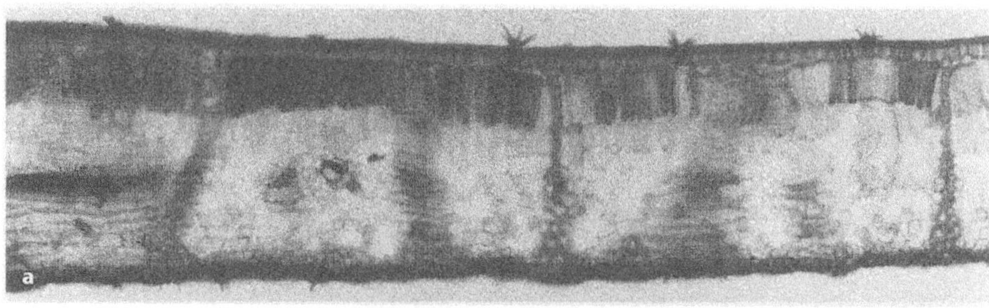

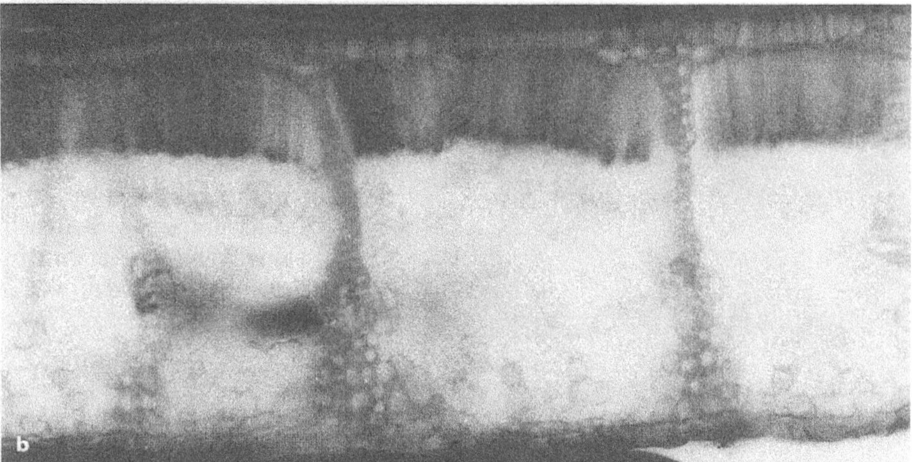

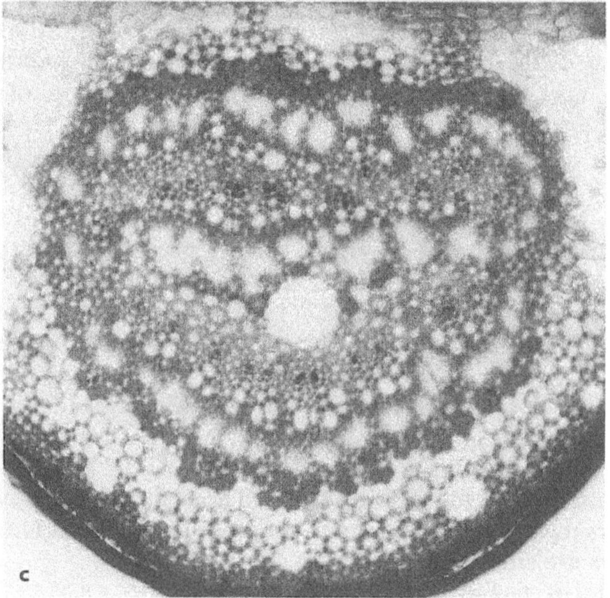

Fig. 143. *Caraipa llanorum*. Leaf. **a, b** T.s. of blade. **c** Midrib. Above and below × 10, center × 20.

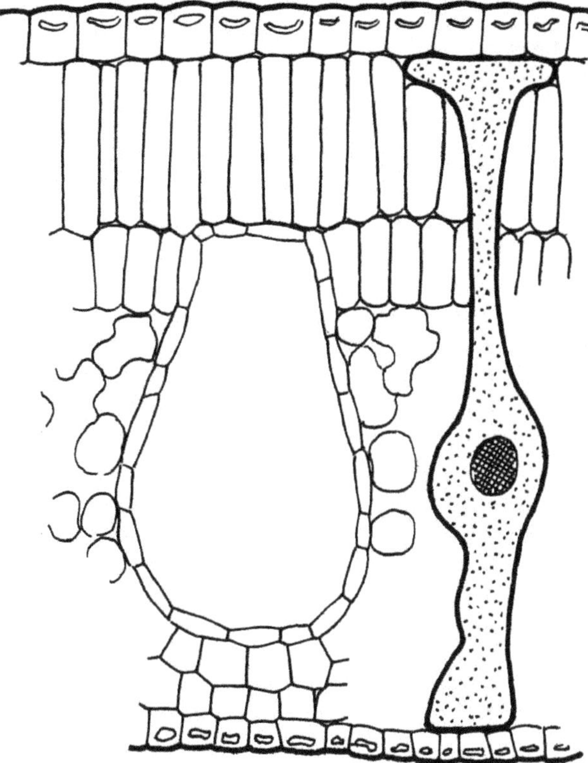

Fig. 144. *Caraipa llanorum*, t.s. of blade with secretory cavity and transcurrent vascular bundle, surrounded by fibers.

Fig. 145. *Clusia rosea*, flowers, **a** female, **b** male.

ers serve for toothache and are antidysenteric. Fruits are diuretic.

Clusia rosea JACQ., synonym: *Clusia maior* (copey, cupey, chuchi, chucúi, isfuque, quiripití, tampaco, tampeque)

Taxonomical description

Clusia rosea (Fig. 145) is a tree, 6–15 m (30 m) high, or an epiphyte which contains a yellow latex. In the juvenile stage, the plant often grows between rocks or germinates on other trees which it slowly destroys; for this reason it is also called matapalo (tree-killer) in some regions. From the branches where it settled first, the plant develops aerial roots which grow in length until they penetrate the soil. The opposed simple leaves have a short petiole and a thick fleshy, somewhat coriaceous blade, which is obovate, 8–19 cm long and 6–16 cm broad, and has an obtuse apex and a obtuse or cuneiform base. The venation is pinnate, but only the midrib is prominent on the abaxial side; the petiole is more or less cylindric and about 1–2 cm long and 0.4–0.9 cm in diameter.

The inflorescences are axillary and terminal, and 3–5 cm long not including the flowers. The flowers are arranged in dichasia of 1, 2 or 3, but the lateral flowers may occasionally develop a further pair of flowers. The 4–6 sepals are whitish-green and up to 2 cm long. The 6–8 petals are obovate or obcordiform and of white or rose colour; they reach a length of 3–4 cm.

The male flowers have numerous stamens; the outer ones are fertile and united at their base to form a cup or ring; the inner sterile stamens are tightly united forming a solid resinous mass.

In the female flowers, the staminodia are united in the form of a cup. The female flowers have 6–9 or more yellow stigmata and are smaller than the male flowers.

The globular fruit is a septicidal capsule, greenish or almost white, 5–8 cm in diameter and with 6–9 (12) locules. The seeds have an aril.

In Venezuela, the tree flowers in September and October.

Origin

Tropical America.

Occurrence

The tree has an ample distribution in Venezuela, growing from 400 m a.s.l. up to 2000 m, particularly in the coastal range (Cordillera de la Costa). In America it is found from Florida to the north of South America, including the Westindies (Venezuela, Ecuador, Colombia, Panama).

Anatomical description

Leaf (Fig. 146, 147 a). The very thick bifacial leaf is hypostomatic. The single-layered small-celled upper epidermis has thickened outer walls and a thick cuticle. As seen in surface view, the cells have straight walls. The underlying water-storing hypodermis is bistratified. It is composed of larger cells than the epidermis. As seen in paradermic section, the cells have also straight walls. The following palisade parenchyma consists of anticlinal rows of anticlinally elongated cells, each row comprising 5–6 cells. The spongy parenchyma comprises about 15–16 cell layers, but the cells are partly arranged in anticlinal rows. The cells have lobules or short arms, leaving intercellular spaces between each other. The spongy parenchyma is larger than the palisade parenchyma, occupying abuot double the space of the palisade parenchyma. The lower hypodermis is single-layered or bistratified, consisting of smaller cells than the upper hypodermis. As seen in paradermic view, the cells have straight walls. The lower epidermis is single-layered and composed of smaller cells than the upper epidermis. The outer walls are strongly thickened and cutinized. As seen in a surface view, the anticlinal walls are slightly wavy. The stomata are without subsidiary cells and are slightly sunk below the surface. The anterior cuticular horns are large.

The midrib projects somewhat above the lower leaf side, but has a depression on the upper side.

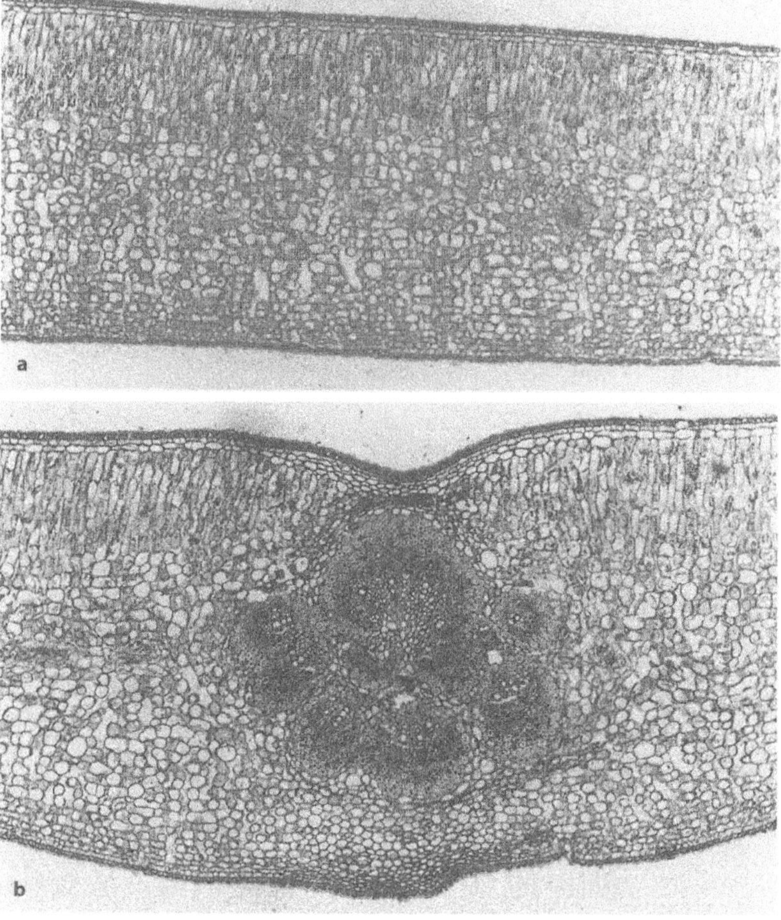

Fig. 146. *Clusia rosea*, leaf. **a** t.s. of blade, × 5. **b** of midrib, × 6.3.

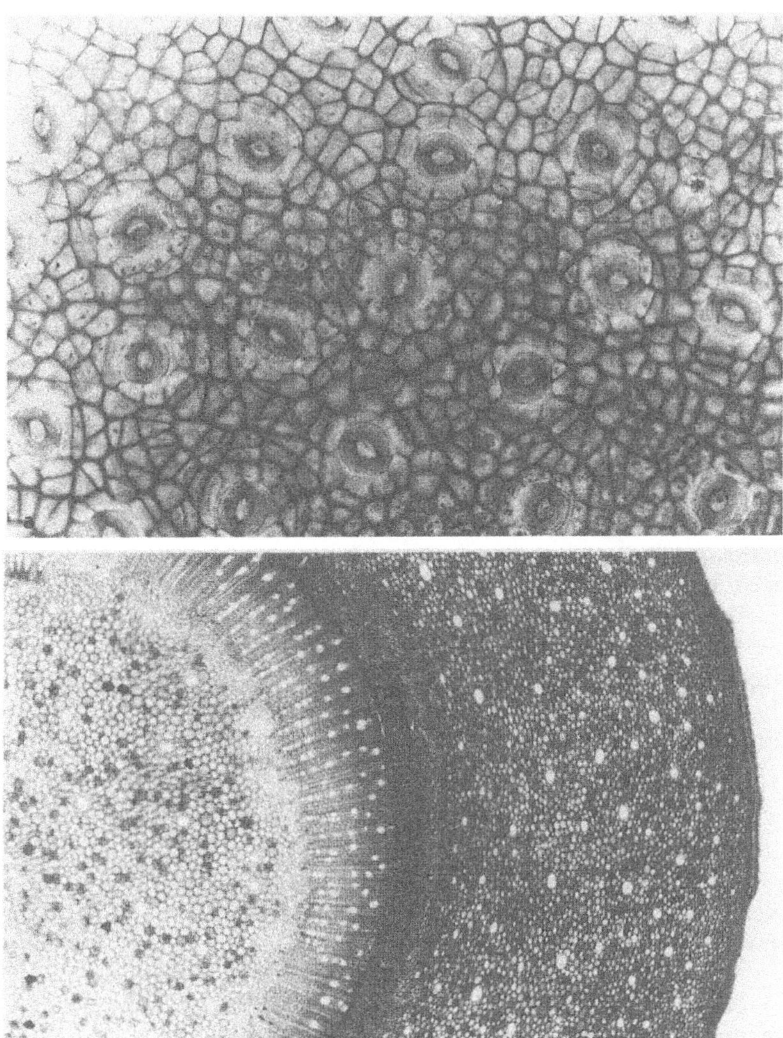

Fig. 147. *Clusia rosea*. **a** Lower epidermis with stomata (× 20). **b** Axis in t.s. × 2.5.

The pattern imposed by the vascular bundles is very characteristic. On the lower side, an arc of 4–7 individual bundles is found, each surrounded by a thick sclerenchymatous sheath. Above, a ring of 8–12 vascular bundles with the xylem pointing towards the center is found. The vascular bundles are surrounded by a common sheath of sclerenchyma which is very thick and partly penetrates into the space between the bundles, partly filling the center of the ring. However, a small passage of parenchyma leads from the center to the bundle arc below. The palisade parenchyma is interrupted in the midrib and substituted by collenchyma; the same tissue is also found on the lower side. The rest of the filling tissue is parenchyma.

The secondary nerves consists of a single vascular bundle which is surrounded by a thick sclerenchymatous sheath. A sclerenchymatous sheath also surrounds the veins of higher order, but is absent from the smallest veins.

Druses of calcium oxalate are very frequent in the mesophyll and in the hypodermis.

Stem (Fig. 147b). In a twig of 8 mm in diameter, the epidermis is single-layered and has strongly thickened and cutinized outer walls which are vaulted above the surface. Cork begins to form in the first subepidermal layer in some places. The primary cortex is very large being composed of more or less spheric parenchyma cells with thickened walls and conspicuous simple pits. Druses of calcium oxalate are often found in the cortex. Besides, secretory canals with a secretory epithelium are frequently dispersed in the parenchyma. The phloem is surrounded by a continuous ring of fibers in which some sclereids are interpersed. The phloem is of the regular type, being arranged in the form of a ring, when seen in transverse section. Some rays are slightly enlarged towards the outside. The xylem, which is also arranged in the form of a ring, is not very abundant. The vessels are found in radial multiples. The axial (paratracheal) parenchyma is scarce and fibers form the basic tissue. The rays are 1–2 seriate. The very ample pith in the center has about the same structure as the cortex. The spheric parenchyma cells are slightly thick-walled and have conspicuous pits. Druses of calcium oxalate are frequently found and secretory canals with a secretory epithelium are frequently dispersed in the ground parenchyma.

A mature fruit is seen in Fig. 148.

Fig. 148. *Clusia*, Guttiferae, fruits. **a** Opened fruit from above. **b** Half-opened fruit from the side. The fruit is a capsule with a complicated dehiscing mechanism (ROTH 1987).

Ethnobotanical and general use

Economical utilization

The tree is used as an ornamental plant in parks, gardens and as an alley tree. It may also be used for afforestation of slopes with a poor soil.

Together with spines of other plants sombreros are plaited of the leaves of *Clusia*, which the golddiggers in Chocó put on.

The wood is of reddish colour, strong and heavy, but it is only used as a combustible.

The aromatic flowers can be used industrially in perfumery.

Medical use

Name of the drug: *Clusia major* L. cortex, fructus. Besides the stem and twigs and the fruits, the leaves are also used in popular medicine.

Leaf. The infusion of the leaves is applied for pectoral diseases.

Stem. The cortex of the stem is astringent. The resin extracted from stem and fruit is applied as a purgative and to cure wounds. The resin of stem and fruit is also used to cure luxations and bone fractures. The latex particularly is the healing component. The latex of *Clusia rosea* is used in Trinidad as a cataplasm to combat aches.

Fruit. A decoction of the fruits is applied for rheumatism. The fruit (resin) is applied as a vulnerary and purgative, as is the cortex.

The oleoresin of this plant is a good vulnerary applied to burns. Mixed with liqueur of cucuy (agave) the oleoresin is applied for frictions of dislocated articulations.

Healing properties

Vulnerary, emollient, antispasmodic, antirheumatic, antineuralgic.

Chemical contents

The oleoresin of the stem contains triterpenoids, eteric oil, pectine and starch.

Related species

There are quite a few species of *Clusia* that have similar contents and similar healing properties (see also CORREA & BERNAL 1993, SCHULTES & RAFFAUF 1990). *Clusia rosea* (synonym: *Clusia major*), however is best known.

Cultivation

The plant is propagated by seeds. Its growth is slow to medium. It has a superficial radical system. The plant is highly resistant and durable.

Observations

The plant is easily identified as a *Clusia* species by its characteristic flowers. Likewise the fleshy-coriaceous leaves are very typical. However, the species of *Clusia* are variable in their anatomical structure (see also METCALFE & CHALK 1950). In our case the palisade cells of the leaves were not a palisade tissue with reticulately thickened walls, almost prosenchymatous in shape, as indicated by METCALFE & CHALK, and the stomata were not of the rubiaceous type, but anomocytic. Variations within a species due to environmental and particularly to soil conditions are not uncommon, as DE CORDEMOY (1911) already emphasized. This may lead to difficulties in distinguishing species.

Rheedia

At least 6 useful species are known.

The species supply balsam (maria balsam), resin and wax. Fruits are edible.

ROTH studied the bark structure 1981, leaf structure 1984, fruit structure and dispersal 1987, leaf venation 1996.

Symphonia

At least 6 useful species are known.

The wood is useful. Fruits are edible. Resin and wax are obtained. Leaves are food for silk worms. The resinous pitch serves for manufacture of torches and dancing masks, for caulking canoes, and as a general purpose glue. Ashes of the bark heal wounds and recalcitrant ulcers. Seed oil is used for skin diseases.

S. globulifera L. Wax and gum are obtained. Ashes of the bark are used for infected wounds and recalcitrant ulcers of abdomen and legs. Seed oil is applied for skin diseases. Boiled leaves applied as a lotion are used for dandruffs.

ROTH studied the bark structure 1981, leaf structure 1984, fruit structure and dispersal 1987, leaf venation 1996.

Tovomita

The genus comprises about 50 species of shrubs or small trees which have their center of distribution in the Amazon basin but extend northward into the West Indies and Central America.

The small flowers are borne in terminal cymes. The capsule dehisces septicidally with 4–6 valves; the seeds are withouth aril, but have a scarlet fleshy testa. Dispersal is endozoochorous. The bark has a yellow latex.

The timber is of good quality and is used locally for staves, shingles, and furniture.

The flowers of *Tovomita* aff. *laurina* are used in the form of a tea to cure diarrhoea.

Leaf anatomy of *Tovomita stigmatosa* PLANCH. & TRIANA has been studied by ROTH 1992.

The bark structure of *Tovomita brevistamina* has been studied by ROTH 1981, leaf anatomy by ROTH 1984, fruit structure and dispersal 1987, leaf venation 1996.

Fruit structure and dispersal of *T.* cf. *calodictyos* has been studied by ROTH 1987.

Vismia

At least 7 useful species are known.

An ointment of the resinous bark heals the craw-craw disease. Bark exudate (a yellow resin) is used for the treatment of wounds and infected sores, of herpes of the lips and fungal infections on the skin. A tea of the leaves is diuretic.

V. guianensis CHOISY. A gum resin is obtained from this species.

ROTH studied the leaf structure 1984, fruit structure and dispersal 1987, leaf venation 1996.

The bark structure of *V. macrophylla* has been studied by ROTH 1981.

Vismia baccifera (L.) TRIANA. The latex is applied for cicatrization. The species is described by CASTILLO (1995) for the state Amazonas.

Vismia ferruginea H. B. K. The latex is used to control skin diseases. Leaves and wood produce a liquid of a brick red colour which is applied externally to the affected region.

Vismia cayenensis (JACQ.) PERS. The bark soaked in water and applied as a footbath cures pains of the feet.

Hernandiaceae

Hernandia

The wood can be used as a substitute for balsa. The sap of the leaves is depilatory and painlessly destroys unwanted hairs on the skin. Fruit structure and dispersal have been described by ROTH (1987).

Hernandia

Saccoglottis

The smoke of the burned bark is inhaled for coughing due to tuberculosis.

The leaf structure was studies by ROTH 1984, bark structure by ROTH 1981.

Iridaceae

I. are monocotyledons, with equitant leaves in distichous position. Flowers trimerous, style endings often petaloid. Flowers arranged in spikes, racemes, panicles, fan-shaped or cincinnal inflorescences with spathaceous subtending floral bracts. Capsules loculicidal.

Sisyrinchieae with a rhizome, *Sisyrinchium* characteristic of South and Central America, in the Andes occurring up to 5000 m a.s.l.

The Iridaceae are not only ornamental plants. *Crocus sativus* supplies saffron (from the flower stigma) used as a condiment and a dye. It contains carotenoids one of which, the crocetin is active against arteriosclerosis.

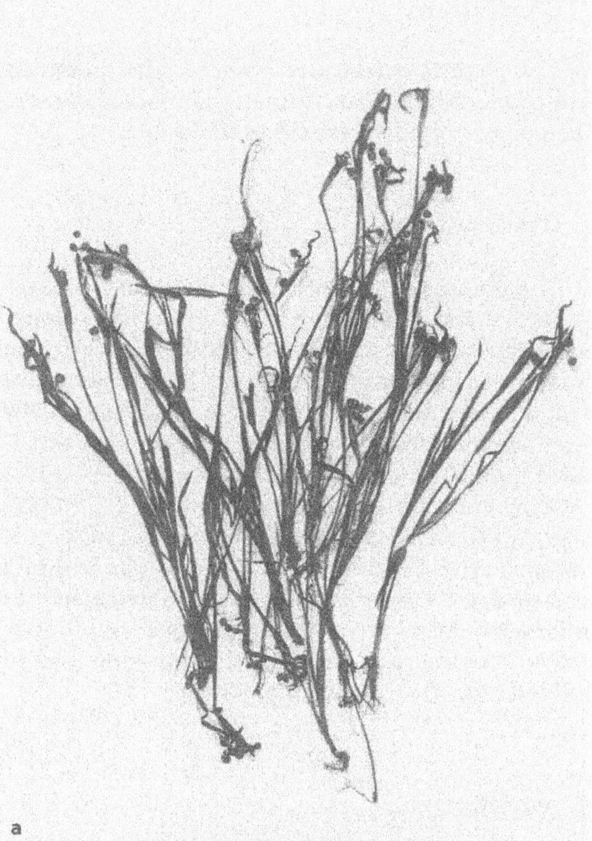

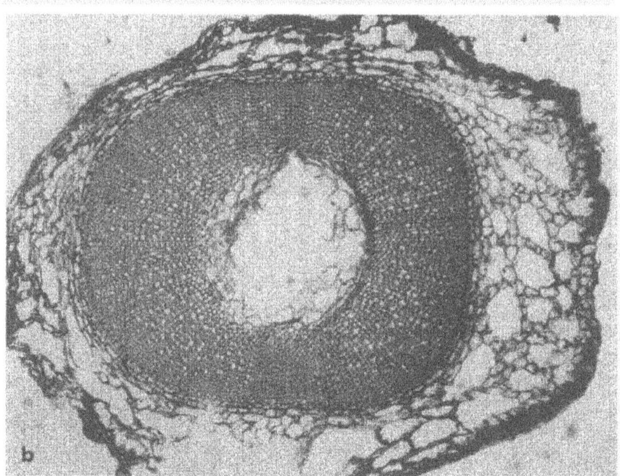

Fig. 149. *Sisyrinchium* sp. **a** Habitus. **b** t.s. of axis.

Sisyrinchium micranthum CAV. (espadilla, espadilla chiquita, little sword grass)

Taxonomical description

The species (Fig. 149 a) is an annual herb, reaching 5–18 (30) cm in height. The grass-like leaves have a long sheath and a linear blade, flattened in the median plane. The distichously arranged leaves reach a length of 7–10 (15) cm and a breadth of 0.2–0.4 cm. The inflorescence axis has acute angles and is slightly winged, it is about 5–20 cm long and is supplied with some leaves shorter than the nomophylls and several spathae. Each spathe bears 2–6 flowers. The perianth is blue or white, the 6 segments are 6 mm long or less. The globular capsule measures 3 mm in diameter.

Occurrence

Andes of Central and South America. In Venezuela, the plant grows in the páramos of the Andes and in the subpáramos of the Cordillera de la Costa between 2000 and 3000 m.

Anatomical description

Leaf. The leaves are more or less vertically oriented on the plant. The teeth, which are dispersed over the entire leaf surface, but become most conspicuous on the false margins (in the median plane), are very characteristic and useful for taxonomic identification. Their tips are directed towards the leaf apex. They are thick-walled and have small pits. They consists of a single epidermis cell of a more or less triangular shape, but are slightly curved towards the leaf tip. They are inserted in the epidermis with a broadened foot piece. The leaf margins are furthermore strengthened by a thicker-walled epidermis and mesophyll cells.

As seen in transverse section, the epidermis of the blade is very large-celled and has a water-storing function. Owing to the formation of teeth the epidermis cells have partly a papillose aspect. The outer walls are only slightly thickened. Stomata are frequent on both leaf sides; they are conspicuously sunk below the epidermis level and are surrounded by large papillose cells. The mesophyll is homogeneous consisting of more or less roundish (globular) cells which are small below the epidermis but become larger towards the leaf center. The larger cells surround the vascular bundles substituting a parenchymatous sheath and may also have a water-storing function; however, a true parenchymatous sheath is not clearly developed. The mesophyll is thin-walled. The vascular bundles are capped by sclerenchyma (fibers) in the xylem side. Their position is characteritic of Iridaceae leaves, so that the xylem alternately points towards the upper or towards the lower leaf side.

The leaf has a clear longitudinal symmetry similar to that of Gramineae, being divided into costal and intercostal fields. The very large papillose epidermis cells occur in the intercostal fields; they are longitudinally enlarged (parallel to the long axis of the leaf) and appear almost balloon-like or inflated, as seen in transverse section. Stomata are likewise found in the intercostal fields and their apperture extends parallel to the longitudinal axis of the leaf. The epidermis cells above the ribs are also longitudinally extended, but are somewhat smaller, as seen in transverse section. All cells of epidermal origin are more or less conspicuously arranged in longitudinal rows.

The leaf has some xeromorphic features, such as the waterstoring cells in the epidermis as well as in the mesophyll, and the stomata sunk below the surface.

Axis (Fig. 149 b, 150 a). In an axis of 1–1.2 mm in diameter, the thick-walled epidermis is maintained. The primary cortex is transformed into an aerenchyma composed of parenchymatous cell chains which leave large intercellular spaces between on another. The cells are thicker-walled. The endodermis with Casparian stripes is conspicuous. The vascular cylinder is completely closed, but its outer border has 4 flattened sides, while the inner border is round (cylindric). The phloem is reduced to a minimum. The xylem is well developed. Vessels are small and inconspicuous. Rays seem to be absent. The cells of the parenchymatous pith increase in size towards the inside. The very center is occupied by a hollow.

Ethnobotanical and general use

Medical use

Drug: Entire plant, root.

In infusion it is said to be a depurative of the blood; it also helps against flu, cough, biliary fever and is antiinflammatory.

Macerated in water it is used as a laxative and febrifuge. Likewise the entire plant in decoction is purgative. It is also considered as a sudorific and helps against irritations.

Taken 3 times a day, the decoction cures dyspepsia and even helps to improve the memory.

Method of use

Infusion, decoction, macerated. The decoction is prepared with 10 g of the plant stuff in 100 g water and taken 3 times a day.

Healing properties

Laxative, sudorific, depurative of the blood, febrifuge, antiinflammatory.

Chemical contents

Leucoanthocyanins, flavonoids, fatty oils in the seeds.

Related species

Sisyrinchium iridifolium H. B. K.

Observations

The leaf is typical of Iridaceae. The teeth dispersed over the entire leaf surface are the most

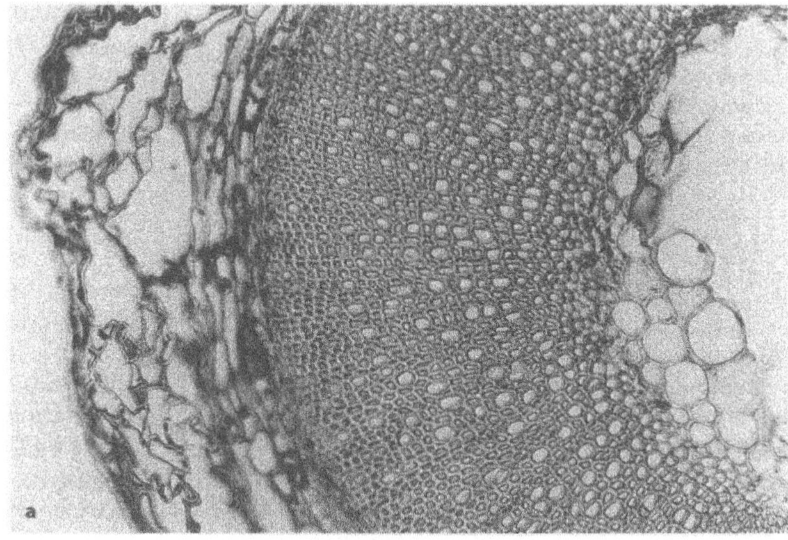

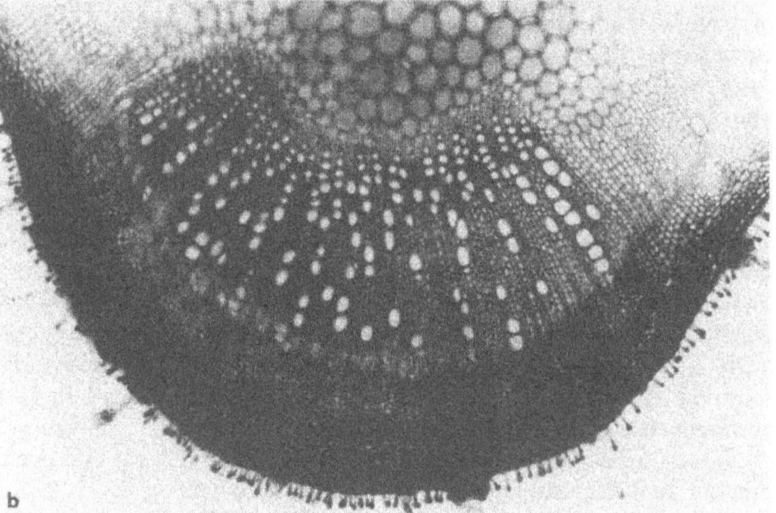

Fig. 150. a *Sisyrinchium* sp. Detail of axis in t.s. with cortical aerenchyma. b *Hyptis suaveolens*. T.s. of axis. × 5.

characteristic feature. Further characteristics are the large-celled water-storing epidermis and the stomata sunk below the surface. In the axis, the cortical aerenchyma is the most conspicuous feature.

Krameriaceae

The family is possibly related to the Leguminosae or Polygalales. The representatives are mainly shrubs, but subshrubs and perennial herbs also occur. The leaves are simple or, more seldom, trilobed.

The flowres are axillary or united in terminal racemes. The perianth is 5-merous, the sepals are free, stamens 4, carpels one with 2 ovules. The globular fruit has prickles, is dry, indehiscent and one-seeded. The seed has thick cotyledons, but no endosperm.

The only genus is Krameria with about 20 species occurring from the south of North America down to Argentina and Chile.

The species are rich in calcium oxalate, mainly in the form of druses.

Krameria lappacea (DOMBEY) BURDET & SIMPOSON, synonym: *K. triandra* RUIZ & PAVON (ratania)

Krameria lappacea is characteristic of the Andes of Bolivia and Peru and supplies the drug Radix Ratanhiae with 8–18% Ratanhia tannic acid. The

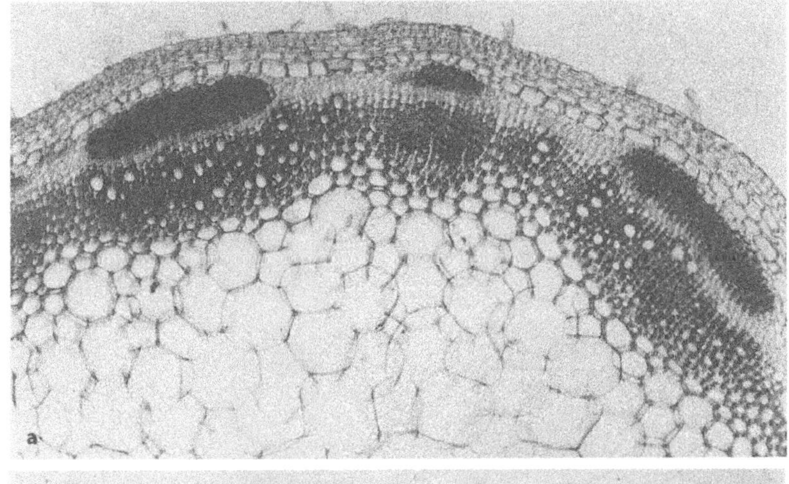

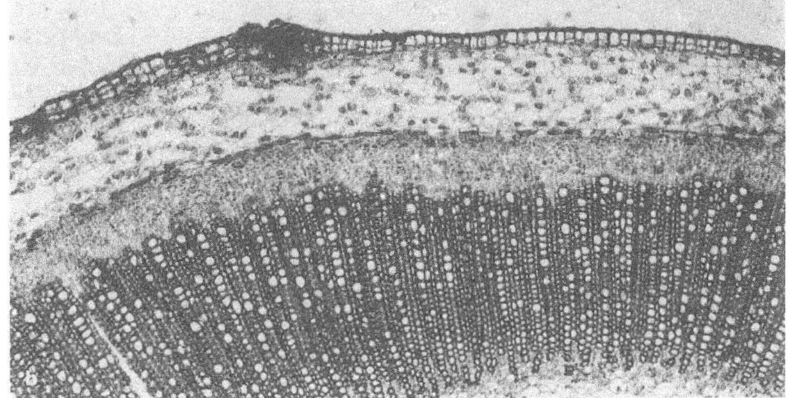

Fig. 151. b T.s. through the axis. Above: **a** *Gnaphalium antennarioides*. (Phloem stained darkly.) **b** *Angelonia salicariaefolia*. Stem with small cork layer, primary cortex in the form of an aerenchyma.

plant is furthermore rich in calcium oxalate which mainly occurs in the form of druses.

The leaves are lanceolate, about 0.8–1 cm long and only 3–3.5 mm broad, with a mucronate tip and a minute acute point. The petiole reaches a maximum length of 1 mm. The leaf arrangement is dispersed. Within the bud, the young leaves point upwards with their tips and are more or less appressed to the axis. Later on, however, the leaves turn downwards at an angle of 120–130°, probably due to an epitonic growth process at the leaf base, so that the upper leaf surface becomes fully exposed to light, while the lower leaf side is turned towards the axis.

The leaf surface is very hairy. Although relatively small, the unicellular hairs are thick-walled and considerably long. The type of indumentum seems to be characteristic of the genus. The hair felt is somewhat denser on the upper than on the lower side.

Anatomical description

Leaf (Fig. 152 a). As seen in a transverse section, the blade has the outlines of a succulent leaf; the midrib is hardly developed, the wings of the blade are reduced; the leaf is therefore thick. Its inner structure is isolateral or centric (METCALFE & CHALK 1950). In the region of the midrib, where the blade is thickest, only a single vascular bundle is found. the leaf margins are rounded so that the palisade parenchyma completely surrounds the blade without interruption.

In contrast to other species of *Krameria*, the upper epidermis cells are of normal size and not very thick-walled, and do not have a well-developed cuticle. The cells of the lower epidermis are only slightly smaller than those of the upper epidermis. Upper and lower epidermis cells have straight anticlinal walls, as seen in a surface view. The lower epidermis cells are more regularly arranged in more or less orderly longitudinal rows. Stomata sunk below the surface are found on both the upper and the lower side but are more frequent on the lower surface. They are surrounded by a variable number of subsidiary cells (anomocytic). A rubiaceous type of stomata, as observed by KUNZ (1913), could not be observed.

Most of the species of Krameria seem to be amphistomatic. The hairs lie closely appressed to the

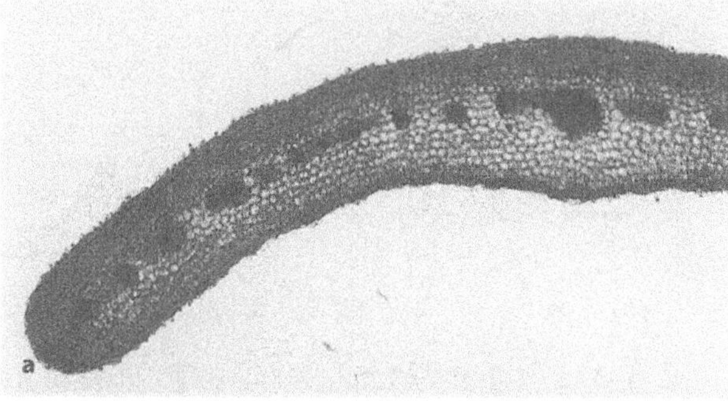

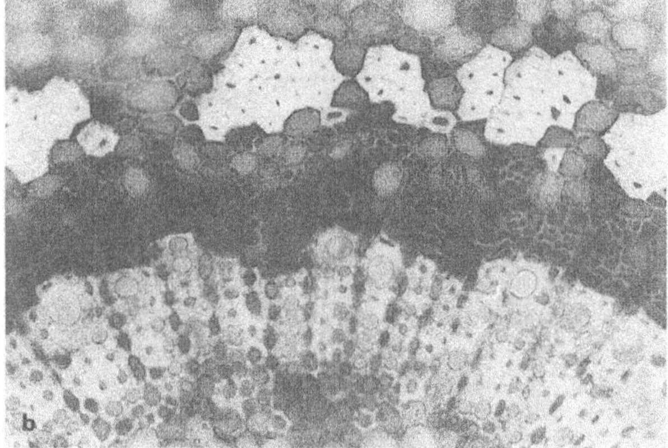

Fig. 152. *Krameria lappacea*. **a** T.s. of leaf: One wing of the leaf and the region of the midrib. **b** T.s. of one year-old stem with interrupted sclerenchymatous ring, xylem and phloem (Roth 1995).

surface; they form a bend at the base above their insertion on the epidermis; as seen in a transverse section, consequently they do not always appear completely on a section.

The differences between palisade parenchyma and spongy parenchyma are not very conspicuous; this is another characteristic feature of the genus. However, on the lower blade side, only one (sometimes 2) palisade-like layer is found, whereas on the upper side 2 (-3) palisade layers are present. The innermost of these 2-3 layers often merges more or less gradually with the spongy parenchyma. The palisade form is not very pronounced and the ratio length/width of the palisade cells oscillates between 2.6 and 3.2. The spongy parenchyma, situated in the middle of the blade, consists of more or less isodiametric, usually rounded, cells leaving only very small intercellular spaces. The chloroplasts of the palisade cells as well as of the spongy parenchyma cells are of medium size. The position of the palisade cells in the blade as well as in the axis is very interesting. Not only are the hairs orientied apically, but also the palisade cells are directed diagonally apically; this becomes most conspicuous in longitudinal sections.

The vascular bundles are slightly closer to the upper leaf surface. The largest bundle ist situated in the blade center. In the blade wings, a variable number of 2-4 or more vascular bundles may be present. Preserved in alcohol, the vascular bundles adopt a dark-brown colour, probably due to the presence of tannins. The vascular bundles are transversely connected by more or less isodiametric cells with thick walls and many large pits. These cells we consider water-storing tracheids and not sclerosed cells (METCALF & CHALK 1950), although they may supply support to the leaf. The larger bundles are accompanied by fibers on their lower side. But some fibers (with strongly light refracting walls) may be found in the xylem side. Smaller and larger druses of calcium oxalate are frequently found in the entire mesophyll.

Leaf petiole. The ventral side of the leaf petiole is flattened so that this appears semicircular in a transverse section. Immediately above its insertion on the axis, the petiole has a single vascular bundle, from which the lateral bundles separate, as soon as the petiole expands in the transverse plane. The parenchyma cells which form the major bulk of the tissue, are thickest-walled at the base of the petiole. But also in the entire petiole, parenchyma cells are, for supporting reasons, more thick-walled than they are in the blade. Where the petiole gains in thickness, the lateral bundles deviate from the main bundle and become arranged more or less in the form of a line, as seen in transverse section, i.e. they remain in the same plane. About 6-7 vascular bundles may pass throug the petiole. The transition between petiole and blade is fluid and more or less gradual. This structure of the petiole is essentially distinguished from that of other species of *Krameria* where a horseshoe-shaped bundle often with inwardly bent margins and surrounded by fiber bundles, is present (e.g. in *K. tomentosa, K. argentea,* according to METCALFE & CHALK 1950).

Axis (Fig. 152b). An abundant parenchymatous pith lies in the center of the axis. The cells are more or less of globular shape and have small, often triangular, intercellular spaces, as seen in transverse section. The cell walls are relatively thin. The largest cells lie in the center, so that cell size diminishes towards the outside, in the direction of the xylem, while the wall thickness increases in this direction. The cells contain small leucoplasts.

A continuius xylem ring, as seen in transverse section, follows towards the outside. The medullary rays are uniseriate, a further peculiarity of the genus *Krameria*. The xylem is surrounded by a smaller phloem ring, which consists mainly of soft bast. Hardbast can hardly be observed, apart from some dispersed fibers. In a one-year-old branch, a single ring of vessels could be observed. The vessels are small with a diameter of less than 30 µm, a further characteristic of the small-leafed *Krameria* species. An interrupted ring of sclerenchyma is found towards the outside; it appears in the form of small groups of fibers which are embedded in the parenchyma. The sclerenchyma in its turn is surrounded by a parenchymatous cortex. The outermost layer of the primary cortex is chlorenchymatous consisting of typical palisade cells. These are, according to METCALFE & CHALK 1950, absent in all species of *Krameria* studied. As mentioned above, the palisade cells are arranged diagonally apically as seen in longitudinal section. Relatively few crystals of calcium oxalate are found in the parenchymatous cortex.

The epidermis is comparatively large-celled and has strongly thickened outer walls in the form of a reverse horseshoe, as the thickenings also extend to the anticlinal walls. A 3–5 fold layering of the outer epidermis walls is very conspicuous. Preserved in alcohol, the epidermis cells as well as the phloem cells adopt a brown colour. The same unicellular and thick-walled long hairs as found in the blade, are also present in the epidermis of the axis, but are less frequent.

In an older, about 3 years old branch, the formation of the first periderm directly beneath the epidermis may be observed; it becomes about 4–6 layered. This peculiarity fundamentally distinguishes *K. lappacea* from other *Krameria* species, in which the first cork is formed more profoundly in the cortex or even in the pericycle (METCALFE & CHALK 1950). The cork cells remain thin-walled in *K. lappacea* and the material studied did not show thickened outer walls, as described by METCALFE & CHALK for other species of *Krameria*.

The branch collected on 07.10.1993 shows abundant storage starch. The parenchymatous pith cells are filled with globular starch grains, much starch is likewise deposited in the uniseriate ray cells, and the parenchymatous cortex cells contain starch in abundance. In the phloem, the rays enlarge slightly towards the outside throug tangential extension of the cells, but no cell divisions take place. The axis thus seemed to be well prepared for the winter.

The xylem shows 3 annual rings which are, however, partly very indistinct, as the vessels are in part dispersed over the section and are not always arranged in the form of conspicuous rings. The wood parenchyma surrounding the vessels is likewise partly distributed in the form of indistinct rings, as seen in transverse section. But radial rows of vessels described by METCALFE & CHALK (1950) as characteristic of *Krameria* were not observed. It may be left to further studies to ascertain whether they only develop in older twigs or whether their absence is a consequence of the more or less homogeneous climate in the greenhouse.

Ethnobotanical and general use

Economical utilization

The plant supplies the drug radix ratanhiae which contains 8–18% tannic acid and is used for tanning.

Medical use

Drug: radix ratanhiae, rich in tannin.

The dried root is used medicinally; it is astringent and tonic. It furthermore contains krameric acid. In former times it was used as a tooth preservative. Powdered root is said to be present in certain tooth-powders of today. The powder is also applied against diarhoea, gonorrhoea and haemorrhage.

Method of use

1–1.5 g root powder against acute and chronic diarrhoea per day. Decoctions of the root in 5% aqueous solution are used as injections against gonorrhoea.

Haemorrhages of the stomach, the intestine, the lungs and the matrix are stopped immediately, when 1 g of the root powder is taken with a spoonful of water.

Healing properties

Astringent, soothing, haemostatic, antidiarrhoeic.

Chemical contents

The species contains much tannin and is used medically as a source of tannic substances. It is well known that polyphenols have important functions in the plant, such as protection against fungus infection or against predation by animals. Recently, it has been proved that polyphenols may also protect the plant against excessive UV radiation. Raxid Ratanhiae is therefore also used in lotions, emulsions and gels which serve as solar protection of the human skin (ALBORNOZ 1980).

Varieties and related species

Species with similar properties are *K. argentea*, Brazil, *K. ixina*, Colombia, *K. tomentosa*, West Indies. These 3 species are also rich in tannins and have similar healing properties and uses as *K. lappacea*.

Observations

Although the various species of Krameria are similar in their inner structure and may easily be confused, *Krameria lappacea* has some specific characteristics which may very well serve for distinction. METCALFE & CHALK (1950) suggest that the transverse section of the leaves may be taken as a good diagnostic feature, as the outlines of the blades differ in the various species of *Krameria*. To distinguish *Krameria lappacea* from other species of *Krameria*, the following anatomical peculiarities may be observed: Besides the outlines of the blade, the upper epidermis cells of *K. lappacea* are not strongly cutinized as compared with the other species. The strongly thickened and layered outer walls of the epidermis of the axis in *K. lappacea*, the position of the first periderm directly beneath the epidermis, the thin-walled cork cells, as well as the lack of radial rows of vessels, the structure of the petiole without a central horseshoe-shaped bundle and the anomocytic stomatal type instead of the rubiaceous type are further peculiarities useful for identification.

Krameria ixina is native of Venezuela and is applied to cure afflictions of the liver and kidney. It is also an abortive. Longterm application, however, may produce cancer of the oesophagus due to the high tannin content.

Labiatae

Flowers bilabiate, fruits 4-celled (through false septs) separating into 4 nut-like mericarps each with a single seed. Herbs or shrubs, seldom trees (*Hyptis*) with tetragonal branches (leaves frequently in decussate position). Inflorescences cymous. Glandular hairs with differently compound etheric oils (aromatic alcohols, phenols, terpenes, ketones, aldehydes etc.) are frequent. The Ocimeae are tropical and subtropical.

The neotropical genus *Hyptis* belongs to the Hyptidinae.

Characteristic chemical contents of the Labiatae are besides essential oils, bitter principles and tannins.

The etheric oils occur in glandular hairs; mono- and sesquiterpenes are the most frequently occurring compounds.

Within the bitter principles, picrosalvin may be mentioned.

Within the polyphenols and tannins, the phenolic compounds called labiatae tannins are characteristic of the family. Many representatives of the Labiatae are used as condiments or as medicinal plants. The tannins of the labiatae are used externally.

Hyptis suaveolens (L.) POT. (mastranto, lavaplatos, jujure)

Taxonomical description

Herb, 1–1.5 cm high. Branches quadrangular, glandular velvety-hairy. Leaves with a long petiole, ovate, 4–10 cm long, with acute tip and a rounded to subcordiform base; margins serrate, surface velvety or hairy.

Flowers arranged in small dense cymes, usually solitary in the axils of the subtending hypsophylls or in spiciform inflorescences. Calyx with 5 long teeth, subulate, rigid, and 4–5 mm long. Corolla blue or purple, bilabiate; tube 4–6 mm long. Fruit separating into 4 nutlets. Very aromatic plant.

Origin

Neotropical

Occurrence

Pantropical (UPHOF 1968). In Venezuela, the species is widely distributed particularly in savannas and other open regions, occurring from sea level up to 900 m a.s.l.

Anatomical description

Leaf. Hyptis suaveolens has a leaf of the sun type (Fig. 153 a, 154 a, b). The leaf is dorsiventral and amphistomatic. The upper epidermis cells are of normal size. Their outer walls are thickened. There is a single row of very long palisade cells with small chloroplasts which have a length/width index of about 9/1. The spongy parenchyma is somewhat smaller than the palisade parenchyma and consists of cells with arms which leave intercellular spaces. Large cells of roundish shape, containing a druse of calcium oxalate are conspicuous at the border line towards the palisade parenchyma and may have a water storing function; they belong to the parenchymatous sheath of the vascular bundles of higher order. The lower epidermis cells are smaller, as seen in t.s., than those of the upper epidermis and the stomata lie at epidermis level.

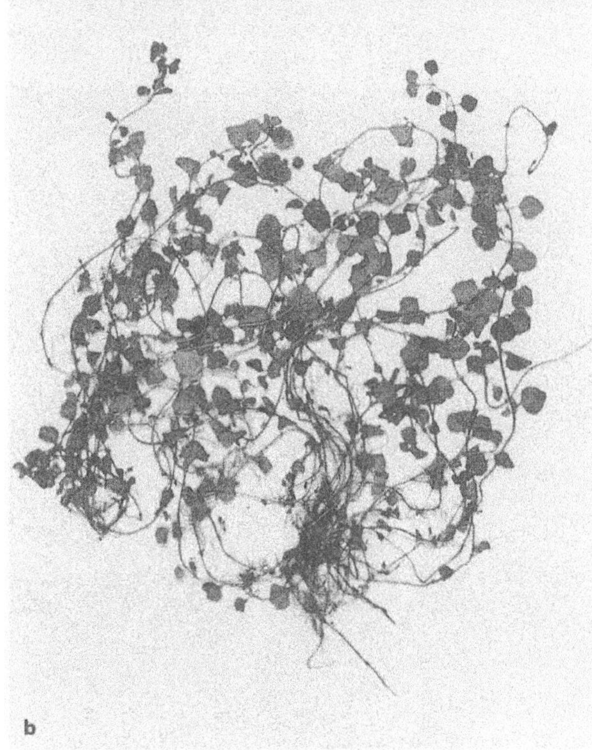

Fig. 153. a *Hyptis suaveolens*. Uniseriate hairs of the lower epidermis (× 5). b *Satureia brownei*. Habitus.

The veins are particularly well developed. The midrib is very prominent on the lower side, but shows a depression on the upper side. Palisade and spongy parenchyma are interrupted in the midrib. The vascular system is arranged in the form of a shallow open arc. The filling tissue is parenchymatous, but may transform into an angular collenchyma in very well developed ribs. The lateral veins of first order have a similar structure to that of the midrib. The smaller bundles of higher order are surrounded by a parenchymatous sheath which is transcurrent to the upper as well as to the lower side; each cell contains a druse of calcium oxalate.

The cells of the upper epidermis have sinuous anticlinal walls, as seen in a surface view, but the walls of the lower epidermis cells are much more undulated. While the stomata of the upper epidermis are relatively numerous, those of the lower epidermis are still more frequent. The stomata belong to the diacytic or caryophyllaceous type, according to METCALFE & CHALK (1950) or to the developmental transversal type ('tipo perpendicular' or 'cruzado' of ROTH & CLAUSNITZER (1969 a) which is also very frequently found in the Rubiaceae. Long and slender uniseriate hairs with a pointed end and a swollen foot cell occur more frequently on the lower than on the upper side. Giant uniseriate hairs occur mainly above the veins; they frequently have a single swollen foot cell, but 2–3 swollen basal cells may also be found. Glandular hairs with a bicellular head and a short stalk are very characteristic and occur more frequently on the lower leaf surface. Above the ribs, their stalk may possibly become somewhat longer (Figs. 153 a, 154 b).

Axis (Figs. 154 c, 155 e). The branches are quadrangular. The epidermis cells are comparatively small. Below follow about 5 layers of photosynthetic tissue which becomes collenchymatous in the 4 angles. Towards the inside follows a more or less continuous vascular cylinder which is closed by the action of the interfasciclular cambium. The vascular cylinder is surrounded by an interrupted pericycle ring composed of arcs or irregular bundles of fibers. Initially, only 4 arcs of vascular tissue are present in the 4 angles of the axis, but later on, separate bundles are formed by the interfascicular cambium so that the ring is finally closed. The vessels in the xylem are relatively small. The rays are inconspicuous and uniseriate; broader rays may however arise in the interfascicular regions. The pith is ample and leaves a big hollow in the center.

Very conspicuous are the glandular hairs with shorter or longer stalks in the epidermis, as well as the uniseriate hairs of different length and sizes and with a varying number of basal cells (see the description of the leaf).

Ethnobotanical and general use

Nutritional use

Hyptis suaveolens, a pantropical aromatic annual, is used mainly in medicine. However, in some parts of West Africa a mintflavoured tea-substitute is prepared as a beverage from the leaves and flowers.

Economical utilization

The foliage is the source of an essential oil extracted from the glandular hairs and used for adultering patchouli (from *Microtanea cymosa* PRAIN.).

Branches are placed under beds and chairs to drive away bedbugs.

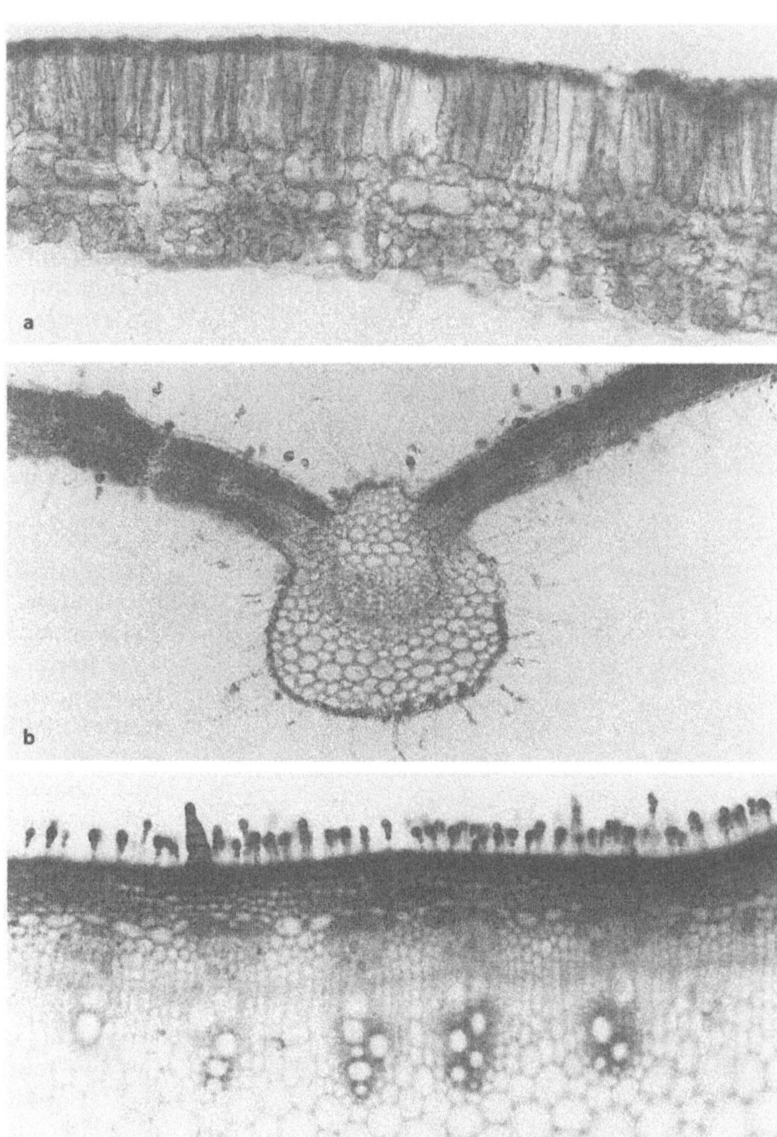

Fig. 154. *Hyptis suaveolens*. **a** T.s. of blade × 20. **b** Midrib × 10. **c** T.s. of outer part of axis with hairs (× 10).

Medical use

Leaves, branches and flowers (inflorescences) are the useful parts.

Leaves and buds are used in antirheumatic baths. Leaves in decoction for hot baths are also applied against muscular weakness and gout. Externally applied, the baths also combat lice and worms. Leaves and flowers in infusion act as a hypotensive. The entire plant in decoction for a bath combats scabies and insect bites. Leaves and flowering parts are used internally against cough and as a digestive. Taken internally the plants have stimulant and sudorific effects. Digestive, emollient, hypotensor, abortive and even aphrodisiac; used against fever, haemorrhoids, hypertension, impotence, diseases of the skin, scabies, cough, muscular debility, rheumatism.

Method of use

It is applied externally as a bath, or internally as decoctions and infusions. To prepare the tea, 1 l of boiling water is poured over a handful flowers and a handful of leaves; a cup of tea with sugar taken before the meals stimulates the appetite, while when taken after meals it is used as a digestive.

Chemical contents

An essential oil is present in the glandular hairs. It is composed mainly of α- and β-pinene, camphene, menthol, cymene and chavicol, as well as malic acid, citric acid, and minerals.

Related species

There are quite a few related species which are used as a food, to flavour food, as a tea, a decoction or infusion against inflammations or as a carminative, sudorific and anticatarrhal. *H. pectinata* supplies a resin used as an incense. The seeds of *H. spicigera* made into a pulp are used as a condiment. Other species mentioned by UPHOF (1968) used in popular medicine are *H. albida, H. breviceps, H. emoryi, H. fasciculata, H. laniflora, H. mutabilis. H. atrorubens* is mentioned by SEAFORTH, ADAMS & SYLVESTER (1983) as a carmiantive (infusion). According to WONG (1976), the leaf juice is used for diarrhoea, dysentery, vomiting; a tea of the leaves is taken for colds, flu, intestinal worms, and a leaf infusion for indigestion. In all these cases, probably the essential oils are the active principles. *H. verticillata* and *H. capitata* are further examples (SCHULTES & RAFFAUF 1990).

Cultivation

It would be easy to cultivate the plant, as it is little exacting and can almost be considered a weed.

Observations

The hairs are the most important characteristic of identification.

Hyptis verticillata JACQ. A decoction of the entire plant is used for headache and fever, applied in the form of a lotion (OCAMPO). Also, the plant shows bactericidal and neoplasm-inhibiting activies (SCHULTES & RAFFAUF 1990, GUPTA 1995).

Satureia brownei (SWARTZ) BRIQUETIER (poleo)

Taxonomical description

The plant (Fig. 153 b) is a delicate herb with procumbent twigs that have long internodes up to 3 cm long; it is creeping on the ground and develops adventicious roots in the nodal regions. The small leaves are suborbicular-deltoid with an obtuse apex and a truncate to almost cordate base. They measure about 5–12 mm in diameter and occupy a surface area of about 1.5 cm^2. The margins of the blade are slightly wavy. The petiole is about 4 mm long. There is a principal nerve and about 2 pronounced laterals in each leaf half which ramify alternately from the midrib. The texture of the blade is herbaceous and delicate.

The filiform flower-stalk is as long as the calyx and at times it is ciliate. The almost cylindric calyx, 4–5 mm long, is hairy in the gorge and has triangular-lanceolate teeth which are also hairy. The crown is 7–8 mm long, of purple colour with white patches; the upper label is at times retuse.

Historical background

Satureia brownei is mentioned as an aromatic garden plant in 1578 in Venezuelan documents (VELEZ & VELEZ 1990).

Occurrence

The plant grows mainly in shady or humid places with a more or less temperate climate. It is regarded as semi-creeping and has an annual life cycle. The species is found in Venezuela, Colombia, Peru and Ecuador.

Anatomical desription

Leaf. The upper epidermis is large-celled and the outer walls which are slightly thickened are vaulted above the surface. The palisade parenchyma is single-layered and consists of relatively short and broad cells. The anticlinal walls of the palisade cells are partly wavy (harmonica-like). The length/width index of the palisade cells amounts to 2, a very low value. The spongy parenchyma is composed of 3 layers of roundish cells and has a more or less compact structure. The lower epidermis is somewhat smaller-celled. The stomata which occur on both leaf surfaces are slightly elevated above the surface by 2 subsidiary cells lying on each side of the stoma, as seen in transverse section. The anticlinal walls of the lower epidermis are more wavy than those of the upper epidermis, as seen in surface view. Druses of calcium oxalate are abundant in the mesophyll, particularly in the palisade parenchyma. The principal nerve is not very strong and none of the nerves has a sclerenchymatous reinforcing sheath, but only a parenchymatous one. The meshes formed by the veins, as seen in a surface view, are medium-sized and show only few ramifications and few free endings. The upper and the lower epidermis are very much alike in their structure, but the upper epidermis has less stomata and in places also forms anticlinal walls which are straight, as seen in surface view. Glands of 2 types are very conspicuous on both leaf sides. There are smaller glands with a short stalk and a unicellular head as well as large glands, intermingled with the small ones, which are sunk slightly below the epidermis and have a very large pluricellular head.

Axis (Fig. 155). As seen in transverse section, the axis has quadrangular outlines with 4 pronounced edges. The epidermis is of regular size and its walls are strongly thickened on the projecting edges. the stomata are equally elevated above the surface as in the leaves. The primary cortex beneath the epidermis is parenchymatous, except the projecting edges which are strengthened by collenchymatous cells. The central cylinder is already closed, although it is obvious that it arose from 4 separate bundles, lying in front of the edges.

The vascular cylinder is surrounded by a very conspicuous large-celled endodermis with Casparian stripes. The phloem forms a continuous ring, as seen in transverse section, and is thinwalled. The vessels in the xylem are arranged in radial rows. The pith in the center is parenchymatous. It con-

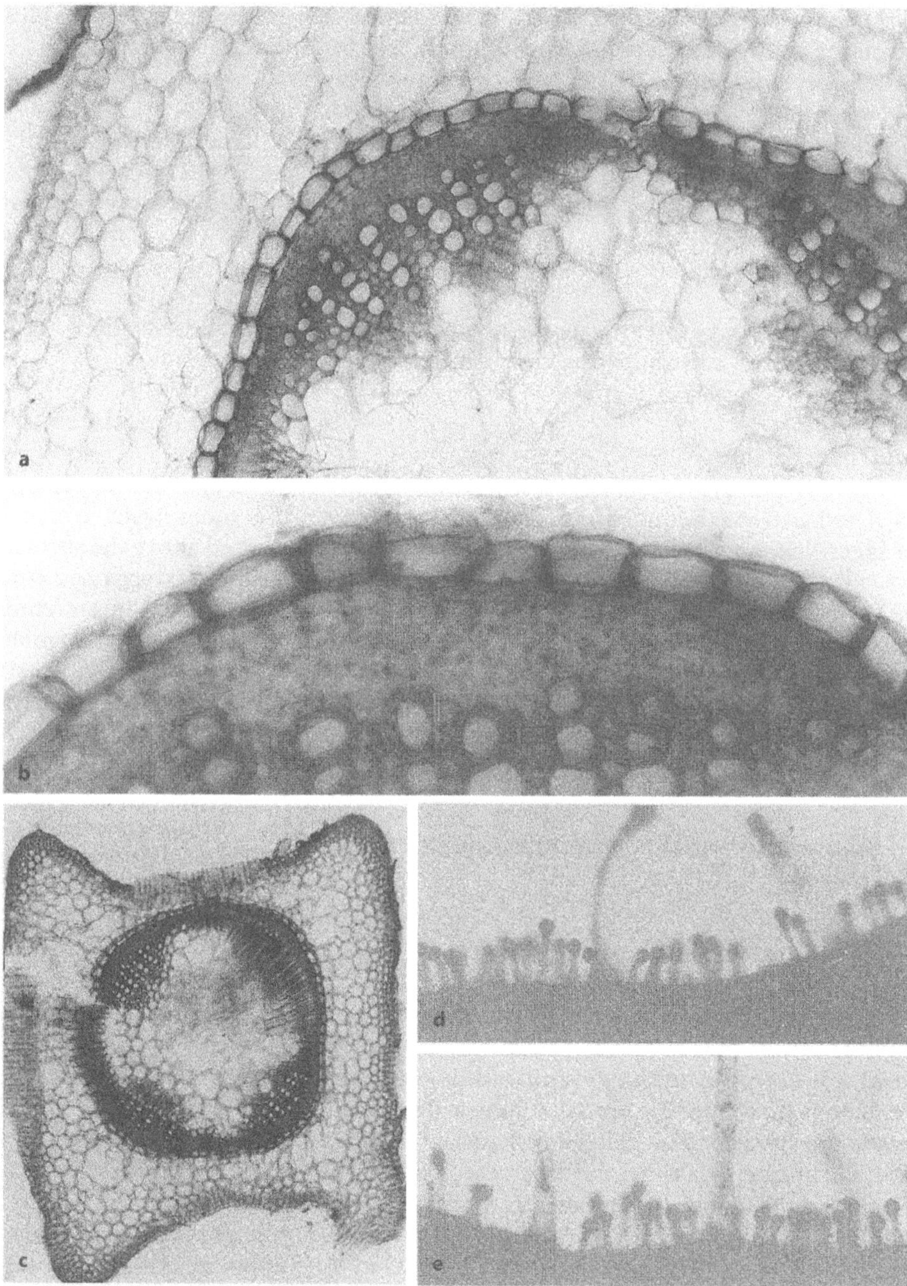

Fig. 155. *Satureia brownei* (a, b, c). **a** T.s. of axis with vascular cylinder. **b** endodermis (× 40). **c** Axis in t.s. (× 5). **d, e** *Hyptis suaveolens*: Hairs of the axis (× 10).

sists of thin-walled cells which increase in size towards the center. The most centreally lying cells may however become ruptured.

Ethnobotanical and general use

Nutritional use
The plant is highly aromatic and used as a condiment, particularly for meat (pounding it before using). The ground plant smells pleasantly of menthol or camphor.

Medical use
The drug is called *Satureia brownei* (SWARTZ) BRIQUETIER folia, rama.
It is applied as a pectoral and emmenagogum and for problems of the liver.

Method of use
The plant is usually prepared as an infusion or decoction. To prepare it as a carmiantive and for the expulsion of wind and worms, 250 ml hot water have to be poured over 10 g leaves. The liquid is drunk during the day.

As a stimulant or carminative and against bronchitis, 40 g leaves are boiled for five minutes in one liter of water; a small cup is taken 3 times a day; honey may be added to the decoction.

People who suffer from gastric ulcer are recommended to take an infusion of Poleo, pouring half a bottle of hot water over 10 g of poleo; this can be taken with sugar in case of hoarseness.

To stimulate digestion after heavy meals, half a bottle of hot water is poured over 5 g poleo and 5 g chamomile; after cooling, the liquid is taken immediately after the meal.

A decoction of poleo is used for inhalations against sinusitis.

Against toothache and mouth ulcers, poleo is boiled together with tobacco, some salt is added and the mouth rinsed with it. The boiled plants can also be rubbed directly in the affected site.

Healing properties
Stomachic, carminative, against flu and catarrh, against debility of the bladder and nocturnal incontinence (decoction or infusion of the entire plant).

Antispasmodic, against stomach ache, colics and flatulence, as an expectorant, against palpitations, trembling and nervous vomiting, very effective against hoarseness.

As an emmenagogum, for problems of the liver, sedative, aphrodisiac.

Chemical contents
Gum and resin. Menthol, limonene, cineol, carvacrol, menthone, dipentene.

Observations

The plant is best recocnized by the shape of the leaves, the quadrangular transection of the axis, and the glandular hairs, which emanate a smell of menthol or camphor. The glands which are abundant on the leaf contain the 'healing substances'.

Lauraceae

Trees or shrubs; small flowers, frequently arranged in racemose inflorescences. Typical representatives of the rain forest, although the leaves are often leathery.

Anthers opening with 4 or 2 valves. Carpels 3. Ovary 1-locular with one hanging ovule. Fruits berries or drupes. Receptacle frequently transformed into a fleshy or woody cupule.

Slime and oil cells typical, as well as bitter substances. Mainly Central and South America.

Cinnamon is obtained from barks of *Cinnamomum* species; the main constituent of the volatile cinnamon oil is cinnamyl aldehyde. Eugenol and camphor are the predominant volatile oils of the leaves and roots. The volatile olis of the root of *Sassafras albidum* var. *molle* have a high content of safrole, which was formerly used for flavouring, but has a carcinogenic effect.

Persea americana, the avocado, supplies fleshy edible fruits.

Aniba

Wood is useful. Wood is source of the essential oil 'essence de bois rose' used in perfumery. Bark is fragrant and has a scent of cinnamon and roses, used for a stimulating tea. Powdered bark is used for perfuming. Bark is stimulant, digestive, antispasmodic, pectoral and helps to treat anemia. Seeds are used for dysentery.

ROTH studied the bark structure 1981, leaf structure 1984, fruit structure and dispersal 1987, leaf venation 1996.

Beilschmiedia

At least 13 useful species are known. Wood is useful. Leaves are used for a body pomade. Fruits are edible. ROTH studied the bark structure 1981, leaf structure 1984, fruit structure and dispersal 1987, leaf venation 1996.

Beilschmiedia mexicana (MEZ) KOSTERM. is a tree up to 22 m high, characteristic of the transitional cloud forest. The simple leaves are elliptic and have a short tip. The coriaceous blade measures 13x17 cm up to 28x13 cm The black berries are edible like those of *B. anay* that reach a length of 10 cm and have a flavour similar to that of avocado.

The leaf anatomy has been briefly described by ROTH (1992).

Nectandra

All species of the genus Nectandra produce timber of good quality for carpentry and general construction. The woods are so much like some of the species of *Ocotea, Aniba* and *Phoebe* that differentiation is very difficult. Taxonomists have the same difficulties, particularly if flowers are lacking. We therefore suppose that the specimen identified as *Nectandra pichurim* in the transitional cloud forest ROTH (1990) is not identical with the specimen described here, as it has a different leaf structure.

The genus Nectandra is well known for its alkaloids. Terpenes are found in the essential oils. Small amounts of tannins occur in the wood. Neolignans have been isolated from some species.

The seeds, important in popular medicine, are very astringent and are used to cure diarrhoea, dysentery and nervous disorders. The wood is useful. Flower, calyx and bark are used as a spice.

The volatile oil is used as a substitute for kerosene, and to treat eczema, psoriasis and to kill nits and lice.

The tea of the bark is febrifugal.

Pichurim or puchury beans are used in medicine.

ROTH studied the bark structure 1981, leaf structure 1984, fruit structure and dispersal 1987.

Nectandra pichurim (H.B.K.) MEZ (laurel capuchino, laurel canelo, capuchino, cobalonga, pucheri)

Taxonomical description

Nectandra pichurim is a tree, 10–20 m high. The alternating leathery leaves are lanceolate, about 10–20 cm long and 3–5 cm wide, largely acuminate, and silky on the lower surface. The petiole reaches 10–20 mm in length. The small flowers are arranged in compound racemes. The inflorescence is of the same size as the leaves or shorter. The flowers are white and measure 3–4 mm in diameter. The pedicels are 1–4 mm long. The berry is egg-shaped and 15–25 mm long.

Origin

From Panama to Argentina.

Occurrence

The species is found in the hot regions of Venezuela.

Anatomical description

Leaf (Fig. 168 a, b). The leaf is equifacial and hypostomatic. As seen in transverse section, the upper epidermis cells are comparatively large and possibly water-storing. There are 2 layers of palisade parenchyma, the upper layer consisting of long and slender cells, while the cells of the second layer are shorter. The spongy parenchyma is small; its cells have irregular outlines forming short arms towards each other so that intercellular spaces arise. Above the lower epidermis, there is again a palisade layer, but of a looser aspect and with shorter cells. Dispersed in the spongy parenchyma as well as in the palisade parenchyma are very large sphaerical cells with an oily content; they are surrounded by a slightly thickened and probably cutinized wall. The lower epidermis is of regular aspect and the stomata lie at epidermis level.

As seen in surface view, the upper epidermis cells are polyhedric and have straight walls. The lower epidermis cells are also polyhedric, but are more irregular and occasionally have slightly bent walls. The stomata are small and have a subsidiary cell on each side (parallel-celled type).

The midrib is vaulted above both surfaces. Palisade parenchyma and spongy parenchyma are interrupted in this region. Most of the space is occupied by a large vascular bundle which is laterally extended and surrounded by a strong sclerenchymatous sheath, this is interrupted on the phloem side by rays. The rest consists of parenchymatous filling tissue and a few sclerenchymatous cells below the lower epidermis. Both epidermal layers are very small-celled and thick-walled. Oil cells are dispersed in the midrib.

The secondary nerves have a structure which is similar to that of the midrib and the veins of higher order are also surrounded by sclerenchyma which is transcurrent to the upper and lower epidermis. The sclerenchyma cells in contact with the lower epidermis have horseshoe-shaped wall thickenings, the thin walls pointing towards the epidermis.

Ethnobotanical and general use

Economical utilization

The wood is valuable. The volatile oil can be used as a substitute of kerosene. It is also used to kill nits and lice.

Medical use

Twigs are used for a bath. The volatile oil is applied to treat eczema and psoriasis.

The seeds are also used in medicine. Pichurim beans are used for diarrhoea, dysentery and nervous irritations. The seeds are astringent and the fruits contain volatile oil.

Chemical contents

The genus contains mainly alkaloids, terpenes may be found in the essential oil. There are several species which are very similar, even in their content and use. Some barks contain alkaloids and a febrifugal effect is ascribed to them.

Observations

The leaves are very characteristic. Most typical are the oil idioblasts and first of all, the horseshoe-shaped thickenings on the sclerenchymatous sheath of the vascular bundles. OSSOWSKI (1929) has shown that these are of considerable diagnostic value in *Laurus nobilis*, even in powdered drugs. The same is true of *N. pichurim*.

Ocotea

At least 13 useful species are known. The wood is useful. The bark is astringent and tonic and used for tanning. Root bark is antirheumatic. Leaves are used for domestic medicine. Fruits are used as an arrow poison (alkaloids). Essential oils contain myristic aldehyde. ROTH studied the bark structure 1981, leaf structure 1984, fruit structure and dispersal 1987, leaf venation 1996.

Persea americana MILL. (aguacate)

Taxonomical description

The avocado is a medium-sized tree of 10–20 m height with a strong root system. The coriaceous leaves are simple oblong, elliptic-lanceolate to ovate 8–40 cm long with an acute or truncate base. Young leaves are pubescent. The bisexual flowers develop in huge quantities on panicles; they have whitish tepals, 9–12 stamens and a short pistil.

The pear-shaped oviform or globular fruit is a berry with a very large and hard kernel (seed). The pear-shaped fruits may attain a length of 16–22 cm and a weight of half a kilo or more.

Different species of *Persea* have been described from Mexico to Chile (BRÜCHER 1989).

Origin and historical background

The cultivated forms must have developed in Mexico and Guatemala. From prehistoric finds it is concluded that avocados were being cultivated 10 000 years before Christ.

In precolombian times, avocado was cultivated from Mexico to Peru. The Aztecs called the tree Ahuacatl from which the name Aguacate is derived. Many more names exist which were given to the tree by other Indian tribes.

Anatomical description

Fruit (Fig. 156, 157, 158). Development and anatomy of the avocado pear have been studied by HECCTOR LOPEZ-NARANJO, a student of the senior author, in his thesis (ROTH 1977).

The main bulk of the pulp consists of parenchyma, whereby the cells of the periphery are smaller increasing in size towards the middle region of the pericarp. The most peripheral layers are rich in chloroplasts and add the green colour to the fruit. The outermost 10–15 layers are usually free from stone cells (Fig. 158). Inside from these layers, groups of stone cells are found; middle and inner

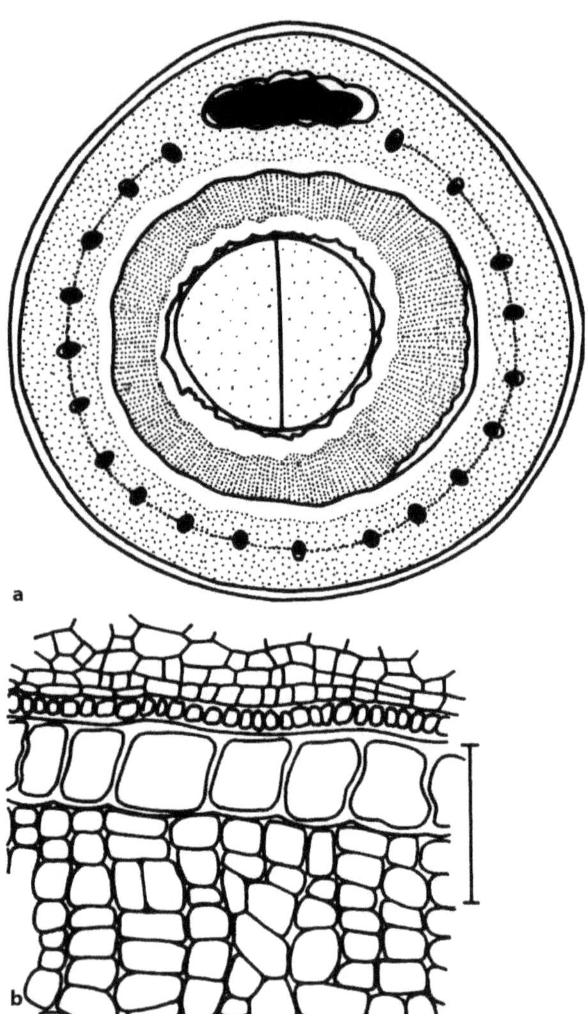

Fig. 156. *Persea americana*, Lauraceae. **a** T.s. of seed. Outer integument with a ring of vascular bundles, inner integument surrounding the 2 cotyledons (Roth 1977). **b** T.s. of young fruit at the border of endocarp and outer integument. The large-celled layer belongs to the outer epidermis of the outer integument. The small-celled layer above belongs to the inner fruit epidermis (ROTH 1977).

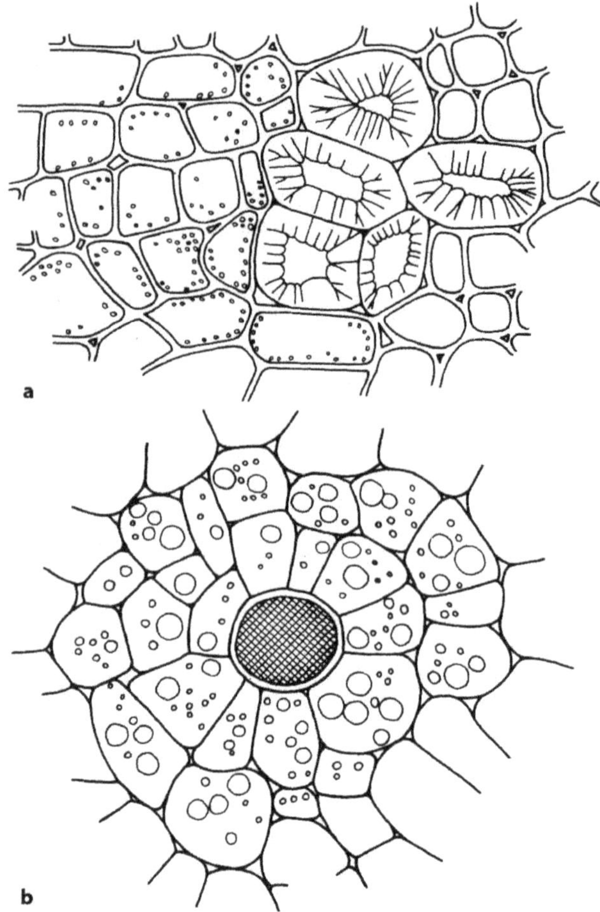

Fig. 157. *Persea americana*. T.s. of ripe fruit. **a** Peripheral part of flesh with parenchyma and stone cells. **b** Central part of flesh with oil idioblast (hatched). The parenchyma cells contain many oil drops (ROTH 1977).

parts of the pericarp are free from stone cells. The amount of stone cells differs according to the variety: less valuable varieties show large quantities of stone cells in the pericarp periphery, whereas the better the quality, the fewer the stone cells so that the skin becomes thinner. At the inner border line of the skin, where stone cell aggregations cease to develop, the formation of oil idioblasts begins; these are polyhedral to globular cells with slightly thickened and suberized cell walls filled with only a single oil drop. They are surrounded by a ring of parenchyma cells. Oil idioblasts are irregularly dispersed all over the fleshy mesocarp. Besides the oil idioblasts, all other parenchyma cells contain numerous oil drops of different sizes which stain orange with Sudan III. In the fruit periphery, the oil drops are smaller than in the deeper lying regions. The inner epidermis (endocarp) of the pericarp consists of very small cells and seems not to be cutinized.

The exocarp thus comprises more than the outer epidermis. The skin which can be peeled off the ripe fruit corresponds to the outer epidermis together with the peripheral fruit part in which the stone cell aggregations lie. In very hard-shelled types, this rind contains large quantities of stone cells which exercise a pressure over the neighbouring parenchyma cells so that these partly disintegrate. In addition, the peripheral parenchyma may develop very thick cell walls resembling a collenchyma. The outer walls of the outer epidermis are likewise heavily thickened and cutinized. The skin receives much strength in this way. Stomata may be observed only in the young fruit, but are often

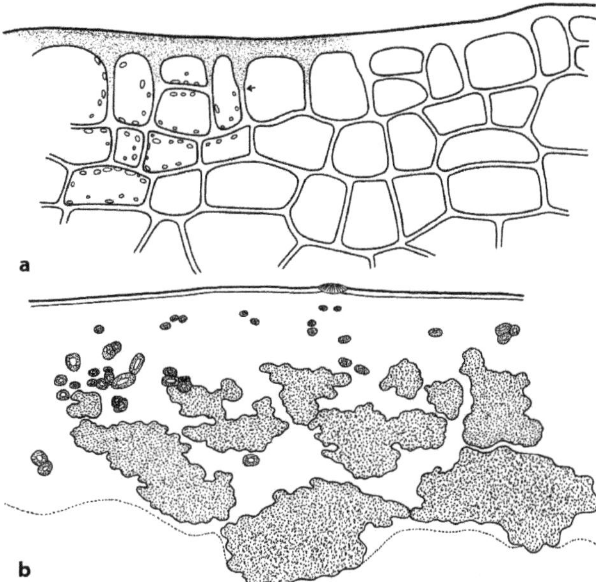

Fig. 158. *Persea americana*. T.s. of ripe fruit. **a** Outer part of skin with outer epidermis. **b** Complete skin of a very hard-shelled fruit with large groups of stone cells (dotted) (ROTH 1977).

replaced later by lenticels which appear scattered over the fruit surface as white or gray dots.

In contrast to the fuerte and other smooth-skinned varieties, the rind of carlsbad and nimlioh are rough in texture due to the cork which forms in the lenticels. The epidermis is ruptured through lenticel development and several lenticels may coalesce to form one large, rough and corky patch.

In the attachment zone of the pedicel and continuing into the fruit base, the vascular system consists of a solid cylinder which divides further upwards into 6 major strands; these divide and redivide in their turn anastomosing until the entire mesocarp is finally penetraded by a network of vascular bundles.

SCOTT et al. (1963) studied the structure of the oil idioblasts under the electron microscope. The idioblasts are distinguished from the other parenchyma cells by their larger size and the thicker cell walls, as well as by the presence of oil sacs which are stalked; the sac membrane and the stalk consist of cellulose, but are impregnated with a suberin-like material. The membrane is sheathed by cytoplasm from which strands radiate between vacuoles. The content of the sacs is a mixture of saturated and unsaturated fatty acids with slight traces of terpenes. In the full grown idioblast, the oil sacs fill almost the entire cell cavity; the wall is impermeable and, besides the impregnation with suberin, may occasionally contain traces of lignin. In addition, all intercellular spaces throughout the pericarp tissue are lined with a suberin film.

The fruit does not mature while attached to the tree, hence it must be harvested to soften and to reach edible maturity.

Woody avocado fruits occur as abnormalities, produced by insect damage, and show the structure of a woody stem; the larger part of these fruits consists of vessels, fibers and woody parenchyma; stone cells and oil idioblasts form in the cortex.

Ethnobotanical and general use

Nutritional use

In the regions of production, the fruits are eaten raw as a vegetable, as a salad prepared with salt and vinegar, or with a filling of meat or lobster; and ice cream is even made of them. In the USA, the fruits are canned, because they are of very short durability. The fruits have a high nutritive value due to the high content of saturated and unsaturated fatty acids (14–30%).

Medical use

Leaves, fruit and seed are mainly applied. Fruits and seeds are used to cure infections of the digestive apparatus and of the skin. They also help as a vermifuge and have antibacterical activities.

For centuries the oil extracted from the seeds is applied to treat the dry scalp. An ointment is also prepared to relieve pains and to soften the skin of wounds.

The dried and pulverized fruit shell is applied as an antidysenteric; so is an infusion of the leaves, which is also used to treat infected and inflamed wounds. The same infusion is also helpful to cure infectious diarrhoea and indigestion. The infusion of the fruit shell helps in the treatment of intestinal parasites.

A decoction of the seeds taken monthly during menstruation is considered an contraceptive.

Crushed seeds dissolved in brandy, are used to treat snakebites.

The fruit pulp is said to regulate menstruation and to be helpful as an aphrodisiac.

Leaves are crushed for a cataplasm to alleviate haematoma.

The antimicrobial activity of the fruit and seed is worthy to note. Extracts of the seeds show antimicrobial activity against *Escherichia coli*, *Micrococcus pyogenes*, *Sarcina lutea*, and *Staphylococcus aureus*.

The long-chained aliphatic compounds of the fruit shell have bactericidal activity on *Bacillus subtilis*, *Bacillus cereus*, *Salmonella typhi*, *Shigella dysenteriae*, *Staphylococcus aureus*. Extracts of fresh leaves and shoots have shown anticancerous activities, as well as cytotoxic properties. They inhibit the growth of gram-negative microorganisms.

Further diseases which are cured with avocado are: scabies, ulcers, vesical incontinence (seed); it helps as an antihelmintic and as a vermifuge (fruit), for rheumatic pains, malaria, as an emmenagogue and an abortive (seeds); it is pectoral, stomachic, and is used as an emmenagogue (leaves and bark).

The fruit is recommended as food for diabetics, as it contains neither sugar nor starch.

The oil of the seed may be applied as a cosmetic to soften the skin and to accelerate cicatrization.

An infusion of the seed used as a bath is recommended in case of blennorrhagia.

Chemical contents

Fruit and seed are rich in fatty acids such as oleic, linoleic, linolenic, palmitic, stearic, capric, and myristic acid which form 80 % of the total fatty content of the fruit. The seed oil is rich in tocopherol. Further substances present in the fruit are squalene, saturated aliphatic hydrocarbons and aliphatic alcohols, terpenes, aspartic and glutamic acid as well as many other substances (GUPTA 1995).

The leaves principally contain a yellow-greenish essential oil composed of estragole, pinene, cineol, transanethole, camphor and traces of other substances (GUPTA 1995).

Aqueous extracts of the leaves contain a high percentage of essential oil, dopamine, serotonin, flavonoids, a bitter principle and traces of other substances (GUPTA 1995).

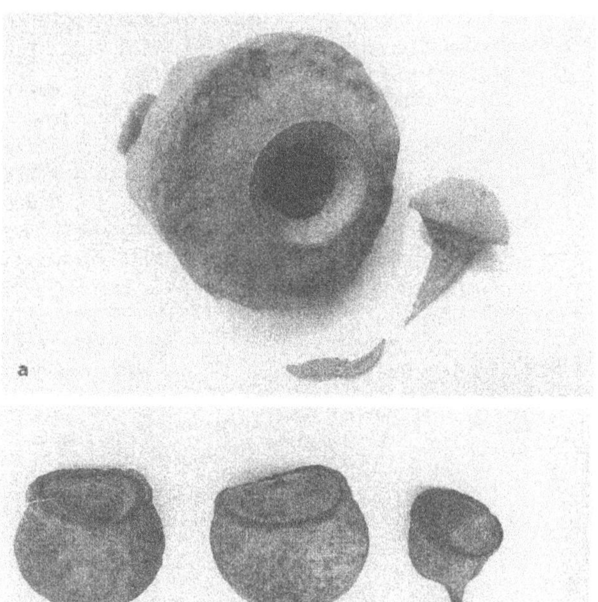

Fig. 159. Fruits of Lecythidaceae. **a** *Lecythis davisii*. **b** *Gustavia augusta*. The fruits of *Gustavia* are still closed by a lid. In Lecythis the lid with the 'thunderbolt' separated already from the fruit and a seed was released (ROTH 1987).

Lecythidaceae

Trees or shrubs with entire leaves, only found in the tropics, and particularly rich in the rain forest of tropical America.

Flowers large, stamens numerous often at their bases united in an androphore or staminodial tubus (pseudocorolla). A disc covers the ovary. Carpels 2–6; many to one seed.

Fruit fleshy or leathery, but frequently a woody pyxidium with lid (Lecythidoideae) (Fig. 160) Seeds without endosperm, but winged samaras, e.g. in *Couratari pulchra*.

The brazil nut (*Bertholletia excelsa*) consists almost exclusively of the enlarged hypocotyl which is rich in oil (50–60 %) and in proteins.

Couroupita guianensis AUBL. (cannon-ball tree) has large rose-coloured to yellowish flowers which arise in racemes from the stem and older branches (cauliflory); the woody subglobular fruits (cannon-balls) reach an enormous size of 15–20 cm in diameter or more and are quite heavy. The tree reaches up to 40 m in height and is a fascinating ornamental. The species is also the source of an excellent timber.

Another South American member with edible nuts is *Lecythis zabucajo*, the paradise nut.

Commercially important is also the West Indian anchovy pear *Grias cauliflora* L.

The bark structure of the Lecythidaceae is very uniform and regular. Layers of fibers regularly alternate with layers of soft bast (ROTH 1981). The fibrous bark can be used for several purposes, e.g. for halter, ropes and cables (in Colonia Tovar, *Eschweilera fendleriana* is used). The fiber of the bark of *Couratari* is used for cloth.

Eschweilera

About 80 species of medium-sized to very large trees distributed from eastern Brazil through the Amazon basin to Trinidad and Costa Rica. The leaves are leathery and the fruit opens with a large operculum; the seeds are not winged and in many species have very bitter kernels; some cause stomach cramps when eaten.

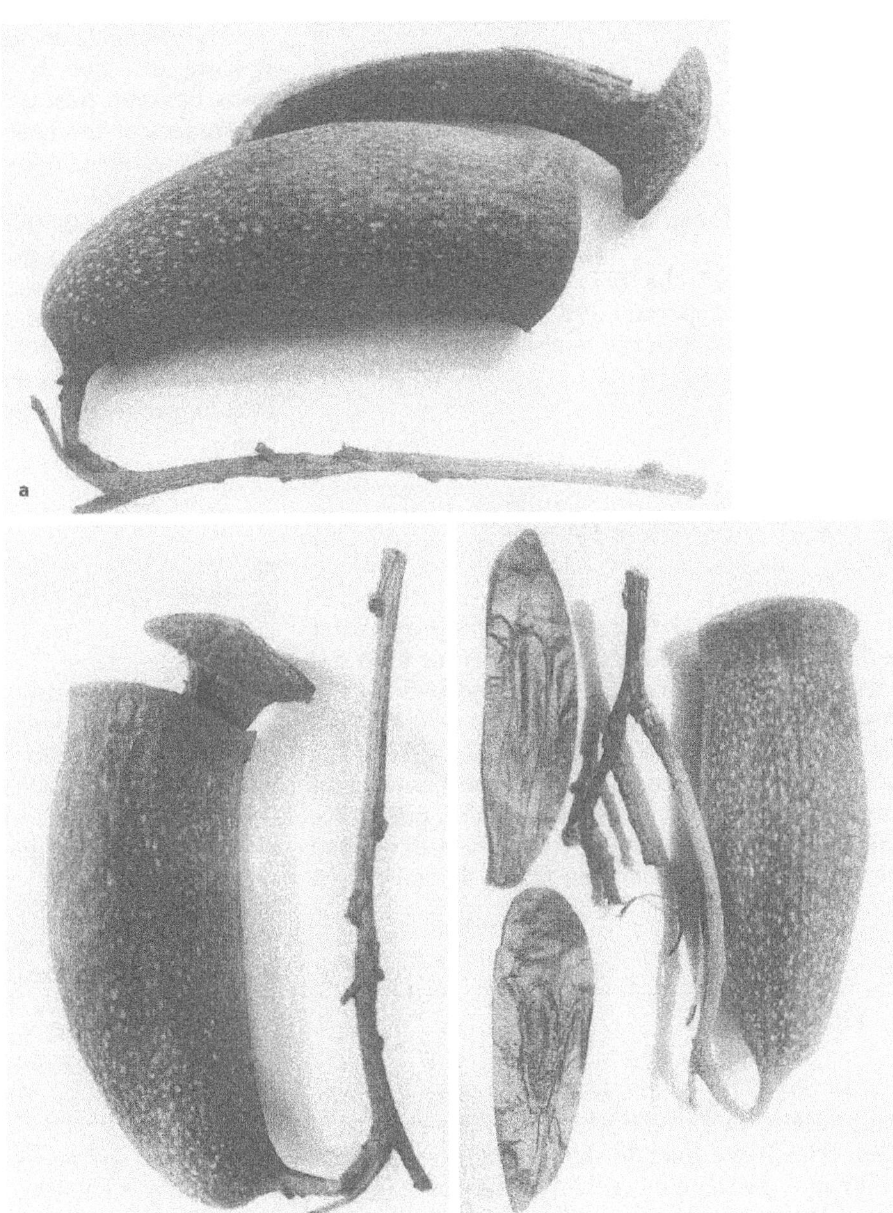

Fig. 160. *Couratari pulchra*, Lecythidaceae, fruits and seeds. Note the very long 'thunderbolt' (ROTH 1987).

E. is recognized as a difficult genus and many of the species are imperfectly known.

The timber is of commercial use, particularly that of *E. subglandulosa* and *E. longipes*, and to a lesser degree that of *E. odora* and *E. corrugata*. The timber is highly resistant to marine borers.

The bark of the following species was studied by ROTH (1981): *E.chartacea, E. corrugata, E. grata, E. odorata, E. subglandulosa, E. cf. trinitensis, E. perumbonata, E. sp.* (Majaguillo erizado). The seeds of *E. odorata* MIERS. are edible and very nutritious and a healthy food, rich on oils according to the natives. The species deserves genetic improvement (BRÜCHER 1989).

The ashes of certain species of *Eschweilera* are vermifugal. Leaf venation has been studied by ROTH 1996, fruit structure and dispersal by ROTH 1987.

The chemistry of the genus is not known.

Gustavia

(Fig. 159 b) G. has a useful wood. The bitter root is purgative. Curare is prepared from the bark. A tea of the fruit is emetic. The seed is purgative. ROTH studied the bark structure 1981, leaf structure 1984, fruit structure and dispersal 1987.

Lecythis

(Fig. 159 a) At least 9 useful species are known.

Wood is useful. Burned bark mixed with clay is used in pottery. The seeds are a source of oil for illumination and soap manufacture. The nuts are edible.

ROTH studied the bark structure 1981, fruit structure and disperesal 1987, leaf venation 1996. BRÜCHER 1989, SCHULTES & RAFFAUF 1990, UPHOF 1968.

Loranthaceae

The representatives are shrubby half-parasites with berry or drupelike fruits. The ovary does not develop well-structured placentae and ovules. The inner layer of the floral axis forms the so-called viscid layer which may contain natural rubber. The fruits dispersed by birds stick to the branches of the host by the viscid substance. The plants are chlorophyll-bearing and have well-developed leaves. The structure of this family is considered very derived.

Phoradendron

The ovary is inferior and is surrounded by the receptacle or flower cup which provides the viscid layer. The tissue outside the viscid layer is often fleshy and parenchymatous. At fruit maturity, the viscid layer consists of anticlinally elongated palisade cells with thin walls. They separate along their anticlinal walls and transform into long threads. These may be subdivided by transverse walls. The palisade cells release a juicy yellow-greenish content in the form of drops or irregular masses, which stain orange with Sudan III, and possibly correspond to fatty oils; in addition, gums, resins, sugar, viscotoxins and other substances have been indentified in the viscid layer (ROTH 1977).

Phoratoxin and flavone glycosides have been isolated from *P. tomentosum*.

Crushed leaves of *P. crassifolium* are used by the natives as a poultice for wounds.

A tea of the leaves of *P. piperoides* alleviates poor condition after an ill-balanced diet. Branches in decoction are used by Venezuelans as a bath to cure dermatitis and fungal infections of the feet.

The chemical constituents of the family are not well-studied, due to the intimate metabolic relations between host and parasite so that chemical substances of the host are also found in the commensal. A thorough study would however, be worthwhile.

It is well known that *Viscum* is used for the control of cancer. The cancerostatic activity is ascribed to basic glycoproteids; they seem to control transcription in a similar way as histines. Likewise the peptid viscotoxin has tumor necrotizing effects. *Phoradendron* and other genera of the family may thus be promising representatives of the Loranthaceae.

Malpighiaceae

The Malpighiaceae are particularly well developed in South America. Staminodia and prolongations of the anthers or of the connectives are frequently occurring characteristics. Hairs on stem, leaf or inflorescence are of a special type (malpighian hairs). Large glands are often found at the leaf margins. Many species are lianas with an anomalous secondary growth of the stem. The fruits of certain species are fleshy and edible. *Bunchosia costaricensis*, for example, is a fruit tree of Costa Rica. The bark of certain species of *Byrsonima* is used for tanning. Although some species of *Byrsonima* have an attractively coloured timber, the trees are too small to be used commercially.

Certain species contain beta-carboline and tryptamine alkaloids, hiptagen, polyphenols and saponins. 'The chemistry of the family is known primarily through the study of the hallucinogenic drinks prepared from a few South American genera and species...' (SCHULTES & RAFFAUF 1990).

To date, about 50 species of *Bunchosia* native of tropical America are known.

About 120 species of *Byrsonima* are known in tropical America.

Bunchosia argentea DC (ciruelo de fraile, ciruela, ciruelita)

Altogether 4 species of *Bunchosia* occur in Venezuela that are known as ciruela or similar names, refering to the similarity of the fruits with a cherry.

Fig. 161. *Bunchosia argentea.* **a** Twig. **b** Flowers and fruits.

Taxonomical descriptions

Buchosia argentea (Fig. 161) is a shrub or small tree and is about 6–12 m high, but is said to attain a height of up to 20 m. The short cylindric stem which reaches a diameter up to 60 cm has a white woolly pubescence on all its young parts; this consists of trichomes 0.5–1 mm long. The branches are abundantly covered with brown lenticels. The simple leaves are opposed and have a papery consistency. The elliptic, oblong or ovate blade is pubescent on both, upper and lower side; it is 14–20 cm long and 7–9 cm broad, has an acuminate tip, entire margins and an obtuse base. The venation is pinnate, broquidodromous, with 7–10 pairs of secondary nerves. The middle nerve is prominent on the abaxial side, but flat on the adaxial side. The petiole is 0.8–1.2 cm long, ridged and pubescent. The axillary inflorescences are racemose. The 5 sepals are ovalate, and have greenish glands. The 5 ovate petals are yellow and 6–8 mm long. The globose to oviform drupe is 2.5–3.5 cm long.

Origin

Tropical America.

Occurrence

The species is distributed in the north of South America. In Venezuela it occurs especially in the Cordillera de la Costa and in the Andes between 1000 and 1500 m a.s.l. In Caracas it may be found in some gardens and parks of the east of the city (HOYOS 1976).

Anatomical description

Anatomical details can be found in Fig. 162, 163, 164, 165, 166, 167 a, b, 168 c.

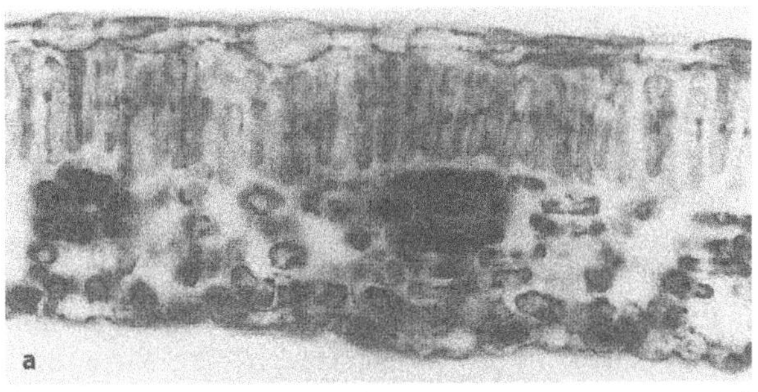

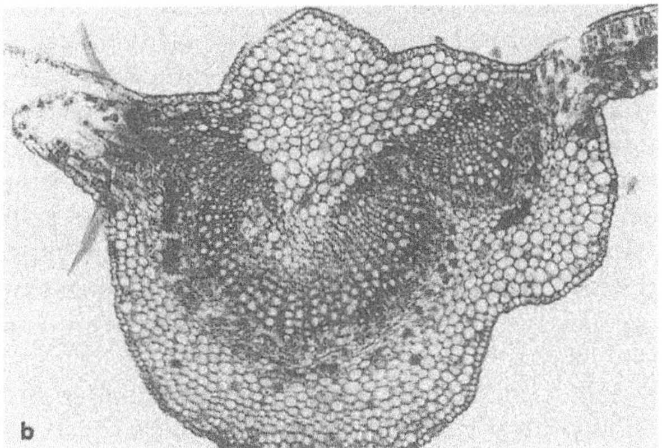

Fig. 162. *Bunchosia argentea.* Leaf. **a** T.s. of blade, × 16. **b** of midrib × 5.

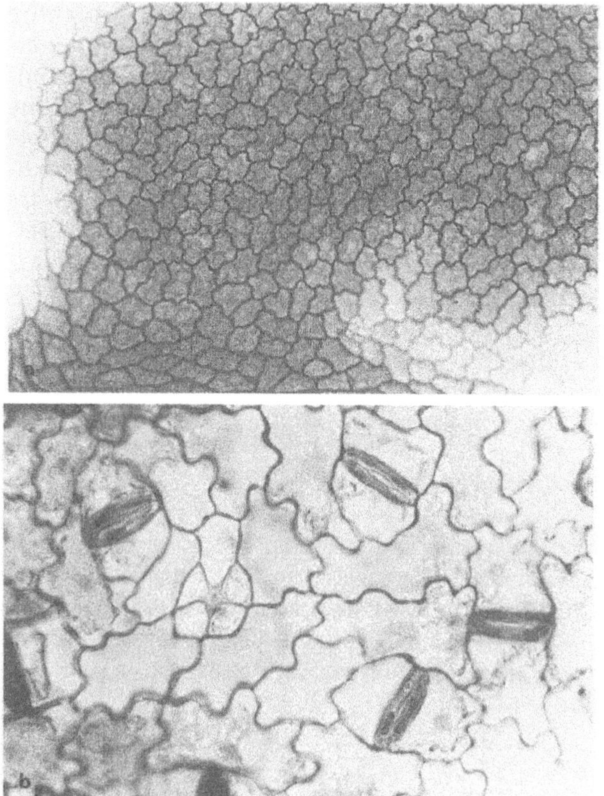

Fig. 163. *Bunchosia argentea*. Above: Upper epidermis (× 6.3). Below: Lower epidermis (× 25).

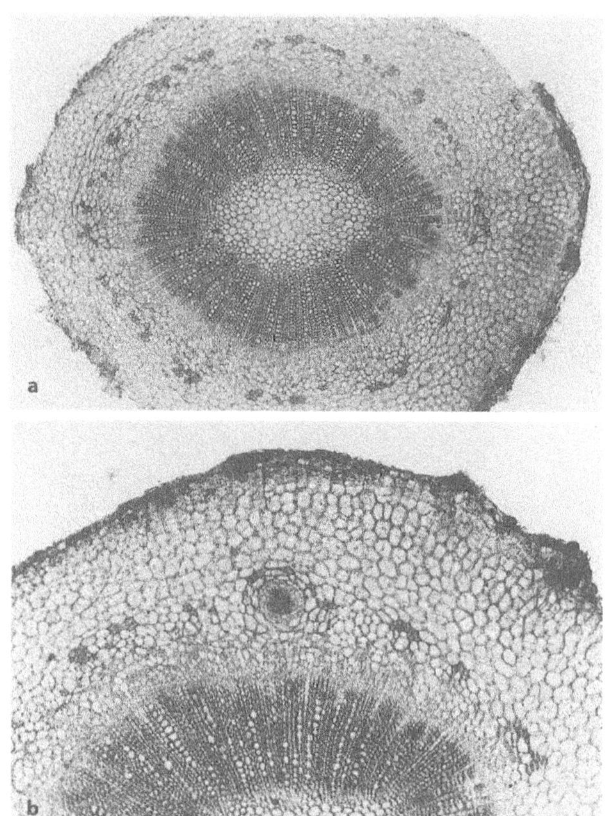

Fig. 164. *Bunchosia argentea*. **a, b** T.s. of stem. Note the cortical vascular bundle in b.

Leaf. The leaf is bifacial and hypostomatic. The upper epidermis consists of cells of a regular size with relatively thin outer walls. As seen in surface view, the anticlinal walls are wavy. The typical malpighian two-armed hairs are scarce on the upper leaf side. The cuticle is delicate. The cells of the lower epidermis are only slightly smaller than those of the upper side, but their outer walls are somewhat vaulted above the surface in the form of a watchglass. The two-armed hairs are more frequent on the lower leaf side. They correspond to unicellular trichomes, usually with 2 more or less horizontal arms attached to the plant by a short vertical stalk. The stomata have 2 subsidiary cells arranged parallel to the pore and are of the rubiaceous type (according to METCALFE & CHALK). The guard cells stain intensely with methylene or toluidine blue, contrasting with the regular epidermis cells in this way (Fig. 163 b, 167 a). Glandular hairs with 5-6 head cells are infrequent (Fig. 167 b).

The midrib is raised above both the upper and the lower leaf side, but more conspicuously so on the lower side (Fig. 162). Its vascular tissue has the shape of a horseshoe, as seen in transverse section, which is open towards the upper side. The extremities of the horseshoe are bent towards the wings of the leaf blade where they give off branches to the side nerves. The phloem is surrounded by a horseshoe-shaped fiber cap.

The palisade parenchyma consists of a single layer of cells which are about 4 times longer than broad. The loser spongy parenchyma comprises 4-5 cell layers reaching about the same proportions as the palisade parenchyma, in transeverse section (Fig. 162). The cells have lobed outlines with short arms.

Stem. The single-layered epidermis is substituted here and there by a cork which is of subepidermal origin (Fig. 164). The primary cortex is composed of 10-12 layers of more or less spherical parenchyma cells, some of which are dilated tangentially, while others are radially compressed. Towards the inside, small fiber bundles are arranged in the form of a ring; the fibers are of the gelatinous type. These isolated fiber strands pertain to the pericycle. Very conspicuous are hadrocentric cortical bundles with a central xylem surrounded by phloem (Fig. 164). In the central cylinder, the medullary rays are uniseriate and the small vessels are

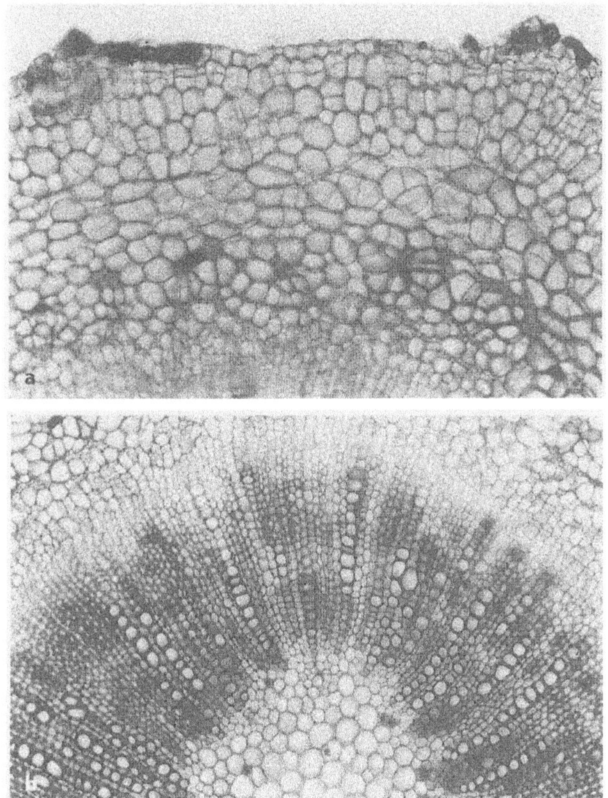

Fig. 165. *Bunchosia argentea*. **a** Cork formation below the epidermis. **b** Xylem and pith of stem.

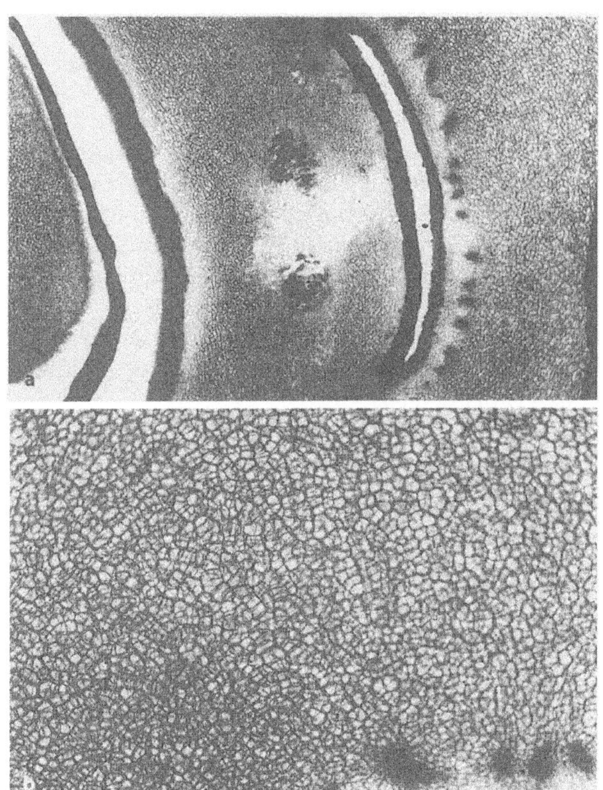

Fig. 166. *Bunchosia argentea*. **a** Sector of fruit with fleshy mesocarp, fibrous endocarp (dark) and vascular bundles near the endocarp. **b** Fruit flesh (× 6.3).

mostly arranged in the form of radial rows. The pith in the center consists of spherical parenchymatous cells.

Clustered crystals in the form of druses occur in the mesophyll as well as in the parenchyma of the stem (Fig. 162, 164).

Fruit. The fruit is a drupe. The exocarp corresponding to the epidermis is composed of large water storing cells. The major part of the drupe consists of parenchyma cells which constitute the fleshy mesocarp. The endocarp is composed of crossing fiber layers. Clustered crystals (druses) are found in the mesocarp parenchyma. The vascular bundles lie close to the endocarp.

Ethnobotanical and general use

Nutritional use

The edible yellowish-red drupe is fleshy and of globular shape, when ripe. It has a fleshy somewhat mealy red pulp with a sweet agreable taste. The species could thus be cultivated as a fruit tree.

Economical utilization

The species is cultivated as an ornamental tree. *Bunchosia argentea* attains a height of 50–65 feet with a trunk diameter of 12–16 inches in Colombia and has a rich darkbrown heartwood which is in sharp contrast to the nearly white sapwood; the wood would thus be good for furniture if available in sufficient quantity and large enough dimensions (RECORD & HESS 1943).

Medicinal use

A gum is extracted from the stem which is used in popular medicine as a remedy for icterus.

Related species

According to SCHULTES & RAFFAUF (1990), *Bunchosia* sp. (wan-ee-koo-soo) is used by the Tikunas who prepare a tea of the reddish fruits to treat diarrhea. *Bunchosia costaricensis* is a fruit tree.

Cultivation

The plant is propagated by seeds, however it does not develop well in a hot climate.

Observations

The chemistry of the species is not yet known (SCHULTES & RAFFAUF 1990).

The species is easily recognized as a Malpighiacea by the presence of the typical malpighian two-armed hairs. The species could well be recognized by its cortical hadrocentric vascular bundles which were not previonsly reported for the Malpighiaceae.

Bunchosia glandulifera. The leaves in the form of a cataplasm cure erysipelas. The gum is efficacious against icterus.

Byrsonima

At least 7 useful species are known. The wood is recommended for fancy furniture. The fruits (drupes) are edible. The bark is applied to hasten wound healing. A tea of the leaves is antidiarrhoeic and vermifuge. Fruit structure and dispersal were studied by ROTH 1987.

Malpighia glabra L. (cerezo, cemeruco, cemiruco, cereza de monte, semeruco, barbados cherry)

Taxonomical description

Malpighia glabra is a shrub or small tree, 2–6 m high with a globular crown (Fig. 169). The opposite leaves are glabrous, ovate-elliptic to lanceolate, sometimes elliptic, 3–8 cm long, 1–3.5 cm broad, with an acute or acuminate tip which seldom becomes obtuse; petiole 5–12 mm long.

3–8 flowers arranged in corymbs or umbels; pedicels 5–15 mm long; calyx with 6–10 glands; sepals oblong to ovate-oblong, glabrous, 2,5 mm long. Petals 5, rose-coloured; one petal larger than all the others and 2 smaller than the others. Limbs of the petals rounded or elliptic with dentate margins, slightly larger than the claws; total length of the petals 8 mm; 3 styles; 10 stamens.

Fruit a scarlet drupe, globose-trilobulate, about 1–2 cm in diameter, with 3 seeds, each with a crest. The fruit is edible.

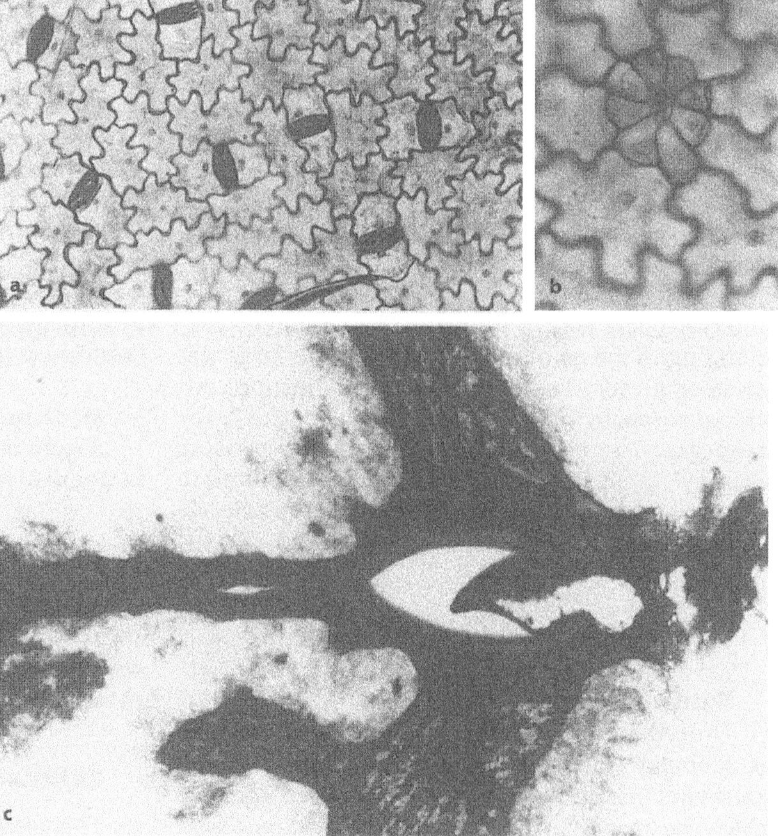

Fig. 167. *Bunchosia argentea.* **a** Lower epidermis. **b** Gland. **c** *Malpighia glabra*, drupe (detail) in longitudinal section (× 2.5).

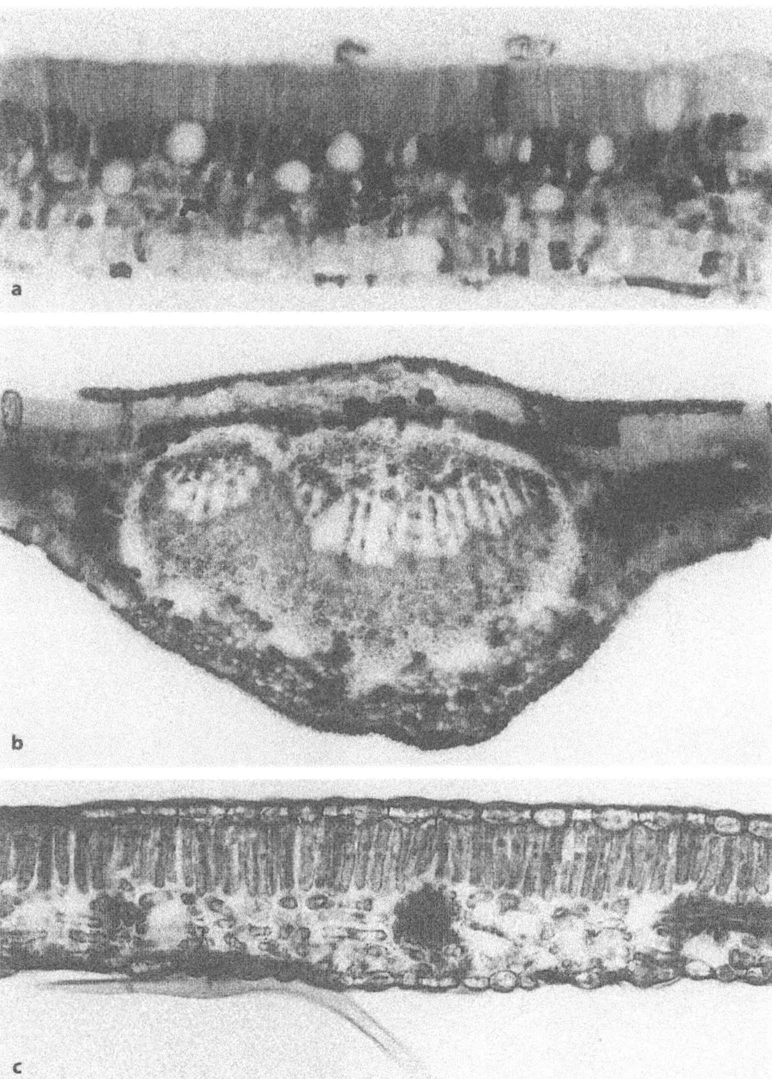

Fig. 168. a, b *Nectandra pichurim*, t.s. of leaf blade (× 20) and midrib (× 10). c *Bunchosia argentea*, leaf blade.

Origin

Tropical America (South America and Caribbean region).

Historical background

The cemirucos were first mentioned by the Spaniards in 1548 as fruits resembling cherries (F. DE OVIEDO: IN VELEZ & VELEZ 1990).

Occurrence

Tropical America, West Indies, South of Mexico and Texas. In Venezuela it is amply distributed in the hot regions. It is now cultivated all over the tropics. The Semeruco is the emblematic plant of the State Lara in Venezuela.

Anatomical description

Fruit (Fig. 167 c, 169, 170). Although a drupe, the fruit is divided into 3 compartments by 3 septs, each compartment with a seed.

The exocarp or outer epidermis consists of polygonal cells with thickened walls and pits. This part forms the skin of the fruit.

The outer fleshy and juicy part (mesocarp) is composed of a very largecelled parenchyma with cells of a variable and irregular shape. The septs are formed by smaller parenchymatous cells of roundish shape and with short arms so that the tissue appears slightly spongy. Interspersed within the parenchyma are cells which intensely stain with toluidine blue.

The endocarp (Fig. 170 a) is hard being composed of sclereids which cross each other more or less at right angles, the outer sclereids running par-

Fig. 169. *Malpighia glabra.* **a** Habitus. **b** Fruits.

allel to the locule, the inner ones transversely. Rhombic crystals can frequently be observed in the sclereids.

Leaf. The upper epidermis serves as a water-storing tissue being composed of very large turgescent cells with thin walls; only the outer walls are thickened. The palisade parenchyma originates from a single layer of palisade cells which divide by one (sometimes 2) periclinal walls. The cells are comparatively short and broad. The length/width index of the entire palisade parenchyma, composed of 2–3 cell layers , amounts to 3.2–4.5. Some palisade cells remain undivided, others remain very short, so that an irregular structure of the palisade parenchyma results. The limit between palisade and spongy parenchyma is thus nonhomogeneous. The spongy parenchyma comprises about 5–6 layers. The cells are more or less of globular shape and partly leave large intercellular spaces. In the center of the mesophyll, a large proportion of the cells are water-storing, becoming conspicuous by their large size. This is a structural characteristic of the Malpighiaceae. Crystals in the form of druses are abundant in the mesophyll, in the palisade as well as in the spongy parenchyma. Some druses in the water-storing cells reach a considerable size. The lower epidermis cells are smaller than those of the upper

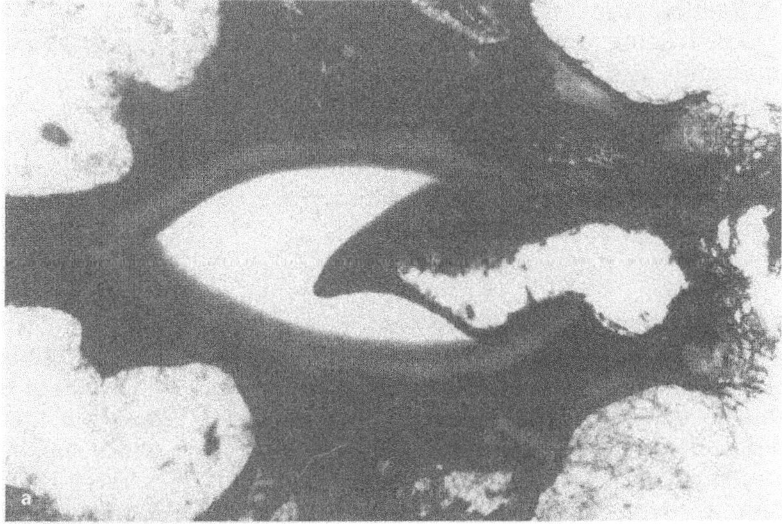

Fig. 170. **a** *Malpighia glabra*, sector of fruit with locule and seed; endocarp dark, surrounded by fleshy mesocarp × 5. **b** *Gossypium barbadense*. Leaves, flowers and fruits.

epidermis and have thin walls. Stomata are confined to the lower epidermis and occur at epidermis level. Vascular bundles are infrequent. This leaf has only a single xeromorphic characteristic: the water-storing cells.

Ethnobotanical and general use

Nutritional use

The juicy sour-sweet yellow pulp (mesocarp) is edible and is eaten either raw or used in jellies or preserves.

The red, cherry-like fruit has recently attained much importance as a source of vitamin C. It has a vitamin C content which exceeds that of oranges more than 100-fold. 100 g of fruit pulp yield 1.7–4.0 mg ascorbic acid. The seeds are also edible.

Economical utilization

In pharmaceutical industry, the fruit is used to produce vitamin C tablets.

The wood is used as a dye for a red colour.

The shrub is furthermore a beautiful ornamental with its brilliant dark green foliage and the contrasting scarlet fruits; it is planted in gardens and as a hedge.

Medical use

The drug is called *Malpighia glabra* L. FRUCTUS. Root, bark and green immature fruits have astringent properties. Unripe fruits are eaten to prevent flu, as they contain a large amount of ascorbic acid (vitamin C).

Powdered seeds mixed with the resin of the plant help against bronchitis.

Healing properties

The healing properties are due to the presence of ascorbic acid and possibly of tannins (astringent properties).

Chemical contents

Ascorbic acid and malic acid, which give a taste of apple to the fruit.

Varieties and related species

There are many varieties which show considerable fluctuations in their vitamin C content (chemical races?) (BRÜCHER 1989).

Malpighia emarginata SESSE & MOC. is a very close relative of *M. glabra* which also occurs in Venezuela. The 2 species are morphologically very similar and they have many characteristics in common. They may be distinguished by the shape of their leaves and fruits. *M. glabra* has leaves with an acute or acuminate tip with 5–12 mm long petioles and globular fruits; *M. emarginata*, on the contrary, has obtuse or emarginate leaf tips, petioles less than 5 mm long and fruits having 4–9 lobules.

Cultivation

The plant may be propagated by seeds, but as it shows a large array of varieties which fluctuate in their aroma and vitamin C content, it is advisable to propagate the plant vegetatively with cuttings. As the plants are easily attacked by nematodes, they could be grafted on more resistant stock. But there are also resistant varieties, such as Florida Sweet cultivated in Florida. After 8 years, this variety yields 25 tons of sweet and very agreeable fruit on 4.047 square meters of land. The plant needs minerals in the soil and regular irrigation, particularly during the flowering period (in May and June), otherwise the flowers and young fruits may drop down. For commercial production, regular pruning is recommended. Temperature and water availability influence the content of vitamin C in the fruit.

The fruits mature mainly during the rainy season.

Observations

The fruit is easily recognized by its outer and inner structure. Although similar to a cherry (barbados cherry), it contains 3 seeds.

Malvaceae

The Malvaceae are a family which shows its richest development in the tropics. The leaves have caducous stipules. The flowers which are usually conspicuous, often have an outer calyx and numerous stamens. The fruits are frequently loculicidal capsules or schizocarps. Slime cells are characteristic of bark and pith. Stellate hairs occur in the epidermis.

The Hibisceae include the ornamental shrubs and trees of the genus *Hibiscus* which are cultivated in green houses, as well as the genus *Gossypium* which supplies cotton and oil.

Very characterisatic of the Malvales are the cyclopropenoid fatty acids which mainly occur in the seed oil (cotton oil). Slime occurs in slime cells or cavities not only of the Malvaceae, but also of the Tiliaceae, Bombacaceae and Sterculiaceae, which are closely related.

Some species of the genus *Sida* supply fibers for textiles.

The family is widely distributed in temperate and tropical regions, reaching high elevations only in the Andes.

Gossypium barbadense L.
synonym: *G. vitifolium*
(algodon de las islas, sea-island cotton)

Taxonomic description

Cultivated perennial shrub 1–2.5 m high, velvety hairy leaves alternate, cordiform to suborbicular, 3–5 lobate or entire, lobules triangular ovate to ovate-lanceolate, 5–15 cm long and 7.5–20 cm broad. Petiole 5–6 cm long. Solitary flowers are pedunculate and subtended by 3 broad-cordiform fringed bracts (outer calyx) that are shorter than the petals (Fig. 170 b).

Calyx truncate or 5-lobate. The contorted petals are 8–10 cm long, much longer than the bracts, of yellow or rose colour with an orangered base. Staminal column monadelphous. Ovary 3-5-locular with numerous ovules in each locule and 5 stigmas.

The ovoid fruit is a coriaceous loculicidal capsule opening with 3-(4)-5 valves, 3–5 cm long. The long white seed hairs called lint are easily separated from the free seeds. Endosperm delicate, cotyledons folded into plaits.

The plant has a well developed tap root reaching 2–4 in in depth.

Origin

The genus has pantropical distribution and cultivars derived from wild forms are found in Asia, Africa and America.

Gossypium barbadense or *vitifolium* is of neotropical origin. According to BRÜCHER (1989), many thousands of years ago, a unique phylogenetical event occurred in South America: a combination, followed by polyploidy of the Old World A-genome with a New World D-genome. The tetraploid cotton species *G. barbadense* is therefore considered neotropical.

Historical background

Cotton, the world's most important and oldest fiber plant, has been planted, gathered and used on various continents for thousands of years. Remains of well-preserved cotton tissues have been recovered from the ruins of Mojendjo-Daro, Pakistan, which date back to 2500 years B.C., and in abundance from ancient Egypt.

The vernacular name cotton is a misspelling of the Arabian word katun. As already mentioned above, many thousands of years ago, a combination followed by polyploidy of the Old World A-genome with a New World D-genome took place. The tetraploid species are considered neotropical. Cotton was already cultivated in Mexico 3400 B.C. and in central Peru 3600 B.C. In central America, *G. hirsutum* was cultivated. Along the South American Pacific coast line, *G. vitifolium* (Synonym: *G. barbadense*) was grown. Thousands of years ago, the inhabitants of both Americas had innumerable applications for cotton fiber, besides spinning and weaving. The techniques of dyeing and weaving were improved in the course of thousands of years. In the Chavin culture spectacular developments are observed which culminated in the Inca time. Although the Spaniards destroyed most of the marvellous cloth and artefacts, some of the remains are well preserved in local museums in Bolivia, Ecuador and Peru.

Archeological excavations of Ica and Huaca Prieta indicate that the colour of the fibers was brown (not white), similar to the existing primitive algodon pardo which grows in West-Peru and Ecuador (BRÜCHER 1989).

Occurrence

Of the 2 most important American species, *G. hirsutum* and *G. vitifolium*, the latter is better adapted to a humid and hot environment. Today, the species is cultivated in all tropical regions.

Anotomical description

Leaf (Fig. 171). The leaf is dorsiventral and amphistomatic. The upper epidermis cells are medium-sized to comparatively large and have strongly convex outer walls, almost partly papillary. Cells with a purple sap are scattered between the regular epidermis cells. Stomata are found at epidermis level. Stellate hairs with long arms and a short stalk are somewhat less frequent than on the lower side. The cuticle of the epidermis cells is slightly ribbed. Pluricellular headed (capitate) glandular hairs are found on the upper as well as on the lower side. As seen in a surface view, the anticlinal walls of the epidermis cells are only very slightly bent.

The palisade parenchyma consists of a single layer of long and slender cells, which occasionally show a periclinal division wall.

The spongy parenchyma is proportionately slightly smaller than the palisade parenchyma and consists of relatively large cells with arms extended in various directions. It proceeds from about 3–4 cell layers. Small and large crystal druses of calcium oxalate occur in the whole mesophyll.

The lower epidermis is similar to the upper epidermis, but is slightly smaller-celled. The stomata which are more frequent in the lower than in the upper epidermis occur at epidermis level. Cells with a purple content (maybe anthocyanine?) are scattered between the regular epidermis cells, in the same way as in the upper epidermis; the cuticle of the epidermis cells is slightly granular. As seen in surface view, the anticlinal walls of the lower epidermis cells are more curved or wavier than on the upper side. Stomata, stellate hairs with very long arms and pluricellular headed glandular hairs are much more frequent in the lower than in the upper epidermis. The stomata are surrounded by 3–4 epidermis cells following the developmental type of *Sedum* or the irregular type of cell division (ROTH & CLAUSNITZER 1969).

Very large circular dots covered by smaller epidermis cells indicate the position of the large secretory cavities (with a brown content) in the mesophyll; they are surrounded by an epithelium of secretory cells.

The midrib is prominent on both surfaces, on the upper side in the form of a bulge or more or less conical protrusion composed of parenchyma. This protuberance proceeds from a ventral meristem which acts by periclinal cell divisions so that anti-

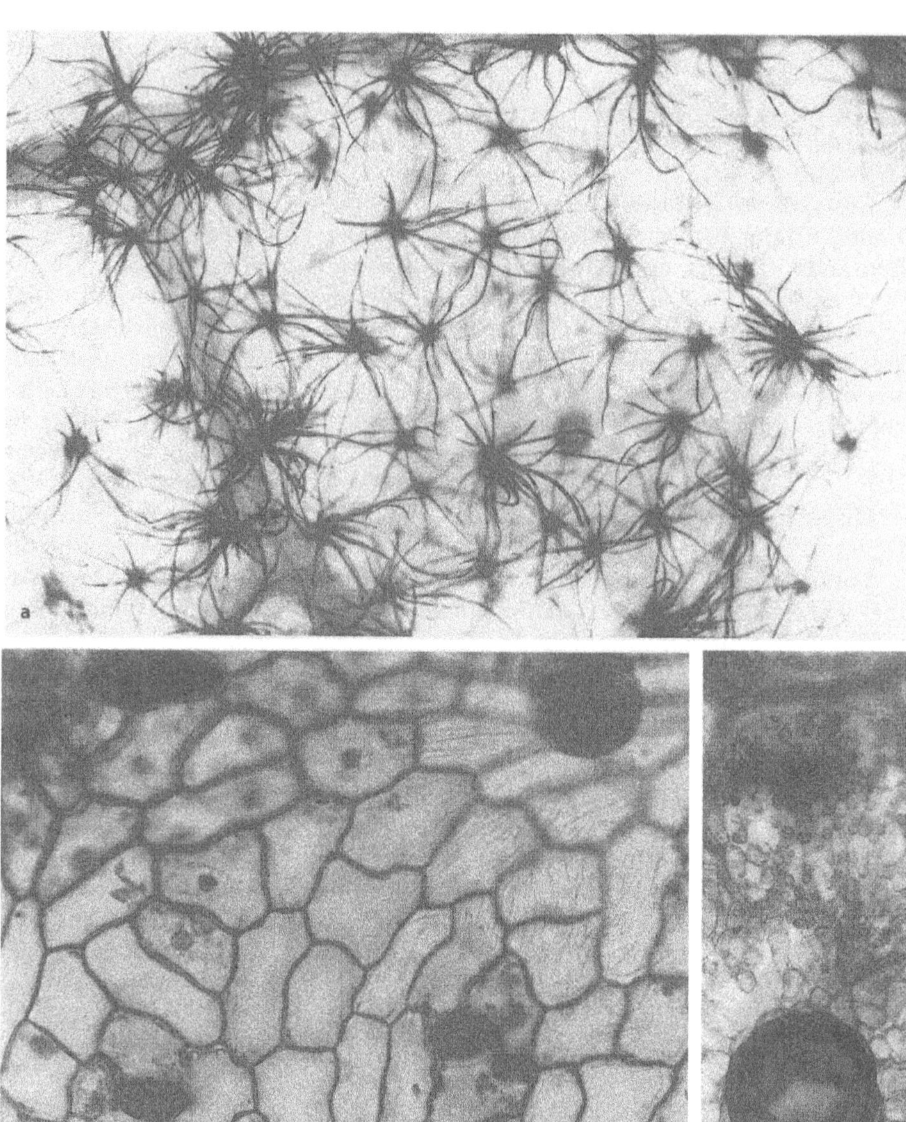

Fig. 171. *Gossypium barbadense.* **a** Lower epidermis with stellate hairs. **b** Stomata of the upper epidermis in surface view. **c** Secretory cavity of the mesophyll with brown content.

clinal cell rows arise. The vascular system forms an open arc with inwardly turned rims from which (2) vascular bundles laterally deviate into the secondary nerves. A collenchyma below the lower epidermis reinforces the midrib whereas the palisade parenchyma is absent in this region.

Collenchyma may also develop on the upper side of very strong midribs. Secretory cavities with a brown content are present in the parenchyma of the upper and lower side. Large colourless cells may be aquiferous. Stellate hairs as well as glandular hairs occur in the epidermis of both sides of the midrib.

The larger secondary veins are transcurrent to the upper and lower epidermis by parenchyma. They also project above the lower leaf side.

Axis. In a twig, about 5 mm in diameter, the superficially originating cork is thin-walled and only several layers thick. The cortex is mainly parenchymatous and its cells partially tangentially extended. Cartilaginous collenchyma is found here and there beneath the cork. The presence of stone cells was not observed. The pericycle is found in the form of small groups of fibers. Druses of calcium oxalate occur in parenchymatous cortex cells. Very conspicuous are large secretory cavities with a redbrown content.

The rays are of 2 sizes in the phloem: Uniseriate rays abundantly contain crystal druses, while strong pluricellular rays considerably enlarge towards the outside in the funnel form, similar to those of many Tiliaceae, Sterculiaceae, Bombacaceae etc. (see ROTH 1981).

The rays enlarge by tangential extension of the cells and following anticlinal cell division. The phloem is therefore divided into triangular strands being separated by the enlarging rays. The phloem is slightly stratified into tangential bands of fibers with a large cell lumen, alternating with softbast; however, the pattern is not very regular.

As seen in transection, the xylem forms a continuous ring. The vessels – irregularly scattered over the section – are solitary or occur in small irregular groups of 2 or more or in short radial multiples. The apotracheal parenchyma is arranged in the form of small tangential bands. The rays are uni or pluriseriate.

The pith is composed of more or less spherical parenchyma cells, partly with thick walls and pits, particularly so towards the center of the axis. Druses are frequent in the cells of the pith. Starch grains are mostly compound.

Root. The secondary xylem of the root constitutes a large proportion of the entire root. As in the stem, the vessels are solitary or occur in groups of 2–4; they frequently contain conspicuous tyloses. The wood fibers have thick walls, tapering ends and bordered pits. The rays are often 4 cells wide and have the height of several rows of cells. Secretory cavities with a brown content, similar to those of stem and leaf, are present in the phloem rays (GORE & TAUBENHAUS 1931).

Fruit. As to the fruit structure, the fruit of *Gossypium hirsutum* was studied by ROTH (1977). The ripe capsule consists of an outer epidermis or exocarp, a parenchymatous mesocarp and a lignified endocarp; secretory (oil) cavities are scattered throughout the parenchyma. The parenchyma cells are radially elongated and dead at maturity. The vascular system is divided into an adaxial and an abaxial part, the latter consisting of 14–16 bundles per carpel which run parallel to the fruit surface and form a vascular network. The endocarp develops from the inner epidermis by periclinal cell divisions; each cell divides about 3–4 times periclinally. The elements proceeding from these cell divisions elongate tangentially and transform into fibers. In the median plane of each carpel, a dehiscence tissue is formed which extends to the center of the fruit through a false septum. On both sides of this very small-celled and thin-walled separation tissue, two trapezoid fiber plates traverse the false septum longitudinally. The dehiscence is loculicidal. The separation tissue is first meristematic, but later on transforms into a lacunar tissue (DE COENE 1950).

The fruit structure of *Gossypium barbadense* slightly differs from that of *G. hirsutum*. The fruit surface is sculptured or warty. To give this outer structuration more strength, the outer epidermis as well as some subepidermal layers have thickened cell walls. The ground tissue is parenchymatous, but the cell walls are slightly thickened. Deeply stained cells are scattered in the parenchyma; they probably contain anthocyanins. Very large secretory cavities with a brown content are more frequent in the outer half of the pericarp. Embedded in the parenchyma are numerous fiber bundles and vascular bundles, the latter are possibly more concentrated towards the inside. The fiber bundles in particular give strength to the pericarp. The surface sculpturing of the fruit (warty outgrowths) is due to the bulging out of fiber bundles in this region. The endocarp occurs in the form of a thin membrane which only consists of fiber bundles; the fibers are however very short and have a large cell lumen, but a comparatively thin wall.

The syncarpous gynoecium carries the seeds in central position along the central column (in so-called axile placentation). A separation tissue also develops on the ventral suture of the carpels, terminating between the 2 funicles of the seeds of 2 adjacent locules (Fig. 37 and 38 in ROTH 1977). The capsule thus opens loculicidallly and ventricidally at the same time (see also ROTH 1977). The separation tissue at the ventral suture is likewise small-celled and thin-walled (see above).

Seed (Fig. 172). The anatropous seeds, which reach a length of up to 12 mm, are pear-shaped and have a brown, red or black colour. The hilum and the micropyle are at the pointed end of the seeds, the chalaza at the other end. A slight ridge along the seeds indicates the position of the raphe. The seed-coat hairs are of 2 sorts: the long hairs or lint, and the short hairs, called linters or fuzz. The lint hairs reach a length of up to 5 cm, by a thickness of only 0.14 mm. They consist of a single cell and are ribbonlike flattened and usually helicoidally twisted. The white hairs originate from the outer epidermis of the outer integument and originally served for the dispersal of the seeds by wind. When removed from the seed, the hairs are open on the base. They have only a thin cuticle which bursts, when the seeds are soaked in water, so that only here and there a ring-shaped constriction becomes visible on the hair. When swelling, the delicate layering of the cell wall becomes conspicuous.

The hairs consist of 91 % cellulose, 7 % water, 0.4 % fat, and 0.1–0.3 % ash.

The outer epidermis of the testa consists of thick-walled cells with dark contents. The hairs which arise from the outer epidermis have an

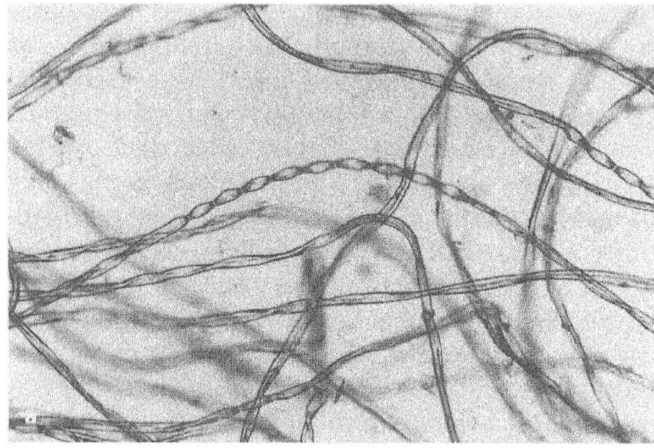

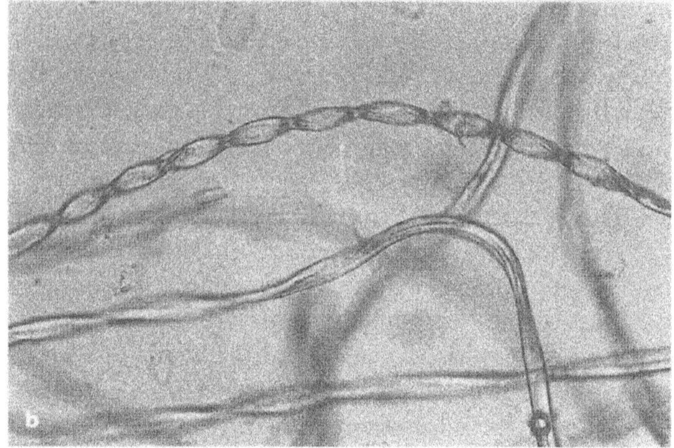

Fig. 172. *Gossypium barbadense.* Seedhairs. **a** ×10; **b** ×20. Note the torsion!

expanded base and their walls are thinner than the lumen. They are characteristically twisted. Stomata are occasionally found in the epidermis. Below the epidermis are 2–3 layers of parenchyma cells with brown contents (pigmented zone).Within this zone is a single layer of clear cells some of which contain small calcium oxalate crystals. The most distinctive region of the testa is the layer of palisade cells, which reach a radial length of 150 μm by a diameter of 15 μm. This layer is characteristic of Malvaceae and Bombacaceae.

The palisade layer develops from the outermost layer of the inner integument. A light line can usually be seen in this palisade layer. An interesting microchemical peculiarity of the palisade cells is that different parts of the cell wall give either a lignin or a cellulose reaction, possibily depending on the species or variety. The rest of the testa consists of rather flattened parenchyma cells, many of which contain brown pigments. The innermost layer of the testa or inner epidermis of the inner integument is composed of cells with curiously pitted walls, called the fringe layer.

The endosperm consists of a single layer of cells with minute aleurone grains.

The embryo has a long radicle and stronly folded cotyledons with marked resin or gossypol cavities.

Cell contents of the cotyledons are oil drops, occasionally clustered crystals of calcium oxalate, and aleurone grains. Small quantities of starch may be present. A characteristic feature of the cotton embryo is the presence of gossypol or resin cavities, up to 120 μm in diameter which are surrounded by secretory cells.

When ripe, the fruit opens with 3–5 valves to expose the seeds surrounded by their hairs. WETTSTEIN (1901/1935) emphasizes that the hairs stick very tightly together so that in his opinion, they can hardly be useful for dissemination.

Ethnobotanical and general use

Nutritional use

The seeds have been ground and eaten or used as a cooking oil by the ancient Indians (BRÜCHER 1989).

Nowadays, cotton-seed oil is a by-product of fiber winning. Cotton seed contains 15–24% oil which, when freshly extracted, is deep red. When refined, it is used mainly for edible purposes, as salad oil, shortening or margarine and, to a lesser extent, in packing of fish and cured meats.

Cotton seed contains 16–28% protein, and the cake or meal is one of the best protein supplements for dairy cows, beef cattle, and sheep, but must be controlled in pig and poultry rations because of the presence of gossypol. In India, cotton seed is fed directly to cattle, and the cake has been used as a human food. Cotton seed could be an important source of vegetable protein for supplementary nourishment of humans in developing countries (VAUGHAN 1970).

The oil as well as the gossypol cavities are located in the embryo (cotyledons).

The large flowers are the source of a mild honey.

Economical utilization

Besides the culinary utilization, the plant also serves other purposes. The fibers are certainly the most important part of the plant. *Gossypium barbadense* has the longest fiber of any cotton species (up to 5 cm long) which is also strong and of excellent quality (UPHOF 1968). About 22% of the seed is crude fiber. Raw cotton consists of 91% cellulose, 7% water, 0.4% fat, and 0.3% ash. To produce a good and absorbent cotton wool quality, the fat has first to be removed from the surface of the fiber.

As already mentioned, two types of hair are distinguished: the long hair which supplies the lint, and the short hair, called fuzz or linters. The long hair is the valuable product. The short fuzz supplies the raw material for the fabrication of felt, paper and synthetic silk, for the production of pure cellulose and of nitrocotton, for rubber-tire fabrics, stuffing cushions and pillows, manufacture of twine and ropes.

The lint is used in numerous cotton goods, such as carpets, mercerized cotton, rayon, any kind of cotton cloth, yarn or textile fabric.

Absorbent cotton which has been cleaned from oily covering substances, being almost pure cellulose, is used, when sterilized, in surgial dressings, for wadding, gauze dressings and sanitary napkins.

The cotton-seed oil is also used for manufacture of soap and soap powders.

Oil cake is not only used as fodder for cattle, but also as fertilizer and dyestuff.

Hulls of the seeds are furthermore used for lining oil-wells, and production of xylose which can be converted into alcohol or explosives.

Stalks of the plant can be employed in paper manufacture or used as fuel.

The petals of the flowers are the source of a yellow or brown dye in some parts of India.

Medical use

Leaves, flowers, seeds and roots are used medically. The leaves are used for flu, fever, cough, consumption, cold in the chest and against constipation. The leaf juice is applied against ear ache. Shoots in decoction are used for prostatitis. ALBORNOZ (1993) states that the drugs are anticolic, emmenagogic, galactogogic and abortive.

Leaf. The leaves are also used against diarrhoea, dysentery, haemorrhoids, cough, bronchitis, colics of the uterus, calculus and strangury. Shoots in decoction are used for prostatitis; 2–3 doses are taken a day.

Flower. The flowers are used against diarrhoea and dysentery. According to WONG (1976), flowers in teas are used for flu and colds.

Seed. The seeds contain up to 25% lipids (predominantly olein), 20% protein, 23% nitrogenated substances, 7% gossypol and phenol. The oil extracted from the seeds is nourishing and has laxative, expectorant, antidysenteric, emollient, abortive, aphrodisiac, tonic, and galactogogue effects.

Root. The root in general contains gossypol (polyphenols), olein, salicylic acid, betaine, and sugar.

The cortex of the root contains hydroxybenzoic acid, salicylic acid, ceryl alcohol and phytosteroles. It has abortive effects.

Method of use

Leaves, flowers and seeds are applied in the form of a 1% infusion against cough and menstrual disorders. A bath with an infusion of leaves, flowers and seeds is used against diarrhoea, dysentery and strangury

1/4 ounces of the sap of leaves or the infusion of the leaves mixed with lemonade is taken against diarrhoea, dysentery, haemorrhoids, strangury and calculus.

A 1 g infusion of the leaves is taken against cold in the chest, and menstrual disorders.

A hot infusion is used as a hip-bath against uterine colic. The leaves are used as a cataplasm for wounds.

A 4 g infusion of leaves or flowers in 360 g of boiling water is also used in a bath for diarrhoea and dysentery.

The seeds contain an oil which is nourishing and laxative at the same time.

To cure cough, 25 g of the shredded seeds together with 30 g of sugar and 20 g of arabic gum are mixed with 175 ml water, and strained. A spoon full (or 1/2 spoonful for children) is taken every 3 hours.

In case of deficient lactation, 30 g of pulverized seeds are macerated for 15 days in 100 g of pure alcohol. 10 drops are taken 3 times a day after meals.

A 1% decoction of the root cures strangury.

Against disorders of the uterus, 20–60 g of the powdered root cortex or 1–2 ounces of the decoction are taken every half hour.

Healing properties

The seeds are said to be laxative, expectorant, antidysenteric, aphrodisiac, emollient, tonic (for nervous tonicity: 2–8 dry seeds without endosperm), a galactogogue, and are abortive.

The root cortex is an emmenagogue and a haemostatic due to the presence of ergotine.

Chemical contents

After being destilled the extract of the cortex supplies a volatile oil which contains furfurol and acetovanillone; the active principle of the cortex is a resinous phenolic substance, gossypol, of a yellowish colour which turns red through oxidation, losing its activity at the same time. The cortex contains 8% gossypol.

The following substances are obtained from the alcoholic extract of the cortex: Salicylic acid, hydroxybenzoic acid, as well as 2 phenolic substances, such as betaine and phytosterol, ceryl alcohol and fatty acids.

The seeds contain 20–25 % lipids, 19–29 % proteins, 23 % non-nitrogenous substances, and gossypol which occurs in the secretory cavities and which corresponds to a mixture of toxic polyphenols.

Gossypol has been associated with antifertility activity in men and also with antiviral activity (SEAFORTH, ADAMS & SYLVESTER 1983).

The pressed seed cake (after the oil is removed) is a rich source of L-glutamic acid (SEAFORTH, ADAMS & SYLVESTER 1983).

Varieties and related species

As already pointed out, the tetraploid cotton species *G. hirsutum* and *G. vitifolium* (*G. barbadense*) are of neotropical orgin. According to its occurrence, *G. vitifolium* is called Sea-Island-Cotton (Caribbean islands, Lowlands of tropical South America), while *G. hirsutum*, the Upland-Cotton, can be cultivated at higher levels.

G. arboreum and *G. herbaceum* are cultivated in India, Asia, and Africa.

Different research groups in the world are working on the improvement of *Gossypium* varieties with high fiber strength, long fibers with a low degree of attachment of the fibers to the seed coat and other peculiarities of better technological value (BRÜCHER 1989).

Cultivation

Cotton, unlike other major crops, suffers from an extremely large quantity of pests and diseases. Pesticides are thus used in large quantities and insects, fungi or bacteria rapidly develop resistance to the pesticides.

Observations

Most characteristic of the species are the fringy bracts of the fruits and the hairy seeds. The capsules are easily recognized by their method of dehiscence and the mass of hairs emerging from the fruit inside. Also very peculiar are the secretory cavities with brown contents which occur in pericarp and embryo.

Sida acuta BURM.
(escoba, escoba amarilla, escobilla, escoba dulce; anu kuraka in Guajira dialect)

Taxonomical description

The plant is an erect shrub, subshrub or suffrutex, about 1 m (0.2–1.8 m) high with a woody stem. The hirsute to glabrate leaves are distichously arranged. The leaf blades are 3–10 cm long and 0.8–3.5 cm broad (about 2–4 times as long as wide), of lanceolate, elliptic to ovate shape, truncate at the base, with an acute tip and serrate margins (at least distally). The short petioles are 1–5–8 mm long. The 7–15 mm long lateral stipules often exceeding the petiole in length, are broadly falcate or lanceolate, several-veined, and persistent. The palmatinerved venation of the blade is brochidodromous with 8–10 pairs of secondary nerves, and a prominent midrib on the lower side. Leaves pubescent, particularly on the lower side. Young twigs slightly pilous.

Flowers solitary or paired in the leaf axils, with a diameter of 9–10 (20) mm. Calyx 6–8 mm long, often ciliate, about half-divided, basally 10-costate. Petals 7–10 mm long, white, yellow, or yellow-orange (often polymorphic for colour in a single population). Stamens numerous; staminal column 2–2.5 mm long, glabrous or pubescent; styles 8–10.

Fruits glabrous schizocarps with 8–10 (12) mericarps; the mericarps laterally reticulate, the apical spines variably developed; total length of the mericarps 2.5–4 mm. Seeds trigonous, 2–2.5 mm long.

The plant is weedy and polymorphic.

Occurrence

Sida acuta occurs nearly throughout Mexico below 1500 m in deciduous forest, evergreen forest, and at roadsides and disturbed sites including urban habitats. It is pantropical in distribution and flowers throughout the year.

It is a weedy polymorhic plant.

The shrub is frequent in Venezuela in deciduous forests, evergreen forests, savannas, disturbed areas and in deserted former cultivated regions (fallows).

Anatomical description

Leaf (Fig. 173, 174 a). The leaf is dorsiventral and amphistomatic. The upper epidermis cells are thin-walled and slightly vaulted above the surface. As seen in surface view, the anticlinal walls are

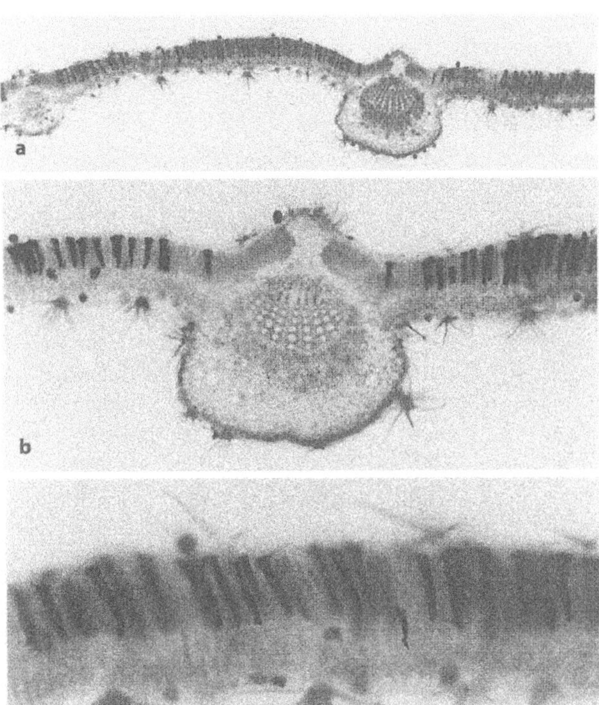

Fig. 173. *Sida acuta*, leaf, midrib. a ×5; b midrib, ×10; c blade ×20.

slightly bent. Stomata are less numerous than in the lower epidermis; they occur at epidermis level. Stellate hairs, glandular hairs with a pluricellular head and flask-shaped glands (Fig. 175) are less frequent than in the lower epidermis. The palisade parenchyma is very well developed in the form of anticlinal rows of about 6–10 cells. The cell rows developed from a single cell layer which repeatedly divided periclinally. Most conspicuous are the sclereids which give support to the palisade parenchyma; they are quite numerous and reach the length of the entire palisade parenchyma; their walls are thickened and their top (directed towards the upper epidermis) is enlarged. As seen in surface view, they appear like dots, leaving about 2–5–10 palisade cells between one another. They stain very intensely with toluidine blue.

The spongy parenchyma is considerably smaller than the palisade parenchyma and reaches about 1/3 or a little more in proportion. There are about 6 layers of very small cells. Very large mucilage cells which may reach an enormous size are found mainly in the spongy parenchyma, but also occur in the palisade parenchyma. Clustered crystals in the form of druses are likewise more frequent in the spongy parenchyma.

As seen in transverse section, depressions occur on the lower leaf side in more or less regular distances (produced by the protruding lateral veins). Stellate hairs as well as glandular hairs (Fig. 175) are found in the depressions. These are furthermore more numerous in the lower epidermis. The lower epidermis cells are smaller than those of the upper epidermis. As seen in surface view, their anticlinal walls are more curved or bent than those of the upper side. Stomata which are more numerous on the lower side, are anomocytic (without distinct subsidiary cells) or anisocytic.

The stellate hairs have about 8–10 long and slender arms. The glandular hairs are of 2 different types: The smaller ones have a foot cell, a stalk and a pluricellular head. The larger ones have a foot, a pluricellular bottle-shaped lower part, a narrower neck composed of about 4 cells and a globular unicellular head. This flask-shaped glandular type (Fig. 175) seems to be characteristic of several species of *Sida*. The small glandular hairs and the large flask-shaped ones are also distinguished by their reaction to toluidine blue: the small ones stain intensely blue, the large ones do not stain.

The midrib is very prominent on the lower side, but forms only a small protrusion on the upper side in which the palisade parenchyma is replaced by angular collenchyma. A single large vascular bundle with a sclerenchymatous cap on top and a sclerenchymatous arc on the lower side, lies in the center. Large-celled parenchyma in which mucilage cells are embedded separates the lower epidermis from the bundle; immediately above the lower epidermis some collenchymatous cells may also be found.

The lateral nerves of secondary order are similar to the midrib, but smaller.

The leaf is characteristically of the sun type.

Axis (Fig. 176, 174b). A young twig of a diameter of 3.5 mm has been studied. The small-celled epidermis is still present. Cork is already developing, but proceeds from the first subepidermal layer, where enlarging cells adopt thin walls and divide periclinally. The 2–4 cell layers underneath are slightly collenchymatous and represent the photosynthetic tissue. Further inwards, the cells enlarge and become parenchymatous. A ring of tangentially extended and partly compressed cells may be observed outside the endodermis. Druses of calcium oxalate are found in the entire cortex, which is, however, not very ample.

The phloem consists of hardbast in the form of fibers and of softbast portions. The hardbast is arranged in more or less irregular layers. Conspicuous are the radial cell rows in the phloem which

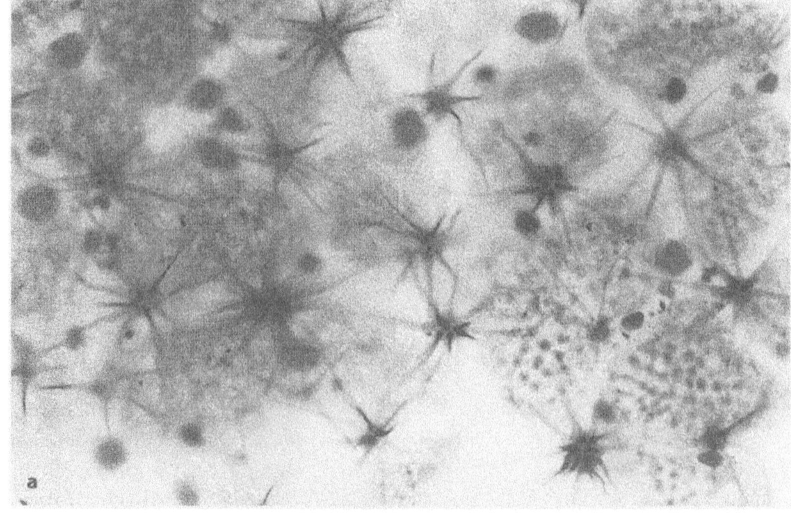

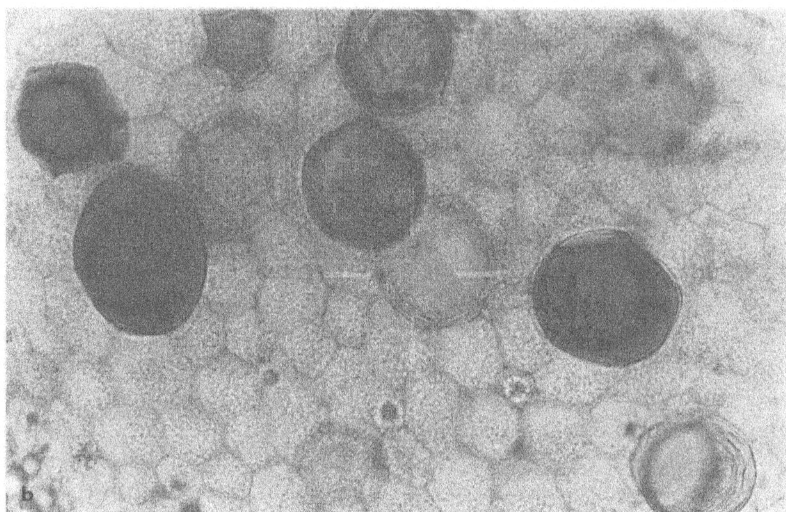

Fig. 174. *Sida acuta.* **a** Upper epidermis with stellata hairs. Note the arrangement of the chloroplasts in the underlying palisade parenchyma (× 20). **b** Pith of the axis in t.s. with slime cells (dark). Note the small druses of calcium oxalate and the small starch grains × 20.

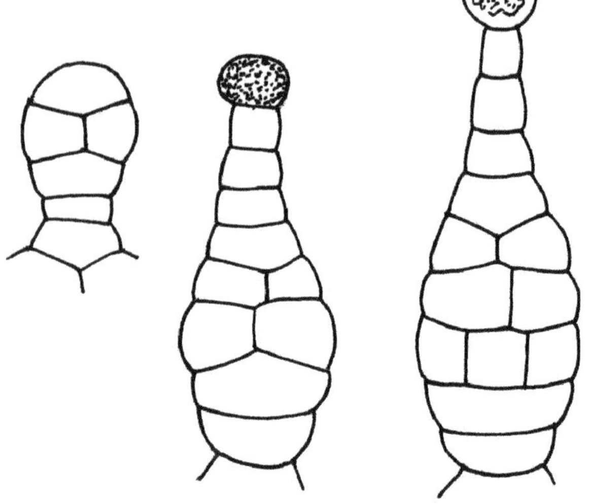

Fig. 175. *Sida acuta.* Small gland (left) and long bottle-shaped glands with belly, neck, and head. Head intensely staining with toluidine blue; discharched head (right).

contain a druse in each cell and which correspond to medullary rays. The rays are 1–3 seriate and some of them dilate towards the outside in a somewhat irregular wedge-shaped form. The phloem is comparatively well developed.

Large irregular groups of fibers outside the phloem may possibly be interpreted as a pericycle.

The vessels of the xylem become very small towards the inside, but enlarge towards the outside. They occur solitarily or are arranged in twins or short radial rows. The parenchymatous pith is well developed and contains very large slime cells which stain violet with toluidine blue. The other medullary cells contain starch or druses. In contrast to the phloem, the rays of the xylem do not contain crystals. The starch grains seem to be of more or less globular to elliptic shape with a cleft of dryness in the center.

Some stellate hairs are only found on very young twigs.

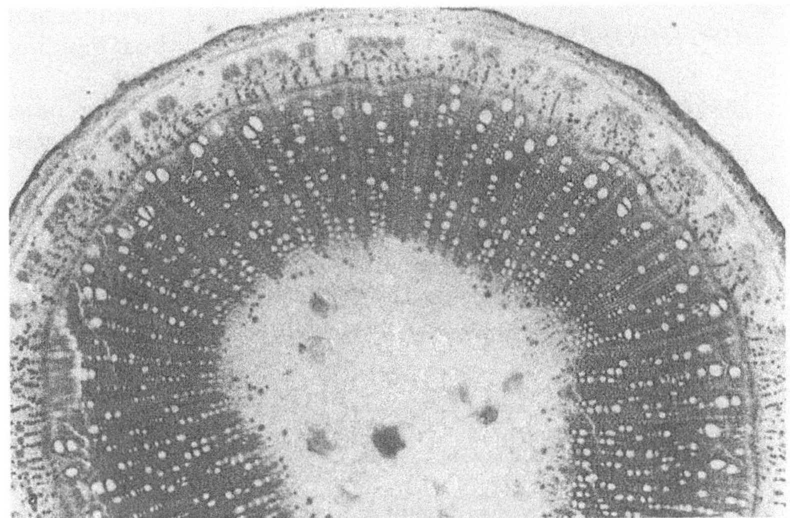

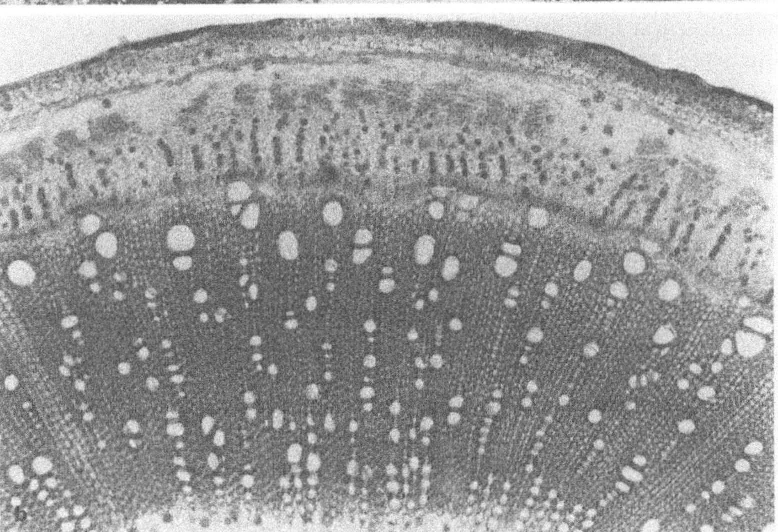

Fig. 176 *Sida acuta*. **a, b** T.s. of axis. **a** Note the slime cells in the pith. Note the dark radial cell rows in the phloem corresponding to medullary rays each cell containing a druse of calcium oxalate. The hardbast of the phloem is arranged in irregular layers (gray)

Ethnobotanical and general use

Economical utilization

Fibers of the stem are used for making cordage. Bundles of the tough branches replace ordinary brooms and brushes.

Medical use

Entire plants and twigs with fruits are mainly used.

Entire plant. A decoction of the plant is used as an expectorant and against inflammations.

Twigs with fruits are sold in herbolarios (herb stands or stalls) of Barquisimeto. They are applied against irritations of the eyes and conjunctivitis.

Healing properties

The slime in the leaves and the stem may help as an expectorant. The plant also contains alkaloids.

Chemical contents

The alkaloids of *Sida acuta* were studied by PRAKASH, VARMA & GHOSAL (1981).

Observations

The plant is very well characterized by the stellate hairs, the bottle-shaped glands, and the large mucilage cells. The very frequently found bacilliform sclereids in the leaf (palisade parenchyma) are an excellent characteristic for identification; these are very rarely found in plants.

Thespesia populnea (L.) SOLAND.

Taxonomical description

The plant is a shrub or small tree (4–8 m in height). The leaves – 5–20 cm long – are heart-shaped with an acuminate tip. The large flowers are solitary and axillary. The petals are 5–7 cm long and of yellowish-purple colour. The indehiscent fruits are globular and have 5 locules, each locule with 3 seeds (1 cm long).

Occurrence

The palaeotropic species is cultivated or grows wild in tropical and subtropical regions (forests and thickets), and is particularly frequent along the sandy littoral.

Anatomical description

Leaf. Upper and lower epidermis are relatively large-celled and probably water-storing. Mucilage cells are absent from the epidermis, although they are characteristic of the family. A layer of very long palisade cells with a length/width index of 7.5–10.7 is found below the upper epidermis. The palisade cells are occasionally divided by periclinal walls. Below follow about 2 layers of arm palisade cells which leave intercellular spaces between one another. In this spongy tissue, the vascular bundles are embedded. A layer of palisade cells is likewise found above the lower epidermis, however, the cells are shorter and broader than on the upper side. The leaf structure may thus be considered isolateral. Of special interest are the transcurrent vascular bundles which are connected with the upper and the lower epidermis by pillars of very large rounded parenchymatous cells; these are colourless, very turgescent, and contain druses of calcium oxalate; their function is water stoarge. Stomata occur only on the lower side. Very characteristic of the genus are peltate (scaly) hairs on the lower epidermis which reduce transpiration, as well as headed glands which may function as hydathodes. The leaf is thick and somewhat fleshy.

Ethnobotanical and general use

Economical utilization

The wood is very hard, light brown with black streaks, and keeps well under water. It is suitable for furniture and cabinet work, and recommended for boat building, wheel wrights, gun stocks and fancy work.

The inner bark of the branches and young stems contains a tough and fine fiber useful for cordage.

The seeds supply a yellow dye.

The plant is also used as an ornamental.

Medical use

Leaf. The leaves have an emollient effect.

Roots. Roots and seeds in decoction are efficient against diseases of the skin.

Bark. A cutaneous bath can be made with the bark.

Fruit. The fruit is likewise medicinal.

Cultivation

The species can be reproduced by seeds. It has a rapid growth. The root system is very profound. The plant is resistant to drought. It grows in plain sun and is little exacting concerning soils.

Observations

The isolateral leaf structure, the vascular bundles which are trancurrent to both epidermal layers by water-storing parenchymatous pillars, together with the peltate hairs and the headed glands are properties sufficiently characteristic to identify the species.

Urena lobata L.
(cadillo de perro, cadillo, pata de perro, cadillo pata de perro, cadillo blanco)

Taxonomical description

Erect strong-stemmed shrubby herb up to 1.5 (2) m high. Leaves spirally arranged, 3–10 (12) cm long and broad, very variable in shape, typically with more or less orbicular slightly angled blades, otherwise more or less deeply divided into five lobes, or ovate to lanceolate, angular, sinuate or lobate, with unequally serrate margins, densely pubescent on both sides with minute stellate hairs and fewer large simple hairs. Rose-coloured flowers axillary. Outer and inner calyx about 5–7 mm long.

Petals pink, darker at the base, about 15 mm long, Staminal column 15–18 mm long. 5 carpels splitting into mericarps, about 6 mm long, indehiscent and covered on the outer surface with numerous hooked spines.

Occurrence

Florida, West Indies, South America, Africa, Asia, tropical and subtropical regions. In Venezuela it is very common, mainly in hot regions. The plant is common along roadsides, in thickets and waste places.

Anatomical decription

Leaf (Fig. 177, 178). As seen in transverse section, the leaf is dorsiventral and amphistomatic. The upper epidermis cells are large, transparent and vaulted above the surface in the watchglass form. There is a single layer of long and slender palisade cells which may occasionally be subdivided by periclinal walls. The spongy parenchyma is proportionately less than the palisade parenchyma and consists of cells of a varying shape with arms which leave intercellular spaces. Dispersed in the mesophyll are numerous more or less globular cells with druses of calcium oxalate which reach an enormous size in the palisade parenchyma. Also very conspicuous are the long unicellular hairs with thick walls as well as the stellate hairs with about 6–8 long and slender arms on the upper and the lower leaf side. The stellate hairs are less frequent on the upper than on the lower side; those of the upper side have fewer (2–3–4, more seldom 6) but stronger arms; those of the lower epidermis have longer more slender and more numerous arms (5–8). However, stronger stellate hairs are intermingled with weaker developed hairs on the lower epidermis. The anticlinical walls of the upper epidermis cells are somewhat wavy, as seen in surface view, while those of the lower epidermis are considerably more sinuous. The stomata are of the *Sedum* developmental type with 3 subsidiary cells surrounding them (ROTH & CLAUSNITZER 1969), imitating a three-edged apical cell dividing at 3 surfaces. The stomata are thus not ranunculaceous or anomocytic, as is usually the case in Malvaceae (METCALFE & CHALK 1950).

The slime is possibly situated in the upper epidermis cells.

The midrib is exceedingly vaulted above the upper and particularly above the lower leaf side. The palisade parenchyma is interrupted in the midrib

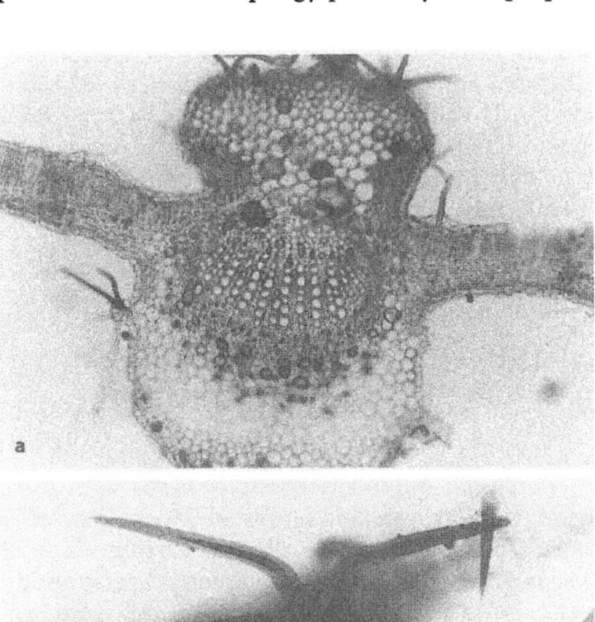

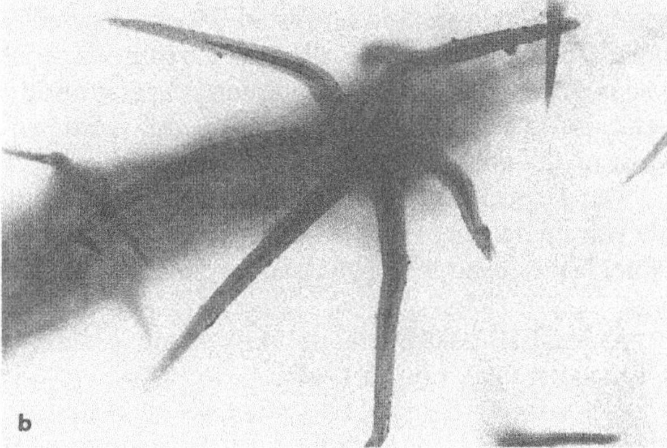

Fig. 177. *Urena lobata.* **a** Midrib of leaf (× 10). **b** Stellate hair on a rib of lower order (× 20).

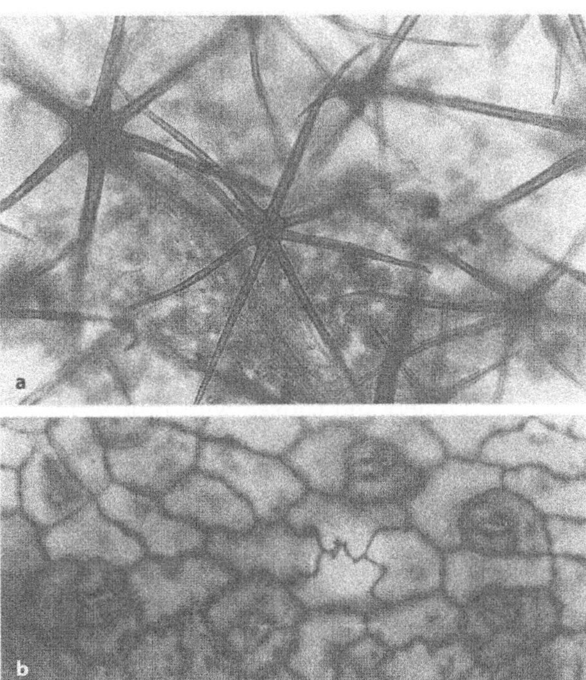

Fig. 178. *Urena lobata.* Upper epidermis: **a** stellate haors (× 40), **b** stomata (× 40).

and replaced by a weak angular collenchyma. The center of the midrib is occupied by the slightly horseshoeshaped vascular bundle. This is surrounded by parenchyma. A weakly developed angular collenchyma is also present on the lower leaf side. Dispersed within the parenchyma and collenchyma are cells whose content easily oxidizes and which are possibly tanniferous. Clustered crystals (mainly small ones) in the form of druses also occur. Hairs of 2 different kinds (see above) are found on both epidermal layers.

The secondary nerves of the first order are transcurrent by waterstoring tracheids.

The leaf is of the sun-type.

Ethnobotanical and general use

Economical utilization

Urena lobata supplies the so-called Aramina fiber which is fine, lustrous and soft, and is comparable to jute. The stem (pericycle and phloem) is the source of the fiber. The small shrub is cultivated for its stem in Florida, the West Indies, South America (Brazil), Africa and Asia. The fiber is used for sacking, cordage, coarse fabrics, ropes, hammocks, fishing tackle and the like. It is said to resist termites and water.

Medical use

Leaf, root and flower are applied medically, and all three are used as a vermifuge.

Leaf. An infusion is taken for the treatment of urinary burning and inflamed kidneys. Leaf infusions are also taken against infections of the urinary tract, gallstones, kidney stones, hepatitis, dysentery, pleurisy, affections of the lungs, and against heat, hangover and gastritis.

Teas are taken for flu and stomachache.

Externally, leaves are applied against erysipelas and eruptions of the skin.

Further applications are in the form of a cataplasm, as an emollient, and refreshing tonic or for lavations.

Leaves ground and cooked with little water are used in cataplasms as a galactogogue. Very hot cataplasms are put on the abdomen against dysmenorrhoea.

Flower. Flowers are used in gargles, against a sore throat and against aphthae.

Method of use

Externally in cataplasms, for lavage, as an emollient and refreshing tonic, and for gargles.

Internally as a tea or infusion.

Healing properties

Antidiarrhoeic, antiseptic, diuretic, emollient, vermifuge. It is furthermore considerded as a colagogue, hepatoprotector and emmenagogue.

Chemical contents

The plant contains a bitter principle, mucilage, a sesquiterpene, oleanolic acid, ferulic and cinnamic acid, sitosterol, sitostenone, dehydrotectol.

Varieties and related species

Variations in the degree of lobing of the leaves are sometimes regarded as formations of distinct species, but intermediates make this supposition questionable. Several different species are united under the vernacular name Cadillo de perro, which pertain to the genus *Urena*, mainly *U. lobata* L. and *U. sinuata* L.

Observations

Large globular cells with druses are conspicuous in the mesophyll. Very long thick-walled hairs as well as stellate hairs with 6–8 arms are characteristic of the upper and lower epidermis. The midrib is extremely vaulted on both, the upper and the lower side; it contains cells with a rapidly oxidizing content which possibly correspond to tanniferous cells. The veins of a higher order are transcurrent by water-storing tracheids.

Marantaceae

Perennial monocotyledonous herbs with distichous pinnately-nerved leaves which are divided in sheath, petiole and blade. Flowers asymmetric with 3 sepals, petals united in a 3-lobed tube, staminodia petaloid. Ovary 3-locular, seeds usually with aril. Most of the genera occur in tropical America.

Many species are ornamentals; others have edible starchy tubers, flowers which are eaten as vegetable, leaves used for roofing or baskets, some yield wax.

At least 13 useful species are known (SCHULTES & RAFFAUF 1990; UPHOF 1968).

Calathea sp. is a high herb of the undergrowth in the Venezuelan transitional and real Cloud forest.

The leaf anatomy has been described by ROTH (1990). There is an upper and lower water storing hypodermis; the mesophyll is not differentiated in palisade and spongy parenchyma; the stomata are of the Rhoeo type (ROTH & CLAUSNITZER 1969).

Calathea lutea (AUBL.) MEY. (casup, platanillo, casigua)

Taxonomical description

Perennial cespitose and robust herb, 1–5 m high. Leaves elliptic or ovate, 50–150 cm long and 25–60 cm broad, with a rounded and apiculate apex. Upper leaf side pale-green, lower side whitish waxy. Petiole about 1.5 m long. Pseudopinnate venation with numerous parallel veins diverging obliquely from the pseudo-principal nerve. These secondary nerves are connected by numerous tertiary nerves. The inflorescences are spikes of 15–20 cm length, with 4–14 coiraceous bracts, distichously arranged, of brown-red colour with intermingled green zones. Flowers yellow. Sepals linear, obtuse, 1 cm long. Crown pale-yellowish, with a tube about 3 cm long and silky lobules of 1.5 cm in length. Stamens pale-yellowish. Ovary pubescent at the base. Fruit a capsule, usually only with a single seed which has a white aril.

Origin and occurrence

Calathea comprises about 130 species in the Americas. *C. lutea* occurs in the West Indies, in Central America and South America down to the Amazonas region. In Venezuela, the species grows spontaneously in the humid forests (Cloud forests) of the Cordillera de la Costa and of the Andes.

Anatomical description

Leaf. Leaf anatomy was studieds by LINDORF 1980.

The leaf is bifical and amphistomatic. Both epidermal coverings are single-layered, with a delicate and smooth cuticle; the lower epidermis is papillose. As seen in surface view, the epidermal cells are of varying shape, their walls are wavy; they show a tendency to become arranged in parallel rows, as often observed in monocotyledons. Below the upper as well as the lower epidermis, there is a single-layered water-storing hypodermis. The upper hypodermis is composed of large hexagonal cells, as seen in paradermic view. The lower hypodermis consists of polygonal cells. The palisade parenchyma forms 1–2 layers. The spongy parenchyma comprises 3–4 layers of cells with irregular lobulate outlines which leave larger intercellular spaces between one another. The stomata found in the intercostal fields are arranged parallel to the major veins and occur at epidermis level. Stomata are accompanied on either side by a subsidiary cell.

The larger vascular bundles extend from the upper to the lower epidermis, the smaller bundles lie in the spongy parenchyma. The larger bundles are surrounded by 2 types of vascular sheath. The outer sheath is parenchymatous and has few chloroplasts of the same type and size as those of the mesophyll; the inner one consists of fibers which surround the phloem and the xylem. The smaller vascular bundles have only an outer vascular sheath on the adaxial side, and an inner sheath on the abaxial side. Above the fibers of the inner sheath, silicious cells or stegmata, arranged in longitudinal rows, may be observed.

Ethnobotanical and general use

The plant is of ornamental value due to its large leaves which resemble those of *Heliconia*; it is suitable for shady sites in parks and gardens. The leaves are used by some indigenous tribes for roofing of their houses.

Varieties and related species

Some species have very interesting patterns on their leaves which make them even more ornamental.

Calathea allouia (AUBL.) LINDL. is cultivated in America for its edible tubercles called lairenes. They are small, about 3–4 cm large, outside of a creamy colour, with a rugose skin which is delicate but strong; inside, the tubercles are white and soft. They are eaten after being boiled in salted water and could even be used as 'pasapalos' (snacks). The soft inflorescences are also edible as a vegetable; the leaves have diuretic effects.

Cultivation

Calathea lutea can be propagated vegetatively by cuttings or by new shoots sprouting from the base of the plant. The plant needs a rich soil and humid shady sites.

Melastomaceae

This family has several structural peculiarities (Fig. 179).

The connective is supplied with spur-like, club-shaped or bifurcate appendices. The anthers open with pores. The fruits are capsules or berries. The seeds are small and without endosperm. The leaves have 3–9 main veins originating at the blade base and running in the form of arcs to the leaf tip.

Myrmecophily is found at times. Pollination is by insects, humming birds or bats.

Mostly shrubs or herbs, sometimes lianas or epiphytes, but rarely trees. Main distribution in tropical America, principally in the rain forest, but also ascending in the Andes.

The Melastomaceae show the highest accumulation of aluminium of almost all angiosperms.

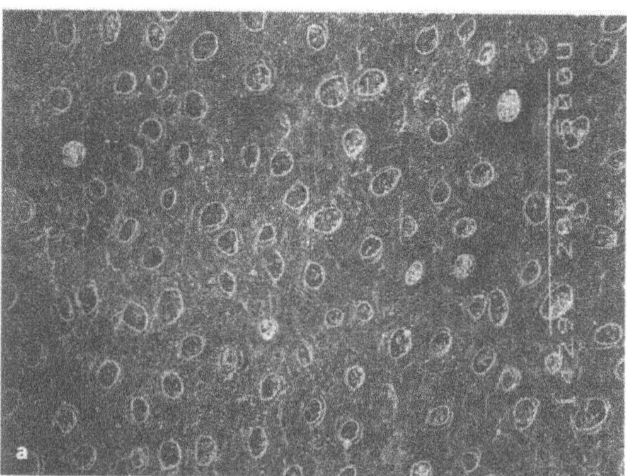

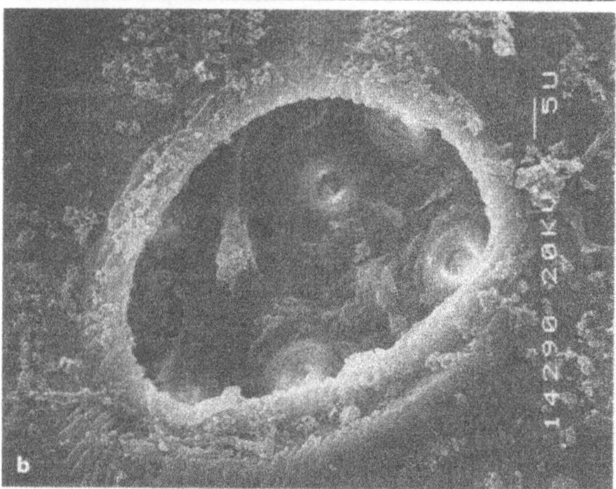

Fig. 179. *Mouriria sideroxylon*, Melastomaceae. **a, b** Lower leaf surface with holes in which the stomata are hidden. Several stomata are present in a crypt (b). Scanning microscope (ROTH 1984).

Polyphenols, cyanogenic compounds, sterols and a few alkaloids have been found; but little is known of the chemistry.

Some of the plants are ornamental, some have edible fruits, and a few yield dyes, tannins, and medicinal products. Only a few attain a large size with timber of economic value which is used for charcoal, fuel, small construction, packing cases, and carpentry.

Clidemia

With more than 100 species of hairy shrubs and a few little trees, C. is widely distributed throughout tropical America and the West Indies.

The fruits are sweet and edible (e.g. of *C. hirta* and *C. ulei*).

A tea of the leaves of *C. heterophylla* is gargled to relieve hoarsness of the throat.

Clidemia plumosa (DESR.) DC is a shrub up to 4 m in height which occurs in the undergrowth of the cloud forest proper. The leaves have an indumentum on both sides.

The leaf anatomy has been studied by ROTH (1990).

Miconia

Miconia is a large genus with nearly 800 species, almost all of which are native to tropical America. Many have beautiful flowers and are cultivated as ornamentals.

A substitutive quinone with antimicrobial and antineoplastic activity has been reported for *M. amazonica* TRIANA.

The fruit of *M. fissa* GLEASON is used as a diuretic.

The fruits of *M. albicans* (SWARTZ.) Triana and of *M.liebmanii* COGN. are used as a food.

The wood of *M. argentea* (SWARTZ.) DC is a source of charcoal.

The bark of *M. willdenowii* KLOTZSCH. is used for swamp fever. The leaves are a substitue for black tea; they contain 0.22 % caffeine.

The leaves of *M. impetiolaris* (SW.) D. DON. are applied for an aromatic bath. A blue-black dye can be obtained from some fruits.

The leaf anatomy of *M. sancti-philippi* NAUD., of *M. serrulata* (DC.) NAUD. and *M. centrodesma* NAUD. has been studied by ROTH 1990. *M. serrulata* is characterized by stellate and glandular hairs.

Leaf venation is very characteristic in the Melastomaceae (see ROTH 1996).

Miconia argyrophylla DC. is a shrub or small tree, up to 9 m high. The simple leaves are elliptic and have a short tip. The globular purplish-black fruit measures 0.4 × 0.5 (0.8) cm and is edible. The wood is used locally for interior construction. The tree grows in the Venezuelan cloud forest. The leaf anatomy has been studied by ROTH (1984).

Mouriria

At least 6–7 useful species are known.

Leaf ashes are used for pottery. The bark is astringent; a tea is used for sores of the mouth. A leaf decoction serves for washes after child birth. A powder of the leathery leaves is applied to clean the teeth. A leaf decoction is used for fungal infections of the skin. Fruits are edible.

M. huberi and *M. sideroxylon*. A decoction of the leaves is used for infections of the skin. ROTH studied the bark structure 1981, leaf structure 1984, leaf venation 1996, fruit structure and dispersal 1987 (Fig. 179).

Tococa

With about 60 species of shrubs or small trees is widely distributed throughout continental tropical America.

The leaf petioles usually bear inflated vesicles which harbor small ants. The common name of *Tococa* is therefore 'ant's plant'.

Tococa juruensis PILGER. An infusion of the leaves with their basal swellings for ants is taken by pregnant women to influence the sex of the unborn child; it assures the birth of a male child.

Tococa aff. longisepala COGN. The bark of the root is astringent and after softening in water it is chewed by the natives to treat sores in the mucuous membranes of the mouth.

The inner structure of the basal swellings of the petiole inhabited by ants, and called domatia was studied by ROTH (1976). Two species, *T. guianensis* var. *orinocensis* and *Tococa macrophysca* were investigated. The swellings form as double structures and are hollow; the cavities are lined with a special tissue in which glandular hairs as well as ramified and sclerified hairs prevail. The content of the glandular hairs gives the reaction of vanillina. The glands are apparently eaten by the larvae of the ants.

The chemistry of the genus is only little studied. Some of the species are very ornamental.

Meliaceae

Carapa

Durable wood. Edible fruit. The seed oil is used for soap manufacture, illumination, protection from insect bites; on the skin it is applied for burns and yaws.

Bitter bark and leaves are tonic, febrifugal and vermifugal.

Bark is applied for tanning.

The peels of the fruit serve as a vegetable and appetiser.

Bark structure has been studied by ROTH 1981, leaf structure 1984, fruit structure and dispersal 1987.

C. guianensis AUBL. Useful wood. The bitter seed oil is used for illumination, soapmanufacture, and as a protection against insect bites. Externally applied it cures skin problems. Internally (drop by drop) it is taken for diseases of the chest and for rheumatic pain. Bark and leaves are applied for a febrifugal and vermifugal tea. Externally applied, they are used for ulcer and skin trouble. The leaves have antidiarrhoeic properties. Fruit structure and dispersal have been studied by ROTH 1987, leaf veneation 1996.

Cedrela

At least 9 useful species are known.

Shade tree. Flowers supply a yellow and red dye.

The aromatic and resinous wood is used for fever and epilepsy.

The bark is astringent, emetic, and helps for leucorrhoea.

The mucilaginous seeds are vermifuge.

C. odorata L. Wood used for moth-proof chests.

Bitter root bark is a febrifuge.

The seeds are vermifuge.

Bark structure has been studied by ROTH 1981, fruit structure and dispersal 1987.

Guarea

At least 7 useful species.

Useful wood. Sandal oil is obtained from wood.

Gum resin of leaves is haemostatic.

Bark is emetic, haemostatic, depurative, expectorant.

Fruit structure and dispersal have been studied by ROTH 1987.

Swietenia

S. the source of the original or true Mahagony, supplies the premier cabinetwood of the world (RECORD & HESS (1943). The genus was described by JACQUIN IN 1760 and the single species was *Swietenia mahagoni*. However, Linnaeus had already named the species a year before as *Cedrela mahagoni*.

The species occurs in Southern Florida, the West Indies, Mexico, Central America, Colombia, Venezuela and the upper Amazonian region.

The fruit is characterized by a strong and hard exocarp and winged seeds (Fig. 180, 181).

Trichilia

At least 13 useful species are known. Useful wood.

Tannins, saponins and limonoids (insect repellents) have been identified.

Seed oil is applied for soap and candle manufacture, as well as for hair dressing.

Leaves serve for pulmonary ailments and fever.

Bark is astringent and febrifuge, it helps for malaria and fever. Scrapings of stem and root bark are vermifuge.

Roots are applied for snake bites.

Bark structure has been studied by ROTH 1981, leaf structure 1984, leaf venation 1996.

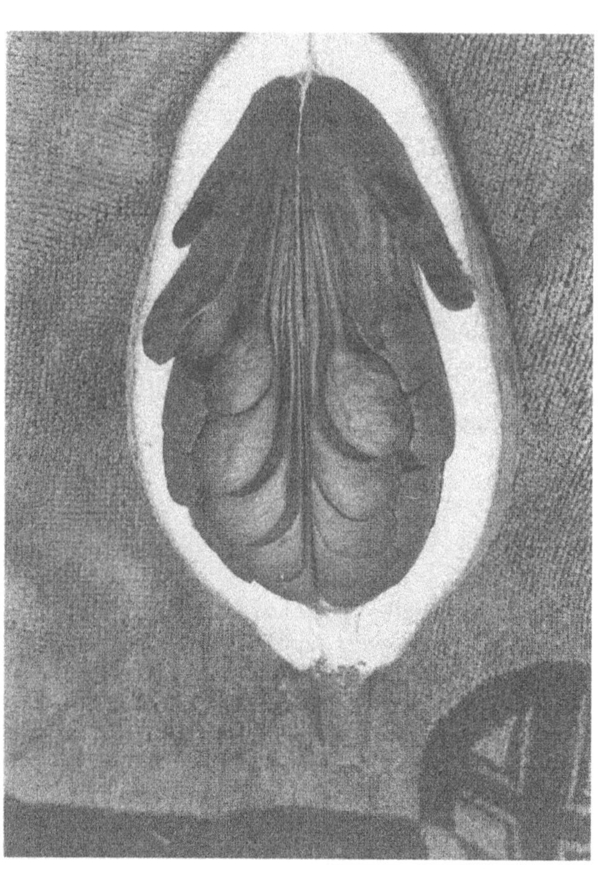

Fig. 180. *Swietenia macrophylla*, Meliaceae. Fruit cut into halves exposing the winged seeds.

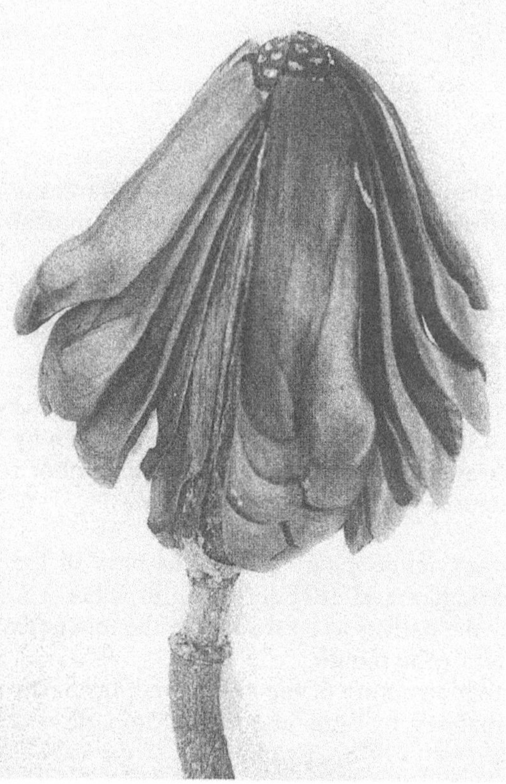

Fig. 181. *Swietenia macrophylla*, Meliaceae. Opened fruit with winged seeds attached to the placenta in the form of superposed rows (ROTH 1984).

Mimosaceae

The Mimosaceae are mostly trees and shrubs with pinnate leaves (simple or double), which may sometimes be modified into phyllodia (expanded blade-like petioles). Flowers actinomorphic with numerous stamens which have long attractively coloured filaments. Pollen united in cohering groups of variable numbers and sizes (several to pluricellular pollinia). A sarcotesta may be present. The flowers are arranged in dense heads, spikes or racemes. The fruit is a legume.

The Mimosaceae occur almost exclusively in the Tropics and Subtropicas.

Mimosa pudica is a pantropical weed with sensitive leaf cushions which are stimulated by the touch. Leaf movements can however also be oberseved on many other tropical Mimosaceae.

Tannins and gum-like slimes, such as gummi arabicum of *Acacia senegal* which is used as a glue and thickening agent, are frequent, mainly in the barks.

Enterolobium

The bark is used as a fish poison and as a source of curare. On the skin it has a fungicidal effect. Roots and seed coats are anthelmintic. Saponins are used as a soap substitute.

E. cyclocarpum GRISEB. Wood is useful. Young pods are edible. Bark and fruit are a substitute for soap. A syrup made of the bark is used for colds. A gum is extracted called 'goma de caro'. Fruits are food for cattle.

ROTH studied the bark structure 1981, fruit structure and dispersal 1987 (Fig. 182).

Inga

At least 30 useful species are known.

Leaves are used for earache and for smoking. Leaves and roots are applied for colouration of the skin (pinto).

Flowers are likewise used for earache and for nasal congestions. Flowers also serve to strengthen the hair.

Some pods are edible, in other cases the sweet fruit pulp around the seeds is edible. Fruit pulp is used for flatulence, to wipe infected eyes and to clean the teeth.

Gum is used as a mordant, the reddish exudate as a dye: *Inga heterophylla* WILLD., *I. rubiginosa* DE CANDOLLE (flower powder as a sniff for nasal congestions), *I. ruiziana* G. DON, *I. scabriuscula* BENTH.

Inga edulis MARTUIS. is a shade plant and has edible fruits. A decoction of the leaves is used for diarrhoea (CASTILLO 1995).

ROTH studied the bark structure 1981, leaf structure 1984, fruit structure and dispersal 1987, leaf venation 1996.

Fig. 182. *Enterolobium cyclocarpum*, Mimosaceae, helicoidally bent fruit pods with a shiny surface (ROTH 1987).

Inga edulis (guama, guamo) is a small, up to 7 m high tree which occurs in the Amazonas basin, but is cultivated everywhere in the Neotropics. It grows in sandy, acid soils with a high aluminium toxicity and supports precipitations between 2000 mm and 5000 mm yearly. The indehiscent fruit contains yellow-orange to black seeds with a white, sweet and edible aril. The optimal temperature for cultivation is between 17 °C and 25 °C. The fruit can be harvested all the year round. Of the aril or pulp refreshing drinks and preseves are made; it is also suitable for alcohol production and as a food for animals. The plant which flowers permanently also supplies a good honey. The seeds contain about 20 % protein and 65 % carbohydrates. Leaves and seeds are medically used as antidiarrhoeic and antirheumatic (NARBAIZA 1999).

Mimosa

The 500 species comprise herbs, shrubs, vines or small trees, mainly distributed in South America. Many species are armed with stipular thorns. The bipinnate leaves have usually numerous small leaflets, which are often sensitive. The legumes are thin, flat and wingless and break up into few to many joints.

Some of the species are ornamentals.

Psychotomimetic constituents have been identified; norepinephrine occurs in *M. pudica.*

Mimosa albida H. & B. is a herb or small shrub with paripinnate leaves. The plant reaches 1.50 m in height; its flowers are pink. It occurs at altitudes of 900–1000 m in the Venezuelan cloud forest, but is more frequent in the savanna and the Mattorales (thorny forests) of Venezuela.

The leaf anatomy has been studied by ROTH (1990).

Simple hairs and glandular hairs are present in the lower epidermis.

Mimosa pudica L. is not only interesting for its sensitive leaves, but is also employed as a medicine by the natives. Leaves, roots and seeds are applied. It is said to be antidysenteric, aphrodisiac, sedative, antiseptic, antispasmodic and emetic. It is also used for epilepsy, problems of the kidney, as a bitter tonic, and a restoring medicine.

Caution is however necessary in the application, because in higher doses the plant is toxic and may cause loss of hair.

Parkia

At least 9 useful species are known.

Useful wood. Bark is rich in tannin and helps for dysentery and for toothache. Fruits are edible and seeds are used as a condiment, as a substitute for coffee, for sauces and cakes. Some pods have a garlic-like taste and are used as condiment.

The sweet mucilage around the seeds is used for sore gums. Seeds serve against colics. *P. speciosa* contains antifungal and antibacterial cyclic polysulfides.

P. oppositifolia SPRUCE EX BENTH. A drink is made of scraped bark for dysentery. ROTH studied the leaf structure 1984, fruit structure and dispersal 1987, leaf venation 1996.

P. pendula BENTH. EX WALPERS. A decoction of the plant serves to treat haemorrhages. Bark boiled is used for stomach ache. The timber is useful.

ROTH studied the bark structure 1981, leaf structure 1984, fruit structure and dispersal 1987, leaf venation 1996.

Pentaclethra

Useful wood. Lotion of bark for sores. Seeds with a poisonous alkaloid used as a vomitory, for ulcers and against snake bites. Seeds are source of a semi-solid fat. Seeds of other species contain oil manufactured into candles and soap, and are also made into bread and used as a food.

ROTH studied the bark structure 1981, leaf structure 1984, fruit structure and dispersal 1987, leaf venation 1996.

Piptadenia

At least 7 useful species are known.

Wood is useful. Bark is source if mucilage. Bark for tanning. Bark astringent and applied for diseases of the lungs.

Seeds narcotic (snuff). Gum with 80 % arabin.

ROTH studied the bark structure 1981, leaf structure 1984, fruit structure and dispersal 1987, leaf venation 1996.

Pithecellobium

At least 25 species are useful.

Useful wood. Wood for dysentery and diarrhoea (in the form of a tea). Yellow dye and tannin from bark. Bark as fish poison. Bark for cramps. Bark purgative. Bitter root bark after malarial attacks. Mucilage from bark. Leaves vermifuge. Leaflets as a substitute for soap and to remove rubber stains. Leaf decoction to promote growth of hair. Leaves, fruits and flowers as vegetable. Twigs as forage.

Excellent honey from flowers. Flowers for headache.

Edible fruits. Edible seeds. Edible arillus. Seed coat as a coffee substitute. Seeds vermifuge. Seeds for diabetes.

Gum from stem.

P. jupunba URB. Useful wood. Decoction of wood for dysentery. Bark to stupefy fish. Leaflets as a soap substitute.

ROTH studied the bark structure 1981, leaf structure 1984, fruit structure and dispersal 1987, leaf venation 1996.

P. claviflorum SPRUCE EX BENTH. Seeds a strong vermifuge. ROTH studied the bark structure 1981, leaf structure 1984, fruit structure and dispersal 1987, leaf venation 1996.

Prosopis juliflora DC (cují-yaque, yaque, yaque blanco, yaque negro, cují, cují negro, cují carora)

Taxonomical description

Prosopis juliflora is a spiny shrub or tree, 5-12 m high, with a short somewhat twisted trunk and a DBH up to 45 cm. It has an extensive umbrella-shaped strongly ramified crown which is often 'wind-shaped'. The more or less pendulous branches are armed with spines. The leaves are bipinnate with 1-3 pairs of pinnae of the first order each of which develops 10-30 pairs of linear oblong leaflets, about 5-25 cm long and 2-8 mm broad.

The small flowers are of a yellowish-white colour arranged in dense spikes 5-10 cm long.

The fruit is a linear brown legume, straight or bent, 5-20 cm long and 6-15 mm broad. The brown, 5 mm long seeds are embedded in a whitish slightly sweet and edible pulp (endocarp).

The tree flowers and bears fruit most of the year.

Occurrence

The tree grows in western Texas and eastern New Mexico, throughout most of Mexico, Central America and the West-Indies.

In Venezuela it is characteristic of the State Falcón where it belongs to the 'flora emblemática' as a symbol of this State (HOYOS 1985). In South America, the tree grows as far down as Argentina.

Anatomical description

Leaf (Fig. 183.1). The leaflet, although very small, is very thick due to the extensive mesophyll it develops. The upper and lower epidermis are large-celled and water-storing. Palisade parenchyma occurs on the upper as well as on the lower side, but is better developed on the upper side, where it comprises 3-4 layers. Part of the palisade cells are very long with a length/width index of 6-10,4. The palisade cells are arranged in anticlinal rows, originating from a single mother cell which repeatedly divided periclinally. The palisade parenchyma on the lower side comprises 2-3 layers of shorter and broader cells.

The tissue in the leaf center between the upper and the lower palisade parenchyma is composed of more or less globular cells which leave intercellular spaces between one another and may be considered as a spongy parenchyma. Embedded in this parenchyma are the vascular bundles, which occur very frequently. They are surrounded by a parenchymatous sheath composed of large cells. Colourless cells - partly palisade-like and partly globular - which

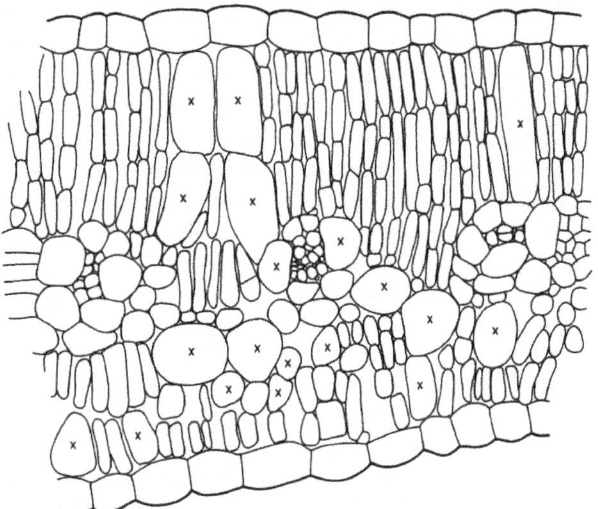

Fig. 183.1. *Prosopis juliflora*, t. s. of leaf. Note the large cells marked with X which are colourless and waterstoring (ROTH 1992).

are interspersed between the palisade as well as the spongy parenchyma are particularly interesting. Their function is to store water. Stomata occur on both surfaces, but are more frequent in the upper leaf side.

The leaf is isolateral and has a fleshy consistence due to its water-toring elements (colourless cells, epidermis cells).

Ethnobotanical and general use

Nutritional use

The flowers are eaten by man (nectar). The flowers, which are much frequented by bees, yield a good grade of a light-coloured honey.

The sweet pods are an important source of forage for stock of all kinds. They are also eaten by man. Non-alcoholic and alcoholic drinks are prepared from the pulp. Ground pods are also used as a meal to bake bread, cakes and cookies.

The gum of the bark 'mesquite gum' is used for the preparation of sweets.

Economical utilization

The inner bark of the root supplies fiber for cordage and coarse fabric.

An amber-coloured translucent gum similar to gummi-arabicum exudes from the trunk through incisions in the bark and is employed industrially for alimentary purposes and for preparation of sweets. The gum is also used in textile industries.

Fruit, bark and wood contain tannins which are used for tanning and dyeing. Roots contain 6–7 % tannin.

The wood is not only valuable for firewood and fuel, but also serves for railway cross ties, vehicle construction, fence posts, and for paving blocks. The heartwood is dark brown, with fine vessel lines. The sapwood is yellow, rather sharply demarcated, and has an aromatic taste. The wood is hard, heavy and strong. It is coarse-textured, mostly with irregular grain, easy to work, finishing smoothly and assuming a high natural polish. It is also very resistant to decay probably due to its tannin content.

The species is also appropriate for afforestation (see below).

Medical use

Drugs: folia, cortex, radix, fructus.

The plant is purgative and emetic, helps against diarrhoea, bronchitis, catarrh and cough, diseases of the eyes, cures dysentery and asthma, regulates the intestinal pH, absorbs toxic substances and regulates hyperperistalsis (mucilage) and is used as a vulnerary.

Method of use

Gum of the bark dissolved in water is used against dysentery and as a gargle for infections of the throat. The bark has a bitter taste.

A decoction of the shoots mixed with milk is used as an eye bath and for conjunctivitis.

Together with *Saccharum officinarum* the drug is applied as an abortive.

A decoction of the bark cures the flu.

Healing properties

Antipyretic disinfectant, pectoral, ophtalmic, vulnerary, antibacterial antispasmodic, diuretic, antimycotic, hypothermic.

Chemical contents

The plant contains flavonoid glycosides, tryptamine; the leaves contain alkaloids such as juliprosopine, juliprosine, juliflorinine, julifloricine; the bark contains tannins, the fruit coumarin and kaempferol.

Cultivation

The plant is very resistant to permanent dryness of the soil, to heat and strong insolation. The tree is even able to survive in such inhospitable regions as the shifting dunes of Coro (State Falcón). Shifted by the wind, the dunes bury all vegetation; only *Prosopis juliflora* defends the territory against these forces stopping the advancing sandy avalanches. The species is also able to grow in the mountains up to an altitude of 1,600 m. It furthermore grows in very different soils, including the poorest.

The species is very appropriate for afforestation of arid regions. It may also be planted as an ornamental and shade tree. It can be planted at sites where other forest species do not grow. The plant can be propagated by seeds which have a high germinating power or by slips or cuttings. It has also a great regenerating power when damaged by goats or sheep. The costs for maintenance reach a minimum from the 2nd year on. Production and harvest begin in the third year. *P. juliflora* is a nitrogen-fixing species which regenerates soils and controls erosion by its profound radical system. It has however to be planted at the appropriate sites, because it may become a 'weed' where it is not wanted.

Observations

The leaf structure is very characteristic with its large-celled upper and lower epidermis, the large colourless water-storing cells in the mesophyll, which consists almost solely of palisade cells (isolateral structure).

Stryphnodendron

Bark is haemostatic, used for diarrhoea, for uterine haemorrhages. Bark is used for tanning. Leaves are tonic. Wood is useful.

ROTH studied the bark structure 1981, fruit structure and dispersal 1987.

Moraceae

The Moraceae are usually woody plants, and only seldom herbs. The leaves are supplied with stipules (median stipule in *Ficus*). The flowers are small and united in cymose or racemose inflorescences. The perigone is often fleshy, only one of the 2 carpels is fertile and develops an ovule. Solitary flowers may be united in a fleshy receptacle or syncarpium. Syncarpia may be disc-shaped, cup-shaped or globular and hollow inside (*Ficus, Dorstenia* etc.).

Very characteristic are the non-articulated laticifers which contain a milky latex, and the cystoliths, which are frequent.

The Moraceae mainly occur in warmer regions. *Morus alba* is very well known; the leaves of this plant supply the food for silk-words,

The latex of the Moraceae contains polyisoprenes which are used for the production of natural rubber (e.g. from *Castilloa elastica*, Central America). The protein fraction of the latex sometimes contains up to 20 % of papain-like proteinases. The toxicity of certain milky secretions is due to the content of cardenolides, cardiac glycosides, which may be used as an arrow poison.

This information of useful chemical contents of the Moraceae taken from FROHNE-JENSEN (1979) is rather modest, as compared with the information available from only one single species of *Cecropia*, completely unknown in Europe (see below). This is not intended as a criticism of the very valuable work of the above cited authors; it only shows what an immense quantity of information has still to be gained and how much further investigation is required of the tropical flora.

With about 75 genera and some 3000 species, the Moraceae is one of the largest families of flowering plants. It is primarily tropical and subtropical. Subdivision into tribes is mainly based on the types of inflorescences. The family is economically very important as a source of paper, of poisons, of edible fruits and useful timbers. A wide assortment of chemical compounds with potential biodynamic activity has been isolated from this family (SCHULTES & RAFFAUF 1990).

Artocarpus

About 20 useful species are mentioned by UPHOF (1968). Most of them have a good durable wood and some are even used as a substitute for teak. The fruits of most species are edible, as are their seeds, but are of different quality. The latex of some species is recommended or used as a source of chewing gum.

All species cited by Uphof are of Asiactic origin (Malayan Archipelago, Philippine Islands etc.) with one exception: *Artocarpus brasiliensis* Gomez. However, the bread fruit is cultivated everywhere in the tropics.

Artocarpus altilis (PARK.) FOSBERG, synonym: *A. communis* FORST., *Artocarpus incia* L. F. (arbol de pan, fruta de pan, ñame de pan)

The breadfruit tree is a small tree of 8–20 m height with evergreen leaves, which are large and glossy on the upper side and have deep incisions; the stipules reach up to 13 cm in length.

The tree is monoecious with male and female flowers on the same individual. The female flowers aggregate in large globular inflorescences. Many inflorescences with a complex structure develop in the Moraceae family. In *Artocarpus altilis*, the countless female flowers, consisting of the ovary and the tubular perigone, are arranged on a more or less elongated to globular and fleshy spadiceous inflorescence axis. The entire inflorescence transforms into the ‘fruit’. The inflorescence axis together with the connate fleshy perigone tubes forms the edible part of the infrutescence, whereas the areoles – covering the surface of the infrutescence and corresponding to the uppermost free parts of the perigone – represent the rind. The border lines between the individual fruits are well marked externally by depressions, each areole attaining a more or less pentagonal shape, although the outlines are very irregular; the small dot in the center of the areole corresponds to the stigma. In the mature stage, the infruitescence has a yellowbrownish colour. The warty ring encloses a fibrous starchy flesh rich in carbohydrates; this also contains proteins, minerals and vitamins (A and C) and has a distinct smell.

Both seedless and seeded types are known. In the latter type, the pulp is poorly developed and of low quality, but the large seeds are eaten boiled or roasted. the seedless variety is used as a vegetable.

Although native to Asia, the breadfruit is cultivated all over the tropics and is frequently planted in the Caribbean room.

Anatomical description

The bulk of the mature infrutescence is composed of parenchyma in which large intercellular spaces occur so that a kind of aerenchyma arises; the cell walls of this tissue are comparatively thick. The cells of the peripheral layers are smaller than those of the inner parts. Due to periclinal cell divisions in the outer epidermis, a multilayered epidermis is formed, about 2-3 cell layers deep. Stomata are relatively rare. The walls of the epidermis cells become thickened and a thick cuticle develops on the outer tangential walls. Likewise subepidermal cells divide periclinally to form anticlinal cell rows; these later transform into a thick-walled collenchyma. During development, starch grains and druses of calcium oxalate appear; laticifers, which are often ramified as well as idioblasts with a yellow content are present in the parenchymatous flesh.

Unicellular hairs are abundant on the outer surface of the infrutescence and have a very characteristic shape. In the seedless form, relatively thin-walled cylindric hairs occur, whereas in the seeded type, hairs are found with an enlarged base, an acute tip and a very thick stratified wall. The infrutescence of *Artocarpus altilis* thus appears as a highly specialized form.

Ethnobotanical and general use

As mentioned already, of the seeded form the seeds are eaten either roasted or boiled.

The seedless form is used as a vegetable and substitute for potatoes. Rich in starch, it can be boiled or fried or baked.

In some regions, the male flowers are used to prepare sweets.

The leaves sometimes serve as a food for livestock.

The bark abundantly produces latex with a high tannin content. Likewise fibers are obtained from the bark.

Brosimum

At least 9 useful species are known.

Useful wood. Edible seeds, roasted a substitute for coffee. Edible 'milk'. Latex as a beverage and base for chewing gum and as an adulterant of rubber. Latex is tonic, purgative, used for indigestion, for asthma and pulmonary ailments, also applied for menstrual pain and pain in child birth.

Leaves are a fodder for livestock. Fiber of bark is used for cloth, blankets, sails.

The Venezuelan 'cow tree' abundantly yields drinkable 'milk' (*B. galactodendron*).

ROTH studied the bark structure 1981, leaf structure 1984, leaf venation 1996.

Cecropia (yagrumo, guarumo)

The genus *Cecropia* comprises about 80 species of neotropical origin (VELASQUEZ 1971). The height of the plants varies between 8 and 25 m. They have a rapid growth and often show a chandelier-like branching. The trees are dioecious and produce latex. According to VELÁSQUEZ, the leaves are palmatifid, compound-digitate or even simple; the number of lobules is variable (7-16). The petiole is long. The stipules are ochrea-like and caducous.

The female twin inflorescences with 4-6 sessile or pedicellate aments, covered by caducous spathes, have flowers with a tubular, pentangular perianth, a unilocular ovary and an orthotropous ovule.

The male twin inflorescences have a variable number of aments (4-20) and are covered by caducous spathes. The male flowers have perianths of a different shape and 2 stamens.

The fruit is a small monospermic nutlet surrounded by the persistant perianth.

In young stages of the tree, the leaves of the species studied are entire and lanceolate, about 5-8 cm long, and 2-3 cm broad, with serrate margins. The upper leaf surface is pilous, the lower side is white tomentose. During the development of the tree, the leaves form first 3 segments or lobules and only later 7 to 16 segments, according to the species in question. Furthermore, the leaf consistency changes and becomes coriaceous and the leaves become much larger, reaching a size of 25-90 cm in length (or in width) in the adult stage of the tree (ROTH 1995).

VELÁSQUEZ comments that the species are difficult to separate, because of their similarity; many species only differ in the characteristics of one or both sexes. To distinguish the species of *Cecropia* it is therefore necessary to collect female and male samples of the same age which come from the same stand.

80–100 species are known from tropical America, the West Indies, Mexico and Brazil.

VELÁSQUEZ (1971) studied 12 of them.

The genus *Cecropia* comprises at least 8 useful species.

Wood is used for paper pulp. The hollow trunk is used for water conduction. Bark fiber is applied for sails. The spongy wood serves as tinder. Rubber is obtained from the latex.

The caustic latex is applied to remove warts; it is also used for dysentery. Leaves are applied for liver ailments and oedema. Young buds are eaten as a vegetable. Inflorescences heal sores of the tongue and the mouth. The brownish gummy exudate is used for bleeding gums. The sap of the root is diuretic and increases the energy of the cardiac muscle.

ROTH studied the bark structure 1981, leaf structure 1984, fruit structure and dispersal 1987.

Cecropia peltata L. The apical meristem of the tree and/or macerated leaves are applied for cicatrization (CASTILLO 1995).

Cecropia palmatisecta CUATR. (yagrumo)

Taxonomical description

The tree is 10–12 m high with a black-brown bark. The subcoriaceous leaves are deeply palmatisect, hence the species name; the 7–14 sections (lobes or leaflets) leave an undissected common part in the center so that the leaf results peltate; during development, the leaf forms a so-called transverse zone which connects the original margins on the ventral side. The disposition of the pinnae follows the palmate type, i.e. all pinnae radiate from the very center. Well developed leaves usually have 10–12 leaflets. The leaflets are elliptic-oblong, rounded or obtuse at the apex, and cuneiform at the base; the largest leaflet is 35–50 cm long and 11–15 cm broad; the entire central part has a diameter of 4–60 mm. The upper leaf side is brilliant green, glabrous or rough and provided with microscopically small tubercles dispersed over the surface, corresponding to the cystoliths. The lower leaf side is silvery white due to the dense indumentum. The veins are reddish; the secondary nerves are 6–12 mm apart.

The petiole is about 50–60 cm long, longitudinally striated, has a scarce pilosity and a hairy basal protuberance (pulvinus) of 2–3 cm. The stipule is 16–25 cm long, acuminate, externally more or less glabrous, but inside silky pubescent.

The male twin inflorescences have a peduncle of 5–8 cm in length, provided with a minute pilosity, a spathe 6–11 cm long, acuminate, covered on its outside with a minute pilosity, inside partially silky pubescent. The male aments, in number 18–20, are 4–9 cm long and 5–6 mm broad, the perianth is glabrous, the pedicels 5–10 mm long.

The female inflorescences with a penduncle of 1.5 cm, have a 11 cm long spathe which is acuminate, externally with a minute pilosity, inside partially silky pubescent; 4–5 aments, 4–5 cm long and 5–15 cm broad, enlarged. Perianth scabridous, claviform.

The inflorescences are of a creamy colour, being surrounded by conspicuous spathes of a brick-red colour.

The leaves are the preferred food of the sloth.

The stems are hollow between the nodes and are inhabited by ants of the genus Azteca which preferentially live in the upper stem parts and in the branches. It is supposed that they defend the plants against enemies which they attack very aggressively.

The tree has a rapid growth and the species is therefore very common. It is the first pioneer in clearings. In young stages, the plant gives the impression of a giant herb.

The species is found in the transitional forest as well as in the cloud forest of the Cordillera de la Costa and of the Andes. It grows there between 900 m and 2000 m a.s.l. Ascending to the Pico Bolívar, the whitish-green crowns of *Cecropia palmatisecta* are very conspicuous. The crowns are easily identified within the forest, even from very far, due to the whitish colour of the lower side of the large leaves (white indumentum).

Anatomical description

The leaves of a young tree and those of an adult one are structurally distinct (ROTH 1995).

The leaf anatomy of a grown-up tree differs considerably from that of a young tree. The blade of the former is thicker. While the upper epidermis is single-layered in the leaves of a young tree, it is multi (2–3) layered in the leaves of the adult trees. The outermost epidermis cells are small, but the inner ones are large and function as a water reservoir. The outer walls of the upper epidermis cells measure 2 μm in the young plant, they increase to a width of 6.75 μm in the adult form. As seen in surface view, the epidermis cells are smaller in the adult form. The palisade parenchyma remains single-layered, but the cells become much longer and their length/width index increases. The somewhat loose spongy parenchyma of the young tree becomes compact in the adult form. It consists of roundish cells with short lateral arms so that only small intercellular spaces arise. The single-layered lower epidermis of the immature plant becomes small-celled and partly multi-layered. The wavy anticlinal walls of the young tree (as seen in surface view) become straight in the adult form. Stomata density is considerably increased, as compared with the juvenile form, stomata size is, however, more or less maintained. Druses of calcium oxalate occur in the mesophyll. In the stronger veins, laticifers are arranged in the form of a ring, as seen in transverse section. The vascular network becomes very dense. While the leaf of the young tree fits into the category between the medium and the suntype, the leaf of the adult form is of the very sun type. While the leaf of youth is classified as mesomorphic with some xeromorphic features, the leaf of the adult form is very xeromorphic.

The structural peculiarities maintained in the leaf of the adult form are the following:

Stomata are conspicuously elevated above the lower epidermis surface, but are covered with a very dense indumentum of filamentous hairs. This indumentum produces the white aspect of the lower leaf side. The upper surface is rough due to cystolith hairs; these are transparent and may function as ocelli. Cystolith hairs also occur above the nerves of the lower leaf side. The vascular bundles are transcurrent to the upper and lower epidermis by sclerenchymatous pillars. A more or less regular pattern of squares is produced by the vascular bundles, larger meshes being subdivided into smaller ones. No bundle endings occur in the small meshes so that the vascular system is completely closed, since all bundles anastomose (see also ROTH 1996, ROLLET, HÖGERMANN & ROTH 1990, HÖGERMANN 1987) – this may be a further xeromorphic characteristic.

Ethnobotanical and general use

The genus of *Cecropia* is economically important and the distinct species have more or less similar uses. The plants contain latex.

Nutritional use

Young buds are occasionally eaten as pot herb (*Cecropia peltata*). The fruits are edible (*C. pachystachya*).

Economical utilization

The plant is also used as an ornamental. The hollow trunks are used as tubes for water conduction. Trumpets and drums are likewise made of the trunks. The soft and spongy wood with a fine fiber is easily inflammable and is employed as tinder. The rough leaf surface is used like sand paper for scraping and cleaning objects. The Faculty of Forestry Sciences of the University in Mérida has started experiments on the use of the wood as a pulp for paper. In Brazil wood is used for paper pulp. The latex has been recommended as a source of rubber (*C. palmata*). In some parts of Brazil the bark is a source of fiber for the manufacture of sails (*C. peltata* and *C. pachystachya*). The bark is also used for tanning. Poles are made of the wood. *C. peltata* is the host of a caterpillar which supplies a red colour for dyeing.

Medical use

Used parts: Entire plant, leaf, bud, bark, root, inflorescence, latex.

Name of the drug: *Cecropia peltata* L. latex, folia, cortex.

Inflorescences are chewed by native Indians to treat sores of the tongue and the mouth.

Latex is caustic and used for the removal of warts and against herpes, ulcers, gangrene, cancer, dysentery. Latex of the branches is usually applied locally.

Bark is used as an emmenagogue, an antidiarrhoeic, against sunstroke. The juice of the bark (latex?) is applied against snake bites and scorpion stings (external application).

The trunk (bark?) cures diseases of the liver and diabetes.

Buds are toasted, crushed and pulverized as a wound healing and cicatrizing remedy.

Root sap is applied against eczemas and for cicatrization. A decoction is tonic, a cardiac stimulant, and helps against oedema, biliary conditions, chorea and St. Vitus dance. The sap of the roots is used against snake bites and scorpion stings. The sap also has diuretic properties and increases the energy of the cardiac muscle without increasing the heart beat.

Infusions help against hepatitis, icterus (jaundice).

Leaf. The leaf is the most used organ of the plant. Cataplasms cure swellings. A decoction of young leaves is used for liver ailments and oedema; it stimulates the cardiovascular system, increass the energy of the cardiac muscles, acts diuretically. Extracts, infusions and tinctures of the leaves cure acute affections of the respiratory ducts, pneumonia, bronchitis, asthma, pulmonary congestions, bronchopneumonia. A decoction is used for cough, flu, fever. Dry leaves as a tea are used against hypertension. A tea or syrup of a mixture of *Cecropia* leaves, leaves of *Cordia curassavica* and tomato leaves is taken for colds and cough.

Leaves in general are recommended against cough, shortness of breath, dyspnoea, asthma, pneumonia, and as a heart tonic.

Excellent results with a rapid improvement were moreover observed in the treatment of Parkinson's disease (see below).

A decoction of leaves is a biliar remedy and helps against chorea and St. Vitus dance.

Leaves toasted, crushed or pulverized are cicatrizing.

Any part of the plant. In decoction is good for the spleen and cures oedema.

Method of use

Fresh leaves as a cataplasm.

Dry leaves crushed or pulverized or in the form of a tea. Decoctions and infusions.

A decoction of the leaves or roots is used against nervous cough, asthma, oedema, diseases of the spleen.

To cure asthma, a single leaf is put into a bottle of water; the drink is taken during the day.

The best results were obtained with Parkinson's disease; to cure Parkinson's disease and chorea, 1–2 leaves per liter of water are boiled for 2 minutes; the drink is taken cup after cup during the day. The treatment has to be followed for 3 months. Asthma is more rapidly cured, but the treatment has to be prolonged for several months (MANFRED 1982).

Three cups a day of a decoction or syrup of the leaves are taken against cough, shortness of breath, dyspnoea, pneumonia and asthma. MANFRED (1977) recommends to eat 2 dry figs softened in white wine before drinking the first cup in the morning. the above mentioned diseases are very rapidly cured (LOPEZ PALACIOS (1987).

30 g bark in a liter of water as a decoction taken 5 times a day are used to combat sunstroke, and as an emmenagogue and antidiarrhoeic (ARIAS 1982).

A decoction of roots and leaves sweetened with honey is effective against chorea and St. Vitus dance.

Any part of the plant can be used against oedema and pain in the spleen (LOPEZ PALACIOS 1987).

Against whooping cough, a 5 % infusion of *Cecropia* is recommended. Against hypertrophy of the heart, 10 drops of *Cecropia* in water are taken. A decoction of 30 g leaves in 300 gr water, 3 times a day helps against nervous asthma.

Syrup of *Cecropia* is recommended against pectoral and asthmatic diseases; the syrup is prepared with 2 leaves of 'yagrumo' in 250 g water, after boiling (decoction), the liquid is sweetened according to preference. This dose is administered within 24 hours.

The latex used against warts, herpes and corns, is pharmaceutically prepared as a liquid.

HERNANDEZ (1992) mentions the following homoeopathic application of *C. peltata*: the mother tincture is prepared with one part of fresh leaves and 3 parts of alcohol 40 %; the dose of 5 drops every 2 hours is prescribed against asthma, bronchitis, oedema, whooping cough, chorea, nervous afflictions, and to facilitate menstruation or to regulate the function of the heart.

In 1889, the illustrious botanist of Capanema prepared a very effective syrup against asthma of bronchial or cardiac origin which he called cecropina: 100 g leaves of *Cecropia* in 750 g water are boiled until reduced by half; then 1/4 kg sugar is added, and the liquid is boiled again and filtered. A soupspoonfull is taken every 2 hours.

An infusion of dry leaves acts as an antiinflammatory, and is recommended for cough and diabetes.

Leaves are crushed with little water and a small cup is taken 3 times a day: to calm the nerves, against stomachache, arthritis, rheumatism, as an expectorant and for the kidneys (as a diuretic).

A bath with a leaf decoction is made and additonally 3–4 cups of the decoction are taken a day internally to cure arthritis.

Against cough, 3 cups of a tea made of the leaves are recommended daily.

For the kidneys, crushed bark (cáscara) is used.

In veterinary medicine, salted leaves of yagrumo are given to the cows for rapid expulsion of the placenta.

Healing properties

Tonic, cardiotonic, stimulating the cardiovascular system, diuretic, antidiarrhoeic, astringent, mucolytic, choleretic, abortive, spasmolytic, wound healing, febrifuge, against hypertension.

Leaf. Expectorant, antispasmodic, diuretic, cardiotonic, antiasthmatic, antiinflammatory. Cures cough, stomach ache, diabetes, arthritis, rheumatism and calms the nerves.

Leaf, bud and bark. Astringent. As a heart tonic it is preferred to *Digitalis*, although its effect is less strong, because it has no cummulative effects.

Extracts of bark and leaves show an inhibitory effect on *Escherichia coli* and *Staphylococcus aureus*; furthermore, effects on blood sugar, and a hypotensive activity are observed.

Latex. The latex is caustic (against warts and herpes) and is applied to bleeding gums.

Chemical contents

Bark. Contains cecropine and tannic acid.

Leaf. Contains the alkaloids ambaine, ambainine, cecropine, cecropinine, arachidic acid, and other rare acids such as behenic, lignoceric, margaric, heneicosanoic, tricosanoic, pentacosanoic, nonadecanoic, cerotinic, stearic, fumaric, myristic, caffeic and gallic acid; β-sitosterol, sigma-4-en-3-one, alpha- and beta-amirine.

Leaf and bark contain alkaloids, cardiotonic glycosides, flavonoids, tannin, triterpenes, saponinic glycosides, sterols etc.

Use in religious rites

For ritual uses, ashes of leaves are mixed with powdered coca. The leaves are gathered and burned the day of use. The ashes supply the alkaline mixture necessary to release the alkaloids from the coca in the normal acidity of the mouth. According to SCHULTES & RAFFAUF (1990), this preparation is a daily custom of the Indians.

Varieties and related species

As indicated above, all Venezuelan (South American) species of *Cecropia* are more or less used in the same way.

C. palmatisecta, *Cecropia peltata* L., *C. palmata* WILLD. and *C. pachystachya* MART. are very well known.

Veslasquez mentions the following species for Venezuela: *C. sciadophylla*, *C. orinocensis*, *C. angulata*, *C. palmatisecta*, *C. kavanayensis*, *C. auyantepuiana*, *C. santanderensis*, *C. telenitida*, *C. libradensis*, *C. metensis*, *C. peltata*, *C. peltata* var. *candida*.

Cultivation

The tree grows quickly and the species is therefore frequently found. It is the first pioneer in clearings.

C. palmatisecta grows between 800 and 1000 (even up to 2000) m a.s.l. and prefers humid sites, e.g. cloud forests.

The species of *Cecropia* grow along river banks and highways, around gorges, bordering precipices, and appear as pioneers of burned land which is much exposed to insolation, in disturbed areas, in abandoned small settlements, and were therefore considered arboreous weeds. The invasory plants thus grow abundantly in deforested regions.

Species of *Cecropia* are found between 50 m and 2700 m a.s.l. They often form dense secondary colonies.

It thus seems not to be difficult to cultivate the plants at a higher range.

Observations

A precise taxonomical description of the 12 species studied in Venezuela is found in Velasquez 1971.

Presence of cystoliths, laticifers around the stronger veins, a dense indumentum on the lower leaf side, transcurrent vascular bundles with sclerenchymatous pillars and a closed system of vascular bundles in the form of a pattern of squares are very important characteristics of the species palmatisecta.

The latex seems to be the most important chemical content of the plant; it is present in the leaves, the bark and probably also in the roots.

An interesting peculiarity of the yagrumos is the division of the hollow stems by transverse septs or partition walls forming a great number of compartments which are used as a shelter by ants; these find at the same time their food in large glands or pulvini at the base of the petioles. In compensation, the ants defend the tree against any enemy so that the sloth seems to be a creature which is uniquely indifferent to these attacks.

Chlorophora

C. tinctoria (L.) GAUDICHAUD is used for ondontological purposes. The wood is useful. The bark is tonic, astringent and purgative. ROTH studied the fruit structure and dispersal 1987.

The northern form of *Chlorophora tinctoria*, growing in Venezuela, southern Mexico, the West Indies, Central America and northern South America, yields the well-known dyewood, 'fustic'. The coloring principle, maclurin, imparts a dull yellowish brown or khaki. The southern form (Argentina,

Paraguay and southern Brazil) is a timber tree. It is suited for fence posts, props and fuel.

Ficus

The genus is represented by small to very large trees and shrubs, many of which are epiphytic, at least in their youth; distributed over the warmer regions of the world, they are most abundant in the East Indies and in Africa.

The best known plants are the fig, *Ficus carica* L., and the rubber plant, *Ficus elastica* ROXB. *Ficus bengalensis* L. with its numerous prop roots, may spread over an acre of ground. RECORD & HESS 1943 estimate that about 50 species occur in South America, but they emphasize the very variable characteristics of some species. Some peculiarities which have been used to separate species are now found to be unreliable. The authors thus think that the number of species probably has to be reduced.

The Aztecs of Mexico already used the bark of fig trees for the preparation of paper. Large trunks are possibly used for making canoes. Where timber is scarce, *Ficus* wood is sometimes used for making boxes and for light construction.

All species are laticiferous. Alkaloids, sterols, triterpenes, coumarins and flavonoids have been found in the genus.

Ficus sp., a tree with latex and simple elliptic leaves, has been found in the transitional cloud forest of Venezuela. Its leaf anatomy is described by ROTH (1990). The leaf has a large-celled upper epidermis which is water storing and possesses lithocysts.

At least 82 useful species of *Ficus* are known in the whole world.

The wood is useful. Bark fiber is used for cloth, cordage and ropes. Bark is pounded into cloth; pounded bark is used for painted mats in folkloric handicraft (artesania) see frontispiece. Bark is also used for paper pulp. Paper was made of *Ficus* bark by Indians in precolumbian times. Likewise, aerial roots are applied for ropes and suspension bridges. Bark is also used for tanning.

Bark is diaphoretic and used against fever, and is even taken in cases of malaria in the form of a juice.

Leaves are used as sand paper (rough surface) and as food for a specific silk worm. Leaves and shoots may also be eaten as a vegetable.

Leaves are used for diseases of the liver and as an antidote after having eaten tainted fish or meat; leaves are also applied for ant stings. The juice of the leaves is used in cases of malaria.

A decoction of the root fiber helps against gonorrhoea.

Fruits are edible. Figs are eaten raw, dried, in preserves and are used for pastries, cookies, canned or prepared as fig wine etc.

A scarlet dye is extracted from the fruits.

Fruits are also used as an aphrodisiac and to improve the memory.

Wax is used for batik work and in the candle industry.

Latex is a source of rubber and Gutta-Percha, and is also used as an adulterant of balatá; it is likewise a base of chewing gum. It is furthermore used as a bird lime and for oxidation of copper. Latex has many medicinal applications.

Natives use latex to make curare and boneset, for cracks in the feet, for herpes sores, to treat bleeding ulcers of the skin, as a vermifuge, for fungal infections and itching and as a protection for wounds – internally for a sore stomach, as a sedative for indigestion, as a purgative and anthelmintic or for diarrhoea and even for malaria. Latex is also used as a glue and disinfectant for wounds, as an astringent and – externally – for rheumatism, lumbago and ant stings. The latex of branches is applied for stomach pain. It is even used as a filling for molar cavities. Latex with powdered bark is applied for wounds and bruises, the gum is used for hernias and broken bones. ROTH studied the bark structure 1981 leaf structure 1984, fruit structure and dispersal 1987, leaf venation 1996.

Helicostylis

H. comprises about 8 species of unarmed, laticiferous medium-sized to large trees, the best-known of which is *H. tomentosa* (Fig. 183.2). However, the wood could be used for the same purposes as that of *Carya* (RECORD & HESS 1943).

SCHULTES & RAFFAUF (1990) mention that the latex of *H. tomentosa* and H. scabra is toxic. It is however used by certain tribes as a vermifuge for intestinal paraistes and as an antifungal agent when repeatedly put on the affected parts of the skin. The bark of *H. tomentosa* is the source of an hallucinogen. ROTH studied the bark structure 1981, leaf structure 1984, fruit structure and dispersal 1987, leaf venation 1996.

Helicostylis tovarensis (KLOTZSCH. & KARST.) C. C. BERG has been studied by ROTH (1992) concerning its leaf structure (Fig. 183.2). The species mainly occurs in the cloud forest. For further details see ROTH (1992).

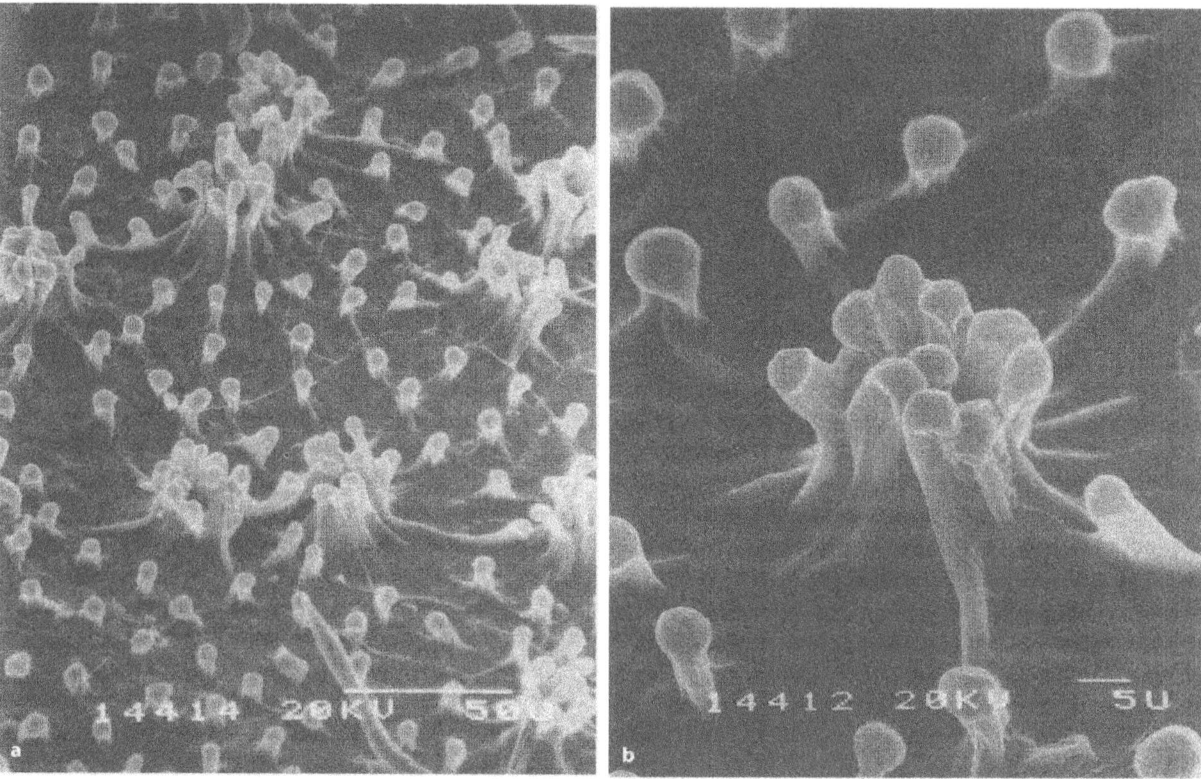

Fig. 183.2. *Heliocostylis tomentosa*, Moraceae. **a, b** Lower leaf surface with papillas. Papillas are more elongated where they surround a stoma (ROTH 1984).

Poulsenia armata (MIG.) STANDLEY

This plant is the only species of the genus. It is a medium-sized to large tree, 12–25 m high and with a DBH up to 60 cm. It is found from Mexico through Central America up to the north of South America. In Venezuela it grows in the transitional cloud forest of the Cordillera de la Costa at altitutes between 1000 and 1300 m. The simple leaves are almost rotund, have a short tip and measure 10 × 7.5 cm up to 30 × 24 cm. The flowers are greenish and measure 1.5–2 mm in diameter. The inflorescence is of a globular shape. The species is readily distinguished from all other species of this family by the numerous prickles on the stipules and twigs. The leaf anatomy has been described by ROTH (1992). Lithocysts are characteristic of both epidermal layers.

Ethnobotanical and general use

The ripe fruit heads are edible and sometimes sold at the markets.

The timber is preferred for construction because of its reputed resistance to fire.

The inner bark of mature trees is very thick and composed of many layers of strongly interlaced fibers. It has long been used by the aborigines for making hammocks, blankets, mats, and clothing. Bark cloth is still made by the Sumu Indians of Honduras and Nicaragua.

The bark is soaked in water for a few days after which the sticky gum or milk adhering to it is scraped off. The bark is then dried in the sun. As it becomes hard and shrinks, it has to be submerged in water for a short time before the pounding begins. This is performed on a smaller log with the aid of a wooden mallet made from the stems of 2 different species of palms. The bark extends gradually upon being pounded and becomes soft and flexible. After being washed and dried it is ready for use and has a brownish colour.

A similar, almost white, cloth of superior quality is obtained by the same processes from the inner bark of a species of *Ficus* and likewise from the rubber tree (*Castilloa*).

Pourouma

Fruits are edible. Roots are taken by women to cause permanent sterility (*P. cecropiaefolia*).

Bark is used for aching joints and rheuma, the ashes of the bark for sores and ulcers.

Musaceae

The representatives are large perennial herbs (giants herbs) with a rhizome and often with pseudostems composed of the leaf sheath. The large inflorescences are furnished with large, often very showy, coloured bracts. Ornithogamy prevails. The ovary is 3-locular, the fruit a woody capsule or becomes fleshy and berry-like.

The most important species are *Musa* (banana, plantain, manila hemp), *Ravenala, Strelitzia.*

Alcohols, aldehydes, alkaloids, phenolic acids and triterpenoids were reported.

ARISTEGUIETA (1961) erstimates that about 30 species of *Heliconia* exist in Venezuela.

Heliconia hirsuta LINNAEUS FIL. is a giant herb in the undergrowth of the transitional and real cloud forest of Venezuela.

The inflorescence is showy with 7-11 red or red-orange coloured bracts; it reaches 20-35 cm in length.

Most of the *Helicinias* are ornamental or could be used as ornamentals.

The leaf anatomy of *H. hirsuta* has been studied by ROTH (1990). The upper epidermis cells are lenticular and possibly have the function of a lens for light absorption. The upper hypodermis consists of transparent cells, while the lower one is small-celled. The upper hypodermis serves as a water reservoir. A similar leaf anatomy is found in several species of Heliconia. LINDORF (1980) studied the leaf anatomy of *H. hirsuta* and *H. revoluta.*

H. hirsuta is widely distributed in Venezuelan humid forests and together with *H. caribaea* is the commonest species of the genus in Venezuela.

It is a resistant plant which is easily reproduced by rhizomes. It is a species which is very appropriate as an ornamental in gardens.

The rhizome is edible and a fermented drink can be prepared from it.

Musa acuminata and *M. balbisiana* – *ciltivars* (cambur and plátano)

According to SIMMONDS (1962), most of the cultivated types of banana have been derived from the above mentioned species. The Linnean names, *Musa paradisiaca* and *Musa sapientium* do not correspond to the concept of a species in any biologically reasonable meaning of the word (SIMMONDS 1962). Most of the cultivated bananas are triploid. The fruits are commonly regarded as berries, although a minority of wild species has dehiscent fruits with a pericarp splitting lengthwise in the form of segments which curl backwards at maturity to expose a mass of seeds.

Although the banana is of Asian (Indomalaysian) origin, it is now cultivted all over the tropical world being one of the most important crops of the tropics. In approximately 1516, it was introduced from the Canarian islands to Santo Domingo/Haiti.

Anatomical description

Fruit development and anatomy have been described by ROTH (1977), based on the thesis results of her student H. LOPEZ-NARANJO.

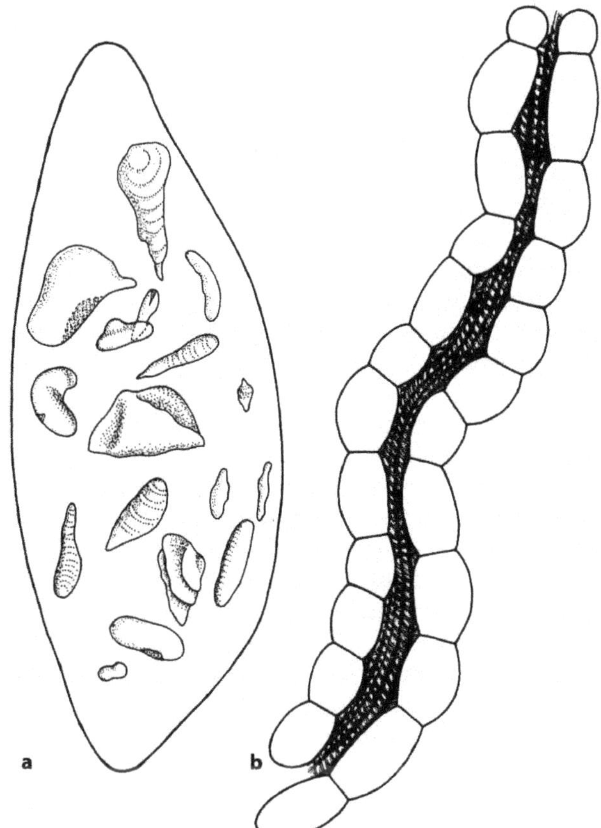

Fig. 184. Banana. Musaceae. **a** Parenchyma cell of overripe fruit with erodet starch grains. **b** Vascular bundle of the fruit flesh with. The tracheids are surrounded by laticifers in the form of a 'Knackwurst' chain (ROTH 1977).

The 3-carpellary ovule is 3-locular, resulting from the fusion and infolding of the carpels. The 3 edges on the outside of the shell correspond to the midrib regions of the carpels. The placentae arise from the fused carpel margins, but only in seed-bearing fruits. The cells from which the pulp develops are situated below the inner epidermis of the pericarp wall and septs. In the parthenocarpic bananas, these cells begin to proliferate 2 weeks after the emergence of the inflorescence from the crown of leaves and their meristematic activity lasts for about 2 weeks; thereafter, the growth of the pulp proceeds by increase of cell size, mainly cell elongation in a radial direction. A subepidermal meristem lying beneath the inner fruit eoidermis is thus responsible for pulp initiation by periclinal divisions, followed by radial enlargement of the derivatives so that long chains originate which, for reasons of limited space, are bent. Filamentous elongation of pulp cells is also observed in cocoa (ROTH & LINDORF 1971, ROTH 1977).

The pulp is composed of thin-walled parenchyma cells which are loosely arranged. All cells of immature fruits are densely filled with large starch grains which disappear during the ripening processes. By enzymatic processes, starch is converted into sugar so that the starch grains gradually disappear by corrosion (Fig. 184). Characteristic figures of erosion of starch grains can be observed in ripening bananas. Starch deposition in the pulp commences some 4 weeks after emergency of the inflorescence. Starch grains develop first in those cells of the pulp periphery which are in the vicinity of vascular bundles and from there starch accumulation proceeds centripetally. In the outer part of the pulp, the cells are more or less isodiametric, but towards the inside they elongate gradually in a radial direction.

The vascular bundles which pass through the pulp in a transverse direction show a very characteristic pattern, particularly in the ripe, partly macerated flesh. They are accompanied by large articulated laticifers composed of short cells which are constricted on both ends, thus giving the impression of a sausage chain (Fig. 184). Their transverse walls are perforated. The vascular bundles together with their laticifers form the most characteristic feature of banana anatomy. They can pass the digestive tract of humans undigested and may then be identified in the excrements by their particular shape, facilitating the medical diagnosis of certain types of diarrhoea.

In the unripe banana, the latex is under pressure and is released when the fruit is cut. Besides tannin it contains globular bodies which stain orange with Sudan III. but are said to correspond to chicle rubber. In the ripe fruit, the latex having lost water is solid and lies in the center of the cell. The pulp is finally macerated. The pectin of the cell walls undergoes a process of hydrolysis during fruit ripening and appears in a slimy condition, while the cells gradually separate from one another, particularly along their longitudinal walls. In this way, the pulp becomes soft.

In the green banana, the cells of the shell or skin and even more those of the pulp are densely filled with starch grains of irregular shape and size (Fig. 184), whereas in the completely ripe banana the cells are to a large extent empty or contain only a few 'eroded' grains, particularly in the center of the fruit. These processes of erosion which lead to the very characteristic grain shape occur by enzymatic action that can be observed in the very late stages of maturation. The length of the grains varies between 5 and 100 μm, the larger grains are found in the pulp. The cone-shaped, bacilliform, oblated and club-shaped forms described by SEIDEMANN (1966) seem partly to correspond to products of erosion, as in young fruits the oval, elliptic forms generally prevail.

The shell or skin consists of the outer fruit epidermis together with quite a few underlying cell layers. The epidermis is small-celled and covered by a cuticle. the stomata are of the developmental type of *Rhoeo* with 4 subsidiary cells surrounding them (ROTH & CLAUSNITZER 1969). The main bulk of the skin is parenchyma. The subepidermal layers close to the epidermis are small-celled, however, further inwards cell size increases and the tissue becomes looser, as the intercellular spaces enlarge. The vascular bundles are irregularly dispersed over the section, as usual in monocotyledons.

The photosynthetic tissue together with the vascular bundles produces a very characteristic pattern in the skin; the chlorophyll-bearing photosynthetic cells form a network of large meshes around the bundles; these consist of a few xylem and phloem elements crowned by an outer collenchymatous cap and surrounded by a variable amount of more or less colourless parenchyma in which large laticifers are interspersed, often in the form of a ring around the larger bundles, while laticifers arranged in pairs or even singly accompany the smaller bundles (Fig. 184). The laticifers are thus always associated with the vascular bundles. The vascular bundles scattered within the skin run from the fruit pedicel longitudinally through the fruit periphery and pass into the perianth members and the stamens.

Ethnobotanical and general use

Nutritional use

The commonly known edible banana is eaten raw, prepared as a desert in puddings and creams, in milk shakes, ice cream, preserves, cakes and drinks. It is also dried or processed as flour. Banana bread is made of a mixture of ripe bananas and wheat flour. Banana is also a source of coffee surrogate, alcohol, and oil.

The plantain (plátano), on the other hand, is a cooking banana, usually not sold on European markets; it has a very hard consistency and must either be cooked or boiled in water or fried. It makes a good substitute for potatoes or other starchy vegetables.

Economical utilization

Wax is obtained from the leaves of certain varieties, yielding an ounce of wax to every 3–4 leaves (Dacca of India). Leaves are also used to wrap up certain dishes in Venezuela. Fibers of the leaves could be used commercially.

Medical use

The green plantain is astringent due to the high tanning content and is applied against diarrhoea.

Cultivation and varieties

The banana is a large herbaceous perennial with a rhizome. The edible banana is a cultivar and is often triploid. Edible bananas are usually seedless and therefore cannot be proparaget by seeds. They are vegetatively parthenocarpic and the fruits develop without the need of any stimulus by pollination. But the plants can easily be vegetatively propagated. Aerial shoots of the rhizomes are used for propagation. There exists a large number of varieties in all parts of the world. Varieties differ mainly in fruit size, colour of skin and flesh, and in aroma. Bananas are sweetest when overripe having many brown dots on the skin (tannin). At this stage, most of the starch is converted into sugar. Plantains, on the other hand, are more mealy due to their high starch content.

Myricaceae

Trees or shrubs with simple alternating leaves. Flowers small and inconspicuous, greenish or yellowish, unisexual, and united in small aments. The achlamydeous male and female flowers are either on the same plant (monoecious) or on different plants (dioecious). Male and female flowers occur in the axil of a bract. Male flowers usually have 4–8 stamens, female flowers a unilocular ovary composed of 2 carpels. The fruit is a drupe with a very hard endocarp and a wax-secreting exocarp. The aromatic leaves are supplied with glandular hairs which contain a fragrant resin.

Myrica gale is rich in volatile oils. The family is characterized by its abundance of polyphenols and triterpenes.

Leaves and fruits are used for liqueur manufacture. The bark is used for tanning and the flower buds for dyeing.

Myrica caracasana H.B.K. (encinillo, palomero, torcaz)

SCHNEE (1960) describes *M. arguta* H.B.K. and *M. pubescens* H. & B. Both have the same vernacular names as *M. caracasana*. Both are found in the Cordillera de la Costa, roughly between 2000 and 3000 m. STEYERMARK & HUBER (1978) found *M. caracasana* in the subparamo between 2300 and 2450 m. The species is a small tree or shrub, 1.5–3 m high. The simple leaves are glossy on the upper side, but pubescent underneath. They are of lanceolate shape, 4–5 cm long and have slightly serrate margins. The petiole is 1 cm long. When triturated, the leaves emit a very pleasant aromatic odour.

Anatomical description

Leaf (Fig. 185). The cells of the upper epidermis have extremely thick cutinized outer walls and thick radial walls.

The leaf has a more or less isolateral structure, as the mesophyll is mostly compound of elongated palisade-like cells. On the lower leaf side, the palisade cells become more loosely arranged. The veins are very frequent and the larger ones are transcurrent to both epidermal layers by a colourless parenchyma with thickerwalled pitted cells. This tissue has a water-storing and supporting function.

The stomata are confined to the lower leaf side. The lower epidermis is thick-walled and cutinized on its outside.

The glands found in the lower epidermis are peltate and scale-like with a pluricellular stalk and a pluricellular head; they are sunk in depressions below the surface and resemble the absorbing scales

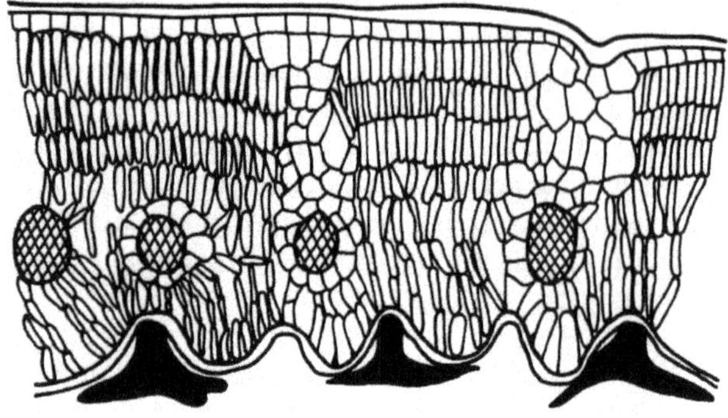

Fig. 185. *Myrica caracasana*, Myricaceae. T.s. of leaf. Note the mesophyll in the form of palisade cells, the vascular bundles which are transcurrent to the upper and lower epidermis by colourless parenchyma cells and the scale-like glands in the lower epidermis (ROTH 1995).

of *Tillandsia*. Their heads are irregularly divided into cells of different shape and the outermost cells may possibly elongate so that star-like structures arise (Fig. 185). Others have smaller heads without star-like extensions. And a third type has just a bicellular head. The glands are certainly responsible for the aromatic fragrance of the leaves.

The leaf is of the suntype and has several important xeromorphic characteristics. It has a leathery consistence.

Ethnobotanical and general use

The species of *Myrica* all have a very similar use. UPHOF (1968) mentions 9 species. Most of them are a source of wax derived from the surface of the fruit, possibly used for manufacture of candles. *M. pubescens* WILLD. of the cool Andean belt also growing in Venezuela produces a wax of excellent quality.

The wax is prepared from the fruits by placing the branches in boiling water so that the wax is skimmed off.

Bark of the roots is astringent and tonic.

Fruits or seeds are eaten.

Medicinally the plants are used against dysentery, as antiparasiticum (against moths), for skin diseases, for jaundice and diarrhoea. A decoction of rootbark of *M. mexicana* is supposed to be acrid, astringent, and in large doses emetic.

Myrica gale is sometimes put in beer to increase foaming (saponin). The family is rich in polyphenols and triterpenes.

Myristicaceae

Stamens are almost always united in a column; anthers are extrors. Only one carpel with a single seed is developed. Fruit is fleshy or woody, dehiscing ventri- and dorsicidally. Seeds with fleshy aril. Flowers are small.

The representatives are trees, more seldom shrubs.

Ethereal oils (eugenol), phenols (lignans and tannins) are common, sterols and triterpenes are also found.

The most useful plant of the Myristicaceae is *Myristica fragrans* with a ruminate endosperm. Terpene hydrocarbons are the main compounds of the ethereal oil, while phenylpropan derivatives such as myristicin, elemicin and safrol produce the toxic and narcotic effects.

Besides *Myristica*, species of *Virola* and *Iryanthera* are of importance.

Dialyanthera

Shrubs or tall trees native to Costa Rica, the Western Amazon and Venezuela. The most widely distributed species is *D. otoba*. The fat obtained from the seeds by boiling and pressing is employed as a remedy for parasites in animals. The timber is very easy to work and utilized locally for boxes and interior construction.

Dialyanthera parvifolia MARKGRAF. The bark is crushed and rubbed on the skin for treating infections caused by mites and fungi. The Indians believe that the tree has wound healing properties.

The leaf anatomy has been studied by ROTH (1992).

Dialyanthera otoba (H. & B.) WARBURG is the best known species. The tree has a large very oily seed with an aril. PITTIER (1926, 1970) gives the following information, citing GUMILLA ('Orinoco ilustrado...', Madrid 1741): 'the oil obtained from the seeds is as soft as butter and is rolled into globules of one pound in weight and sold for 8 reales of silver each. It is a much wanted remedy against itching and an excellent preventive for sand-flea and mosquitos. It is furthermore a stomachic: a globule of the size of a hazelnut taken with 2 mouthful of water releases from stomach ache. Two or three globules of the same size taken with some water have a purgative effect'.

GUMILLA thinks that many other virtues of this species will be discovered with time; perhaps now is the time for these discoveries.

Iryanthera

At least 10 useful species are known. The resin-like bark exudate is used for fungal infections, to kill mites; internally taken it serves against diarrhoea and food poisoning due to consumption of bad fish or meat. Resin, bark and leaves are used for the manufacture of perfume. Bark is also used for pottery.

ROTH studied the bark structure 1981, leaf venation 1996.

Myristica fragrans

Although this tree is native of the Moluccas, it is cultivated all over the tropics. It is the source of nutmeg and mace, used as condiments.

The fruit anatomy has been studied by ROTH & LINDORF 1974. The follicular fruit is unicarpellary and dehisces into 2 valves along the ventral and dorsal sutures. The mature fruit consists mainly of parenchyma. A ring of stone cells of very irregular shape develops, however, in the periphery. The outer epidermis is composed of small and thin-walled cells. Stellate hairs with two elongated apical cells crossing one another characterize the epidermis. The hairs have lignified walls. Beneath the epidermis a hypodermis, possibly with crystals of calcium oxalate, is present. Towards the inside follow several layers of stone cells. Small groups of stone cells may also occur in deeper lying tissues.

Characteristic of the species are idioblasts of spherical or polygonal shape with slightly suberized walls which contain a volatile oil of yellow colour, adding fragrance to the fruit.

Besides the aromatic substances, the presence of tannin can be proven with $FeCl_3$ in the cell walls of young fruits. Later on, the tannin concentrates in special elongate and somewhat ramified cells; these cells have a light-brown content and additionally rodlike starch grains. The regular parenchyma cells, however, do not contain starch grains. In the nearly ripe fruit, small globular starch grains can only be observed in the 2 epidermal layers of the ventral suture.

Small intercellular spaces may be observed in the inner epidermis. For more details see ROTH (1977) or ROTH & LINDORF (1974).

The fruits are globular or pyriform, about 3–6 cm long, of a yellowish colour, slightly resembling apricots. They hang down from the tree and, when opening, contain a large single seed of brown colour which is surrounded by a red laciniate aril, the socalled mace. The seed reaches 1–4 (5) cm in diameter and is very showy.

Nutmeg and mace are used as condiments.

The fruit flesh has a slightly sour taste. Fresh husks of ripe fruits are components of jellies, mixed pickles, preserves and sweets.

The mace contains 7.5–25 % essential oil (oleum myristicae) which includes the toxic myristicin besides eugenol and a variety of terpenes. It is used as a condiment for savoury dishes, meat, pickles, sausages, sauces, ketchup, pastries, pudding and beverages.

The nut contains 20–30 % fatty oils (myristic acid and ethereal oil). It is used in a similar way to the mace, namely as a condiment. However, larger amounts are toxic; consumption of only half a nut can lead to death.

Nutmeg butter extracted from the seeds that are unfit for the spice trade is applied in ointments and candles, liniments, plasters and soaps. This fatty oil is obtained by pressing.

The ethereal oil extracted by distillation from seed or mace is used in condiments, liquors, chocolate and the foodstuff industry, as well as in perfumery, cosmetic and soap manufacture.

Oleum myristicae is also used in medicine as an aromatic and carminative, for dentifrices, perfumes and in the tobacco industry.

In medicine, the oil is furthermore used to cure flu and diarrhoea. Besides, the nutmeg contains antoxidants.

An excessive consumption of the fruit may provoke optic and acoustic hallucinations.

Virola

At least 23 useful species are known.

Useful wood. Bark is narcotic, used for fungal infections, malaria.

Ashes of bark are put on festering wounds.

Leaves and bark are applied to a skin disease causing spots of discoloration.

Leaves and twigs are used for arthritis and skin infections.

Leaves serve for colics and dyspepsia. Leaves are rubbed on the gum of teething children. Leaves are also used for mites and skin infections. Pulverized leaves are applied as an insect repellent. Leaves are also the source of hallucinogens.

The fat extracted from the seeds contains myristic acid. It is used for rheumatism as well as for the manufacture of soap and candles. The bark exudate, a resin, stops bleeding from cuts, it is furthermore wound healing and used for haemorrhoids. It helps also against fungal infections.

V. sebifera AUBL. Seeds are the source of *Virola* fat with a scent of nutmeg which is used in candle and soap manufacture, transmitting a pleasant scent to candles and soap.

ROTH studied the bark structure 1981, leaf structure 1984, fruit structure and dispersal 1987.

V. surinamensis WARB. Tea of leaves for colics and dyspepsia. Bark exudate for erysipelas. Infusion of bark for cleansing and healing of wounds, also applied to haemorrhoids. Seeds are the source of a butter rich in myristic acid.

The species contains large oil cells in the bark and in the leaf. ROTH studied the bark structure 1981, leaf structure 1984, fruit structure and dispersal 1987.

Myrsinaceae

This family comprises mainly shrubs, small trees and a few herbs distributed throughout tropical and subtropical regions. Some species are epiphytic or climbing.

The crown is wheel or plate-shaped with valvate or contorte aestivation. Stamens 5; ovary with central placenta. Flowers are snall, united in inflorescences. The fruit is a berry or drupe. The leaves are frequently evergreen.

Schizogeneous resin canals as well as glandular hairs are frequent in leaf, bark, pith and flower.

The timber is of good quality and has an attractive oak-like figure.

Cybianthus fendleri MEZ. is a shrub or small tree in the transitional and real cloud forest of Venezuela. The species occurs at altitudes between 1390 and 1500 m.

The leaf anatomy has been studied by ROTH (1990). Glandular hairs with an 8-celled head, probably functioning as hydathodes, occur in the lower epidermis; the regular epidermis cells have a ribbed surface structure.

Myrtaceae

The representatives of the Myrtaceae are trees, shrubs or subshrubs, occasionally climbing. The leaves are mostly simple, evergreen and leathery with reduced stipules. The stamens are numerous and frequently coloured. Central placentation and an ovary with 2–3 locules are common. The flowers are often arranged in racemes. The fruits are loculicidal capsules, berries or drupes with one to few seeds which are occasionally winged.

Secretory cavities localized in different tissue types are characteristic of all species and add the aromatic smell to the plant. Oil glands in flowers and leaves supply ethereal oils (*Eucalyptus*), fruits of some species are edible (*Eugenia, Jambolana, Psidium*).

The family is not only rich in ethereal oils, it occasionally also contains polyphenols, alkaloids, tannins, sterols, triterpenes, cyanogenic compounds and triketones.

Genera as important as *Eucalyptus, Callistemon, Eugenia, Calycorectes, Myrciaria, Myrcia, Psidium* and *Syzygium* (clove) belong to the family.

The family is almost exclusively tropical and principally occurs in South America and Australia.

The various species of a genus are not seldom difficult to differentiate and to identify (gregarious species).

Eugenia

The genus Eugenia comprises at least 53 useful species.

It supplies useful wood, edible fruits, commercial cloves for flavouring food and candles; oleum caryophylli, a stimulant, carminative, and antiseptic, with 84–90 % eugenol, also a source of vanillin. Bark supplies fibers, tannins and terpenes, and is a source of black or brown dye. Leaves may be used as an insect repellent.

Besides edible flowers there are edible fruits made into beverages, wine, vinegar, preserves, jellies, jam, compote, sherbet, pies and syrup.

Medically the leaves are used as a tonic, expectorant, diuretic, a remedy for catarrhal disorders of the respiratory organs, for diarrhoea, for earache, pain in the chest and cough.

Bark is used for chronic chest ailments and for diarrhoea. Seeds are astringent, diuretic and are used for diabetes.

This pantropical genus with numerous useful species also has some neotropical representatives. Their often edible fruits have a high vitamin A and C content. Unfortunately, they are still underexploited in the Americas. Ethereal oils are characteristic of the genus.

Eugenia caryophyllata THUNB., synonym: *Syzygium aromaticum* (clove)

The clove although not indigenous, is also cultivated in the Neotropics. It is therefore briefly mentioned here.

Anatomical description

Flower bud (Fig. 186, 187). As seen in a transverse section cut below the locules, the West Indian clove may roughly be subdivided into 5 sections. The outer epidermis forms the exocarp together with some layers of very small subepidermal cells which represent a certain type of hypodermis. The outline of the receptacle is wavy and stomata are situated on the crests of the waves. The outer tangential walls of the epidermis cells are extremely thickened and cutinized so that the cell lumen is compressed and the tangential walls exceed the cell lumen about 6 times in width. The peripheral part of the mesocarp consists of parenchyma in which oil cavities mainly of an elliptic shape are densely interspersed. They contain volatile oil and can reach a diameter of up to 200 μm. In this zone, the radially elongated parenchymatous cells are arranged in the form of conspicuous radial rows. The rows extend from the outer epidermis to the collenchymatous region, which comes after the parenchyma towards the inside. Oil cavities only occur in the parenchyma, but not in the collenchymatous

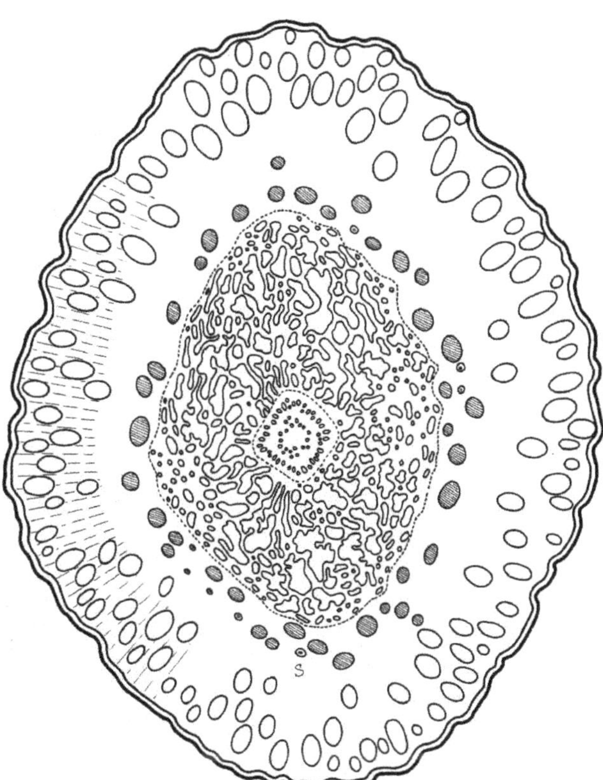

Fig. 186. *Eugenia caryophyllata* (clove), Myrtaceae. T.s. of the receptacle with peripheral oil canals, a ring of vascular bundles (shaded) and the spongy parenchyma, the central column with 2 rings of vascular bundles surrounded by abundant spongy parenchyma (ROTH 1977).

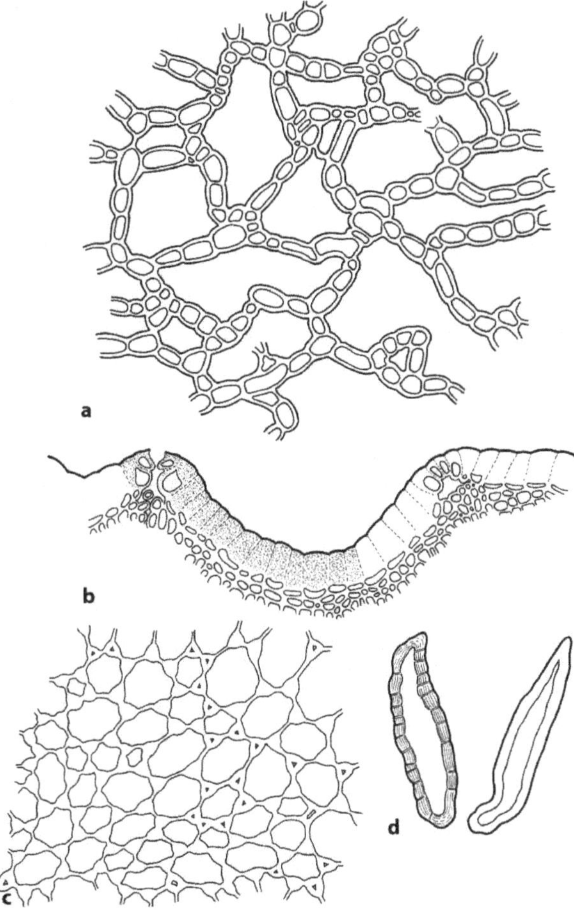

Fig. 187. *Eugenia caryophyllata* clove), Myrtaceae. T.s. of receptacle. **a** Inner spongy parenchyma with thick cell walls. **b** Outer epidermis with very thick outer cell walls and stoma. **c** Angular collenchyma of the region between the inner border line of the oil ducts and the outer ring of vascular bundles. **d** Large sclereids found in the neighbourhood of the vascular bundles (ROTH 1977).

zone. This is typically angular with thickened angles in the center of which small intercellular spaces occasionally form. The collenchyma is bordered by a ring of large vascular bundles of a bicollateral structure. Interspersed between the bundles are large elongate sclereids of irregular shape. Towards the inside there is a large region of lose spongy aerenchyma composed of relatively thick-walled cells which form an irregular network around large intercellular spaces. The central column is composed of small cells rich in protoplasm which occasionally contain druses of calcium oxalate. Two concentric rings of vascular bundles can be distinguished here.

In contrast to the flower bud, in the fruit or in the so-called mother clove, all elements are more strongly developed and the bundle elements are larger; in the outer parenchyma, cells have thick wavy walls (ROTH 1977).

Ethnobotanical and general use

Nutritional use

The interior ovary of the unopened flower bud with its 4 calyx lobes is about 1 cm long and 3 mm in diameter. It is the source of the spice and the ethereal oil. It contains up to 23 % of ethereal oil, fatty oils and tannins. 90 % of the ethereeal oil is eugenol; further contents are vanillin, sesquiterpenes, α and β caryophyllene.

The dried flower buds are either used whole or are ground and pulverized as a spice in bakery, in the manufacture of chocolate, sausages, sauces, ketchup, candies, and for flavouring of sweet and savoury dishes, pickles or liqueur.

Economical utilization

Oil of cloves is not only used in the fabrication of perfumes, cosmetics and soap, but also as a flavouring of cigarettes.

It is technically applied in painting on china.

In histological work it is used as a clearing agent.

Medical use

Medically, the oil is appreciated as a stimulant, aromatic, antiseptic (in dentistry), carminative, stomachic and as a remedy against toothache. It is furthermore applied for flavouring and as a taste corrective.

Stems, leaves, flowers and fruits contain the oil. Buds, leaves, and stalks of inflorescences are used as drugs. The extracted oil of the leaves is cheaper and is mainly used for extraction of vanillin.

E. patriisii VAHL. A tea of the leaves, twigs and fruits is used as a remedy for repeated coughs and other respiratory problems.

ROTH studied the bark structure 1981, fruit structure and dispersal 1987.

Myrcia

M. comprises about 500 species of shrubs and small to medium-sized trees distributed in tropical South America and the West Indies.

Some species supply edible fruits, dyes and tannin from the bark, and timber for fuel and miscellaneous purposes.

The timber is similar in appearance and structure to that of *Eugenia*. Some species have insect resistant wood.

Terpenes, sesquiterpenes, amyrin and eucalyptin have been identified and a patent covering the antitumour constituents of *M. fallax* has been issued.

The leaf anatomy of *M.* sp. growing in the Venezuelan cloud forest has been described by ROTH (1990). The leaf contains oil idioblasts in the mesophyll; this species is perhaps a promising plant.

The bark of *M. splendens* supplies tannin and a black dye. ROTH studied the bark structure 1981, fruit structure and dispersal 1987.

Myrciaria has edible fruits of high quality.

ROTH studied the bark structure 1981, fruit structure and dispersal 1987.

Plinia pinnata L.

Taxonomical description

The shrub or tree, up to 9 m high, occurs in riverine and marsh forests, and is fairly rare.

The flowers are subsessile. The calyx splits irregularly into 4 segments. The ovary is 2-locular. The fruit is a globose berry crowned by remnants of the sepals and contains a single seed; its surface is yellowish to orange and slightly hairy, smooth or 12-ribbed. The pulp is edible. The only seed is large and subglobose (ROOSMALEN 1985).

Occurrence

Mainly Guiana and Surinam. Along rivers and in marsh forests.

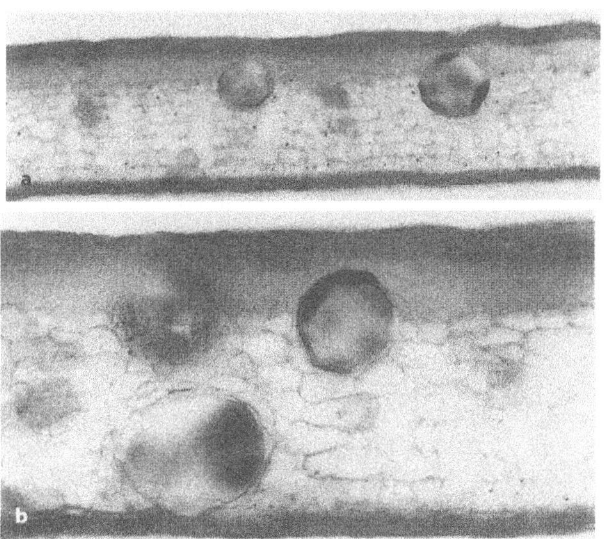

Fig. 188. *Plinia pinnata*, t.s. of leaf blade (a ×20, b ×40). Note the secretory cavities with an oily content.

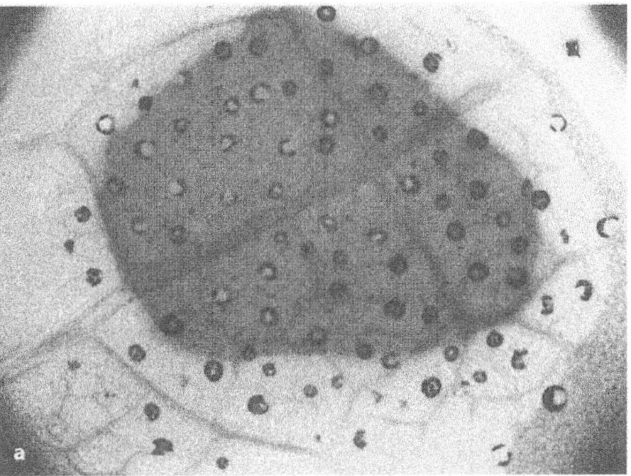

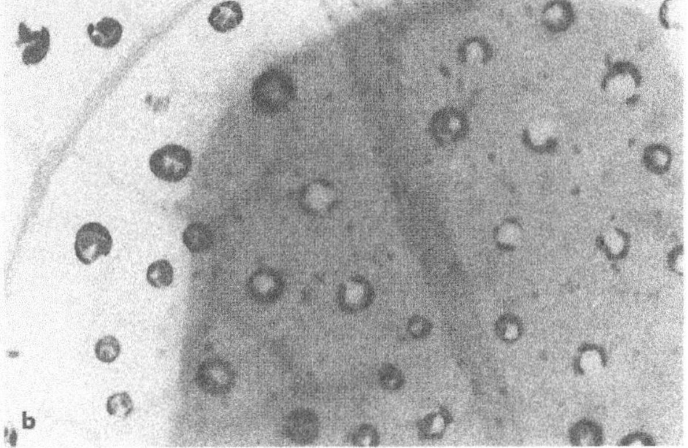

Fig. 189. a, b *Plinia pinnata*, lower epidermis and oil cavities in the mesophyll, as seen in a surface view. (a ×5, b ×10).

Anatomical description

Leaf (Figs. 188, 189, 190 d). The leaf is dorsiventral and hypostomatic with a tendency towards isobilaterality. The upper epidermis cells are more or less of regular size, but have enormously thickened and cutinized outer tangential walls so that the cell lumen is much reduced. As seen in surface view, the upper epidermis cells are small and have straight or slightly bent anticlinal walls. The cuticle has an irregular partly ribbed surface.

There is only one compact layer of palisade cells. The spongy parenchyma is well developed with about 8 layers of irregularly-shaped cells with short arms. The intercellular spaces are of medium size. The lowest mesophyll layer, above the lower epidermis, has the aspect of a reduced palisade parenchyma. The cells are oriented perpendicularly to the surface and are slightly elongated in this direction, but are relatively broad.

Frequently dispersed in the mesophyll are very large globular secretory cavities with an oily content (Sudan III positive) surrounded by a wreath of secretory cells. Clustered crystals in the form of druses occur here and there in the mesophyll, but are particularly frequent below the upper epidermis.

The vascular bundles of higher order are very much reduced and without a fibrous sheath. The larger bundles of higher order, on the contrary, have strong fibrous caps on both upper and lower side.

The lower epidermis cells likewise have extremely thick and cutinized outer tangential walls. Stomata lie at epidermis level, but their cuticular horns project slightly over the epidermis surface. Stomata are frequent on the lower leaf side. Scars of hairs are also present. As seen in surface view, the anticlinal walls of the lower epidermis cells are slightly bent. Their cuticle is ribbed.

Medical use

The plant induces hypotension.

Observations

The leaf is characterized by the enormously thickened outer epidermis walls, the large globular oil cavities, the shape of the stomata, as seen in t.s., with their cuticular horns and the ribbed cuticular surface. The lower layer of palisade-like cells may however be subject to variations, according to climatic and environmental conditions and to the position of the leaves.

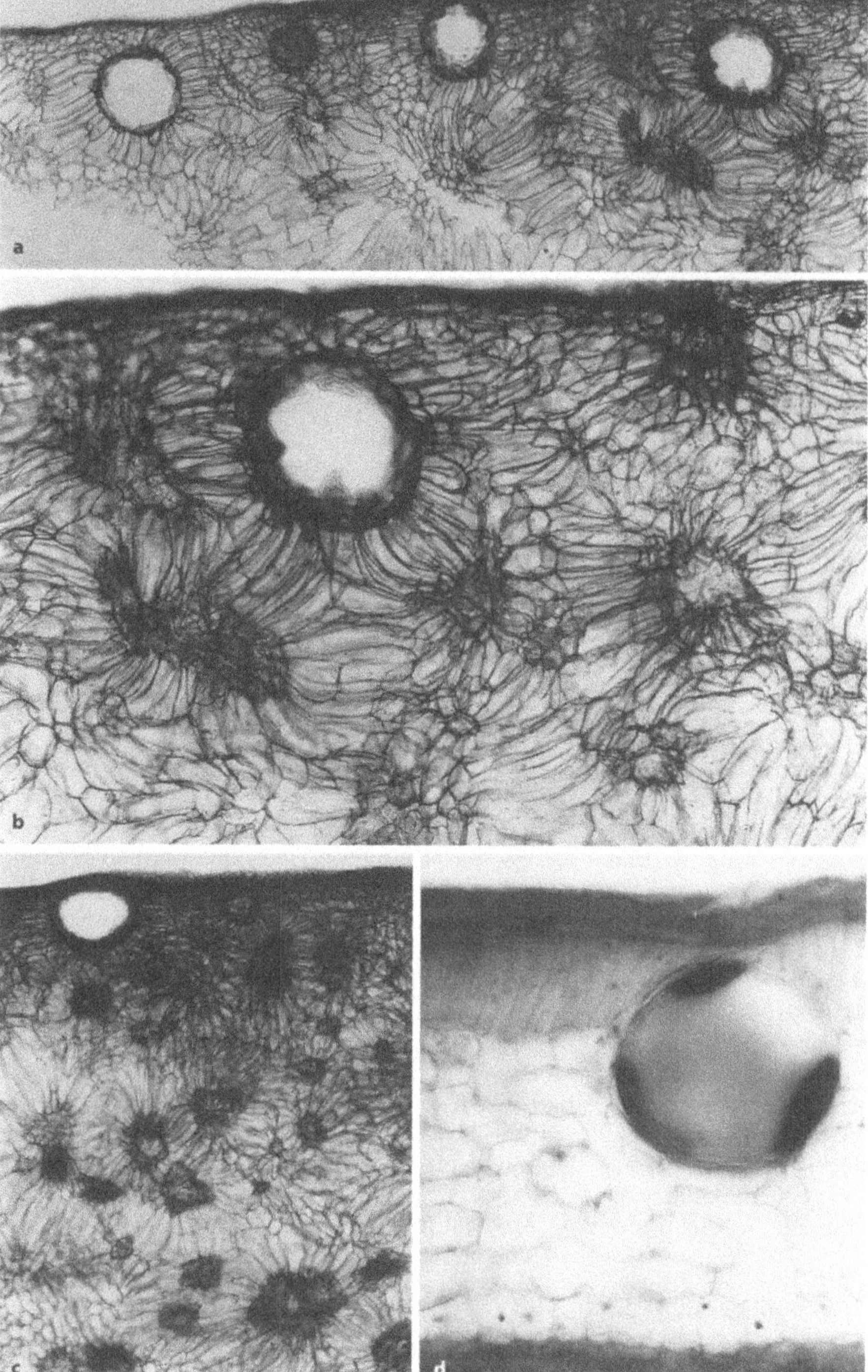

Fig. 190. *Psidium guajava.* **a–c** Fruit. Note the secretory cavities and the small groups of stone cells surrounded by radially extended parenchyma cells radiating from the stone cell groups. **d** *Plinia pinnata*, oil cavitiy (× 40).

Psidium guajava L. (guayaba)

Taxonomical description

Psidium guajava (Fig. 191) is a shrub or small tree, 3–6 (8) m high with a diameter of 20 cm; the young branches are quadrangular and slightly winged; they are pubescent through simple trichomes (Fig. 191). The bark is of a light chocolate colour and dehisces in the form of delicate scales which are in contrast with the green or reddish stem surface. The wood is chocolate brown or reddish. The simple leaves are opposite and have a short petiole. The leathery blades are glabrous on

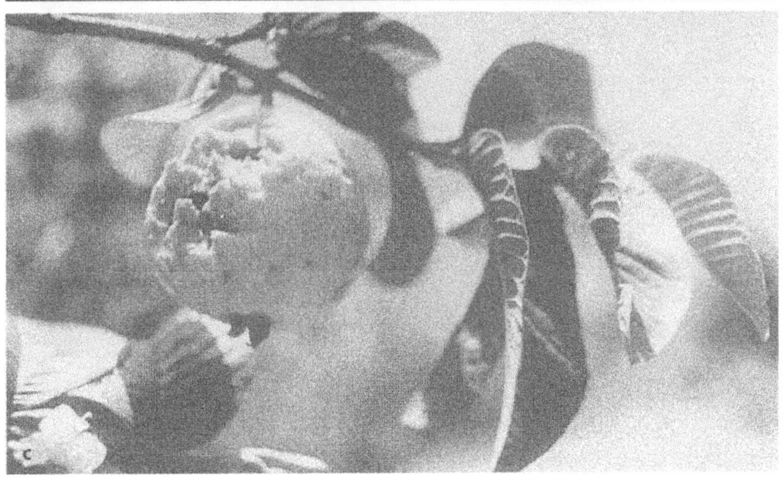

Fig. 191. *Psidium guajava*. **a** Tree. **b** Flowers. **c** Fruit.

the upper and pubescent on the lower side, of elliptic, oblong or lanceolate shape and measure 5–18 cm in length and 3–6.5 cm in width; the leaf tip is obtuse, rounded or retuse and the margins are entire; the leaf base is obtuse or rounded; the petiole measures 0.3–0.8 cm in length, is pubescent and ribbed. The venation is pinnate, brochidodromous, and shows 13–19 pairs of secondary nerves. The midrib is prominent on the abaxial side.

The inflorescences are axillary, the flowers actinomorphous, the calyx is 5-lobed, the crown white, the stamens are numerous.

The fruit is a globular or pear-shaped berry with a diameter of 3–6–12 cm; the pulp is rose-coloured or yellow. The seeds are numerous and hard. The reniform seeds are 3–5 mm long. The calyx is persistent on the fruit top.

Origin

Tropical America: South of Mexico or Amazonas.

Occurrence

In Venezuela, the species is very frequently cultivated.

Historical background

The species is mentioned by the Spanish chroniclers in the first years of colonization as a plant which supplies food and medicine (GUPTA 1995). In Venezuelan documents the plant has already been mentioned in 1578 (VELEZ & VELEZ 1990).

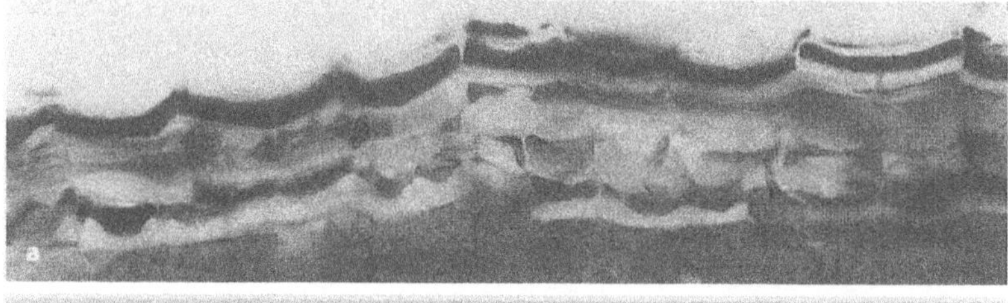

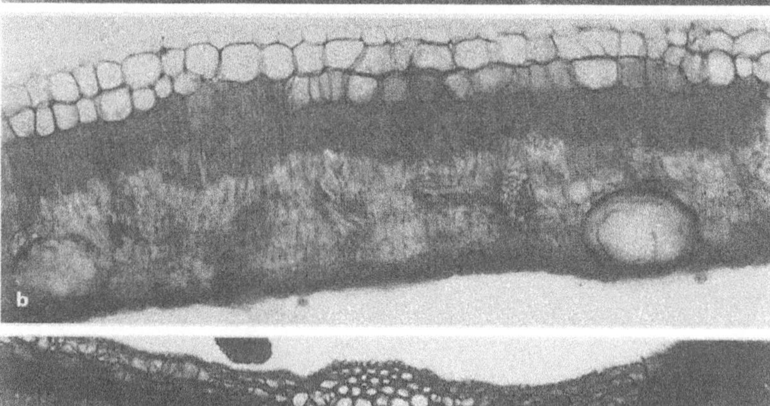

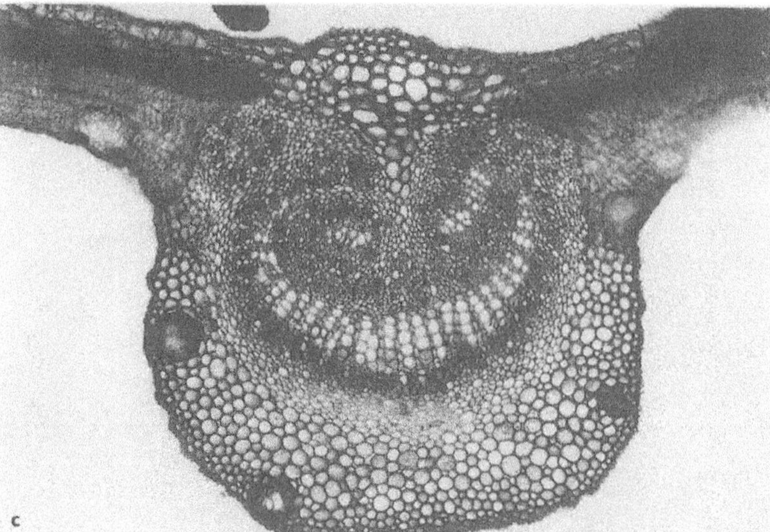

Fig. 192. *Psidium guajava.* **a** Cork of stem (× 25). **b** Leaf blade (× 16). **c** Midrib.

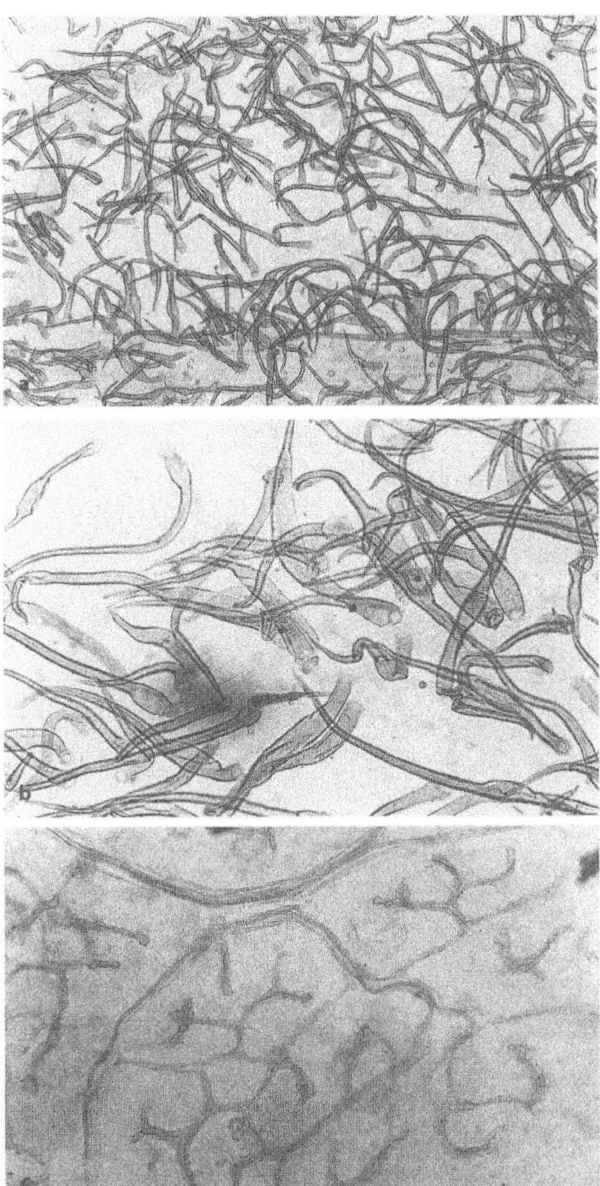

Fig. 193. *Psidium guajava*. **a, b** Lower epidermis with hairs. **c** Venation pattern of leaf.

Anatomical description

Leaf (Figs. 192 b c, 193). The leaf is almost isolateral (isobilateral). As seen in transverse section, the upper epidermis is very small-celled and slightly papillose above the midrib. As seen in surface view, the cells are polyhedric with straight walls. The scars of the hair bases are very conspicuous. Beneath follows a very large-celled hypodermis comprising 2 layers. This extensive tissue with large pits in the slightly thickened walls is water-storing. The following 2-layered palisade parenchyma occupies about the same space (as seen in transverse section) as epi and hypodermis together. Its cells are palisade-like and represent the photosynthetic tissue proper. The following 4–5 layers are also composed of anticlinally elongated cells, but the length of their anticlinal axis diminishes towards the lower epidermis. This tissue is only slightly less compact than the 2 uppermost palisade layers. The lower epidermis is small-celled and its walls are straight, as seen in surface view. The long simple and unicellular hairs with an acute tip are more numerous than on the upper side. At times, they show a constriction above the hair base. Stomata which are small and very numerous only occur on the lower side; they are anomocytic or have 2 lateral subsidiary cells. They are very slightly elevated above the surface. Most conspicuous are the very large secretory cavities below the upper hypodermis and, more frequently, below the lower epidermis. They are surrounded by a secretory epithelium.

The midrib is very prominent on the lower leaf side. The main vascular strand is U-shaped with somewhat incurved ends and with 2 additional strands at both ends. It thus resembles the petiolar structure (METCALFE & CHALK 1950). It is only very slightly bicollateral. The surrounding tissue is parenchymatous and becomes somewhat collenchymatous towards the upper and lower epidermis. The large hypodermis-like parenchyma cells on the upper side seem to also be water-storing. There is a varying number of secretory cavities in the midrib.

The secondary veins are partly transcurrent to the upper hypodermis as well as to the lower epidermis by parenchymatous extensions. The cells composing the secondary veins are comparatively thin-walled and fibers are absent.

The leathery consistency of the leaf is mainly due to the 2-layered hypodermis and to the compactness of the mesophyll.

Axis (Figs. 194, 195, 192 a). A transverse section through a twig of about 8 mm in diameter shows a few layers of cork composed of cells with U-shaped wall thickenings. The phelloderm consists of a few layers (4–6) of thinwalled isodiametric cells. A radial arrangement of cell rows is very conspicuous in phloem and xylem. The rays are mostly 1–2 seriate, enlarging slightly towards the outside through cell enlargement only. The phloem consists mainly of soft bast, being interrupted by a single more or less continuous ring of stone cells, about 1–3 cells high.

The vessels of the xylem are small and remain mostly solitary or are arranged in short radial rows. A certain rhythmical growth is perceptible through formation of smaller thicker-walled cells at border

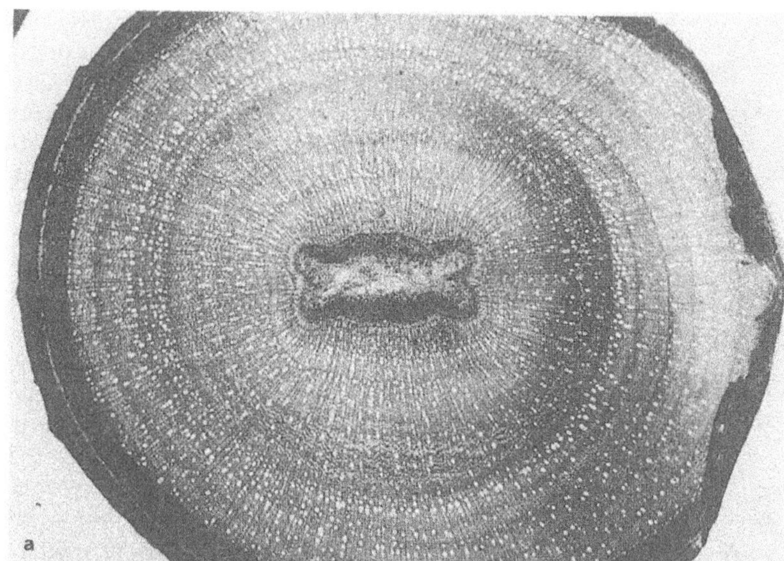

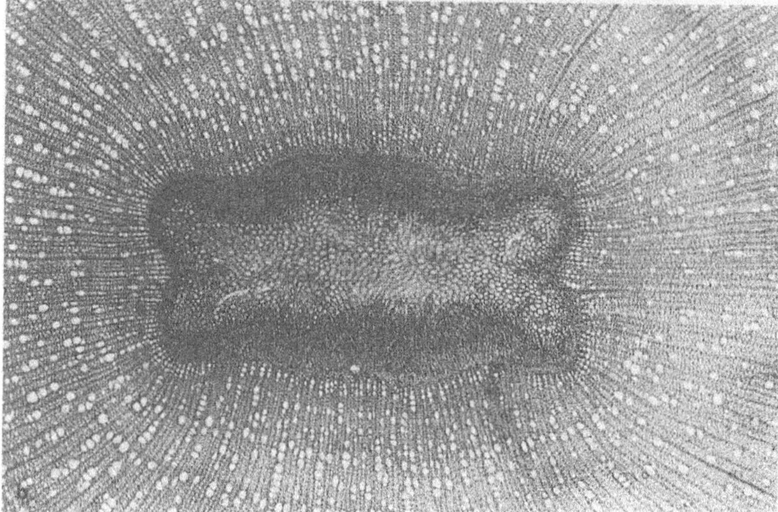

Fig. 194. *Psidium guajava.* **a** T.s. of axis. **b** Pith surrounded by xylem.

lines, but this formation of rings is very irregular and less conspicuons (growth rings).

The pith in the center adapts in its shape to the quadrangular form of the twig. It extends more vigorously towards 2 opposite sides forming 4 lobes. The intraxylary phloem on the xylem inside forms a comparatively broad band. The pith itself consists of roundish parenchyma cells which frequently contain crystals.

Bark (Fig. 195). A bark about 2 mm in thickness shows several cork layers with U-shaped wall thickenings (Fig. 192 a) and a few layered phelloderm. The cell arrangement in radial rows becomes very conspicuous in the phloem. Small rays (1–2 seriate) are very frequent. They form a more or less parenchymatous network together with the axial parenchyma which often occurs in the form of tangential bands, a single cell row thick. Rays and parenchyma stain intensely with toluidine blue and contain abundant starch. Sieve tubes with their companion cells alternate with the parenchymatous bands.

The hardbast is very scarce and only 2 rings of stone cells could be observed, leaving large distances between one another in a radial direction. As seen in longitudinal section, parenchymatous strands of cells, each with a rhombic crystal, occur in the soft bast.

The structure of the phloem is very homogeneous.

Fruit (Fig. 190 a–c, 191 c). The fruit is an edible berry and consequently consists mainly of parenchyma. The epidermis surrounding the fruit is very small-celled and has somewhat thickened outer walls. The following 3–5 parenchymatous layers beneath the epidermis are likewise small-celled. Embedded in this tissue are numerous large secretory

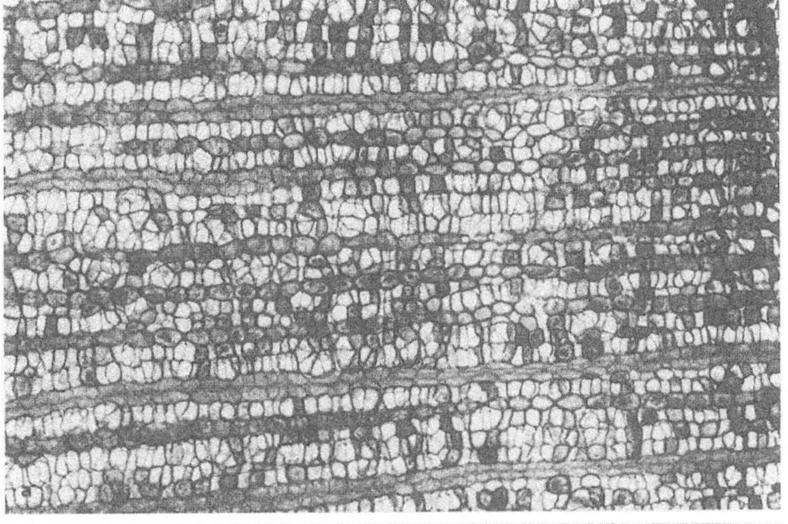

Fig. 195. *Psidium guajava*, bark. **b** Entire section (cambium to the right) with 2 fiber layers (× 6.3). **a** Detail (× 16). Note the medullary rays (1–2 seriate).

cavities surrounded by a secretory epithehum. The principal pattern imposed on the fruit section are groups of stone cells surrounded by radially elongated parenchyma cells which radiate from the stone cell groups as the center of resistance. This star-like arrangement of parenchymatous cells around obstacles such as stone cell groups is found in many fruits. Stone cell aggregations diffusely dispersed between parenchymatous cells behave like foreign bodies, because they cannot follow the growth of the fruit in thickness and in circumference due to their rigid walls. The neighbouring cells are thus exposed to strong tension and elongate in different directions. Groups of stone cells scattered between the pericarp parenchyma are for example also found in the avocado pear (see also ROTH 1977, and Fig. 157).

Ethobotanical and general use

Nutritional use

The edible part of the plant is the fruit. Eaten raw it has a sour-sweet and aromatic taste due to the etheric oil which is localized in the secretory cavities beneath the epidermis. More frequently the fruit is eaten cooked in the form of preserves, jam, marmalade or jelly. In this form, the guava has a taste and consistency similar to that of the European quince whose flesh also contains stone cell groups giving it a granular texture. The stone cell groups can also be recognized in the marmalade under the microscope, thus serving as a characteristic of identification (ROTH 1973: Vida y uso de los frutos, p. 141, Fig. 35). As both fruits, the guava and the quince, are rich in pectic substances, a form of cheese can be made of them; this cheese is solid and can be moulded. Refreshing drinks as well as cookies and cakes can also be made out of the pulp (see also ROTH 1973 'Vida y uso de los frutos', Fig.35 and the recipe p. 118).

Economical utilization

The yellowish bark is applied to tan skins and to dye cotton and silk.

The wood is of a reddish-yellow colour being compact, heavy, strong and durable; the fiber is fine. However due to its small diameter (20 cm), it is used only locally.

The species is furthermore grown in coffee plantations as a shade tree. It is likewise planted as an ornamental tree.

The leaves put under the bed serve as an insecticide.

Medical use

Leaf. The leaf has a high tannin and Vitamin C content.

It is said to be astringent (tannin), antibacterial, antiinflammatory, anthelmintic, antiseptic, antidiarrhoeic, antiemetic, carminative, spasmolytic and tonic.

Locally applied, it cures afflictions of the skin, asthma, inflammations of the tongue, and scrofula.

Leaves are used as a cataplasm against swellings and obstruction of the spleen.

As a bath, it cures scabies, itching, cutaneous diseases and ulcers.

As a footbath it is used for swollen feet.

A handful of leaves, boiled for 10 minutes in 1 l water is used for compresses against varices.

An infusion has stimulating and antispasmodic effects. It is also taken against indigestion, dysentery, diarrhoea and liver disorders.

A tea of the leaves is taken for heat, diarrhoea and dysentery. Tea also helps against diabetes.

A decoction of the leaves (and bark) is taken orally against diarrhoea, dysentery, colics, vomiting, diseases of the skin (fistula, ulcer, pyodermia, abrasion, tinea), leucorrhoea, diabetes, haemorrhage, swellings and urethritis. 30 g leaf, fruit or bark are boiled in 1 l water to prepare the decoction.

Young shoots. Young shoots are applied as a tonic for the hair to combat alopecia and to cure colds.

Leaf buds. Leaf buds help against fever and diarrhoea in the form of a tea.

Bark. Bark is rich in tannins and is therefore astringent. It has antiseptic, antiinflammatory, anthelmintic, antimicrobial, spasmolytic, carminative and tonic effects.

A tea or decoction of the bark helps against feverish diseases, dysentery, stomach aches, gastritis, diarrhoea, leucorrhoea and cough. A decoction is also taken against amoebas and as an emmenagogue.

Externally, the decoction is used as a vaginal irrigation, a lavage of ulcers, a gargle for infections of the mouth and throat.

Wood. Wood is also occasionally used against diarrhoea.

Root. The root is astringent and helps against dysentery, it is antispasmodic, antiparasitic and resolving.

Against oedema, a decoction is taken twice a day for a prolonged time.

As a mouthwash the root is good for the gums. It is also recommended for stomach and skin.

The cortex of the root is given to children to cure diarrhoea.

Flowers. A decoction is used as a mouthwash against aphthous ulcer in the mouth.

An infusion helps against irregular menstruation.

A decoction taken internally is applied as an emmenagogue, against amoebas, and for measles (WILBERT 1996).

Fruit. The fruit has a high tannin and vitamin C content.

Fresh fruit (one red fruit a day) is taken for Basedow's illness. Fresh fruit which is almost ripe has antidiarrhoeic and antidysenteric effects.

Ripe fruit is laxative.

Green fruit in decoction cures diseases of the urinary tract and haematuria. Fresh green fruit cures tonsillitis.

A tea of the fruit is taken for dysentery.

Dry fruit is used for haemorrhoids.

The green fruit is astringent (tannin), acts as a digestive, is antipyretic, cures inflammations and alleviates respiratory congestion.

Seeds. Pulverized, the seeds help against diabetes.

Method of use

The distinct organs can be used in the most varied ways. Fruit is eaten raw, either unripe (antidiarrhoeic and antidysenteric) or ripe (laxative). Green fruit cures haematuria and diseases of the urinary tract.

Seeds are pulverized to cure diabetes.

Flowers as an infusion or decoction are taken internally as an emmenagogue and against amoebas. A decoction serves as a mouthwash to cure ulcers.

Roots in decoction are used against oedema, dysentery and diarrhoea. As a mouthwash the decoction is applied to cure the gums.

Bark in decoction is taken orally against fever, dysentery, diarrhoea, leucorrhoea, cough etc. Externally, the decoction serves as a vaginal irrigation, a gargle or a lavage.

Leaves are applied locally as a compress or cataplasm to cure diseases of the skin, swellings, inflammation, asthma etc.

A bath cures scabies, itches and the like.

A footbath is used for swollen feet.

An infusion has stimulating and antispasmodic effects.

A tea is taken against diarrhoea and dysentery.

A decoction taken orally cures all diseases which can be controlled by the healing forces of the plant.

Against diarrhoea, leaves, flowers, unripe green fruits, bark and root are ground together, mixed with water and taken fresh.

Healing properties

The following curative effects are ascribed to the plant: Astringent, antibacterial, antimycotic, antiinflammatory, spasmolytic, antiscorbutic (vitamin C), antibiotic (against *Trichomanes vaginalis*), anthelmintic, antidiarrhoeal (unripe fruits), antileucorrhoeic, aphrodisiac , liver protecting during infections, against ulcers, vomiting, stomach aches, and dysentery.

Chemical contents

Among many other substances, the following contents were found: coumarins, alkaloids, pectins, vitamins C, A, B1 and B2, iron, phosphorus, calcium oxalate, β-sitosterol, guaiol, eugenol.

In leaves and bark, there are essential oils (in the secretory canals), flavonoids, sesquiterpenic alcohols, triterpene acids, complex tannins, polyphenols, eugenol.

The tannins give the astringent taste. Vitamin C has antiscorbutic effects, a narcotic-like principle is responsible for the spasmolytic effects.

Varieties and related species

According to BRÜCHER (1989), numerous local races and selections exist in America. Wild-growing guavas have mostly fruit of poor quality and contain many seeds; their mesocarp is brittle or sandy due to the presence of many stone cell grains. This property has been eliminated in better cultivars. Seedless varieties now exist; these have a triploid genome structure and must be propagated vegetatively. Other selected varieties have a high content of ascorbic acid (vitamin C): an average of 1000 mg/100 g fruit pulp. Seedless varieties could be very profitable for the fruit industry.Chemical fruit analyses are given by BRÜCHER (1989) and VELEZ & VELEZ (1990).

HOYOS (1989) gives detailed descriptions of the two most important varieties in Venezuela: Some varieties have a hard and thin exocarp, while in others the exocarp is thick and soft; the latter selection is used for the famous Venezuelan national dish 'cascos de guayaba' in which guava shells are cooked in syrup and eaten together with cottage-cheese (quesillo).

There is an off white (yellowish) variety with a higher vitamin C content, and a rose-coloured variety with less vitamin C and more vitamin A.

Cultivation

Propagation is carried out either by seeds or vegetatively. The seedlings serve as stock on to which the selected varieties are grafted. The seeds have to be sown immediately after gathering from the fruit. After the permanence in a seedbed for 110 days, they are put into a nursery where the grafting is accomplished. When the plant has reached a height of about 1 m, the treelets are put on the definitive site at distances of about 7 m. The grafts are obtained from young twigs which have already lost the green colour of the bark.

The plant can be grown between 0 and 1800 m a.s.l. at temperatures between 18 and 28 °C and an annual precipitation between 1000 and 3000 mm. Excessive rain during the fruiting time damages the fruits.

The wild growing varieties are very resistant and tolerant. Cultivated, the tree grows best in a soil of pH 5–6.

The fruits can be harvested twice a year (April/July and November/February). (For more details see HOYOS 1989). Fruit production starts after 2–3 years.

Observations

The leaf is easy to recognize by its isolateral structure, the secretory canals, the water-storing hypodermis and the long unicellular hairs. It is also well characterized within the family by having stomata only on the lower side.

The axis is characterized by the U-shaped wall thickenings of the cork, the quadrangular pith adapted in its shape to the quadrangular circumference of the twig, the homogeneous structure of the phloem with only little hardbast in the form of widely-spaced rings of stone cells and the 1–2 seriate rays.

The fruit which consists mainly of parenchymatous cells is characterized by the secretory canals beneath the epidermis and the abundant stone cell groups in the flesh. These 2 properties allow the identification of industrial products under the microscope.

Psidium guineense SW. Leaves, bark and fruit in decoction are used for dysentery and diarrhoea. Pulverized seeds are applied for diabetes (CASTILLO 1995).

Punica granatum, the pomegranate, is now cultivated everywhere in the tropics and subtropics. The fruit is an irregularly dehiscing capsule (Fig. 196) which exposes a great quantity of juicy seeds at maturity. The peculiar arrangement of the carpels in the form of two stories, the upper one 4–6 celled and the lower one 2–5 celled, has claimed the attention of several authors. At the beginning, the carpels are arranged in two – or possibly three – concentric whorls; by favoured growth of the peripheral ovary parts, however, the outer whorl is lifted up and the originally central placentae are shifted to a parietal position. Both central and parietal placentation therefore seem to be realized in one and the same ovary. The ovary is inferior and in the mature fruit stage crowned by the calyx. The floral development has been studied in detail by TUNG (1935).

Development and anatomy of the fruit have been studied by ROTH & LINDORF (1972), ROTH 1977).

The plant is not only used as a fruit tree, but also as an ornamental; the bark of stem and root which is rich in tannins has medicinal application.

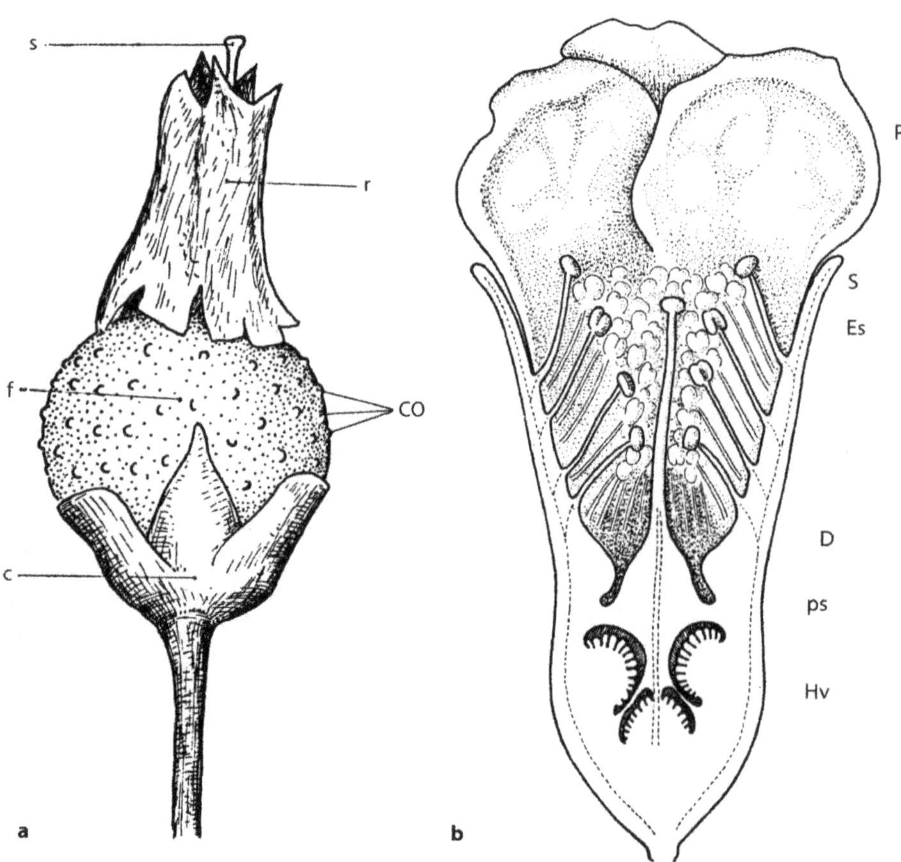

Fig. 196. a Young fruit of *Achras zapota*, Sapotaceae with rests of the corolla around the stigma in the form of a 'calyptra', fruit surface covered with corky scales (CO). **b** *Punica granatum*, Punicaceae, longitudinal section of flower with 2 storeys of carpels in the ovary (ROTH & LINDORF 1972 b).

Nyctaginaceae

Shrubs, trees or herbs. Lower part of the perigone tube transforming into an anthocarp which encloses and protects the fruit; this is interpreted as a nut. One carpel.

Vascular bundles in the pith frequent. Anomalous growth in thickness in all woody stems and partly also in roots with successive rings of vascular bundles. Oxalate crystals in the form of raphides.

The plants are mainly tropical and mostly occur in America.

Mirabilis

M. has about 60 species in America.

Mirabilis jalapa (jasmin de tarde, buenas tardes, good evening) is a perennial herb, 30–90 cm high, with red, yellow or white flowers.

The anthocarp consisting of the lower part of the perigone tube completely substitutes the real pericarp which obliterates during development. Already at anthesis the perigone divides into a lower part and an upper attractive part which is shed shortly after anthesis by a separation tissue. The lower persistent part increases considerably and transforms into a hard shell which surrounds and protects the real fruit. Most conspicuous in the anthocarp of the Mirabileae is the mucilage production by subepidermal palisade cells which prevail in ledges, warts or other kinds of protuberances producing the special sculpturing of the anthocarp surface. Absorbing water, the palisade cells swell so much that the epidermis bursts and the slime is released to fasten the fruit on the substrate (ROTH 1977).

The plant is very ornamental and planted in gardens. The flowers open in the afternoon, therefore the common name Buenas tardes. The root is thick and tuberous and has purgative effects. With 1–2 spoonful of the crushed root and half a bottle boiling water the purgative is prepared. Roots are also used for oedema.

Ear ache is cured with the squeezed flowers (slime).

Powdered seeds are sometimes used as a cosmetic.

A crimson dye is obtained from flowers steeped in water to color jellies or cakes.

Ochnaceae

Ouratea

Bitter roots and leaves serve as a tonic, a stomachic, anthelmintic and antiemetic, as well as for dysentery.

The wood is hard and heavy.

The kernels of the fruit supply an oil for soap manufacture.

A fast drying oil of the fruit husks is applied in varnishes.

ROTH 1984 studied the leaf structure and 1996 the venation pattern of the leaves.

Olacaceae

Heisteria

Leaves and bark are used for boils to reduce swelling and pain.

A scopolamine with psychoactive properties has been found in the genus.

Bark structure has been studied by ROTH 1981.

Opiliaceae

Agonandra brasiliensis BENTH. & HOOK. The ivory wood seed oil can be manufactured into a rubber substitute.

Bark structure was studied by ROTH 1981, fruit morphology and dispersal by ROTH 1987, leaf structure by ROTH 1984.

Orchidaceae

The Vanillinae comprise 5 genera. 100 species are known of the genus Vanilla which are indigenous to tropical America. *Vanilla planifolia* (*V. fragrans*) (Mexico), much cultivated now in the tropics, supplies the vanilla of commerce (Fructus Vanillae); it contains 1–2.9 % of vanilline, furthermore glucovanilline etc.

Fig. 197. *Vanilla planifolia*, Orchidaceae, climbing on a wall; note the distichous phyllotaxis of the succulent leaves. This plant was growing in the garden of the Senior author in Caracas.

Vanilla pompona (Mexico down to Guiana) is used in perfumery, supplying the vanillons.

Vanilla planifolia is easily cultivated. Fig. 197 shows the plant climbing on a wall in the garden of the senior author in Caracas. The succulent leaves are distichiously arranged.

Palmae (Arecaceae)

With more than 2000 species, the palmae represent one of the largest taxonomic units of the Monocotyledoneae. In their economic importance they are second only to the Gramineae.

The richest diversity of palm species is found in Brazil and Venezuela. Nearly 180 species have been reported from Venezuela (BRAUN 1970).

The ovary of the Arecaceae usually consists of 3 carpels. The fruits are berries or drupes, the seeds large and often joined to the endocarp.

The Palmae are usually woody trees, shrubs or lianas, but are devoid of a secondary growth in thickness of the stem. The stems are only seldom ramified. The leaf blade is generally divided (palmate and pinnate leaves).

The palms are pantropic-subtropical.

They supply oil and fat, starch, wax, fiber, vegetables, beverages, wood. Fibers are used for clothing, fishing nets, mats, and leaves for roofing and building of shelter. Palm oils are non-drying oils and generally correspond to tasteless triglycerides which remain liquid when exposed to the air.

BRÜCHER (1989) points to the danger of extermination of certain palms the hearts of which now represent delicacies in the industrialised countries; he emphasizes that there is a remarkable trend in the European fancy-food market for expensive tropical food. For this reason he recommends the cultivation of the most frequently requested species to prevent the destruction of rain forests in search of such palm species.

To date, almost nothing is known about the chemistry of South American tropical palm species.

NARVAEZ CORDOBA & STAUFFER (1999) studied the products of palms which are available at the markets of Puerto Ayachucho, Edo. Amazonas, Venezuela. They observed that the products are used for food, medicine and handicraft. Altogether nine species are useful (see Table 9 below).

Table 9. Uses and common names of studied species.

Species	Uses	Common names
Astrocaryum jauari	In the Amazonas basin, the fruit is used as bait for fishing (Stauffer et al. 334, 476–VEN). The Pumé Indians use the fiber from the leaves to weave baskets and bags (Gragson 1992).	Kiajuara (Gentry and Stein 47325–NY)
Attalea butyracea	In Cojedes, the oil extracted from the fruits is used to fortify the hair of men and animals and the stems to build corrals and bridges (Delascio 11710–VEN). In Barinas, the pulp of fruits is eaten by hogs and the leaves used for roof (Ramia 1669–VEN). In the upper Orinoco River, the Yanomamis eat the fruit after boiling it.	Yagua, corozo (Cojedes State) (Delacio and López 111710–VEN)

Table 9 (Continued). Uses and common names of studies species.

Species	Uses	Common names
Attalea maripa	In Bolivar, the fruit is eaten boiled and with the flour, a flat bread (*arepa*) is made (Stauffer et al. 274–VEN). In Amazonas, from the crushed and boiled seed semi-solid whitish grease is obtained which is used for medicines. Mixed with vegetable pigments or minerals, it serves as a base for paint the indians use in their bodies. The oil also serves as a fuel or burners In the upper Orinoco a starch from the mesocarp is obtained to make cakes (Civrieux 1957: Braun und Delascio 1987). The Yanomamis eat the raw fruit (Fuentes 1980). The big peduncular bracts serves as toys for children (Civrieux 1957).	Careshi (Yanomami) (Stauffer et al. 274–VEN) Wasai (Yekuana, Bolívar state) (Goldstein and Salas 328–VEN) Mabaco (Piapoco) (Braun and Delascio 1987)
Bactris gasipaes	The wood from the trunk is used when the palm becomes sterile and the shoot is edible (Braun and Delascio 1987). The Yekuanas use the wood to make bows (*haia*) and the epicarp is used to sweeten the *yarake* (fermented drink made with *Manihot esculenta*) (Delascio 1992). The Venezuelan and Brazilian Yanomamis celebrate the harvest of Pijiguao with dances and a thick drink made from the fruit (Patiño 1992). In the Venezuelan Amazonas, the liquid extracted from the fruits is used as an astringent (Braun and Delascio 1987). When there is no pottery for cooking, the fruits are roasted (Civrieux 1957).	Pijiwao (Wessels-Boer 2252–NY) Jijiri, fhi-hidi. Lasha (Yekuana) Braun and Delascio 1987; Delascio 1992) Raxa (Yanomami) (Fuentes 1980).
Euterpe precatoria var. *precatoria*	Occasionally boards are obtained from the trunk. The fruits are used to prepare a refreshing drink (Stauffer et al. 286, 306, 328, 475–VEN). With the ashes of the leaves the Pumé Indians from the southwest Venezuela, in Apure, prepare *paramán* a substance used to seal, tie and make many products waterproof (Gragson 1992).	Nenea (Piaroa) (Stauffer et al. 286, 306, 328, 475–VEN) Wa-hu (Steyermark 107188–NY) Wahima (Yanomami) (Salaroli and Rucci 7–NY).
Leopoldinia piassaba	The fiber is used to make handicrafts (Stauffer et al. 391–VEN). The threads from the petioles are very resistant to water, flexible, and long-lasting; because of this, they are used to make ropes, brooms, collars, and brushes to clean budares, bracelets, hammocks, carpets and, mats (Braun and Delascio 1987). The leaves are also used for thatching roofs (Henderson 1997).	Marama, madama (Yekuana) (Delascio 1992)
Mauritia flexuosa	The leaves are used for thatching, the fiber resulting fromt he first partly grown leaf, is used to make handicrafts. From the fruit a paste is made to produce a refreshing drink. Old stems shelter the beetle larvae of (*Rynchophorus palmarum*) which is eaten by the Yanomamis due to ist high fat content. The fruits are eaten raw or cooked. When they are too ripe the Waraos Indians of the Orinoco Delta crush the fruit to obtain a paste, *ojiguari*, a kind of cheese that lasts a few days (Braun and Delascio 1987). Starting from the seeds a thick drink or *carato* is prepared (Braun and Delascio 1987). The heart of this palm is consumed by the Pumé Indians from Apure State (Gragson 1992).	Kuia (Yekuana), eteweshi (Yanomami) (Braun and Delascio 1987) Cuhuai (Yekuana) (Delascio 1992)
Oenocarpus bacaba	The epidermis of the leaf sheath is used to make cigars (Braun and Delascio 1987). The leaves are used for thatching (Henderson 1997). From the fruits alcoholic and refreshing drinks are prepared, and also a good edible oil (Berry 791–VEN): Braun and Delascio 1987). The pulp from the fruit crushed and mixed with water, produces an emulsion which is drunk with coffee instead of milk (Braun 1997).	Palma de vino, macaba (Yanomami) (Braun and Delascio 1987) Kujedi (Yekuana, Bolívar State) (Goldstein and Salas 343–VEN)

Table 9 (Continued). Uses and common names of studies species.

Species	Uses	Common names
Oenocarpus bataua var. *bataua*	The leaf is used to make arrows for hunting and the trunk is used to make bows (Braun and Delascio 1987). The oil that comes from the fruit is used against asthma and to cure skin irritation and diseases (Stauffer et al. 382-VEN) and Guánchez (1996) points out that oil is used as an antidote to the bite of poisonous animals. The medicinal oil also has culinary uses (Williams 14345-VEN). The fruits soaked in water for several days produce a very palatable strong drink (Braun and Delascio 1987) that is nutritious (Berry 2133-VEN). From the fruit, a starch is obtained which is used in confectionery. The seeds in decoction, crushed and emulsified in water are used to prepare a *carato* which is very nutritious. The seeds can be eaten raw (Civrieux 1957).	Kudai (Yekuana, Bolívar State) (Goldstein and Salas 342-VEN)

NARVÁEZ, STAUFFER and GERTSCH studied the utility of palms in the State Amazonas. Besides the already mentioned palms, the following species are used by the native people.

The flexible shoots of *Desmoncus* are used for the manufacture of ropes and baskets. The fruits of *D. orthocanthus* are occasionally consumed.

A refreshing drink is made from the boiled fruits of *Euterpe*. *E. oleracea* is cultivated in the State Amazonas, due to the qualitiy of the fruits which are manufactured in drinks and in ice-cream. The trunks serve for the construction of huts.

The leaves of *Geonoma* supply the material for the roofing of houses (e. g. of *G. deversa*, *G. baculifera* and *G. maxima*).

The native people use the shoots of *Iriartella* (*I. setigera*) for the manufacture of tubes which mainly serve for hunting and for the manufacture of traps to catch fish.

In the following a few examples of useful palm species are added (mentioned by Braun).

Euterpe. The root is used for malarian fever. *E. edulis* has edible fruits; the heart, which contains phenols, is used as a vegetable.

Jessenia. The fruits are edible, the seeds supply oil; the oil is antitubercular at the same time. The adventilious roots serve to expel worms and to treat diarrhoea, headache and stomach trouble.

Manicaria saccifera (Fig. 199). Supplies oil in the seeds. Leaves are used for roof covering, fibers of the spathe for mats.

Maximiliana. The leaf petiole is used for blow gun darts; the fruit is edible and the source of an oil; it is also used to cure colds. Leaves serve for thatching; seeds are a source of fat.

Sabal. Strips of leaves are made into hats, baskets, mats. Leaves are used for thatching, trunks for construction, leaf buds as vegetable. Flowers are a source of honey, fruits are edible, cross sections of stems serve as tables.

Mauritia flexuosa. (Fig. 198 b). Supplies starch.

Other palm trees are rich in oil or have edible fruits (e. g. *Cocos nucifera*).

Bactris

The genus comprises low to medium sized very spiny monoecious palms with solitary or clustered stems and pinnate leaves. Althoug many species are extremely interesting and of graceful aspect, they are rarely cultivated.

Some furnish valuable wood for carpentry. The straight stems of severeal species are worked up into walking sticks. The natives use the fleshy fruits of certain species to make vinegar.

B. gasipaës is the most important species. Its fruits are cooked in salted water and eaten. A kind of flour is made of roasted fruits. The fruits are also the source of a fermented drink. Fruits of certain varieties contain no kernels. The seeds are the source of an oil, similar to coconut oil. The wood is used for building purposes and also made into bows. Spines are used for tattooing (BRAUN 1968, UPHOF 1968).

Fig. 198. **a** *Chamaedorea pinnatifrons* (leaf and fruits). **b** *Mauritia flexuosa*, ripe fruits. Note the spiral arrangement of scales.

Bactris setulosa KARSTEN (albarico, corozo, macanilla)

Bactris setulosa KARSTEN is a palm tree characteristic of the transitional cloud forest as well as of the cloud forest proper. It reaches 10–12 m in height, with a stem diameter of 10 cm. The large pinnate leaves have about 50 pairs of leaflets which are linear-lanceolate, 40–60 cm long and 3–4.5 cm broad. Reproduction takes place in the second half of the rainy season. The fruit is almost globular and has a mealy-fleshy pericarp. According to VARESCHI (1986) it is the most frequently found palm species in the transitional and real cloud forest (ROTH 1992, SCHNEE 1960).

Anatomical description

Leaf. The leaf is comparatively thin; it has a small-celled and thick-walled upper epidermis and a large-celled water-storing hypodermis. The mesophyll is only poorly differentaited, consisting of more or less roundish cells leaving small intercellular spaces between one another. The stomata are restricted to the lower epidermis. For further details see ROTH 1992.

Ethnobotanical and general use

Clubs, tips of arrows and bows are made of the wood.

A decoction of the fruits of *B. guineensis* (L.) H.E. MOORE is applied to remove parasites from the intestinal tract.

Chamaedorea pinnatifrons (JACQ.) OERSTED. (caña molinillo, molinillo).

Taxonomical description

Solitary palm, inerm and dioecious (Fig. 198 a). Stem erect, occasionally arched, 1.5–2 m high by a stem diameter of 2–3 cm. Roots inerm, fulcrate, in number 12–15 forming a cone at the base of the stem. Leaves 3–6, slightly arched, pinnate, not persistent when old. Sheath 25–30 cm long, with a ventral cleft, rhachis of the leaf 60 cm long. The leaflets rhomboid, arranged in 7–8 pairs, in the same plane, usually opposed, rarely alternating, with a median rib and 2 marginal ones and numerous parallel nerves.

Male and female inflorescences similar, solitary, intrafoliar and pendulous. Flowers very small. Fruit with a single seed (possibly a drupe) globose-ellipsoid and black, when mature.

Origin and occurrence

From Costa Rica to Brazil and Bolivia, low regions in Amazonia, but ascending up to 2700 a.s.l.) (STAUFFER 1994).

The plant prefers shade or half-shade; it is common in the cloud forests of the Cordillera de la Costa of Venezuela, between 1000 m and 1850 m a.s.l. (BRAUN & DELASCIO CHITTY 1987).

Anatomical description

The leaf is equifacial and amphistomatic. The epidermis of the upper and lower side is similar comprising a single layer of cells with delicate walls. As seen in surface view, a distinction can be made in costal regions with 4–6 longitudinal rows of rectangular cells, and in broader intercostal regions with rhomboid or fusiform cells, the longer axis of which is obliquely oriented to the long axis of the leaf.

The stomata are found in the intercostal regions of the upper and lower epidermis, being distributed in longitudinal rows, but are more frequently found on the lower side. Four subsidiary cells accompany the stomata, 2 in lateral and 2 in terminal positions (possibly of the *Rhoeo* type, as frequently found in monocotyledons: (ROTH & CLAUSNITZER 1969). The stomata are oriented parallel to the long axis of the leaf.

The mesophyll comprises 5–6 layers of more or less homogenous cells. The adaxial cells are isodiametric and form a more compact layer, the cells of the central layers have more irregular forms, extending short arms in several directions so that small intercellular spaces arise; the abaxial cells are of more globular shape, being arranged in a single compact layer. The chloroplasts which are found in all layers of the mesophyll, appear large. In the adaxial mesophyll layers, non-lignified fibers are oriented parallel to the longitudinal axis of the leaf; they are solitary or are arranged in groups of 2–3. In the central mesophyll layers, there are cylindric idioblasts with raphides, running parallel to the veins; as seen in t. s., they appear as large circles between the vascular bundles. Silicia cells (stegmata) in the form of an umbrella are disposed in longitudinal rows, accompanying the fibers and the vascular bundles. Occasionally, siliceous crystals are found in the idioblasts with raphides of calcium oxalate. The vascular bundles are embedded in the central part of the mesophyll and are not transcurrent to the epidermal layers. The larger bundles are completely surrounded by a fibrous sheath, while the smaller bundles are crowned by cascades of fibers on the phloem and xylem side.

Ethnobotanical and general use

Like many palm trees, this species is ornamental and can be planted in the interior of houses.

According to PITTIER (1926), the lateral roots are radially arranged on the stem base which permits their use as 'molinillos' (little mills or grinders) to mix chocolate etc. For this reason, the palm received its vernacular name.

Other species of this genus have edible fruits or spathes, which are eaten fresh or cooked.

Varieties and related species

Many species of this genus are ornamental. Best known is *Chamaedorea elegans*, native to Mexico, which is extensively used to adorn the interior of houses and apartments.

Cultivation

The plant is propagated by seeds which germinate within 3–4 months. It needs shady places, rich soils and abundant watering (HOYOS & BRAUN 1984).

Cocos nucifera L. (coco, cocotero)

Although the etymology of the word 'cocos' remains doubtful, it may be deduced from the Spanish word 'coco', applied to a monkey's grotesque face (CHILD 1964).

Origin and historic background

Just as doubtful as the etymology of *Cocos* is the origin of the plant. The case of a Central American origin has been debated, but an Eastern origin is more likely, as in the East a far larger range of varieties exists and fossil nuts of a Cocos species were found in the Pliocene deposits at Mangonui, New Zealand. The coconut palm was probably introduced to Puerto Rico by the Spaniards, but it seems certain that the coconut preceded the white man on the Pacific shores of Central America. Whatever may have been the original center of coconut evolution, a subsequent question is, how did it become disseminated all over the tropical regions. It is supposed that it was partly brought by man, but that also the ocean current cooperated in dissemination and that coconuts are able to float in sea water for

Fig. 199. *Manicaria saccifera*, Palmae (Arecaceae). Bilocular 'double fruit' (ROTH 1987).

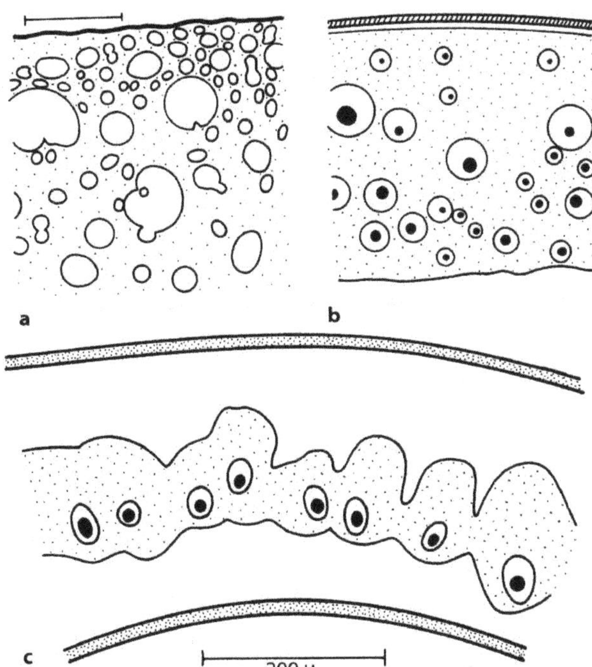

Fig. 200. *Cocos nucifera*. Diagrams of cross sections **a** Peripheral mesocarp part of a young fruit with irregularly dispersed vascular bundles. **b** Perigone leaf of a female flower. Vascular bundles (black) surrounded by a strong sclerenchymatous sheath. **c** Perigone leaf of a male flower Vascular bundles embedded in a sclerenchymatous band (dotted). (ROTH 1977).

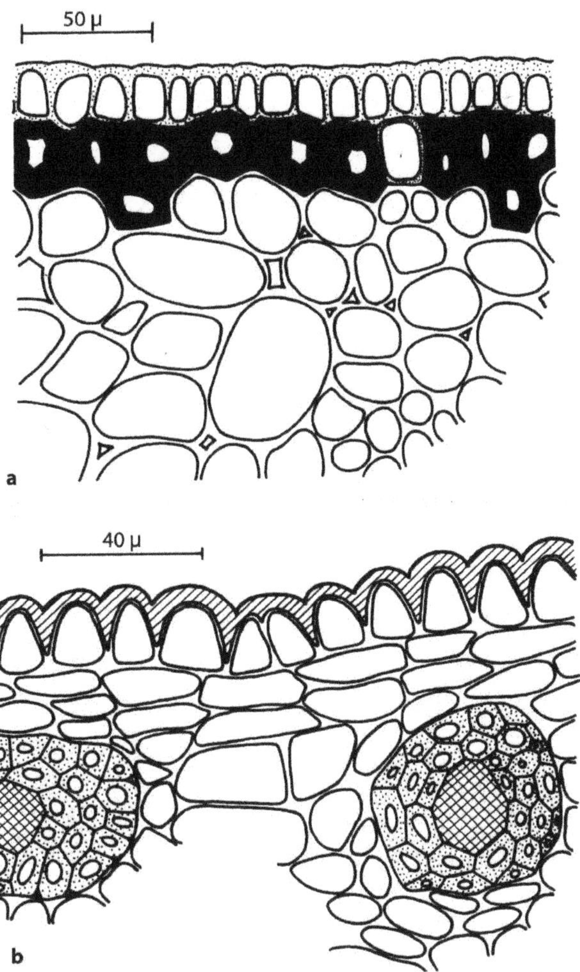

Fig. 201. *Cocos nucifera*. **a** Detail of perigone leaf periphery in t.s. Sclerenchyma cells in black. **b** T.s. of exocarp of young fruit with heavily cutinized outer epidermis and 2 vascular bundles surrounded by a sclerenchymatous sheath (ROTH 1977).

more than 100 days without losing their germination capacity (CHILD 1964).

Although it is improbable that the coconut originated in America we shall give a short survey of its structure and uses in the following passages.

Anatomical description

The coconut (Figs. 200, 201, 202) is regarded as a drupe, although it develops from an originally tricarpellary ovary with originally 3 ovules, 2 of which degenerate. Three germination holes, the 'eyes', are still visible on the mature fruit. The 3 eyes may have been the reason for the name coco (grotesque face). Each hole pertains to one carpel. As only one ovule develops, the 2 other ovules become compressed together with their locules.

In the following, the results of the thesis of HECTOR LOPEZ-NARANJO directed by the senior author will be presented briefly.

The pericarp is divided into a single-layered exocarp-originating from the outer epidermis – with thick cutinized outer and radial walls, a parenchymatous mesocarp densely penetrated by vascular bundles, and a hard sclerenchymatous and lignified endocarp which protects the only seed together with its endosperm. Fruit development is described by ROTH (1977).

In the mature fruit state, the outer epidermis cells which form the exocarp are anticlinally elongated and have strongly thickened and completely cutinized walls. As seen in surface view (Fig. 202 b), the arrangement of the cells in longitudinal rows has been lost and stomata were partly elongated tangentially through dilatation growth.

The ground tissue of the mesocarp is parenchymatous consisting of relatively large cells which become elongated in various directions due to the growth in thickness and in circumference of the pericarp. Simultaneously with the enlargement of the parenchyma cells, tanniferous cells begin to differentiate. Several layers beneath the outer epidermis become thick-walled and tangetially elongated.

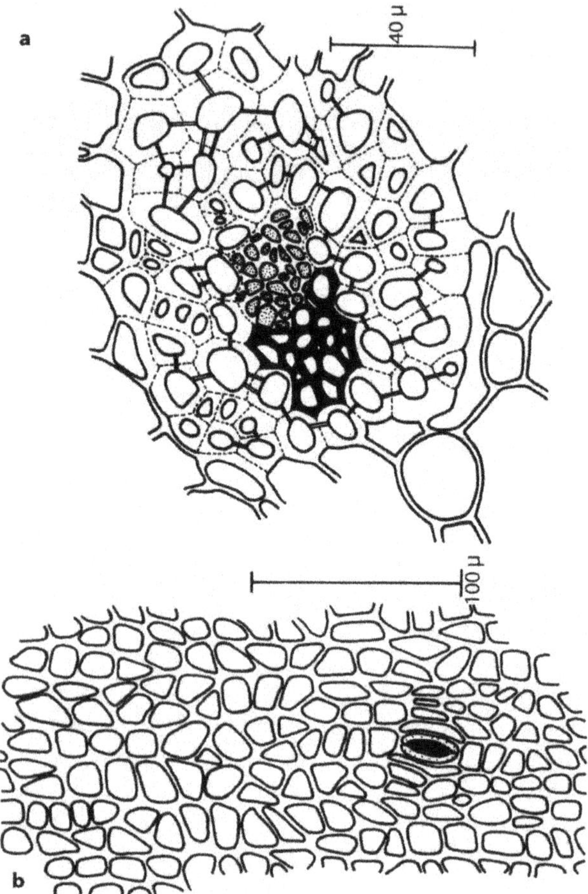

Fig. 202. *Cocos nucifera*. **a** Vascular bundle of perigone leaf of female flower. **b** Surface view of outer fruit epidermis with stoma: The epidermis cells are still more or less arranged in the form of longitudinal rows (ROTH 1977).

The vascular bundles are irregularly dispersed in the mesocarp, as usual in monocotyledons, and are surrounded by an enormous fibrous sheath; the bundles are smaller in the periphery, but enlarge considerably towards the center. Each bundle is collateral with the xylem pointing towards the fruit center, being composed of vessels and tracheids with reticulate, scalariform or helicoidal wall thickenings. In larger strands, two or more bundles may be enclosed in a single sheath. In the periphery, on the contrary, vascular elements may become reduced or are completely absent so that the strands consist only of fibers.

The vascular strands surrounded by their fibrous sheath form the coir of commerce. The original fibers are short, reaching 0.4–1.0 mm in length and 15–18 µm in width. But the technically used coir fibers correspond to entire fibrovascular strands of 15–35 cm length with a diamter of 0.1–1.5 mm.

During development, the fibrovascular strands increase in number by vigorous branching and finally form a dense network in the mature pericarp. Lastly, the mesocarp dries out and the soft phloem as well as parenchyma cells are partly disintegrated. In this way, the fibrovascular strands become separated from one another remaining surrounded by dead parenchyma cells. Sclerenchyma cells enclosing siliceous bodies with wart-like protuberances (the so-called stegmata) accompany the fibers.

The cells of the future endocarp also enlarge in different directions so that their axes become crossed, and some cells elongate so much that they have to bend several times: an interlocking structure results in this way, giving the endocarp more strength for tension and pressure. In the end, the endocarp cells differentiate into thick-walled sclereids and lignify.

The 'eyes' are also covered by endocarp, but the tissue is softer and thinner in this region. The functional eye through which the embryo later germinates, has only a spongy covering combined with a very thin plate of lignified cells, proceeding from the inner epidermis of the endocarp and from the outer integument. The outer integument of the seed and the endocarp come into close contact, whereas the inner integument adheres to the endosperm. The outer integument is thick-walled and stone cells are scattered in it, whereas the inner integument is much thicker and consists of parenchyma.

Ethnobotanical and general use

Nutritional use

Toddy is obtained from the unopened spathe of the inflorescence when wounded; it yields a sap rich in sugars and vitamin B and is used fresh as a drink; syrups and jaggery sugar may also be manufactured from it. When fermented, the sap becomes an alcoholic drink which can be distilled to arrack. By exposure of the alcoholic sap to air, the alcohol is converted into acetic acid and vinegar is thus obtained.

Coconut milk obtained from unripe fruits is a refreshing drink in the tropics. It can be condensed into a very pleasant substitute for cream. Coconut milk is also used for tissue culture in the laboratory. The fresh milk contains sugars and a small percentage of oil and protein, as well as vitamins (B, C) and minerals (K, P, Mg).

The dried solid endosperm is used to make sweets, biscuits, pastries, and confectionary in general. The grated material is similar to copra, but has a lower moisture content and a higher oil content and is of a finer quality.

'Cabbage' and 'apple' are luxury foods. The cabbage is derived from young leaf buds which are consumed as vegetable; the apple is the common name of the haustorium of the embryo.

'Copra' is the dried solid endosperm used in commerce. Coconut oil is expressed from copra. Good copra has an oil content of 60–65 %. Cooking fat, margarine, soap and other cosmetics are manufactured from the oil.

Economical utilization

Leaves are made into mats, hats, baskets and are used for thatching. Midribs of leaves are made into brushes. Trunks (stems) are used for building material. Wood is employed for cabinet work, construction, building and as a fire wood. Leaves, inflorescences, shells and other parts of the plant are used as fuel. Shells (entire pericarp of ripe fruit) are not a good manure, because they break down very slowly, but they are used in some regions for this purpose. Shells however make a good charcoal. Shells are also used as cups, bowls, spoons or for fuel. Commercial coire obtained from the husks (mesocarp) has many applications. Brushes, cordage, and ropes resistant to salty sea water are made of it. It is also used for matting (longer fibers) and filling material; the ropes are particularly elastic.

Coir, as a light and elastic material, is also used for making yarn, cables, rugs, carpets, bags, bristle fiber, mattress fiber and rigging.

Coir waste, the binding material separated fromt the fiber during processing can be used for manufacture of fiber boards, insulating material, rubberised flooring etc.

'Coconut cake' obtained by expressing the oil from the copra, still contains oil, protein, carbohydrates and fiber and makes an excellent food for cattle and poultry. Sometimes, it is also employed as a fertilizer.

As mentioned above, not only food is manufactured from the oil, but also cosmetics, such as soap, shaving creams, and shampoos. Stearin is separated from the oil and used in manufacture of candles.

The beautiful trees are also appreciated as ornamentals along the beaches and in gardens.

Medical use

In certain tropical regions, roots, bark, wax, flowers and fruits are used medically. The bark may be used as a dentifrice and and antiseptic. The ash is employed for scabies. The soft, downy substance from the lower leaf surface is used as a styptic. The astringent roots are employed in dysentery and intestinal ailments. Coconut water is used as a diuretic and an antihelmintic.

All parts of the plant are thus used in one way or another, several parts have many applications.

Commercially the fruit is however the most important organ of the plant. When unripe, the fruit has a smooth shining green skin which becomes yellow at maturity. Unripe fruits yield coconut water or milk, originating from the fluid endosperm. In this stage, a thin white and fleshy layer surrounds a cavity filled with sap. But as the fruit matures the soft and milky endosperm becomes solid.

Coir of good quality is obtained from unripe fruit, being harvested when the fruit is still green. The fruit mesocarp is first subject to a rotting process lasting for about one year during which the parenchymatous connecting tissue is broken down by fermentation. The material is then beaten to remove the remnants mechanically. The waste of coir, mainly composed of parenchyma, is thus separated from the fiber which corresponds to entire vascular bundles or to fiber bundles.

In the mature fruit, the endosperm is solid and the commercial copra is obtained from it. The main product of copra is the coconut oil; as the oil is expressed from the endosperm, the residues i. e. the remaining cell material forms the coconut cake.

Geonoma

G. comprises about 150 species of shrubs and treelets, distributed in tropical America and the West Indies. They are mostly shade-loving palms of montane forests, but also grow in hot and damp lowlands. The leaves are used for roofing by the natives.

Young flower clusters of *G. binervia* OERST. serve as a food when cooked.

The leaf anatomy of *Geonoma* sp. of the cloud forest of Venezuela is characterized by a non-differentiated mesophyll and by sclereids dispersed in the mesophyll. The stomata are of the *Rhoeo* type of development (ROTH & CLAUSNITZER 1969, ROTH 1992).

Geonoma simplicifrons WILLD. is a very common palm tree in the undergrowth of the Venezuelan transitional and real cloud forest. The treelet reaches 2–3 (5) m in height. Young plants have bipartite leaves. The leaves reach a length of 0.4–0.7 m.

The leaf anatomy has been studied by ROTH (1990). The mesophyll is not differentiated in palisade and spongy parenchyma, but contains fibers as well as raphides in some cells. The leaf is of the shade type.

Mauritia flexuosa L.

Mauritia flexuosa L. belongs to the Lepidocaryeae or Mauritieae with palmate (fan-shaped) leaves.

M. flexuosa is a tall dioecious palm with a branchless stem, about 15–20 m (30 m) high and 60 cm in diameter. It is usually called moriche in Venezuela, but has many other synonyms such as muriche, morichi, gábe, gábi, uara, iséol, ité, quiteve.

The leaves are deeply palmatifid. The blade is about 80–120 cm long, the petiole (with prickles) reaches 2–3 m. The segments reach 2–4 cm in width. The 2–3 m long spadix is pendant and has about 30 branches. The fruit is globular with a depression at the apex and measures about 4–6 cm in diameter. The pericarp is scaly on its outside and is red when ripe (Fig. 198 b).

Origin

Equatorial Brazil and adjoining regions.

Historical background

The plant was first mentioned by PIMENTEL in 1597. The vernacular name is deduced from the Tupí, 'mbur' meaning food and 'iti' meaning high tree.

Occurence

In Venezuela the palm occurs from the Orinoco Delta crossing the eastern Llanos up to the Alto Orinoco. It prefers a hot and humid climate. It is also found in Guiana and Western Amazonia. It forms large stands called morichales.

Anatomical description

Fruit. The fruit is about the size of a chicken egg. It is externally covered with brilliant-brown scales of rhomboidal shape. The scales are spirally arranged on the fruit, the spirals turning to the left as well as to the right. The yellow edible mesocarp consists of an outer small-celled and thick-walled parenchymatous part and an inner large-celled and thin-walled parenchymatous part. Dispersed in the mesocarp there are small vascular bundles and very conspicious large oil cells which are however mainly restricted to the outer mesocarp part, where they are quite numerous.

The scaly exocarp as well as the hard endocarp are sclerenymatous. The scales are interpreted as emergences by GUÉRIN (1949). The same author refers to the weakly developed vascular bundles as being naked because they are devoid of sclerenchymatous elements.

The storage substances in the mesocarp are oil and sugar. The pulp (mesocarp) is yellow, sweet and edible.

Ethnobotanical and general use

Nutritional use

The plant is highly appreciated by the indigenous Warao Indians, because it supplies them with many useful objects.

Young leaf buds 'cogollo' are eaten as a vegetable delicacy.

From the pith of the stem a sago-like starch can be extracted. The pith is also dried and toasted to be eaten as a kind of bread. The natives prepare wine from the fermented sap of the stem. The sweet fruits which have a taste of mango are eaten raw or refreshing drinks and a brandy are prepared from them. From fruit and seed a yellow oil is extracted.

Economical utilization.

Adult leaves are used as a covering for roofs and also are a source of fiber. Leaf-sheets are made into sandals.

The stems are used as poles in construction and for floating bridges.

The hard 'seeds' (endocarp) are used industrially to make ivory-like buttons.

Medical use

The fruit is digestive and laxative. An alcoholic beverage of the fruit is made for colds, influenza and to strengthen old and weak people. Leaves and shoots are used for cough. Young sprouts help for fever, vomiting, eye trouble, stomach ache. The cambium of the stem is masticated to cure whooping cough.

The Warao Indians use leaves and trunk for coughs, young sprouts for fever, vomiting, ear and stomachache, and the pith of the trunk for haemorrhage and whooping cough (WILBERT 1996).

Varieties

M. flexuosa var. *venezuelana* STEYERMARK has fruits 3.5–4 cm in diameter and occurs in the savannas around the Rio Caroní and Ciudad Bolivar.

Cultivation

The tree is easily propagated by seeds, which germinate after 2 months. The growth is rapid. The plant requires a hot and humid climate.

In nature, the tree grows along river beds and in boggy zones of the savanna where it forms the so-called Morichales.

Syagrus

Species of this genus are monoecious medium-sized or dwarf palms with ringed stems and pinnate leaves. The fruit is a drupe. The genus is restricted to South America. About 4 species occur in Venezuela of which

Syagrus orinocensis (Spruce) BURRET is cultivated. It has a slender stem, 5–8 m high. It is a very graceful palm which is widespread in the Amazon territory. The palm is extremely resistant to drought and even grows on rocky crevices of granite massifs. It endures hot winds for months. The seeds germinate quickly. Cultivation is however difficult. The seeds have to be sown directly in their final location or habitat. The palm is recommended as an ornamental (BRAUN 1968).

Syagrus sp. has been described for the Cloud forest of Venezuela. Its leaf anatomy has been described by ROTH (1992). The mesophyll is not differentiated in palisade and spongy parenchyma.

Papilionaceae

The zygomorphic crown is butterfly-shaped having a vexillum, 2 alae and a carina: there are usually 10 stamens. The leaves are simple-pinnate, the fruit is a legume. The Papilionaceae are mostly herbs, shrubs, and more seldom lianas or trees. Many species supply ornamentals.

The family is of great economic importance, supplying proteins, starch and oil (e. g. *Pisum, Lathyrus, Lens* and *Phaseolus*); further important species are *Arachis hypogaea* (peanut) and *Glycine max* (soya). *Lupinus, Trifolium, Melilotus, Medicago* etc. are essential fodder plants. Flowers are a source of honey. Many species are medical plants supplying gums, resins, saponins, alkaloids, lectins (phytohaemagglutinins), or sweet glycyrrhizin. The isoflavon rotenone, specific of the Papilionaceae, is used as a fish poison and insecticide. Several species of *Astragalus* supply the polysaccharide tragacanth. Balsam of Peru obtained from *Myroxylon* is antiseptic and wound healing.

Valuable timber such as Palisander or Rosewood (*Dalbergia, Machaerium, Pterocarpus*) is obtained from this family. *Machaerium* is well known by its anomalous secondary growth with intraphhloematic xylem (ROTH 1981).

Andira

At least 6 species are useful. The wood is useful. Chrysarobin from wood is irritant and antiparasitic. Bark and seeds are vermifuge, febrifuge, anthelmintic, purgative and emetic. bark is narcotic.

ROTH studied the bark structure 1981, leaf structure 1984, fruit structure and dispersal 1987, leaf venation 1996.

Andira surinamensis (BONDT.) SPLITG. EX SPRUCE. Buds and leaves ground and toasted are used for cicatrization. The grated fruit applied on the skin cures scabies (CASTILLO 1995).

Centrolobium

See Fig. 203. The wood is useful.

ROTH studied the bark structure 1981, fruit structure and dispersal 1987, leaf venation 1996.

Diplotropis

Wood is useful. Bark is vermifuge. Leaves help when blood is found in the stool.

D. purpurea AMSHOFF is ichthyotoxic, medicinal.

ROTH studied the bark structure 1981, leaf structure 1984, fruit structure and dispersal 1987, leaf venation 1996.

Dipteryx

Bark is febrifugal. Seeds are edible. Flowers are rubbed on the skin to cure rough, dry hands.

D. odorata WILLD. (tonka tree). Seeds are source of tonka beans producing, after fermen-

Fig. 203. *Centrolobium paraense*, Papilionaceae. Winged fruit (samara) with a large wing and spines on the base (ROTH 1987).

tation, a coumarin, used for flavoring and perfumery which is supposed to be narcotic and stimulant.

ROTH studied the bark structure 1981, leaf structure 1984, fruit structure and dispersal 1987, leaf venation 1996.

Lonchocarpus

At least 12 useful species are known. Wood is useful. Wood is antipyretic. Wood and bark are a fish poison; both are also used to destroy leaf-cutting ants. The Bark is the source of a drink.

Roots are used for leprosy. Roots are a fish and arrow poison, and are also applied against leaf-cutting ants. Young roots and leaves supply a blue dye. Roots are source of rotenone, used as an insecticide.

L. sericeus H.B.K. Wood is useful, bark is a laxative. ROTH studied the bark structure 1981, leaf structure 1984, fruit structure and dispersal 1987, leaf venation 1996.

Machaerium

Fruits are diuretic. Bark is the source of a gum-resin used as an antidote for snake bites.

The genus is characterized by intraphloematic xylem in the bark and the rhythmic formation of secretory cells in the form of continous rings in the bark. The secretory cells are very large being extended in a longitudinal direction (parallel to the axis); they have one large pit in the transverse wall and several large pit holes in the side walls (Figs. 29 and 32 in ROTH 1981).

ROTH studied the bark structure 1981, fruit structure and dispersal 1987.

Ormosia

Wood is useful. Bark is used for infected sores with pus. Seeds relieve cramps. An arrow poison is obtained from the seeds.

ROTH studied the bark structure 1981, fruit structure and dispersal 1987.

Platymiscium

P. pinnatum DUGAND. Wood is useful. ROTH studied the bark structure in 1981.

Pterocarpus

At least 12 species are useful. Wood is useful. Wood is used instead of cork because of its light specific weight. A red dye is obtained from it. West-Indian 'Dragon's blood' is extracted from the bark. A gum kino (astringent resin) is used to treat wounds. The kino contains kinotannic acid. Wood powder is emetic. Roots are used as an aphrodisiac.

P. rohrii VAHL. A leaf tea is an antipyretic. ROTH studied the bark structure in 1981, leaf structure 1984, fruit structure and dispersal in 1987, leaf venation in 1996.

P. officinalis JACQ. A gum exudates from the bark called Sangredrago (dragon-blood) used as an astringent and to harden and strengthen the teeth. An infusion of the bark is phosphorescent. The bark exudate is haemostatic and was formerly exported to Europe (PITTIER 1970). ROTH studied the bark structure (1981) and found tannin idioblasts of a giant size in the bark. ROTH studied fruit structure and dispersal (1987).

Passifloraceae

The representatives are herbs or shrubs and often climbers with inflorescences modified to tendrils.

The flowers are frequently large and showy (eg: passion flower). They usually have an androgynophore and a second corolla. Stamens mostly 5, carpels 3. The fruit is a capsule or berry. Seeds often with arillus and a fleshy endosperm.

Leaves are frequently lobed or pinnate, stipules are occasionally foliar; nectaries often on the blade base or leaf petiole.

Pollination by insects or humming-birds. Vessels frequently large and with simple perforations.

About 12 genera with 600 species, mostly in America and Africa.

Some species are ornamental (*Passiflora*) or supply edible fruits (granadilla, maracuja).

Extracts of the drug (herba passiflorae) are considered mild sedatives.

Cyanogenic compounds with a cyclopentenoid ring system (gynocardine type) are of taxonomic interest, as they have been only found in the Passifloraceae and Flacourtiaceae.

Passiflora

Historical background

CAULIN mentions *P. quadrangularis* with its fruits called 'parchas' in 1760. Between 1786 and 1789, ALCEDO described the same species and mentions the heavy load of the fruits which are of a lustrous yellow colour and have a sweet fragrance: it has a subtle peel and a delicate flesh, about 2–3 fingers thick and the hollow in the center is filled with an orange juice, even sweeter and more fragrant, with seeds covered by a delicious flesh (aril) which is also edible. In 1826, ANDRÉS BELLO dedicated some kind of poem to the 'Parcha Granadina'. ABBAD and HUMBOLDT mention also the plant.

The 4 species of *Passiflora* described in the following pages were collected by the senior author in the Cordillera de la Costa, pico El Avila. STEYERMARK & HUBER (1978) mention 17 species of *Passiflora* in their Flora del Avila, the majority of which are herbaceous and climbing.

The first botanical collection at the Avila was carried out by the German gardener BREDEMEYER in 1786. 14 years later, ALEXANDER VON HUMBOLDT and AIMÉÉ BONPLAND ascended the peak of the Silla (close to the Avila), and there noticed Passiflora.

Leaves. The leaves of Passiflora may be simple, compound, lobate or even divided, and their shape can vary considerably. Glands (extrafloral nectaries) either sessile or shortly stalked are common on the leaves (METCALFE & CHALK 1950) and frequently occur on the leaf petiole, while glandular spots are found less frequently on the lower side of the foliar lamina. On the petiole, the glands often form pairs, whereas glandular pustules on the blade are relatively rare. Glands serving as extrafloral nectaries were also found by ROTH (1969) on the involucral bracts and vegetative leaves of *Passiflora foltida* L. where they appear as 'emergences'. The bracts of *P. foetida* are divided into very delicate segments each of which terminates in a gland that has the shape of a tentacle, resembling that of *Drosera*.

Leaf shape. The leaf morphology of the Passifloraceae was studied by TROLL (1939). He considers the compound leaf with 3 well-developed leaflets (e.g. of *P. caerulea* and *P. racemosa*) as a fundamental type from which he deduces the leaves of *P. trifasciata* and *P. hahnei* in which the central part is favoured, while the lateral leaflets are reduced. However in another developmental series he observes a reduction of the terminal leaflet, as in *P. coriacea* and *P. capsularis*, where the blade adopts a butterfly shape similar to the leaf of *Bauhinia*.

P. kalbreyeri MAST. has a simple leaf with 3 strong principal veins, one middle vein and 2 laterals; at the end of each a slight indentation is formed so that 4 shallow lobules develop. The submarginal veins, on the contrary, are not continuous, but correspond to a complex system, i.e. a union of many vein endings so that a compound submarginal vein originates. *P. suberosa* L., on the contrary, has a compound leaf with 3 long leaflets and with 3 strong veins each of which supplies a leaflet. However, this species shows a great variability of leaf shape, as leaves with only 3 lobes in which the central part is very well developed have also been observed. In a young leaf stage, the lobes are less developed, while the central part is increased in surface. In the adult leaf, the terminal leaflet is better developed than the lateral ones. The leaves of the juvenile plant have a leaf shape in which the central part is better developed, whereas the successive leaves of the adult plant develop the more incised type with well-developed leaflets. The proportions between central part/free leaflets thus change during the development of the plant. A change of proportions can be observed not only in *P. suberosa*, but in the entire genus *Passiflora*, when different species are compared with one another.

In *Passiflora cuneata* WILLD. (Fig. 204), on the other hand, a complete reduction of the terminal leaflet takes place, while the 2 lateral leaflets develop very well. The terminal leaflet is reduced to a short tip. In spite of the reduction of the terminal leaflet and the favoured development of the 2 lateral leaflets, 3 strong principal veins are present. However, the leaves of this species are very variable. In other leaf forms, the central part of the blade is favoured and increases in length, while the lateral leaflets are reduced; in this leaf type, the central part of the limb develops first, whereas the lateral leaflets are delayed in their development. Thus the definite leaf shape of *P. cuneata* does not seem to have been established yet. Comparing the leaves of *P. kalbeyeri* with those of *P. cuneata*, the leaves of the former seem to remain in the juvenile stage.

The flowers of *Passiflora* are very characteristic. The name is deduced from the passion of Christ with the implements of his crucifixion: The hammers are the fertile stamens with their anthers (usually five), the nails correspond to the styles (usually 3) and the crown is composed of the sterile filaments, often present in several circles and beautifully coloured (additional crown).

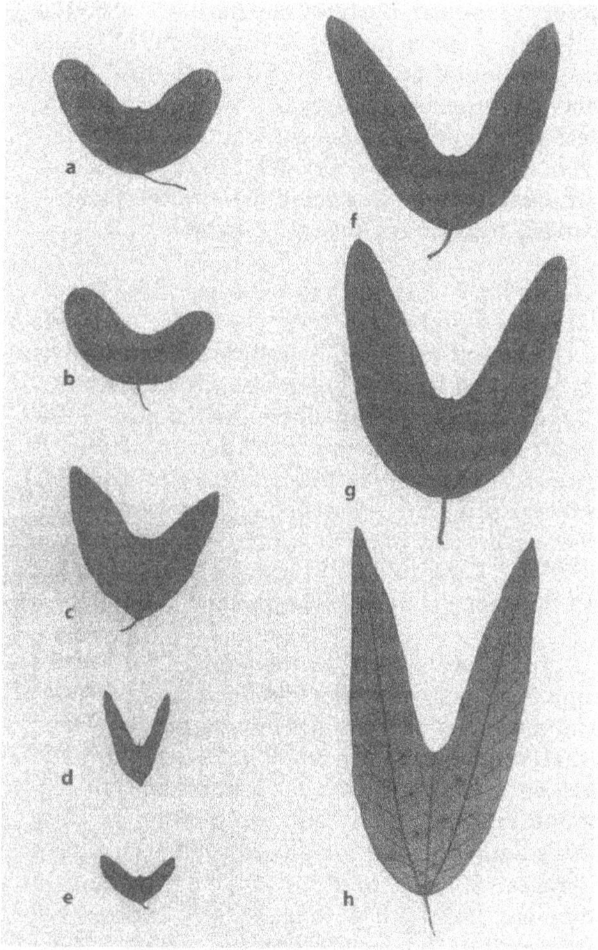

Fig. 204. *Passiflora cuneata*, Passifloraceae. Left: Series of successive stages of development of leaves, the lowest leaf is the youngest, the 2 following leaves are older. All the other leaves are adult and show the variation possibilities of shape (ROTH 1995).

Passiflora foetida L. (parchita de montaña, taguatagua, parcha de culebra, Fig. 205, 206) is a climbing herb, viscous to the touch and provided

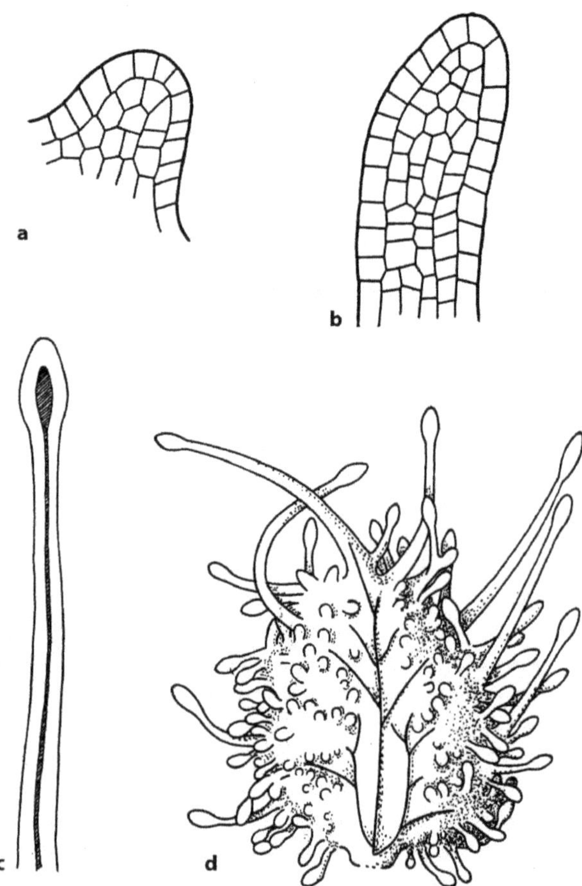

Fig. 205. *Passiflora foetida*, Passifloraceae. Developmental stages **a, b, c** of the glands of a young bract **d** (ROTH 1995).

with a dense indumentum of yellowish-brown colour. The leaves are more or less heart-shaped, 4–7 cm long and 3–4.5 cm broad, generally glanduloso-ciliate and more or less tightly hirsute. The terminal lobe is lanceolate or ovate-lanceolate, acuminate; the lateral lobes are suborbicular and often apiculate; the petiole reaches 1–4 cm in length. The flower peduncles arise solitary in the leaf axils and are 2–3 cm long. The bracts have ovate outlines, are 2–3 cm long, bi-tripinnatisect, the last segments are filiform. The calyx tube is shortly bell-shaped. Sepals and petals are oblong or ovate-oblong, 1–2 cm long, of white colour often mixed with lila. The filaments of the additional crown occur in several circles, in the 2 outer circles they are filiform, 5–10 mm long, violet and white or purple and white; the filaments of the further circles are capillaceous, 1–2 mm long and of violet or purple colour. The ovary is more or less globose and densely covered with white or brownish hairs.

The fruit is globular, 2–2.5 cm in diamter, yellow and more or less densely pilous.

The species is found in almost the entire tropical America. In Venezuela it is amply distributed from sea level up to 1500 m.

Passiflora suberosa L. (parchita de culebra)

Taxonomical description

The species is a climbing plant, glabrous or densely pubescent. The lower part of the axis is slightly woody and suberized hence the species name. The stipules are linear-subulate, 6–8 mm long, 1 mm broad, and inconspicuous. The leaves are very variable in shape: entire to profoundly trilobate, the lobules linear to broad-ovate, acute or obtuse, the blade base is rounded or at times peltate. Petiole 0.5–4 cm long with 2 glands at the base of the blade.

Flowers 0.8–3 cm in diameter, solitary or in pairs in the axils of the leaves or sometimes arranged in axillary racemes. Bracts minute, fine and caducous. Sepals ovate-lanceolate, slightly obtuse, of greenish-yellow colour, petals absent. An inner crown is formed by the filaments which are white with a yellow top and a red-purple base. Fruit globose to oviform, of dark purple or black colour, 6–15 mm in diameter. Common weed in almost all of tropical America. In Venezuela it is frequent in the Cordillera de la Costa between 900 and 1500 m, in the deciduous forest and cultivated in 'cafetales'.

Anatomical description

Leaf. The upper epidermis of *Passiflora cuneata* is composed of comparatively large cells with thickened outer walls and a cuticle. Below follows a single layer of palisade cells; the cells are relatively short and some of them appear funnel-shaped becoming constricted towards the base; they contain large chloroplasts arranged along the lateral walls and the cell bottom. Four to five layers of loose spongy parenchyma composed of roundish cells with large chloroplasts lie beneath; the cells leave larger intercellular spaces between one another. The lower epidermis is distinguished by cells with convex outer walls which are vaulted above the surface in the form of a watch glass. The cell walls are delicate. The stomata are restricted to the lower epidermis. The veins develop only little mechanical tissue; a parenchymatous sheath surrounds the veins which is rich in druses of calcium oxalate. The entire leaf is thus thin and delicate and more of the shade type.

Foliar glands. Glands, probably extrafloral nectaries, are found in pairs on the leaf petiole of *P. suberosa*. In *P. kalbreyery* and *P. cuneata*, on the contrary, glands are found on the lower side of the limb, along both sides of the midvein. Frequently, the glands occur in pairs on the limb, but the uppermost gland can remain single. At the leaf base, the glands are closer together, while they separate continuously from one another towards the leaf tip. This is due to the fact that the limb elongates more in its upper part than in the basal part, although the distance between the pairs, as seen in a longitudinal direction, is approximately equal in very young stages of the leaf. The glands appear on the limb as transparent spots like small windows. The blade is thicker where the glands develop.

The glands of *P. cuneata* arise from the lower epidermis which becomes pluristratified by periclinal cell divisions so that anticlinal cell rows of up to 10 cells are formed. A pluristratified epidermis often occurs in the Passifloraceae and is a family characteristic. But in *P. cuneata*, the epidermis becomes only pluristratified where the glands develop.

Likewise, the first epidermal layer beneath the lower epidermis divides periclinally. The subepidermal cells are distinguished by the presence of crystal druses. Tracheids with helicoidal wall thickenings may be observed above the subepidermal layers. Between the tracheids and the upper epidermis, there are few parenchymatous layers. The first layer below the upper epidermis is not transformed into a palisade parenchyma, but is composed of more or less roundish cells with few chloroplasts. The glandular zone is free from chloroplasts in its peripherical part, while few chloroplasts are found towards the inside of the gland.

The outermost cells of the gland on the lower surface which are in contact with the environment have thickened outer walls covered by a cuticle. The glandular cells maintain their disposition in the form of anticlinal rows even in the adult leaf; they have delicate walls, are rich in protoplasm and accumulate substances of secretion, but do not leave intercellular spaces between one another. In old leaves, the glands are frequently attacked by fungi (see also BRÜCHER 1989).

The glands of *P. kalbreyeri* develop in the same way and have more or less the same structure.

Extrafloral nectaries of P. foetida. (Fig. 205, 206) Below the showy flower, there are 3 involucral bracts which are divided into very fine filamentous segments each of which terminates in a gland (Fig. 205, 206). The bracts are viscid to the touch due to the secretions of the glands.

Nectar secreting glands are present in almost all species of *Passiflora*, either in the form of protuberances on the leaf petiole or along the margins of the bracts (KILLIP 1938).

Presence or absence of glands as well as their shape, position and number constitute important differences for the identification of species or groups of species.

The glands of the adult bract of *P. foetida* are very similar in their structure to the tentacles of *Drosera* (Fig. 206). A pluricellular stalk and a head are distinguished; tracheids pass through the center of the stalk and only very slightly enlarge in the head of the gland of *Passiflora*. Three distinct layers surround the tracheids: The outermost layer of the head – corresponding to the modified epidermis – is composed of anticlinally elongated cells. The underlying first subepidermal layer is also very much alike in both *Passiflora* and *Drosera*. Only the third layer shows some differences with respect to the thickening of the cell walls and the tangential enlargement of the cells in *Drosera*. This intermediate layer which is also called a physiological sheath, is of particular importance in secretory glands, as the entire walls or only the anticlinal walls are cutinized and possibly have a selective function in the passage of substances. The secretory epithelium in *Drosera* as well as that of *Passiflora* corresponds to the outermost (epidermal) layer. This layer secretes slime and digestive enzymes in *Drosera*, while the glands of *Passiflora* secrete nectar. Many small oil drops which stain orange with Sudan III are present in the cells of the secretory layer of *Passiflora*.

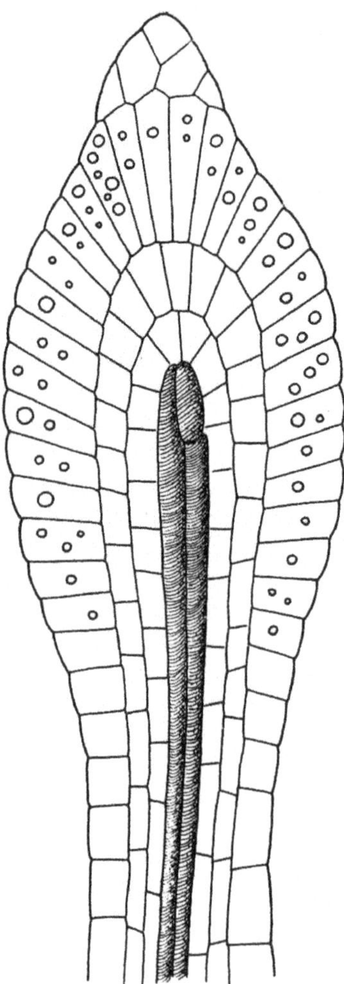

Fig. 206. *Passiflora foetida.* Head of an adult tentacle with 2 rows of tracheids in the center; the oil drops in the epidermis are conspicuous (ROTH 1995).

Although the glands of the bracts of *Passiflora* and the tentacles of the leaves of *Drosera* have a very similar structure, they serve different functions. The physiological function of the glands in *Passiflora* is probably that of attracting insects for fecundation of the ovule.

The same tentacles as observed on the bracts of *P. foetida* are also found on the marginal teeth and on the upper side of the vegetative leaves of *P. foetida*, while glands are rare on the lower leaf side. The same glands also cover the surface of the petiole of the vegetative leaves. Certainly, the glands on the bracts are larger and more attractive, whereas those of the vegetative leaves are hidden between simple hairs.

Three different types of glands may thus be distinguished in the genus *Passiflora*: The most common type is represented by the glands of the leaf petiole which usually appear in the form of protuberances and in pairs. A less frequent type are the glands in the form of transparent windows on the limb which may also form in pairs. The third type is found on the bracts in the form of tentacles that arise as true emergences.

The glands are only furnished with tracheids with helicoidal wall thickenings which supply the glands with water and solutes. For this reason, the glands seem to be a transitional form between nectaries which usually are in contact with the phloem, and hydrathodes, which are usually provided with tracheids.

Ethnobotanical and general use

Nutritional use

More than 380 taxa of the neotropical genus *Passiflora* have been described since LINNÉ (1753). It is said that 50 of them have edible fruits (BRÜCHER 1989). Some of them are mentioned in the following: *P. alata* DRYAND, *P. amethystina* MIKAN., *P. cearensis* BARB., *P. cincinnata* MASTERS, *P. caerulea* L., *P. edulis* SIMS, *P. incarnata* L., *P. laurifolia* L., *P. ligularis* JUSS., *P. maliformis* L., *P. mollissima* (H. B. K.) BAILEY, *P. mooreana* HOOKER, *P. organensis* GARDN., *P. pinnatistipula* CAV., *P. popenovii* KILLIP, *P. psilantha* (SODIRO) KILLIP, *P. quadrangularis* L., *P. salvadorensis* DON SMITH, *P. tripartita* (JUSS.) POIR., *P. van-volxemii* TRIANA & PLANCH., *P. vitifolia* H. B. K., *P. warmingii* MAST.

This enumeration does not however imply that the other species are not edible, but they may be less tasty or are used medically (UPHOFF 1968).

The pulp of the fruit as well as the aril of the seeds which can be made into a juice at fruit maturity, are edible in many cases. A great variety of uses of the fruit is known: The fruit may be eaten raw, because it has an aromatic taste, or it may be consumed with sugar, when it is sour-sweet. It is used for flavoring sherbets, for icing, cakes, confectionary, trifles (e.g. *P. edulis*). This species also supplies the maracuya juice, which is prepared on a commercial scale. In Venezuela, it is called 'parchita maracuya', a name which comes from the Quechua 'murukúya' or 'murukkóya'. The Granadilla corresponds to *P. ligularis* which is now less cultivated in Venezuela, because the Maracuya is preferred; the fruit of the Granadilla contains numerous seeds with a mucilaginous aril which is the edible part. However, the pulp of many other species of *Passiflora* is also edible and is converted into refreshing drinks. *P. ligularis* is also used as a breakfast fruit and is sold on markets. Jelly is made of *Passiflora* and ice-cream. Crema de Curuba is made of *P. mollissima*. *Passiflora* is also used for a variety of pastries, sauces, and a chutney is made from unripe fruits. Further preparations are gelatinous sweets, mousse, milk shakes; the fruit is preserved in syrup or wine. Unfortunately its durability is short and it has to be consumed rapidly or immediately made into a preserve.

VELEZ & VELEZ (1990) mention the following edible species for Venezuela: *P. quadrangularis*, *P. ligularis*, *P. edulis* SIMS var. *flavicarpa* DEGENER. and *P. edulis* SIMS. var. *edulis*, *P. manicata* JUSS. and *P. mixta* L. They also mention a poisonous species: *P. adenopoda* DC. which looses its toxicity when the skin is completely purple; in the unripe state however it contains a cyanogenic glycoside. Some other species contain this glycoside in the roots, in the leaves, the fruit shell and in the flowers.

The pulp of the edible fruits is usually rich in vitamin C, niacin, vitamin A. Maracuya contains 1–4 mg/ml serotonin in 100 g. Further contents are sugar (8–10 %), fruit acid (2–5 %), pectin (0.2–0.3 %) and minerals (0.4–0–6 %), e.g. phosphate and potassium (*P. edulis*: BRÜCHER 1989). The same author also cultivated a giant passion fruit from *P. quadrangularis*.

Due to the presence of a very large number of seeds in the fruit, it is preferably made into a juice, particularly for commercial purposes.

Economical utilization

Many species of *Passiflora* are beautiful ornamentals with their 'passion flowers' in blue, red or purple colour and with the symbols of the passion of Christ: hammer (anthers), nail (stigma) and crown, e.g. *P. caerulea* L., *P. racemosa*.

Medical use

Certain species have sedative and narcotic properties, e.g. *P. ciliata* DRYAND. It is said to produce a deep and restful sleep. It is therefore used as a remedy for insomnia, hysteria and convulsions in children. Dried flowering and fruit tops of *P. incarnata* are used medicinally in neuralgia, insomnia, diarhoea and dysmenorrhoea. *P. salvadorensis* DON SMITH has diuretic properties. *P. quadrangularis* L. supplies pharmaceutical preparations made in Venezuela from the alkaloids which have sedative, antispasmodic and hypnotic effects. The infusion of the leaves of *P. foetida* L. is used as an emmenagogue; at the same time it has a sedative effect (UPHOF 1968, VELEZ & VELEZ 1990).

P. edulis SIMS is taken orally against bronchitis, asthma and as a diuretic. Externally it is applied against haermorrhoidal inflammations. Fresh leaves are used in the treatment of hypertension and to induce diuresis. An aqueous extract of all aerial parts of the plant is used against tetanus, epilepsy, insomnia, nervous tension and as an antihypertensive (GUPTA 1995).

The leaves of *P. suberosa* L. are applied against bucal infections with *Candida* (candidiasis), as a 'purifyer' of the blood and as a diaphoretic (GUPTA 1995).

The entire plant of *P. foetida* L. has diuretic effects. The infusion of the leaves is used as an emmengogue and to tranquillize during hysteria. It is also used against inflammations of the face and to cure diarrhoea in hares (in veterinary medicine). An infusion of the flowers is applied to cure high fever and to preparare enemas. An infusion also regulates the menstrual flow (RODRIGUEZ 1983, DELASCIO CHITTY 1995).

P. mixta L. F. A decoction of the leaves is applied as a tranquillizer of the nerves and helps to cure fever (DELASCIO CHITTY 1995).

An infusion of the entire plant of *P. mollissima* (H. B. K.) Bailey is taken to combat insomnia, depression, anxiety and any other nervous disease (DELASCIO CHITTY 1995).

Leaf teas of *P. quadrangularis* are taken for high blood pressure. Tea of leaves helps against heat, hypertenseion and diabetes. The plant has sedative and antihypertensive properties (SEAFORTH, ADAMS & SYLVESTER 1983).

The roots as well as the leaves and the flowers of *P. rubra* have a slightly narcotic property, similar to that of opium (MANFRED 1982).

Leaves and flowers of *P. foetida* L. have antipyretic and sedative properties and are used as an emmenagogue. They are used against fever, hypertension, insomnia, problems of menstruation and cough (ALBORNOZ 1993).

Leaves, flowers and fruits of *P. mixta* L. have diuretic, antipyretic, hypotensive, diaphoretic, sedative, refreshing and analgesic properties. They are applied against cystitis, depressions, headache, fever, irregularities of menstruation, nervousness, neuralgia, parasitosis and parotitis (mumps). The root is a powerful cumulative narcotic (ALBORNOZ 1993).

The leaves of *Passiflora ligularis* and of other species of *Passiflora* are used as a cataplasm for liver disorder. Leaves with almond oil and brandy applied as a cataplasm help against head ache. A decoction of the leaves is used as a bath against ulcers and sores. The fresh sap of the leaves taken together with sweetened water has antipyretic, antispasmodic, diaphoretic and anthelmintic properties. The leaves of *P. edulis* seem to produce hallucinogenic effects.

The flowers in infusion are taken against epilepsy. A decoction of the flowers only or flowers together with the leaves, have sedative effects in case of insomnia, restlessness, hysteria, nervous headache, depression: A decoctionm of 1–2 spoonfuls of the flowers in a cup of water is taken 2–3 times a day or a tincture of 50–60 drops is taken according to requirements.

Ground seeds have anthelmintic effects. The fruit shell as a decoction is used against cough. LOPEZ-PALACIOS (1987) considers passiflorine as the most active substance.

Method of use

See above. Cataplasm, infusion, decoction, or fresh, in the form of a bath.

Healing properties

Mostly sedative, diaphoretic, anthelmintic, against infecitons with *Candida*, antipyretic, analgesic, against depressions, headache, irregular menstruation, antihypertensive etc. The main application is against insomnia and nervousness.

Chemical contents

Passiflorine, the alkaloids harman and harmine, harmol, tryptamine, several sesquiterpenes, coumarin, caffeic acid, hesperidin, vitexin, quercetin, β-sitosterol, stigmasterol and many other substances (GUPTA 1995) such as, gynocardin, suberin A, passisuberosin, epipassisuberosin, epivolkenin and serotonin are present.

Toxicity

As mentioned above, some species of *P.*, or some of their organs (e.g. roots) are poisonous. The poisonous principles are, according to BLOHM (1962), prussic acid (in leaves, pericarp and immature seeds) and passiflorine found in *P. quadrangularis*. When the pulp is eaten in large quantities, it has a somniferous effect; this may be attributed to passiflorine which possesses lethargic properties. The root is anthelmintic; it is regarded as a narcotic and poisonous and in a powdered form, mixed with oil, it is esteemed as an emollient poultice.

VELEZ & VELEZ (1990) ascribe the toxicity of *P. adenopoda* and of other species of *Passiflora* to cyanogenic glycosides.

Varieties and related species

The species with similar properties were mentioned above. *P. edulis* is the commercially most interesting species. Many varieties of *P. foetida* L., exist in Venezuela (VELEZ & VELEZ (1990). The best known variety of *P. edulis* in Venezuela is var. *flavicarpa* DEGENER (Maracuyá) (HOYOS 1989). Hybrids are known of *P. mollissima* x *P. mixta*.

Cultivation

The species of *Passiflora* are usually propagated by seeds which germinate after a few weeks. The seeds of *P. edulis* can be stored for 3 months or more after having been washed at room temperature. In commercial culture, the young seedlings are planted at distances of 3–5 m. The plant must be supported by wire or wooden ledges. The plant originates from the hill region of Brazil and the most suitable ecological conditions are therefore altitudes above 1500 m in a cool tropical mountain climate. Commercial plantations have failed when undertaken in the hot humid tropics (BRÜCHER 1989). The growth of the plant is rapid and vigorous. The best conditions for this plant are sandy, non-calcareous soils with some clay and much organic matter, a good drainage and a pH between 4.5 and 5.5. The first fruits can be harvested after 6–8 months. The harvest is almost continuous, but is most abundant in the first months of the year. An annual yield of 6–10 t/ha can be considered satisfactory. The economic life span of the plant is 3–8 years. At present, there is a considerable increase in the production of passion fruit in many tropical and subtropical regions of Africa, Australia, Oceania and America, but self-fertile highly productive selections are still lacking (BRÜCHER 1989, HOYOS 1989).

HOYOS describes conditions for cultivation of *P. cincinnata* MASTERS, *P. edulis* SIMS, *P. laurifolia* L., *P. ligularis* Juss., *P. mollissima* (H. B. K.) BAILEY, *P. mixta* L., *P. nitida* H. B. K., *P. quadrangularis* L., and *P. seemanni* GRISEB.

Passiflora quadrangularis has the largest fruits, measuring 20–30 cm in length and 10–18 cm in diameter.

Observations

The following species are described by BRÜCHER (1989): *P. cincinnata* MASTERS, *P. caerulea* L., *P. ligularis* JUSS., *P. laurifolia* L., *P. mollissima* (H. B. K.) BAILEY, *P. mooreana* HOOKER, *P. vitifolia* H. B. K. *P. edulis* SIMS, and *P. quadrangularis*.

HOYOS describes (1989) *P. cincinnata* MASTERS, *P. edulis* SIMS, *P. laurifolia* L., *P. ligularis* JUSS., *P. mollissima* (H. B. K. BAILEY, *P. mixta* L., *P. nitida* (H. B. K., *P. quadrangularis* L., and *P. seemanni* GRISEB.

VELEZ & VELEZ (1990) describe *P. quadrangularis* L., *P. ligularis* JUSS., *P. foetida* L., *P. edulis* SIMS var. *flavicarpa* DEGENER and var. *edulis*, and *P. adenopoda* DC.

Some species show a great variability in leaf shape which causes confusion in identification, e.g. *P. cincinnata* and *P. suberosa*.

As indicated above, the presence or absence of glands as well as their type, shape, position on the plant, arrangement (in pairs) and number constitute important characteristics for the distinction of species or groups of species.

Phytolaccaceae

This family is mostly haplochlamydeous, actinomorphous or very rarely zygomorphous (*Petiveria*), generally hypogynous and only seldom perigynous (*Petiveria*); tepals 4–5, mostly inconspicuous, herbaceous or membranaceous. Androecium and gynoecium are very variable in form and number of members, each carpel has only one basal campylotropous ovary. Flowers in racemes, spikes and seldom in panicles. Embryo always bent. Fruits are berries, nuts or loculicidal capsules.

Trees, shrubs, herbs or frutescent. Anomalous secondary growth and persistent subterraneous rapes are occasionally present. Pantropical, mainly American.

Crystals of calcium oxalate in the form of raphides and styloids, druses are rare. Chemical peculiarities are betalains (betaxanthins) and cyclit (pinit: mesembrin, mescaline).

Petiveria alliacea L. (anamú, mapurite)

Taxonomical description

Herb or undershrub (30) 50–150 cm high, ramified (Fig. 207 a). Stalk ribbed, with a white pubescence consisting of simple trichomes in the young parts. Leaves simple, alternating, blades of a papery consistency, glabrous or only slightly pubescent. Blades are small-elliptic, 6–18 cm long and 2.5–6 cm broad, with an acuminate tip, entire margins and an attenuate base. Venation pinnate, brochidodromous with 10–13 pairs of secondary nerves. Midrib prominent on the abaxial side, but only slightly elevated above the upper leaf side. Petiole 0.7–1.2 cm long with a reduced pubescence. Stipules linear-lanceolate, 0.1–0.2 cm long.

Flowers in axillary or terminal spikes, which are 10–40 cm long: the spikes arise solitarily or in pairs in the leaf axils. Perigone greenish-white, consisting of 4 segments, about 4 mm long; stamens 4–8, perigynous. Ovary is ellipsoid, tomentous, originating

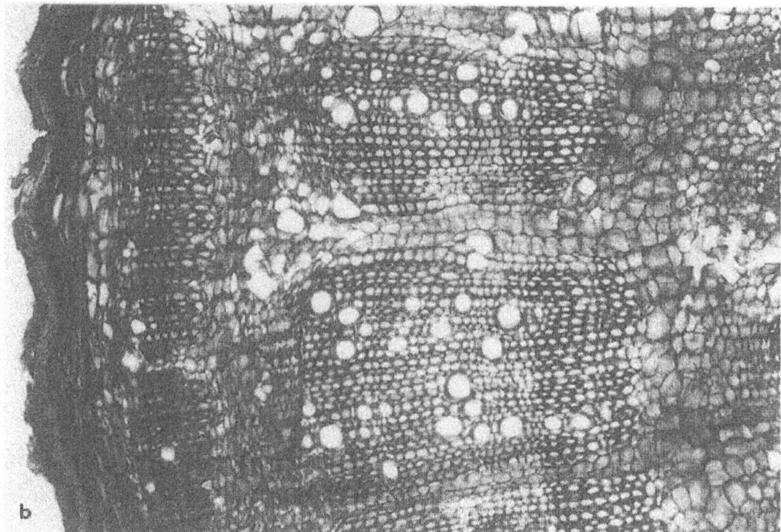

Fig. 207. *Petiveria alliaceae.* **a** Habitus. **b** Section of root in t.s.

from a single carpel and with 4 hook-shaped prolongations.

The fruit is elongated - cuneiform, 0.8–1 cm long, and provided with 4 hooks bent downwards.

The garlic smell of this plant is very strong and even lasts many years after the plant has been prepared and dried for a herbarium specimen.

Origin

Neotropical. From Florida and Mexico, throughout the West-Indies down to Argentina.

Historical background

Principally the Mayas used this plant medicinally and in magic art. In Brazil, it is applied in religious rituals of Afro-Brazilians. At present, it is one of the plants with the most ample medicinal application in the area.

Occurrence

In Venezuela, the species occurs in temperate as well as in hot regions ascending up to 1200 m. It is now grown in some parts of Asia and Africa.

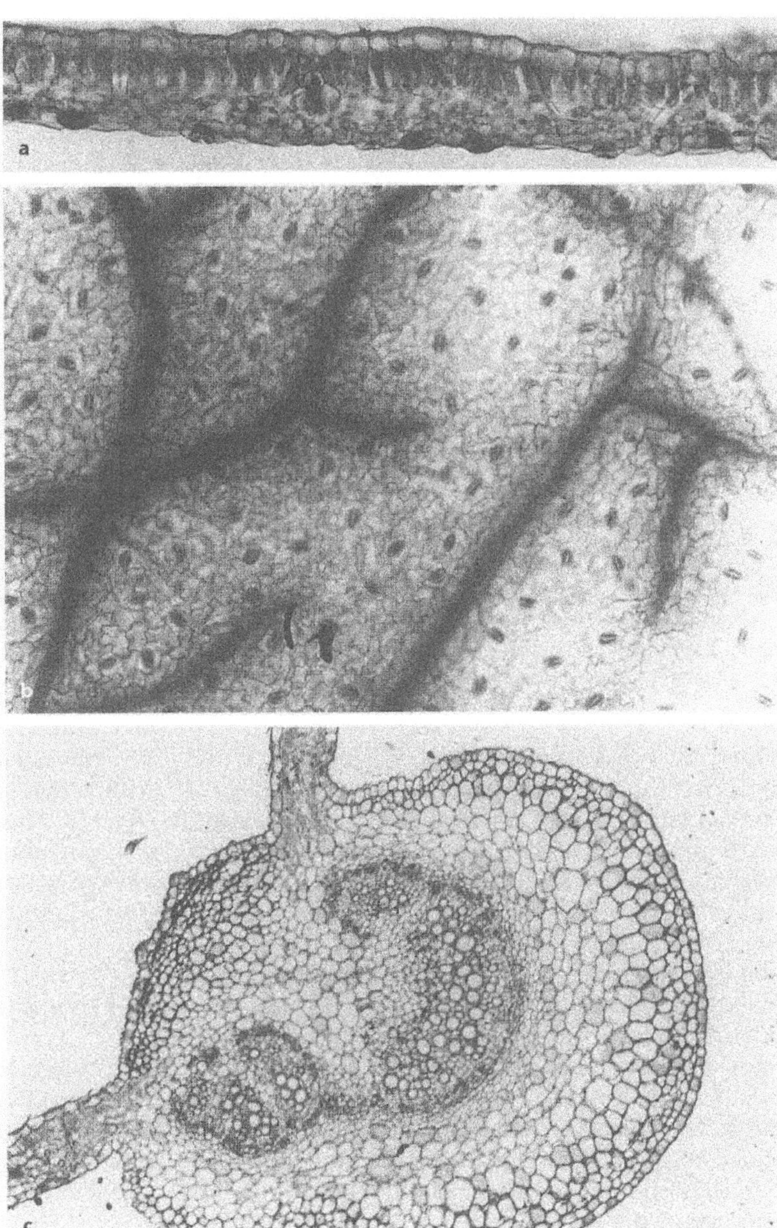

Fig. 208. *Petiveria alliaceae.* **a** T.s. of leaf blade. **b** Lower epidermis in surface view with stomata (note the vascular bundles). **c** Midrib (× 10).

Anatomical description

Leaf (Fig. 208, 209). The leaf is dorsiventral and amphistomatic, however, the stomata are very rarely found on the upper side. As seen in a surface view, the anticlinal epidermis walls are very wavy on both, upper and lower surface. As seen in t.s. the cells of upper and lower epidermis are of a regular size. The palisade parenchyma forms a single layer of comparatively short cells. The spongy parenchyma occupies about the same proportion as the palisade parenchyma or is only slightly better developed. Its cells are more or less isodiametric or globular and form arms towards one another; the spongy parenchyma is shomewhat looser than the palisade parenchyma. Very conspicuous are the very large crystals in the form of styloids (Fig. 209) which may traverse the mesophyll in an anticlinal direction reaching from the upper to the lower epidermis. They attain an enormous size, when they extend obliquely parallel to the leaf surface, occasionally excceding 10 or more epidermis cells in length.

Simple uniseriate hairs and glandular hairs with a multicellular pedicel and a unicellular head are scarce in the upper as well as in the lower epidermis. The stomata are slightly elevated above the surface and have a subsidiary cell on each side, one

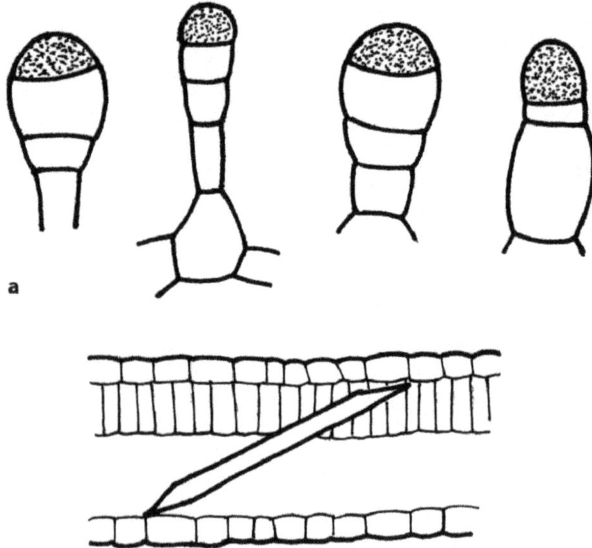

Fig. 209. *Petiveria alliaceae.* **a** Glandular hairs (head intensely stained with toluidine blue). **b** T.s. of leaf blade, indicating upper and lower epidermis, palisade parenchyma and a large styloid, oriented diagonally in the mesophyll.

larger, the other smaller, which are parallel to the long axis of the pore (rubiaceous) according to METCALFE and CHALK, or parallel type, according to ROTH & CLAUSNITZER (1969). As both subsidiary cells are developed one after the other in the same way as the segments of a 2-edged apical cell, the different sizes of the subsidiary cells are easily explained. METCALFE & CHALK (1950) do not report glandular hairs for the entire family of Phytolaccaceae, which is a very astonishing fact. However, we often observed club-shaped hairs or hairs with a swollen end, the apical cell of which was intensely dyed with toluidine blue, and which may correspond to glandular hairs; they may probably contain slime and the strong smell of the plant possibly emanates from these glands.

The midrib is elevated above the lower and the upper side, but projects more conspicuously on the lower one. Palisade and spongy parenchyma are interrupted in the midrib region and replaced by collenchyma. The vascular system is arranged in the form of an open arc, but where bundles deviate to the secondary nerves, the extremity of the arcs is transformed into a ring, as seen in transverse section.

The secondary nerves and those of a higher order contain a single vascular bundle which is surrounded by a parenchymatous vascular sheath; this is transcurrent to both epidermal laysers in the larger bundles. The leaf shows a tendency towards the shade type.

Axis. In a twig, about 3 mm in diameter, the epidermis is still maintained. The primary cortex consists of parenchyma in which large styloids are embedded. Fiber bundles of varying size and shape unite in an interrupted pericycle ring, as seen in transverse section. The vascular system forms a continuous ring (cylinder). The rays are inconspicuous, varying in size from 1 to pluriseriate. The parenchymatous pith is very extensive and quite a few cells contain large styloids.

Root. (Figs. 207 b, 210). A root, about 7–8 mm in diameter, has a multi-layered cork, a small parenchymatous primary cortex and a well-developed vascular system of anomalous structure similar to that observed in *Phytolacca*. As seen in transverse section, 5 superposed rings or stories of xylem and phloem are separated from one another by 'connective' parenchyma which is comparatively extensive. In each ring, individual bundles with a well-developed xylem and a cap of phloem on top are tangentially separated from one another by broad multiseriate (2–12 seriate) rays. The xylem is dispersed-porous and the vessels are mainly solitary. Some type of growth rings may be observed in the xylem. In the center, there is the primary xylem with about 3–5 xylem rays. All types of parenchyma cells may contain styloids. The connective parenchyma cells are stuffed with starch grains, as are the rays; this region thus functions as storage parenchyma.

Ethnobotanical and general use

Economical utilization

The natives use the plant as an insecticide. Powdered roots are put between woolen articles to protect them against moths. The plant is also applied as a disinfectant.

Furthermore, the plant is used as a poison for the preparation of a certain type of curare and as a fish poison.

Medicinal use

Used parts: Entire plant, root, leaf.

Entire plant. The entire plant is boiled in wine when used as a cataplasm; to conbat arthritis, the entire plant is boiled in water, some urine and Epsom salt are added, and the liquid is applied lukewarm on a cloth to bandage the painful part.

A decoction of the aerial parts of the plant is used every day in the form of a mouthwash to avoid caries and to fortify the gums.

Leaf. The leaves are applied externally as well as internally. The sap of the entire plant or of the

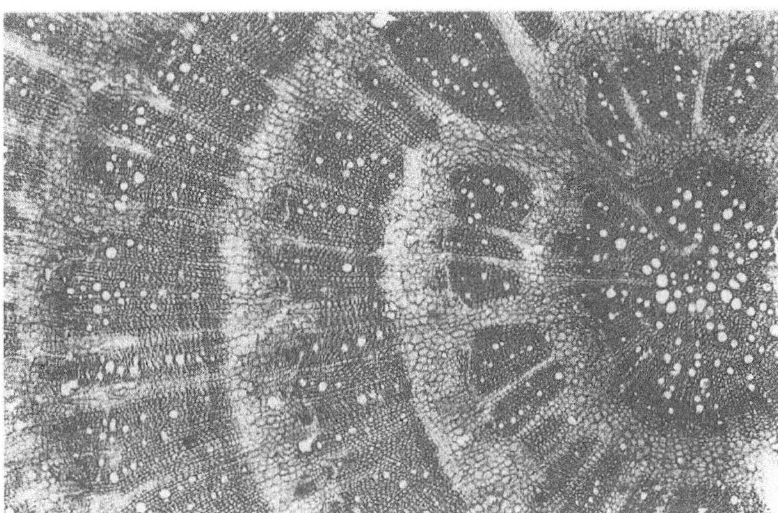

Fig. 210. *Petiveria alliacea*. Root. Note the stories of xylem, phloem and connective parenchyma (light) (× 6.3).

leaves alone applied externally cures skin diseases, arthritis and tooth ache.

A decoction of the leaves is used against colds, cramps, inflammation of the bladder and asthma. A decoction of 15–20 leaves in one liter of water is prepared and a cup of the decoction is taken 3 times a day. The liquid is used for a mouthwash against toothache.

The infusion of the leaves is applied 'to alleviate the labourpains of women and cows'. Appranetly there is not much of a difference between women and cows according to the indigenous people.

The leaves are furthermore used as an antipyretic and against disorders of the lungs. They are also applied as an aphrodisiac, as an abortive and against snake bites.

To cure inflammation of the bladder, the leaves in decoction are used as a footh bath.

An infusion of the leaves mixed with milk helps against intestinal parasites. The leaves are said to be sudorific and depurative. A bath prepared with the leaves is used as an antihysteric. Leaves in decoction are also used against spasms.

Leaves and branches in decoction fortify the gums and avoid caries, if used every day.

A decoction of leaves and roots is diuretic and lowers the fever. An infusion of the leaves mixed with milk is a vermifuge.

The anticancerous properties ascribed to the plant seem to be interesting. It is assumed that the plant has antineoplasmic and oncolytic properties. The first publication on the anticarcinogenic effects of the plant was probably that of Toyos Alcalá in 1975. The Anamú Foundation of America Inc., Miami, Florida, was founded to study the healing properties of the plant in 1979. In 1983, the plant was presented as a healing marvel at the VII. Botanical Congress of Venezuela. Some cases of healing of leukaemia were reported.

Root. The roots are mainly used against the same sicknesses and in the same way as the leaves. Root teas and infusions are used in cases of flu, venereal diseases, cystitis, bleb, dysmenorrhoea, womb inflammation, and as an abortifacient. A root bath is used for heat and a poultice is applied for colds in the head.

The root is said to have antispasmodic, vermifuge and abortive properties. Crushed root put on a sick tooth destroys the nerve and alleviates pain.

The root is used as a sudorific, diuretic, antirheumatic, as a antipyretic and against venereal diseases.

Method of use

To cure cancer, 24–30 fresh and green leaves are put in a liter of cold water; this dose is taken in 3 portions a day in the morning, at noon and at night. The treatment is continued for several months until the symptoms of the sickness disappear. Some patients take it for years. When fresh leaves are not available, a decoction of dried leaves can also be used.

For further applications, see also the recipes described above.

Healing properties

Diuretic, sudorific, expectorant, antispasmodic, depurative, abortifacient, vermifuge, antiinflammatory, antitumoural.

The plant is furthermore used against arthritis, bronchitis, asthma, flu, hysteria and nervous diseases, to treat scorpion stings, toothache and fever;

it helps also as an emmenagogue, against venereal diseases and disorders of the uterus.

The plant extract shows antimicrobial activity.

Chemical contents

The milk of cows which feed on the plant is tainted with sulphur compounds (flavour of garlic).

Leaf extracts give a positive test for alkaloids.

The toxic principle of the plant is unknown. It is used as a fish poison and for the preparation of a certain type of curare. Among other substances the presence of tannins, saponins, glycosides, alkaloids and β-sitosterol have been proved.

An antibacterial activity of ethanolic extracts of the plant was reported for Gram-negative organisms and an antimycotic activity against several pathogenic fungi in aqueous extracts was found as well.

The watery extract of leaves and shoots shows a stimulant effect in the uterus. It is used as a stimulant of menstruation. Activities against *Mycobacterium tuberculosis* and *Candida albicans* were observed. Another effect is the stimulation of phagocytic activity.

A decoction of the leaves shows effects against *Epipdermophyton floccosum.*

Pharmacologically, it could be demonstrated that a decoction of the leaves has antiinflammatory and analgesic effects. It inhibits oedema of the feet.

The essential oil of the leaves exercises a suppressive action of the nutrition of certain pathogeneous insect larvas (*Attagenus piceus*), and acts as an insecticide against adult insects (*Cimex lectularius, Musca domestica*, mosquitos); it is furthermore a repellent of clothes moths. The alcoholic extract of the leaves acts as a nematicide for *Melodiogyne* spp. possibly due to the presence of thiophene derivatives (trithiolane?).

Abortive properties in humans are attributed to the root, of which curare and a fish poison are prepared. Ethanolic extract of the roots applied locally and orally showed an inhibitory effect on dermatitis caused by croton oil (*Croton tiglium* L.) and on granuloma induced by cotton pellet; the activity was stronger when applied locally than when taken orally.

Antiproliferative effects of *Petiveria alliacea* on several tumor cell lines could furthermore be shown. A macrolide is said to have antitumoural activity.

Other contents are triterpenes, a sesquiterpenic lactone, β-sitosterol, uronic acid.

Cultivation

The plant is a weed and can easily be cultivated.

Observations

The glandular hairs, not described as such up to date and which apparently are responsible for the penetrating smell of garlic of the plant, are the most characteristic feature.

Piperaceae

The Piperaceae are small trees, shrubs or more or less woody vines. More than 2000 species are distinguished and classified into a few genera. Economically important is *Piper nigrum* (white and black pepper), *P. betle* (aromatic leaves are chewed), and *P. cubeba* (cubeb berries).

Many species entered into the folk-medicine in tropical America.

Ethereal oils, pungent amides, mono- and sesquiterpenes, phenyl propanoids, pyrones, polyphenols, lignans and alkaloids have been found in the family.

Piper

Although a large number of species is known of this genus, they are distinguished by only very small differences. The leaves are entire and stipulate. The small greenish flowers occur in dense spikes. The fruit is a small berry.

The plants are aromatic and the source of medicines and condiments. Many contain alkaloids, others have antifertility effects or insecticidal activities.

Piper marginatum JACQ.

P. marginatum is a small shrub, up to 3 m high. The cordate leaves have an elongated tip (possibly a drip tip). Their consistency is membranaceous. The inflorescences are 10–25 cm long. The fruit is a drupe, 0.4–0.7 mm in diameter. The plant occurs from 0 to 1200 a.s.l.

The leaf anatomy has been studied by ROTH (1992). The upper epidermis is large-celled and serves as a water reservoir. The lower epidermis is usually 2-layered and functions likewise as a water reservoir. Hydathodes occur in the lower epidermis. The leaf consistency is slightly succulent due to the extense water-storage tissues.

Ethnobotanical and general use

The plant is applied for amoebiasis: 3 leaves are boiled in 2 liter of water; 3 cups a day are taken of this liquid for 13 days. Squeezed leaves are put on the forhead to cure headache. Ground leaves are put into the nose for nasal haemorrhage.

A spoonful of the leaf sap is taken orally for menorrhagia. The sap of the leaves also cures irritations of the eyes.

A decoction of the leaves is taken to stimulate the appetite, or for use as an emmenagogue and diuretic. It is furthermore useful for colds, headaches and as an external haemostatic.

The fresh entire plant is used to protect the teeth which adopt a black colour when masticating the plant.

Root in decoction is used for malaria, fever and toothache.

Healing properties

Water saturated with essential oil is molluscicidal. The entire plant has immunostimulating properties, antibacterial activity (*Staphylococcus aureus* and others), and relaxing effects on the smooth muscular system.

Piper sp. is a small shrub of the undergrowth in the Venezuelan cloud forest. The leaves are cordate and have a short tip.

The upper epidermis is large-celled. Three layers of hypodermis form, together with the epidermis, a water storing tissue. Below follows one layer of very small and short palisade cells. The spongy parenchyma, consisting of larger cells, occupies several layers. The leaf is typical of succulent Piperaceae (ROTH 1990).

Plantaginaceae

The Plantaginaceae are either represented by annual or perennial herbs or by undershrubs with parallel-nerved leaves - morphologically interpreted as enlarged midribs.

The inconspicuous gamopetalous flowers have originally 4 sepais, 4 petals, 4 stamens and 2 carpels. The anemogamous flowers may unite in reduced spikes.

The fruits are either capsules which open transversely or single-seeded nuts.

The cosmopolitan *Plantago* however develops a pyxidium with a lid.

The seed coat often transforms into a slime epidermis, as in *P. psyllium* (semen Psyllii: mucilaginous flea-seeds).

Plantago major L. (llanten)

Taxonomical description

Perennial rosette plant with reduced internodes (Fig. 211 a). Leaves 5–33 cm long and 2–9 cm broad, simple with long petioles. The membranaceous blades are elliptic-ovate to obovate, with an acute to obtuse apex; the margins are irregularly dentate or sinuate; blade base decurrent so that the petiole may become slightly alate.

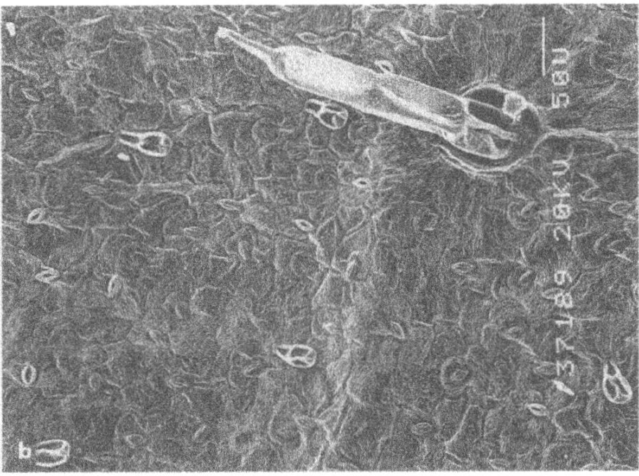

Fig. 211. *Plantago major.* **a** Habitus. **b** Lower epidermis with hairs and striated cuticle and stomata.

Venation parallel-acrodromous, usually consisting of 2 pairs of secondary nerves, deviating at the blade base, and a principal nerve, prominent at the abaxial side. Petiole, 4–20 cm long, convex at the abaxial and plane at the adaxial side.

The small green flowers are arranged in spikes of 10–25 cm length. Floral bracts are shorter than the calyx. The persistent sepals are broadly elliptic to ovate and conspicuously carinate; crown with 4 small obtuse lobules, 4 stamens, ovary bilocular. Capsule oviform or globular, 0.2 cm long, with 6 to numerous seeds.

Plant introduced to Venezuela, preferring the Andean regions.

Origin

Europe, temperate Asia.

Occurrence

Naturalized in America. In Venezuela frequently found in the Andes.

Anatomical descrption

Leaf (Fig. 211 b, 212 b, 214 a). The leaf is amphistomatic, but only slightly dorsiventral (FUCHS 1932, FISCHER 1937). The epidermis cells are large on the upper as well as on the lower leaf side, but are slightly larger on the upper side. As seen in transverse section, they are of globular shape. As seen in surface view, the anticlinal walls of the upper and lower epidermis cells are straight or only very slightly bent. The cuticle is slightly striated. The stomata are comparatively numerous on the upper side and lie at epidermis level. On the upper

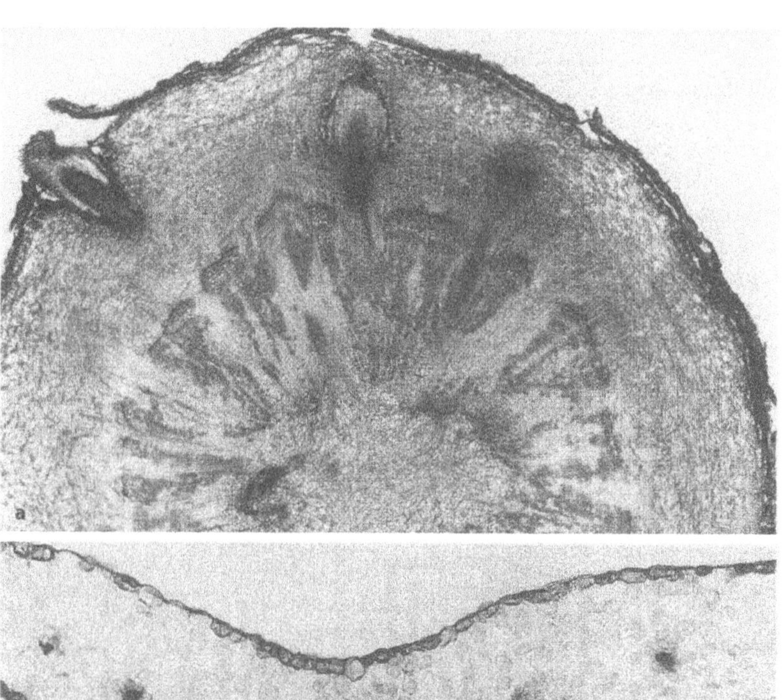

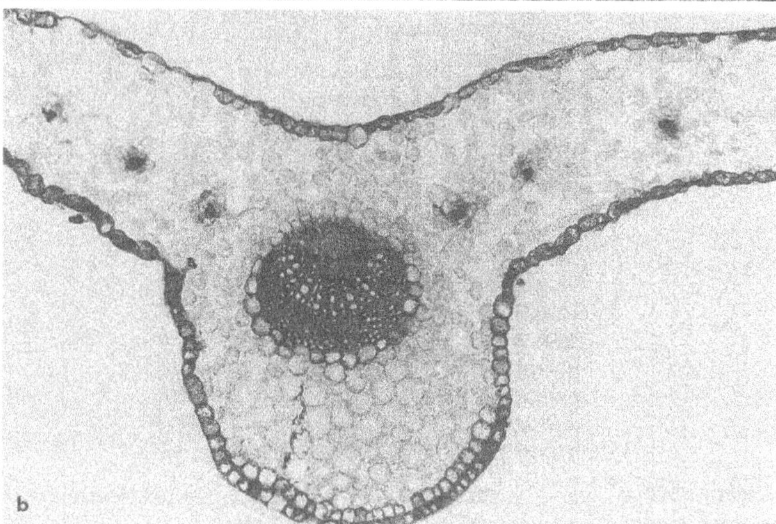

Fig. 212. *Plantago major.* **a** T.s. of shoot with formation of adventicious roots (rosette plant!). **b** Midrib of leaf.

side, there are more or less 2 (1–3) layers of anticlinally somewhat elongated cells which are only seldom palisade-like, whereas a loser spongy tissue composed of more isodiametric cells is present on the lower side. The lower epidermis is similar to that of the upper side, but the cuticle is slightly more striated. The stomata are anomocytic, not having any subsidiary cells, or are associated with a single cell, at times even with 3 different epidermis cells. There are uniseriate hairs of giant size as well as glandular hairs with 2 head cells in the epidermis.

The midrib as well as the large secondary nerves are prominent on the lower leaf surface. They have bicollateral vascular bundles with a fibrous cap on the abaxial side. Midrib and secondary nerves are surrounded by an endodermic sheath with Casparian stripes. The veins of higher order are collateral, but also surrounded by a parenchymatous sheath.

Shoot (Fig. 212 a). The shoot axis studied measured about 5–6 mm in diameter. The axis is surrounded by the leaf bases which are tightly appressed to it. A thin irregular periderm forms superficially (originating from the epidermis, according to PILGER 1937). Long uniseriate trichomes are conspicuous at the surface of the axis. They may help to retain water by capillary forces between the axis and the leaf bases.

The primary cortex is very ample, consisting of thin-walled parenchymatous cells which decrease in size towards the inside. Leaf traces abundantly deviate from the central cylinder, crossing the cortex on their way. The parenchyma forms partially tangential cell chains due to the growth in thickness of the axis.

The vascular system forms a continuous cylinder in which the cells are disposed in radial rows, more conspicuously so in the xylem where the very small vessels form radial rows of small multiples. Rays are absent, but the xylem parenchyma is extremely abundant.

The thin-walled parenchymatous pith is likewise very abundant. Bundles of cambiform tissue are evident in the pith, consisting of elongated narrow cells which contain abundant protoplasm. In better developed samples short tracheids become apparent in the center of the bundles, but phloem seems to be absent. The bundles are of secondary origin forming an intercommunicating system, which is however not connected with the central cylinder. They are interpreted as reduced vascular strands by SOLEREDER (1908, p. 644), and may even transform into complete bundles in very well developed samples. They are of different sizes and seem to continue into the adventitious roots.

Adventitious root (Fig. 212 a). The rhizodermis is comparatively large-celled. The parenchymatous cortex cells, which are arranged in radial rows, radiate from the endodermis; the globular cells enlarge in size towards the outside. The endodermis shows Casparian stripes; the pericycle is small-celled. The xylem has 5–6 or more rays.

The axis of the inflorescence (flowering scape) is also characterized by an endodermis.

The anatomical structure is variable in relation to the environment. *Plantago* is a 'plastic' genus, according to METCALFE & CHALK 1950.

Fruit and seed (Fig. 213, 214 b). The fruit wall is thin. The outer epidermis cells are elongated parallel to the fruit axis and are thin-walled. The inner epidermis cells, however, have very wavy walls, as seen in a surface view (Fig. 214 b).

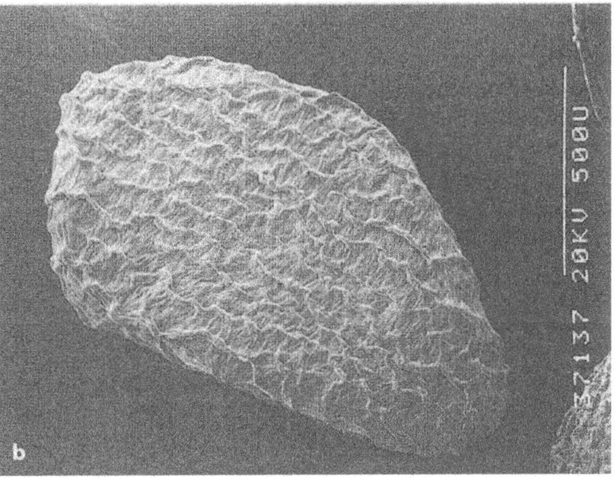

Fig. 213. *Plantago major.* **a** Fruit, opened, with seeds. **b** Seed.

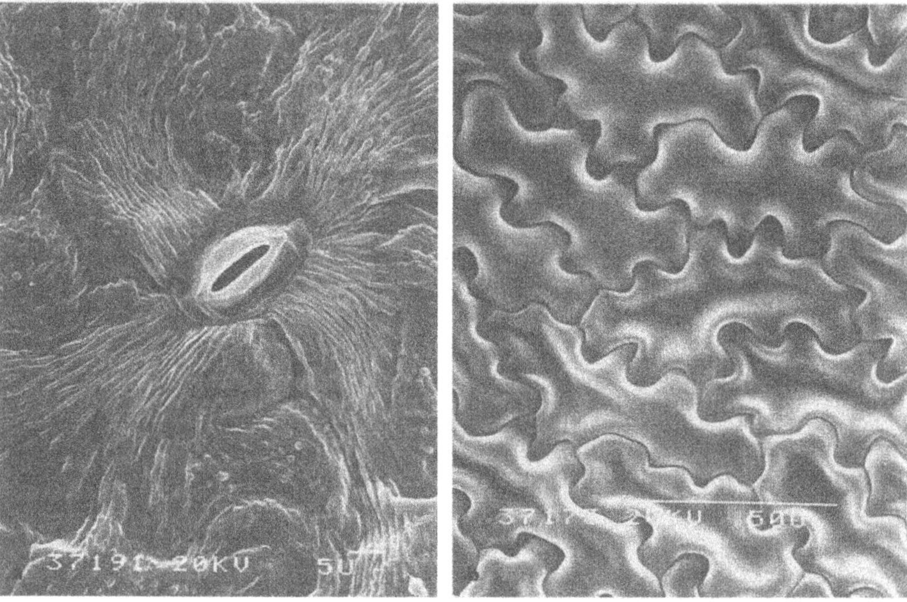

Fig. 214. *Plantago major.* **a** Stoma of lower leaf epidermis. **b** Inner fruit epidermis.

In a dry state, the outer epidermis of the seed coat is strongly sculptured (Fig. 213 b). When soaked in water, a slimy layer surrounds the seed.

Plantago psyllium

P. psyllium, semen psyllii (Fig. 215). M indicates the mucilaginous layers of the seed coat, when soaked with water and treated with toludine blue. Beneath follows the hard sclerotesta (Sc).

Ethnobotanical and general use

Medical use

Name of the drug: Folia, Semen, Planta.

Leaf. The leaves are slightly astringent. Leaves are said to have cooling, diuretic and alterative effects. Leaves are used against liver and kidney diseases, cancer, inflammations, ulcers, haemorrhoids, fever, malaria, chronic gastritis and diarrhoea. They are also used as a laxative, against cold and to cure bladder ailments.

Fresh leaves rubbed on a bee sting bring relief. Indians mix crushed leaves with a raw egg and take 2 teaspoonfuls twice a day to cure fever and bronchitis. Crushed leaves are put on the wound to accelerate the haemostatic action. Leaves are also used for cicatrization.

An infusion of the leaves is used to cure diarrhoea, liver conditions, ulcers and bruises.

Leaf tea, dew on leaf and leaf juice is used for eye-washes in ophthalmia. Leaf juice is used for eyedrops.

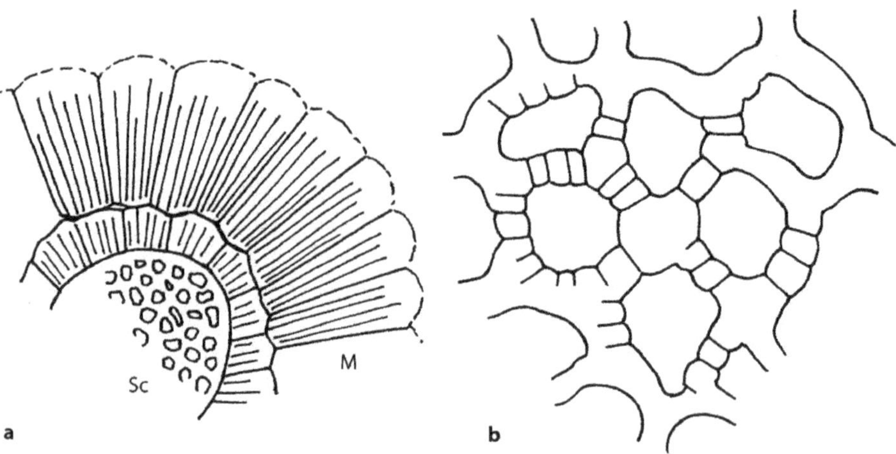

Fig. 215. *Plantago psyllium*, Semen psyllii. **a** T.s. of seed coat. The mucilage layer (M) soaked with water surrounds the seed. Sc = Sclerenchyma. **b** Sclerotesta. Seed treated with toluidine blue.

Root. A decoction of roots or leaves is used to treat liver problems. It is also applied as an emmenagogue when irregularities of the menstruation are observed.

Entire plant. A decoction of the plant is applied as an eye lotion.

Tea made of the plant is taken for dysmenorrhoea and for cooling. Mixed with bath water treats skin rashes in babies.

A decoction is used for gargles, internally taken it alleviates heart diseases.

Seed. The mucilaginous layer of the seeds has a laxative effect.

Method of use

Leaves and entire plant are used fresh, crushed (sap) and in the form of a decoction, tea or infusion. Roots are used in decoction. Seeds are taken untreated or dried. There are also commercial products for sale.

The leaf juice is extracted by placing a leaf over the top of a clean cup, and covering it with hot iron; the juice that drops down is collected in the cup. This juice is used for eyedrops. Gargles, cataplasms and compresses are also common.

Crushed leaves with oil of *Lilium candidum* (azucena) are applied externally to cure pimples.

The pure sap of the leaves, (3 or more drops a day), are used against pain in the ear canal. Sap with hot vinegar, taken in the morning, helps against intermittent fever. Sap mixed with honey of roses is used as a mouth rinse to cure aphthae and inflammation of the tonsils. Sap diluted with water or mixed with some starch, wine and sugar can be used as an eyewash (3–4 times a day) to cure cataracts.

Sap in the form of an ointment calms the pain of haemorrhoids; it is applied against eczema in the form of a bandage.

Taken internally, the sap is said to cure ulcers, chronical gastritis, and diseases of the liver.

Infusions of the leaves are taken as a febrifuge, against malaria and heart diseases.

A decoction of leaves taken internally is good for the stomach, the liver and kidneys, helps against kidney stones, as a haemostatic and as a vaginal lotion. A weak decoction of leaves is given to children for indisposition, against cough and to relieve gout.

In a proportion of 60/1000 (leaves/water) the decoction helps against dysentery, intestinal diseases, as a diuretic, a refresher during diseases of kidney, liver and bladder: 3 cups are taken daily.

Adding oil or fat, a cream is made of the decoction to cure haemorrhoids.

Gargles or mouth rinses prepared with the leaves help against angina, and inflammations of mouth and throat.

A cataplasm has maturing and resolving effects, when used for wound healing and against ulcers.

A compress has emollient effects.

A special recipe helps against herpes:

60 g leaves of *Plantago*, 15 g leaves of *Pilea microphylla* (yedra), 15 g roots of *Beta vulgaris*, 10 g saltpetre.

Root boiled in beef-tea is given to children against diarrhoea.

The entire plant boiled and mixed with toasted rice is given to children to cure diarrhoea.

Seeds are used in suppositories for haemorrhoids.

Seeds are taken internally against diarrhoea and dysentery.

Four grams of pulverized seeds are taken with milk to cure haemorrhoids.

Seeds in decoction have diuretic effects.

The plant is almost a panacea (universal remedy), remarks RODRIGUEZ!

Healing properties

The following curative properties are ascribed to the plant: it is astringent, wound healing, purifying, emollient, abortive, refreshing, diuretic, laxative, antiinflammatory, antibacterial, antiviral, antipyretic, antihaemorrhagic, oestrogenic. It also has cholesterol reducing effects, acts as an expectorant, and dissolves kidney stones.

It is used against infections of the urinary tract, prostatitis, acute conjunctivitis, necrotic oedema, hepatitis etc.

Extracts of the plant which contain flavonoids, polyphenols and alkaloids have an antiinflammatory action. Plantamajoside has antibacterial effects (against *Escherichia coli* and *Staphylococcus aureus*). A mixture of polyphenols extracted from *Plantago major* has anticarcinogenic effects. Hydroxycinnamic acids exercise an antiinflammatory effect.

Chemical contents

The seeds contain oil and slime. The following substances were found in the slime: galactose, glucose, xylose, arabinose, rhamnose, galacturonic acid, saccharose and fructose, plantabiose and the trisaccharide planteose, whereas in the subterraneous organs the tetrasaccharide stachyose was found.

Further contents of the plant are: hydroxy cinnamic acid, polyphenolic compounds, caffeic acid sugar ester (plantamajoside), iridoid glycosides (iridoid aucubin), alkaloids, flavonoids, triterpenes, coumaric acid, β-sitosterol, vitamins A and C and others.

Varieties and related species

Plantago major L. var. *asiaticum* DCNE. is a perennial herb the seeds of which are used in Chinese medicine as a diuretic.

Plantago psyllium L. supplies Psyllium seeds also called plantain seeds or flea seeds. Dried ripe seeds are used as a laxative, because their mucilaginous epidermal layer swells in the intestine. Seeds of *Plantago ovata* FORSK. are used in a similar way.

The flea seeds of *P. psyllium* are often adulterated with seeds of other *Plantago* species e.g. of *P. indica* L. and *P. lanceolata* L.

Cultivation

The plant is practically a weed and is easy to cultivate.

Observations

The variability of this genus in relation to the environment makes distinction of species difficult. In the leaf, the striation of the cuticle may be a diagnostic feature. Furthermore, the inner epidermis of the fruit with its undulated walls is very characteristic. Shape and surface sculpturing of the seed surface may also be helpful in identification.

The endodermis with Casparian stripes which surrounds the individual bundles of the leaf is extraordinary. However, experiments led TRAPP (1933) to conclude that the possession of a foliar endodermis is a character which is neither beneficial nor prejudicial to the physiological economy of the leaf. METCALFE & CALK (1950) suggest that the endodermis may be regarded as a primitive structure from the phylogenetic point of view which gradually disappears due to lack of function.

Polygonaceae

The Polygonaceae are mostly herbaceous and less frequently woody. The axis has knotty articulations a fact that is reflected in the name of the family. The leaf base is usually supplied with an ochrea. The flowers are small and occur in larger inflorescences. The gynoecium frequently consists of 3 carpels and the fruit is a triangular nut.

Frequently found contents are: Calcium oxalate in druses or solitary crystals, tannins, flavonoids, flavonolglucosides, and anthraglucosides.

Coccoloba uvifera (L.) JACQ. (uvero de playa, uva de playa, uva de costa, uva, uvero macho, mangle de falda, arahueque, cumaro blanco, camare, dreifi).

Taxonomical description

Taxonomic details are found in Fig. 216. Coccoloba uvifera (sea-grape) is a small tree, reaching 6–9 m in height, often with a torsive stem which begins to ramify close to the base and has an extensive crown. The simple alternating leaves are broad-orbicular or suborbicular with an obtuse or retuse apex and a cordiform base; they are 5–20 cm long and 7–25 cm broad, are glabrous and of a coriaceous-fleshy consistency. The venation is pinnate, broquidodromous, with 5–7 pairs of secondary nerves. The middle nerve is prominent on the abaxial leaf side and flat on the adaxial side. The venation is of pink or reddish colour in young leaves which adds a certain ornamental quality to the plant. The petiole is 1.5–2.5 cm long and flat-convex. The persistent ochrea at the base of the petiole is about 0.8–1 cm long.

The terminal pendent and racemose inflorescences are 12–30 cm long and fragrant. The creamy white coloured flowers are composed of 5 tepals, 8 stamens and 3 styles. The globose fruit is about 2.5 cm long and 2 cm in diameter and is truncate at the apex. When ripe, the fruit has a purple-violet colour. The infrutescence resembles a bunch of grapes, for this reason it is called uva de playa (beach grape). During maturation, the perigone increases in size together with the real fruit, a nut, and becomes fleshy. The edible part of the pseudogrape is thus the perigone; it has a sour-sweet taste and is astringent (tannins). The stone of the fruit corresponds to the nut.

Origin

Tropical America. It was probably the first tree Cristobal Colon (Christopher Columbus) saw when he discovered America!

Occurrence

Abundant at sandy and rocky beaches of the Atlantic and Pacific Ocean in Central and eastern South America, down to Peru and in the Caribbean room (Westindies and the north of South America / Venezuela). In Venezuela, it is frequent in the littoral of the Caribbean Sea.

Fig. 216. a. *Coccoloba uvifera*. **a** Shrub. **b** Leaves, flowers and fruits.

It is cultivated along alleys, in parks and gardens close to the shore, due to its adaptation to saline soils. It may also be observed growing half-wild at altitudes up to 1000 m, even in moist woods of Cuba and Jamaica. At the limit of its range and in poor situations, it is only a shrub.

Anatomical description

Leaf (Fig. 217). The leaves are comparatively large with a Length/width index of 0,68 and possibly broader than long and reach a mean value of surface area of 200 cm^2 (ROTH 1992). The leaf is dorsiventral and hypostomatic. Upper and lower epidermis are comparatively small-celled and thick-walled. A thin-walled water-storing hypodermis lies beneath the upper and the lower epidermis. It comprises about 2–3 layers of large turgescent cells. The hypodermal layers are mainly responsible for the leathery-succulent consistency of the leaves. As seen in transverse section, the thickness of the hypodermis exceeds that of the epidermis about 3 times. The mesophyll exclusively consists of palisade parenchyma; the leaf is thus isolateral. The palisade cells are long and slender. Their length/width index amounts to 7–10, particularly in the 2 uppermost layers. The palisade parenchyma comprises about 6–8 layers; however it is easily perceptible that these layers – at least partly – originated

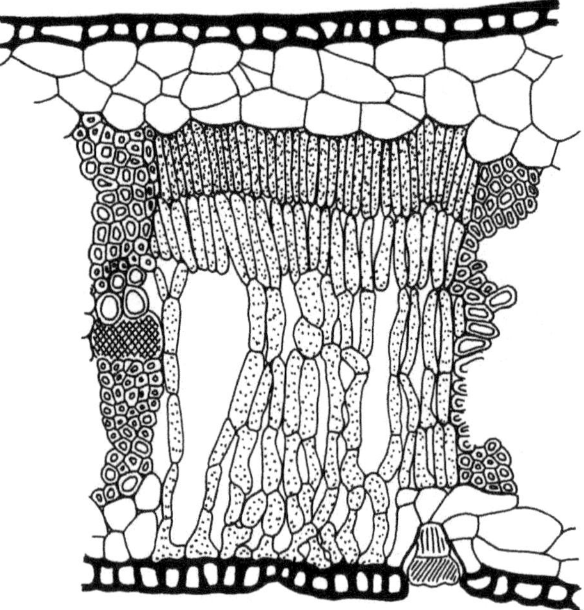

Fig. 217. *Coccoloba uvifera*, Polygonaceae. T.s. of leaf with water storing hypodermis and a mesophyll consisting of palisade parenchyma (Roth 1992).

by periclinal divisions so that anticlinal cell rows developed. In the 2 uppermost layers, the cells are very densely packed, while the underlying palisade parenchyma is looser thus giving rise to intercellular spaces. The vascular bundles are very frequent and are surrounded by abundant sclerenchyma. Upper and lower sclerenchymatous caps may unite to form a continuous sheath around the larger vascular bundles. This sclerenchymatous tissue adds the leathery consistency to the leaf. The stomata – sunk below the epidermis surface – are confined to the lower epidermis; about 122 stomata/mm^2 were counted; the length of the stomata amounts to 31 µm. Glandular hairs are additionally found in the lower epidermis, being hidden in small depressions below the surface. The glands are short-stalked and peltate and probably originally correspond to hydathodes. It has not been investigated as yet whether they are able to absorb water.

The leaf is characteristiv of the sun type and has several xeromorphic peculiarities (description taken from ROTH 1992).

Stem, Bark (Figs. 218, 219). It is interesting to note that, there is a clear distinction between bark

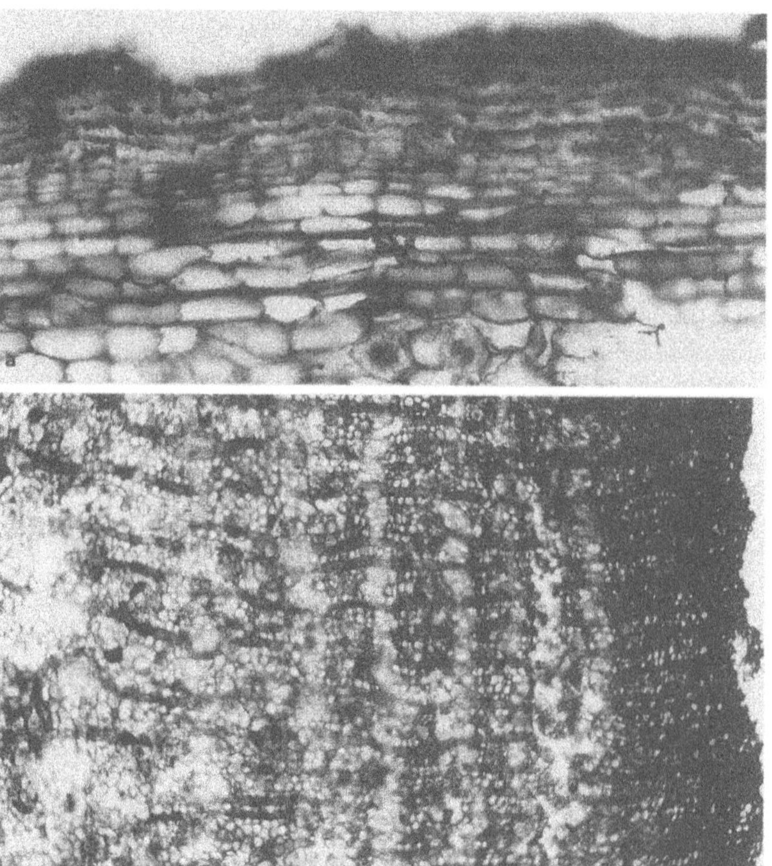

Fig. 218. *Coccoloba uvifera*. **a** Cork: **b** Bark (cambial zone to owards the right). (Above: × 25, below × 6.3).

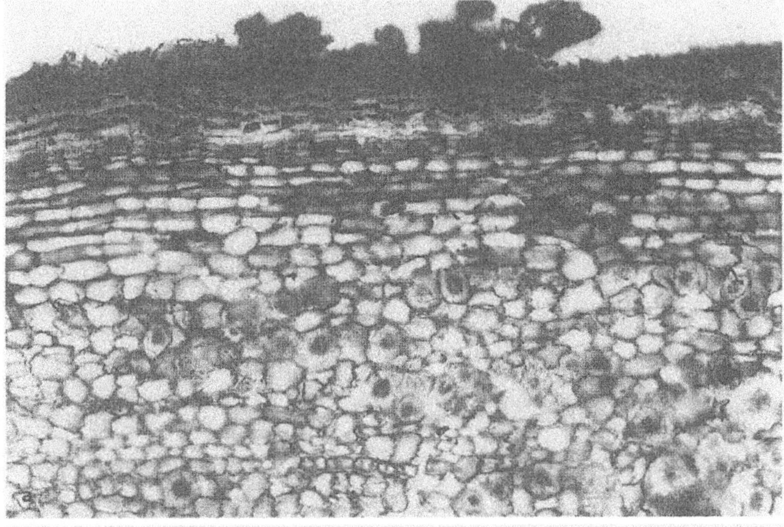

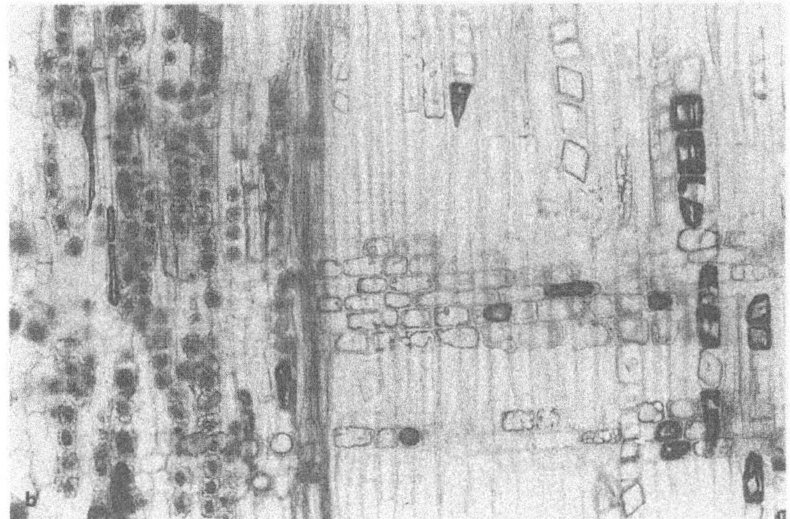

Fig. 219. *Coccoloba uvifera*. **a** Longitudinal section of cork. **b** Longitudinal section of bark and wood. Note the crystals! Druses in the bark and rhomboids in the wood!

and wood in the formation of the crystals; this becomes particularly obvious in a longitudinal section: In the bark exclusively clustered crystals in the form of druses are found (Fig. 223) while in the wood there are exclusively solitary crystals of a rhombic shape; these occur in septate crystal strands (Fig. 219).

The inner functional bark is of a more or less regular size.

In the middle bark, a clear stratification of soft and hard bast is obvious. The hard bast consists of a mixture of elongated fibersclereids and of shorter but larger sclereids with a large cell lumen. The rays are uniseriate. The ray cells are lignified where they cross the hard bast. The frequently occurring crystal druses are dispersed all over the section. They occur mostly in the axial parenchyma in septate longitudinal strands. In the outer bark, the rays partly dilate and a part of the parenchyma cells shows tangential dilatation and formation of anticlinal walls. However, the dilatation growth is very moderate in a bark about 3 mm wide. The phelloderm comprises up to 10 layers of thin-walled cells, while the innermost cork layers are composed of cells with U-shaped wall thickenings; the inner walls are the thickest and are crossed by very delicate pits.

Fatty and possibly resinous substances are deposited in the ray cells and in cells of the axial parenchyma, but no starch could be observed.

Wood. The vessels have simple perforation plates. The pits are small. The rays are uniseriate and homogeneous. The wood parenchyma is diffuse. Septate crystal strands with rhombic crystals are frequent in the axial parenchyma. Growth rings may be present, but are never conspicuous; their intensity probably depends on the environmental conditions.

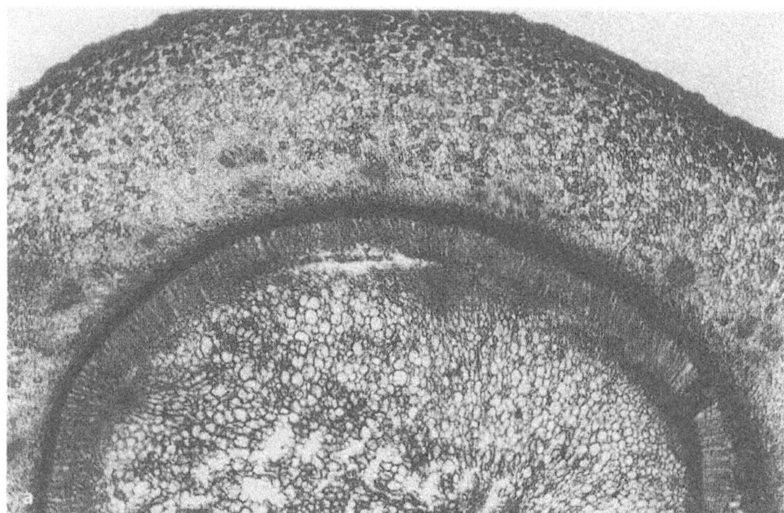

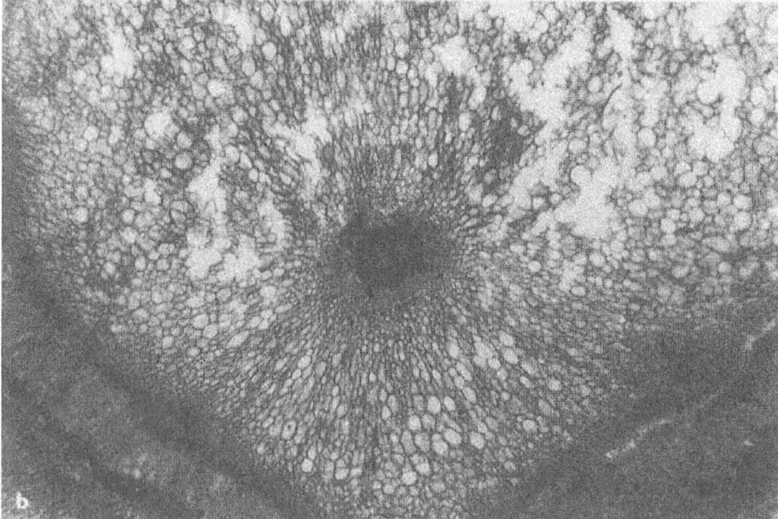

Fig. 220. *Coccoloba uvifera.* **a, b** Pseudofruit composed of a 'perigone ring', as seen in t.s., mainly consisting of parenchyma. In the innermost zone of this 'ring', the vascular bundles are to be found. Towards the inside follows the 'pseudoendocarp' corresponding to the outer epidermis of the real fruit, a nut; this part consists of radially elongated thick-walled cells. Parenchyma follows beneath. The center of the fruit with a vascular cylinder **b** is cut in the basal region, near the peduncle.

Fruit (Figs. 220, 221, 222). The pseudofruit is composed of an outer part which is formed by the united perigone leaves, and of an inner part which is formed by the real fruit, the nut.

The outermost layer of the pseudofruit or the pseudoexocarp consists of epidermal cells united in small groups which form knobs on the outside. The epidermis cells which correspond to the lower epidermis cells of the perigone leaves are radially (perpendicular to the surface) palisade-like elongated and are often divided by periclinal walls. The outer tangential walls of the epidermis cells are heavily thickened. Beneath follows a large parenchymatous zone composed of more or less globular cells with thin walls and large pit areas. Quite a few cells have a granular content which stains intensely with toluidine blue. A ring of vascular bundles embedded in the parenchyma lies in the innermost third of the parenchymatous zone, close to the pseudoendocarp. The vascular bundles are arranged in an irregular way, partly forming small arcs and are partly arranged in a concentric way. The inner epidermis of the perigone leaves is small-celled and inconspicuous.

The hard pseudoendocarp formed by the epidermis cells of the real fruit (nut) lies towards the inside. The cells are palisade-like, radially elongated, partly anticlinally divided, and have thick undulated walls. This layer which is very broad, is intimately united with the tepals (perigone leaves).

Towards the inside follow more or less globular parenchyma cells with relatively thin walls and large pit areas. The size of the cells enlarges notably towards the inside. In the outer part, below the pseudoendocarp, some vascular bundles occur here and there.

In the center of the pseudofruit, there is a vascular cylinder pertaining to the peduncle of the pseudofruit. It is surrounded by small parenchyma cells. The fruit was cut in its basal part. The edible part is

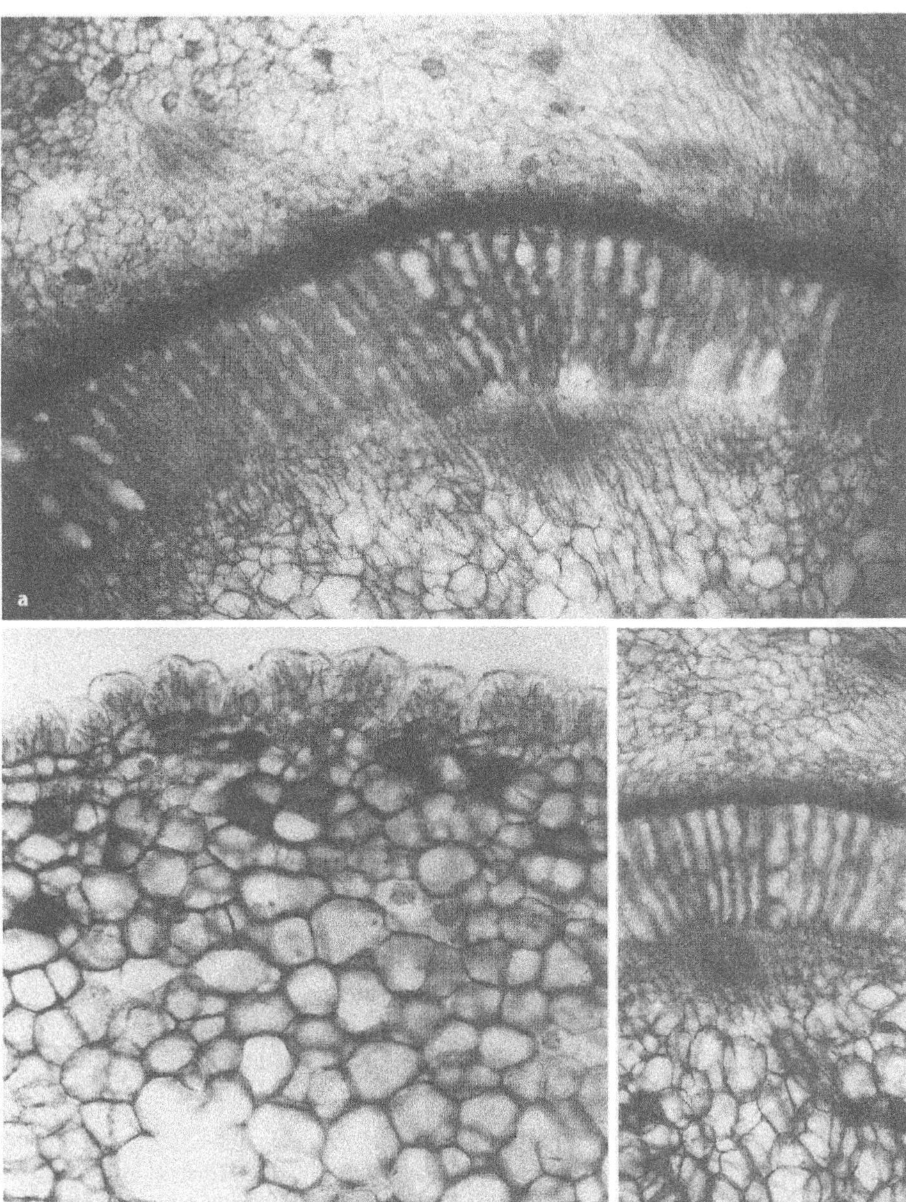

Fig. 221. *Coccoloba uvifera.* **a** 'Pseudoendocarp'. **b** 'Pseudoexocarp' consisting of epidermal cell groups which are wart-like; the cells are radially elongated and their outer walls are thickened. They correspond to the outer epidermis cells of the basal part of the perigone leaves. **c** A vascular bundle beneath the 'pseudoendocarp'.

the juicy parenchyma of the perigone, while the nut supplies the stone.

Ethnobotanical and general use

Nutritional use

The fruit is edible: it has an agreable sour-sweet taste, but is astringent. It is however only eaten by the natives. It contains thiamin, riboflavin, niacin, vitamin C and A. It is made into a jelly and into a fermented drink.

Economical utilization

The timber is of good quality, but because of the small size of the tree, it is used only locally and then mostly for fuel. The heartwood is pinkish or reddish and without distinctive odour or taste. The texture is uniform, medium to fine. The timber is sometimes used for furniture.

The tree is occasionally planted for ornamental purposes in parks or along small roads. According to BRÜCHER (1989), the species deserves domestication, especially for colonization of large sea beaches, as a windbreaker and living fence (hedge).

The bark yields, upon excision, an astringent red juice which is the source of the West Indian

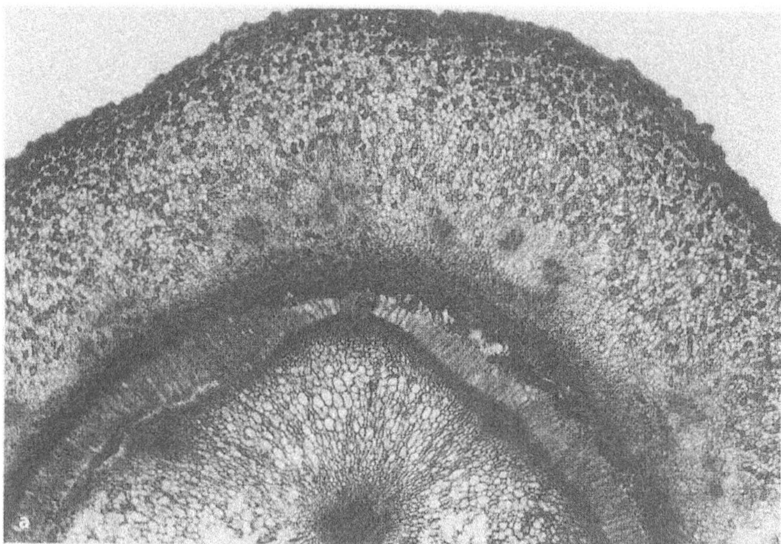

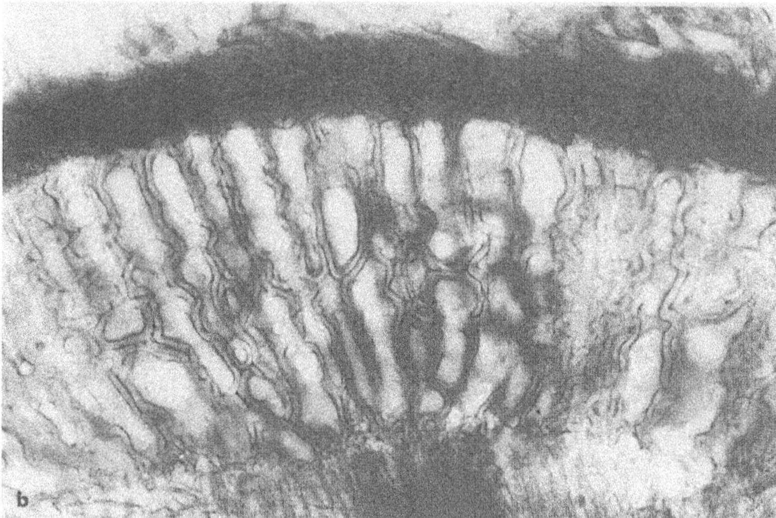

Fig. 222. *Coccoloba uvifera.* **a, b** T.s. of pseudofruit. 'Pseudoendocarp' × 16 **b.**

kino or gum kino. This was formerly an article of trade, but commercial kino is now supplied by West Africa and by some parts of the East Indies.

The root contains up to 25 % tannin and is used for tanning. The anthocyanins of the fruit are used as a colouring matter for tea and wine.

Medical use

Fruit and bark are astringent and contain tannins. A decoction of the bark is used to combat chronic diarrhoea. The fruit favours the dissolution of uric acid and has an effect on muscular debility. A cup of bark tea taken on an empty stomach fortifies the sphincters of the bladders of children with nocturnal enuresis.

Method of use

Fruit raw, bark in decoction or as a tea. For diarrhoea, 35 g bark and 25 g leaves are boiled 15 minutes in 1 l of water. A small cup is taken every 4 hours.

Healing properties

Astringent, tonic, haemostatic, antidiarrhoeic.

Chemical contents

Vitamins of the B and C group, tannin.

Related spcies

There exist more than 125 species of *Coccoloba*, mostly in the neotropics and in subtropical regions. Some are cultivated in green houses for their large ornamental leaves.

BRÜCHER (1989) mentions *Coccoloba venenosa* which is a poisonous plant. Natives of Puerto Rico call it calambrena, due to its nerve-paralyzing effect.

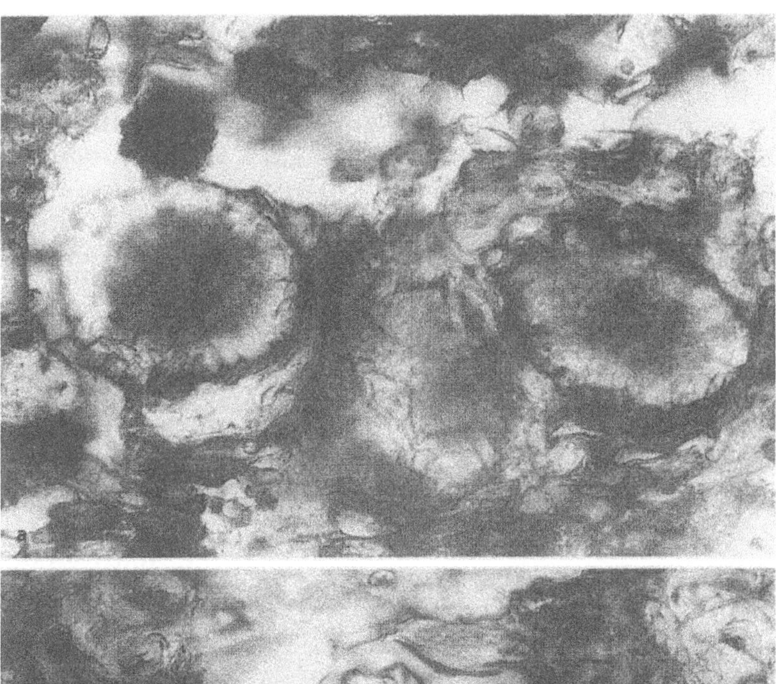
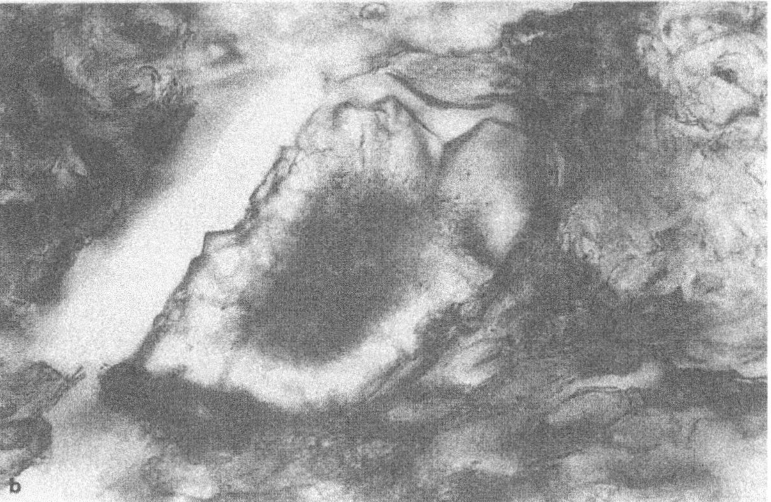

Fig. 223. **a** *Coccoloba*. **a** 'Uvero', **b** 'Arahueque' with 2 different types oif giant druses (Roth 1981).

SCHNEE (1960) mentions *Coccoloba latifolia* LAM., a small tree which occurs in Guiana, Brasil, Trinidad and Tobago. It is also found in Venezuelan Guayana and is called uvero.

ROTH (1981) studied the bark of the uvero in Venezuelan Guayana which possibly corresponds to *Coccoloba latifolia*. This species contains giant size druses of calcium oxalate in the soft bast.

The bark of another *Coccoloba* species of Venezuelan Guayana called Arahueque was likewise studied by ROTH (1981). The bark structure of both, Arahueque and Uvero is very much alike, but the structure of the Arahueque is more regular. Both species show conspicuous growth rings in the wood (annual rings?).

It is not yet known whether these species have some application in popular medicine or have economic potential.

Leaves, fruit and bark of *C.* cf. *marginata* are used against tuberculosis (WILBERT 1996).

Cultivation

The plant can be reproduced by seeds. The growth is medium. Although the radical system is not very profound, the tree is a pioneer plant and resistant to drought and to saline soils. It also resists strong winds and is long-lived. The tree merits greater utilization, as it can also be used as an ornamental. Its commercial cultivation would be lucrative, as it yields a rich harvest. The fruits could be used to make sweets, marmalades and refreshing drinks. The tree flowers twice a year, in May and October. As the tree is dioecious, it may also be propagated by cuttings of female specimens.

Observations

Coccoloba uvifera is a well known plant in Venezuela and easily recognized by its habit, the large leathery leaves with the ochrea, the flowers and edible grape-like fruits.

In drinks, marmalade and sweets, remnants of the pseudoendocarp or outer epidermis of the nut are good criteria for identification.

The two not yet clearly identified Guianan species of *Coccoloba* are well distinguished by their bark structure (ROTH 1981).

Bark structure of *Coccoloba* sp. has been studied by ROTH 1981, leaf structure 1984, leaf venation 1996.

Triplaris surinamensis CHAM. Useful wood. Bark contains tannin and is used for tooth ache.

Leaf structure has been studied by ROTH 1984, fruit structure and dispersal 1987, leaf venation 1996.

Punicaceae

The Punicaceae are a very small family with only one genus: *Punica*. The 9 carpels are arranged in 1–3 stories, one above the other. The fruit is berry-like or a capsule and contains a large number of seeds. The representatives of the Punicaceae are shrubs or small trees. The flowers are handsome having a red hypanthium (axial cup) and calyx.

Punica granatum L. (granada)

Taxonomical description

The pomegranate is a 1.5–4 m high shrub or tree which is slightly spiny. The branches are angular, the foliage is caducuous. The spiny glabrous leaves are opposed, of an ovate to oblong shape and 2–8 cm long; the petiole is short.

The red-orange flowers are furnished with a fleshy cup-shaped hypanthium terminating in 5–7 teeth. The red-orange crown has 5–8 free petals; the stamens are very numerous.

The spheric fruits are crowned by the persistent sepals and have the size of an orange (6–8 cm in diameter). The carpels are usually arranged in 2 stories; the locules are separated by membranaceous septs. The seeds, covered by a fleshy sarcotesta, are very numerous.

Origin and occurrence

The pomegranate comes from Asia, but is now much cultivated or grows wild everywhere in the subtropics and tropics.

Anatomical description

The pomegranate represents an irregularly dehiscing capsule which exposes a great quantity of juicy seeds at maturity. The peculiar arrangement of the carpels in the form of 2 stories (Fig. 196 b), the upper one 4–9 celled and the lower one 2–5 celled, has claimed the attention of several authors. At the beginning of development, the carpels are distributed in 2 (or 3) concentric whorls, but by favored growth of the peripheral ovary parts, the outer whorl is lifted up and the original central placentae are thus shifted to a parietal position (NIEDENZU 1898). The ovary is inferior in position. The floral development of *Punica granatum* has been studied in detail by TUNG (1935). The vascular anatomy is of importance for the interpretation of the ovary position.

From the floral development and the vascular supply of the floral members a partial invagination of the receptacle could eventually be deduced, especially regarding the irregular sequence of the developing organs, i.e. appearance of carpel primordia before stamen primordia, in the sense that a confusion of the normal sequence has taken place. There are, however, not enough ontogenetic and anatomical data available from the studies of TUNG (1935) to give a definite interpretation (see also ROTH 1977).

Fruit anatomy of *Punica granatum* has been studied by ROTH & LINDORF (1972), ROTH (1977). At the beginning, the entire ovary wall is parenchymatous. Only in a young fruit of 1.5 cm in diameter, do the sclereids begin to differentiate in the form of small groups or isolated cells. Differentiation of the brachysclereids starts approximately in the middle part of the pericarp wall and from there progresses towards the periphery. Towards the pericarp inside, almost no sclerenchyma cells are formed. In the mature stage, the sclereids show an irregular shape. The remainder of the fruit wall is parenchymatous. Towards the inside, the parenchyma has a slightly spongy consistencs due to the formation of long cell chains which partly separate from one another at their tangential cell walls.

The peripheral parenchymatous layers are rich in chlorophyll and anthocyanin concentrates mainly in the first subepidermal layer of the fruit wall. The outer epidermis has a fine cuticle and shows only few stomata.

In the very ripe fruit, a partial maceration becomes obvious in the fruit parenchyma. The parenchymatous cell chains which have formed by anticlinal cell division, separate more and more from one another so that large intercellular spaces develop and the pericarp adopts a spongy aspect in this way. In some cells of the fruit periphery, druses of calcium oxalate can be observed. Very characteristic is the large quantity of tannin in the fruit wall which becomes obvious upon oxidation and a brown colouring of the tissue. However, the pericarp shows little tissue differentiation in comparison with the seed coat which supplies the juicy edible part.

Seed anatomy, which is very interesting, has also been studied by ROTH & LINDORF (1972). Ovules are anatropous and have 2 in teguments. The inner integument forms immediately after the formation of the outer one. The inner integument is smaller (with 1–2 cell layers) than the outer one (with 2–3 layers). The inner integument seems to grow with a one-sided apical cell, while the thicker outer integument has a 2-sided apical cell. The juicy layer directly originates from the segments produced by the outer integument towards the outside, while in the segments formed towards the inside, further periclinal divisions take place. In the following, the outer epidermis cells of the outer integument augment in size, whereas the underlying cells divide in various planes so that 3–4 subepidermal layers result. The inner epidermis consists of small cells. The inner integument increases in tickness by formation of periclinal cell divisions.The outer epidermis cells of the outer integument enlarge more and more in an anticlinal direction and small starch grains become visible in them.

In the adult stage, the seed coat comprises the following parts: An outer integument with a juicy outer epidermal layer consisting of anticlinally enlarged palisade-like cells with delicate walls and simple or compound starch grains. The enormously enlarged cells are transparent and contain a usually rose-coloured cell sap due to the presence of anthocyanin. Below this layer lie several parenchymatous layers of the outer integument. Towards the inside the stony part becomes evident which is composed of stone cells with irregular outlines (Cornea) (ENGLER & PRANTL 1898). The small cells of the inner epidermis of the outer integument are also transformed into small stone cells.

The inner integument, on the other side, shows 2 distinct layers: a layer of tracheidal cells with reticulate wall thickenings, and a layer of cells with smooth walls. Some times, rests of the nucellus may be observed towards the inside (ROTH 1977).

Ethnobotanical and general use

Nutritional use

Marmalade or a syrup are made from the sarcotesta of the seeds. Seeds are also eaten fresh. The subacid pulp makes a refreshing drink (grenadine).

Economical utilization

The shrub or tree is often used as an ornamental for its showy red-orange flowers and fruits. It is also sometimes used as a living fence (hedge).

The leathery fruit skin produces up to 26 % tannin used for tanning a fine quality of leather.

Medical use

The drug cortex granati corresponding to the dried bark of stem, branches and roots contains alkaloids (pelletierine and granatonine) and has anthelmintic and taenifuge effects. It is furthermore used for intermittent fever, diarrhoea and night sweat.

The sap of the seeds is applied for afflictions of the throat.

Chemical contents

Alkaloids (Pelletierine and granatonine), ellagic tannins, organic acids, estrone (in seeds).

Cultivation

Propagation is possible by seeds or cuttings. The plant tolerates poor soils and a hot climate.

Quiinaceae

Quiina

Quiina is medicinal.

Touroulia guianensis. The bark anatomy has been studied by ROTH 1981, Fruit structure and dispersal by ROTH 1987, leaf anatomy by ROTH 1984, leaf venation by ROTH 1996.

The venation pattern of the leaves of *Touroulia guianensis* is unique; there is no formation of meshes, only of anastomoses which are undulated (ROTH 1996; Fig. 224)

The fruits form ribs with 2 different tissue types: an inner parenchymatous part and a peripheral sclerenchymatous part. In between the ribs lie large oil pockets densely filled with oil drops (Fig. 225).

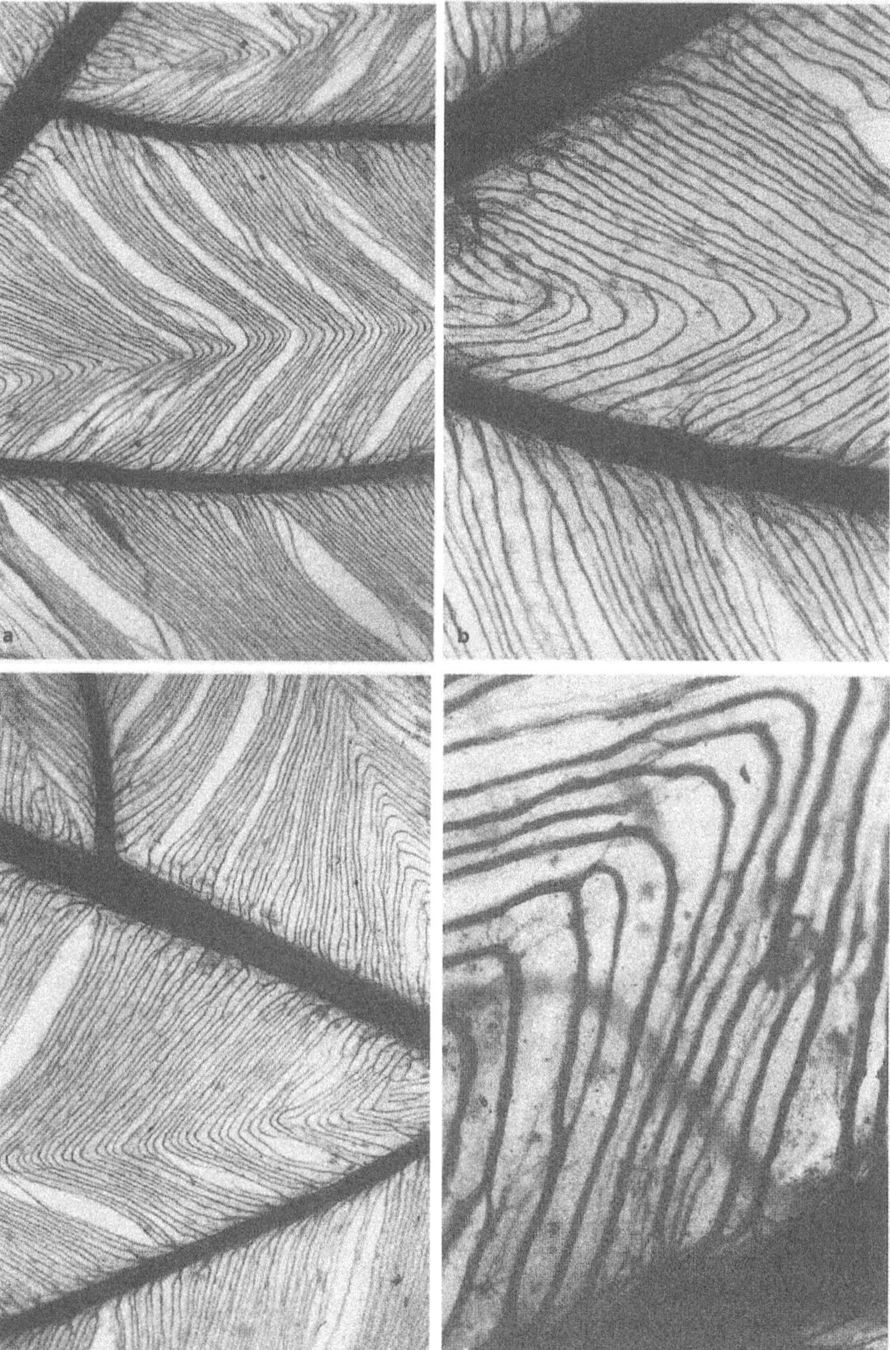

Fig. 224. *Touroulia guianensis*, Quiinaceae. **a–d** Unique venation pattern in the leaf without mesh formation, but only with undulated anastomoses (ROTH 1996).

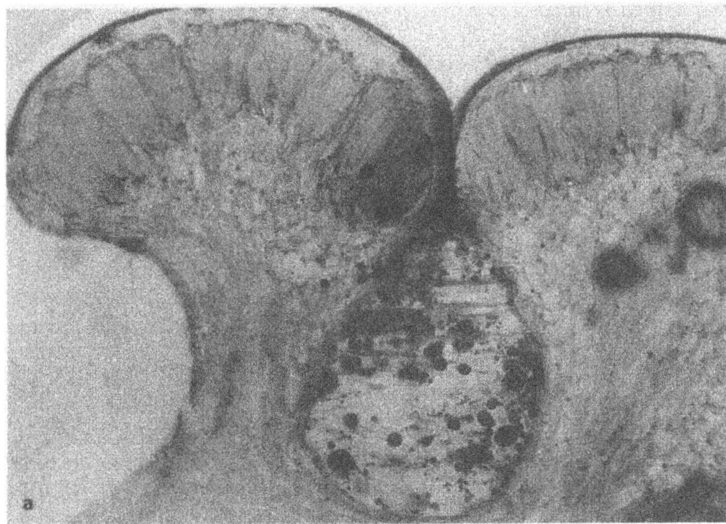

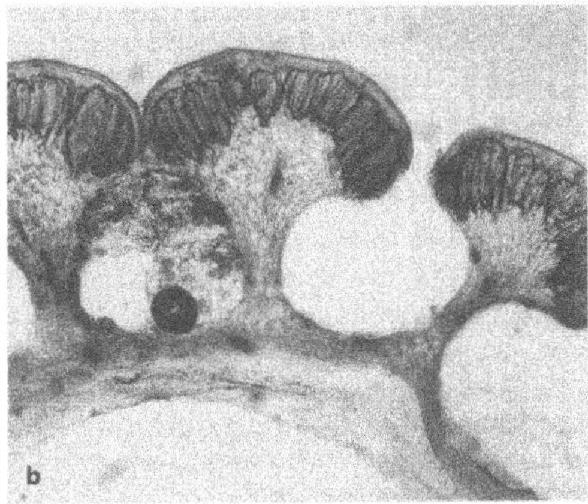

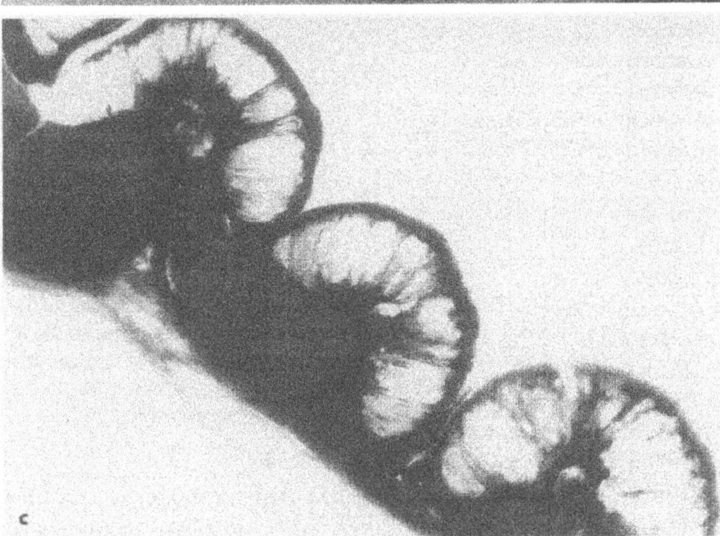

Fig. 225. *Touroulia guianensis.* **a–c** Ribs of the fruit in t.s. with an inner parenchymatous part and a peripheral sclerenchymatous part. In between the ribs lie large oil pockets filled with oil drops (ROTH 1987).

Rhamnaceae

Colubrina

Useful wood. Bark and root are a substitute for soap. Leaves are eaten raw. Fruits are a fish poison and abortifacienst. 'Snake bark' is antidysenteric, antipyretic, antiscorbutic. 'Saguaragy' bark is used for fever. Leaves of some species are anthelmintic and an emmenagogue.

Bark structure bas been studied by ROTH 1981, fruit structure and dispersal, 1987.

Rhizophoraceae

Trees or shrubs mostly with opposite often leathery leaves and lanceolate caducous interpetiolar stipules. Flowers small or medium-sized, solitary or in small inflorescences. The bark is often rich in tannins. A very natural group, secondarily adapted to the mangrove conditions. Vivipary in littoral species. Roots modified to pneumatophores or stiltroots. The large amounts of tannins protect the plants from rotting in their aquatic environment.

Rhizophora mangle L. (mangle, mangle colorado, mangle rojo, purgua)

The tree reaches a height of 10 m (up to 30 m). The opposite leaves are evergreen and coriaceous, entire, elliptic or lanceolate to obovate, obtuse at the apex, glabrous and with a brilliant upper surface. They are 7–15 cm long and 4–6.5 cm broad; the petiole is up to 3 cm long. The angusti-lanceolate, large and 7 cm long stipules are caducous. The axillary inflorescences have 2 to numerous flowers. The pedicels are 4–10 mm long.

The tetramerous flowers are yellowish-green to yellow, being surrounded by 2 ovate and acute bracteoles, measuring 2–10 mm in length and 2 mm in width. The 4 sepals are thick, coriaceous, lanceolate concave, acute at the tip and 7 mm long and 3.5 mm wide. The coriaceous petals are lanceolate, 5.5 mm long and 1.5 mm broad, with an emarginate apex and a woolly interior. The anthers of the 8 stamens are sessile or almost sessile. The disc is fleshy. The semi-inferior ovary is bilocular with 2 ovules in each compartment. The style has 2 lobular stigmas. The conic fruit is about 2 cm long and usually contains a single seed.

The embryo already germinates on the tree (vivipary) with a hypocotyl that may reach 15–25 cm or more in length (see also ROTH 1965). The hypocotyl thus perforates the fruit wall. Seedlings are therefore found on the mother tree in different stages of development. According to ROTH (1965) the hypocotyl may even reach a length of up to 60 cm, until the seedling separates from the mother plant and falls into the mud (fango).

The stem is supported by stilt roots which often form impenetrable thickets along low muddy seashores.

Occurrence

The species occurs in tropical America and in tropical occidental Africa. In Venezuela, the plant is frequent along the northern coast line where it may form small woods. It is the emblematic plant of the state Delta Amacuro where it is called mangle rojo.

Anatomical description

Leaf (Fig. 226). The upper epidermis is single-layered and small-celled. The outer walls are thick and strongly cutinized. Beneath follows a hypodermis of about 4–5 layers, not counting the slime cells. The hypodermis is distinguished in 2 zones: The cells of the upper zone (in direct contact with the epidermis) are smaller and contain a high amount of tannic substances in the form of a brownish content, while the cells of the second zone are larger and much less stained. The innermost (5th or 6th) layer of the hypoderm is however very distinct from the other layers. The cells are funnel-shaped and slimy. The inner slimy cell walls almost entirely fill the cell lumen and are conspicuously layered (Fig. 226). The contiguous palisade cells surround the slime cells in the form of a cup. The funnel-shaped hypodermis cells apparently serve as light collectors for the palisade cells. The palisade cells are very long and slender and contain small chloroplasts. The spongy parenchyma in the form of a loose network is very airy. Additionally, osteosclereids are found in the mesophyll. The mesophyll layer in direct contact with the lower epidermis may be considered as a lower hypodermis; it also has a tanniferous content. The stomata are confined to the lower epidermis; they have well-developed cuticular ledges (horns). The lower epidermis is small-celled.

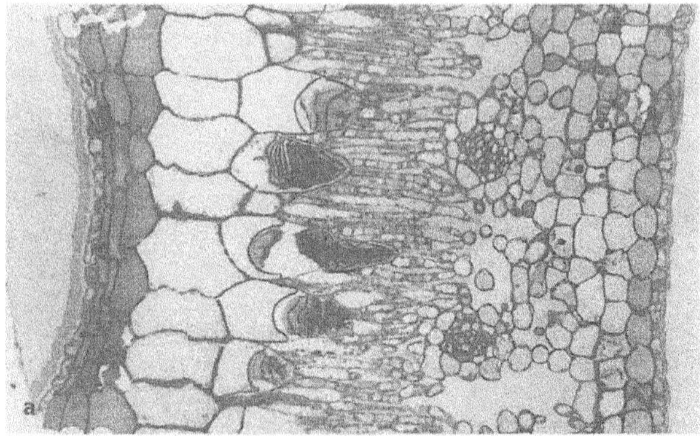

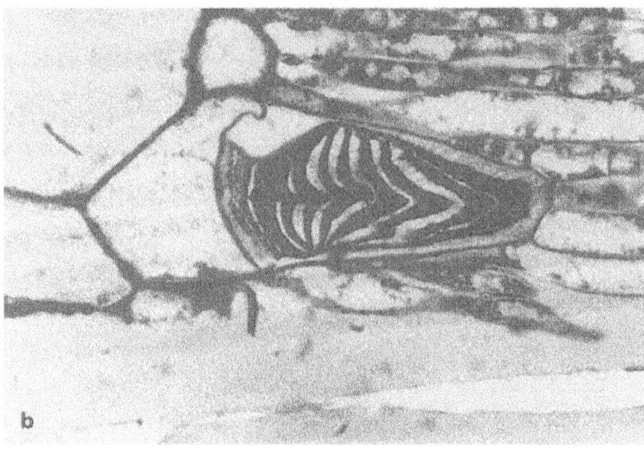

Fig. 226. *Rhizophora mangle*, Rhizophoraceae. **a** T.s. of leaf with thick-walled upper epidermis, 3 layers of tanniferous hypodermis cells, a water-storing hypodermis with slime cells and palisade cells surrounding the slime cells. **b** Slime cell with layered content. Electron microscope (ROTH 1992).

The funnel-shaped slime cells function not only as light collectors (Fig. 226), but also have a very strong water absorbing capacity. The small vascular bundles terminate blindly in the meshes in the form of enlarged tracheids with thickened walls and pits and may also serve as water-storing cells. Cork warts or lenticels are found on the leaf surface; they may serve for aeration and even as hydathodes, secreting water and chloride (ROTH 1992).

Hypocotyl. The 4 sepals are persistent on the ripe fruit, which is perforated at its tip by the developing hypocotyl. This elongates with a basal-intercalary meristem. The collateral vascular bundles form a ring, as seen in t.s. and ramified H-shaped trichoblasts are found in the cortex as well as in the pith. Large intercellular spaces form in the cortex during development. Most interesting are the lenticels which cover a large part of the surface. They arise in basipetal sequence on the hypocotyl and are responsible for aeration. They develop independently of stomata (which are absent on the hypocotyl) and show a certain layering in the adult stage.

No stable epidermis develops, only a surface layer which constantly divides peri and anticlinally producing lenticels (for further information see ROTH 1965).

Ethnobotanical and general use

Economical utilization

The bark contains 20–30% tannin and is employed for dyeing and tanning leather.

The fruits contain 16% tannin.

The wood is hard, heavy, dark reddish-brown, is durable in water, is not attacked by molluscs and teredo. It is not only used for fuel, charcoal, firewood, but also for fences and as construction lumber (for rafters, beams and joints of buildings, for keels and ribs of boats, and also for posts, piling, railway crossties).

The flowers are a source of a commercial honey. Bark and roots supply a reddish dye.

Medical use

Bark and leaf. The bark is used in general medicine for haemorrhage and angina and is also considered as a specific remedy against leprosy. It is also used against tuberculosis.

Bark and leaves are considered to be haemostatic and febrifuge. The Warao Indians use the bark for muscular spasms, skin ulcers, haemorrhagic dysentery and diarrhoea.

Fruit and leaf. The fruit is applied for tooth ache, the leaves for fever (WILBERT 1996).

Observations

Rhizophora mangle L. is a very well known species which was much used for tannig in former times. It is easily recognizable from a distance because of the long green hypocotyls of the seedlings hanging down like candles from the tree. The most characteristic feature of the leaf is the largecelled slime hypodermis; the layering of the slime becomes visible when an aqueous solution of methylene blue is added to the slide (ROTH 1992).

Rosaceae

Prunus

At least 34 useful species are known.

The manifold applications of this genus are well known, e.g. useful wood, edible fruits and seeds. Fungicides and bactericides as well as antitubercular activity and antifertility agents have additionally been found.

ROTH studied the bark structure 1981, leaf structure 1984, fruit structure and dispersal 1987, leaf venation 1996.

Rubiaceae

Trees, shrubs and herbs are found in this family. The entire leaves are usually opposite and have stipules (often interpetiolar i.e. united between 2 neighbouring petioles). Flowers frequently aggregated in conspicuous inflorescences. Gynoecium inferior, usually with 2 carpels. The fruits are capsules, mericarps, berries or drupes.

The family is particularly represented in the tropics. Well-known genera such as *Coffea* belong to this family as well as ornamental plants, species supplying drugs and edible fruits or timber of economic use and high quality.

The family is distinguished by the presence of certain alkaloids and iridoid-like compounds. Of the indolealkaloids, yohimbine of *Pausinystalia yohimba* (cortex yohimbe) is well known for its sympatholytic effect. Yohimbine is also found in *Rauwolfia*. Best known are the alkaloids of *Cinchona* species which contain quinine in their bark (e.g. yellowbark, ledgerbark).

The bitter tasting quinine has antipyretic effects and is applied for malaria; it is furthermore ecbolic.

The family is rich in chemical constituents of potential biodynamic activity, such as iridoids, alkaloids, triterpenes, sterols, anthraquinones, naphthalene derivatives, polyphenols, tannins, cyanogenic compounds.

Chimarrhis

C. comprises about a dozen species of small to medium-sized trees which are widely distributed in tropical South America and Southern Central America.

The wood is not difficult to work and is probably durable. The leaf anatomy of *Chimarrhis microcarpa* Standley var. *speciosa* Steyermark has been described by ROTH (1992). The chemistry of this genus is not known, but may be interesting.

Cinchona

The genus is represented by quite a few species which occur mainly in the montane cloud forests of the Andes of South and Central America, though some are also found in the Cordillera de la Costa (Figs. 227–229). It is difficult to classify the group owing to the facility with which the species hybridize with one another, giving rise to many different forms (natural crosses). Bolivia and Peru are probably the most important countries where Cinchona occurrs. There are about 8 species which are well known and used medicinally for their stem and root bark.

The species are evergreen shrubs or small trees with a rough bark. The simple leaves are oblong-elliptic and have a smooth surface. The inflorescences occur in terminal panicles with small fragrant yellow or pink flowers, depending on the local variety. The flowers have to be cross-pollinated to produce seeds. The capsules are 1–2 cm long.

In the Andes, the plants grow between 500 m and 3000 m a.s.l. The different species have similar climatic requirements. In culture they grow best at altitudes between 1000 m and 3000 m where the average temperature is high with a relatively small range of variation, high atmospheric humidity, high rainfall well distributed throughout the year, a light well-drained soil rich in organic matter, and a sloping ground sheltered from the wind.

Ethnobotanical and general use

Medical use

Stem and root bark. The stem and root bark have become famous through their curative power of malaria. The bark contains a number of alkaloids among which quinine, quinidine, cinchonine and cinchonidine are the most important. They are used as antipyretics against periodic fever, as a tonic and ecbolic.

However, quinine was isolated in a pure state as early as 1820. It acts as a protoplasm poison of the malaria plasmodium. In view of the alarming acquired resistance of the malaria plasmodia to synthetic drugs, the natural *Cinchona* alkaloids have now regained their importance.

In the meantime, the plants have been widely used as tonics and as a stimulant in beverages (Tonic water).

The most important species are: *Cinchona calisaya* WEDD., *C. ledgeriana* MOENS., *C. officinalis* L. and *C. succirubra* PAVON.

Adulterations are made with species of *Remija*, Rubiaceae, Esenbeckia, Rutaceae, and in Venezuela with *Cusparia trifoliata* (Rutaceae) which also contains a bitter principle known as Angostura bitter. It is tonic and antipyretic and less irritating for the stomach than quinine.

Cinchona sp. (quina)

Anatomical description

Bark (Figs. 227, 228, 229). The cork is small and thin-walled. Dilatation growth is present in the primary cortex with abundant formation of relatively thin-walled stone cells. The pericycle is already ruptured and replaced by stone cells. The hard bast, composed of fibers, is arranged in radial rows; these are about 1–3 (seldom 4) cells in width. Most of the rays are uniseriate but become biseriate by forming anticlinal walls; rays enlarge irregularly in this way. Formation of large stone cells can be observed in the rays. Some rays enlarge more towards the outside than others. The hardbast fibers are accompanied by septate crystal strands (Fig. 229). However here and there druses occur instead of rhomboid crystals. Dilatation growth becomes pronounced in the outer bark. Parenchymatous cells as well as stone cells are tangentially extended in this region.

The bark is similar to that of *C. succirubra*, but shows formation of stone cells and has rhomboid crystals instead of crystal sand.

Fig. 227. *Cinchona (officinalis?)*, 'Quina'. **a** T.s. of bark. Note the radial rows of the hardbast (fibers) and the dilatation growth in the peripheral part (cells extended tangentially). **b** Detail with hard and soft bast.

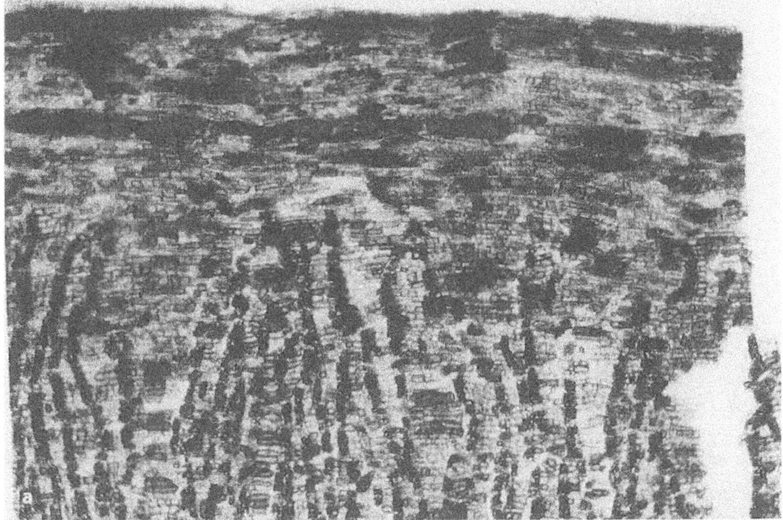

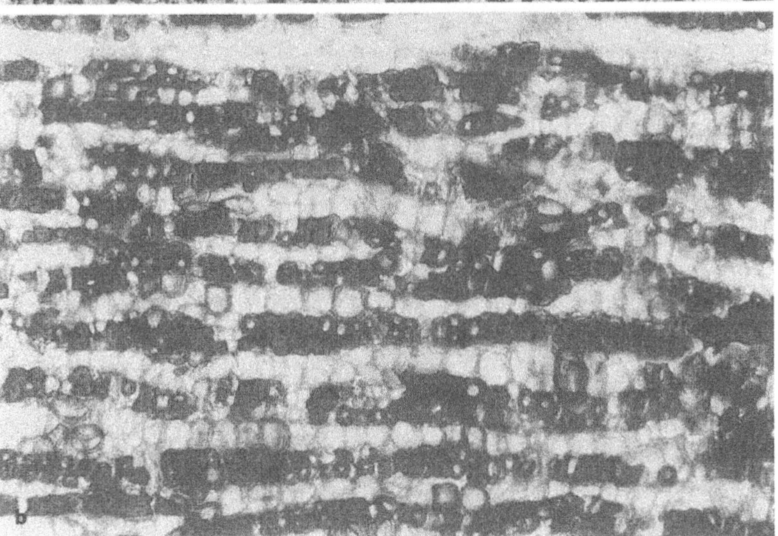

Coffea arabica L. (cafe)

Although the coffee is indigenous to Africa, it is now cultivated in the tropics of the Old and New World and shall therefore be mentioned shortly here. Economically seen, coffee fruits are the most important cultivated fruits in Latin America. In Venezuela, coffee was first introduced in 1748 by a priest. Fruit and seed anatomy and development have been studied by ROTH & LINDORF (1971).

Anatomical description

Fruit. The ovary is inferior and develops into a drupe with two seeds. In an advanced stage, the fruit consists of an exocarp corresponding to the outer epidermis. Beneath, several subepidermal layers with thickwalled cells form the parenchymatous hypodermis. The bulk of the mesocarp is still composed of relatively thin-walled parenchyma cells.

The endocarp, mainly originating from a subepidermal ventral meristem, undergoes periclinal cell divisions so that 3–4 layers are formed. The endocarp mother cells of the inner epidermis and of the 3–4 subepidermal layers are subdivided by parallel division walls; each mother cell divides about 5–20 times. However, as the direction of cell division of neighbouring mother cells differs, a parquetry pattern results, as seen in tangential sections through the ripe endocarp. The daughter cells become thick-walled and lignified so that the endocarp is finally fibrous.

Two to three cell layers of the mesophyll which are in direct contact with the endocarp, maintain their meristematic activity until the endocarp is

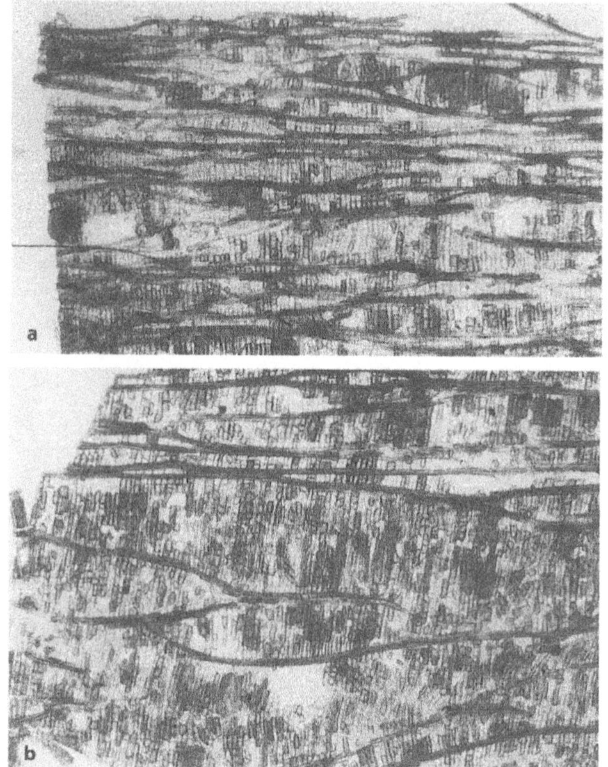

Fig. 228. 'Quina' bark. a, b Longitudinal tangential section. Note the fibers and the rays.

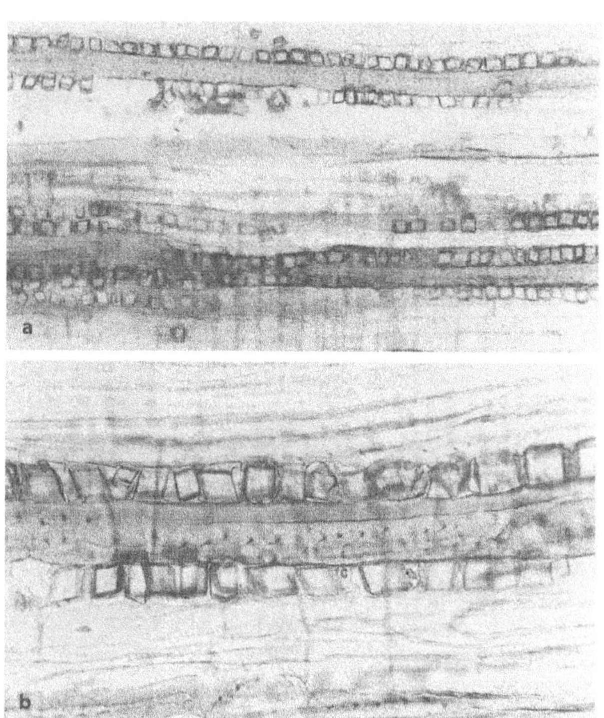

Fig. 229. 'Quina' bark. a, b Longitudinal radial section of bark. Note the fibers and the septate crystal strands accompanying the fibers. Each cell contains a rhomboid crystal.

already differentiated. Now they reassume their activity, principally dividing in an anticlinal direction. During differentiation, they enlarge considerably in a radial (anticlinal) direction so that the pericarp markedly increases in width. The cells in direct contact with the endocarp enlarge most. Finally, these palisade-like cells become slimy.

In the mature stage, the pericarp consists of a single-layered exocarp (outher epidermis), a mesocarp which is divided into an outer parenchymatous region in which the vascular bundles lie, and into an inner slimy palisade layer, and finally of a fibrous endo or sclerocarp which has originated from the inner epidermis together with some subepidermal layers (ROTH 1977, ROTH & LINDORF 1971a).

Seed. The so-called silver skin in contact with the endocarp, originates from the seed coat and in the mature state is composed of an interrupted layer of sclereids, several layers of thin-walled parenchyma cells and a layer of completely compressed and obliterated cells. The layer of sclereids corresponds to the outer epidermis of the seed coat. The epidermis cells cannot adapt to the growth of the seed in circumference in later stages, either owing to cell division or to further cell enlargement, but partly separate from each other leaving large intercellular spaces.

In this way, a discontinuous layer results. The sclereids of the silver skin become lignified, although to a lesser degree than the fibers of the endocarp. As seen in a surface view, the sclereids adopt very different forms and part of them are very irregular in shape; their walls are partly bent and less thick than those of the endocarp; their pits are also larger.

Under the microscope, pulverized coffee may be distinguished by the presence of endosperm cells and rests of the silver skin. The endosperm cells are easily recognized by their cell walls in the form of a string of pearls; the walls are thickened, but repeatealy strangulated by the formation of pits; the cells contain oil drops. The lignified sclereids of the silver skin are larger and stain red with phloroglucine and hydrochloric acid: lignin. The embryo itself is so small that almost no rests of it are recognizable in ground and roasted coffee powder.

As coffee is one of the most popular drinks in the world, but has a relatively high price, it is frequently adulterated, e.g. with ground nut shells or fruit kernels, grape cake or other residues of fruit, leached chips of turnips or coffee grounds.

Likewise, a large variety of substitutes exists which may be detected microscopically in the way decribed above.

Cichorium intybus is used as a substitute as well as an adulterant.

The interested reader can find a long list of adulterants and substitutes of coffee in COOLHAAS, DE FLUITER & KOENIG (1960).

Ethnobotanical and general use

Nutritional use

Coffee is used by at least one third of the world population as a beverage and is one of the most important commercial crops of the tropics. Ninety percent of the coffee used in commerce is derived from *Coffea arabica*. The flavour of coffee is due to the presence of the essential oil caffeol, the stimulating effect is ascribed to the alkaloid caffeine. Fatty oil is mainly stored in the endosperm. There is very little true tannin in Coffee. Besides being used as a beverage, coffee is also used to flavour ice-cream, candies and pastries as well as liqueurs.

Economical utilization

The shrub or treelet is also appreciated as an ornamental. The wood is little used, as the stem does not reach an adequate size.

Medical use

Caffeine is a stimulant, a nervine and diuretic. It acts on the central nervous system, kidneys, heart and muscles. The methylxanthines caffeine, theobromine and theophylline have psychoanaleptical and diuretic effects. Caffeine is present in quite a few medicines.

Coffee is said to be an aphrodisiac and a stimulant. People can become addicted to it, it can increase the cholesterol content in the blood and affect the mucosa of the stomach – but nevertheless it is much appreciated, since it seems to meet the following description:

Negro como el diablo,
caliente como el infierno, y
dulce como un angelo
(Black like the devil,
hot like the hell,
and sweet as an angel).

Coutarea

Copalchi bark is tonic, used for intermittent fevers (malaria) and ailments of the lungs.

ROTH studied the bark structure 1981, fruit structure and dispersal 1987. Further information: UPHOF 1968.

C. hexandra supplies Quina do Pernambuco.

Duroia

Leaf decoction for headache following alcoholic intoxication. Stem infusion for cough. Bark for cicatrization. Pheromones of ants living in the swollen stem parts are used by the Waoranis for ulcers of the mouth. Exudates of roots or ants living in the internodes are probably poisonous for other plants. Bark for cicafrization. ROTH studied the bark structure 1981, leaf structure 1984, fruit structure and dispersal 1987, leaf venation 1996.

Further information: SCHULTES & RAFFAUF 1990.

Faramea

At least 5 species are useful. Wood is useful; The bark is emetic and febrifugal. The roots are febrifugal.

ROTH studied the leaf structure 1984, fruit structure and dispersal 1987, leaf venation 1996.

Genipa americana L. Wood is useful. The fruit is edible and source of jam and marmalade (marmalade box). It is also manufactured into refreshing drinks and alcoholic beverages. It contains a black-blue pigment used as a dye. Immature fruits are applied for toothache and extraction of teeth.

A decoction of the leaves is used as an astringent and for dysentery and gonorrhoea. The flowers are tonic and febrifugal. The fruit is said to be emetic.

ROTH studied the bark structure 1981, leaf structure 1984, fruit structure and dispersal 1987, leaf venation 1996.

Further information: UPHOF 1968, GUPTA 1995, SCHULTE & RAFFAUF 1990.

Gonzalagunia

G., with about 13 species of erect or straggling shrubs and a few trees is widely distributed in tropical America. The wood is light-coloured and fine-textured. It is locally used.

G. dicocca CHAM. & SCHLECHT SSP. *venezuelensis* STEYERMARK is a low shrub, 1–3 m high, occuring in the transitional cloud forest at altitudes of 900 m. The fruits are globular fleshy berries of purple colour.

The leaf anatomy has been discribed by ROTH (1990).

Guettarda

Wood useful. Leaves serve to expel intestinal worms. The root is astringent and vulnerary, used for diarrhoea, particularly in veterinary medicine.

ROTH studied the bark structure 1981, fruit structure and dispersal 1987. Further information: UPHOF 1968, SCHULTES & RAFFAUF 1990.

Palicourea

P. comprises more than 200 species of shrubs and small trees, native to tropical America. All timber is fine-textured, easy to work with and is used locally. About 17 species are medically used (SCHULTES & RAFFAUT 1990).

Palicourea crocea (SW.) ROEM. & SCHULT. is a 1–3 m high shrub. The species is frequent in the Cloud forest of the Andes and of the Cordillera de la Costa. Its common name is Café de monte. It would be interesting to know, whether the plant contains caffeine.

The natives consider a decoction of the leaves to be an effective antirheumatic and an emergency emetic in cases of poisoning from ingestion of spoiled fish or meat.

The leaf anatomy of this species has been described by ROTH (1990).

Posoqueria

With about 15 species of shrubs and small to medium-sized trees, P. is widely distributed in tropical America.

The fine textured wood is used for handles, turnery, and small articles of joinery.

The aromatic flowers of *P. latifolia* are pulverized to repel fleas in clothing and hammocks.

Posoqueria coriacea MART. & GAL. SSP. *formosa* (KARST.) STEYERMARK is a shrub or small tree, 5–12 (20) m high. The thick coriaceous leaves are elliptic, 10–15 cm long and 6–8 cm broad. The very variable species is frequently found in Venezuela. It is characteristic of the cloud forest at an altitude of 1500–2000 m. The flowers are white and fragrant. The globular to ovoid berries are dark-green to yellow, fleshy and 7.5–8 cm long and 5.5 cm broad.

The leaf anatomy has been described by ROTH (1990).

Psychotria

P. is a very large genus with about 1200 species of low to medium-sized shrubs and small trees. However, distinction of the genus from other genera is difficult. *Psychotria* is closely related to *Palicourea*.

There are at least 30 species which can be used. The small berries of some species are edible; other species have highly toxic fruits.

Medicinal plants for rheumatism, snake bites and insect stings or asthma are described as well as are species which are used as an emmenagogue or which have emetic properties. Some fish poisons are also found.

GUPTA (1995) describes sudorific and emetic activities, curative effects of bronchitis and amebiasis of *P. ipecacuanha* STOKES (synonym: *Cephaelis ipecacuanha* (BROTERO) RICH.

Phytosterols and dimethyltryptamine as well as a complex alkaloid, psychotrine, were found.

The leaf anatomy of *P. agostinii*, *P. araguana* STANDLEY, *P. aubletiana* STEYERMARK and *P. botryocephala* (STANDL.) STEYERMARK has been described by ROTH (1990), *P. anceps* HBK and *P. capitata* R. ET P. by LINDORF 1992.

P. agostinii is a frequent and a characteristically tufted treelet of the undergrowth of the Venezuelan cloud forest proper. The leaves are of herbacious-leathery consistence and their upper surface has a bluish brilliant hue. The fleshy berries, eaten by birds, have an attractive red colour. The leaf anatomy has been described by ROTH (1990).

Uncaria guianensis (AUBL.) GMEL (uña de gavilan, bejuco de murcielago, bejuco de gavilan)

Taxonomical description

The species is a climbing shrub with spines which reaches 16–18 m in height. The opposed glabrous leaves are of herbaceous consistency, of elliptic shape and acuminate; they reach 5–9 cm in length and 2.5–4.5 cm in width. The petioles are 5–15 mm long. The lateral stipules are oblong, glabrous and erect.

The flowers are united in dense solitary heads which develop axillary or terminally. The filiform bracts are pubescent and 1.5 mm long. The calyx is pubescent on its outside; the calyx tube measures 2.5–4 mm in length, the lobules 0.5–0.8 mm. The white or yellowish crown is silky-hairy outside; the tube is 4.5–6.5 mm long, the lobules 1.5–3 mm. The stamens have short filaments and sagittate anthers, they reach a length of 1.5–2 mm. The disc shows a ring of silky hairs. The filiform style reaches 8.5–10 mm in length.

The infrutescence develops numerous fusiform brown capsules, 1.8–2.4 (3) cm long and 5–7 mm in diameter.

Occurrence

The species is found wild in the tropical part of South America. In Venezuela it occurs in Guiana.

Anatomical description

Bark. (compare Fig. 230–234). A bark of about 7 mm in width was studied. The bark is scaly and the scales are of medium size. Hardbast is found only in the form of fibers, although fibers are usually rare in the Rubiaceae (METCALFE & CHALK 1950). The hardbast is relatively scarce and occurs in very long alternating bands of irregular shape; solitary fibers are also irregularly dispersed in the parenchyma. The fiber bands are occasionally wavy. The hardbast pattern is very irregular and the distances between the bands, particularly in a radial

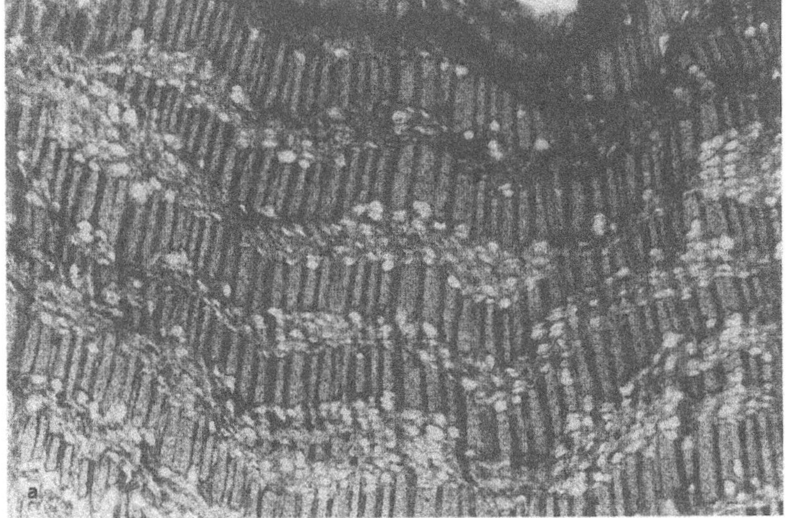

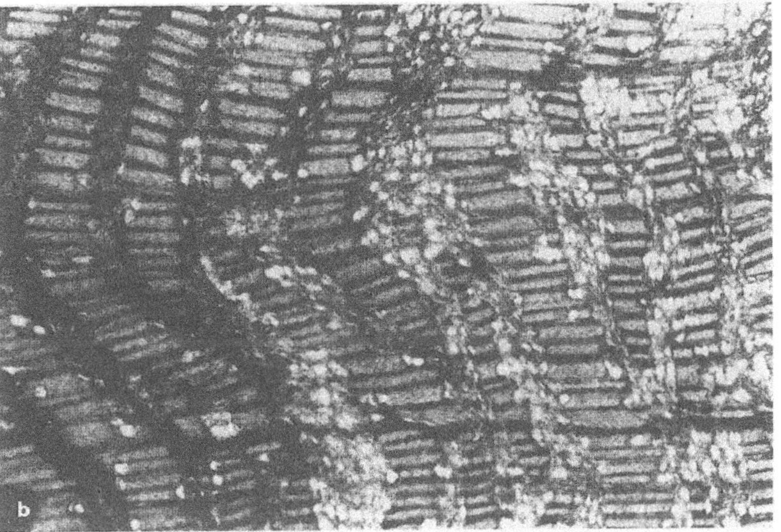

Fig. 230. *Uncaria tomentosa*. **a, b** T.s. of bark (× 5). Note the broad bands of hard bast alternating with bands of soft bast and the small rays (× 5).

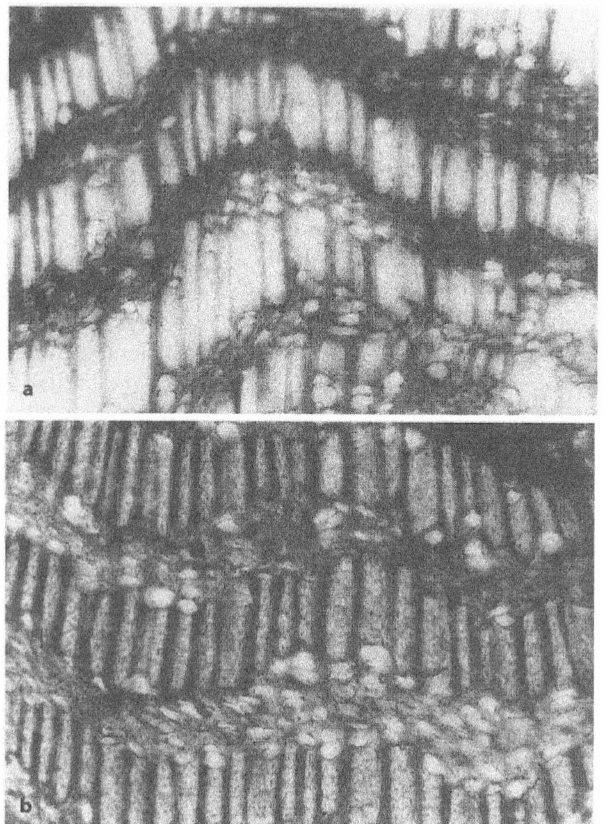

Fig. 231. *Uncaria tomentosa*, **a**, **b** bark × 10. Note the fibers in the hard bast.

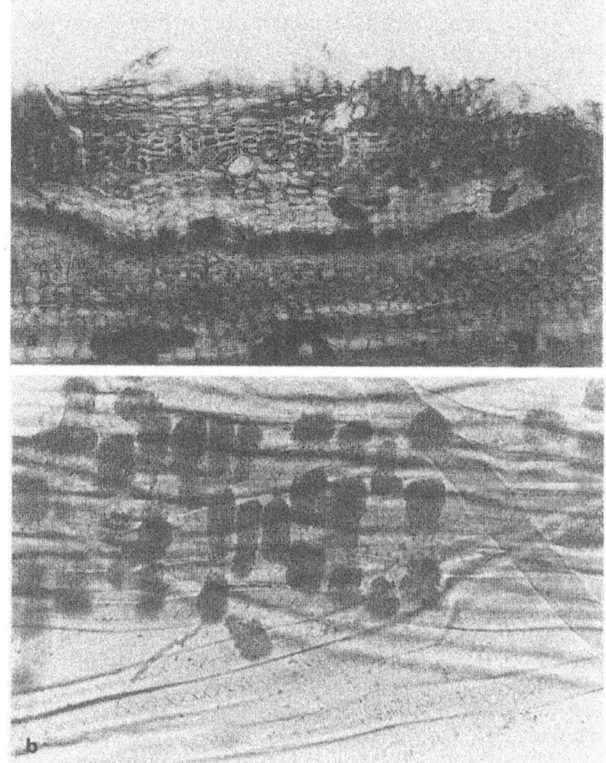

Fig. 232. *Uncaria tomentosa*. **a** Periderm, longitudinal section × 10. **b** Macerated bark with fibers and crystal sand in the ray cells × 20.

direction, are occasionally very large. The rays are narrow, but freqeunt, and only 1–2 cells wide. Cells filled with a brown content (probably tannins) are dispersed in the axial parenchyma. Crystal sand and solitary crystals in the form of molars are abundantly found in isodiametric parenchyma cells.

Secondary formation of stone cells is very scarce and dilatation growth is very moderate, mainly involving the axial parenchyma, and only occasionally the rays. The phelloderm is of medium size and composed of stone cells: it comprises about 6–10 cells in width.

Ethnobotanical and general use

Medical use

In popular medicine, the entire plant or the leaves are used against malignant tumorus, rheumatism, arthritis, diabetes; cirrhosis (GUPTA 1995), and against diarrhoea, dysentery and to cure wounds (PHILLIPSON et al. 1978, uphof 1968).

Healing properties

Febrifuge, antitumoural

Method of use

Two spoonfuls of the entire plant are boiled in 1.5 l water for 30 minutes. Half a glass of the cool liquid is taken 3 times a day before the meals.

Chemical contents

Two flavonoids are found in the bark: kaempferol and dihydrokaempferol, furthermore tannins (cells with brown content of the bark), and 4 glycosides of quinovic acid. Indolic alkaloids are found in the roots and the leaves.

Varieties and related species

Uncaria tomentosa is the other neotropical species which has similar properties and is used medically in a similar way.

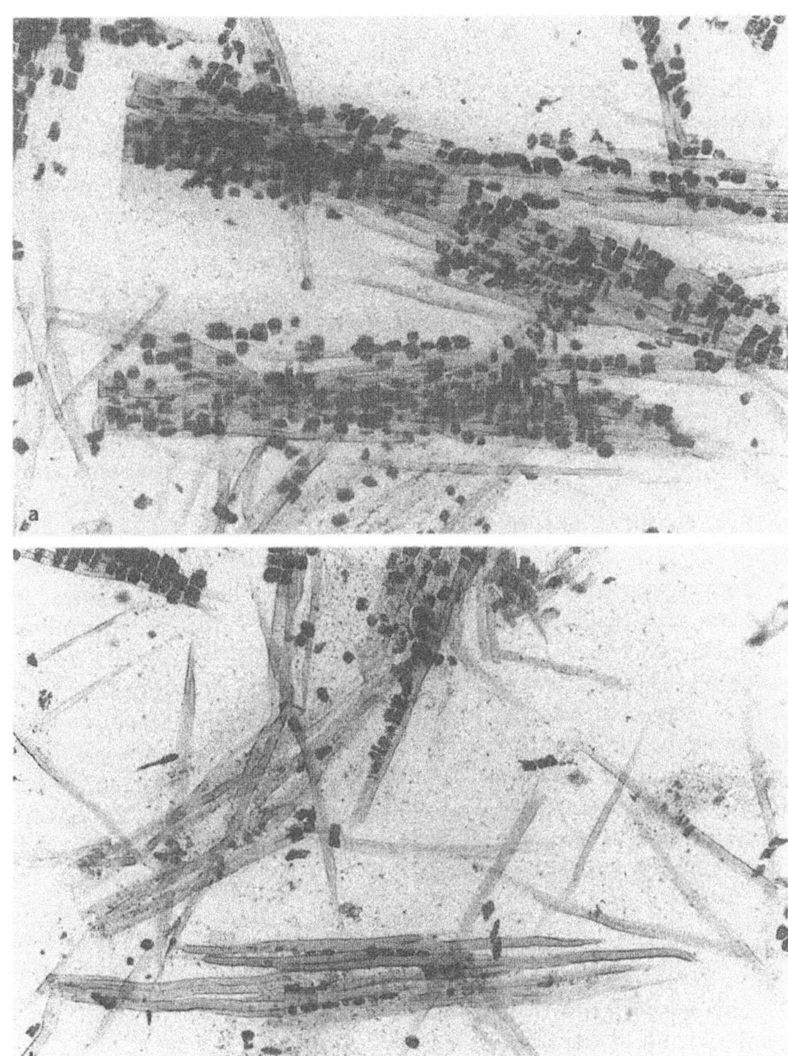

Fig. 233. *Uncaria tomentosa*. **a, b** Macerated bark with fibers and crystal sand in the ray cells × 5.

Fig. 234. *Uncaria guianensis*. Macerated bark with solitary crystals in the form of 'molars' (styloids).

Cultivation

The species grows in high forest with abundant insolation up to a height of 450 m a.s.l.

Observations

See the comparison with *Uncaria tomentosa*. *Uncaria tomentosa* (WILLD.) DC (uña de gato) and *U. guianensis* (AUBL.) GMEL.

Uncaria tomentosa (WILLD.) DC a comparison

Taxonomical description

Both species are vines. *U. tomentosa* reaches 18–19 m in height. The young leaves are of brownish-red colour. In contrast to *U. guianensis* which has glabrous leaves, the lower leaf surface of *U. tomentosa* is pubescent to tomentose. While the hooks of *U. guianensis* are conspicuously curved backwards and are glabrous from the beginning on, those of *U. tomentosa* are only seldom curved, but are tomentose to glabrous.

The brown fruits are pubescent.

Occurrence

The species grow in high forests with abundant insolation up to altitudes of 500–600 m a.s.l. *U. tomentosa* is typical of Venezuela, but also occurs in other countries of South America. *U. guianensis* is characteristic of Guiana.

Anatomical description

Bark (Figs. 230–233). The width of the bark studied was about 5 mm. As compared with *U. guianensis, U. tomentosa* develops much more hardbast. The fiber bands are more numerous and are closer together, as seen in a radial direction. They may comprise up to 11 fibers in a radial row, while in *U. guianensis* there are usually only 3 (2–4) fibers in a radial row. Solitary fibers are rare, in contrast to *U. guianensis* where they occur abundantly. The fiber bands are conspicuously wavy. The bands may reach such a length that they almost touch neighbouring bands to become concentric. The hardbast pattern is thus more regular in *U. tomentosa*.

The rays are mostly uniseriate and only seldom 2- to 4 seriate. They are more frequent in *U. tomentosa*, leaving a distance of 1–2 fibers between one another, as seen in a tangential direction; in U. *guianensis*, on the other hand, there are 1–3 fibers between 2 neighbouring rays. Crystal sand is abundantly found in the ray cells.

The rhytidome is scaly and the periderm is of the same structure as in *U. guianensis* with a phelloderm composed of stone cells.

Ethnobotanical and general use

Medical use

Used parts: Bark, leaves. Bark in pieces or ground. In popular medicine, the plant is used against malignant tumour, rheumatism, arthritis, diabetes and cirrhosis of the liver (GUPTA 1995).

While in *Uncaria guianensis* it is mainly the leaves that are used for healing purposes, in *U. tomentosa* it is the bark which contains the healing properties. In Peru, the bark has been used from time immemorial by the Ashanica Indians to cure a variety of sicknesses, such as immunological and digestive problems (STEINBERG 1994). Intense investigations since 1970 carried out at universities throughout the world have shown that the plant has the following properties: it is immunostimulant, antiinflammatory, antiviral, antiallergic; it even contains substances inhibitory to tumours and stimulates the immune system against aids.

U. tomentosa is therefore a species largely used in popular medicine. *U. guianensis* possesses similar properties, but its effects on the immunological system are less striking as it lacks one of the alkaloids present in *Uncaria tomentosa*. The pharmaceutical industry already produces tablets, capsules and ointment of *U. tomentosa* (LINDORF 2000).

Method of use

2 spoonfuls of the bark are boiled in 1,5 l of water for 30 minutes and left to cool. Half a glass of this liquid is taken 3 times a day before meals (GUPTA 1995).

Healing properties

Antiinflammatory, antiviral, contraceptive; increases phagocytosis, exhibits cytostatic activity, inhibits DNA synthesis in sarcoma, increases the level of immunoglobulin.

Chemical contents

Six alkaloids could be identified. The bark contains campesterol (see also GUPTA 1995).

Observations

The best criteria to distinguish the barks of the 2 species are the styloids in *U. guianensis* and the hardbast structure: A more abundant hardbast in the form of long undulated bands which reach a width of 11 fibers in a radial row and which are closer to each other, characterize *U. tomentosa*.

LINDORF (1997, 2000) compared the barks of the 2 species anatomically and obtained the following results.

Comparison of the 2 species (according to LINDORF 1997, 2000). The 2 species of *Uncaria* occurring in tropical America have many characteristics in common, even in their bark structure: The fibers are 1100–2750 μm long. The sieve tubes are dispersed or occur in the form of radially arranged groups with 5–7 members in a file. The sieve plates are inclined, being composed of approximately 6 sieve areas.

The axial parenchyma is intercalated between the sieve tubes forming long parenchymatous series of 4–6 (11) cells, some of which are sclerosed.

3–6 rays occur per mm, they are uniseriate (or bi to four seriate) with long uniseriate margins. The multiseriate rays often vary considerably in height due to the vertical union of 2 or more rays. The uniseriate rays are composed of erect to square-shaped cells, and often of globular cells. The multiseriate rays are heterogeneous with square-shaped cells and frequently with globular cells in the uniseriate extremities, and with procumbent cells in the main body. Occasionally, the entire ray is composed of globular cells. Dilatation growth mostly acts in the axial parenchyma and only occasionally in the rays. The cells of the dilating zones often become sclerosed with time.

Solitary crystals as well as crystal sand are found in the axial parenchyma and in the rays. The solitary crystals (styloids with bifurcate extremities) however only occur in *U. guianensis*, and only crystalsand is present in *U. tomentosa*. Simple and compound starch grains occur in the rays and in the axial parenchyma.

Cells with a thickened wall and a brown content are dispersed in the parenchyma, in the phelloderm, and in the lenticels.

Alternating fiber bands are also found in other species of the Rubiaceae. Dilatation growth and secondary formation of stone cells in the dilated zones are rare in this family.

Although the 2 species studied are similar in their structure they show certain differences: in *U. tomentosa* the fiber bands are higher, more densely arranged and are more regular.

Crystals are less suitable for a comparison, because their formation depends on fluctuations of the environment, such as differences in insolation, water supply or of nutrients.

Frequency and morphology of the crystals may also change during plant development. To obtain more precise results, a larger quantity of samples coming from different stands and geographic regions should be studied, experiments should be carried out and the entire life cycle of the plants should be investigated.

However, in the case of *Uncaria guianansis*, the dentate styloids are very characteristic and can be used as an important criterion for identification (Fig. 234).

Rutaceae

This family comprises aromatic trees, shrubs and a few herbs. Glands in the bark, in leaves and fruits are common.

The most important genus is *Citrus*.

Many species are sources of essential oils used in perfumery and medicine. Ethereal oils, alkaloids, cyanogenic compounds, coumarins, flavonoids, aromatic acids, tannins, lignans, tetracyclic terpenes, as well as saponins are found. Acronycin, an acridonalkaloid, has oncolytic effects.

Anatomical description

Citrus fruits

Although species of *Citrus* (Figs. 235, 236, 237) have their origin in Asia, they are cultivated in the tropics and subtropics all over the world, and may therefore be briefly mentioned here.

All the commercially interesting species of *Citrus*, like the sweet orange, the lemon and lime, grapefruit, bitter orange, tangerine and mandarine orange, show more or less the same fruit structure and development. There appear however some structural differences of minor importance in the various cultivated varieties. As the fruit is the most important and the most used organ of the *Citrus* species, we only describe fruit structure in the following.

Fruit anatomy and development have mainly been described by POULSEN (1877), PENZIG (1887), BIERMANN (1896), FORD (1942), BARTHOLOMEW & REED (1943), SCOTT & BAKER (1947), ROTH & LINDORF (1972, ROTH 1977).

Besides many other studies dealing with certain structures of the orange fruit have been published.

The fruit, called a hesperidium, represents a berry of special structure with a thick leathery rind proceeding from exo- and mesocarp and a juicy

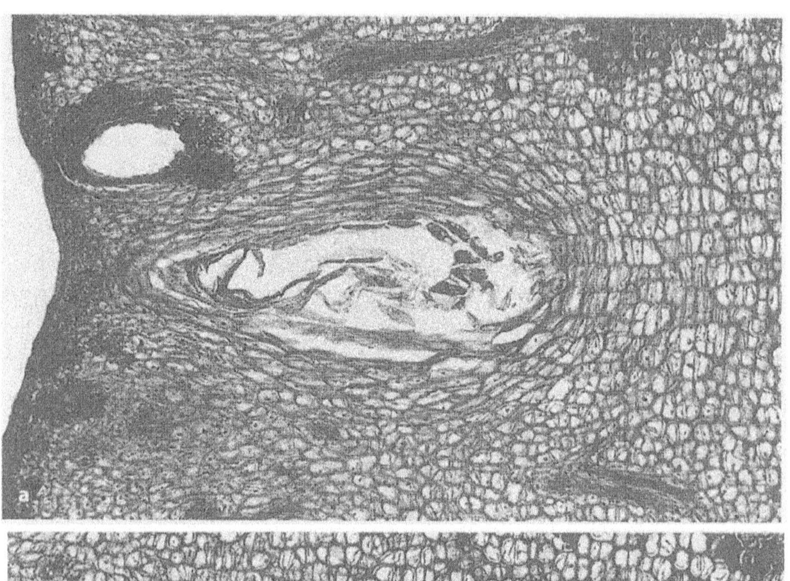

Fig. 235. *Citurs limonia*, Rutaceae. T.s. of young fruit. **a** Oil cavity. **b** Radial rows of inner flavedo (ROTH 1977).

pulp originating from proliferations of the endocarp. The superior ovary is composed of numerous (10–13) carpels.

General structure of the Citrus fruit. In the navel orange the rind is very thin, being about 8 mm thick. The cells of the outer epidermis contain plastids and small oil drops. The outer tangential walls are cutinized and additionally protected by a layer of wax. Stomata are frequent with an approximate density of 13 per square millimeter; they are of the actinocytic type with the subsidiary cells concentrically arranged around them.

Below follows a 1–3 layered hypoderm with collenchymatous wall thickenings. Commonly, no distinction is made between exo- and mesocarp, but the pericarp wall is subdivided into epidermis, hypodermis, flavedo, albedo and endocarp. However ESAU (1967) and ROTH & LINDORF (1972) distinguish an exocarp or flavedo, a mesocarp or albedo (Figs. 237 a + b), and an endocarp (Fig. 235).

The flavedo comprises the oil cavities (Fig. 235 a), vein endings and the zone of chloro- and chromoplasts, whereas towards the inside of the pericarp the plastids diminish in number. The tissue of the flavedo (Fig. 235 b), composed of polygonal cells, is relatively compact. Towards the middle part of the pericarp, cells become larger in size and develop thicker walls. There is a typical transitional zone between flavedo and albedo. The oil cavities are surrounded by concentric rings of relatively small, very thick-walled and characteristically pitted cells which are flattened parallel to the cavity, giving the impression of a collenchyma. Towards the inside of the cavity, the cells become thinner-walled and contain small oil drops; large drops of essential oil are embedded in the cavity itself. In fruits of a lower quality and with a heavy rind, the cells of the flavedo are throughout thick-walled and give the aspect of a collenchyma.

Towards the inside of the pericarp follows the spongy white tissue of the albedo (Fig. 237 a + b)

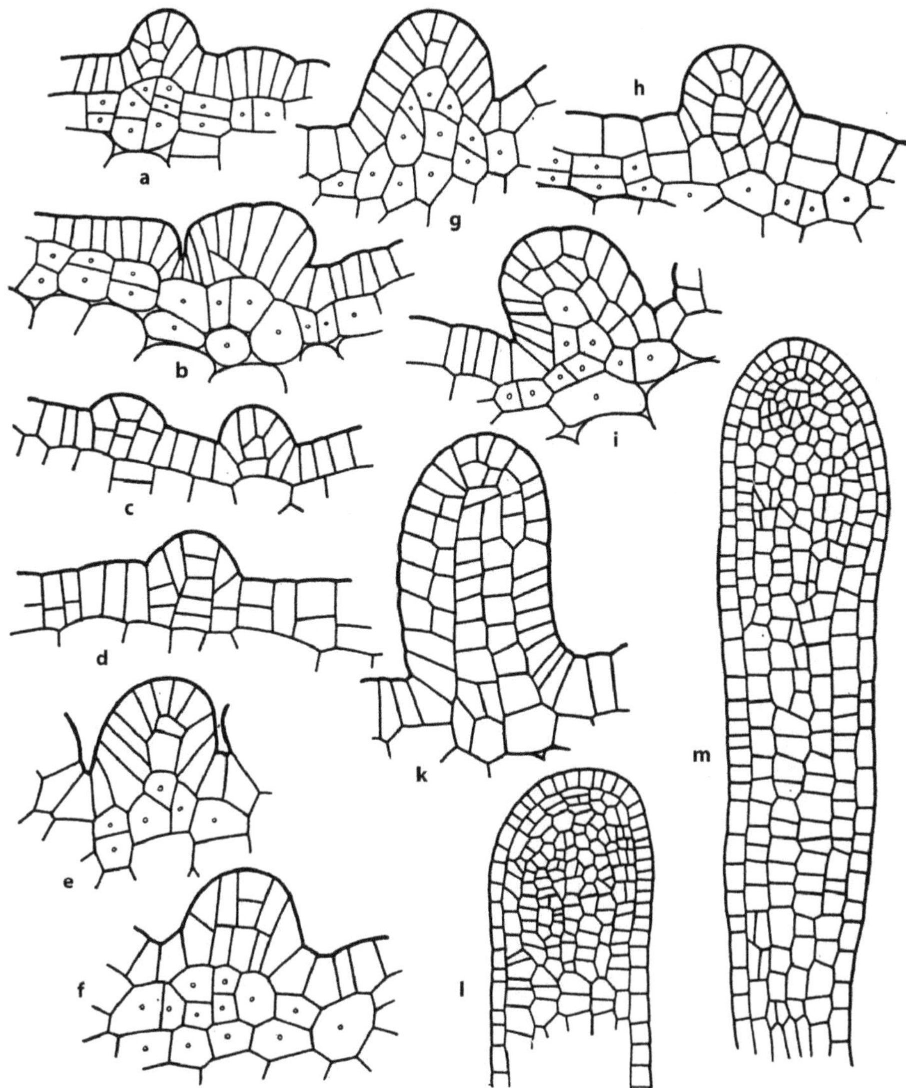

Fig. 236. *Citrus.* **a–m** Cross sections through papillas of the endocarp in successive stages of development (ROTH & LINDORF 1972 c).

which contains large air spaces. The colourless cells are typically 8-armed.

The most outstanding part of the pericarp is the endocarp; the inner epidermis together with some subepidermal layers is involved in its formation. In the mature fruit, these layers may form a resistant membrane around the pulp, due to cell wall thickening. The inner epidermis is covered by a thin cuticle and has no stomata.

The endocarp gives rise to the juicy endocarp sacs which fill the locules entirely at fruit maturity. Each vesicle has a clubshaped body raised above the epidermis by a thin elongate stalk. The epidermis cells of the adult papillae are elongated parallel to the long axis of the vesicle, while the inner cells are much larger, particularly in the center of the juice sacs, and very thin-walled; they contain large vacuoles and small chloroplasts. Schizolysigeneous spherical cavities which contain oil, wax and a granular matter are found in the center of the juice sacs of most Citrus varieties. The plastids in the juice sacs seem to be characteristic of the species: certain tangerines and mandarin oranges have spindle-shaped plastids; isodiametric plastids occur in the king orange and in numerous varieties of the sweet orange. The colourless type of plastids (elaioplasts) is found in the grape fruit, sour orange, and lemon. A possible granulation of the papillae is due to wall thickening and lignification of some interior cells.

Most interesting is the formation of the juice sacs in the fruit endocarp (Fig. 236). At early stages, the inner protodermis is very meristematic dividing anti and periclinally so that almost the entire subepidermal zone originates from the inner protoderm, as the true inner epidermis is established much later. The protuberances emerging from the protoderm could therefore be considered as protodermal derivatives, although subepidermal tissue participates in their formation. They arise exclusively on the inner tangential walls of the locules, but not along the septs (except in some varieties).

Additionally to the juice sacs which form the juicy pulp of the fruit, we find ramified pluricellular papillose hairs on the inner tangential walls of the locules which are said to secrete mucilage; they possess large papillose end cells with thick probably mucilaginous walls. Another type of hair is present in the stylar canals; they are unicellular or – at times – pluricellular.

The spongy parenchyma of the rind (albedo) corresponds to a parenchyma with armed cells of star-like appearance. The meristematic cells of the young albedo are tetrakaidecahedral (14-hedrous). The mature tissue consists of 8-armed cells (SCOTT & BAKER 1947).

Secretory glands or cavities which can reach a diameter of 0.13 mm occur only in the periphery of the rind and have elliptic outlines, as seen in a longitudinal section. They probably originate from one single mother cell which repeatedly divides into daughter cells. The origin of the cavity is believed to be of schizogeneous nature, but its expansion takes place partly in a lysigeneous way. At the mature stage, the essential oil is deposited in the center of the cavity. The mean density of oil glands is about 2.3 per square millimeter in the Washington navel orange.

Chloroplasts are present in the flavedo only at early stages, as long as the parenchyma is photosynthetically active, but later transform into chromoplasts adding the orange colour to the rind. The albedo usually only contains leucoplasts and serves as a storage parenchyma. Starch has been observed mainly in the albedo and only in young fruits, but completely disappears at more advanced stages. The essential oil develops in the oil cavities of the rind and in the juice sacs. A very peculiar substance present in the *Citrus* fruits is the glucoside hesperidin, found in large quantities in young fruits. It occurs in solution in the vacuoles, but crystallizes in the form of fine needles-when alcohol is added-uniting in aggregations of variable shape. It also adopts crystalline form in frozen tissue and may therefore be used as an indicator of frost damage to fruits. Another substance which may occur in crystalline form is naringin which is found in the grapefruit. The characteristic orange and yellow colour of the peel is due to chromoplasts which contain carotene and xanthophyll. The characteristic colour in the rind of mature limes is due to the pigment phlobatannin which occurs in the cell sap. The pink and reddish colours of blood oranges and some grape fruits derive from pigments dissolved in the vacuoles.

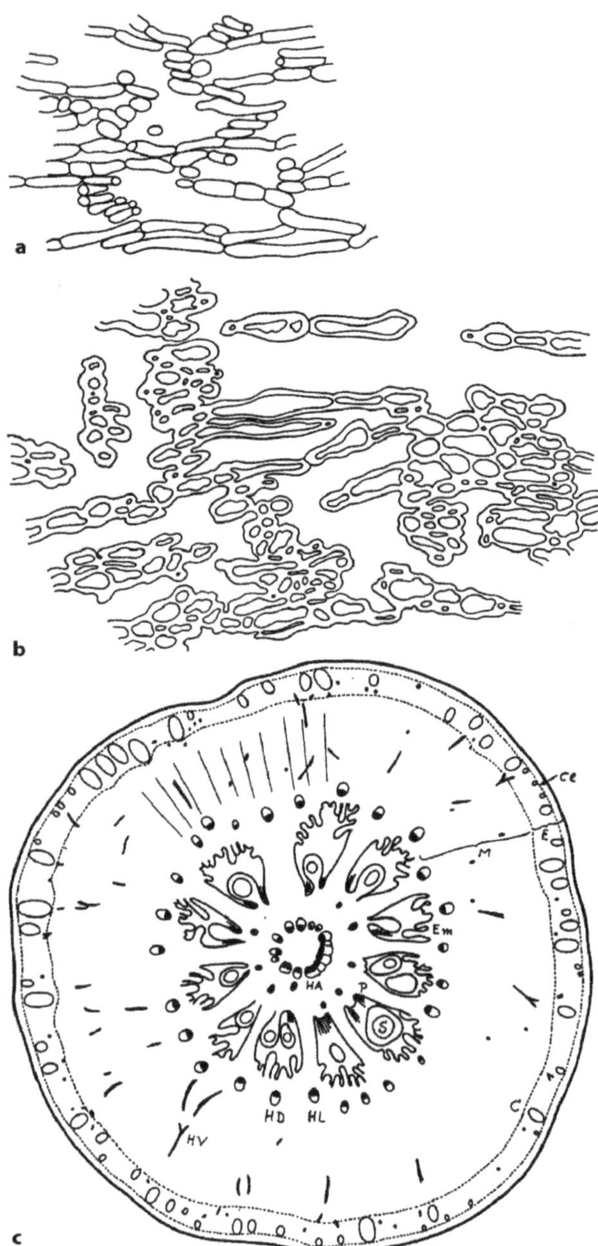

Fig. 237. *Citrus.* **c** T.s. through young fruit. **a** Part of the albedo. **b** Spongy parenchyma of the albedo of a ripe fruit (Roth & Lindorf 1972 c).

Those who are interested in more details about fruit anatomy and development are referred to ROTH (1977) and ROTH & LINDORF (1972) and to the bibliography cited there.

Seed anatomy and development. The ovules are anatropous and have 2 integuments, each composed of various layers of cells.

The outer integument has an outer epidermis the cells of which enlarge anticlinally; the outer walls become thickened and lignified very early. By further enlargement, the cells adopt a palisade-like structure, as seen in transverse section. As wall thickening continues, the outer epidermis cells transform into stone cells, adding the hard texture to the seed. The entire walls now become lignified. The outer epidermis becomes mucilaginous. The rest of the outer integument consists of parenchyma which contains starch grains. The innermost layers of the parenchyma become compressed towards the inner integument, as may be seen in the adult stage.

The inner integument is completely parenchymatous. The cotyledons store many oil drops in their tissues.

Ethnobotanical and general use

Nutritional use

The uses of *Citrus* fruits are well-known and therefore only briefly mentioned here. *Citrus* fruits are eaten raw, consumed in fruit salads, non-alcoholic beverages, for flavoring, or are processed in confectionary, candies, ice-cream, cakes, marmalade, jelly and in quite a few types of liqueurs.

Additionally, citric acid and volatile oils are extracted from fruit and seed (lime seed oil).

Economical utilization

From the volatile oils of fruits, flowers and leaves perfumes, creams, toilet water, powders, cleansing agents and soap are manufactured.

The shrubs or small treelets are also appreciated as ornamentals.

Medical use

The peel is used as a stomachic, stimulant, carminative and tonic.

Cusparia

Useful wood. The plants serve as a fish poison and contain saponins and tannins, particularly the bark. The bark is used as a tonic, antidysenteric, febrifuge and for chronic diarrhoea. The bitter bark acts on the spinal motor nerves and is used for paralysis. Leaf anatomy has been studied by ROTH 1984, fruit structure and dispersal 1987.

Erythrochiton

E. is a curare plant.

Leaf structure has been studied by Roth 1984, fruit structure and dispersal by ROTH 1987.

Fagara

At least 9 useful species are known.

Useful wood. Leaves are used as a condiment. Bark is applied for tooth ache and to kill lice. Roots serve as an aphrodisiac or heal blenorrhagia. Fruits are used as a spice. Seed oil is applied as perfume. ROTH studied the bark structure 1981, fruit structure and dispersal 1987.

Zanthoxylum or Xanthoxylum

Perhaps synonymous with FAGARA, RECORD & HESS (1943).

The bark is aromatic and that of the roots is sometimes used in medicine as a stimulant and tonic, e.g. of *Z. americanum* MILL. the toothache tree.

Bark and leaves of *Z. caribaeum* LAM. contain coumarins and lactones and are said to be antispasmodic, antipyretic and astringent.

The leaf anatomy of *Zanthoxylum ocumarense* (PITTIER) STEYERMARK has been described by Roth (1990). The lower epidermis is supplied with multicellular glands.

Sabiaceae

Meliosma herbertii ROLFE. Attractive useful wood. Bark structure has been studied by ROTH 1981, leaf structure 1984, fruit structure and dispersal 1987.

Sapindaceae

The Sapindaceae are a tropical or subtropical family rich in phenotypes. They are represented by trees or shrubs, more seldom by lianas, often with an anomalous growth in thickness. The flowers are small and occur in paniculiform inflorescences. The seeds are often supplied with a large sugary aril, but have no endosperm. Resinous or latex-like, often poisonous secretions which contain triterpene saponins are frequent. Tannins are also present.

Species of Paullinia contain caffeine and theobromine. Pasta guarana (obtained from *Paullinia cupana*) is the drug richest in caffeine with a 4–8 % content.

Hypoglycaemic active amino acids have been isolated from *Blighia sapida* fruits.

The fruit flesh of *Sapindus saponaria* is rich in saponins which are used for soap production.

Saponins are common in the Sapindaceae and many species are locally employed as a soap substitute.

Besides saponins, sterols, polyphenols, di- and triterpenes, tannins, flavonoids, cyanogenic compounds, ethereal oils, alkaloids and a few amino acids of medicinal/toxological interest have been described.

About 300 species are woody lianas with very interesting types of anomalous secondary growth (ROTH 1966).

Allophylus

The wood of some African species is used for construction work. The leaves of *A. africanus* P. BEAUV. are crushed by the natives and employed for rheumatics and headache. The juice of the leaves is employed as a vulnerary and as a remedy for ailments of the chest.

A. edulis (ST. HIL.) RADLK., the argentine species has juicy drupaceous fruits which are edible. Leaves macerated in water are used for icterus and liver problems; the preparation stimulates the biliary ducts; it is also employed as a colagogum and for diabetes.

The leaf anatomy has been studied by ROTH (1995).

A. occidentalis (SW.) RADLK., a tree of 12–15 m height, occurs in the Venezuelan cloud forest. Its leaf anatomy has been studied by ROTH (1990); the lower leaf epidermis is supplied with glandular hairs which possibly function as hydathodes.

Melicocca bijuga L. (mamon, macao, mauco, maco, muco)

Taxonomical description

Taxonomical details can be found in Fig. 238 a, 239. The species is a tree about (6) 15–30 m high occurring as dioecious or monoecious individuals. The evergreen leaves are alternate, compound paripinnate with 2 pairs of leaflets, opposite or subopposite, sessile or subsessile. The blades are coriaceous, glabrous, elliptic or elliptic-lanceolate, the upper pair is larger than the lower one; about 8–20 cm long and 3–6.5 cm broad, with an acuminate tip, entire margins and an oblique base. The venation is pinnate, brochidodromus with 9–11 pairs of secondary nerves; the midrib is prominent on the abaxial side and slightly vaulted on the adaxial side. The rachis is angular, 8–15 cm long and 0.3–0.8 cm broad. The petiole is 0.5–1 cm long and plane-convex. The inflorescences are racemiform and terminal. The flowers are small and greenish-white. The male flowers occur in ramified racemes, the female flowers in simple racemes; the pedicels are 4–8 mm long. The 4–5 sepals are almost free, 1.5–2 mm long; 4–5 petals; the disc is orbicular and glabrous. Stamens 8–10, stigma bilobed.

The coriaceous fruit is globular and measures 2–3 cm in diameter, it is of green colour outside and encloses an ellipsoid seed, 1–2 cm long. The edible pulp (aril) which surrounds the seed is fleshy gelatinous, of sour-sweet taste and of a salmon-pink colour.

Origin

Tropical America. Central and north of South America, West Indies, Nicaragua, Colombia, Venezuela.

Occurrence

In Venezuela the tree grows wild in the hot regions and is also cultivated for its edible fruits. It occurs from sea level up to 1000 m.

Anatomical description

Leaf (Figs. 238, 239). The blade is very thin. The upper epidermis cells are of regular size and structure. Very characteristic is a large-celled single-layered hypodermis consisting of transparent cells which are probably water-storing. The mesophyll is almost equifacial or centric (METCALFE & CHALK 1950), consisting mostly of palisadelike cells. However, only 2–3 layers show real palisade structure being considerably elongated anticlinally; some of them may however be divided by transverse walls. The longest cells lie directly below the hypodermis, while the long axis of the cells contin-

Fig. 238. *Melicocca bijuga.* **a** Tree. **b** T.s. of leaf. **c** Midrib (× 10).

ually decreases in size towards the lower leaf side. In this region, the cells become very small and partly adopt a globular shape so that intercellular spaces also arise. Most of the cells of this more spongy parenchyma are stained red with toluidine blue so that the spongy parenchyma is distinguished by its colour from the palisade parenchyma. Some cells of the hypodermis and the mesophyll intensely stain blue with the same dye. The lower epidermis is small-celled and the stomata lie at epidermis level.

As seen in surface view, the upper epidermis cells have wavy anticlinal walls. Stomata are rarely found in the upper epidermis, but occur more frequently in the neighbourhood of the veins. The stomata are anomocytic (METCALFE & CHALK 1950). i.e. subsidiary cells can not be distinguished. The lower epidermis cells are polyedric, as seen in a surface view, and have straight anticlinal walls. Wavy anticlinal walls generally occur in the lower leaf epidermis, while straight anticlinal walls are typical of the upper epidermis. The opposite is the case of *Melicocca bijuga*. However, METCALFE & CHALK (1950) report that the walls of the upper epidermis are often sinuous in the leaves of Sapindaceae.

The midrib is vaulted above both leaf surfaces. The disposition of the vascular tissue may be either interpreted as an anticlinally compressed ring surrounded by a continuous sheath of fibrous sclerenchyma – or as 2 superposed horseshoe-shaped arcs with introrse xylem. The epidermis cells covering the midrib are small and their outer walls which are slightly vaulted above the surface are thick-walled. The filling tissue in the center of the vascular ring

Fig. 239. *Melicocca bijuga*. **a** Inflorescences. **b** Fruits.

and that surrounding the fiber sheath is parenchymatous.

Veins of secondary order are completely surrounded by a very well-developed fibrous sheath. The veins of higher order, are however naked, being devoid of a sheath.

Some cells of the outer parenchymatous tissue as well as of the phloem and the rays of the midrib stain intensely blue with toluidine blue (maybe tanniferous cells?). Rhombic crystals frequently appear in cells above the sclerenchymatous sheath of the vascular bundles of secondary order.

Fruit and seed (Figs. 239, 240, 241, 242). The dry indehiscent fruit has a leathery pericarp mainly composed of parenchyma in which large quantities of stone cell groups as well as vascular bundles are embedded. Rhombic crystals are frequently observed in the neighbourhood of the bundles. The ripe pericarp is easy to open.

Towards the inside follows the aril or sarcotesta (Fig. 241 b) with very small and thin-walled epidermis cells. As seen in surface view, a very conspicuous pattern becomes obvious in the epidermis: About 10–30 polydric epidermis cells are united in polyedric groups of higher order corresponding to epidermis mother cells of the outer integument. These have divided many times in different directions so that some kind of a net-like pattern resulted. Exactly below each cell group a very large subepidermal cell becomes visible. The subepidermal cells thus have about the same outlines as the epidermis mother cells. This leads to the conclusion that the underlying subepidermal layer (the palisade layer) originated from the epidermis. In transverse section it becomes obvious that the very large subepidermal cells correspond to the aril. They have an enormous length and are thus palisade cells, but at certain distances they are periclinally divided by thin transverse or oblique walls. The side walls of neighbouring palisade cells are connected by very short arms. The palisade layer can reach a width of up to 5 mm.

The seed coat proper (Fig. 242 a) is sclerenchymatous consisting of groups of elongated sclerenchyma cells crossing each other in different directions. The cells may be considered as a transitional stage between fibers and sclereids; they have thick walls with many pits and partly develop a very large cell lumen and very short arms towards the neighbouring cells. The embryo inside mainly consists of 2 large cotyledons (Fig. 242 b). The outermost cells of the cotyledons are small, but cell size increases towards the inside; larger cells are densely filled with starch grains. Vascular bundles are dispersedly arranged in the cotyledons. As usual in the Sapindaceae, an endosperm is not developed.

The fruit pericarp is thus easy to crack so that dispersers can easily get to the sugary pulp; but the seed developed a very hard coat so that the embryo remains undamaged when the fruit is eaten.

Ethnobotanical and general use

Nutritional use

The plant is cultivated as a fruit-tree due to the large quantities of fruit which it yields. The fruit is eaten raw or as a refreshing drink called leche de mamon or mamonada which is prepared by adding sugar, water and milk. Likewise sweets can be made of the pulp (aril) which adopt a beautiful salmon-red colour. Although the aril is not of high nutritive value, the kernels contain an elevated quantity of protein (more than 6 % of the humid matter). They are eaten toasted by the natives. A kind of meal is also prepared of the seeds. The flowers supply honey.

Economical utilization

The tree is also cultivated as an ornamental for its beautiful appearance, the abundant foliage of its

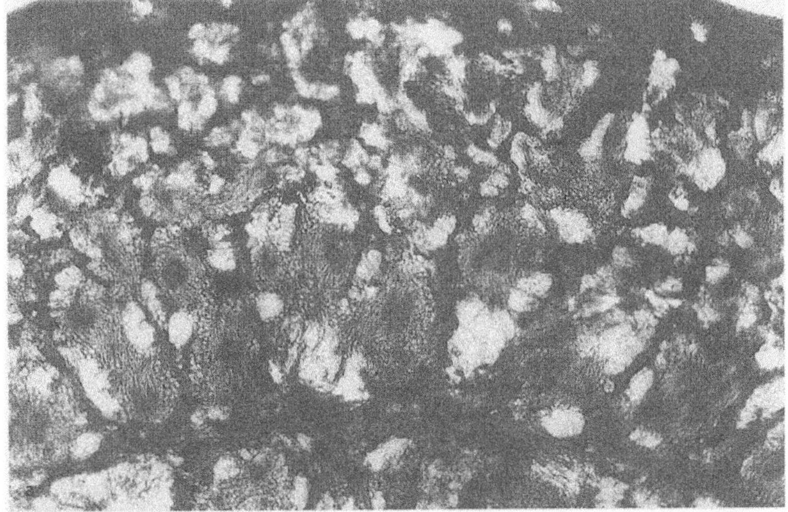

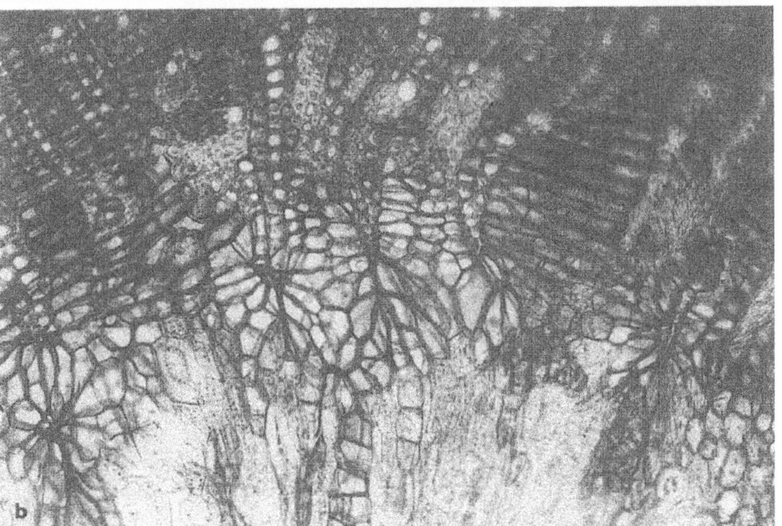

Fig. 240. *Melicocca bijuga*. **a** Fruit periphery. Light spots are groups of stone cells; parenchyma darkly stained (× 6.3). **b** Fruit center × 16.

crown makes it appropiate as a shade tree in parks, gardens and in urban places and squares. Furthermore, the flowers contain a large quantiy of pollen which is a great attraction for bees.

Medical use

When the fruit is not completely ripe, the aril has a very astringent taste. Because of its high content of tannin, the aril is therefore used as an antidiarrhoeic in the form of a refreshing drink. It is particularly given to children and may substitute for the grated pulp of an apple. The seeds may also be pulverized and mixed with honey to control diarrhoea. A decoction of the bark is used for dysentery. Curative properties for the breast and the stomach were also observed.

Cultivation

The plant is easily propagated by seeds, but as it is a dioecious tree with female and male individuals the desired type is developed by grafting only. Seeds germinate rapidly, but the tree grows slowly. It is not demanding as far as soil and climate are concerned and even grows well in arid soils. But it prefers a hot climate and develops best in humid soils which are rich in organic matter. However, it also grows at an altitude of 500–1000 m a.s.l.

The tree gives a rich yield of fruit. lt has a profound radical system and is long-lived.

Besides female and male individuals, there are also monoecious trees. In cultivation, the plant has produced varieties: from originally small and acid fruits towards larger and sweet fruits.

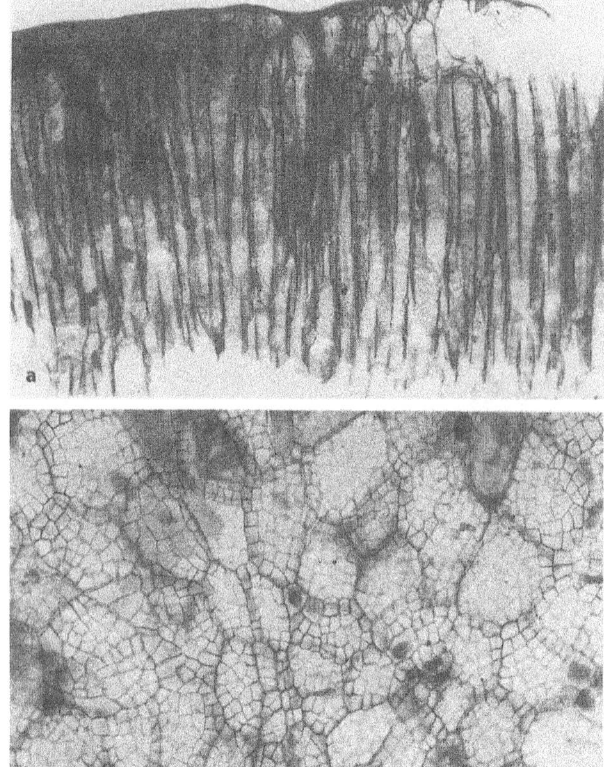

Fig. 241. *Melicocca bijuga*. **a** Fleshy fruit part corresponding to the palisade cells of the aril × 6.3. **b** Aril in surface view. Epidermis and underlying subepidermal palisade layer × 16.

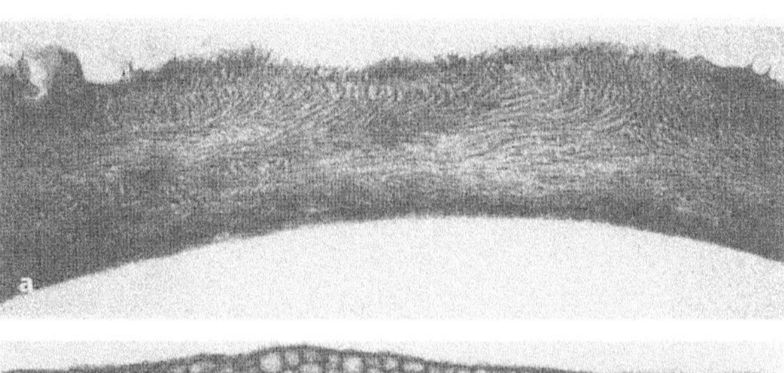

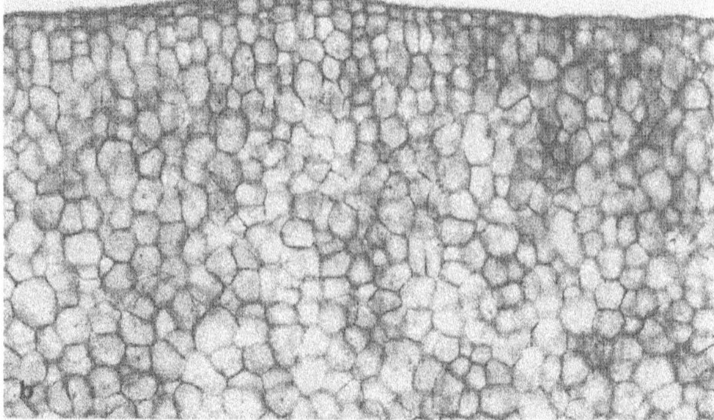

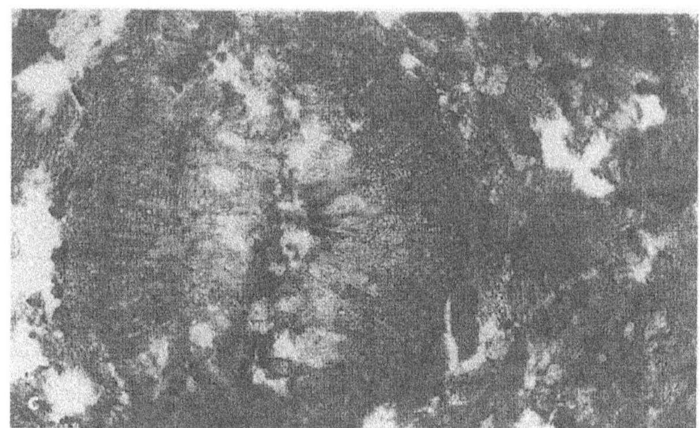

Fig. 242. *Melicocca bijuga*. **a** Sclerenchymatous seed coat × 6.3. **b** Cotyledonary tissue × 16. **c** Fruit center with vascular bundle × 6.3.

Observations

The aril of the seeds with its net-like pattern in the epidermis and the enormously elongated palisade layer of epidermal origin together with the sclerenchymatous seed coat are the most conspicuous features for identification of the tree. The green colour of the ripe fruit and the attractive salmon-red of the aril are additional characteristics.

Paullinia cupana KUNTH is a South American shrub. The Indians prepare an alcoholic drink from the seeds. Crushed seeds are medically used for chronic diarrhoea. The plant contains guaranine, an alkaloid. Seeds are also made into Pasta Guarana (guarana paste) used as a stimulating drink which contains 4.88 % caffeine or even more.

LINDORF (1992) studied the leaf structure of *P. leiocarpa* GRISEB.

Sapindus

At least 6 species are useful. Useful wood. Bark as fish poison. Saponin in fruit pulp. The fruit is a substitute for soap (soapberry) and is used to restore silverware. The fruits can also be applied as an insecticide and as a fish poison. Externally they serve as a detergent. Medically, the fruits are used as a tonic, alexipharmic, expectorant, emetic, nauseant, purgative, for scabies, epilepsy, hysteria, asthma, migraine.

The seeds supply an oil and are cathartic.

Sapindus saponaria L. (prapara, pepo, zapatero, soapberry, savonnier)

Taxonomical description

For taxonomical details see Fig. 243, 244 a. The species is a small tree, about 8–15 m in height. The alternating leaves are usually paripinnate, one leaflet of the uppermost pair can become reduced to

Fig. 243. *Sapindus saponaria.* **a** Branches with inflorescences. **b** Tree.

such an extent that a pseudoterminal leaflet results. The leaves are 15–20 (30) cm long, but 4–8 cm of the length correspond to the slightly hairy and winged petiole, which is thickened at its base. The 3–6 pairs of pinnae are in opposite position, almost opposite or seldom almost alternating; they are sessile and of variable sizes on one and the same leaf. The blade is small-elliptic, lanceolate or lanceolate-ovalate to oblong, 5–12 (18) cm long and 1.5–7 cm broad; the tip is obtuse to acuminate, the base is asymmetric, acute or obtuse, the entire margins are slightly wavy. The rhachis is slightly alate. The venation is prominent on the lower surface. The pinnate venation is brochidodromus with 9–11 pairs of secondary nerves; the midrib is prominent on the abaxial side and slightly vaulted above the adaxial side. The texture of the glabrous blade ranges from papery to subcoriaceous.

The inflorescences are axillary or terminal panicles, about 6–12 cm long. The whitish flowers are actinomorphic and small. They measure about 4 mm in diameter. The petals are 3 mm long.

The reddish globular fruit is fleshy, glabrous and wrinkled on its surface; it measures 1–2 cm in diameter. It is a drupe and not a berry (as written in some books) and has a hard endocarp. It contains black seeds rich in fat.

The species is very variable.

Origin

Tropical America.

Occurrence

In Venezuela it is amply distributed in hot regions. The tree grows frequently in the Orinoco and Amazon region.

The species is either found wild or as an escape from cultivation, occurring from the south of the USA to Argentina.

Anatomical description

Leaf (Fig. 244 b, 245, 246, 247). Shape and size of the leaves vary due to the great variability of the species. The leaves are usually paripinnate, they can however become pseudo-imparipinnate by the re-

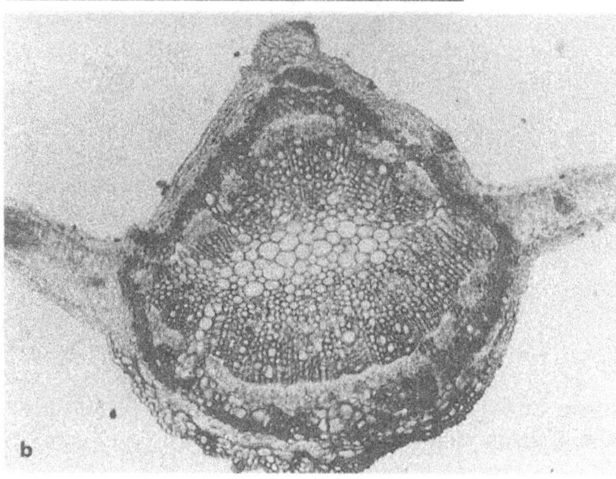

Fig. 244. *Sapindus saponaria.* **a** Leaf with slightly winged rhachis and fruits. **b** Midrib of leaflet in t.s. × 6.3.

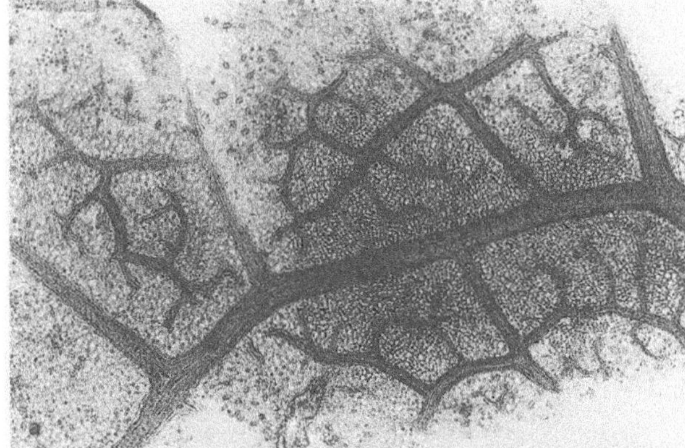

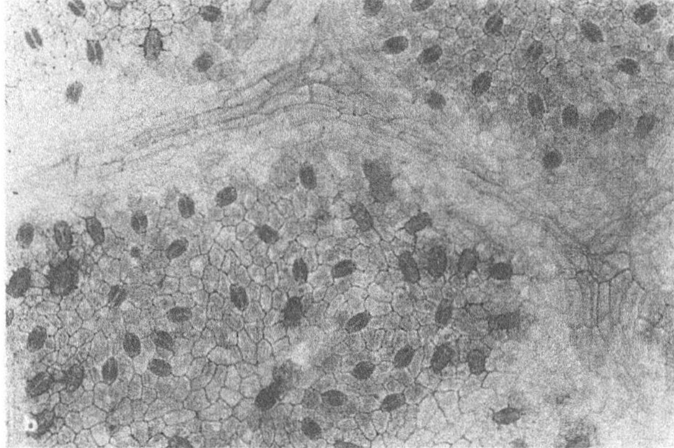

Fig. 245. *Sapindus saponaria.* **a** Venation pattern (cleared preparation) × 6.3. **b** Lower epidermis with stomata in surface view × 16.

duction of one leaflet of the uppermost pair. The number of leaflets is likewise variable. The position of the leaflets on the rhachis is usually opposite, may however oscillate between almost opposite and almost alternating. Leaflets furthermore show variable sizes on one and the same leaf. The leaf consistency varies between papery and coriaceous.

The cells of the upper epidermis are larger than those of the lower one; they are polyedric and have straight or somewhat bent anticlinal walls, as seen in surface view. Their cuticle is ribbed. The palisade parenchyma is one to occasionally 3-layered (in the vicinity of the midrib); the second and third layer seem to proceed from periclinal cell divisions of the original layer. The cells of the second and third layer are usually shorter; the palisade parenchyma remains one-layered above the weaker veins and disappears completely above the stronger veins and in the midrib. The spongy parenchyma is less developed than the palisade parenchyma. The cells have a more or less globular shape and leave small intercellular spaces between one another. The arrangement of the cells in the form of anticlinal rows suggests the development from mother cells by periclinal divisions. Very small druses are dispersed within the mesophyll, but are most frequent in the palisade parenchyma.

The lower epidermis is similar to the upper one. The cells are polyedric and only seldom have bent anticlinal walls. The stomata lie at epidermis level and are devoid of subsidiary cells. This situation seems to be common in the Sapindaceae (METCALFE & CHALK 1950) and is frequently found in neotropical species. Cuticular ledges radiating from the guard cells reinforce the stomata (Fig. 246).

Uniseriate hairs with very thick walls and unicellular hairs occur mainly above the veins on the lower epidermis. Very peculiar are the glandular hairs on the lower epidermis (Figs. 246, 247) which are partly sunk below the surface (SOLEREDER 1908). They are composed of a foot cell, a pluricellular pedicel and a pluricellular head (Fig. 246). Curiously, the glands are found in depressions of the

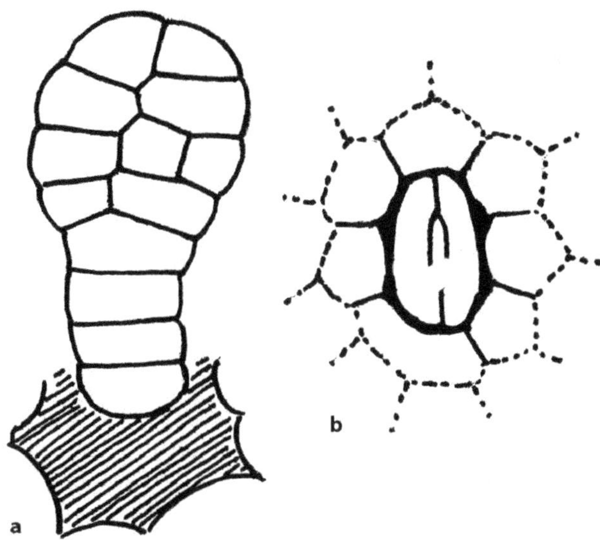

Fig. 246. *Sapindus saponaria.* **a** Gland in depression. **b** Stoma with cuticular ledges, ca. × 40.

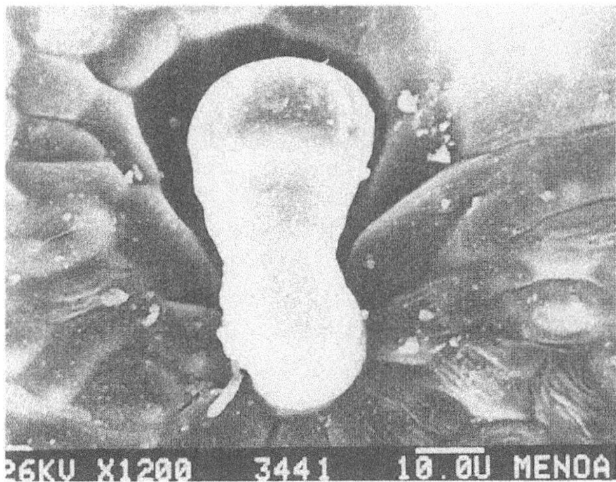

Fig. 247. *Sapindus saponaria.* Gland of the leaf as seen with the scanning microscope.

epidermis. They may be better protected against mechanical injuries in this way.

The smallest veins are devoid of any strengthening tissue. Medium-sized veins are surrounded by sclerenchyma, and this is surrounded by parenchymatous cells with rhombic crystals. Larger veins are transcurrent, reaching from the upper to the lower epidermis by a sclerenchymatous sheath. All veins are accompanied by septate crystal strands with rhombic crystals.

The midrib is strongly vaulted above the adaxial side with a slight protrusion on top, as seen in a transection. The vascular system is nearly arranged in the form of a ring; it can however also be interpreted as being composed of 2 opposite semicircles. The vascular system is surrounded by a sclerenchymatous ring. Large parenchyma cells are found in the center of the midrib. Solitary and clustered crystals occur mainly in the center of the midrib. Collenchyma is developed beneath both epidermal layers.

Fruit. (Fig. 248 a). The outer epidermis of the pericarp has very thick outer walls. Beneath follow several layers of tangentially extended cells with thickened walls and pits. The main bulk of the mesocarp is parenchymatous; a large amount of the cells enlarge enormously to form giant cells which are mainly arranged in large groups. They contain the saponins. Vascular bundles with a fibrous cap are found towards the inside where the parenchyma becomes more compact and the secretory cells disappear. Small clustered crystals are rare. The mesocarp is viscous and has a bitter taste.

At the border line between meso and endocarp, there is a layer of small cells with rhombic crystals in each cell. The endocarp is hard and fibrous. Fiber bundles cross each other, but mainly run longitudinally.

Ethnobotanical and general use

Economical utilization.

The fruits (called soap berries) contain saponins in their fleshy mesocarp and are actually drupes with a hard endocarp. The fruit flesh - when macerated with water-produces suds and foam like soap and is therefore used as a substitute for soap for washing purposes, e.g. clothes. It tastes bitter because of the high tannin content (RECORD & HESS 1943).

Pounded seeds thrown into the water are used as a fish poison because they cause asphyxia. Up to 90 % of the seeds germinate rapidly when released from the pericarp and washed. But only 50 % germinate, when the pericarp is not removed; it is thus suggested that an inhibitory substance of germination is present in the pericarp (PONESSA DE MERCADO, PERSONAL COMMUNICATION). Seeds are furthermore used for ornamental purposes.

The seeds contain 28–30 % fatty substances with a low melting point of 15 °C. The oil extracted from the cotyledons is used in soap manufacture and medicine.

Fruit pulp and roots pounded can paralyse fish. The poisonous seeds may likewise be used as an insecticide.

Not only the fleshy mesocarp of the fruits, but also the leaves, the bark and the roots contain saponin.

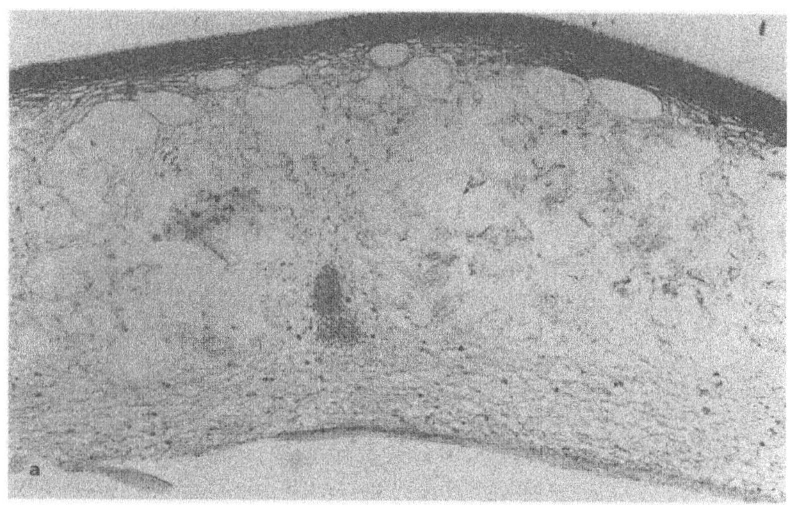

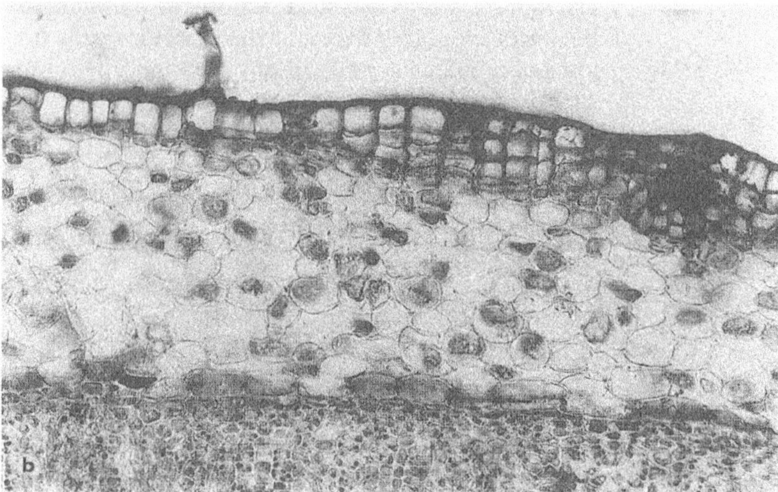

Fig. 248. **a** *Sapindus saponaria*, fruit pericarp × 6.3. **b** *Angelonia salicariaefolia*, cork formation in the bark × 16.

Medical use

Leaves, bark, roots, fruits and seeds are medically used.

Leaf. The leaves in infusion help against snake bites and bites of rays (raya).

Bark. Root and bark in decoction are applied for their astringent and tonic effects.

Root. The root has febrifuge effects. It is also astringent and tonic, when used as a decoction.

Fruit. The fruit has febrifugal activity and is also used against blennorrhagia.

Seeds. A tonic is prepared from the crushed and boiled seeds to combat dandruff. The seeds have febrifugal activities. Seeds soaked in kerosene applied externally help against rheumatism and are also used as an insecticide.

The oil extracted from the cotyledons of the seed is used in medicine and for the manufacture of soap (PITTIER 1926, 1970).

Method of use

Leaves in infusion, root and bark as a decoction, seeds crushed and boiled (externally), seeds soaked in kerosene (externally). The plant has to be used with much caution, particularly the seeds which are very poisonous. The aqueous extract of roots, wood and especially of the fruits is also poisonous.

In some animals it has produced convulsions and paralysis of the nervous system, with consequent asphyxia (BLOHM 1962). The fruit may cause a pronounced rash in some individuals.

Healing properties

Tonic, astringent, scabicide, antirheumatic, febrifuge, against blennorrhagia, and as a detergent. It is also an antidote against snake bites and bites of rays.

Chemical contents

Saponins with strong haemolytic effects, tannins (which are astringent), fatty substances in the seed which could be used industrially.

Toxicity

The plant is toxic, as all parts contain saponins which have a strong haemolytic effect.

Cultivation

The tree requires little care and is easy to cultivate.

Observations

The structure and the shape of the foliar midrib is very characteristic. Most peculiar are the glandular hairs in depressions of the lower leaf epidermis. The fruit mesocarp is distinguished by the presence of giant cells; the saponin content is another important characteristic.

Sapotaceae

The Sapotaceae are mainly a tropical and subtropical family. The evergreen woody plants have simple leaves. Two-armed hairs and laticifers are very characteristic of the family. The fruit is a berry and only seldom a leathery capsule. The seeds are smooth and shiny.

The species are difficult to identify, as they are very variable and have many characteristics in common (gregarious species).

The woods are hard, heavy and durable and some are of commercial value.

Achras zapota, species of *Manilkara*, *Chrysophyllum cainito* and the marmalade plum (*Lucuma mammosa*) have edible fruits.

The miraculous berry of *Synsepalum dulcifolium* gives a sweet taste to even the sourest food.

Shea butter is obtained from the seeds of *Butyrospermum parkii*, whilst a substitute for olive oil is extracted from the seeds of *Argania sideroxylon*.

Gutta-percha is derived from *Palaquium gutta*, and a similar substance, balata, is obtained from *Manilkara bidentata* and other species of *Manilkara*.

Chewing gum is prepared from the latex of a group of variable and closely related species of *Manilkara* (synonym: *Achras*).

Achras zapota L. (nispero, sapodilla)

Taxonomical description

Medium-sized tree (10–20 m high) with a dense globular crown and a brown bark, which releases a white latex when cut. The simple coriaceous leaves are 5–15 cm long and 2.5–6 cm broad with an acute or emarginate tip and an obtuse or truncate base, glabrous on the upper side and with a rust-coloured pubescence on the lower one.

The small whitish flowers have short pedicels. The cylindrical crown is 8–10 mm long and has 6 lobules. The calyx has 5–6 sepals, the 6 stamens are petaloid. The gynoecium is superior being composed of approx. 10 carpels, a number which is subject to variation. Remnants of the crown may remain on the fruit surrounding the style in the form of a calyptra. The calyx is persistent on the ripe fruit.

The fruit is a more or less ovoid to globular berry the size of an apple, measuring about 5–10 cm in diameter, of brownish colour and with an asperous scaly surface. The exocarp is leathery, the pulp or mesocarp is brownish-yellowish, brown-rose coloured or reddish-brown, aromatic and slightly granulose. The shiny black seeds which are 2–3 cm long and hard, vary in number from 0–10 (12).

Origin and occurrence

The tree is native to Central America and the south of Mexico, but is now cultivated not only in the tropics and subtropics of America, but also in tropical regions of the Old World.

Anatomical description

Fruit (Fig. 249, 250, 251). The fruit of the sapodilla which has been studied in its anatomy and develeopment by ROTH & LINDORF (1972) shows strong radial symmetry and develops approximately 10 locules and 10 dorsal carpellary bundles in front of them (Fig. 249). The pericarp is characterized by the early formation of cork on the surface and by the presence of lacticifers in the mesocarp. In the young ovary, the laticifers first appear near the vascular bundles and in the center of the fruit. Bifurcate hairs characteristic of the family of Sapotaceae occur in the outer fruit epidermis (Fig. 250).

In a very young fruit, the laticifers (Fig. 251) are dispersed over the whole transverse section and

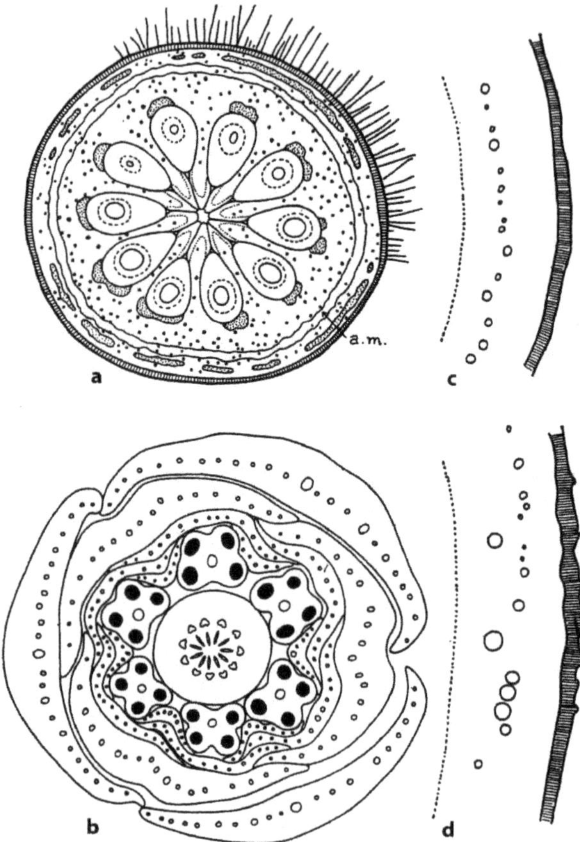

Fig. 249. *Achras zapota*, Sapotaceae. **a** T.s. of ovary with 10 locules. **b** T.s. of flower. **c** and **d** Peripheral part of fruit with cork (shaded) and vascular bundles (ROTH & LINDORF 1972 b).

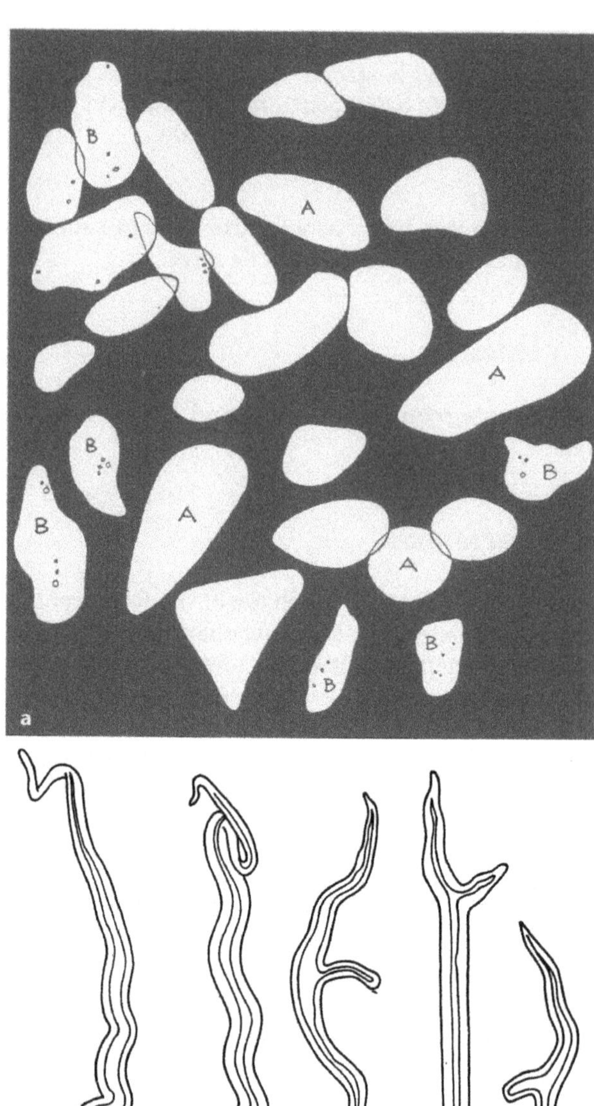

Fig. 250. *Achras zaptoa*. **a** Macerated pulp of the ripe fruit with very turgescent parenchyma cells of yellow colour (A) and colourless cells (B). **b** Bifurcate unicellular hairs from the outer surface of a young ovary (ROTH & LINDORF 1972 b).

some parenchymatous cells in the periphery begin to transform into stone cells; at this early stage, cork formation begins to develop. A meristematic activity may be observed in the first subepidermal layer which however is very irregular; meristematic activity is often restricted to a small cell group, the epidermis breaks above the group and a lenticel develops. The phellogen produces up to 20 cells towards the periphery, but only 1–2 cells towards the inside. The epidermis disappears gradually in this way, the peripheral cells of the cork become suberized and finally disintegrated; they desquamate step by step so that the cell rows of the cork are shorter in the ripe fruit comprising only 5–10 cells. It is surprising, how early in the fruit development the cork is formed, although the fruit – a few millimeters in diameter – is still enclosed within the calyx.

The laticifers begin to form near the center of the fruit and from there cell differentiation progresses towards the periphery. As seen in transverse section, the laticifers are surrounded by a ring of parenchymatous cells. The laticifers are of the articulated non-anastomosing type and appear in a longitudinal section as curved chains of short cells (Fig. 251). Their cell content is somewhat granulated and of grayish colour at early stages, but becomes agglomerated and of yellowish colour in advanced stages. The little droplets in the latex correspond to the chicle rubber extracted from the stem of the plant for chewing gum production.

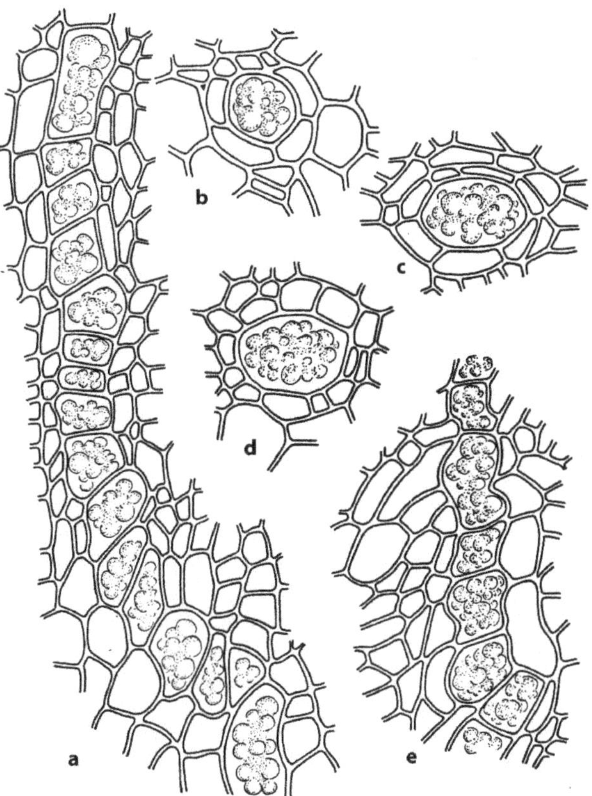

Fig. 251. *Achras zapota.* a–e Laticifers in the fruit flesh (ROTH & LINDORF 1972 b).

Brachysclereids or stone cells, isolated or united in groups, begin to form in the periphery of the young fruit. Certain parenchymatous cells enlarge enormously, often in a radial direction, and then gradually form thick cell walls, occasionally with ramified pit canals. After a more or less continuous ring of stone cells has been formed in the pericarp periphery, large caps of sclereids develop above the locules. Differentiation of stone cells then progresses from the fruit periphery towards the pericarp inside so that at more advanced stages of fruit develpopment small groups of brachysclereids are scattered over the whole pericarp section. These interspersed stone cell groups add the gritty granulous consistence to the mature fruit flesh which resembles that of a pear. The majority of the pericarp consists of parenchymatous cells many of which contain tannin; at maturity this is transformed into an insoluble form so that inclusions develop in the cells. In the mature fruit, these tannin cells adopt a red colour when treated with vanillin and hydrochloric acid; they contain acetaldehyd.

The inner fruit epidermis is composed of smaller cells than the outer epidermis and has no thickened cell walls. The cells have the capacity to enlarge in an anticlinal direction and to divide periclinally so that a multilayered epidermis results, particularly in the median region of the carpels.

The outer epidermis is not only characterized by its early cork formation, but also by the presence of bifurcate unicellular hairs (Fig. 250).

At the mature stage, the fruit flesh is partly macerated (Fig. 250). Turgescent reddish-brown or yellowish cells (tannin-cells?) are distinguished from less turgescent colourless cells (Fig. 250).

Ethnobotanical and general use

Nutritional use

The fruits which mature within 4 months have a soft flesh and a pleasant taste. They have to be gathered when completely ripe, because unripe fruits are very astringent due their high content of tannin. The fruits are eaten fresh or prepared into a drink, as a milk shake or made into ice-cream. Jam and vinegar are also obtained from the fruits.

Chewing gum and chicle gum are prepared from the latex of the trunk, which contains 20–40 % gum. The bark can be tapped every 2–3 years.

Economical utilization

The wood is hard and is used locally. The tree is also used as an ornamental.

Medical use

Bark. The bark is very astringent and is applied for catarrhal conditions.

Fruit. A cataplasm is prepared from the fruits to treat inflammations of the liver.

Seed. The seeds are diuretic. They are said to dissolve calculi of the gall bladder and uroliths.

Varieties and related species

There are quite a few species of Manilkara with similar properties, such as *M. zapota, M. chicle, M. inundata.*

Cultivation

The tree is easily propagated by seeds, which germinate within 4–5 weeks. The growth however is slow. The radical system is deep and the tree is long-lived. It requires a hot climate and strong insolation. It can however be cultivated from 0 to 1200 m a.s.l. It flowers and fruits most of the year.

Observations

The fruit is well characterized by the cork and the bifurcate hairs.

Chrysophyllum

At least 16 useful species are known. Useful wood. Leaves for coagulation. Seed oil for candles and food, as well as for soap manufacture. Edible fruits. Star apple from *C. caimito*. Bark astringent, stomacbic, expectorant, supplying tannins and saponins. Latex of bark for fungal infections and to hasten healing of wounds.

ROTH studied the bark structure 1981, leaf structure 1984, fruit structure and dispersal 1987, leaf venation 1996.

Ecclinusa

E. supplies a substitute of Balata latex.

ROTH studied the bark structure 1981, leaf structure 1984, fruit structure and dispersal 1987, leaf venation 1996.

Manilkara

At least 7 useful species are known. Useful wood. Latex is a source of balata and substitute of chicle. Flowers are a source of essential oil used in perfumery. Roots are used for fever and as a mouth wash. Roots and bark are applied for diarrhoea. Bark is tonic and febrifuge.

Seeds are tonic, anthelmintic, used in ophthalmia, for leprosy and delirium.

M. bidentata CHEV. Latex is a source of balata and a substitute of chicle. ROTH studied the bark structure 1981, leaf structure 1984, fruit structure and dispersal 1987, leaf venation 1996.

Micropholis

M. supplies saponins, phenols and alkaloids. The wood is useful.

M. melinoniana PIERRE supplies a substitute of balata (balata blanc). ROTH studied the bark structure 1981, fruit structure and dispersal 1987.

Oxythece

O. has a useful wood. Drinks are prepared of the fruit. ROTH studied fruit structure and dispersal 1987.

Pouteria

At least 6 useful species are known. Useful wood. Leaves are disinfectant. A tea of the bark is purgative. The fruits are edible and vermifugal. The fruit pulp looses its acidity when boiled. Saponins, phenols, triterpenes and tannins are in the bark. ROTH studied the bark structure 1981, fruit structure and dispersal 1987, leaf venation 1996.

Pradosia

The bark is tonic, astringent and haemostatic. ROTH studied the bark 1981, fruit structure and dispersal 1987, ROTH & LINDORF 1972.

Scrophulariaceae

The Scrophulariaceae usually have zygomorphic flowers which are bilabiate. The 2 carpels form a 2-celled ovary. The flowers occur in racemes, spikes, panicles or false umbels. The fruit is a capsule dehiscing septicidally, at times loculicidally or with pores.

Mostly herbs and shrubs, seldom lianas.

Iridoid-like chemical compounds are characteristic of many representatives of this family. Their presence may become apparent, when leaves adopt a black colour on drying. The genus *Digitalis* contains the well-known digitalis-glycosides which are indispensable in the treatment of heart diseases. Saponins from Flores Verbasci are well known as an expectorant.

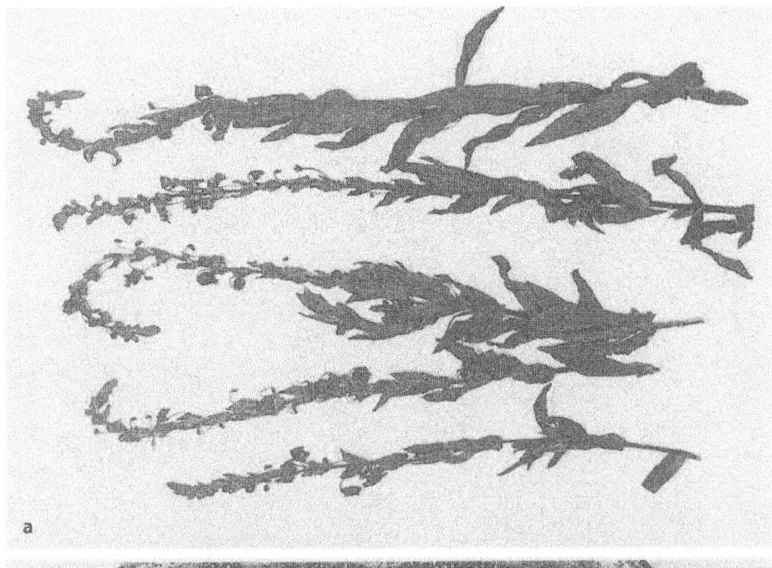

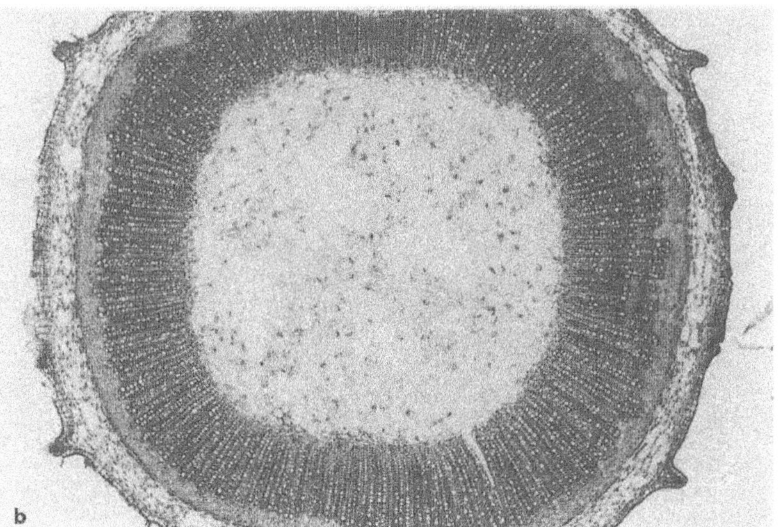

Fig. 252. *Angelonia salicariaefolia.* **a** Twigs with flowers. **b** Axis in t.s. ×2.5.

Angelonia salicariaefolia H & B (angelon)

Taxonomical description

The species is a perennial herb, 30 cm up to 1 m high, and pubescent (Fig. 252 a). The leaves are opposite, the upper leaves may however alternate; they are almost sessile, lanceolate to linear-oblong, 3–8 cm long, 1–1.7 cm broad, and slightly dentate. The blue to violet flowers raised on a pedicel are 2 cm in diameter and are arranged in long racemes (up to 15 cm or more). The calyx is 5-dentate, the crown is bilabiate and the tube is very reduced. The upper labellum is bilobed, the lower one trilobed. The 4 stamens have short filaments. The capsule has a globular shape.

Occurrence

The species is principally found in Venezuela in the Cordillera de la Costa at altitudes between 1000 m and 1620 m.

It occurs from the south of North America and the West-Indies down to the north of South America.

Anatomical descrption

Leaf (Fig. 255a). The leaf is thin and has a very pronounced midrib which is sunk below the upper surface, but strongly projects above the lower surface. The leaf is furthermore dorsiventral and amphistomatic.

As seen in transverse section, the upper epidermis is proportionally very large-celled; the cells are

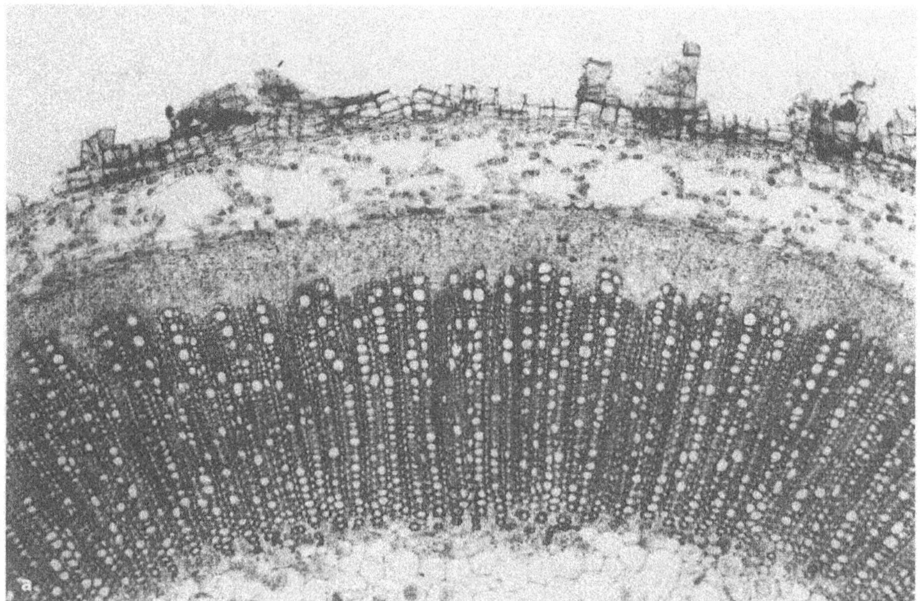

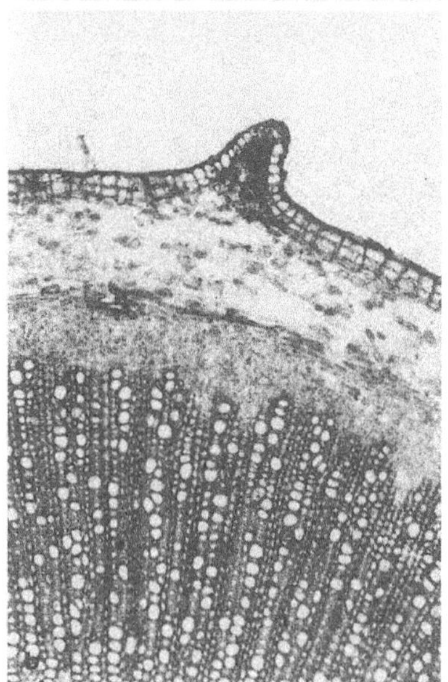

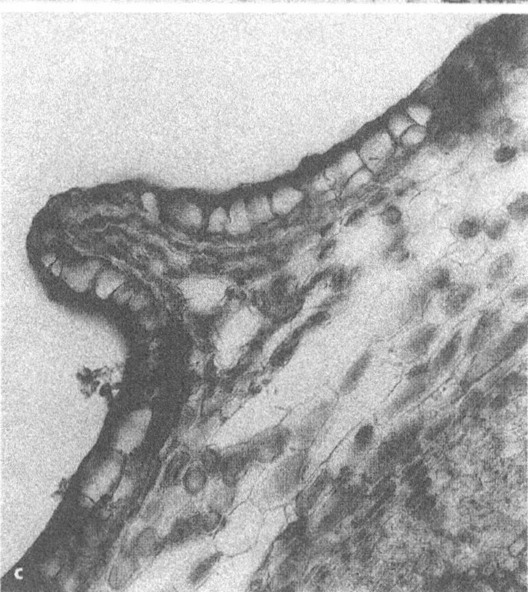

Fig. 253. *Angelonia salicariaefolia.* **a** T.s. of axis. Note the cork, the aerenchyma below (primary cortex), the phloem and xylem part. × 6.3. **b** Cortical outgrowth on leaf base. **b** × 6.3. **c** × 16.

vaulted above the surface and sometimes become papillary. The lower epidermis cells are similar in structure, but smaller. The outer tangential walls, particularly those of the upper epidermis are considerably thickened and the cuticle is striated or ribbed. As seen in a surface view, the anticlinal walls of both epidermal layers are slightly wavy. The leaf surface is densely covered with numerous glandular hairs with an enlarged foot cell (likewise with a striated cuticle) (Fig. 255), a 3–4 celled stalk and a roundish head of 4–8 cells. There are shorter and longer glandular hairs; the length of the stalk of the longer glands exceeds the width of the blade two or even several times. Few sessile glands were also observed.

Stalked glandular hairs accompany the leaf margins. The comparatively large stomata occur on both surfaces, but are more numerous on the lower side, where they are conspicuously elevated above the surface. The single-layered palisade parenchyma is very reduced being composed of short and broad cells with a length/width index of only 2 or less. The spongy parenchyma comprises about 4–5 layers. The midrib pronouncedly projects above the lower surface and contains a single vascular bundle which is of a slightly crescent shape and seems to be bicollateral, having a small portion of phloem on the adaxial side. The midrib is reinforced by thick-walled collenchyma which is particularly well developed on the lower side. The laterals which are

Fig. 254. *Angelonia salicariaefolia.*
a Anomalous cambium producing anomalous secondary growth. × 10.
b Endodermis surrounding the phloem (dark). × 20.

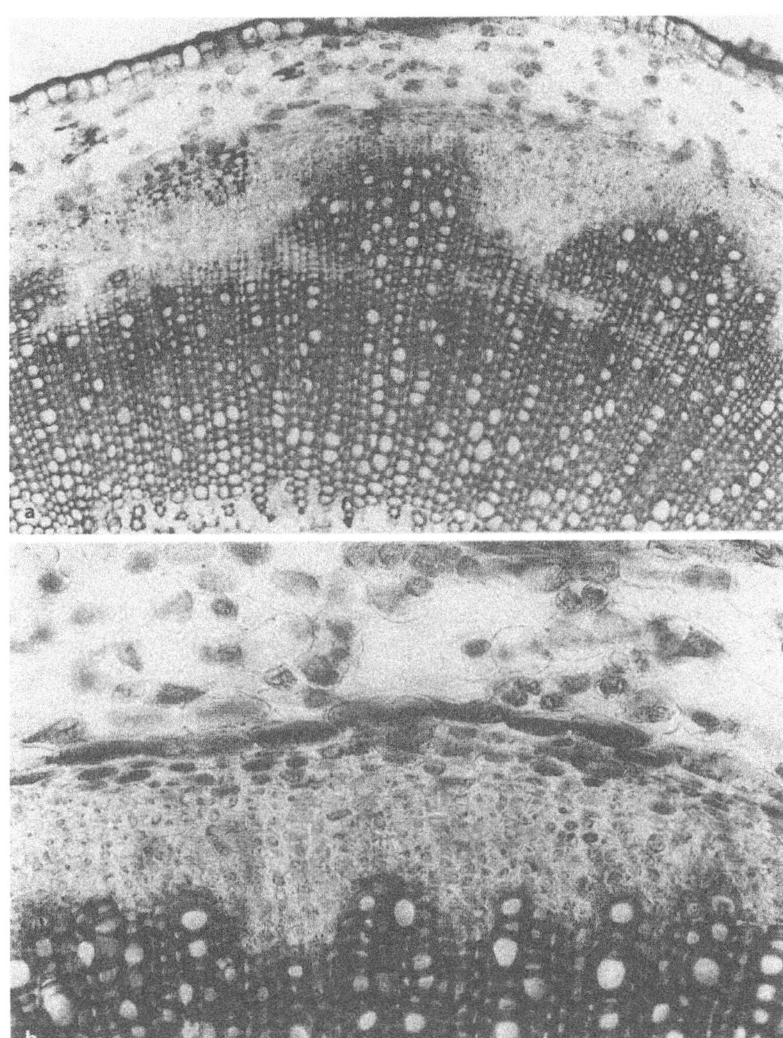

not too frequent seem to be reduced in their structure and are not reinforced by mechanical tissue. The meshwork of the vascular bundles, as seen in surface view, shows only few ramifications and free endings in the meshes.

The strong collenchyma of the midrib, which may also be water-storing, as well as the large-celled water-storing upper epidermis with its thick walls, together with the large thickened foot cells of the hairs give mechanical support to the blade.

Axis (Figs. 252 b, 253, 254). An axis of 5 mm in diameter is in the stage to form the cork which arises owing to periclinal divisions in the epidermis (Fig. 248 b). The primary cortex below is composed of parenchymatous cells which leave large intercellular spaces, so that the tissue appears aerenchymatous. The endodermis is conspicuous by its tangentially elongated cells and the Casparian strips in the radial walls; furthermore it shows a dense granulated content. Rests of the pericycle are found beneath the endodermis in the form of isolated fibers or small fiber groups.

The vascular system forms a continuous cylinder. The phloem consists only of a soft bast. However, the cambium works in a very irregular manner. A case of anomalous secondary growth in thickness is realized here. Only partly, the cambium forms phloem towards the outside and xylem towards the inside, following the regular way. In some places, however, it forms both phloem as well as xylem towards the outside, and then it ceases to grow and a new cambium develops further outside between the recently formed phloem and xylem. The cambium is thus discontinuous making a patchwork: old cambia cease to grow, while new ones arise. This proceeds in a very irregular manner so that the outer border lines of the xylem cylinder become very irregular, just as the inner border lines of the phloem (Fig. 254).

Vessels are relatively frequent and of a smaller size, being mainly arranged in radial rows. Medullary rays are 1-3 seriate. The pith consists of large and thin-walled parenchymatous cells the size of which increases towards the center. Solitary crystals, prisms and needles, are frequent. In the very center, the pith may already be ruptured.

It seems that the secondary growth begins in a regular way and that growth anomalies occur later when the axis has already reached a certain thickness.

Ethnobotanical and general use

Economical utilization

As an ornamental plant.

Medical use

Leaves and flowers are used. Flowers in decoction are used as a pectoral and sudorific. A syrup of the flowers is prepared to combat cough, chronic cacatarrh and is used as a diaphoretic.

Flowers or parts of the entire plant in infusion or decoction serve as an expectorant, antispasmodic, and as a remedy against afflictions of the bronchopulmonary tract; they cure cough and influenza, are diaphoretic and are even used as an abortive.

Method of use

The syrup which is applied as a sudorific and against chronic catarrh and cough is prepared in the following way: *Angelonia* is mixed with violets, borage, elder and sugar in some water. Three spoonfuls are taken in the morning and at night.

Healing properties

Diaphoretic, pectoral, (abortive).

Chemical contents

The floral essence (nectar?) contains iridoid glucosides and fatty oils.

Observations

The extremely long-stalked glands which densely cover leaves and axis are very characteristic. The large-celled, water-storing upper leaf epidermis and the midrib reinforced by abundant collenchyma on the lower side are further particularities of the leaf.

The cork is of epidermal origin, while a subepidermal development is usual in Scrophulariaceae. The cortical aerenchyma and the conspicuous endodermis are further characteristics; however, the formation of the endodermis seems to be subject to variation (METCALFE & CHALK 1950). The presence of 1-3 seriate rays is exceptional in herbaceous species of the Scrophulariaceae and no anomalous secondary growth in thickness of the axis has been reported as yet for this family (METCALFE & CHALK 1950).

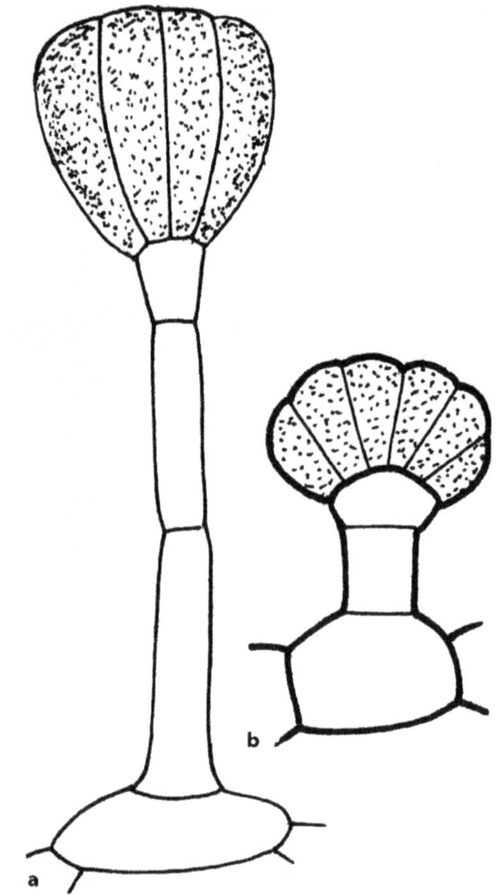

Fig. 255. **a** *Angelonia salicariaefolia*: gland of leaf and axis. **b** Glandular hair of *Capraria biflora* (lower leaf side).

Capraria biflora L. (fregosa)

Taxonomical description

C. is a perennial ramified herb with variable hairiness, attaining a height between 30 cm and 1 m, occasionally 1.5 m (Fig. 256). The alternating leaves are lanceolate, 1-4 cm long, and 0.7-1.3 cm broad with serrate margins in the distal half or two thirds. Flowers axillary, on slender stalks. Occasion-

Fig. 256. *Capraria biflora.* **a** Habitus. **b** Twig with flowers.

ally 2 stalks in an axil. Calyx with 5 narrow linear or lanceolate sepals, 4–6 mm long. The bell-shaped crown is white, 1 cm long and has 5 lanceolate lobes, about of the same length as the tube. Usually 5 stamens. Stigma 2-lobed. The 4–6 mm long ovoid capsule dehisces loculicidally. The numerous seeds, of yellow colour and with a rugose surface, are 0.5 mm long.

Origin

Tropical and subtropical America.

Occurrence

The species has an ample distribution in Venezuela. The northern border line of distribution is Florida. The plant occurs throughout tropical and subtropical America and is very common on disturbed ground and along the roadside.

Anatomical description

Leaf (Figs. 257, 258 a). The leaf is bifacial and amphistomatic. The upper epidermis consists of slightly papillose cells with thin walls. The anticlinal walls are strongly wavy, as seen in a surface view. The stomata do not show special companion cells and are anomocytic according to METCALFE AND CHALK (1950). The glandular hairs with a short stalk and a pluricellular head are more frequent on the lower than on the upper side (Fig. 255, b). They consist of a foot cell, a stalk cell, a middle cell, and a head composed of about 12 cells (see also RADLKOFER 1886). The palisadeparenchyma is single-layered but occasionally individual cells show transverse partition walls. The spongy parenchyma occupies a larger proportion than the palisade parenchyma and comprises cells with short arms.The cells of the palisade parenchyma are comparatively short. The intercellular spaces in the spongy parenchyma are small. The stomata which are slightly elevated above the surface, are more numerous on the lower side, the epidermis cells are smaller and their anticlinal walls form more profound folds, as seen in a surface view, so that teeth arise.

The vascular bundles are surrounded by a parenchymatous sheath. The midrib which projects above the lower surface, has a slightly horse-shoe-shaped vascular bundle. The palisade parenchyma does not penetrate into the middrib.

Stem (Figs. 258 b, 259 a). A stem of about 5–6 cm in diameter shows secondary growth. Several layers of cork are followed by several layers of parenchyma with tangentially extended cells. The pericycle composed of an interrupted ring of fibers lies below. The arrangement of the phloem cells in the form of radial rows is very striking. The xylem is diffuse porous with solitary vessels and radial multiples. The vessels are small. The paratracheal parenchyma is extremely scarce. The rays are uni or

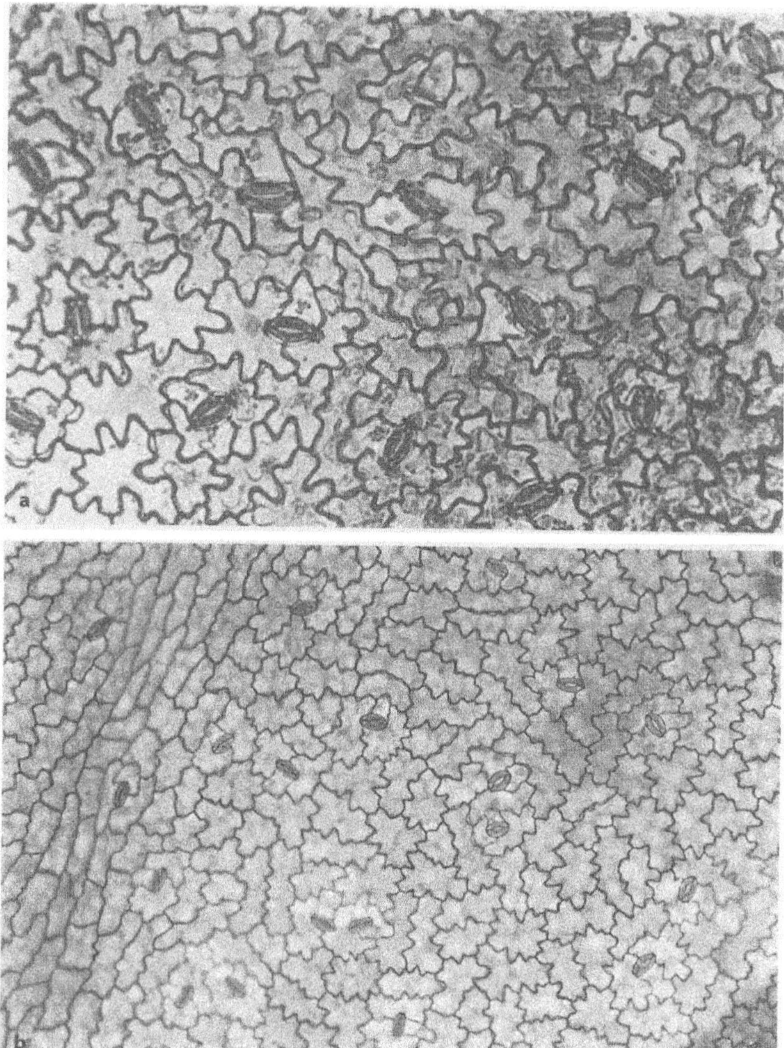

Fig. 257. *Capraria biflora*. **a, b** Surface view of lower epidermis with stomata (×6.3 and ×16).

biseriate. The basic tissue consists of fibers. Instead of a pith there is a large hole in the center.

Root (Figs. 259 b, 260 a). A root of about 8 cm in diameter is in the state of secondary growth. The cork is better developed than in the axis and comprises 15–20 layers of thin-walled cells. The primary cortex is very ample being composed of tangentially dilated cells. The 1–2 seriate rays dilate in the phloem in the funnel form, mainly due to tangential enlargement of the cells, but also through cell divisions. Stone cells, either in very small groups or solitary, are dispersed within the primary cortex and the phloem. The xylem is abundant and shows characteristics which are similar to those of the stem. However, the biseriate rays are more frequent than in the stem. In contrast to other roots, there is no axial parenchyma developed. The center is filled with primary xylem. From there, the lateral roots originate. Furthermore, growth rings may be observed in the xylem of the root; in the case studied, there were 3 growth rings each starting with an accumulation of larger vessels in the ring form, as seen in transverse section. Such distinct growth rings were not observed in the stem.

Ethnobotanical and general use

Medical use

The plant is mainly of medicinal use. The medically mostly used part is the leaf.

Leaf. Teas of the leaf are used as a febrifuge, for flu, dysmenorrhoea, heat, vomiting, diarrhoea, measles, as a postpartum depurant, and as an eyewash. They are also applied to treat the griping effects accompanying the purgative action of Castor oil (Ricinus communis).

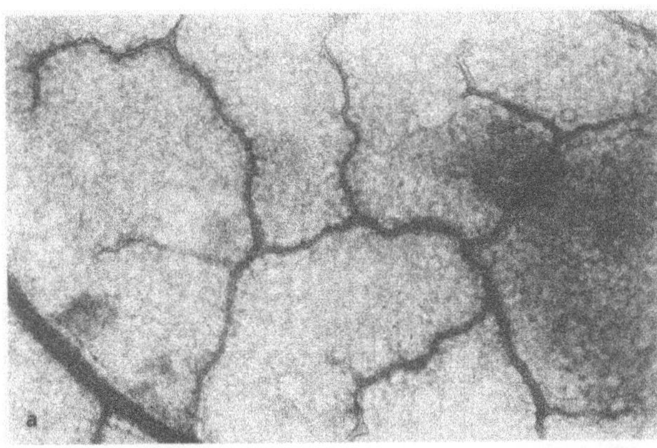

Fig. 258. *Capraria biflora.* **a** Venation pattern of leaf in cleared preparation. × 16. **b** Axis in t.s., mainly consisting of xylem. × 6.3.

Method of use

The leaves are mostly used as a tea.

Leaf tea, dew on the leaves, the juice of fresh leaves or the aqueous extracts of the leaves are used for an eyewash in opthalmia. A decoction of the entire plant cures diarrhoea and problems of blood pressure.

Healing properties

Antispasmodic, carminative, digestive, tonic, antidiarrhoeic, vermifuge, and hypotensor.

Chemical contents

The following substances were identified: meliacine of a similar structure to hirtine, flavones, and phenolic derivatives. Caution is necessary in the application of the dose! In strong doses, the leaves produce paralysis.

Observations

The species is well defined within the Scrophulariaceae by its particular glandular hairs with a pluricellular head, the presence of 1–2 seriate medullary rays in stem and root, the papillose upper epidermis cells of the leaf, the stomata slightly elevated above the surface, and the cork on the stem which is of subepidermal origin.

Simaroubaceae

Picramnia

Leaves are used for a black and golden dye.

A purple dye is obtained from reddish fruits and leaves.

Roots and bark are febrifugal.

The so-called Honduras bark is diuretic and helps for skin diseases. Leaf structure has been studied by ROTH 1984, leaf venation 1996.

Simaba

Cotyledons are used for fever and snake bites.

The bitter seeds are febrifugal, tonic, sedative, antispasmodic, and help for malaria and dyspepsia.

The bark is tonic, vermifugal, febrifugal, anthelmintic, it is used for fever, dysentery and as an appetiser. Bark is also used as a substitute of *Quassia*.

Bark structure has been studied by ROTH 1981, fruit structure and dispersal 1987, leaf structure 1984.

Simarouba

Bark for intestinal worms and snake bites. Powdered it is used as an insecticide.

S. amara AUBL. (orinoco simarouba bark) is used as a bitter tonic.

Bark structure has been studied by ROTH 1981, leaf structure 1984, fruit structure and dispersal 1987.

S. glauca D.C. The bark boiled in water is used as a tonic and vermifuge.

Solanaceae

The most outstanding anatomical feature of the Solanaceae is the presence of bicollateral vascular bundles and of intraxylary phloem in the axis. The

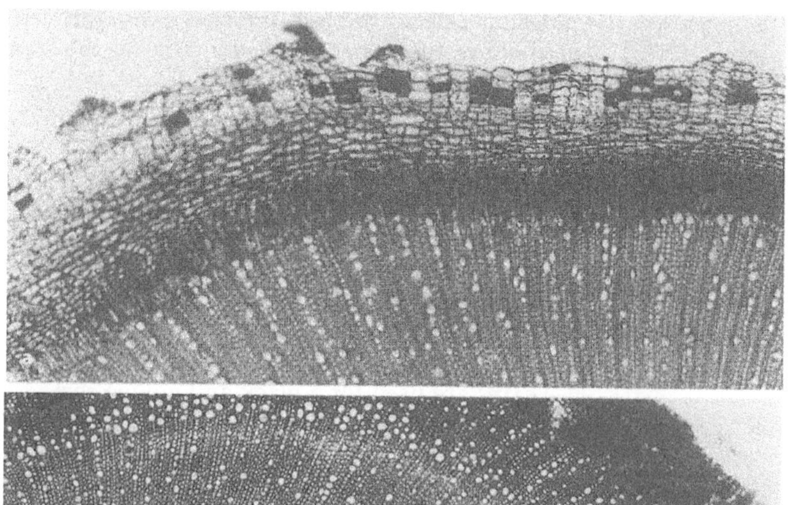

Fig. 259. *Capraria biflora.* **a** Cork of the axis in t.s. × 6.3. **b** T.s. of root. × 5.

family is mainly of tropical and subtropical origin, being very well developed in South and Central America. Representatives which supply very important spices, vegetables, stimulants and medicinal plants are species of *Capsicum* (chillies, red and green pepper), *Solanum tuberosum* (potato), *Solanum melongena* (egg-fruit), *Lycopersicum esculentum* (tomato), *Nicotiana tabacum* (tobacco), *Atropa belladonna* and *Datura stramonium*. The family which is rich in alkaloids is consequently called alkaloid family by some authors. Calcium oxalate in the form of solitary crystals, clusters or as crystal sand is furthermore frequently found in the Solanaceae.

The 2 carpels unite in a bilocular ovary which usually develops into a berry (more seldom into a capsule), occasionally with false septs.

40 genera (trees, shrubs and herbs) are endemic in America.

Capsicum frutescens L. var. BACCATUM (L.) (aji chirel)

Taxonomical description

The genus *Capsicum* is of neotropical origin and occurs in South and Central America (Fig. 260 b).

Capsicum frutescens is a small shrub and produces the hot cayenne pepper or chilli.

Aji is in Venezuela the collective name for species and varieties of *Capsicum*, such as *C. frutescens* L., *C. frutescens* L. var. *longum* BAILEY, *C. frutescens* L. var. *cerasiforme* BAILEY, *C. frutescens* L. var. *microcarpum* (DC) (SCHNEE 1960).

Capsicum frutescens var. *baccatum* reaches a height between 1 and 3 m, abundantly ramifying, and occasionally occurs as a climber. The simple leaves are ovalate to ovalate-oblong or ovalate-lanceolate, 2–5 cm long, acute or acuminate and abruptly attenuate or truncate at the base. However, the leaves are very variable in their shape and size, reaching between 3 and 15 cm in length (SCHNEE 1960). The leaf petiole has half of the length of the

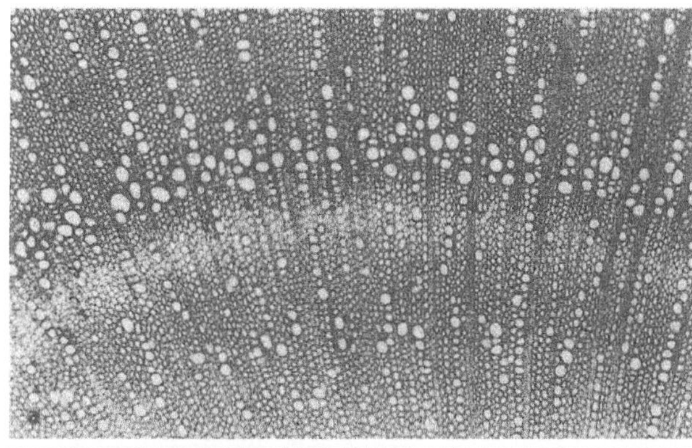

Fig. 260. **a** *Capraria biflora*, root with growth ring. **b** *Capsicum frutescens*. Twig with flowers.

blade or remains shorter. The flowers occur solitary in the axils or form small cymes; their pedicels are 1–2 cm long. The 5-lobed calyx is 2.5–3 mm long; the lobules reach the length of the tube or are shorter. The crown is commonly white and its 5 lobules are imbricate. The 5 stamens are united with the tube. The ovary is 2–5 locular. The elongated elliptic, dorsiventral or even globular berries are 5–15 mm long, of red colour and pungent. The yellow almost circular seeds are flattened and disc-shaped. Equally as the leaves, the flowers and fruits are subject to variations (ROIG & MESA 1953). The calyx persists on the fruit, which contains only few seeds.

Origin

Capsicum frutescens is of South American Origin (BRÜCHER 1989).

Historical background

Archaeological findings of *Capsicum* date back to 2000 B.C. in South America and as much as 7000 B.C. in Central America (BRÜCHER 1989). Domesticated selections, some used for condiment, others for food preservation and as a vegetable already existed in 1590, when Padre Acosta wrote his 'Historia de las Indias'. Since the discovery of the New World by Columbus the *Capsicum* pepper has outstripped the black pepper (*Piper nigrum*) of Asiatic origin in production and trade, and gained the position of the most appreciated condiment of the world (BRÜCHER 1989).

Occurrence

Capsicum species grow best in a warm climate. *Capsicum frutescens* is now cultivated all over Venezuela.

Anatomical description

Leaf (Figs. 261, 262, 264, left). The bifacial leaf is amphistomatic with a higher number of stomata on the lower side. The epidermis cells are slightly papillose on both sides, but the cells are somewhat larger on the upper leaf side. The anticlinal walls are strongly bent, as seen in surface view, but those of the lower epidermis form more profound folds so that deep folds or teeth arise. Upper and lower epidermis are single-layered. The cuticle is delicate on both sides of the leaf. The single layered palisade parenchyma is composed of cells which are longer than broad. The spongy parenchyma consists of 3–4 layers of slightly lobed cells or cells with very short arms. The palisade parenchyma surpasses the spongy parenchyma slightly in proportion. Frequently dispersed within the mesophyll (mainly within the spongy parenchyma) are large globular cells filled with crystal sand and clustered crystals. The stomata on the lower leaf side are somewhat larger than those of the upper one. Distinct subsidiary cells could not be observed, the stomata are thus of the anomocytic type, according to the nomenclature of METCALFE & CHALK (1950). On the upper leaf side, some reduced or degenerated stomata could be observed, some of which had only a single guard cell.

The middle nerve projects over both, the upper and the lower side, but particularly over the upper leaf side, an extraordinary case, because usually its growth is favoured on the lower side. The bicollateral vascular bundle has the shape of a horse-shoe with phloem on both sides. The palisade parenchyma is interrupted in the midrib and a collenchyma is instead developed. Collenchyma also occurs on the lower side below the epidermis. The rest

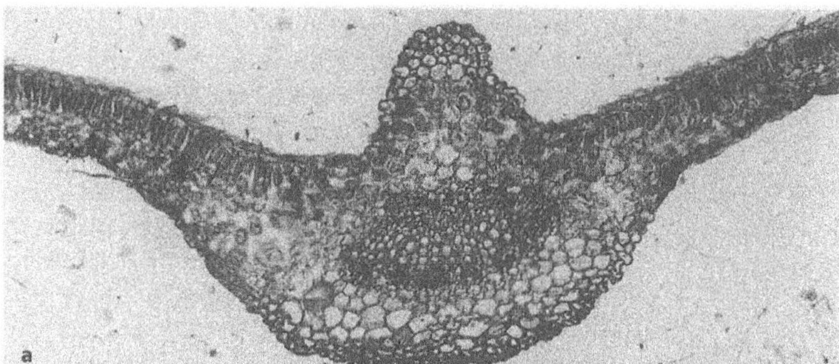

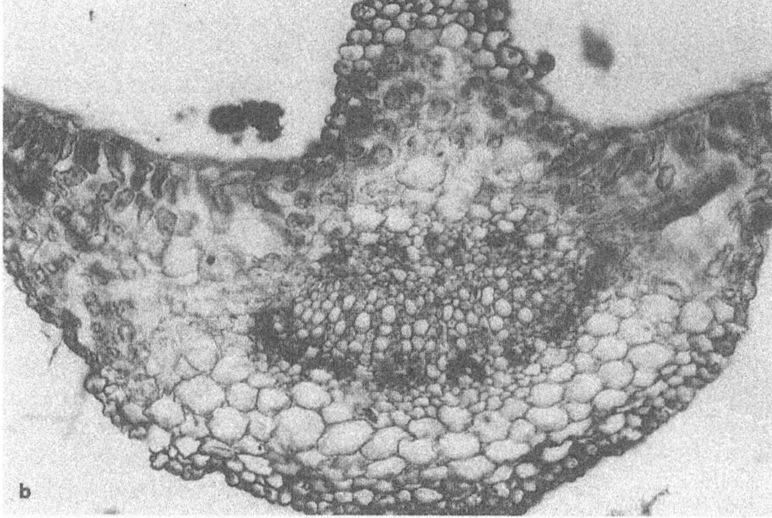

Fig. 261. *Capsicum frutescens.* **a, b** Leaf with midrib in t.s. **a** × 6.3. **b** × 16.

of the tissue is parenchymatous. The strongest secondary nerves are similar to the middle nerve, but project less over the surface and have a normal bicollateral bundle of circular shape, as seen in a transverse section. The less developed bundles of higher order are collateral and do not show an adaxial phloem. They are surrounded by a parenchymatous sheath. Longer or shorter uniseriate pluricellular hairs preferably occur above the veins. They have blunt apices and show the typical shape of Solanaceae hairs. Glandular hairs with a short stalk and a pluricellular head can also be observed (Fig. 264). They have an orange-coloured content. They occur with greater frequency in the furrows on either side of the protruding midrib. Insects also seem to prefer to lay their eggs in the furrows. The pluricellular head of the glands is variable in its structure, but not infrequently consists of 4 cells. Below lies an intermediate cell.

Fruit (Figs. 263, 264 right, 265, 266). The anatomical structure of the fruit of *Capsicum frutescens* var. *baccatum* is somewhat different from that of *Capsicum frutescens (annuum)*. The outer epidermis cells are of a regular size and have strongly thickened outer and anticlinal walls. As seen in surface view, the mediumsized cells have thick, frequently slightly folded, walls. The epidermis cells are partly arranged in longitudinal rows. The outer tangential walls are covered by a relatively thick cuticle. Stomata are extremely rare. It is the outer epidermis that mainly gives the strength to the pericarp. Beneath the outer epidermis there are about 5–10 layers of parenchyma cells. The vascular bundles are embedded in the parenchymatous tissue and the larger ones are bicollateral or amphicribral, while the smaller ones are reduced and collateral. The size of the parenchyma cells continuously enlarges towards the inner part of the fruit. The inner surface of the fruit is blistered by the formation of giant cells (ROTH 1977). The giant cells develop from the innermost parenchyma layer due to inflation of certain cells. Since not all cells of this layer can enlarge, small bridges of very small parenchyma cells develop between the giant cells, when a transverse section is considered (Fig. 263 b). The cells of the bridges usually elongate in a radial direction and form a sort of network around the giant cells. The inner epidermis covers the giant cells. As seen in transverse section, the epidermis cells seem to be

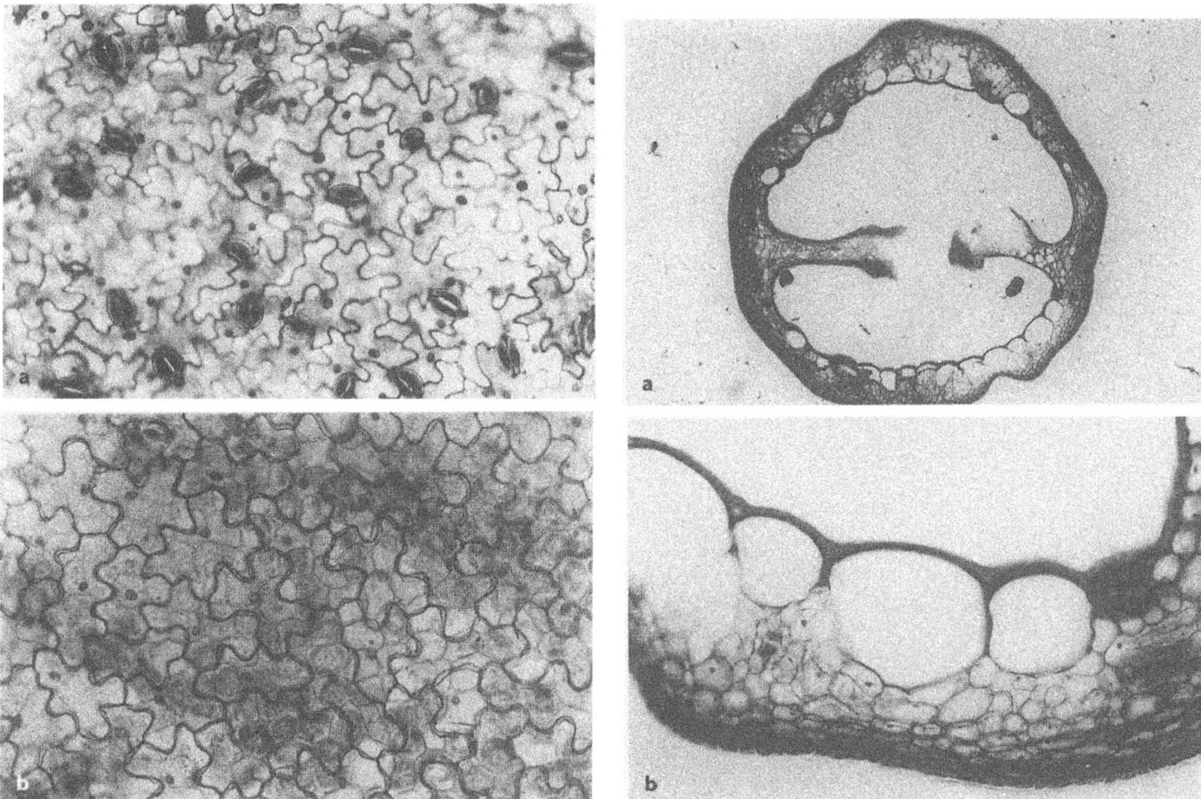

Fig. 262. *Capsicum frutescens.* **a** Lower epidermis with stomata (× 25). **b** Upper epidermis. × 25.

Fig. 263. *Capsicum frutescens.* Fruit. **a** t.s. of entire fruit. × 6.3. **b** Sector of pericarp with 'giant cells'. × 16.

very small; in surface view, however, the cells are more or less of the same size as the upper epidermis cells, but their walls are slightly more undulated and the pits in the walls are very conspicuous. Additionally, their walls are slightly thicker than those of the outer epidermis. The bilocular fruit is divided by a sept which is longitudinally traversed by vascular bundles. The epidermis cells of the sept are thin-walled; dispersed within the epidermis are the capsaicin glands which arise from the epidermis by radial elongation of cells. These cells unite in groups to form the glands. The capsaicin is secreted into the space between the cellulose wall and the cuticle; the cuticle is lifted up by the oily secretion and the oil is accumulated in the free space.

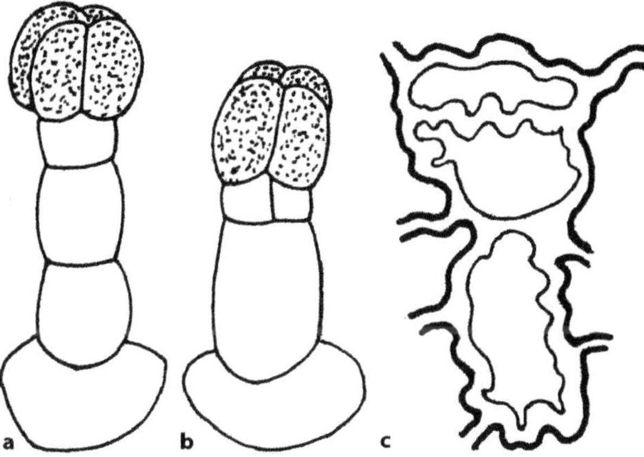

Fig. 264. *Capsicum frutescens.* **a, b** Glandular hairs of leaf. **c** Outer epidermis cells of fruit in surface view.

The complete anatomical structure of the fruit of *C. frutescens* (synonym: *anuum*) is described by ROTH (1977).

Seed (Figs. 265 left, 266 b). The yellow disc-shaped seed which may reach 4 mm in length is surrounded by a flat rim. As seen in transverse section, the seed has a single-layered epidermis with strongly thickened walls. However, on the flat sides, the epidermis cells are small, while their height increases towards the rim. In this zone they adopt a palisade-like shape through radial elongation. The layering of the walls is easily perceptible. Although the outer and inner tangential walls are also thickened, they are much smaller than the anticlinal walls. Where the anticlinal walls unite with the

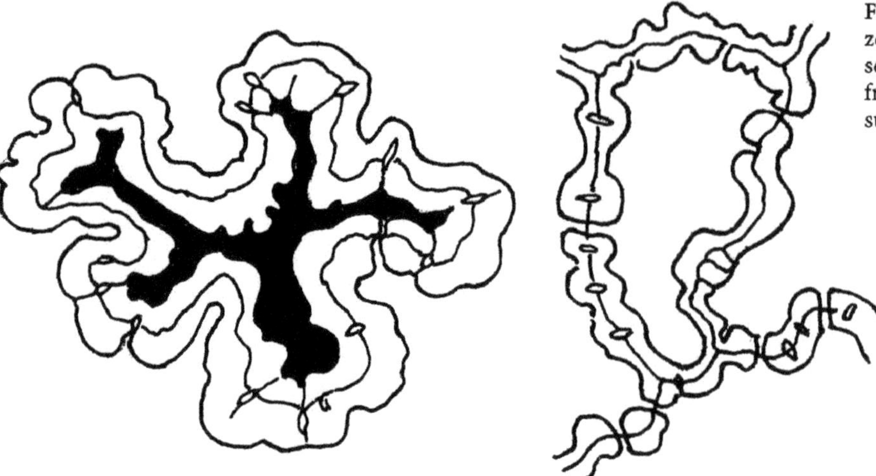

Fig. 265. Left: 'Gekrösezelle' of outer epidermis of seed coat. Right: Inner fruit epidermis. Both in surface view.

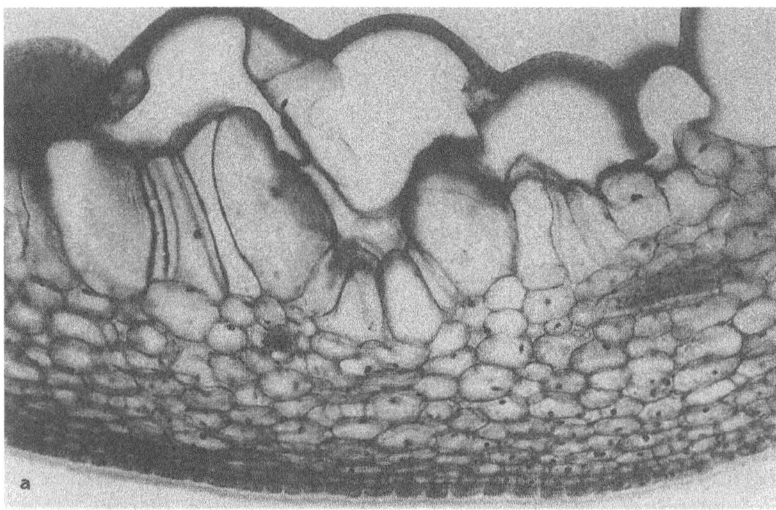

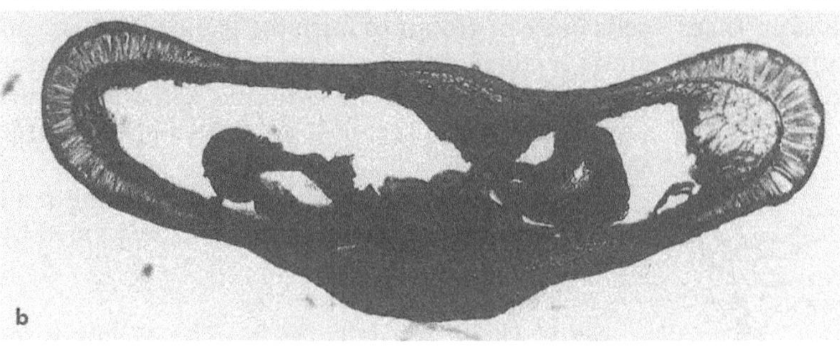

Fig. 266. *Capsicum frutescens*. **a** Fruit wall with blister cells. × 16. **b** Seed cut transversely. × 6.3.

outer tangential walls, there is a large pit, which may be seen as a small hole, when a surface view of the epidermis is observed. As seen in paradermic section (parallel to the seed surface) through the epidermis, it becomes obvious that the cells walls are extremely folded and are thickened to such an extent that the cell lumen is considerably reduced. Due to their very wavy aspect, the epidermis cells are called mesenteric cells, because the wavy walls resemble intestinal loops. The formation of folds and the thickening of the walls makes the seed coat extremly strong.

Below the epidermis, a thin-walled slightly spongy parenchymatous tissue is found. Towards the rim of the seed, the parenchyma is better developed mainly because of the considerable enlargement of the cells. The endosperm surrounds a strongly bent embryo.

Ethnobotanical and general use

Nutritional use

The fruit is used as a condiment. The capsaicin content in the glands of the septs stimulates the appetite and the secretion of the salivary gland in humans. The fruit is also a source of vitamin A and C. The more pungent a variety is, the more it is appreciated by the Indians. Capsaicin is the pungent principle.

Medical use

Name of the drug: fructus capsici. The Indians know a large variety of its medicinal applications. *Capsicum* is first of all used as a stomach remedy e.g. as an appetite stimulant and as a digestive. Natives eat the fruit raw to relieve flatulence. The recently detected vitamin P in *Capsicum* regulates blood circulation. Indians who live at very high altitudes in the Altiplano plateau take the fruit to improve blood circulation at high altitudes (4–5000 m a.s.l.). The alkaloid capsicin has bactericidal effects and is used as an antidysenteric.

Ground fruit as a powder is added to food to relieve abdominal pain. In Ecuador, the fruit is used for odontological purposes. Further applications of the fruit are: accelerating difficult childbirth, against intoxication with the hallucinogenic drink prepared from *Banisteriopsis*, as a snuff to improve breathing, to mask unpleasant tastes of medicinal drinks and even for the weaning of infants (the mothers smear the juice on the nipples to discourage further nursing) (BRÜCHER 1989, SCHULTES & RAFFAUF 1990). The juice of the ingested fruit facilitates evacuation of the bile.

Fruit. Fresh fruit is also used externally: a friction of the skin is used as a cutaneous irritant (rubefacient). Fruits are eaten for indigestion; gargles are for sore throat.

Leaf. Even leaves are applied: together with hot talcum the leaves are put on furuncles to cure them (RODRIGUEZ 1983). In the West Indies leaf decoctions are used to cure asthma, coughs, chest colds, consumption; leaf poultices are put on boils.

Method of use

The fruits are applied fresh or ground as a powder or are used in gargles. A tea and a poultice are prepared of the leaves. To prepare a tincture, 200 g fresh fruit are macerated in 60 % alcohol for 19 days.

Healing properties

Stimulant, tonic, digestive, regulator of blood pressure and circulation, bactericidal and antiseptic, emetic, haemorstatic, rubefacient, skin irritant, dissolving agent, hyperaemic.

Chemical contents

Vitamin A and C, capsaicin (pungent component), vitamin P, the alkaloid capsicin. Overdoses taken internally can affect the digestive tract and the kidneys, externally applied, blisters and dermatitis may be proveked; capsaicin produces hyperaemia and is therefore also applied in rheumatism and arthritis.

Varieties and related species

As already mentioned above, SCHNEE (1960) distinguishes 3 further varieties of *C. frutescens* (besides var. *baccatum*) in Venezuela: var. *longum* BAILEY, var. *cerasiforme* BAILEY, and var. *microcarpum* (DC).

Species names of *Capsicum* have been changed continuously until, at the end of the last century, 100 names were established. To put an end to this inconvenience, BAILEY (1923) concentrated all taxa in only one collective species. TERPO (1966), on the contrary, split the species *Capsicum annuum* in 33 concultas, based on fruit form and size. His nomenclature was however rejected for the simple reason that these fruit characteristics had been selected by man and not by nature. DAVENPORT (1970) and others proposed a monophyletic origin of all domesticates from a single ancestor: *Capsicum frutescens.* But such a suggestion would be in contradiction with archaeological and morphological evidence (BRÜCHER 1989). Varieties are also distinguished by their pungency: The more pungent the variety, the more it is appreciated by the Indians. The more pungent varieties contain more of the alkaloid capsicin which has a bactericidal effect, and the Indians appreciated the plant more for medicinal purposes (see also SCHULTES & RAFFAUF 1990).

According to BRÜCHER (1989), the genus *Capsicum* includes 30 wild-growing species, five of which have been domesticated by distinct Indian tribes in different and very remote places of the American continent. Some experts in Solanaceae are inclined to separate *Capsicum* into 2 phylogenetically distinct lineages: i.e. purple-flowered and white-flowered lineages; *C. frutescens* belongs to the latter group; it was probably domesticated in Costa Rica and Nicaragua (BRÜCHER 1989).

Cultivation

Cultivated *Capsicum* species have 2 n = 24 chromosomes. No casual polyploidy has been observed in our days and artificially induced genome duplications have never yielded practical results in commercial varieties. This is a consequence of the fact that Indians long ago were able to make use of the so-called gigas effect in *Capsicum*. The dramatic increase in fruit size in *Capsicum annuum* of 1:100 (from small bird pepper to the local varieties in Central America) has been achieved by factor combination alone, practiced by the Indians.

As a result of the huge monocultures of *Capsicum* in North America and Europe, the susceptibilities increase each year in a spectacular way, e.g. by fungus, bacteria and particularly virus infections; insects preferentially attack the sweet peppers. Many primitive and wild *Capsicum* strains such as *C. frutescens* have contributed to the genetic improvement of modern varieties. Special conservation centers have been established (Gatersleben/Germany; Leningrad/Russia; Davis/California) to save the gene pools of Indian peppers for the benefit of mankind. *Capsicum* has trespassed widely beyond the natural limits of its neotropical origin: Today Asia and particularly the Indian subcontinent are the main producers of chillies (BRÜCHER 1989).

Observations

Criteria of distinction of the *Capsicum* drug from adulterants are the giant cells in the inner fruit wall, the outer and inner fruit epidermis, as seen in surface view, the outer epidermis of the seed coat, as seen in t.s. and in surface view. In the powder of the drug, the yellowish-green outer epidermis cells of the seed coat are very conspicuous and, more rarely, the inner epidermis cells of the fruit wall, called rosary cells, as well as reddish oil drops of the fruit septs.

In the leaf, the wavy walls of the upper and lower epidermis, the bicollateral vascular bundles, the typical Solanaceae hairs, and the characteristic glandular hairs with their oily content, are good criteria of identification.

The fruit anatomy of *Capsicum frutescens* slightly differs from that of *Capsicum annuum* by the lack of collenchyma in the outer part of the fruit wall and the presence of bicollateral vascular bundles.

Cestrum

The genus includes shrubs and small trees. Several species are cultivated as ornamentals.

Steroidal saponins (with an effect on blood pressure), sterols, catechin, tannins, flavonoids, alkaloids of the nicotine type have been analyzed.

Cestrum bigibbosum PITTIER. This is a small shrub occurring in the Venezuelan cloud forest. The leaf anatomy has been studied by ROTH (1990).

Lycium

The fruits of several species of *Lycium* are edible.

Lycium tweedianum GRISEB. The taxonomic and anatomical descriptions are found in ROTH 1992. Possibly, the fruits of this species are also edible.

Lycopersicon esculentum MILLER. The tomato deduces its name from the Central American Nahuatl word tomatle. *L. esculentum* was already domesticated in pre-Colombian times in Central America, where no wild tomato species is native. However a dozen wild species of Lycopersicon exist in South America. Importation of the tomato to Europe took place from Mexico, in any case before 1544 (TEPPNER 1993). Today the tomato fruit is very much used as a vegetable, in spite of the steroidal alkaloids tomatidin and demissidin which have a poisonous effect on insects, acting as an insect repellent for the plant (potato-beetle etc.). Tomato leaves are thus slightly toxic for humans and animals. However, the content of toxins very much varies according to the genetic background, the state of development of the plant, time of the year and environmental conditions.

Slimy-viscid secretions of the glandular hairs are likewise insect repellent. The volatile oil of the glandular hairs of *Lycopersicon esculentum* may provoke an allergy on the skin in sensitive persons. The potassium content of the fruit is relatively high, but the content of oxalic acid is comparatively low.

Fruit anatomy of *Lycopersicon esculentum* has been described by ROTH (1977). Further information is available in BRÜCHER 1989 and TEPPNER 1993 and the bibliography cited there.

Nicotiana tabacum L. (tabaco, uenaña, pinaji, tabba, yapo, cauai)

Taxonomical description

Annual erect herb, 0.5–1.5 (3) m high (Figs. 267, 268 a). Viscid-pubescent, unramified or only little ramified; leaves alternate simple, oblong to oblong-lanceolate, 20–40 cm long, sessile, acute or acuminate at the apex, narrowed towards the base; the lowest leaves are decurrent on the stalk. The pentamerous flowers are 3–5 cm long; they have a pedicel and occur in terminal panicles. The tubular 5-partite calyx is oblong; the lanceolate segments are unequal, acute and about 12 mm long. The rose or red coloured bell-shaped crown is about 5 cm long; the tube is longer than the limb which is 5-lobed; the lobes are triangular, lanose outside and slightly inflated at the throat. The 5 stamens are attached to the base of the tube inside; the filaments are filiform and the anthers dehisce longitudinally. The ovary is 2-locular. The ovoid, septicid capsule opens with 2 valves; the seeds are small and numerous.

Mature plants may however reach a height of 6 m with a stem 3–6 cm thick and it is questionable whether they are annual.

Origin

The species is probably introduced from Brazil (BRÜCHER 1989, WILBERT 1972), including the Virginia types.

Historical background

N. rustica has a much higher nicotine content that *N. tabacum* and perhaps for this reason was widespread in pre-Columbian times in whole America (BRÜCHER 1989). The North American Seneca Indians were growing *N. rustica* in the region of New York, when the first European settlers arrived.

Occurrence

The species is now cultivated all over the world.

Anatomical description

Leaf (Figs. 268 b, 269). The leaf is dorsiventral and amphistomatic. As seen in transverse section, the upper epidermis cells are larger than those of the lower one. Below the upper epidermis, there is a single layer of palisade parenchyma consisting of long and slender cells with a length/width index of about 1/3 – 1/4. The spongy parenchyma occupies

Fig. 267. *Nicotiana tabacum*. **a** Young plant. **b** Flowers.

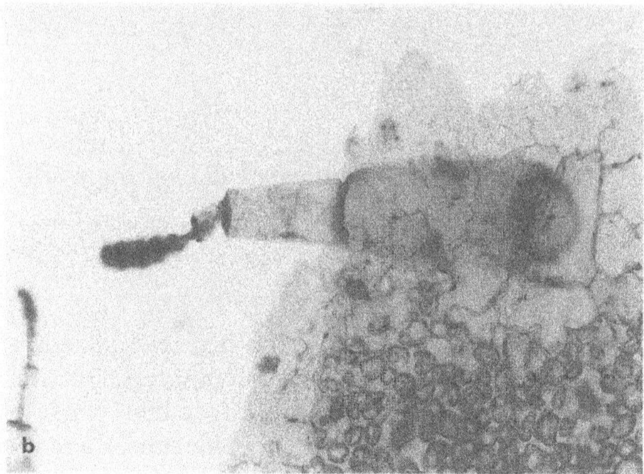

Fig. 268. *Nicotiana tabacum*. **a** Flowers. **b** Hair of the leaf. ×20.

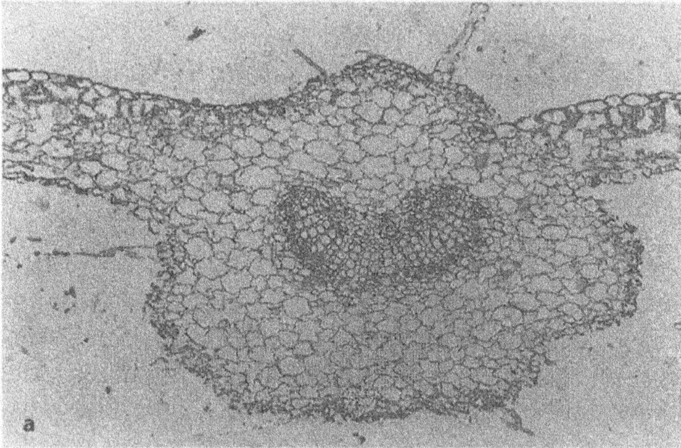

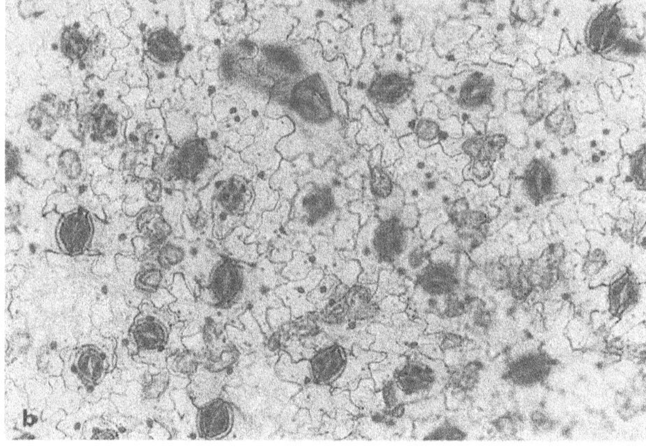

Fig. 269. *Nicotiana tabacum*. **a** Midrib of leaf. ×6.3. **b** Lower epidermis with stomata. ×25.

less space than the palisade parenchyma; the composing cells form arms towards one another leaving intercellular spaces so that the spongy parenchyma becomes relatively loose. Globular cells (idioblasts) which contain crystal sand are dispersed in the spongy parenchyma.

The midrib projects considerably above the lower leaf side and is supplied with a horseshoe-shaped bicollateral vascular bundle. Palisade parenchyma and spongy parenchyma are interrupted in the midrib region and the filling tissue is parenchymatous. The veins of secondary or higher order have a parenchymatous sheath which surrounds a single vascular bundle.

As seen in a surface view, the cells of the upper and lower epidermis have wavy walls, but those of the lower one are much more sinuous. The stomata occur at epidermis level or are slightly elevated; they are anisocytic. Most conspicuous are simple uniseriate glandular giant size hairs with 1–2 or more enormously inflated stalk cells and a pluricellular (storied) head composed of several cells. They occur on both surfaces, are by far the most frequently occurring hairs and are responsible for the viscid leaf surface.

Besides, there are uniseriate hairs with a pointed tip consisting of 2–10 cells, and hairs with unicellular stalks and heads comprising up to 20 cells. However, there are also intermediate types.

Ethnobotanical and general use

Economical utilization

The plant is a source of nicotine, nicotine sulfate and insecticides.

Its major importance however is its use as a narcotic. Dried and fermented leaves are used for manufacture of cigars, cigarettes, as wrapper leaf, binder leaf, filler leaf, for pipe tobacco, chewing tobacco and snuff. It is thus closely related to industrialization and valued as a major source of taxes, although it causes cancer of the lungs.

Medical use

Amazonian tribes use tobacco for many curative rituals. A decoction of the leaves is rubbed over sprains and bruises. Fresh leaves are crushed and poulticed over boils and infected wounds. The crushed leaves mixed with palm oil are rubbed into the hair to prevent baldness. Tobacco juice is taken therapeutically for indisposition, chilis and snakebites. A tobacco sniff may be employed medicinally for a variety of illnesses, particularly to treat pulmonary ailments (SCHULTES & RAFFAUF).

Varieties and related species

The sources of commercial tobacco are two annual herbs: *N. rustica* and *N. tabacum*. American Indians already used as many as 10 different species of *Nicotiana* for narcotic, medicinal, ritualistic and hedonistic purposes.

Until now, 64 distinct taxa have been established. The origin of *N. tabacum* is probably Argentina where crossing of *N. tomentosiformis* with *N. sylvestris* took place. *N. tabacum* as well as *N. rustica* are allo-polyploids which originated by hybridization (BRÜCHER 1989). There exist also Mammoth varieties which are tree-like.

Cultivation

The fruits contain an enormous quantity of minute seeds (about one million seeds per plant) and 1000 seeds weigh only 50–80 g. They maintain their viability for 20 years under adequate storage conditions. Tobacco is cultivated all over the world.

Observations

Tobacco leaves are best identified and recognized by their distinct hairs and the idioblasts of crystal sand. The best characteristics are however the giant size glandular hairs.

Solanum

At least 88 useful species are known.

Stems have a sedative, diaphoretic, diuretic and hypnotic effect.

Bark is emetic. Stem pith is applied for aching ears, to cleanse and gloss the hair.

Roots are stomachic and diuretic, being used for liver, spleen and bladder. Tubers are a source of starch and supply alcohol.

Shoots and leaves are eaten as a vegetable. Leaves of some species are used as a poultice for infected wounds or as an antirheumatic. Leaves are narcotic, diuretic and used for gonorrhoea. Leaves are used as a substitute for tobacco leaves for smoking. Leaves serve for kidney and bladder ailments; they are also applied as a vaginal douche. Leaves furthermore serve for headache, as a vermifuge and for infections through sand fleas. Foliage and fruit are diuretic and used for cystitis.

Fruits and tubers are edible. Fruits are also used as a condiment. Fruits have a sedative and antispasmodic effect. Tumours are treated with fruits. Fruit is used for sore gums. Fruit with 2 aspirins is abortifacient. Berries are also used for cough. Powder of the fruit serves as an insect repellent. Fruit pulp and seed are applied as an insect repellent (against cockroaches); they are also used to cure bites and stings of insects, spiders, scorpions. Fruit is used as a soap substitute. Fruit and bark are purgative.

Seeds are used for curdling milk.

Flavonols and steroidal alkaloids can have anti-inflammatory activities or be toxic, according to the doses being applied.

ROTH studied the leaf structure 1984.

Solanum americanum MILL.
synonym: *Solanum nigrum* L. var. *americanum* O. SCHULZ
(yerba mora, hierba mora)

Taxonomical description

Creeping or erect annual herb, glabrous or slightly pubescent with simple hairs, reaching 30–80 (150) cm in height (Fig. 270 a). Leaves alternate, simple; the thin blades ovate to lanceolate-ovate, slightly asymmetric, 2–8 (15) cm long with an acute or acuminate tip; petiole 2–3 cm long. Margins often sinuate or shallowly toothed.

The axillary inflorescences develop 3–10 flowers in cymose umbels; pedicels 4–14 mm long. Flowers 7–10 mm in diameter. Calyx 5-toothed with oblong to obtuse teeth, persistent on the berry. Crown white or bluey with a short tube and a folded 5-lobular border. Stamens inserted on the tube, filaments short. Ovary bilocular. The globular berry is glabrous, 7–10 mm in diameter, of black colour; it is found on a pendant pedicel with the persistent calyx on its base.

Origin

The species is native to the region between Mexico and Costa Rica and grows in the open field and in mixed forests between 1500 and 3900 m a.s.l. (GUPTA 1995).

Historical background

The species is mentioned in 1654 by the Spanish colonists (COBO 1984). During the Independence war in Cuba in the 19th Century it was used as a medicinal plant by the bearer-corps of the Libertador army.

Occurrence

The plant grows on hills and river banks. It is found in Cuba, the West Indies, Guatemala (GUPTA 1995). In Venezuela it is common in uncultivated regions near houses and cultivated land (SCHNEE 1960).

Anatomical description

Leaf. (Figs. 270, 271) The leaf is dorsiventral and amphistomatic. The upper epidermis cells are larger than those of the lower epidermis; they are papillary. As seen in a surface view, the anticlinal walls are wavy. The stomata, lying at epidermis level, are devoid of subsidiary cells (anomocytic) or are surrounded by 3 subsidiary cells (anisocytic). There is only a single palisade layer and about 2–3 (4) layers of small spongy parenchyma cells. The lower epidermis is similar to the upper one, but the composing cells are smaller. They are likewise papillary and their anticlinal walls are more strongly

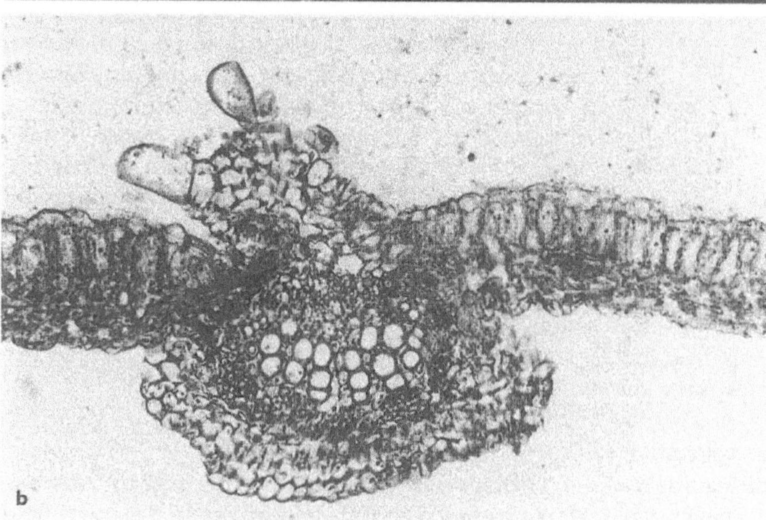

Fig. 270. *Solanum americanum.* **a** Branch with leaves, flowers and fruits. **b** Leaf midrib in t.s. ×6.3.

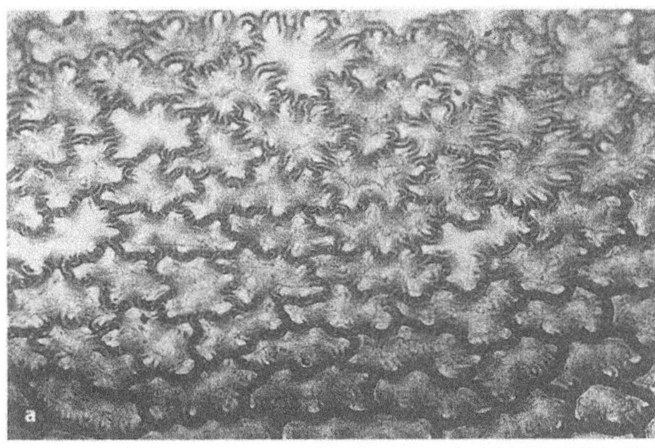

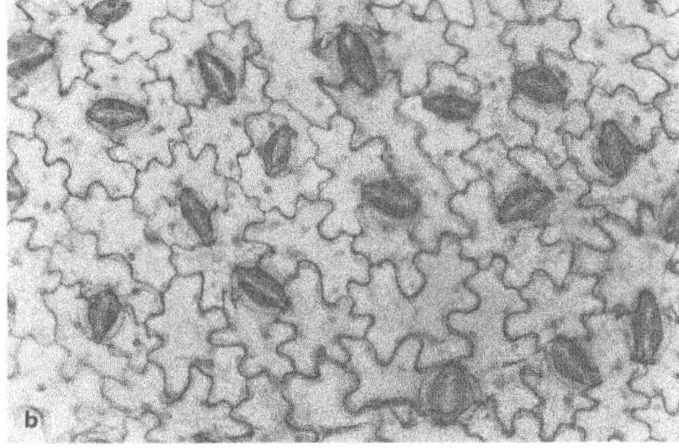

Fig. 271. *Solanum americanum*. **a** Upper epidermis. **b** Lower epidermis. Both × 16.

bent than those of the upper epidermis, as seen in a surface view. The stomata are much more numerous on the lower side and are slightly elevated above the surface.

The midrib is prominent on upper and lower side and has a large bicollateral vascular bundle in the center; angular collenchyma reinforces the midrib on both sides. The vascular bundles of secondary order are likewise bicollateral, but the adaxial phloem is reduced.

Collenchyma is only found on the lower side below the bundles, so that the palisade parenchyma continues on the upper side. The smallest bundles have only collateral structure.

Simple uniseriate hairs consisting of 3–4 cells which have pointed tips and occasionally a swollen base, mainly occur above the veins, but are rare. The cells surrounding the hair base may become elevated above the surface.

The leaf is delicate and has a herbaceous consistency.

Axis (Figs. 272, 273 a). A young stem of hardly 2 mm in diameter has been studied. The epidermis supplied with stomata and hairs is small-celled and has thickened outer walls. Beneath follows a layer of loose chlorenchyma.The following collenchyma is angular and also penetrates into the 3–4 protuberances on the outside of the axis which are due to the longitudinal ribs of the stem. The vascular system is surrounded by a discontinuous ring of fibers, corresponding to the pericycle. The original ring of vascular bundles has already been closed by the activity of the interfascicular cambium. The intraxylary phloem is well developed and the pith is very ample. A few fibers are associated with the intraxylary phloem. Some cells of the pith and the cortex contain clustered crystals or crystal sand. The vessels in the xylem are still restricted to the fascicular zones.

Fruit (Figs. 273 b, 274). The fruit as a berry is composed of parenchyma. The outer epidermis has thickened outer walls. The berry contains a large number of seeds which are embedded in the 2 locules. Vascular bundles occur in the periphery and in the sept; a concentration of bundles is found in the fruit center. The peripheral parenchymatous cells are tangentially extended, while the inner epidermis limiting the locules has radially extended cells.

Seed (Figs. 274–276). The structure of the seed coat is very interesting. The outer epidermis cells show a lower and an upper part. The lower part directed towards the seed inside has strong wall thickenings which appear in the form of a horseshoe, when the seed coat is cut transversely. Furthermore, the walls of the lower part are strongly wavy, when seen in a surface view. The upper part of the cells directed towards the outside, on the other hand, is thin-walled and has only very delicate ledges on the wall inside. The ledges run in anticlinal direction to the seed surface. Furthermore, the 2 parts which are almost equal in size stain differently with toluidine blue; when the lower part stains blue, the upper part adopts a reddish violet (purple) colour. It can therefore be assumed that the wall material of both parts has different chemical composition. Additionally, the walls of the upper part are not sinuous, but straight, and seen in surface view, the cells have a polygonal appearance. Beneath the epidermis follow several layers of thin-walled flattened and compressed cells.

The curved embryo, a characteristic often found in the Solanaceae, is embedded in the endosperm. The division of the seed epidermcies cells into 2 very different halves is most extraordinary.

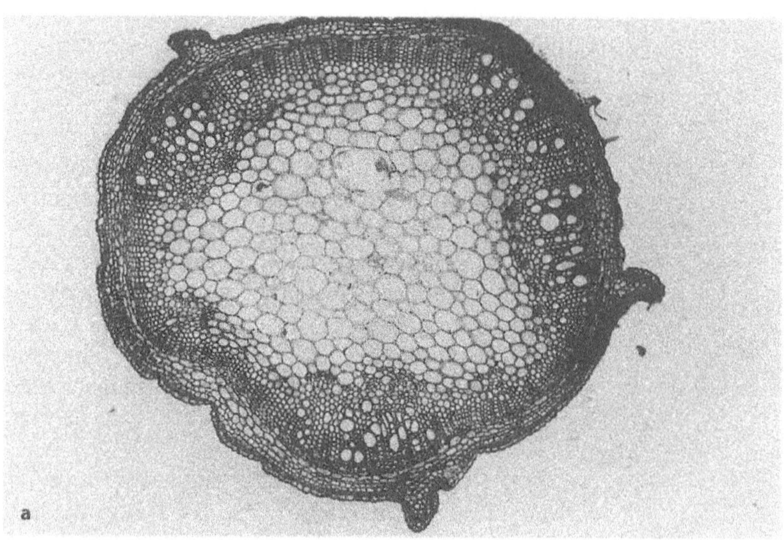

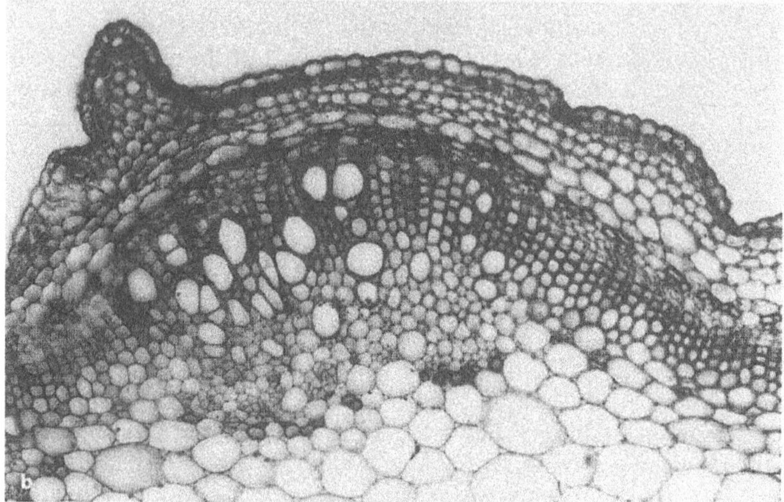

Fig. 272. *Solanum americanum*. Axis. **a** × 6.3; **b** × 16 (detail).

Ethnobotanical and general use

Nutritional use

The leaves are cooked as soups or eaten fried with eggs. The vegetable is found in country markets and is eaten in great quantities. It is particularly consumed during convalescence and during recuperation from diverse diseases.

Medical use

Leaves, roots, fruits and seeds are used.

Leaf. For skin conditions, such as scabies, ringworm, leprosy, pustules, warts, erysipelas, exantheme, acne, dermatitis, eczema, ulcers, and abscesses, a decoction is prepared and is applied locally or in the form of a bath. Arthritis is cured in the same way.

The decoction is taken orally during digestive disorders, against asthma, amygdalitis, anaemia, cirrhosis, colics, diarrhoea, toothache, scorbut, constipation, gastritis, meningitis, malaria, high blood pressure, retention of urine, rheumatism, whooping cough, gastric ulcer, stomachache and nervousness.

Leaf tea is taken for fatigue and heat.

Infusion is ingested for constipation.

A mouthwash is used for boils on lips and tongue.

A vaginal lavage is applied for gynaecological diseases (leucorrhoea).

A sitz-bath of leaves or leaves and roots cures haemorrhoids.

Cataplasms help against burns, tumours, herpes, ulcers. A cataplasm or decoction heals wounds.

Leaf juice is used for asthma. Sap of the leaves applied on the eyes produces dilation of the pupils.

Leaf juice in syrup is used as an expectorant.

In Venezuela it is used as the main remedy against herpes zoster.

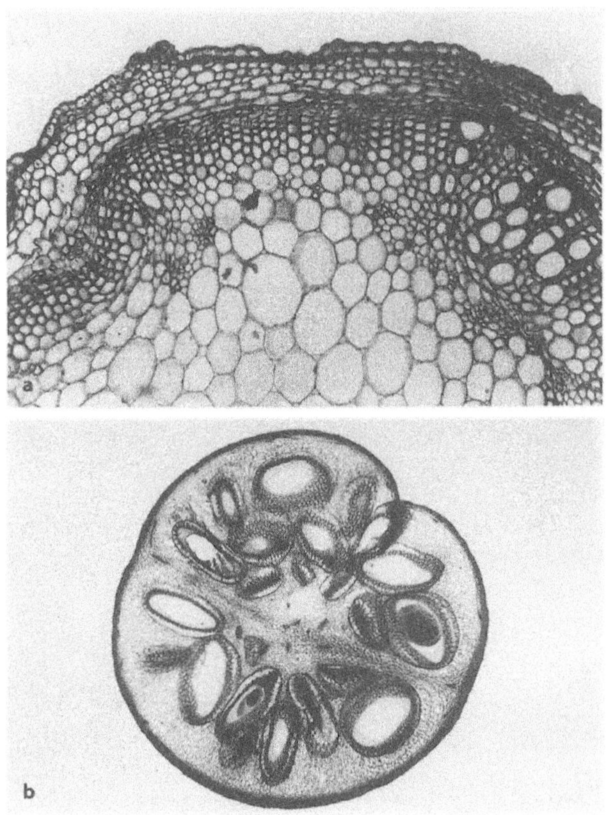

Fig. 273. *Solanum americanum.* **a** T.s. of axis. × 16. **b** T.s. of fruit. × 2.5.

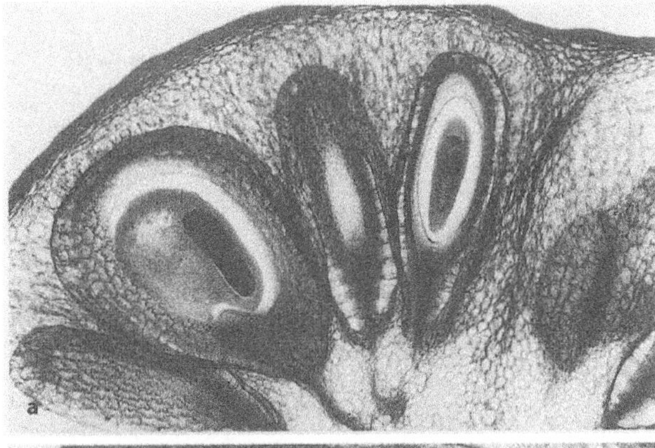

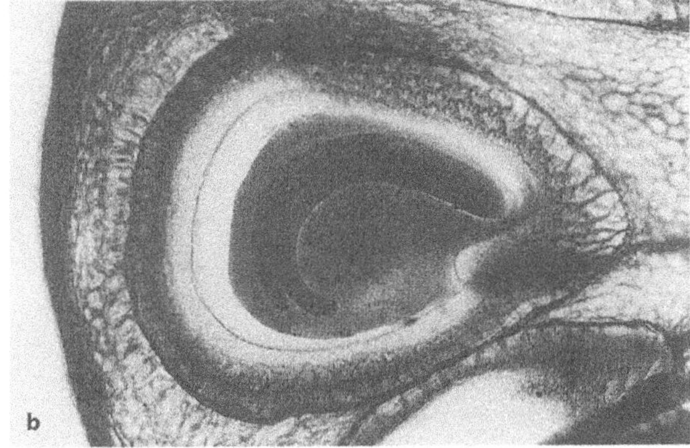

Fig. 274. *Solanum americanum.* **a** Pericarp with locules in t.s. **b** Seed.

Root. Root in decoction and applied as a sitzbath cures haemorrhoids.

Fruit. Fruit is used against skin conditions (see above).

Seed. Seed helps against sinusitis.

Method of use

Leaves are used in decoction, infusion, or as a tea. Decoctions are applied locally, as a cataplasm, in the form of a bath, sitzbath or are taken orally. Leaf juice is taken for asthma, leaf juice in syrup is taken as an expectorant.

The sap put into the eyes can be used as a beautifier, as it dilates the pupils (as does Atropa belladonna) or can also be applied in ophthalmology for eye tests.

A cataplasm of young shoots can be applied against swellings.

To cure gangrene, a lavage is applied with the decoction of 50 g herb in 1 l water.

For sinusitis, 10 g seed are macerated in 30 g of pure alcohol (LOPEZ PALACIOS 1987).

Healing properties

An infusion of the leaves has spasmolytic effects. It inhibits spasm due to the effects of muscarine and musculotropic mechanisms (GUPTA 1995). Leaves have appetite stimulant, sedative, purifying, diuretic, antiinflammatory, emollient, antipyretic, for recovery, wound healing (disinfectant) effects; they have antibiotic activities, being antibacterial (*Staphylococcus aureus*) as well as antimycotic (*Candida albicans*).

The fresh fruit is antiparasitic, diuretic, narcotic, disinfectant and toxic. During ripening, the fruit looses more and more of its toxicity.

Chemical contents

The most important chemical contained is the glycosidal alkaloid solanine. The antibiotic activity of the plant is attributed to this substance.

Toxicity

Leaf and fruit are toxic. However, with ripening, the fruit looses more and more of its toxic character. By boiling, the plant looses much of its toxicity. The plant is therefore not generally regarded as dangerous. Furthermore, the alkaloids are poorly absorbed and rapidly detoxified by the human body (SEAFORTH, ADAMS & SYLVESTER 1983).

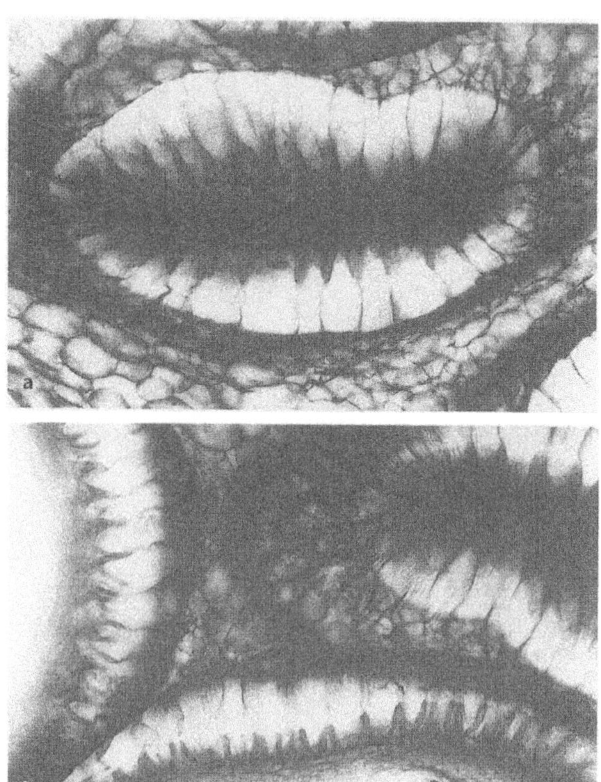

Fig. 275. *Solanum americanum.* **a, b** T.s. of seed. × 16. Note the outer epidermis cells with horse-shoe-shaped wall thickenings (below!).

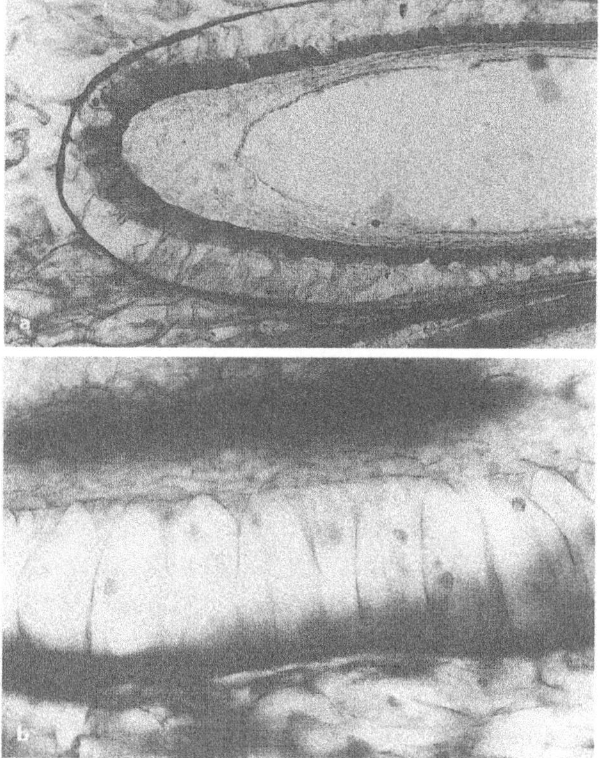

Fig. 276. *Solanum americanum.* **a** Seed in longitudinal section. **b** Outer epidermis of seed coat. Note the anticlinal striation of the cells (× 25).

Varieties and related species

The plant is frequently confused with *Solanum nigrum*. According to BLOHM (1962) contradictory reports exist about the toxicity of this species. BLOHM therefore suggests that there are several different strains which differ somewhat in their properties.

Observations

The Solanaceae are an alkaloid family. The sap of the leaves of *Solanum americanum* dilates the pupils when applied in the eyes, as does atropine in *Atropa belladonna*; the species name refers to this property which beautifies the eyes. Hyoscyamin and atropine are spasmolytics; the same effect is ascribed to the leaves of *Solanum americanum*.

Anatomical characteristics are the bicollateral vascular bundles, the ribbed axis produced by protuberances of subepidermal tissue (collenchyma) and the very peculiar structure of the epidermis cells of the seed coat which are divided in a lower and an upper part. The lower half has strongly thickened walls which are horseshoe-shaped when seen in transverse section, the walls stain blue with toludine blue and are wavy, as seen in surface view. The upper half of the cells, on the contrary, is thin-walled and has only very delicate ledges, the walls are straight, as seen in surface view and stain purple with toludine blue.

The plant is thus very well characterized.

Solanine is the toxic principle of the plant, but it is poorly absorbed and rapidly detoxified in the human body. Unripe fruits are toxic, but as the fruit ripens, the alkaloid content decreases, similarly as in the leaf; adult leaves are used as a vegetable.

Sterculiaceae

The representatives of this tropical family are mostly trees or shrubs, more seldom herbs, with lobed or pinnate leaves and stipules on the leaves. The ovary is frequently 5-carpellary and syncarpous. The fruit often splits into mericarps. Very characteristic are slime cells or cavities in all organs, and stellate hairs. Apopetaly (absence of petals) is observed in certain Sterculieae. The Sterculiaceae are considered the most primitive family of the Malvales.

Some species of *Sterculia* supply fibers (hard-bast) for ropes or for paper fabrication, others supply gum. The existence of *Sterculia* in cretaceous and tertiary formations has been proved.

The family is rich in methylxanthin derivatives. Theobromin prevails in cocoa, caffeine in cola. *Theobroma cacao* (Central and South America) as well as *Cola nitida* (West Africa) are cauliflorous. In both cases, the seeds are commercially used, more exactly: in cola the cotyledons.

Resemblance in anatomical characteristics suggests a close relationship between Sterculiaceae, Tiliaceae, Bombacaceae and Malvaceae.

Guazuma ulmifolia LAM., synonym: *Guazuma tomentosa* H. B. K. (guácimo, guácimo dulce, guácimo cimarrón, cuahulote are only a few of the 47 vernacular names)

Taxonomical description

The tree is 5–15 (occasionally 20) m (Figs. 277b, 278) high and reaches 60 cm in diameter. The trunk is short and generally tortuous and has a grayish-brown bark which is profoundly fissured, dehiscing in the form of small plates; it is fibrous and tastes slightly bitter. The crown is extensive and the branches pendulous.

The simple leaves are alternating and have a short petiole, about 1.5–2.5 cm long, and are pubescent. The blades are coriaceous and of very variable shape, generally ovate to oblong or lanceolate with an acute tip, serrate margins and a truncate asymmetric base; they are about (5) 10–20 cm long and 4–8 cm broad, pubescent on both sides, but more densely so on the lower side. The venation is pinnate, brochidodromus-craspedodromus with 6–8 pairs of secondary nerves. The midrib is very prominent on the lower side, but less so on the upper side.

The flowers are arranged in panicles of 2–5 cm length, they are axillary or cauliflorous. The flowers are small, not showy, white to yellowish, and of a fine fragrance. The calyx is trilobate, woolly outside, 2.5–3 mm long. The yellowish petals are pubescent and have 4–4.5 mm long bipartite appendages in their upper part, while the lower part, about 3.5–4 mm long, is concave. The stamens form a bell-shaped tube with 5 angles, being united in groups of 2–3. The styles are also united.

The globular or slightly elongated black fruit is woody, 2–5 cm in diameter and has a tuberculate surface. It opens either at the apex or irregularly on the wall with pores and is therefore considered a capsule. It contains many small seeds.

Origin

The species probably comes originally from the Caribbean Islands (West Indies).

Occurrence

Central America, West Indies and South America. In Venezuela, the species is amply distributed in all hot and temperate regions.

Anatomical description

Leaf (Figs. 279–281). The leaf is dorsiventral and hypostomatic. The upper epidermis is very irregular, as seen in transverse section, and consists partly of very large cells. Frequently, the cells are periclinally divided; some cells reach an enormous size. They are slime cells, a characteristic of the Sterculiaceae. As seen in surface view, the anticlinal walls are straight. Thick-walled lignified unicellular hairs and thick-walled lignified stellate hairs with 6–8 arms are less frequently found on the upper than on the lower side. Glandular hairs with a foot cell, a stalk cell, and a pluricellular head also occur here and there (Fig. 279). The palisade parenchyma occurs in the form of anticlinal rows of 2–4 or more cells; the cells are short and slender. Large slime cavities and solitary rhombic crystals are dispersed in the parenchyma. The palisade parenchyma is not very clearly distinguished from the spongy parenchyma. The latter is more loose and has a more irregular structure, but its cells are also partly arranged in the form of anticlinal rows. This particularity is characteristic of many Sterculiaceae.

The lower epidermis cells are smaller than those of the upper epidermis, the outer tangential walls of the cells are delicate and the anticlinal walls are somewhat bent, as seen in surface view. The stomata are comparatively small and have no distinct subsidiarys cells, they are thus anomocytic; they occur at epidermis level or are only slightly elevated above it. Stellate hairs with up to 10 arms, unicellular hairs and glandular hairs very frequently occur on the lower surface. The glandular hairs are of different sizes, the larger ones show more head cells,

Fig. 277. a Leaves and flowers of *Melochia tomentosa*. b Habitus of *Guazuma ulmifolia*.

Fig. 278. *Guazuma ulmifolia*. a Twig with leaves. b Fruits.

and their pedicel may consist of several cells instead of only one (Fig. 279, 281.1, 281.2).

The midrib projects considerably above the lower and the upper side. On the upper side, the epidermis is small-celled and the palisade parenchyma is interrupted; instead, a regular parenchyma is developed, in the center of which a large slime cavity is usually situated. The vascular system is horse-shoe shaped and the phloem is surrounded by fiber bundles on its outside. On the upper side of the xylem, very small-celled tissue similar to the phloem tissue as well as some fiber bundles occur. The tissue below the phloem is parenchymatous and contains several slime cavities. The lower epidermis is also small-celled in the midrib region. Solitary crystals are found in the parenchyma of the entire midrib.

The venation is dense and the secondary nerves are transcurrent to the upper and to the lower side mostly by sclerenchyma, but parenchyma may also participate, possibly in the form of a not very well defined sheath in the larger bundles. Septate crystal

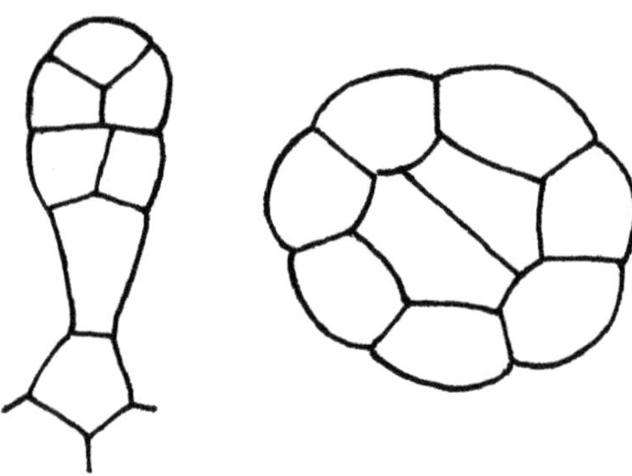

Fig. 279. *Guazuma ulmifolia*. Left: small glandular hair, typical of the Sterculiaceae family. Right: T.s. of the head of a larger glandular hair.

Fig. 280. *Guazuma ulmifolia*. **a, b** Leaf midrib.

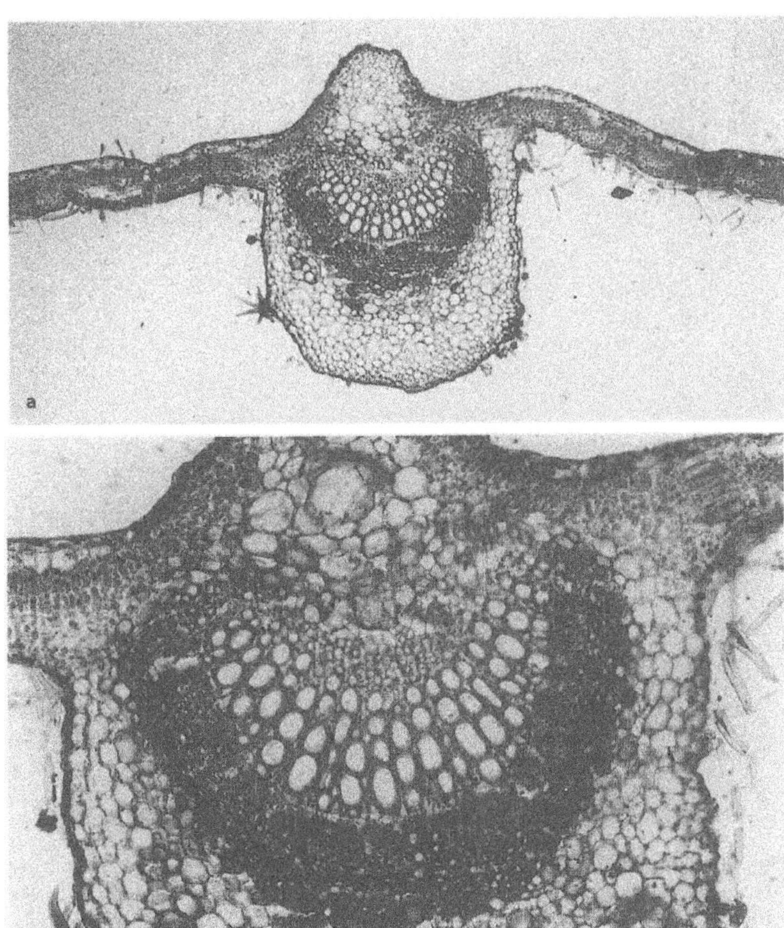

strands accompany the vascular bundles; they become best visible in paradermic sections. The vascular bundles terminate blindly in the mesophyll, 'open meshes' (according to ROTH 1996) with irregularly enlarged tracheids which have a water-storing function.

Bark (Figs. 282, 283 a). The inner conducting bark is of medium size. Following the usual pattern in the Sterculiaceae, the phloem is stratified into soft and hardbast plates. As seen in transverse section, the plates are 2–3 cells high and only 2–6 (9) cells wide. They are accompanied on both upper and lower side by septate crystal strands with rhombic crystals. The soft bast portions reach about twice or three times the height of the fiber plates. The rays which occur frequently are pluriseriate, up to 5-seriate and their course is slightly undulated due to the sieve tube collapse in the form of tangential bands. Some rays enlarge towards the outside in the funnel form, a further peculiarity of the Sterculiaceae. In this case, the ray cells enlarge tangentially and divide anticlinally so that tangential cell rows arise. Compound starch grains are present in the ray cells as well as in the axial parenchyma cells of the phloem. Very large slime cavities are situated in the rays and partly originate at the same tangential level forming tangential rows. As seen in a longitudinal section, the rays are comparatively high and more or less consist of cells of the same type except some marginal cells which are more square-shaped than the others.

The medium-sized periderm takes a typically undulated course. The phelloderm is thin-walled and locally well-developed, the cork cells are slightly thickened. Periderms run parallel to each other for longer distances. In the rhytidome, ray cells and axial parenchyma cells frequently show a brown content which intensely stains with toluidine blue; they probably contain tannins.

Wood. The pinkish wood varies from rather light to moderately heavy, but is firm and strong and used locally for general carpentry and interior construction, slack cooperage, boxes and crates, tool handles and for fuel and fine charcoal.

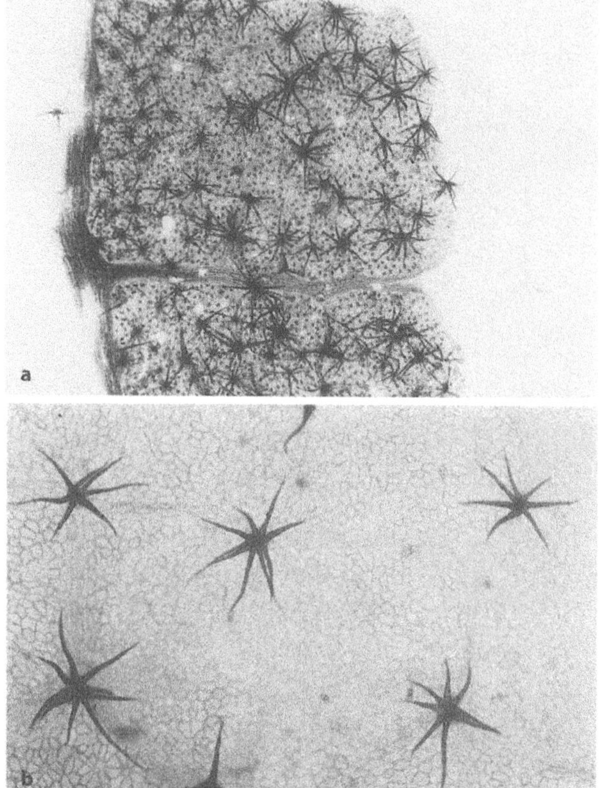

Fig. 281.1. *Guazuma ulmifolia*. **a** Lower epidermis with stellate hairs. **b** Upper epidermis with much less stellate hairs (× 16).

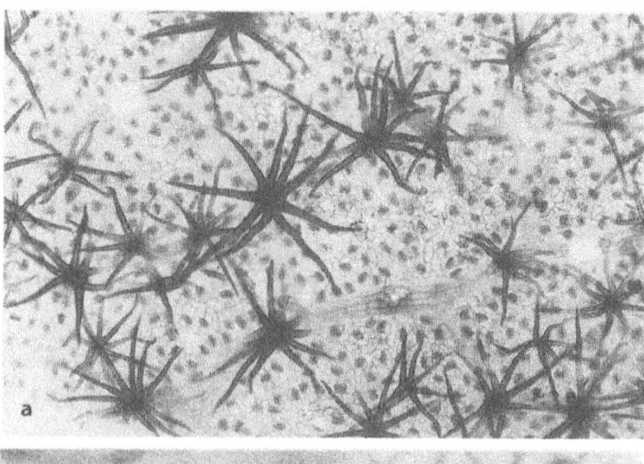

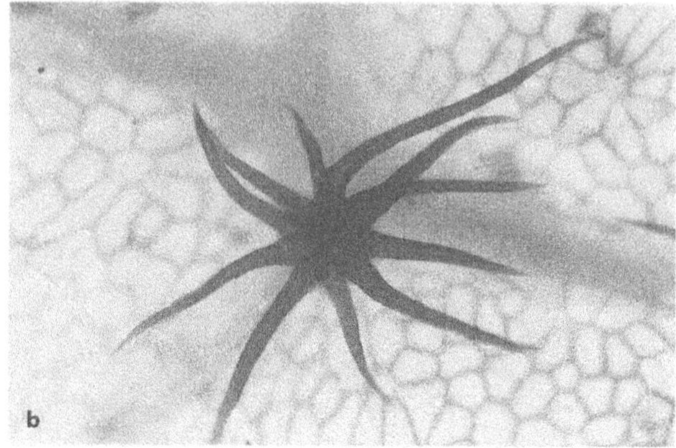

Fig. 281.2. *Guazuma ulmifolia*. **a** Lower epidermis with stomata and stellate hairs. **b** Stellate hair of the upper epidermis, on top of a rib.

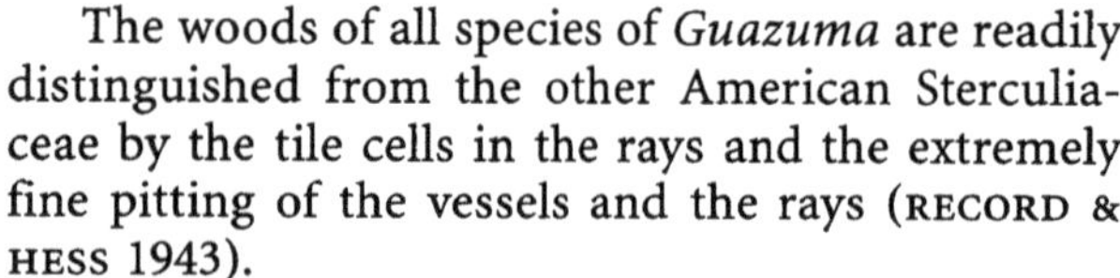

The woods of all species of *Guazuma* are readily distinguished from the other American Sterculiaceae by the tile cells in the rays and the extremely fine pitting of the vessels and the rays (RECORD & HESS 1943).

The wood of *Guazuma ulmifolia* is diffuse-porous. The vessels, solitary or arranged in radial multiples of 2 to several, are small but frequent. The rays are mostly small.

Ethnobotanical and general use

The plant is said to have 36 different applications.

Nutritional use

The fruit is edible; it encloses a slimy and astringent pulp which is enjoyed by cattle and is therefore used as a fodder.

The tree is appropriate for afforestation on steep slopes and on poor soils. It can also be planted in gardens and parks.

The bast fiber of the young stem is employed in making rope. The grayish fibrous wood is coarse-grained and light, but firm and strong, and is used locally for general carpentry, interior construction and furniture, for ribs of small boats, barrel staves, shoe lasts, paneling, slack cooperage, boxes and crates, tool handles and for fuel and fine charcoal to make gun powder.

Medical use

The parts used are leaves, bark, roots, flowers, fruits and seeds.

Leaf. Leaves and roots are said to be antidysenteric. Macerated leaves applied on the skin are used against elephantiasis and cutaneous diseases.

Bark. A gum is prepared from the slime of the bark for promoting hair growth; a decoction of the bark or the bark macerated without heating prevents loss of hair. The slime of the bark is also used against diseases of the skin. A syrup or a decoction of the bark is used to cure elephantiasis and to clarify syrup in sugar manufacture.

A decoction (30 g in 150 ml water) is applied as a depurative against bubo, to heal burns, inflammations, and first symptoms of blindness. A decoction

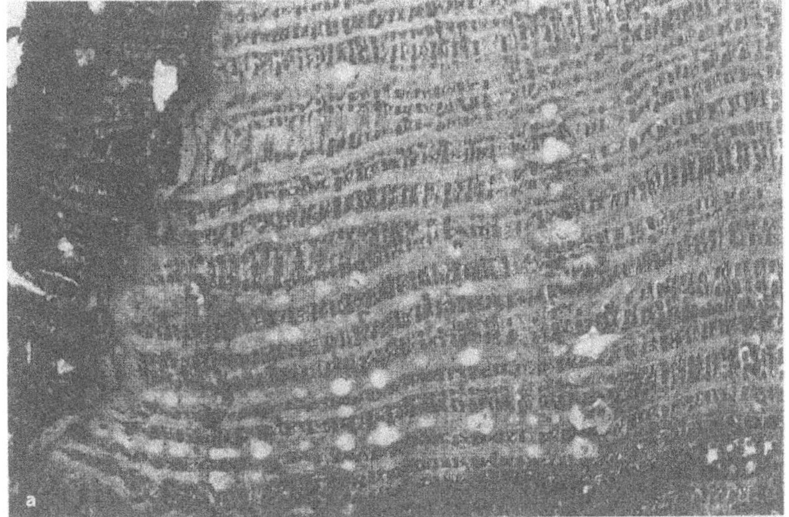
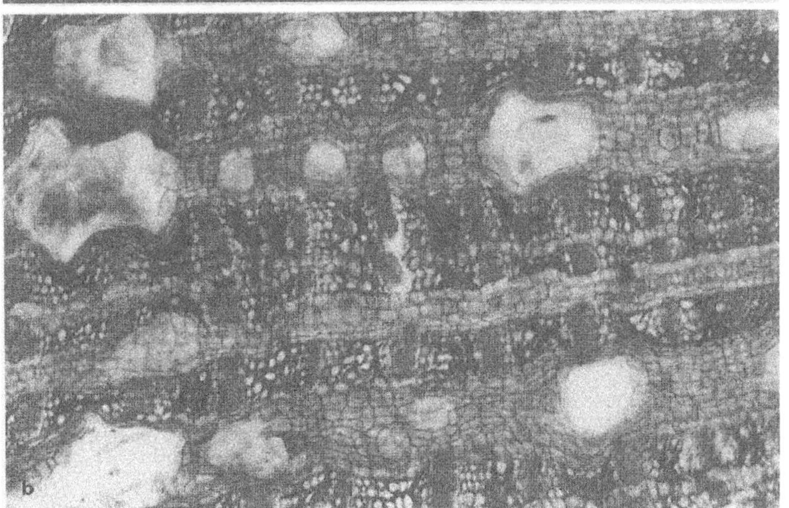

Fig. 282. *Guazuma ulmifolia*. Bark. **a** Note the pluriseriate rays and the undulated periderm towards the outside. **b** Detail with large slime cavities (light). 6.3 × 10.

is also used in a bath against sunstroke and as a contractor of the uterus. An infusion calms liver, spleen and kidney, macerated in decoction it is used for urethral and intestinal affections and for haemorrhoids. A lavation is applied against dysentery.

In a bath, the bark cures the allergy caused by manzanillo (*Hippomane mancinella* L., Euphorbiaceae). The bark is also applied against asthma, to facilitate delivery at birth and to cure diseases of the liver.

Root. The root is antidysenteric, antidiarrhoeic, diuretic and cures measles. It is considered to be the most active part of the plant, and is also applied for baldness.

Fruit. The fruit is used against flu and as an infusion with syrup against irritations of the stomach and against uterine cramps.

Method of use

Seed. Mainly used for respiratory and gastrointestinal diseases.

The bark is used in decoction or infusion, in the form of a lavation or bath or as a syrup; or the bark is macerated and the liquid is taken to treat the mucosa of the intestine. Externally a decoction or maceration is used to stop loss of hair.

The fruit is applied in infusion or decoction or is macerated.

Leaves are macerated on the skin to cure diseases of the skin.

The shoot is used in the form of a tea.

Flowers are applied in infusion.

Seeds are applied in decoction.

Healing properties

Antidysenteric, antidiarrhoeic, astringent, refreshing, diuretic, emollient, febrifugal, against asthma, elephantiasis, irritations of stomach and

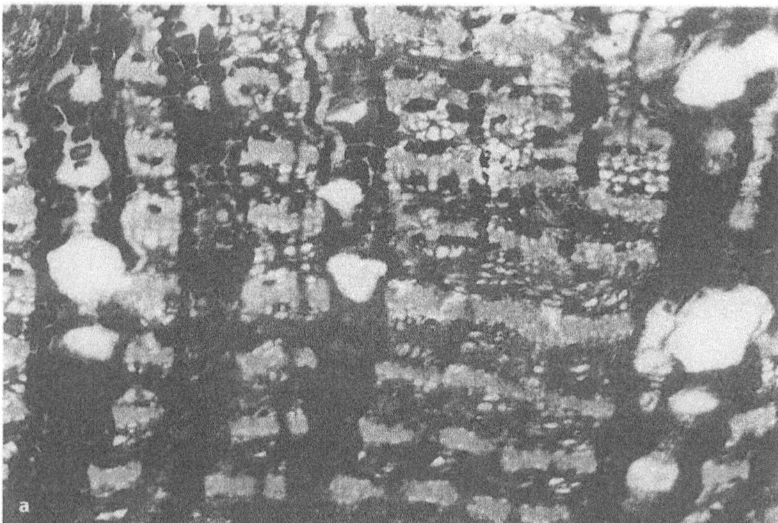

Fig. 283. **a** *Guazuma ulmifolia*. Bark. Note the stratified phloem with layers of hard bast in the form of fibers (gray), parenchyma and obliterated sieve tubes with their companion cells. 6.3 × 5. **b** *Melochia tomentosa*. Young plant.

liver, against haemorrhoids, as a vulnerary and against burns, against irritations of the intestine and the urethra, to soothe the contractions of the uterus and as a pectoral medicine.

Chemical contents

Starch, tannins, saponins, alkaloids; caffeine (leaf). Betulin, β-sitosterol, friedelin, unsaturated esters, cardenolides, bufadienolides, flavonoids, anthocyanins (bark). Flavonoids, kaempferol, kaempferitin and quercetin (flowers). Shoot and root have bactericidal activities.

Cultivation

The tree is propagated by seeds. Its growth is rapid. The radical system is superficial. The species is of medium duration and very resistant to drought and tolerates most soils.

Observations

Special characteristics of the Sterculiaceae are the stellate hairs, slime cells and slime cavities, stomata without distinct subsidiary cells, scattered tannin cells in leaf and stem, septate crystal strands accopmanying the vascular bundles, a stratified phloem with fiber plates and medullary rays which partly enlarge in the funnelform towards the outside of the phloem.

Fig. 284. *Melochia tomentosa*. **a** T.s. of leaf blade (× 16). **b** Midrib (× 16). **c** Midrib (× 20). Note the stellate hairs.

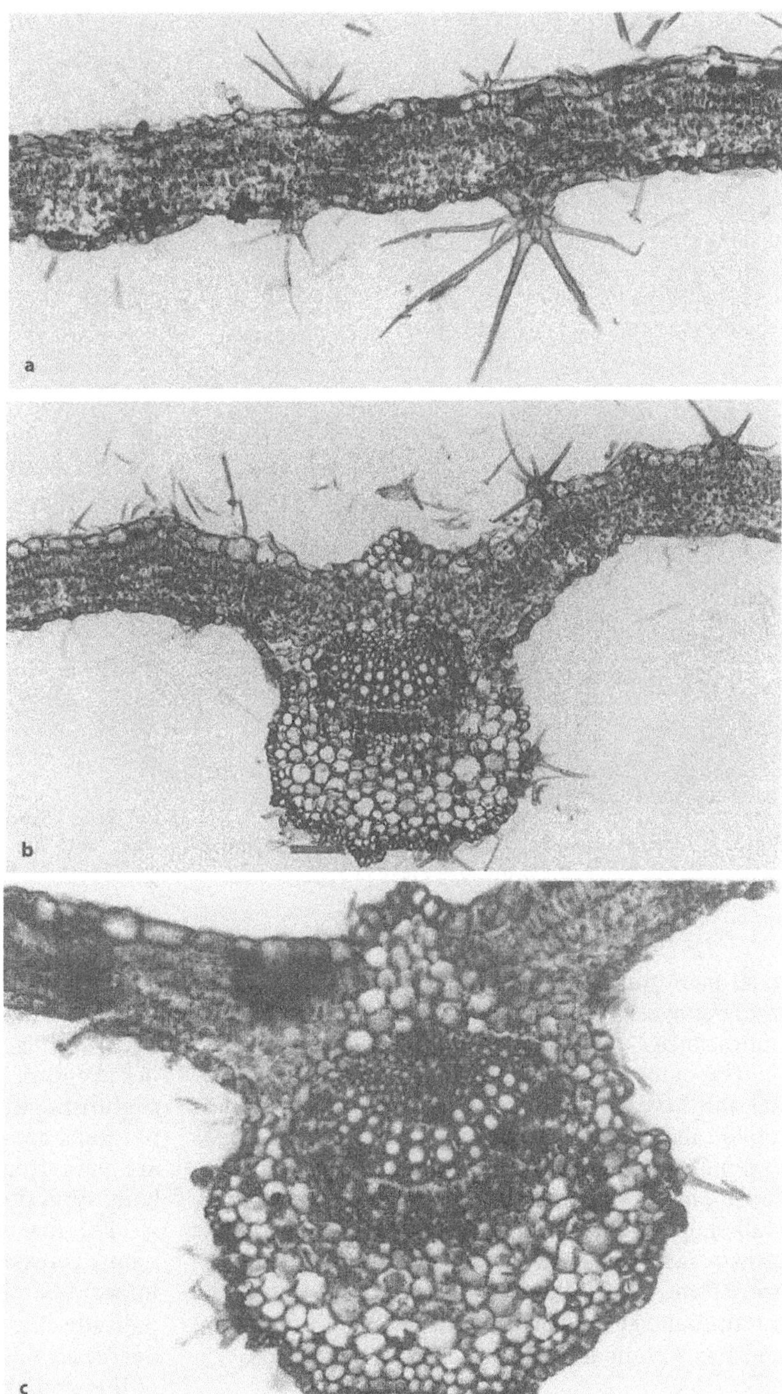

***Melochia tomentosa* L.**
(bretónica morada, totón, cariaquito, tostón, cariaquito morado, yerba de pujo, lava plato, shüüüi (Guajira)

Taxonomical description

Taxonomical details can be found in Fig. 277 a, 283 b. The species is a subshrab or shrub, 1-3 m high, with erect or inclined branches. Stem cylindric, pubescent. Leaves alernate, variable in size and shape; blades covered with woolly white hairs on both surfaces, of ovate, oblong or lanceolate shape, 1-5 (7) cm long and 1-3 cm broad, with an acute to obtuse tip, serrate margins and a subcordate base. The venation is pinnate, craspedodromus, simple, with 6-8 pairs of secondary nerves. Midrib and secondary nerves prominent on the ab-

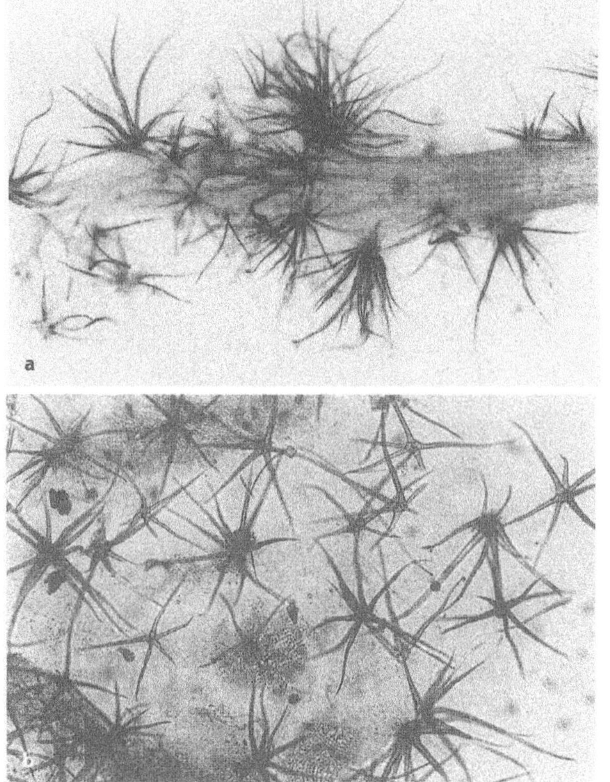

Fig. 285. *Melochia tomentosa*. Stellate hairs. **a** On the midrib of the lower leaf side. **b** On the lower leaf surface.

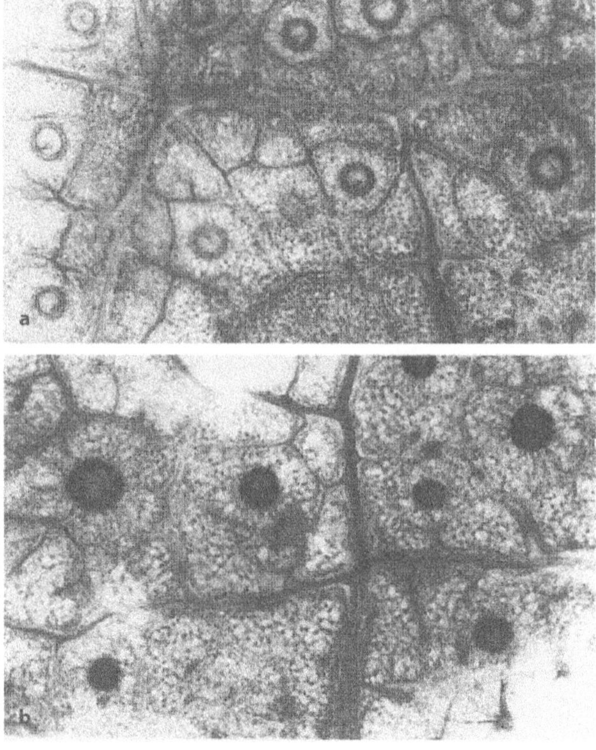

Fig. 286. *Melochia tomentosa*, **a, b** surface view of lower leaf side. Note the large sphaerical slime cavities, one beneath each mesh of the vascular network.

axial side and sunk below the upper side. Petiole plane-convex, 0.4–1.5 cm long, covered with a white pubescence.

The cymose inflorescences are axillary or terminal and have 2–10 flowers, 1.5–2.5 cm long. The 5-lobed calyx is 0.6–0.7 cm long, the lobules are linear and pubescent on the outside; they are covered with stellate hairs and glandular hairs with a globular or club-shaped head. the rose-coloured or purple crown has 5 petals, 0.1–1.3 cm long. The stamens are 0.5 cm long. The ovary is also pubescent. The pyramidal capsule is 5-angular, 0.7–0.9 cm long, and has 5 compartments, each with 1–2 seeds.

Origin

Tropical and subtropical America.

Occurrence

In Venezuela, the plant is amply distributed in arid and semi-arid regions, in xerophilous forests, savannas, in disturbed areas, and on calcareous soils.

Anatomical description

Leaf (Figs. 284, 285, 286). The leaf is thin and dorsiventral. The upper epidermis is large-celled and slightly papillose. As seen in surface view, the anticlinal walls are more or less straight und stellate hairs are very frequently found. Additionally, glandular hairs typical of the Sterculiaceae, are observed.

The mesophyll is partly more or less homogeneous consisting of about 4–5 layers of cells. In the uppermost layer, the cells are elongated and palisade-like, while the length of the anticlinal walls decreases towards the lower leaf side. In the region of the stomata, the mesophyll is looser and the cells are more or less spherical, while the cells in between may maintain their elongated shape. In certain places, 2 layers of palisade parenchyma are very conspicuous, whereas the underlying tissue is spongy, being composed of approximately spherical cells which leave intercellular spaces. The structure of the mesophyll may thus vary in one and the same leaf. As in certain other Sterculiaceae, palisade and spongy parenchyma are not clearly differentiated. But generally, there are about 1–2 layers of palisade parenchyma.

The lower epidermis cells are smaller than those of the upper epidermis. The stomata are slightly elevated above the surface. Very conspicuous are slime cavities below the lower epidermis. They are surrounded by a ring of parenchymatous cells in the form of 1–2 cell layers. Several irregular slimy layers are observed surrounding the parenchymatous ring. The cavity seems to be densely filled with slime. The surrounding parenchyma cells may possibly be involved in the slime secretion.

As seen in surface view, a slime cavity (Fig. 288) comes to lie in the center of each mesh of the venation network (Fig. 286). The stellate hairs are nearly as frequent as on the upper leaf side, but may have more arms. Club-shaped glandular hairs with a puricellular head are possibly more frequent in the lower than in the upper epidermis. The stomata do not show particular subsidiary cells ('anomocytic' type).

The most interesting characteristic of this species are the styloids (Fig. 288), crystals of enormous size with pointed tips and a quadrangular transection which form a network below the upper epidermis, following the network of the veins in the foliar limb.

The upper epidermis cells adopt a yellowish-orange or dark-red colour when treated with Sudan III and may contain oily or fatty substances. Also present are conspicuously dark staining cells when dyed with toluidine blue; they may correspond to tanniferous cells and mainly occur in the epidermis and in subepidermal laysers.

Not only the mesophyll is variable in its structure, but also the midrib. The xylem forms either an open arc in the midrib which is completely surrounded by phloem – or a small xylem portion is developed above the phloem on the adaxial side, while the lateral parts of the phloem are reduced so that the impression of 2 superposed vascular units is created: a larger horseshoe-shaped open arc on top of which a vascular bundle with a horseshoe-shaped phloem and an adaxially oriented xylem arises. The midrib forms a conical outgrowth on the adaxial side and the cells between the upper epidermis and the vascular system are large and parenchymatous. Parenchyma is also found beneath the vascular bundle on the lower side, which is considerably vaulted above the surface. However, towards the lower epidermis the cells become smaller and slightly collenchymatous. Palisade parenchyma as well as spongy parenchyma are thus interrupted in the midrib. Hairs and glands and possibly also stomata are found in the epidermis. Sclerenchyma is however absent, a feature frequently found in the Sterculiaceae.

The veins of second order which are also prominent on the lower leaf side, have a single collateral vascular bundle which is connected with both epidermal layers by parenchyma. The veins of higher order are embedded in parenchyma.

Fig. 287. *Melochia tomentosa*. Xylem of the axis. (**a** ×5, **b** ×10).

Axis (Fig. 287). A twig of about 1 cm in diameter has a superficial cork with abundant lenticels. Dark-brown layers of cells with a very reduced radial width alternate with colourless layers of regular brick-shaped cells. In the dark-brown layers, the cell wall is light-brown, while the cell content has a dark-brown colour. All cork cells are thin-walled. The phelloderm is small comprising up to 7 layers of more or less isodiametric cells. In the lenticels the cork is likewise stratified in large layers of a spongy tissue and in very small layers of cells with radially shortened walls. The loose filling tissue consists of more or less globular cells and some very large partly rhexigeneous intercellular spaces. At places, a scaly rhytidome develops in the form of small scales. A pericycle composed of small fiber groups arranged in the form of a ring (as seen in transverse section) follows below. The small parenchymatous zone of the primary cortex shows dilatation growth in the form of periclinal cell chains which originated by anticlinal cell division.

The phloem is stratified in layers of softbast with sieve tubes, companion cells and axial parenchyma and in layers of hardbast consisting of fibers. Collapsed sieve tubes outside the conducting bark are partly arranged in the form of layers. The 1–3 seriate rays are small, but some develop a vigorous dilatation growth so that funnel- or wedge-shaped (triangular) (METCALFE & CHALK 1950) rays well known of the Sterculiaceae, Tiliaceae, Bombacaceae etc., arise. The dilated rays thus consist of periclinal rows of parenchymatous cells which originated from anticlinal cell divison.

The well-developed xylem shows very irregular growth rings. The vessels are small and occur either solitary or in radial multiples (Fig. 287). The larger part of the tissue is parenchymatous, while fibers are accumulated towards the end of the growth rings. The pith in the center is parenchymatous and extremely small. The cells contain crystals either in the form of druses of Ca-oxalate or in the form of styloids. Styloids (Fig. 288) are also present in the dilated rays, in the parenchyma of the primary cortex, as well as in the axial parenchyma of the softbast, partly accompanying the hardbast fibers.

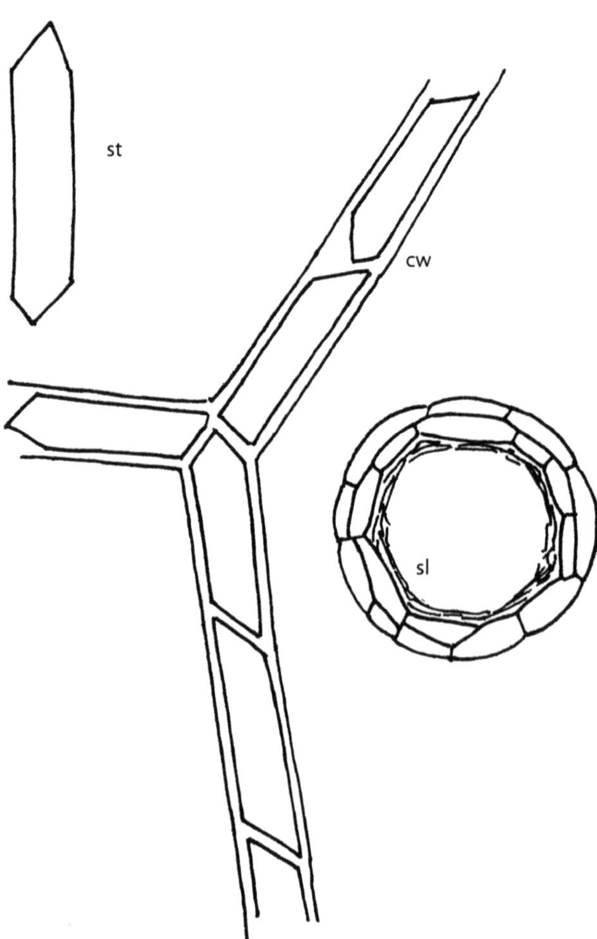

Fig. 288. *Melochia tomentosa*. St styloid. Cw cell wall of styloid cell. Sl slime cavity.

Ethnobotanical and general use

Medical use

The used parts are leaves, bark, shoots, roots, entire plant, flowers, branches with flowers.

Leaf. The triturated leaves taken with water cure vaginal haemorrhage and ulcer.

Bark. The bark is said to have emollient effects.

To cure wounds, the entire plant triturated is used locally.

Root. The root is used against pruritus and aches of the gum.

Flower. The flowers or branches with flowers are applied in infusion to cure flu and liver problems. In the form of drops, the infusion is applied against blindness.

Entire plant. The plant has antiseptic and bacteriostatic properties. It contains phenolic derivatives (tannin), slime and zygosporin (a cytochalasin).

Most of the Sterculiaceae contain tannins and slime. The tannins are probably responsible for the antiseptic effects.

The drugs are sold at stands in Caracas and Maracaibo, where herbal medicine is offered.

Varieties and related species

Melochia pilosa (MILL.) FAWC. & RENDLE which also occurs in Venezuela, where it is called Bretónica amarilla, is equally a medicnal plant. Branches with flowers are sold at stands in Caracas. The bark has emollient effects.

Observations

The plant is still little known and poorly studied. It is however easy to recognize by the chains of styloids which accompany the veins of the leaf blade and by the inner structure of the midrib. As far as the authors are aware, styloids in strands accompanying the veins of the leaf have not yet been reported. They are an extraordinary feature.

The glandular hairs and stellate hairs, as well as the funnel-shaped rays in the phloem and the stratified structure of soft and harbast reveal the relationship with the Sterculiaceae.

Sterculia

At least 20 useful species are known. Useful wood.

Leaves for sores and infected pimples on the head. Stem diuretic. Bark for fiber. Flowers for boils. Edible fruits. Edible seeds (maranháo nuts), source of oil for paints and lubrication. Source of tartar gum. Gum as masticatory with tonic properties. Gum as adulterant of gummi arabicum. Gum for dyeing. Seeds for seasoning and beverages. Bitter gelatinous seeds used in medicine as febrifugal.

S. pruriens (AUB.) SCHUMANN. Leaves for sores and infected pimples on the head. Tea of bark diuretic. Timber of good quality.

ROTH studied the bark structure 1981, leaf structure 1984, fruit structure and dispersal 1987, leaf venation 1996.

Theobroma cacao L. (cacao, cacao criollo, cacao dulce)

Taxonomical description

T. cacao is a small tree, 3–10 m high, with scarce ramification (Figs. 289, 290). The stem has sympodial growth. The tree grows in periodical shoves which explains the arrangement of lateral branches in the form of 'stories'. Young branches are pubescent (with star-shaped hairs). The alternating leaves have a petiole. The leaf blades are leathery and glabrous on the abaxial side, but are scarcely pubescent on the veins of the adaxial side (with simple and star-shaped hairs). The blades are oblong (15)-60 cm long and (6)-18 cm broad, have an elongated drip tip and entire to slightly indented margins and a rounded base. The venation is pinnate, brochidodromus, with 11–12 pairs of seondary nerves; the middle nerve is prominent on the abaxial side and only slightly vaulted on the adaxial side. The petiole is 2–4 cm long, cylindrical and has a scarce pubescence of simple and stellate hairs. Young leaves are distinguished from the adult ones by their reddish-brown colour (presence of anthocyanins), while older leaves are dark green.

The tree is cauliflorous; the inflorescences sprout directly from the stem or from old branches; the flowers, furnished with peduncles, are arranged in cymose glomeruli. The rose-coloured calyx is 5-lobed and about 6 mm long. The crown (about 8 mm long) is of a whitish yellow colour. The 5 petals are divided into a lower cup-shaped fleshy part and an upper petaloid part. There are 5 stamens and 5 staminodes.

The oviform-oblong fruit is acuminate, about 15–20 cm long and of yellow, orange, red or purple colour. The fruit is not a capsule, as indicated in the bibliography, but remains closed and is considered a berry by ROTH & LINDORF (1971A). The ovary is superior and composed of 5 syncarpic carpels with marginal central placentation; the numerous seeds (16–60) are arranged in 2 rows in each locule. The pericarp of the ripe fruit is characteristically divided into an outer hard region and an inner mucilaginous white or pink pulp (berry) in which the seeds are embedded. This part is edible and has a sour-sweet taste. The embryo with its large cotyledons almost fills the entire ripe seed. The seeds are flattened on their sides by the neighbouring seeds which are arranged in the form of piles. They are oviform and about 2–3 cm long. In the adult stage, they seem to occur in a single row in each locule, but are alternately attached to the left or right margin of the carpels. The ripe fruit has either 5 or 10 ribs or has circular outlines without ribs, according to the variety.

Origin

The original center of radiation is probably the Upper Orinoco-Amazonas region from where the species spread to the Pacific coast and to Central America and Mexico (BRÜCHER 1989), as well as to Guiana and Brazil. CUATRECASAS (1964) writes that 'a natural population had been spread throughout the central part of Amazonia-Guiana westwards and northwards to the south of Mexico....'. This means that Central America is not the gene center of cacao, but a secondary region of diversification and domestication under the guidance of the highly developed Oltec and Mayas (BRÜCHER 1989, LATHRAP 1973).

Historical background

Curiously, the tree and the product are distinguished by 2 different names both of which are deduced from Indian names.

The term cacao is generally used to design the tree and the crop, while cocoa means the seed kernel and the beverage which is prepared from the roasted and ground beans. The term cacao is derived from the maya 'kakau', this from 'kaj-kaj', which means bitter and from 'kab' which means juice. The Aztecs called the tree cacahuatl and considered it as a 'food of the gods'. Basing on the leg-

Fig. 289. *Theobroma cacao*. **a** Treelet. **b** Ripe fruit. **c** Flower coming out of a twig. **d** Fruits on the stem of atree (cauliflory/cauli-carpy).

Fig. 290. *Theobroma cacao.* **a, b** Caulicarpy. **c** Fruit cut longitudinally, with seeds embedded in fruit pulp (edible).

ends of the Indians, botanists gave the tree the name Theobroma cacao, because Linné objected to the generic name 'cacao' which he considered a Nomen barbarum.

The name chocolate, designating the solid product of the cacao tree is likewise derived from an aztec expression 'xocatl' which gave rise to the Mexican expression 'chocoatl'.

The Spanish conquerors '(conquistadores)' found cacao already cultivated in Mexico. It was not only used as food, but also as a currency. According to OVIEDO (BRÜCHER 1989), a slave could be bought in Nicaragua for 100 seeds, a rabbit was sold for 10 seeds.

The first cacao fruits were taken to Europe by Hernán Cortés in 1582. Chocolate consumption became more popular when C. J. van Houten developed a new process of chocolate manufacture by eliminating the excess of fat and by adding sugar.

Occurrence

Theobroma is an exclusively neotropical genus. *Theobroma cacao* is now in cultivation everywhere in tropical countries. Due to cultivation and easy hybridization, differentiation of varieties is difficult.

In Venezuela, 2 varieties, the Criollo, and the Forastero-Amazónico are known. This latter type grows wild in the surroundings of Amazonas and Orinoco, according to HOYOS (1989). For further information concerning varieties in Venezuela see HOYOS 1989.

Anatomical description

Leaf (Fig. 291). The leaf is dorsiventral and hypostomatic. The upper epidermis is very large-celled and mucilaginous. Some of the cells enlarge considerably and stain dark-blue with toluidine blue. As seen in a surface view, the cells are polygo-

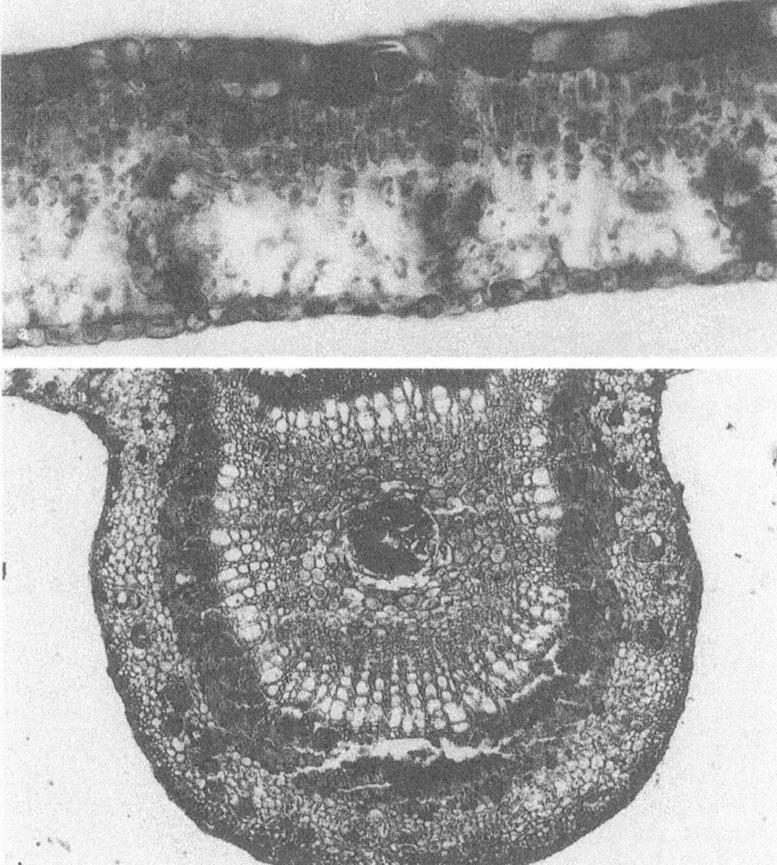

Fig. 291. *Theobroma cacao*. **a** T.s. of leaf. × 25. **b** Midrib. × 6.3. Note the slime cell in the center, the vascular ring, flattened on the upper side, and the sclerenchyma surrounding the vascular cylinder (× 6.3.).

nal and have straight walls. The palisade parenchyma consists of cells which are arranged in anticlinal rows, comprising up to 5 cells in a row. The spongy parenchyma is loose. The vascular bundles of higher order are transcurrent by a parenchymatous sheath in which slimy cells may occasionally occur. Rhombic crystals are found in the parenchyma cells directly surrounding the vascular bundles which therefore may be interpreted as septate crystals strands. The lower epidermis is small-celled and slime cells may possibly occur here and there. The stomata are anomocytic and occur at epidermis level. As seen in a surface view, the cells are polygonal but of a more irregular shape than in the upper epidermis. Trichomes are rare in the mature leaves. Glandular hairs with a several-celled stalk and a pluricellular head as well as long unicellular hairs with an acute tip and stellate hairs occur mainly on the midrib of the lower side and on the secondary nerves, but are occasionally also found on the upper side. Scattered cells in the mesophyll staining deeply are probably tanniferous. Druses of calcium oxalate also occur in the mesophyll.

The midrib has a very characteristic structure (Fig. 291 b). The center is occupied by a very large slime cell which is surrounded by parenchyma. The vascular system forms a continuous ring which is however flattened on the adaxial side. The palisade parenchyma is interrupted in the midrib. The vascular cylinder is surrounded by sclerenchyma. Parenchyma with scattered slime cells surrounds the sclerenchyma. The upper and the lower side of the midrib are reinforced by small collenchymatous cells. The lateral nerves of secondary order have a similar structure as the midrib. The veins of higher order are more or less transcurrent as indicated above.

Fruit (Figs. 292, 293, 294, 295 a). Structure and development of the fruit have been studied by ROTH & LINDORF (1971 A). The mature fruit is divided by a slerenchymatous ring into an outer part of small parenchymatous cells which give rise to small intercellular spaces (mucilaginous cells are interspersed in this tissue) and into an inner part which comprises the edible part surrounding the locules. As often found in fruits (berries), the cells of the outer

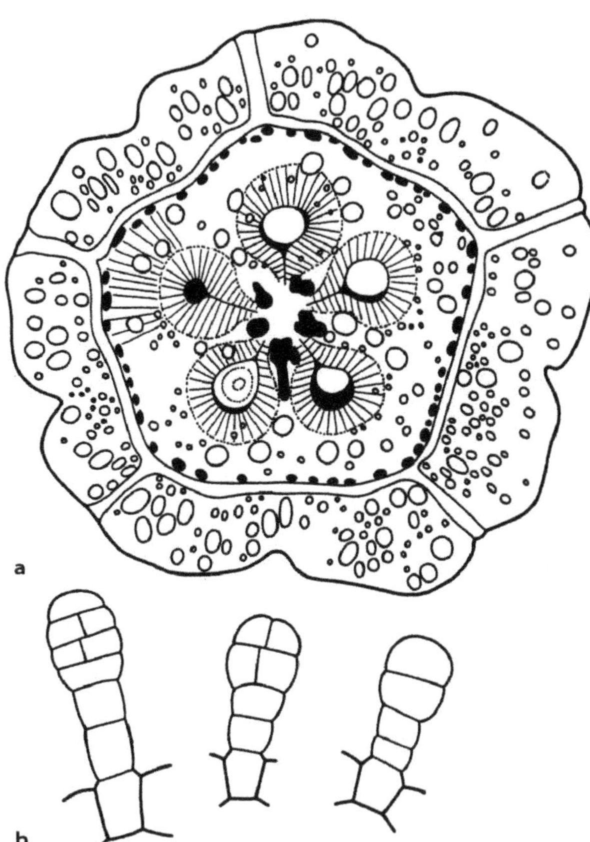

Fig. 292. a. *Theobroma cacao*, Sterculiaceae. **a** T.s. of fruit with meristematic ring from which 5 protrusions extend towards the outer epidermis; a ring of vascular bundles inside the meristematic ring (black); slime cells (circles), and 5 locules surrounded by radiating cell rows. **b** Glandular hairs of the outer fruit epidermis (ROTH & LINDORF 1971 a).

subepidermal laysers are much smaller than those of the deeper lying ones; these small-celled layers add more strength to the exocarp (sensu lato). The parenchymatous cells near the sclerenchymatous ring are often tangentially elongated and later divided anticlinally so that tangential cell rows may arise.

The sclereids composing the sclerenchymatous ring (Fig. 294) have thick walls and pits. They are arranged in the form of crisscross bundles. It is however left to the decision of the reader whether he considers these sclerenchymatous elements as sclereids or short fibers. There are transitional cases between both types which cannot be precisely distinguished, because: transitional cases instead of strong borderlines are found everywhere in nature! Beneath the sclerenchymatous ring, numerous vascular bundles are likewise arranged in ring form. The outermost cells of the inner pericarp section near the vascular bundles are also tangentially extended, due to tensions of the dilatation growth in this region. The edible part surrounding the seeds is completely macerated in the mature state and consists of strongly elongated and secondarily deformed cells. A characteristic feature of the cacao fruit are the glandular hairs on the outer surface, composed of a several-celled stalk and a pluricellular head (Fig. 292).

A transverse section through a young ovary reveals a perfect radial symmetry with 5 locules, each of which accommodates a pair of seed primordia. Inside the placentae, there appear 5 pairs of vascular bundles, corresponding to the marginal bundles of the carpels, each of which fuses – at a higher level – with a neighbouring bundle into one single bundle. At the beginning, only one rib is found over each locule, which however later transforms into 2 more conspicuous ribs, as a result of an inhibitory growth in the center of the original rib. In a more advanced stage, the fruit therefore has 10 ribs.

As in other syncarpic ovaries (e.g. *Achras zapota* or *Citrus*) the ovary wall increases first in width, whereas the locules maintain their original size. Consequently, the pericarp wall attains its final width first, while the development of the locules is retarded. The last stage of fruit development therefore corresponds to the increase in size of the locules and simultaneously to the development of the seed primordia.

Even at early stages, the ovary wall is divided into an outer and an inner part by the presence of a meristematic ring in the middle of the pericarp. Initially, the outer part increases more rapidly in width, while the inner part is delayed in its development. In agreement with the radial symmetry of the ovary, radial cell rows arise in the outer part of the pericarp wall by periclinal cell divisions; these rows contribute to the growth in thickness of the pericarp, whereas the action of growth in circumference is only slightly conspicuous at that time.

In the inner part of the pericarp, a subepidermal meristem develops beneath the inner fruit epidermis from which long cell rows originate, contributing in this way most of the inner pericap wall. This meristematic tissue arising from the first subepidermal layer completely surrounds the locules and consequently the originating cell rows radiate from the locules in all directions. Due to its position in the ovary, it can be intrpeted as a ventral meristem of the carpels. The arrangement of the radial cell rows is later disturbed by the action of growth in circumference in the periphery of the inner pericarp part, where the cells are tangentially extended and then subdivided by anticlinal walls. The cell rows in the immediate vicinity of the locules are however maintained.

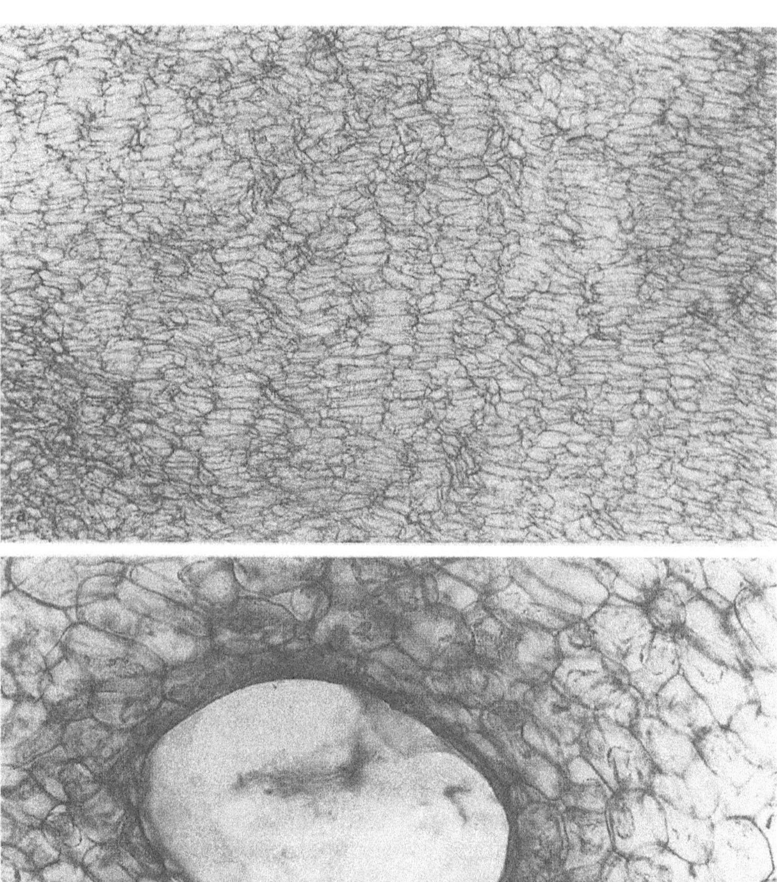

Fig. 293. *Theobroma cacao.* **a** Fruit flesh. × 6.3. **b** Slime cell in the ripe pericarp. × 16.

In the middle zone of the pericarp, there arises a girdle of very small thin-walled cells rich in protoplasm corresponding to remnants of the original embryonic ovary meristem; this 'residual meristem' is preserved for a longer time and reveals itself as a complex structure during the following differentiation processes. From the innermost layers of the meristematic ring many vascular bundles with exoscopic xylem arise; these numerous small bundles are arranged in the form of a ring, as seen in transverse section.

The larger outer part of the meristematic ring is later transformed into sclerenchyma which gives the strength to the pericarp. The thin-walled sclereid mother cells begin to elongate in different directions and to divide parallel to their long axes; finally, the walls of the daughter cells become thickened. Individual bundles of elongate sclereids differentiate in this way into the future sclerenchymatous ring, where cell groups elongate alternately parallel or perpendicularly to the long axis of the fruit. Furthermore, the sclerenchymatous bundles in the perpendicular plane, do not develop in a parallel arrangement to each another, but incline towards one another at an angle of approximately 90°. The pattern consequently follows a more or less zigzag line: one bundle is orientied parallel to the long axis of the fruit; the next one runs perpendicular to it, departing to the right; the following bundle again follows the long axis of the fruit, and the neighbouring bundle turns perpendicular to it, but to the left. The sclerenchymatous elements become interwoven in this way, imitating a plaiting of fibers. A tissue of this type gives more strength to the pericarp.

The pulp which in the mature stage surrounds the seeds like a paste, originates exclusively from the ventral meristem of the carpels. As mentioned above, the cell rows raditing from the locules are maintained in the innermost region of the pericarp

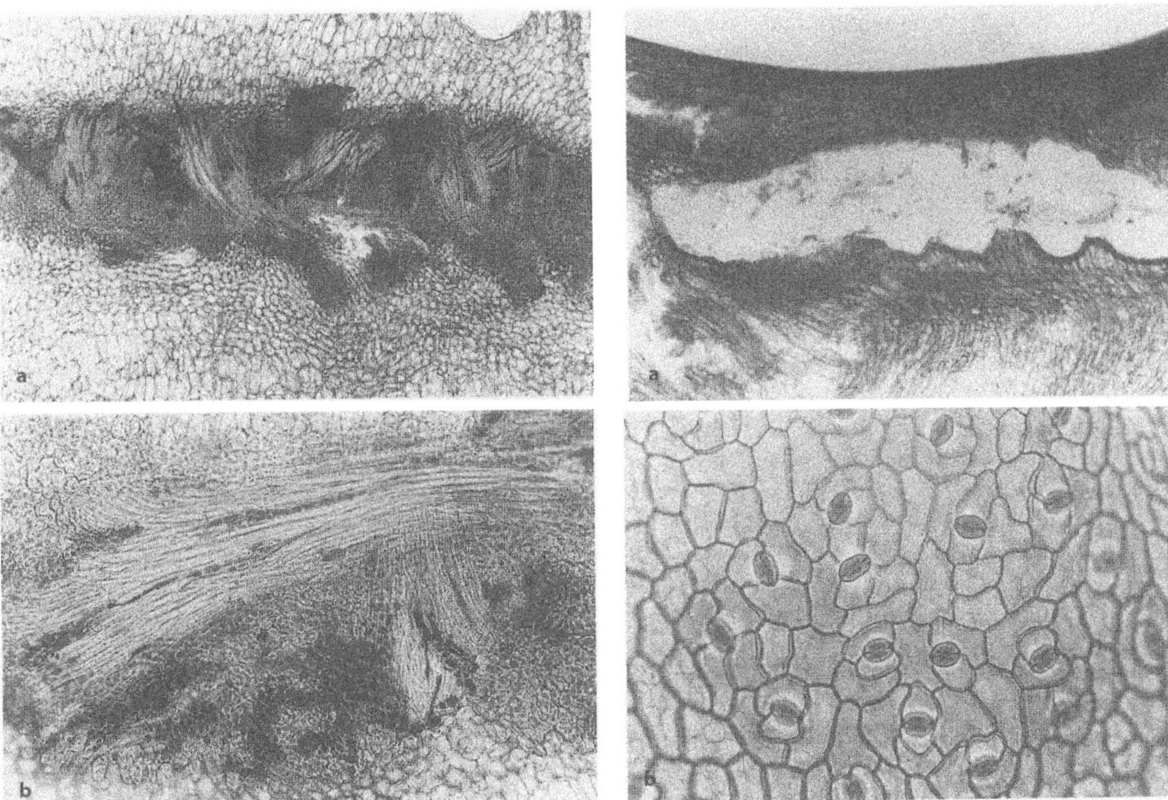

Fig. 294. *Theobroma cacao*. **a, b** Fibers of the sclerenchymatous ring of the fruit pericarp. The fiber bundles run in different directions (× 6.3.).

Fig. 295. **a** *Theobroma cacao*. Seed coat with large slime cells. **b** *Eryngium foetidum*. Lower epidermis with stomata. × 25.

around the locules; later on, the cells elongate considerably parallel to the long axis of the rows transforming themselves into filaments comparable to those of the banana pulp. This filamentous tissue subsequently undergoes a process of maceration; the cells first separate along their longitudinal walls, but may finally become completely separated from one another; they lose their turgor pressure and consequently suffer wall deformations. In the macerated pulp of the ripe fruit, elongated cells are still recognizable. The solid mature pericarp wall therefore appears thinner than at its young stages, because the inner part has completely been transformed into the creamy pulp.

A very characteritic peculiarity of the fruit are the mucilaginous cells which have already begun to develop in a young ovary of 0.8 mm in diameter. They usually appear first in the septs and from there differentition proceeds centrifugally into the outer pericarp part. In the inner region of the pericarp, the ventral meristem is still active continuously adding new cells in a centripetal direction by periclinal cell divisions in the first subepidermal layer; the outermost derivtives of this meristem are therefore the most advanced ones and differentiate first; differentiation of mucilaginous cells consequently proceeds here in a centripetal direction. The large slime cells (Fig. 293 b) are scattered over the entire mature pericarp; the slimy texture of the creamy pulp is therefore due to the presence of numerous mucilaginous cells.

The development of the cacao fruit is somewhat comparable to that of *Achras zapota* because of the radial symmetry, the formation of radial cell rows and the presence of a meristematic ring in the middle region of the pericarp which arises from a rest meristem in the cacao fruit, though in the case of the sapote it originates from a secondary cambium formation.

Seed (Fig. 295 a). Seeds collected from ripe fruits are usually still surrounded by remnants of the inner pulpy fruit flesh. The outer epidermis of the testa is comparatively large-celled with slightly thickened and cutinized outer walls and frequently with folded anticlinal walls. The mass of the testa consists of spongy parenchyma in which large mucilage cells (Fig. 295 a) are embedded; these occur mostly in the form of groups below the outer epidermis. The vascular bundles of the raphe also run through the parenchyma. Further inwards, there follows a single, often interrupted, layer of small stone cells with horseshoe-shaped wall thickenings (inner tangential and radial walls). The innermost zone consists of flattened parenchyma. The inner epidermis is small-celled. Remnants of the endosperm in the form of a small 'silver skin' surround the manifoldly folded cotyledons.

Pluricellular glandular hairs are found in the epidermis of the cotyledons. Most of the cotyledonary tissue consists of isodiametric parenchymatous cells, which abundantly contain needle-shaped crystals, starch grains and aleuron grains; some clustered crystals may also be found. The cacao starch grains are either simple irregularly spheric or elongated or composed in the form of twins or triplets; a layering is absent, but a central cavity may be present. Dispersed in the parenchyma are pigment cells which vary from violet to red-brown, according to the variety. The coloured content is soluble in water. Tannin cells which additionally contain anthocyanins and theobromine are dispersed in the seed coat.

The radicle is not fully developed yet, but rhizodermis, cortex and a central cylinder are already present. The starch-free tissue of the radicle, which contains oil drops and a dense protoplasm, is easily distinguished from the tissue of the cotyledons.

Ethnobotanical and general use

Nutritional use

Mexican Aztecs who appreciated cacao as a beverage, planted and selected it long before the Spaniards came. Brazilian Indians, used the fruit pulp for culinary purposes only. It appears that only the elite class of the Mayas and Aztecs were allowed to drink the precious beverage, mixed with maize flour, vanilla and pepper, and to enjoy its high content of stimulating alkaloids. It is reported by the Spaniards that the emperor Moctezuma drank daily 400 cups of hot xocolaltl.

Today, the pericarp is split open to remove the seeds from the pulp and to subject them to a fermentation process. The purpose of fermentation is to free the beans from the pulp and to prevent germiantion. Besides during fermentation, the bitter taste is reduced and the aroma is enhanced. The testa changes to a pale brown and the beans become plumper, due to absorption of water by the cotyledons. At the same time, the colour of the seed is altered due to disintegration of the tannins into glucose and into a brown pigment (cacao red). The fermented beans are then washed, dried and roasted. These processes make the shell brittle so that it can be removed. The cotyledonous matter known as the nib is ground to prepare the cocoa mass from which much of the fat is extracted. Cacao powder is obtained through a further grinding.

Besides the cocoa beverage and the further processed solid chocolate, the by-products can also be used. The cocoa shells are sometimes used to produce cocoa tea. They can furthermore be used as a fodder for animals, but must be fed to cattle with care because of the presence of theobromine. The shells contain about 3 % fat and 14–16 % protein. Ground shells are occasionally added to the cocoa powder as an adulterant. They are however easily recognized anatomically under the microscope, principally owing to the presence of slime cells and the sclerenchymatous layer with u-shaped thickenings.

Further adulterants, such as starch and flour, are easily recognized by the different shape of the starch grains. The following adulterations are known: wheat, rice, oats, barley, and banana starch, as well as oak and chestnut starch.

Further admixtures are possible: condiments such as cinnamon, cloves, nutmeg, macis, coriander and vanilla, even cola and coffee are added.

Jellies are prepared from the pulp of the fruit which contains theobromine and caffeine or – by fermentation – alcoholic drinks and vinegar.

Economical utilization

Maya merchants and other Indians used cacao beans as currency. The yearly tribute of lowland Indios to the Aztec ruler was paid in cacao. As mentioned above, with 100 beans one could buy a slave, with 10 beans a rabbit. Even falsification of this money occurred (as always with money, adds BRÜCHER (1989) by filling empty bean shells with earth. This cacao bean money was widely used in Central America and seems to have survived as a currency in Guatemala until the last century! Cacao seeds were the basis of the financial system in Mexico when the Spaniards arrived.

Scruple (scrupulo, skrupel) is a very old standard weight which has the value of 20–24 grains and is equivalent to about 1.296–1.198 g. It was not a very exact weight and values indicated in the bibliography therefore vary.

Although the wood is not a source of commercial timber, it is suitable for purposes requiring toughness and strength rather than attractive appearance or resistance to decay. It is also used as fire wood by the natives. BRÜCHER (1989) was impressed by the way Indians made fire, in spite of the humidity of the wood; matches were useless. But spinning of soft wood and dry fungus against the hard wood at high speed produced results.

The fruit shells are not only used as a fodder for cattle, but also as a fertilizer or fuel.

Medical use

Leaf and seed are mainly used.

Leaf. An infusion of the leaves is applied as a heart tonic, a diuretic, and to stimulate blood circulation.

Bark. A decoction of the bark is used as a wash for certain cutaneous diseases, such as scabies.

A decoction of the fruit pericarp helps against whooping cough and dysentery and favours growth.

Seed. Not only is a stimulating drink prepared from the seeds, but cocoa butter which is applied medically is also extracted. Cocoa butter is used as an excipient for ointments. Cocoa butter is also applied as an ointment to improve cicatrization. Externally it is used for raw or chapped lips, for burns on lips, for chafed skin of babies, for chaps of the nipples, abrasions and scratches of the skin, against sunburn and haemorrhoids.

Internally, cocoa butter is taken against whooping cough, bronchitis and catarrh.

Oleum cacao (cocoa butter) has a melting point only slightly below body temperature and is therefore suitable as a basic substance for suppositories and vaginal globuli.

Toasted seeds, boiled in water reached from *Manihot* pressing, relieve eczemas.

A refreshment for feverish people is prepared of the slime in the seed coat.

Method of use

Leaves in an infusion, fruit pericarp in decoction, oleum cacao externally and internally, as indicated above.

The recipe for an ointment to promote better hair growth is recommended by LOPEZ PALACIOS (1987): 4 spoonfuls of cocoa butter, 4 spoonfuls of oil of sesame or almonds, 1 scruple of cinnamon essence, 8 drops of clove essence.

With this mixture the hair is oiled.

Healing properties

Astringent (tannins), digestive, stimulating (theobromine), emollient, febrifuge, and tonic (theobromine) (for nerves and heart). Theobromine dilates the coronary arteries and alleviates angina pectoris.

Chemical contents

The seeds are very rich in fat, but also contain carbohydrates and proteins, as well as potassium, copper, phosphorus, magnesium, calcium, iron, and the vitamins A, C, thiamine, riboflavine and niacin. The fruit is very rich in mucilage. Further contents are polyphenols, tannins, cyanogenic compounds, and the purines caffeine and theobromine.

The alkaloids theobromine and caffeine have stimulant (heart and nerves), vasodilatatory as well as diuretic effects. Chocolate is constipating.

Flavonoids of *Theobroma leiocarpa* have antineoplastic activity.

Varieties and related species

About 30 species of Theobroma are native to tropical America. According to BRÜCHER (1989), *Theobroma cacao* has been divided into 2 subspecies and 3 cultivated groups: 1) Criollo or Central American, 2) Amazonian Forastero and 3) Trinitario type.

A large number of hybrids and varieties complicates a detailed survey.

There are many related species with edible fruits and seeds, such as *T. angustifolia* DC, *T. bernouilli* PITTIER, *T. bicolor* HUMB. ET BONPL., *T. grandiflorum* WILLD., *T. simiarum* DON., *T. speciosum* WILLD., *T. subincanum* MART. etc.

Theobroma mariae and *T. speciosum* contain edible fat in the seeds.

Cultivation

The plant is generally propagated by seeds; in certain regions, however, vegetative propagation by scions is adopted. Young plants have to be planted at distances of 3–4 m during the rainy season (from May to October). Cultivation is only possible in the narrow tropical zone between 15 degrees northern latitude and 18 degrees southern latitude and at an altitude below 300 m. Cocoa needs a hot climate with average temperatures of 24–28 °C; temperatures below 15 °C do great harm to the plants. For good development, the plants need between 1.500 and 2000 mm of precipitation a year. The soils should be deep, rich in nutritive substances, humid, but with a well drained superficial horizon. In the upper layer, the plants tolerate a pH of 4–7.4. To obtain a good vegetative development and a good crop later, the plants need periodic pruning.

Almost all varieties of cocoa need shade; special shade trees are therefore planted in the cultivations, e.g. *Erythrina* or *Inga*. In recent years, however, a more profitable variety not requiring shade was obtained.

The trees begin to fruit reaching the 4th year and can reach a total age of 80 years or more. The ripening of the fruits takes 5 to 8 months. Harvest is possible during the entire year, but the best production occurs from November to April. To avoid germination of seeds within the fruits, the farmers open the fruits immediately, remove the seeds and prepare them for fermentation.

Observations

The most characteristic features of the cocoa plant are the stellate hairs in the epidermis and the large mucilage cells in the leaf, the fruit pericarp and the seed coat.

The presence of adulterations must be studied under the microscope and only after a certain preliminary treatment of the material. Cocoa products are usually pulverized to the size of grains of 20–30 µm so that only fragments of tissues or cells can be recognized. In good quality cocoa, only cotyledonary tissue should be present. The examination of food containing cocoa such as ice cream, liquor or pudding is therefore difficult. Starch grains and fragments of the pigment cells as well as parts of the silver skin may be preserved.

Adulterations with cocoa shells are easily recognized by the stone cell layer having the horseshoe-shaped wall thickenings and the mucilage cells. Adulterants in the form of flour are evident owing to the different shape of the starch grains.

Tepuianthaceae

The family has been created in 1981, when MAGUIRE & STEYERMARK decided to introduce a new family name for the only genus *Tepuianthus*. The genus is endemic in the mesetas of the Roraima sandstone formation in the highlands of Venezuelan Guiana. The name of the genus and family is derived from the Indian word tepui which means the mesetas or 'houses of the gods'. These are table-shaped mountains, rising up to a height of about 3000 m. For geological reasons, endemisms of plants and animals are accumulated in this region, and even differ from Tepui to Tepui.

A certain relationship of the Tepuianthaceae with the Sapindaceae and Rutaceae is obvious (MAGUIRE & STEYERMARK 1981). For a more detailed description see ROTH & LINDORF (1990). Leaf and bark structure of *T. auyantepuiensis* were studied by ROTH & LINDORF (1990).

Tepuianthus auyantepuiensis MAGUIRE & STEYERMARK

Tepuianthus auyantepuiensis (sp. nov) is a low shrub which grows at altitudes between 2000 and 2300 m on the tableland of the Auyantepui. The species ist found at more open rocky sites and is thus exposed to strong insolation and UV radiation. The strength of the irradiation increases with height.

The leaves of this species are simple, have entire margins and are of obovate shape with a mucronate tip. The spiny apex is better developed than in other species of the genus. The leaves are small, of a very coriaceous texture and are hard to the touch. A dense silky indumentum is found on the lower leaf side, while the upper surface is glabrous. The leaves are of the microphyllous size class (lower limit) of VARESCHI (1980) and have a total surface area of about 5 cm^2. The specific leaf area is very low, because the blade is proportionally very thick.

Leaf and bark structure were studied by ROTH and LINDORF (1990).

Anatomical description

Leaf (Figs. 296, 297). The leaf blade reaches a thickness of 600–1000 µm between the veins, and more than 1000 µm in the vicinity of the midrib. The upper epidermis is multilayered, generally 2–3 layered, but may develop up to 4 layers. The outer cell walls are thick and cutinized. The wall thickness oscillates between 30 and 70 µm. The cuticle proper measures 7–20 µm in thickness. Due to repeated periclinal cell divisions, the upper multilayered epidermis forms anticlinal cell rows; these penetrate deeply into the leaf inside, occupying up to 3/4 of the entire transverse section. The outermost cells of the rows are short, while cell length increases towards the inside so that the innermost cells may reach a length of 500–600 (up to 850) µm by a width of 87–120 µm. Each cell row has the shape of a test-tube, tapering towards the inside. The walls of the innermost cells are slimy to such a degree that they almost fill the entire cell lumen. The slimy walls are under mutual pressure. They almost explode when water is added, expanding in the direction of the lowest resistance, i.e. they swell up to their maximal capacity. All epidermis cells are

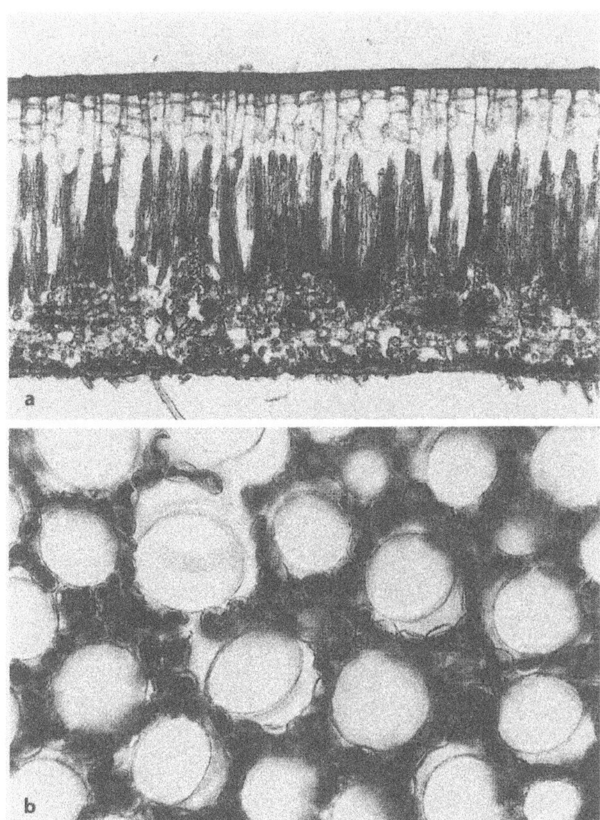

Fig. 296. *Tepuianthus auyantepuiensis*, Tepuianthaceae. **a** T.s. of leaf. **b** paradermic section at the height of the slime cells (large transparent cells) surrounded by palisade cells (dark). (ROTH & LINDORF 1990).

transparent, do not contain chloroplasts or other inclusions and are pervious to rays to a high degree, except the small outermost epidermis cells which contain a brownish or amber-coloured granular and gummy substance that probably corresponds to resins.

The photosynthetic parenchyma covers about two thirds to three quaters (and even up to 80 %) of the anticlinal walls of the epidermal cell rows, so that the anticlinally elongated palisade cells which are likewise arranged in anticlinal rows, surround the slime cells in the form of a mould. Each epidermal cell row is thus surrounded by a shell of palisade cells in such a way that neighbouring epidermal cell rows have palisade rows in common (Fig. 296). As the epidermal rows become reduced in width towards the inside of the leaf, they are joined by several palisade rows at their bottom. The palisade rows are thus arranged around the base of the epidermal rows in a funnel-form so that each epidermal row is placed in the palisade parenchyma like a Chianti-bottle in its basket. Each palisade row is composed of 4–5 (maximum 6) cells. In this case, the palisade parenchyma is not arranged in the usual form of layers parallel to the leaf surface. The total length of the palisade rows may amount to 410–540 μm. The longest (often the innermost) palisade cells reach a length of 54–168 μm. The length/width index of the palisade cells is 4–10, the cells are thus long and slim. About 9–18 palisade cells surround an epidermal slime cell (when seen in paradermal section) Fig. 296.

The spongy parenchyma with a width of 200–236 μm is small in comparison with the palisade parenchyma and the proportion of palisade parenchyma/spongy parenchyma amounts to 2.9–4.6. The walls of the spongy parenchyma cells are slightly thickened. The cells develop short arms in various directions, but the intercellular spaces remain small, so that the spongy parenchyma adopts a more or less solid consistency. The chloroplasts of the mesophyll are small. Resiniferous cells are dispersed in the spongy parenchyma.

The lower epidermis is small-celled. The cells appear small and polyedric, as seen in surface view; the anticlinal walls are straight, the cuticle is smooth. The outer cell walls reach 20–40 μm in thickness. They are completely cutinized. A thin cuticle covers the outside of the walls of the epidermis cells as well as the outside of the hairs. The stomata are variable in size. They reach a length of 34–61 μm and a width of 27–54 μm. The guard cells are at epidermis level, but the cutinized ledges project considerably above the surface. Stomata density oscillates between 75 and 127/mm^2. The lower leaf side has a dense indumentum, composed of 300–1000 μm long, simple and unicellular hairs; the walls of the hairs are thick, but not lignified. Hair density is 386–590/mm^2.

The vascular network is relatively dense (Fig. 297). The distances between the vascular bundles are 70–1300 μm. The vascular bundles are comparatively large, but show fusions and ramifications with free endings of variable number. As seen in transverse section, the oval vascular bundles have an upper and a lower sclerenchymatous cap. The larger bundles are however completely surrounded by sclerenchyma. Curiously, a layer of sclerenchyma may separate phloem and xylem from each other. This layer is composed of fibers with bordered pits. Phloem and xylem of the midrib are likewise separated by a sclerenchymatous layer. The midrib is strengthened on its lower side by collenchyma. The midrib as well as the larger lateral veins maintain connections with the slime cells of the epidermis by tracheidal cells and lignified parenchyma cells. These vascular extensions penetrate between the basal ends of the slime cell rows and may supply them directly with water. The smallest

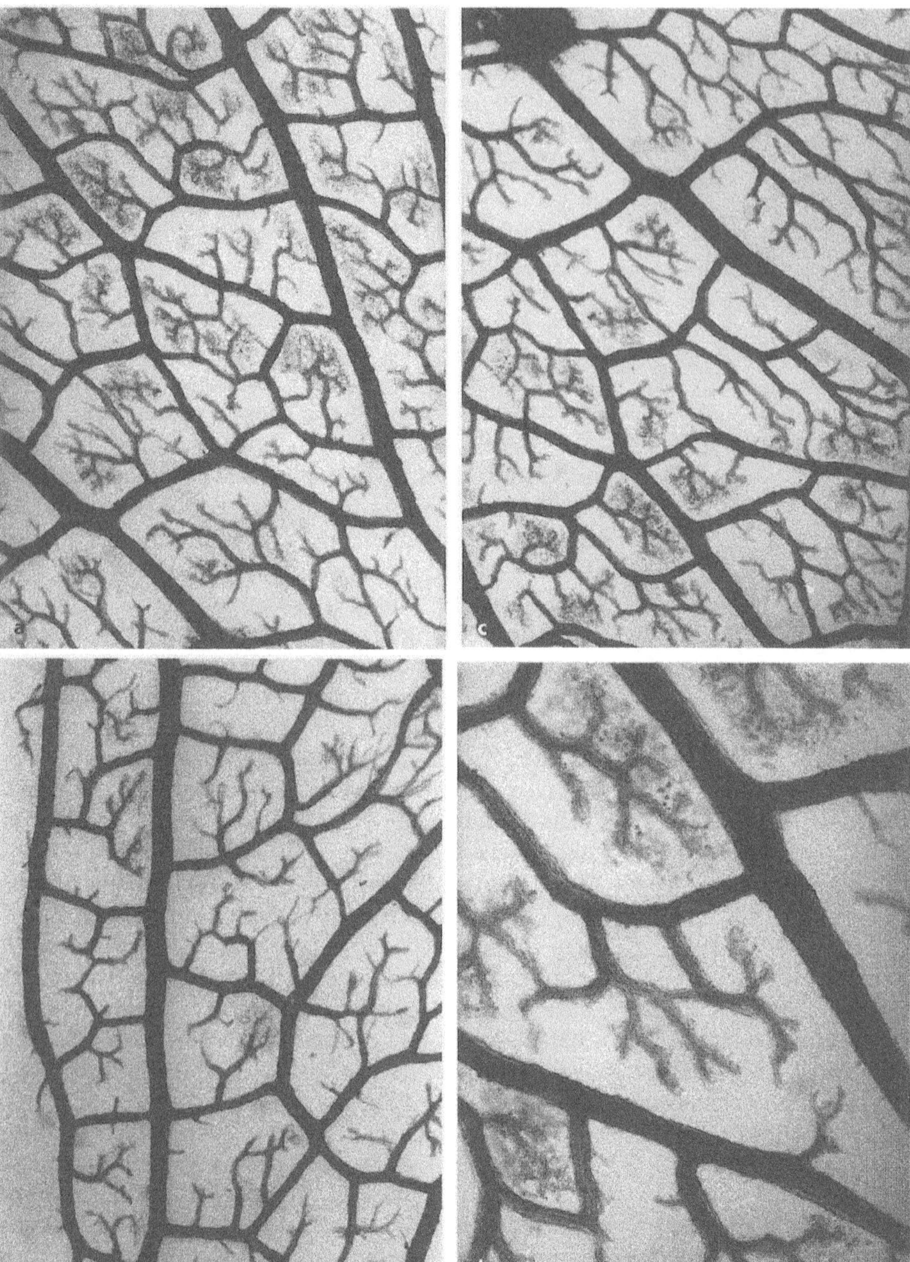

Fig. 297. *Tepuianthus auyantepuiensis.* **a–d** Leaf venation: Meshes with free endings (Roth 1995).

vascular bundles only consist of xylem (tracheids), which is likewise in contact with the upper epidermis through parenchyma or tracheidal cells. The slime cells have the capacity of absorbing large amounts of water and of storing it. Lateral veins as well as the midrib are sporadically surrounded by resiniferous cells. The vein length of the blade amounts to 51.75 cm/cm^2.

Small druses of calcium oxalate are dispersed in the palisade parenchyma. The leaf is xeromorphic and of the sun type.

Bark (Fig. 298). Although the plant is exposed to strong wind and insolation, the bark of the shrub is very small. The total bark width including the cork measures about 2 mm. The cork is proportionally broad with 0.17–0.25 mm. The inner functioning bark or the conducting phloem is likewise relatively broad measuring about 0.28–0.3 mm. A phelloderm is absent.

The bark has a very simple structure. The cork has a layered aspect, as flattened cells alternate with more or less square shaped cells in successive lay-

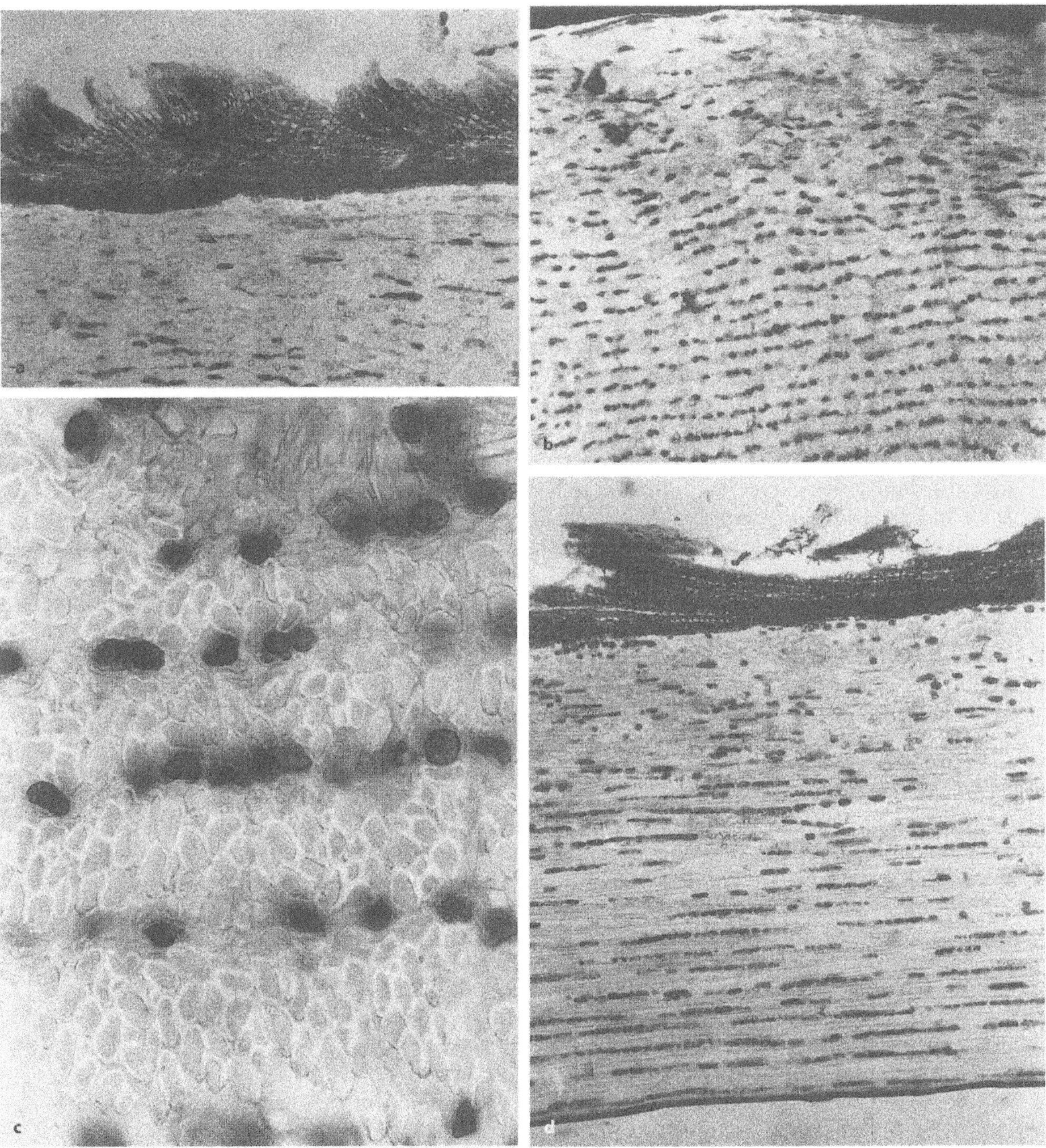

Fig. 298. *Tepuianthus auyantepuiensis.* T.s. of bark. **a** Cork. **b** Detail of inner bark with secretory cells (black). **c** Tangential rows of secretory cells. d) Longitudinal section through entire bark (ROTH 1995).

ers. This layering is particularly well developed in the lenticels. The entire bark is very soft and consists almost entirely of softbast. Fibers are only found sporadically. Chambered crystalliferous parenchyma strands with rhombic crystals may be observed here and there. Most conspicupus are brown cells, arranged in tangential rows, which may be considered secretory cells. As seen in longitudinal section, they appear longitudinally extended and form almost continuous longitudinal rows. In the outer bark, the cells seem to be somewhat tangentially extended, but dilatation growth is very moderate.

Intact sieve elements lie only in the inner bark. They become compressed and crushed towards the outside and partly adopt a grotesque shape with depressions and undulated walls. Thin-walled elements, extending in a longitudinal direction and with oblique transverse walls (as seen in a longitudinal section) occupy the major part of the bark.

They have a parenchymatous character. The rays are generally uniseriate and only rarely biseriate. The distance between 2 consecutive rays (as seen in a tangential direction) is small with about 5–7 cells.

In its entire structure, the bark is very simple and poorly specialized. The brown content in the secretory cells is probably just as resinous as in the leaves or is of a tannic origin. The bitter taste of the bark could be due to the content of the secretory cells. The bark easily exfoliates in the form of longitudinal stripes.

Observations

Tepuianthus is a new genus of a plant family unknown up to the present. MAGUIRE & STEYERMARK (1981) gave the family the name 'Tepuianthaceae'. 'Tepui' is an Indian word for 'meseta', a mountain of tabular shape, which the Indians considered as 'houses of the Gods'. The Tepuianthanceae comprise only a single genus, *Tepuianthus*. It is represented in Venezuela by 6 taxa. MAGUIRE & STEYERMARK (1981) observe a relationship with the Sapindaceae or Rutaceae and localize the family in the Sapindales. About 100 mesetas, 1500–3000 m high, occur in an almost inaccessible area of 1/2 million sqaure kilometers, 65 of which never have been entered by human beings, as far as we know. One of the best known mesetas is the Auyantepui. Endemisms have developed abundantly on the mesetas; about 50–70 % of the plants are endemic. This is due mainly to the characteristic geographical and geological structure of the region. Even from Tepui to Tepui, the plants are distinguished from one another. During each expedition, dozens of new plant and animal species are discovered.

Tepuianthus is a very promising plant because its chemical compounds have not yet been studied. But it contains a large quantity of slime, tannins, resins and a bitter principle.

The multilayered slime epidermis as well as the characteristic venation pattern of the blade facilitate identification.

Theophrastaceae

Is a family of shrubs and small trees which mainly occur in Central America, the North of South America and the West Indies.

Some members supply soap and fish poisons.

Kaempferol and coumaric acid have been reported.

Some species of *Jacquinia* contain saponins and alkaloids.

Jacquinia

Four Venezuelan species of *Jacquinia* have been studied thoroughly by a former studend of the senior author: NORMA CARRASQUEL (1969). She studied the species *aristata, barbasco, loefflingii* and *revoluta* concerning their distribution in Venezuela, their taxonomy, morphology and anatomy (axis, leaf, fruit), palynology and germination capacity (Fig. 299).

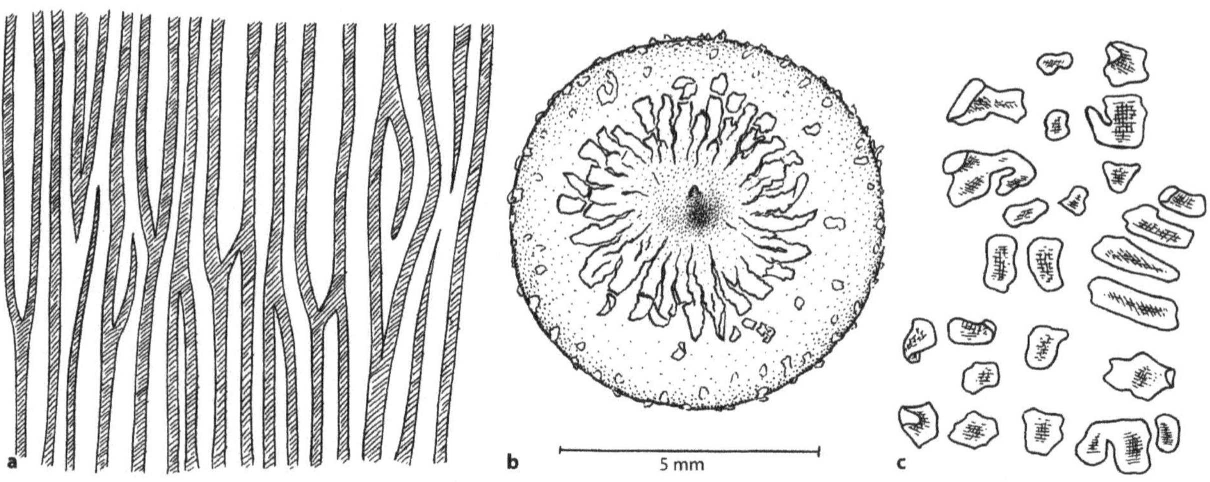

Fig. 299. *Jacquinia aristata*, Theophrastaceae. **a** Surface view of mature fruit with ribs and depressins (Shaded). **b** *J. loeflingii*. Fruit seen from above with a wreath of scales around the style. **c** *J. loeflingii*. Scales desquamated from mature fruit (ROTH 1977).

Two species, *J. barbasco*, and *J. revoluta*, are used as fish poisons; the fruit makes the fish 'sleepy' so that they can easily be catched.

Jacquinia caracasana H.B.K. The taxonomic description and the description of the leaf anatomy are found in ROTH (1992), the multilayered epidermis of Jacquinia is described in ROTH (1977) and in ROTH & CARRASQUEL (1969).

Tiliaceae

Apeiba

The wood is useful. The fruits resemble sea urchins. A 'sniff' of the fruit is good for asthma.

A. tibourbou AUBL. The apeiba oil from seeds is used for rheumatic ailments. ROTH 1984, has studied leaf structure, fruit structure and dispersal 1987, leaf venation 1996.

Luehea

L. has a useful wood. Leaf structure has been studied by ROTH 1984.s

Umbelliferae (Apiaceae)

The Umbelliferae comprise mainly herbaceous plants with a taproot or a rhizome. They usually grow in more temperate regions, occurring from the coast line up to the high mountains. Their inflorescences are umbelliform (hence the name of the family). The inferior bipartite ovary develops a central carpophore and 2 mericarps, each usually with 5 main ribs and alternating valleys; the oil cavities often alternate with the ribs, lying in the valleys or below the secondary ribs.

Although the characteristics of the family are very homogenous and uniform, remaining more or less unchanged, a great richness of species has developed, due to slight alterations of minor characteristics, particularly in the fruit domain. The fruit characteristics therefore are the main base of the natural system of this family.

Volatile oils are frequent in roots, foliage and fruits. Many species supply condiments or medical drugs.

Coumarins occur mainly in roots, but also in fruits. Carminativa, spasmolytica, stomachica, secretolytica, expectorantia, and diuretica are found within the officinal herbs.

Well-known species such as *Coriandrum sativum, Carum carvi, Foeniculum vulgare, Pimpinella anisum, Petroselinum crispum, Levisticum officinale,* and *Anthriscus cerefolium* which supply condiments and/or vegetables belong to the Umbelliferae.

Arracacia esculenta DC, synonym: A. xanthorrhiza BANCR. (apio, apio criollo, arrakacha, arracacia, arracacha, arracate)

Taxonomical description

The species is a perennial herb with thick fleshy and edible roots which are of golden yellow or red colour. The leaves are pinnatisect. The small flowers are yellow. The petals are broad, acuminate and curved upwards. The fruit is ovate-oblong, more or less attenuate at the apex and laterally compressed.

Origin

Probably Peru. Andean regions of Peru and Ecuador.

Historical background

Archaeological finds in Peruvian tombs (BRÜCHER 1989) testify to its early use as a root vegetable in Pre-Columbian times. Its cultivation is mentioned in 1786 (ALTOLAGUIRRE). In 1659, 4 pounds of the tuber were sold for one 'real' (VELEZ & VELEZ 1990).

Occurrence

It is mainly cultivted in the Andean valleys of the Cloud forest regions in Colombia and Venezuela, as well as in the Cordillera de la Costa. Outside of Venezuela it is cultivated in the West-Indies, Central America, Africa, Sri Lanka, and on an industrialized range in the south of Brazil.

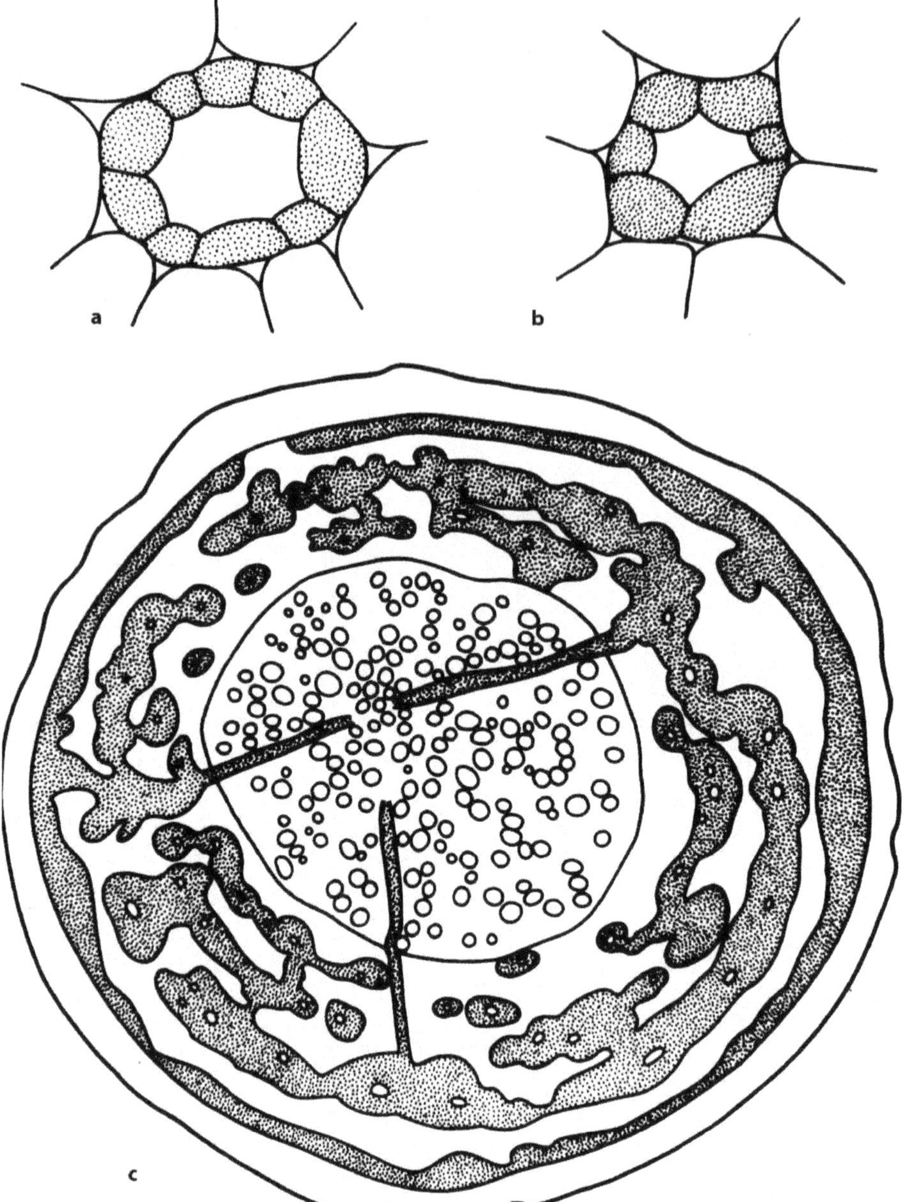

Fig. 300. *Arracacia esculenta.* **a, b.** Secretory canals with secretory cells (dotted) of the tuber. **c** Cross section of the tuber (edible root). Note the numerous vessels in the xylem (circles), the pluriseriate rays and the layered amyliferous parenchyma in the phloem (dotted) with secretory canals (ROTH 1996).

Anatomical description

Anatomical details can be found in Fig. 300–303. *Root.* The normal roots are filamentous and show a few vessels in the center, when they are young. Coalescence or fasciation between several roots may be observed. In their secondary state, the roots have the usual aspect. The phloem consists of sieve tubes with their companion cells and of parenchyma, but sclerenchyma is absent. Secretory canals are dispersed in the amyliferous parenchyma. The protective tissue towards the outside is brought about by some cork layers.

The tubers, however, may proceed from normal filamentous roots through growth in thickness. The development was observed by ROTH (1977 b AND 1995). A filamentous root of 1/2 mm in diameter is already in the state of secondary growth and consists of much xylem and little phloem. In the next stage, the cork gains in numbers of layers. Secretory canals appear here and there in the cork (Fig. 302). Starch grains are still rare. In a root of 2 mm in diameter the medullary rays begin to broaden and the phloem increases in thickness by the formation of starchy layers; secretory canals also begin to form in these layers. In a root of 3 mm in diameter, the xylem remained more or less in its original state, while the phloem very much increased in diameter and additionally became stratified in amyliferous

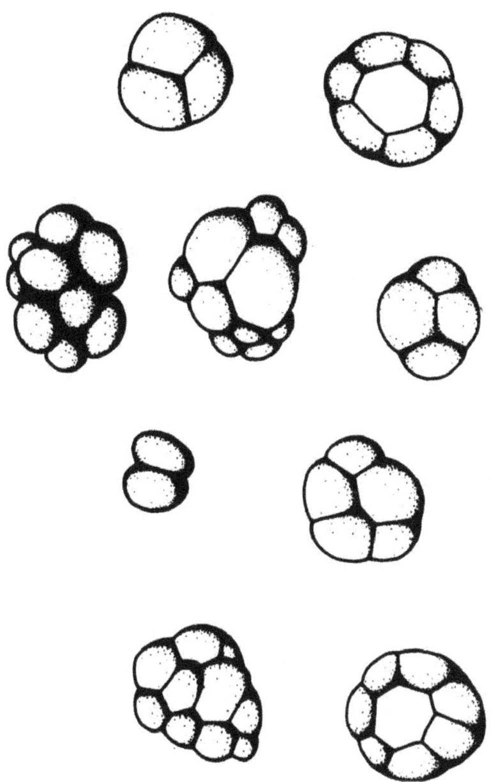
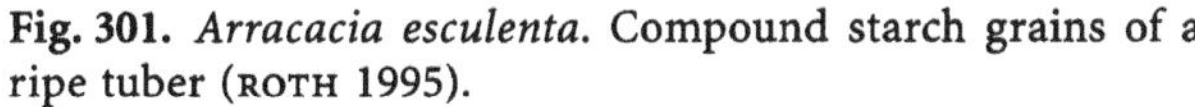

Fig. 301. *Arracacia esculenta.* Compound starch grains of a ripe tuber (ROTH 1995).

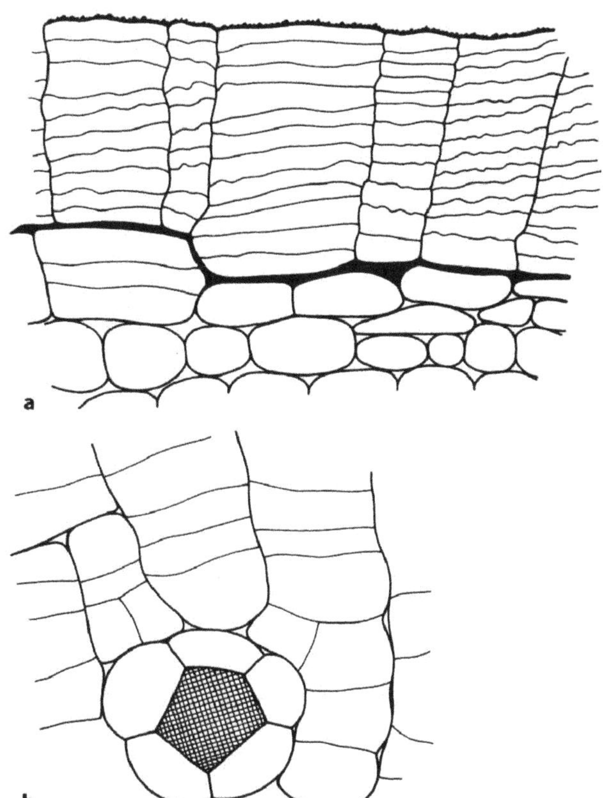

Fig. 302. *Arracacia esculenta.* **a** Storied cork of the tuber in t.s. **b** Secretory canal in the cork (ROTH 1995).

layers full of starch and in layers of normal phloem (Fig. 303). Secretory canals are now abundant in the amyliferous layers. Furthermore, the medullary rays increased in a tangential direction and are now pluriseriate. The medullary rays, also rich in starch, now form a continuous network of starchy tissue together with the amyliferous layers, as seen in transverse section. Even the starch grains gained in size and now appear mostly in the form of compound groups (Fig. 301). In young stages, the secretory canals consist of 3–4 cells in the center of which the intercellular space arises; in the adult stage, on the contrary, the secretory canal is surrounded by about 8–10 secretory cells (Fig. 300). Frequently, the radial symmetry of the tuber is disturbed by a more or less unilateral growth so that only one side is favoured in its decelopment. In a tuber of 4 mm in diameter, the amyliferous layers have augmented in number and the rays have increased in tangential width, while their cells increased in size towards the outside. The root now contains less oil drops. In the following stages, the number of rays augments towards the outside and the cork begins to desquamate. In a root of 2.5 cm in diameter, the major part of the xylem is already parenchymatous either in the form of axial xylem parenchyma or in the form of pluriseriate rays. All parenchymatous cells are filled with starch. The grains are partly very large, compound and of different types. Oil drops also occur in the parenchymatous cells. The stratification of the phloem has almost disappeared, and the entire phloem now seems to consist of amyliferous parenchyma. This is due to the compression and obliteration of the layers with sieve tubes and companion cells which are not functional any more. In a still more advanced stage of a tuber of 3 cm in diameter, many intercellular spaces developed in the root center so that the core of the root adopts a white aspect (through total reflection of light) and assumes a mealy texture (through the formation of intercellular spaces).

In the xylem, the outermost layers are the richest in starch, while in the phloem the innermost layers are those which contain most starch grains. All parenchymatous cells maintained their delicate walls, in the xylem as well as in the phloem. The cork is arranged in the form of 'stories', as individual cells divide periclinally several times, a type of cork formation more characteristic of monocotyle-

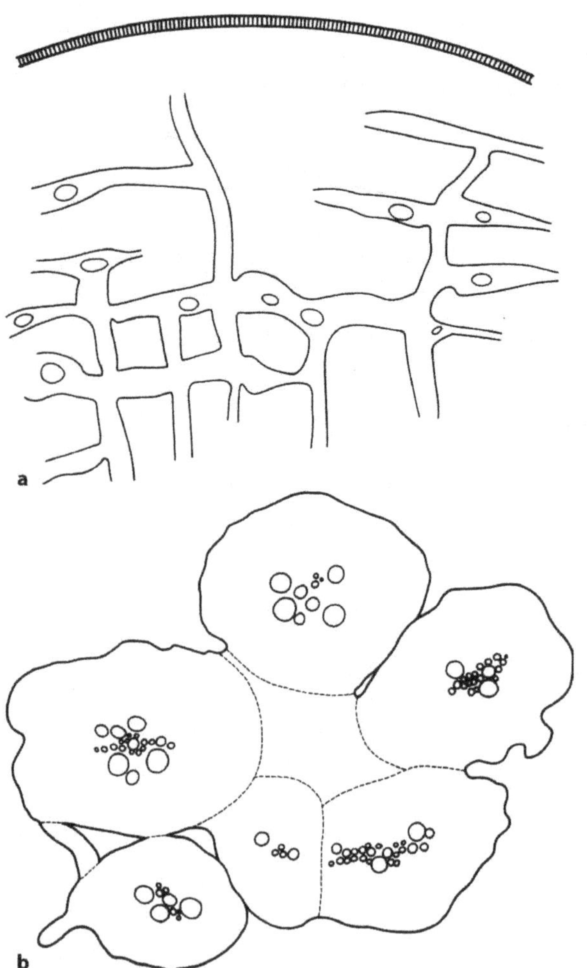

Fig. 303. *Arracacia esculenta.* **a** Part of the phloem of the tuber as seen in t.s., with layers of amyliferous parenchyma. **b** 6 roots coalesced to form a 'fasciation'. (ROTH 1995).

dons in general. Therefore no phellogen develops. The cork formation is somewhat irregular, perhaps because the plant has been in cultivation for hundreds, if not thousands of years.

The most important feature of the tuber is thus the pronounced growth in thickness which continues due to the activity of the original cambium. Abnormal growth in thickness does not occur, except for the abundance of parenchyma. The tissue which is produced most and which prevails, is the amyliferous parenchyma. It supplies the bulk of the tissue and makes the tuber edible, as the cell walls remain thin. A further feature of the tuber is the formation of secretory canals in the cork. The content of the canals certainly makes the tuber more resistant to fungi and microorganisms and also adds special flavour to the vegetable (Fig. 302).

A comparison between the normal filamentous roots and the tubers shows the following differences:

In the tuber, the growth in thickness is much favoured; dilatation growth and tangetial extention of cells is a consequence.

The xylem is principally composed of parenchyma and has few vessels; likewise the functional phloem is reduced in its structure. The phloem parenchyma which is normally differentiated in regular phloem parenchyma and in amyliferous parenchyma, becomes completely amyliferous. Starch becomes abundant.

The secretory canals become less frequent and are possibly compressed in the periphery of the tuber.

The tuber thus has much more parenchyma which is filled with starch grains so that it becomes fleshy and soft, on the one hand; and more nutritious, on the other, while the very pungent taste is diminished because of the reduction of the secretory canals.

Ethnobotanical and general use

The roots which are mainly used for food reach 8–20 cm in length and 6–10 cm in thickness.

Nutritional use

The tubers, which are eaten, correspond to modified roots and serve as a food reservoir for the plant; the parenchyma functions as a storage tissue, mainly of starch (about 10–25 %).

The tubers have an aromatic taste, due to the content of volatile oils in the secretory canals (e.g. carvone).

They are eaten as a vegetable like potatoes, being used in stews and soups, boiled or fried, mashed as purée and even as a sweet cake.

The plant is much used in the Venezuelan cuisine and recommended as palatable, tasty, wholesome, aromatic and abundant.

The famous botanist DECANDOLLE considered the plant as a genuine enrichment of French cuisine! Besides, roots and leaves are a food for cattle and other animals.

Medical use

The leaves of the plant have antiinflammatory properties and are often used in the treatment of dental abscesses. The root has a slightly laxative effect.

Chemical contents

Volatile oils (carvone). The tuber is rich in carbohydrates (starch 25 %), calcium, phosphorus, niacin, vit. C and A., potash and sodium.

Varieties and related species

There is a golden yellow variety and a white one with some purple as well as a purple variety. They slightly differ in their chemical contents (VELEZ & VELEZ 1990).

Cultivation

The plant needs well drained sandy soils and temperatures between 15–20 °C; it does not need much watering. The cuttings are planted in files at distances of 80 cm. The right time of planting is between August and October. The plant grows better at altitudes above 1000 m a.s.l. The harvest has to be made before the flowers appear, i.e. after about 10–11 months. The plant needs short-day conditions. It is cultivated between 1500 and 2500 m in the Andes.

Observations

The tuber has the disadvantage of having a high enzymatic activity in its respiratory system (twice that of other edible roots) and therefore has a short storage life. BRÜCHER (1989) consequently suggests the creation of biotypes which are physiologically indifferent to day-light and have a reduced enzymatic activity.

From the economic point of view, the leaves are too large in comparison with the edible roots. This could also be improved by breeding.

Eryngium foetidum L. (culantro, culantro de monte, culantro de burro, cilantro)

Taxonomical description

Taxonomical details can be found in Fig. 304. Herbaceous rosette plant, 20–80 cm high, shoots with elongate internodes only supporting the inflorescences. Rosette leaves glabrous, papyraceous, lanceolate, oblanceolate to obovate, up to 25 cm long and up to 4 cm broad with an obtuse tip and a decurrent base which narrows into a kind of petiole; the blade has strongly serrate margins. Venation pinnate, semi-craspedodromus, midrib prominent on the abaxial side and almost flat on the adaxial side, blade with 11–13 pairs of secondary nerves.

Lower bracts similar to the rosette leaves or with a 3–5 partite tip and with deeper teeth, divisions acuminate and spiny. The upper bracts incised to lacerate, 3–5 parted. Flowers in cylindric multiflorous umbels, 5–20 mm long and 2–5 mm in diameter. The 5–6 involucral bracts are foliaceous, free and lanceolate to linear. The minute flowers have lanceolate or ovate sepals, 0.5–1 mm long and with

Fig. 304. *Eryngium foetidum*. **a** Complete plant with leaf rosette. **b** Habitus in nature.

acuminate tips. The petals are elliptic-oblong, 0.5–0.8 mm long with the tip almost as long as the limb. The filaments of the stamens are longer than the sepals. The subglobose fruit is 0.1–0.2 mm long, and densely covered with vesicular scales; each schizocarpium has 5 ribs and 5 oil cavities.

Origin

The species is indigenous to tropical America. 'Culantro' was already mentioned in 1578 by PIMENTEL in his 'Relación'.

Occurrence

In Venezuela, the plant is amply distributed in the savanna and the tropopophilous forests as well as in cooler areas and in disturbed places.

Anatomical description

Leaf (Figs. 305, 295 b). The leaf is bifacial and amphistomatic. The upper and the lower epidermis (Fig. 295 b) are single-layered and have a delicate cuticle which is slightly ribbed. Occasionally, certain cells are enlarged and papillose. As seen in surface view, the anticlinal walls are slightly wavy, but more so in the lower epidermis. The very small stomata are more numerous in the lower epidermis and are found at the same level as the epidermis cells (Fig. 295 b). The subsidiary cells are variously oriented, and a tendency of the stomata to become arranged in the form of rows is obvious. The palisade parenchyma consists of 1–2 (3) layers of short cells which are densely filled with small chloroplasts. The spongy parenchyma is more or less compact being composed of more or less spheric or roundish cells with very short arms; about 4–6 layers may be observed. Druses are found in some cells of the mesophyll.

The midrib shows a slight protuberance on the ventral side which is reinforced by collenchyma. Collenchyma is also found below the lower epidermis on the opposite side. Collenchymatous strands interrupting the palisade parenchyma occur additionally on the upper and lower side of the midrib. The collateral vascular bundles are arranged in the form of an open arc. A single vascular bundle with introrse xylem lies on the upper side. The secretory canals are mainly associated with the vascular bundles. A rhexigeneous intercellular space (lacuna) occurs in the center of the midrib. The filling tissue of the midrib is parenchymatous.

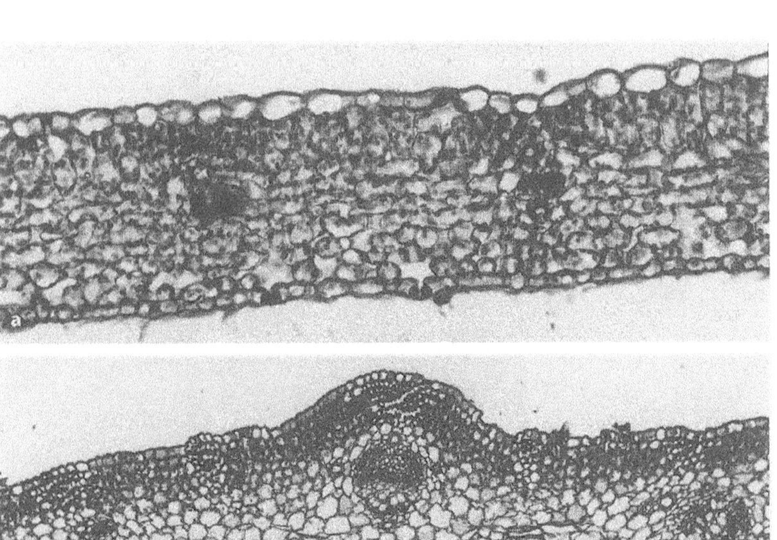

Fig. 305. *Eryngium foetidum*. **a** T.s. of leaf blade. × 16. **b** T.s. of midrib. × 6.5.

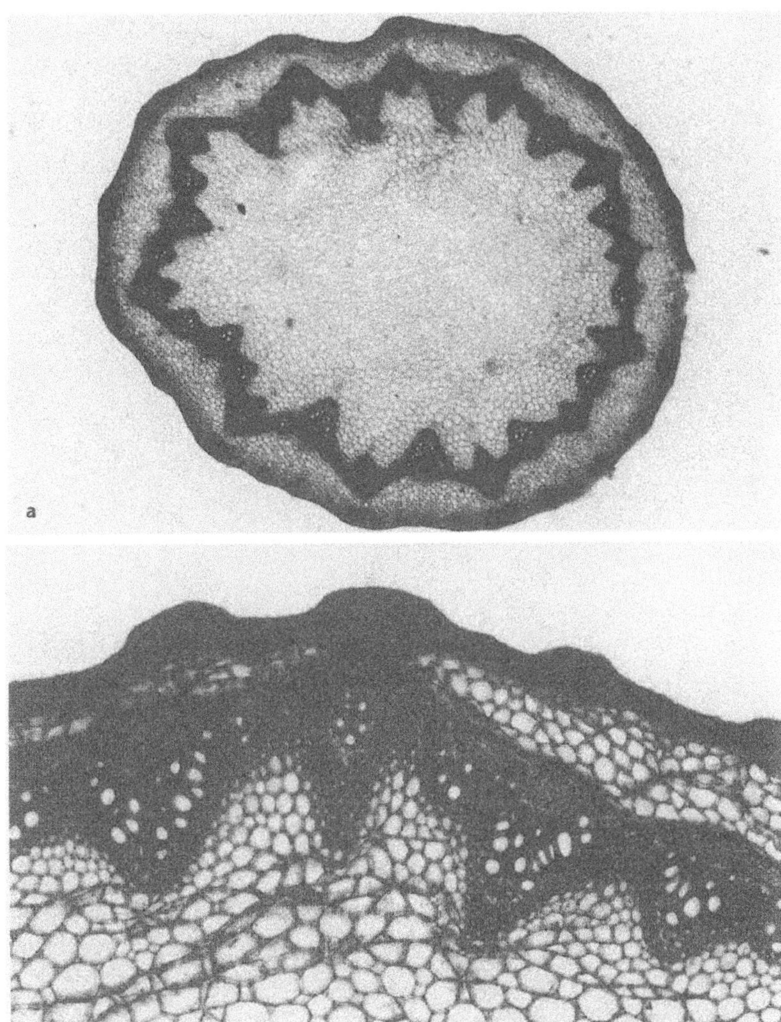

Fig. 306. *Eryngium foetidum.* Axis. **a** Entire t.s. × 6.3. **b** Detail of peripheral part with vascular cylinder.

The secondary veins are surrounded by a somewhat irregular parenchymatous sheath, possibly reinforced by fibers, particularly towards the upper leaf side. The secretory canals are generally associated with the vascular bundles and are at times reinforced by fibers. Collenchyma and possibly also fibers reinforce the leaf margins.

Shoot (Figs. 306, 307 d). An axis of 3–4 mm in diameter is slightly angular; the angles are reinforced by an angular collenchyma. The epidermis is small-celled and the cuticle sculptured. The collenchymatous strands are separated from one another by photosynthetic tissue. The larger part of the cortex consists of colour-less roundish parenchymatous cells; dispersed within the parenchyma are secretory canals. As seen in transverse section, the vascular cylinder has zig-zag outlines where the highest points coincide with the collenchymatous angles.

The vascular cylinder is composed of collateral bundles which are connected with one another by sclerosed and lignified parenchyma, mainly on the xylem side. Only a fascicular cambium is present. The ample pith is composed of more or less globular parenchyma cells; secretory canals are interspersed within the parenchyma, while in the center of the pith a large rhexigeneous lacuna is usually observed.

Root (Fig. 307 a + b). The root is surrounded by a relatively thin layer of compressed cork cells. The parenchymatous cells of the cortex are mainly extended tangentially and partly divided anticlinally. The root studied was of a diamter of about 5 mm and was in the secondary state of growth. The xylem is polyarch, but the rays are very short in comparison with the very long radial phloem sectors which narrow down towards the outside. The pluriseriate medullary rays enlarge towards the outside,

Fig. 307. *Eryngium foetidum*. **a, b** T.s. of root. **d** Detail of vascular bundle of axis. × 16. **c** *Lantana trifolia*. Twig with inflorescences. **a** × 2.5; **b** × 6.3.

mainly owing to cell enlargement. The radial arrangement of the phloem and ray cells is very evident. The cambial zone is relatively broad. Secretory canals are present in the cortex and in the phloem, and some of them occur in the rays and in the parenchymatous areas of the xylem.

Ethnobotanical and general use

Nutritional use

The leaves, the roots and the entire plant are used as a condiment in soups and meat dishes or in other cooked dishes. Although the plant has an offensive smell, as a condiment it imparts a very agreeable flavour to food. Odour and flavour are similar to those of *Coriandrum sativum*, which emanate from the content of the secretory canals. However, the spice must not be used in excess.

Medicinal use

In popular medicine, roots, shoots and leaves are used.

An infusion of the root is used as a stimulant and a febrifuge. In Trinidad and Tobago, a leaf decoction is taken for colds and in cases of pneumonia.

Method of use

The root is used as an infusion (with or without leaves and shoot) or as a decoction. The leaves are used as an infusion (also in combination with the root) or as a decoction. For oral consumption, honey or wine can be added.

Roots, possibly in combination with leaves and shoots are also applied as cataplasms.

Healing properties

Taken as a medicine, the plant has diuretic, diaphoretic, carminative, stimulant and antipyretic properties. It is also taken as an emmenagogue, against cough, as a laxative and to stop haemorrhage of the uterus, to relieve rheumatism, oedema, nervous diseases, or diseases of the liver. The medicine is indicated against stomach ache, against calcifications of the bladder, against small pox and gonorrhoea. In some regions of Venezuela the plant is used as an aphrodisiac and as an abortive.

According to WONG (1976) leaf teas are used for fever, flu, diabetes; leaf baths for cough and heat; urinary infusion baths for fever, flu, pneumonia. Root decoctions are applied for fever, cold, cough, pneumonia, malarial fever and constipation. An infusion of the root in rum and vermouth helps against worms and bles.

Chemical contents

The odour and flavour of the species is certainly due to the etheric oils in the secretory cavities of leaves, shoots, roots and fruits. The toxic effect may be due to toxic acetylene-compounds or to toxic alcaloids, such as coniin. The plant is toxic to live stock and even to man, when used in excess or for a prolonged time. The species also contains saponins and sesquiterpenes (essence of *Eryngium*).

Related species

The Umbelliferae supply a large variety of spices and medicinal plants, some of which are used as expectorants and carminatives. Others are diuretic or have dilating effects on the coronary vessels of the heart.

Cultivation

The plant is easy to cultivate and often appears spontaneously.

Observations

The plant is easily recognized by its characteristic habitus as well as by the anatomy of the foliar midrib and of the axis.

Urticaceae

The family comprises mostly herbs and more seldom woody plants. The leaves have stipules. The small flowers occur in inflorescences. Long fibers, cystoliths and slime canals are characteristic. The family is divided into species with or without stingy hairs.

Alkaloids, phenols, acetylcholine and hydroxytryptamine were reported.

Urera caracasana (JACQ.) GRISEB. (Guaritoto) is a shrub or small tree, 5–8 m high. The ovalate leaves have an acuminate tip and a more or less cordate base; the margins are dentate. The fruit is red or orange. The species is very frequent in Venezuela, occurring in the cloud forest at altitudes between 900 and 1000 m.

The leaf anatomy has been studied by ROTH (1990). The upper epidermis is composed of very large cells which serve as a water reservoir. The lower epidermis has ocelli and hairs. The leaf blade is very delicate.

Ethnobotanical and general use

The plant was formerly used as a remedy for syphilis.

The root is considered to have antihaemorrhagic properties.

An infusion of the leaves is said to alleviate erysipelas.

The natives beat the stinging leaves and inflorescences on aching joints as a counterirritant.

Verbenaceae

This family is intimately related to the Labiatae or Lamiaceae. The mostly simple leaves are generally opposite. The inflorescences are cymose or racemose. The gynoecium is composed of 2 carpels. The fruits are drupes or divide septicidally into mericarps (nutlets); capsules are rare. The Verbenaceae occur mainly in tropical and subtropical regions.

Pollination occurs by Colibris. The inforescences are cymose in the Avicennioideae. The fruit opens with 2 valves. Vivipary occurs: the embryo already germinates on the tree (*Rhizophora mangle*).

Tectona grandis L. is the economically most important species of the family supplying the famous teak timber which is very resistant, although comparatively lightweight. Its resistance to animal predation and fungal diseases is due to the content of silicic acid as well as to the presence of anthraquinone derivatives, such as tectoquinone, and to naphthoquinones. Ethereal oils are also found in this family.

Besides valuable timber, the family supplies edible fruits, oils (ethereal oils), gums, tannins, flavonoids, quinones, sterols, triterpenes, saponins, iridoid glycosides and pseudoindican derivatives.

Lippia triphylla contains the real Verbena oil, *Vitex agnuscastus* agnusid.

Avicennia germinans (L.) STEARN, synonym: _A. nitida_ JACQ. (mangle, mangle negro, mangle rosado, mangle amarillo, mangle prieto)

Taxonomical description

Shrub or tree about 5–15 m high with tetragonal young branches. Evergreen leaves opposite, coriaceous, oblong, lanceolate, elliptic or obovate, 5–15 cm long, base cuneate, margins entire, petiole 2–25 mm long; upper blade side glabrous, lower side pubescent.

The inflorescences are axillary or terminal spikes, 1.5–6.5 cm long and 1–1.5 cm in diameter, with 1–15 pairs of flowers; flowers opposite. Bracteoles ovate or oblong, sessile, and densely pubescent. The lobules of the calyx are ovate, 3–5 mm long and 2–3 mm broad and densely pubescent at the outside. The actinomorphic crown is 4-parted, white or yellow and 12–20 mm long; the tube is as long or shorter than the calyx. The corollary lobes are oblong or almost square-shaped, 2–2.4 mm long, and pubescent on the outside. Stamens 4.

Fruit pyriform or oviform and asymmetric, 1.2–2 cm long and 7–12 mm in diameter, gray-pubescent and with a single seed.

The species grows along the northern coast line of Venezuela in brackish swamps or river banks.

Anatomical description

Leaf (Fig. 308, 96 b and ROTH 1992). Although the upper epidermis cells are small, their outer walls are heavily thickened and cutinized. Glandular hairs sunk in depressions occur very frequently in the upper epidermis so that glandular density is generally higher on the upper than on the lower leaf side. The hypodermis consists of large cells which are waterstoring and comprises about 4 layers. The cells of this tissue are almost devoid of chloroplasts. The palisade parenchyma is composed of 2–4 layers. The palisade cells of the uppermost layer are broader than the underlying palisade cells, which are however shorter. The chloroplasts of the palisade cells are of medium size. The cells of the uppermost layer contain fewer starch grains as a rule than those of the other layers. Oil drops form transverse rows in the distinct palisade layers; usually only one oil drop is found per cell. The spongy parenchyma is looser than the palisade parenchyma and has small intercellular spaces; each cell contains an oil drop.

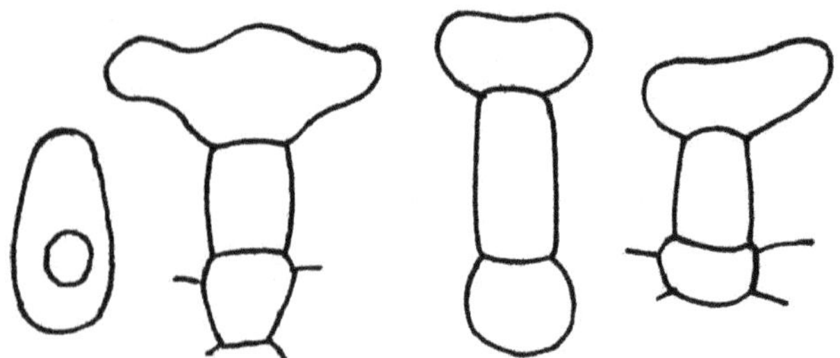

Fig. 308. *Avicennia germinans.* 'Hammer hairs' of leaf. Left: Head cell as seen from above and stalk cell.

The vascular bundles are very frequent; the vein meshes are small and the endings frequently ramify within the meshes; the terminations correspond to short enlarged tracheids with thickened walls and many pits; they are very characteristic and seem to serve as water reservoirs. The ramifications within the meshes are of the dendroid type. The last mesophyll layer above the lower epidermis may be considered a lower hypodermis, being distinguished from the spongy parenchyma by the cell shape. The very characteristic hammer hairs (Fig. 96 b, 308) occur abundantly on the lower leaf side together with stomata and glandular hairs. The hammer hairs are composed of a basal cell, a usually unicellular stalk and an asymmetric head cell.

The glandular hairs have a 4-celled head, a short intermediate cell below and a pluricellular base. They occur at epidermis level in the lower epidermis, while they are sunk in depressions of the upper epidermis. The 4 head cells are the secretory cells; they are covered with a thick cuticle with pores at the top of the gland. The intermediate cell is incorrectly called a basal cell and the basal cells (2–4) are considered as collecting cells. A detailed description of the glands and their aspect as seen in the elctron microscope is found in ROTH 1992. The glands of the upper as well as of the lower side usually lie directly above a vascular bundle. Special mesophyll cells connect the collecting cells with the corresponding vascular bundle by numerous plasmodesmata.

The secretory glands of *A. germinans* coincide in their structure with those of *A. marina*. In both species, the glands serve for salt excretion.

Ethnobotanical and general use

Economical utilization

The astringent bark is used for tanning.

The wood is locally used for fuel. Experiments have demonstrated that it can be pulped successfully by the soda process, but because of the shortness of fiber it must be mixed with coniferous pulp.

Medical use

A decoction of the bark is applied as a gargle to combat pain of the throat and to remove mouth ulcers.

Observations

The hammer hairs and the glandular hairs of the leaves are most characteristic.

Citharexylum

C., with about 115 species in tropical and subtropical America, reaching from the USA through Central America down to Argentina is a small tree or shrub. The flowers are small and often fragrant. The fruit is a drupe with juicy flesh which is in some species edible.

the timber is well known for its strength and toughness. It is called 'fiddlewood' because stringed instruments such as guitars are made of it.

The leaves are reported to be febrifugal when used in the form of a tea (*C. poeppigii*).

Citharexylum subthyrsoideum PITTIER is called in Venezuela 'palo guitarro' because its wood is used for the fabrication of guitars and stringed instruments of other kind.

The leaves of *C. spinosum* L. in decoction are used in Venezuela to cure colds.

The bark of *C. subserratum* contains beta-sitosterol, glucosides, ursolic acid and its acetate along with n-octacosanol; the leaves include several cinnamic acid derivatives, beta-sitosterol and its glucoside, as well as ursolic acid.

Citharexylum macrphyllum POIR. is a tree up to 20 m high. Its leaf anatomy has been described by ROTH (1984) and its bark structure 1981. It is possibly another promising plant.

Lantana camara L. which is a small shrub, native to the West Indies, likewise has medicinal properties, being used for flu, fever, and diseases of the respiratory organs (aerial parts) as well as for diseases of the skin (leaf). The aromatic leaves contain the toxic principle lantadeno which produces hypotension and hypothermia; they have however, antispasmodic and antibiotic activities and are occasionally used as a tonic and stimulant; in this case, a decoction of the entire plant is used.

MORTON (1994) mentions that a decoction of the entire plant is appreciated as an intestinal stimulant (MANFRED 1947). A tea of the leaves is used as an emmenagogue, stomachic and diuretic, and to relieve rheumatism, colds, fever, hypertension and diarrhoea, when taken internally; externally it is applied for measles and chickenpox. A flower decoction serves as a sudorific, a decoction of the root as a pectoral, antiasthmatic and for the treatment of venereal diseases. The leaves are externally applied to cuts, ulcers and bruises (MORTON 1981). HERBERT ET AL. (1991) isolated from lantana leaves the substance verbascoside (or acetocide) with antimicrobial, immunosuppressive and antitumour activity. According to AHMED et al. (1971, 1972) the leaves lower blood pressure and accelerated deep respiration, stimulate intestinal movement and inhibit uterine motility. Toxic species of *Lantana* contain the tetracyclic triterpenes A and B.

The plants are rich in manganese, potassium and nitrogen and are therefore used as a fertilizer and to suppress weeds. Spiny forms are planted as protective hedges. The roughest types of *Lantana* leaves serve to polish wood. Paper can be made of the stems. Twigs are made into toothbrushes. *Lantana* leaves are also used to scent tobacco. The essential oil of the leaves is of interest for perfume manufacture (MORTON 1994).

Lantana trifolia L. (cariaquito morado)

Taxonomical description

Taxonomical details can be found in Fig. 307 c, 310. Shrub, 1–3 m high. Smaller branches angular, pubescent; trichomes simple, white, 0.5–1.5 mm long. Leaves generally in whorls of 3 on a node, sometimes 2 or 4. Leaf blades of papery consistency, pubescent on both sides, oblong-lanceolate to elliptic (3) 5–12 cm long, and 1.5–4 cm broad, apex acuminate (or acute), base attenuated and decurrent, margin serrate or crenate. Venation pinnate, craspedodromus, 6–8 secondary nerves, midrib prominent on the abaxial side and flat on the upper side; petiole short, 0.2–0.5 cm long, pubescent.

Inflorescences capitate when young, elongating into a dense spike of 3–4.5 cm length. Flowers subtended by foliaceous bracts with a broad base and an attenuate tip, pubescent, 0.3–0.8 cm long. Crown of rose colour, blue or purple, 0.8–1 cm long, tube 0.5–0.6 cm long. 4 stamens. Ovary bilocular, each compartment with a single ovule. The globose fruit is a purple or bluish drupe, 0.2–0.4 cm in diameter.

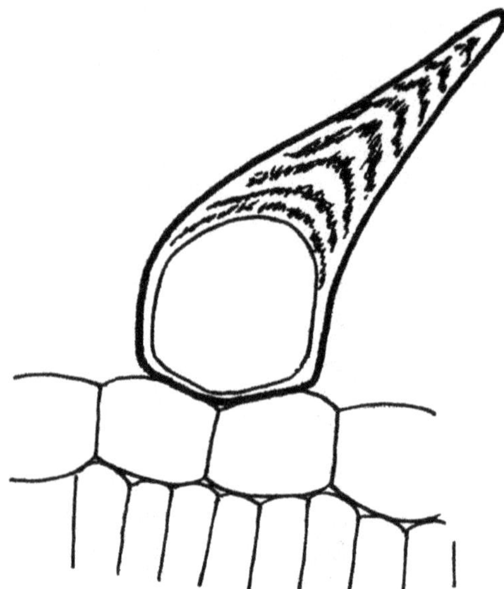

Fig. 309. *Lantana trifolia.* Large hair of leaf (upper epidermis).

Origin

Tropical America.

Occurrence

In Venezuela common from the Andes to Guiana, between 200 to 1.500 meters above sea level.

Anatomical description

Leaf (Figs. 309, 310). The leaf is dorsiventral and hypostomatic. The upper epidermis cells are slightly enlarged and possibly water-storing. There is only one layer of palisade cells with a length/width index of about 1/3. The spongy parenchyma is of about the same size as the palisade parenchyma or slightly larger, relatively loose and with intercellular spaces of medium size. The lower epidermis cells are smaller than those of the upper epidermis and the stomata are conspicuously elevated above the surface; they are of the hygromorphic type. The stomata are comparatively large and numerous. They are without conspicuous subsidiary cells. As seen in a surface view, the anticlinal walls of the upper and lower epidermis cells are conspicuously wavy. Hairs of different types are found in both epidermal layers, but their densitiy is much higher on the lower than on the upper side. Very conspicuous are large unicellular hairs with a broad enlarged base and an acute thick-walled tip (Fig. 309). There are also longer and slender uniseriate, mostly bicellular hairs, likewise with an acute tip, and thirdly, there are glandular hairs with a foot cell, an intermediate (pedicel) cell and a large unicellular head.

The midrib projects considerably above the lower leaf surface. The palisade parenchyma is interrupted in the midrib region and the vascular tissue is arranged in the form of an open arc. However, individual irregular bundles on the upper side above the arc complicate the picture. This pattern renders the midrib very characteristic and useful for identification. Several cell layers below the upper and lower epidermis are slightly collenchymatous; the filling tissue is parenchymatous.

The lateral veins of first order also project above the lower side. They have a smaller arc-shaped vascular system than the midrib and a general structure similar to that of the midrib, but in a reduced form. The palisade parenchyma is also interrupted in the lateral veins of first order.

The veins of higher order consists of a single vascular bundle surrounded by a parenchymatous sheath which is trancurrent to the upper epidermis.

Axis (Fig. 311). In a twig about 7 mm in diameter, the periderm which comprises several cell layers forms the outer limit. The primary cortex is small and consists of parenchyma in which fiber bundles (probably of the pericycle) are embedded. The vascular tissue is conspicuously arranged in radial rows. The rays are small and 1–2 seriate. The outermost phloem cells are compressed. All phloem cells are thin-walled.

An irregular formation of growth rings is obvious in the xylem. The vessels are small, but numerous, mostly occurring solitary or in radial rows.

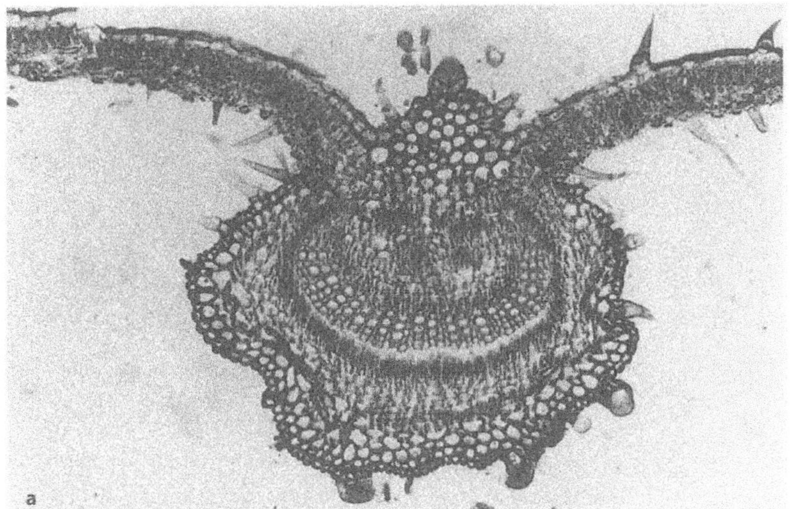

Fig. 310. *Lantana trifolia.* **a** T.s. of leaf with midrib. × 6.3. **b** Twigs with leaves, flowers and fruits.

The growth rings are delimited by cells with radially shortened walls.

The pith is comparatively ample and consists of large thin-walled cells.

Root (Fig. 312). The root structure is similar to that of the stem. The outer part of the bark consists of bands of compressed cells (sieve tubes and companion cells) alternating with bands of uncompressed cells (parenchyma). The conducting bark follows towards the inside. In the phloem as well as in the xylem, the regular radial cell rows are conspicuous. The xylem is well developed. The vessels are small; they occur solitary or in shorter or longer radial multiples. The rays are uni or pluriseriate. The primary xylem in the center from which lateral roots ramify is very conspicuous. Growth ring formation is completely absent or very irregular.

Ethnobotanical and general use

Medical use

The plant is mainly used medicinally. The useful parts are leaves, young shoots, roots, flowers, fruits, seeds and the entire plant. Like other species of the genus, the plant has tonic and stimulant properties and is used against depression, as a tranquilizer

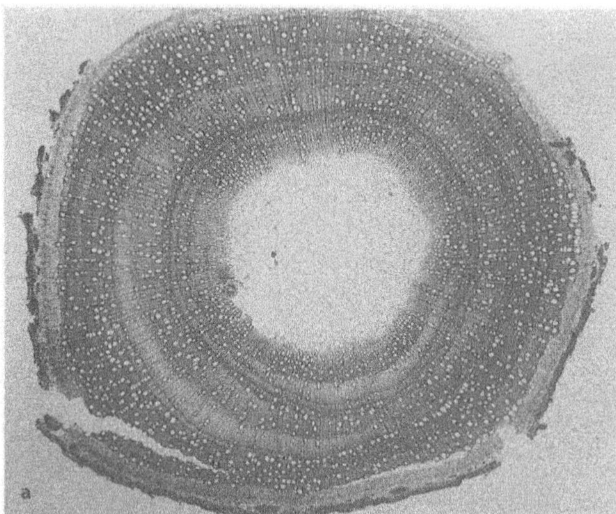

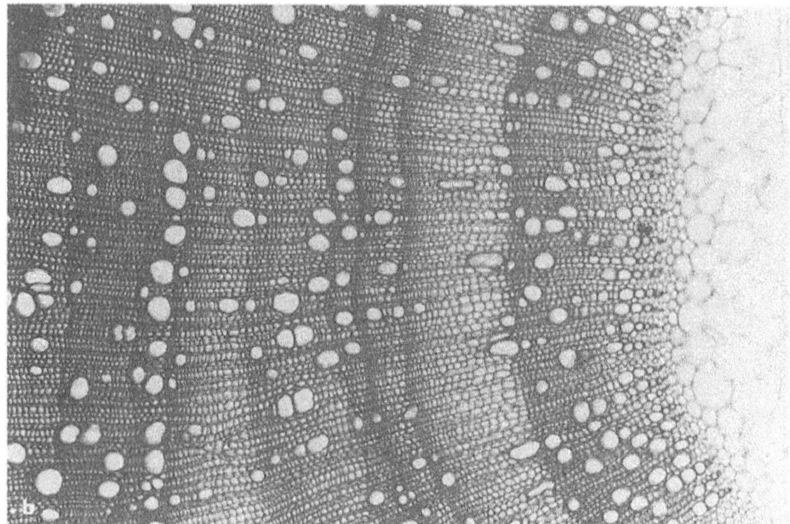

Fig. 311. *Lantana trifolia*. Axis. **a** Complete t.s. × 2.5. **b** Detail of xylem with growth rings. × 6.3.

and an emmenagogue, it is applied against catarrh and bronchitis. It has digestive and antispasmodic properties.

According to RODRÍGUEZ (1983), the leaves in decoction help against catarrh and bronchitis; flowers and leaves in decoction are an emmenagogue and a uterine tonic, roots and flowers in infusion are used against chicken pox. The entire plant and the fruit in decoction is applied as an ophthalmic and against rheumatism. Twigs and flowers in infusion are used as a bath and antispasmodic.

There are quite a few species of *Lantana* which are medicinally used and have similar applications. The best known species is *Lantana camara* L. In Venezuela this species is very popular and it is recommended in baths as a protection against bad luck.

Method of use

The remedy is usually prepared as a tea, putting a handful leaves into a liter of boiling water.

For a bath, a much larger quantity of leaves is used. It is suggested that a bath helps against debility, deformations and tuberculosis of the bones.

Healing properties

The most important healing properties are tonic, stimulant, digestive and antispasmodic effects.

Chemical contents

The plant contains the alkaloid lantanin which has effects similar to those of quinine, and is antispasmodic. A furanonaphthoquinone has been identified in *L. trifolia*. The seeds of *L. camara* contain an essential oil which is anthelmintic. The aromatic substances are found in the glands.

Toxicity

The polycyclic triterpenoid principle called lantaden A is toxic. The species of *Lantana* have to be handled with care because of their toxicity

Fig. 312. *Lantana trifolia*. Root. **a** Central part. **b** Peripheral part. ×6.3 both.

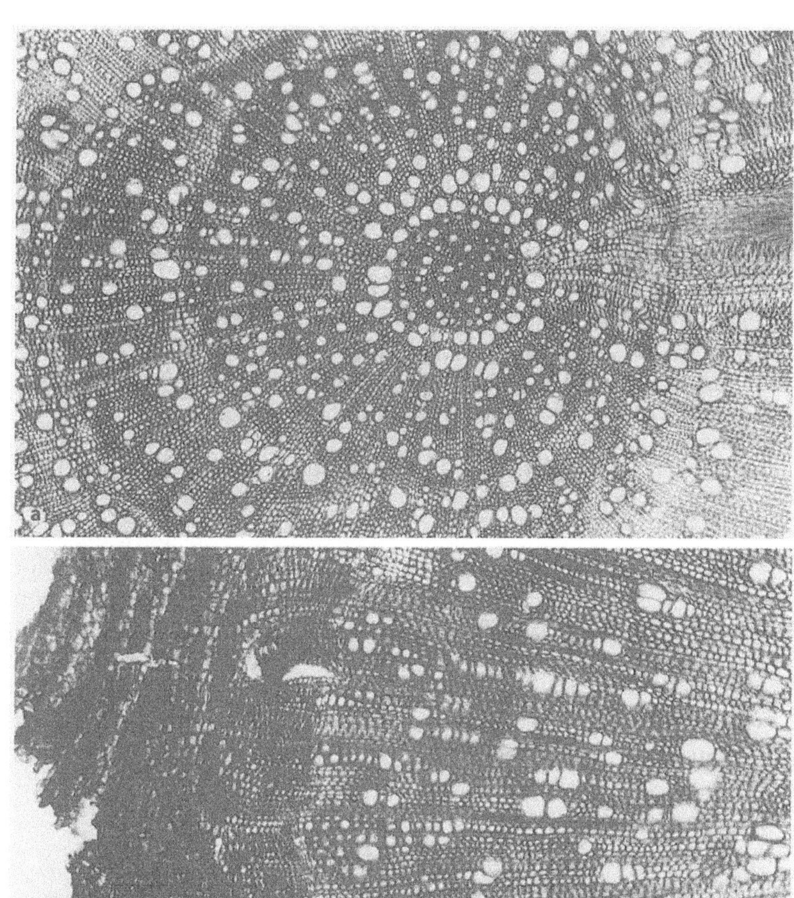

(GUJRAL & VASUDEVAN (1983) and AVADHOOT et al. (1980).

Varieties and related species

The species is very variable and polymorphic.

Closely related species are *Lantana camara* L., and *Lantana canescens*; they have similar medical applications.

Cultivation

The plants should be easy to cultivate, as they develop regionally like weeds.

Observations

The plant may be recognized by its habitus: Serrate leaves in whorls of 3 (or 4).

The 3 types of hairs are very characteristic, particularly those with the large bulbous base and the acute thick-walled tip (Fig. 309) as well as the glandular hairs with unicellular heads. Also pecuiliar is the midrib with an open arc of vascular tissue corwned by several individual bundles of irregular shape and arrangement.

In the axis, the small rays, the uniform phloem and xylem and the small vessels which are solitary or arranged in radial rows are important for identification.

Vitex

At least 14 useful species are known. The plant is a symbol of chastity; it is an antiaphrodisiac (vitex agnus castus L.). It is a source of volatile oil. Twigs are used for basketry. A gum is used for ink and as an antidote for arrow poison. Fruits are a substitute for pepper and are edible. Leaves are rich in tannin, they are made into a tea. Leaves are used for bruises, for beri-beri and oedma. Fresh leaves are applied on wounds. Leaves and fruits serve to combat diarrhoea. The bark is aromatic and tonic. The wood is also used.

Bark structure has been studied by ROTH 1981, leaf venation 1996.

Violaceae

Paypayrola

Decoction of flowers for anemia.

Leaf structure has been studied by ROTH 1984, fruit structure and dispersal 1987.

Rinorea

Leaves as a vegetable.

R. riana KUNTZE. Decoction of leaves dissolves hardened wax in the ear. Leaf structure has been studied by ROTH 1984, fruit structure and dispersal 1987.

Vochysiaceae

Erisma

Seed 'butter' for candle and soap manufacture. Seed oil for itching due to eczema. Seeds have a calmative effect.

Bark structure has been studied by ROTH (1981) leaf structure 1984, leaf venation 1996.

Qualea

Phenolic bark taenifuge. Young leaves might be a cure for leprosy. Bark as styptic, haemostatic, wound healing.

ROTH studied bark structure 1981, leaf venation 1996, leaf structure 1984.

Vochysia

At least 6–7 useful species are known. Useful wood. Alkaloid vochysine. Curare from bark. Bark smoke inhaled for asthma and respiratory ailments; bark for sweating and as a februfuge, as a wash for ulcer, and for skin sores. Bark and leaves abortifacient and contraceptive. Leaves for sores of the mucous membrane of the mouth. Leaves diuretic. Flowers for sores of lips (possibly for herpes).

ROTH studied the bark structure 1981, leaf structure 1984, leaf venation 1996, fruit structure and dispersal 1987.

Zygophyllaceae

Representatives of this family are mainly shrubs and subshrubs, seldom trees or annual herbs. The alternate leaves have persisting stipules. The floral axis forms discs or a gynophore. The obdiplostemonous stamens occur mainly in groups of 10–8 and occasionally have ligula-like appendices. The fruit is a capsule, seldom a berry or drupe or splits into mericarps (cocci). Bur-like fruits which stick to the fur of animals are not seldom.

Many representatives are xerophytes or halophytes typical of deserts and salt steppes.

The Zygophyllaceae are rich in lignans.

From the wood of *Guajacum* species (Lignum guajaci), a resin is extracted which is used to test for oxidizing substances (blue colouring).

From the resin of *Larrea tridentata* the nordihydroguajaret acid (NDGA) is extracted which is used as an antioxidant for the stabilization of easily oxidizing substances, e.g. vitamins.

Besides triterpene saponins, steroid saponins are also found which are normally restricted to the monocotyledons.

Bulnesia arborea (JACQ.) ENGLER. Seed anatomy was studied by LINDORF (1978). The fruit is a hemispheric capsule which contains a single seed in each locule. Dehiscence is septicidal and ventricidal, not releasing the seeds. The reniform seeds are about 15 × 15 mm. The seed coat is crustaceous, the outer part being composed of parenchymatous cells with reticular thickenings leaving larger intercellular spaces between one another so that a spongy appearance results. The inner more obliterated part has a subepidermal crystal layer with rectangular crystals.

The wood of the tree which follows the coastal zone of Venezuela is used for the same purposes as 'lignum vitae' (*Guaiacum*) being made into collars of water turbines and brush backs.

Bulnesia sarmientoi LORENTZ. Bark and wood structure of *Bulnesia sarmientoi* have been studied by ROTH and GIMÉNEZ DE BOLZÓN (1997). The plant supplies a balsam used in perfumery (guaiac oil). See also the paragraph dealing with Argentinian economically important plants.

Tribulus cistoides L. (abrojo, flor amarilla)

Taxonomical description

The plant is a perennial herb reaching a height of 20–30 cm. It is hairy, particularly over the nodules. The paripinnate leaves are 2.5–6 cm long and have (4) 6–9 pairs of leaflets. These are oblong or elliptic, asymmetric, 7–17 mm long and 3–6 mm broad, having a subacute and often mucronate tip and a silky indumentum on the lower side. The flowers arise solitary in the leaf axils.

The sepals are lanceolate, acuminate, 7–9 mm long, pubescent and caducous. The 5 yellow petals are obovate, 2–2.5 cm long and 1–2 cm broad, and have a rounded tip. The 10 stamens are occasionally unequal and some are sterile.

The sessile ovary, covered with white hairs is 2-locular and has 5 stigmas which are more or less linked with one another.

The fruit has a pentagonal shape, dehiscing septicidally at maturity and liberating 5 partial fruits (mericarps or cocci) of hard consistency. These are tuberculate on their outside being beset with knobby excrescences, each of which is provided with 2 horizontally divergent larger spines and 2 smaller spines attached close to the base.

The species is a weed which abundantly occurs in dry and hot regions, particularly in the vicinity of the sea. The spiny mericarps are dispersed by epizoochory.

Occurrence

Tropical America up to the south of the USA. In Venezuela it is found in hot and dry regions on dry spots and frequently along the sea shore.

Anatomical description

See Fig. 313. The leaf is isolateral. A layer of palisade parenchyma is found on the uper as well as on the lower leaf side. However, the palisade cells of the upper side are slightly more elongated than those of the lower side. In the leaf center, the vascular bundles occur at short distances, leaving only 3–4 cells between one another. They are embedded in a layer of shorter palisade cells. The palisade cells contain small chloroplasts. The vascular bundles are surrounded by a sheath of large parenchymatous cells which contain larger chloroplasts than the palisade cells; the chloroplasts aggregate in the horseshoe shape along the inner and radial walls of the sheath cells; they contain larger starch grains. This is the typical 'Kranz' structure of leaves particularly found in plants of dry and hot regions. The plant thus shows the so-called C4 type of photosynthesis.

A hypodermis with few chloroplasts serving as a water-storing tissue is found in contact with the lower epidermis. Both epidermal layers are small-celled and have stomata, but stomata are more numerous on the lower side. Unicellular somewhat curled hairs are likewise present on both surfaces (ROTH 1992).

For further information of distribution and movement of chloroplasts see ROTH 1992. P.70–73.

Seed anatomy has been studied by LINDORF (1978).

The fruit is a globose schizocarpium with 5 mericarps, each of triangular shape, hard consistency and with a spiny dorsal side originating from the ovary wall. 2 large spines in the upper part and 2 smaller ones in the lower one with several small spines in between facilitate epizoochorous dissemination. The 3–4 seeds in each mericarp are separated by transverse walls which originate late in carpel development. The seeds are about 3.6 mm long and 1.6 mm broad. The seed coat is membranaceous. In the outer epidermis, some cells have helicoidal or parallel wall thickenings. Below follow 3–4 layers of thin-walled parenchyma cells with simple and compound starch grains and with oil drops. Then follows a crystal layer, the radial and inner tangential walls of which are thickened, resulting in a U-shape, as seen in t.s.

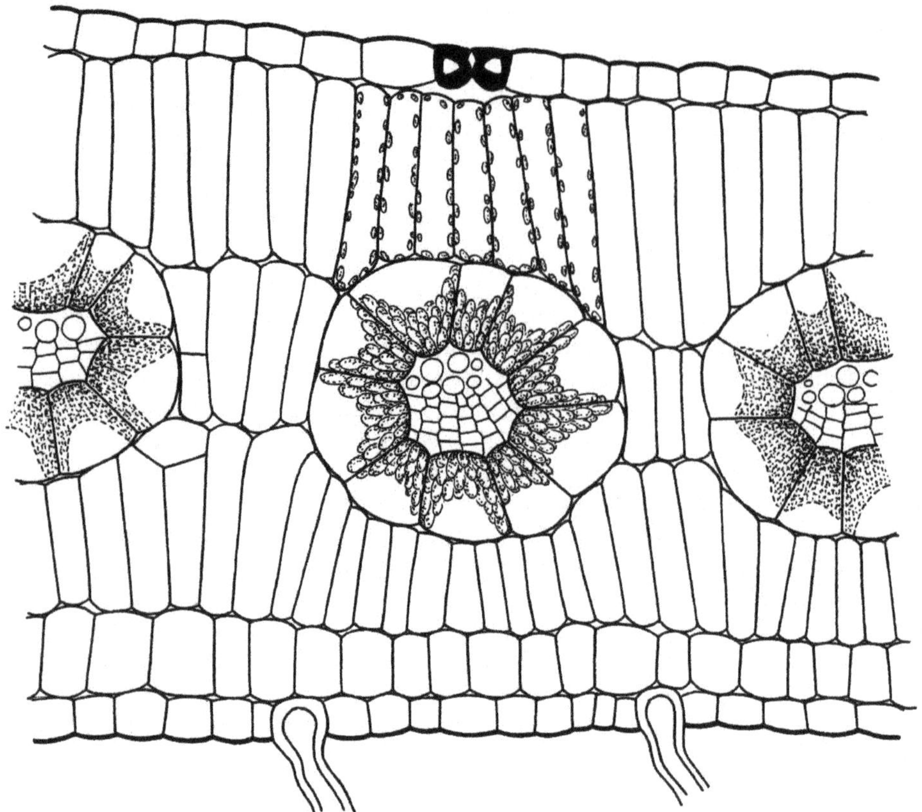

Fig. 313. *Tribulus cistoides*, Zygophyllaceae. T.s. of leaf of the 'Kranz' type and C4 type of photosynthesis. Note the vascular bundles surrounded by a parenchymatous sheath with large chloroplasts, while chloroplasts of the palisade parenchyma are much smaller (ROTH 1992).

Ethnobotanical and general use

Medical use

Leaves, roots, fruits and seeds are medicinally applied.

Ground and cooked leaves are applied as an emollient cataplasm for abscesses and tumors. A decoction is also used to cleanse wounds.

Roots in decoction are tonic and diuretic.

Seeds are astringent. A decoction of the seeds is used for nasal and intestinal haemorrhage, as a gargle and mouth wash for buccal affections, for a sore throat and swollen gums.

Method of use

As a decoction or cataplasm. Abscesses are treated several times a day with some drops of a 2 or 10 % decoction, according to whether it is used internally or externally.

Healing properties

Astringent, antiinflammatory, diuretic, haemostatic, antispasmodic, disinfectant, as an emmenagogue.

The fruits of *T. terrestris* L. are used as a diuretic.

Toxicity

In Australia it is said that the plant has caused the death of stock when eaten on an empty stomach. Likewise in Colombia it is believed to be poisonous to animals.

Observations

The species is easily recognized by its characteristic leaf structure with the typical 'Kranz' structure of the vascular bundles and by its fruit structure.

Epilog

We hope to have demonstrated that a large number of tropical plants have not only a single, but a variety of uses. Most medicinal plants can be applied not only for a single sickness, but frequently cure quite a few diseases. Additionally, these plants also have other uses at the same time. They may have edible organs or parts, they may be used as ornamentals, they may supply fibers, cork, waxes, fats and oils, starch, proteins, they may contain soap substitutes and sugar substitutes (without nutritional value), tannins for tanning, colouring matter for dyeing, gummy and resinous substances, or the plants may be used for charcoal or may serve only for firewood. Trees prevail in the tropics and mainly trees are officinal, in contrast to the European officinal herbs. We did not mention trees which supply timber, because we think that these have been known well enough for a long time, but we may recommend Record & Hess for the study of Timbers of the New World.

Useful plants could thus be cultivated at the same sites where they grow wild and be processed in small factories right away at the same places. Nutmeg and mace are, for example, gathered and produced in this way on Caribben islands. Chocolate, cloves, cinnamon bark are obtained by similar methods. Black women usually work in these factories. Even artesania (handicraft) can be developed at these small productive centers, e. g. wood carving or painting of pressed *Ficus* bark with folkloric designs, weaving of fibers from indigenous plants and many other activities of this kind.

In cultivating useful plants near the regions of their origin, disturbance or alteration to the metabolism of these plants could be avoided. It is well known that the content of secondary compounds of plants cultivated in temperate regions may change quantitatively as well as qualitatively; certain valuable seconday compounds may even disappear completely.

Furthermore, the South American tropical flora which is still unexplored to a large extent, should be intensely studied. More taxonomists should be prepared to study the tropical flora so that identification of plants progresses more rapidly.

Before all tropical forests are destroyed completely, at least all useful tropical plants – and which plants are not useful? – should be studied in the same way as has been done in Europe and in other countries. It would also be necessary to elaborate a complete pharmacognosy of tropical plants which would help to identify and to distinguish species and to separate true drugs from adulterations. Unfortunately, we are only at the beginning of these investigations. Later on, the chemistry of the plants may be studied.

The immense wealth of secondary compounds, which can help to cure many diseases, not to speak of the many other uses of plants, obliges us to protect the tropical forests more and better.

When we show developing countries what treasures they possess in their forests, and when we help them to cultivate the plants and to process them in small factories in the regions of their origin, the natives will treat the forests with more care.

The knowledge of the native population and the local bibliography concerning medicinal and useful plants should be more widely consulted.

European pharmaceutical companies still have very little interest or opportunity to experiment with new promising plants, but large North American companies such as Merck, Glaxo, Montesanto and Shaman are already expecting the great business. They consider the tropical forests as living pharmacies, sending out science brokers to the forests for plant collection. They have already started a multimillion dollar business on this expectation. Europe – as usual – remains behind!

Bibliography

Abbad, Fr. Iñigo (1781) Viage a la America. Banco Nacional de ahorro y préstamo. Edic. facsímil. Graf. Armitano. Caracas 1974.

Abstracts of lectures and poster presentations: 44th Ann Congr Soc med plt res Prague, Sept. 1996. Ed. T. Vaněek & Valterova.

Achiote, el (*Bixa orellana* L.). Boletin Técnico 1, 1985. Talamanca, Costa Rica.

Acosta de Obando, Rosa María (1979) Estudio anatomofoliar del género *Carramboa*. Thesis, Univ. Los Andes, Mérida.

Acosta Saignes, M (1961) Estudios de Etnologia antigua de Venezuela. EBUC, Caracas.

Ahmad, MU, Husain SK et al. (1979) Studies on herbaceous seed oils. XII. Journl. oil Technol. Assoc India 11 (3): 70–72.

Akinsinde KA & Olukoya DK (1995) Vibriocidal activities of some local herbs. J Diaechoeal Dis Res 13: 127–129.

Alban S, Kiefl B & Franz G (1996) Storage stability of essential oils in plant drugs. Abstr. 44th Ann Congr Soc Medicin Plt Res Prague: 153.

Albornoz AM (1980) Productos naturales (sustancias y drogas extraidas de las plantas). Publ UCV, Caracas

Albornoz MA (1993) Medicina tradicional herbaria (Guia de fitoterapia). Instituto Farmacoterápico Latino SA, Caracas. pp 564.

Alcedo A de (1988) Diccionario Geográfico Histórico de las Indias occidentales o América. Ed Ex-libris, Caracas.

Alegría C (1977) Historia de la farmacología e inventarios de farmacias en la época colonial. Rev Fac Farm UCV 24 (38): 144–159, Caracas.

Allain Arbe B (1994) Untersuchungen zur diagnostischen Differenzierung von Bignoniaceen-Rinden der Gattungen Jacaranda, Tabebuia und Tecoma. Dipl. Arb., Univ. Göttingen

Amaro-Feruandez MI, Orfila de Garcia L. Colman-Saizarbitorria I (1999) Citotoxicidad de algunas especies de la familia Annonaceae usadas en la medicina popular en Venezuela. Mem Inst Biol Exper 2: 175–178.

Angehr G, Coley P & Worthington A (1984) Guía de los árboles comunes del Parque Nacional Soberanía Panama. Coed. Smithonian Trop Res Inst – Dirección Nac. de Recursos Naturales Renovables, p. 19. Imp. Impretex S.A.

Anonymus (1979) Tropical legumes. Resources for the future. Nat Acad Sciences, Washington DC, USA.

Arias AE (1982) *Cecropia* ssp. In: Plantas medicinales. 17th edit. Medellín, Colombia: pp. 208.

Aristeguieta L (1961) El género *Heliconia* en Venezuela. Ministerio de Agric. & Cría, Caracas.

Aristeguieta L (1962) Arboles ornamentales de Caracas. UCV, Consejo de Desarrollo Científico y Humanustico, Caracas.

Aristeguieta L (1964) Compositae. In: Flora de Venezuela, Vol. X., Caracas.

Aristeguieta L (1976) Presencia de alcaloides en plantas de Venezuela. Bol Soc Venez. Ciencias Naturales 32 (132–133): 23–34, Caracas.

Armas Alfonzo A (1991) Un arcano: Las Yerbas del Venezolano. Revista M (Corimon), Caracas.

Ayensu E (1982) Medicinal plants of the West Indies. Algonac. Michigan Reference Public Inc. pp. 282.

Bader G, Plohmann B, Franz G & Hiller K (1996) Saponins from *Solidago virgaurea* L. Possible agents for therapy of cancer? Abstr 44th Ann Congr Soc Medicin Plt Res. p. 21.

Baldwa VS & col. (1977) Clinical trial in patients with Diabetes mellitus of an insulin-like compound obtained from plant source. Uppsala, J Med Sci 82: 39–41.

Balick M (1985) Useful plants of Amazonia: A resource of global importance. In: Key enviroments: Amazonia (GT Prance and TE Lovejoy, eds): 339–368. Pergamon Press, Oxford.

Balick MK & Cox PA (1996) Drogen, Kräuter und Kulturen. Wiss Buchges Darmstadt.

Balick MJ, Elilsabetsky E & Laird SA (1996) Medical resources of the tropical forest. Biodiversity and its importance to human health. Biology and Resource Management in the Tropics Series. pp. 464.

Bartholomew ET & Reed HS (1943) In: HJ Webber & LD Batcheler (eds). The Citrus industry, Vol I. History, Botany and Breeding: 669–717. Univ. of Calif. Press, Berkely & Los Angeles.

Baudi JC (1987) Plantas medicinales existentes en Venezuela y Latinoamérica. Edit. America, Caracas.

Bazzochi IL, Alvarenga NL, Ravelo AG & Moujir L (1996) A new bioactive nor-quinone-methide triterpene from *Maytenus scutioides*. Abstr. 44th Ann Congr Soc Medicin Plt Res. Prague p 49.

Beard FS (1944) Key for the identification of the more important trees of Tobago in characters of bark and blaze. Empire Forestry Journal 23: 34–36.

Becco HJ (1991) Crónicas de la naturaleza del Nuevo Mundo. Cuadernos Lagoven, Edit. Arte CA, Caracas.

Bello A (1952) Obras completas. Ministerio de Educación, Caracas.

Benítez de Rojas CE & Ruiz TZ (1982) Las Annonáceas comestibles en Venezuela. Natura 72: 10–13.

Berger I (1949) Handbuch der Drogenkunde, Bd. 1. W. Maudrich, Wien.

Bermúdez A (1999) Enfoques metodológicos para la investigación etnobotánica sobre plantas medicinales. Memorias Inst Biol Exp (MIBE) 2: 3–6.

Bermúdez A & Velásquez D (1999) Plantas medicinales que se venden en los herbolarios del Estado Trujillo. Memorias Inst Biol Exp (MIBE) 2: 137–140.

Bernal HY & Correa JE (1989) Especies vegetales promisorias de los países del Convenio Andrés Bello. SECAB, Bogota.

Bernal HY & Jiménez LC (1990) Haba Criolla, *Canavalia ensiformis* (L.) DC. SECAB, Bogota.

Bernardi AL (1957) Estudio botánico-forestal de las selvas pluviales del rio Apacara, Reg. Urimán, Edo. Bolívar. Publ Dir Cult Uni. Los Andes 63: 149 pp. Mérida, Venezuela.

Bertoni MS (1940) Diccionario botánico latino-guaraní y guaraní-latino. Asunción Reprint 1989. 146 pp.

Bhalsingh SK & Maheswari VL (1997) Importance of secondary plant constituents as drugs. Pkt. Structure & Morphogenesis, 79–85, India (eds. TV Ramana, LL Rao, I.L. Kothari).

Bhat K (1981) Herbolario Tropical. Industria Gráfica Oriental, Cumaná.

Biermann M (1896) Beiträge zur Kenntnis der Entwicklungsgeschichte der Früchte von *Citrus vulgaris* Risso und anderer *Citrus*arten. Diss Univ Bern (Minden, CC Bruns).

Bisset NG (1991) One man's poison, another man's medicine? Jour Ethnopharmacol 32: 71–81.

Blohm H (1962) Poisonous plants of Venezuela. Wiss Verlagsges, Stuttgart.

Borges J, Manresa MT, Martin JL & Vásquez CPYP (1978) Altamisin, a new sesquiterpene lactone from *Ambrosia cumanensis*. Tetrahedron Lett 1978: 1513.

Borges del Castillo J, Manresa Ferrero MT, Martin Ramion JL, Rodríguez F, Vásquez Bueno R (1980) Two new psilostachyins from *Ambrosia cumanensis*. An Quim Ser C 76: 277–280.

Borges del Castillo J, Bradley Delso A, Manresa Ferrero MT, Vasquez Bueno P, Rodriguez Luis F (1983) Salvadorian Compositae 4. A sesquiterpenoid lactone from *Ambrosia cumanensis*. Phytochemistry 22: 782–783.

Borges del Castillo J, Ferrero MTM, Ramon JLM, Luis FR, Vásquez Bueno P (1983) Sesquiterpene lactones from *Ambrosia cumanensis*. An Quim Ser C 77: 52–55.

Bork PM, Schmitz ML, Weinmann C, Castro V, Mora G, Merfort I & Heinrich M (1996) Antiinflammatory activity of Nahua Indian Medicinal Plants (Mexico): Effects of plant extracts and purified compounds on the transcription factor NF-kB. Abstr. 44th. Ann Congr Soc Medicin Plt Res. Prague, p. 24.

Braun A (1968) Cultivated palms of Venezuela. Principes (Journal of the Palm Society) 12, Nos. 2, 3 & 4.

Braun A (1970) Palmas cultivadas de Venezuela. Acta Bot Venez. 5 (1–4): 7–94.

Braun A (1996) El Chaguaramo – sus afinidades, sus características y su cultivo. Litho-Tip, CA, Caracas.

Braun A (1997) La utilidad de las palmas en Venezuela. Fund. Th. Merle, Litopar, Caracas, 67 pp.

Braun A & Delascio Chitty FD (1987) Palmas autóctonas de Venezuela y de los países adyacentes. Inparques – Jardín Botánico de Caracas. Impr. Litopar, Caracas, 156 pp.

Breitwieser K (1942) Pharmakognostische Untersuchungen über Verbenacee. Habilitationsschrift, Gatzer & Hahn, Schramberg/Schwarzwald.

Brown GB, Deakin JR & Wood MB (1969) Identification of *Cucumis* species by paper chromatography of flavonoids. Jour Amer Soc Hort Sci. 94: 231–234.

Brücher H (1955)n Origen y filogenia de los cereales. Min educ nac univ nac Cuyo. ACME, Buenis Aires.

Brücher H (1968) Genetische Reserven Südamerikas für die Kulturpflanzenzüchtung. Theor Appl Genet 38: 9–12.

Brücher H (1969) Gibt es Gen-Zentren? Naturwissenschaften 56: 77–84.

Brücher H (1977) Tropische Nutzpflanzen. Ursprung, Evolution und Domestikation. Springer, Berlin Heidelberg New York.

Brücher H (1982) Die sieben Säulen der Welternährung. Senckenberg Buch 59. Kramer, Frankfurt.

Brücher H (1985) The South American gene pool of economical useful plants. Plant Res Dev 16: 109–120.

Brücher H (1988) Migration and Dispersion amerikanischer Nutzpflanzen über die Landenge von Darién (Panama). Naturwissenschaften 75: 18.

Brücher H (1989) Useful plants of neotropical origin and their wild relatives. Springer, Berlin Heidelberg New York London Paris Toyko Hongkong.

Burger J Die Wächter der Erde (über Sanema Indianer Venezuelas).

Caceres A, Menendez H, Menendez E, Cohobon E, Samoya BE, Jaureguí E, Peralta E & Carrillo G (1995) Antigonorrhoeal activity of plants used in Guatemala for treatment of sexually transmitted diseases. J Ethnopharmacology 48: 85–88.

Caesar W (1992) Kolumbus und die Folgen. Deutsche Apotheker Zeitung (132. Jahrgg.) No. 40: 2107–2113.

Campos Y & Casale I edit (1997) Algunas plantas usadas en la industria artesanal en el Edo. Sucre, Venez. Mem Congr Internat Etnobotánica '97, Mérida, Mexico.

Cannap: The Caribbean network for natural products (investig. of medicinal plants). Coord. OEA St. Augustine, Trinidad, Dr. C.E. Seaforth.

Cannon J & Cannon M (1994) Dye plants and dyeing. Portland OR (Timber Press).

Cano ALM & Hernández AC (1984) El 'Piñoncillo' (*Jatropha curcas*), recurso biotico silvestre del trópico. Cuaderno de divulgación INIREB, Xalapa-Veracruz (Mexico) 14: 1–16.

Carrasquel N (1969) Estudios anatomo-morfológicos de las especies del género *Jacquinia* en Venezuela para su interpretación taxonómica. Acta Bot Venez 4: 307–357

Carrizosa SM (1984) Contribución al estudio fitoquímico de *Gnaphalium antennarioides*. Tesis, Univ. Bogota.

Casale I & Campos I (1997) Plantas medicinales de venta en el mercado principal de Cumaná, Edo. Sucre, Venezuela. II. Congr Internac Etnobotánica 1997, Mérida, Mexico, p. 51.

Castillo Suárez A (1999) Etnobotánica medicinal 'Piaroa' al Norte del Edo. Amazonas. Mem Inst Biol Exper (MIBE) 2: 141–144.

Castillo-Suárez A (1999) (1992) El uso medicinal de los árboles del bosque húmedo del rio Cataniapo, Edo. Amazonas, Venezuela. Acta Biol Venez 14 (1): 7–25.

Castillo Suárez A (1999) Etnobotánica medicinal 'Piaroa' al norte del Estado Amazonas. Mem Inst Biol Exp (MIBE) 2: 141–144.

Castillo A (1994) Aspectos floristicos, fisionomónicos y dendrológicas del bosque húmedo del rio Cataniapo. Part I, Thesis, Caracas.

Castillo A (1995) El uso medicinal de los arbustos del bosque húmedo del rio Cataniapo, Estado Amazonas, Venezuela. Acta Biol Venez 15 (3–4): 41–54.

Castillo A (1997) El uso medicinal de los árboles del bosque humedo del rio Cataniapo, Edo. Amazonas, Venezuela. In: P. Naranjo & A. Crespo (eds.), Etnomedicina Progresos.

Castillo A (1998) Estudio flirsticos, dendroogicos y etnobnotaucos de ecosistemas boscosos Venezolauous. Mem Inst Biol Exper 1: 217–220.

Castillo A, Lopez M, Rodriguez L & Jimenez S (1997) Arboles de uso medicinal de los bosques riberenos del área Sipapo-Cuao, Edo. Amazonas, Venezuela.

Castillo A, Gómez S & Moreno O (1992) Aspectos florísticos y fisionómicos de un ecosistema semiárido del litoral central, Municipio Vargas, Dstro. Federal. Acta Biol Venez 13: 97–115.

Caulin Fr Antonio (1760) Historia corográfica, natural y evangélica de la Nueva Andalucia.

Centeno JC (1993) Amazonia 2000. Dimensiones políticas del manejo sostenido del Amazonas. Caracas: World wide fund for nature. 56 pp.

Chagnon NA, Le Quesne P & Cook JA (1970) Algunos aspectos de drogas, comercio y domesticación de plantas entre los indígenas yanomami de Venezuela y Brasil. Acta Cient Venez 21 (5): 186–193. Caracas

Child R (1964) Coconuts – Tropical Agriculture Series. Longmans, London.

Chinou I, Roussis V, Tzakou O, Karapati C, Mazomenos BE, Chinou E & Golemati-Persidou P (1996) Antibacterial spectrum of Greek bee-honeys. Preliminary studies on their chemical composition. Abstr. 44th. Ann Congr Soc Medicin Plt Res, Prague: 139.

Choi OB, Yoon TJ, Drees M, Scheer R & Kim JB (1996) Comparison of antitumoral activities of extracts of Korean mistletoe with extracts of European mistletoe (*Viscum album* L.). Abstr. 44th. Ann Congr Soc Medicin Plt Res. Prague: 40.

Cimanga K, de Bruyne T, Pieters L, Claeys M, van den Berghe D & Vliethinck AJ (1996) Neocryptolepine and biscryptolepine, Two antibacterial alkaloids from *Cryptolepis sanguinolenta*. Abstr. 44th. Ann Congr Soc Medicin Plt Res. Prague: 45.

Cobley LS (1956) An introduction to the botany of tropical crops. 4th Impression 1963. Longmans, London.

Coene, R de (1950) Anatomie, histologie et histogénèse de la capsule de cotonnier, *Gossypium hirsutum* L. La Cellul3 53: 137–150.

Collin PH, Schick E & Livesey R (1995) Pons – Fremdwörterbuch Medizin (Englisch–Deutsch, Deutsch–Englisch). Klett, Stuttgart Dresden.

Collins JL (1960) The pineapple (botany, cultivation, and utilization). World Crops Books (ed. N. Pilunin). L. Hill, London New York.

Coolhaas C, de Fluiter MJ & Koenig HP (1960) Tropische und subtropische Weltwirtschaftspflanzen III. Genusspflanzen. 2. Bd. Kaffee. 2. Aufl. Enke, Stuttgart.

Copeland HF (1955) The reproductive structure of *Pistacia chinensis* (Anacardiaceae). Phytomorphology 5: 440–449.

Corner EJH (1976) The seeds of dicotyledons. Cambridge University Press, Cambridge.

Corothie H (1948) Maderas de Venezuela. Imprenta Nacional, Caracas.

Correll SD (1953) Vanilla. Its botany, history, cultivation and economic import. Economic Botany 7: 291–358.

Cortes S (1917) *Espeletia*. In: Flora de Colombia. 2nd. ed. Bogota, 312 pp.

Costa de AF & Cavalcanti JJ (1958) Hypoglicenic action of inner bark (bast) of cashew tree (*Anacardium occidentale* L.). Action of a decoction on normal rats. Anais Fac Med Univ Recife (Brasil) 18: 193–197.

Cronquist A (1988) The evolution and classification of flowering plants. 2nd ed. New York Botanical Garden, Bronx, USA.

CURARE. Zeitschrift für Ethnomedizin und transkulturelle Psychiatrie. Herausgeg. Arbeitsgemeinschaft Ethnomedizin E.V. Verlg Vieweg, Wiesbaden, Postfach 58 29.

CURARE. Zeitschrift für Ethnomedizin und transkulturelle Psychiatrie. Arbeitsgem. Ethnomedizin e.V. - Vieweg, Braunschweig, ISSN 0344-8622.

Delascio Chitty F (1978) Aportes al conocimiento de la etnobotánica del Estado Cojedes (Venezuela). Fundac. La Salle de Cienc. Nat. Caracas. 127 pp.

Delascio Chitty F (1984) Datos etnobotánicos de la región de San Carlos de Rio Negro, Terr Fed Amazonas, Venez-Bol Soc Ven Cienc Nat 142: 273-293.

Delascio Chitty F (1985) Algunas plantas usadas en la medicina empírica Venezolana. Dir Investig Biol, Divis. Vegetación, Jardín Botánico, Caracas.

Delascio Chitty F (1989) Algunas plantas útiles de los indios Cariñas de Caíco Seco, Edo. Anzoátegui, Venez. Acta Bot Venez 15: 25-40.

Delgado R, Sanabria M, González R & Cumaná L (1997) Plantas medicinales de Macuro, Edo Sucre, Venezuela. I. Congr Internac, Etnobotánica 1992, Córdoba, Espana.

Dell B & McComb AJ (1978) Plant resins - their formation, secretion and possible function. In: HW Woolhouse (Ed): Advances in Botanical research 275-316. Academic Press, New York.

Den Outer R (1972) Tentative determination key to 600 trees, shrubs and climbers from the Ivory Coast, Africa, mainly based on characters of the living bark, besides the rhytidome and the leaf. Med Landb Wag 72: 18-19-20-21; 73 (41, 52, 60).

Deutsche Apotheker Zeitung. Deutscher Apotheker Verlag, Stuttgart, Germany.

Drago T. A las puertas del descubrimiento. El Universal, 24.1.1992. Caracas, Venezuela.

Duke JA (1988) Handbook of medicinal herbs. CRC Press, Boca Ratón, Florida, USA.

Dunsterville GCK & Garay LA (1959-1976) Venezuelan Orchids Illustrated. T. I-VI. London.

Eckey EW (1954) Vegetable fats and oils. Reinhold, New York.

El Achiote (*Bixa orellana* L.) (1984/85): Proy. Agrofor. ANAI-CINDE. Talamanca, Limón, Costa Rica. Bol. Técn. I, 1985.

Elisabetsky E & Addison D (1989) Use of contraceptive and related plants by the Kayapó Indians (Brazil). Journal of Ethnopharmacology 26: 299-316.

Engler A (1964) Syllabus der Pflanzenfamilien. Bd. II. Angiospermen. Borntraeger, Berlin.

Engler A & Prantl K (1897) Die natürlichen Pflanzenfamilien. Leipzig.

Ernst A (1986) Obras completas (Tomo I. Botánica). Caracas, Ed. Presidencia de la República. 601 pp.

Esau K (1967) Plant Anatomy. 2nd. ed. John Wiley, New York London Sydney.

Esau K (1969) The phloem. Encyclopedia of plant anatomy. V. Part 2, 514 pp. Borntraeger, Berlin.

Esau K (1977) Anatomy of seed plants. 2nd. ed. J. Wiley, USA.

Esdorn I & Pirson H (1973) Die Nutzpflanzen der Tropen und Subtropen in der Weltwirtschaft. 2. Aufl. Fischer, Stuttgart.

Especies Vegetales Promisorias de los países del Convenio Andrés Bello. Tomos I-XII (1989-1998). SECAB, Bogotá (Eds. JE Correa & HY Bernal).

Espinosa Salas AJ (1985) Plantas medicinales de la Hrasteca Hidalguense. Tesis Fac Cienc, Univ Nac Autón, Mexico.

Etnobotánica - Boletín informativo del Grupo Ethnobotánico Latinoamericano. Dr. J Canballero, Jardín Botánico UNAM, Apto. 70-614, Mexico D.F. 04510. Mexico.

Etnomedicina (Unicef, Ecuador) (1995) Memorias del IV. Congr. Italolatinamericano. Eds. P Naranjo & A Crespo.

Ettinghausen CR v (1854) Über die Nervatur der Blätter und blattartigen Organe bei den Euphorbiaceen, mit bes. Rücksicht auf die vorweltlichen Formen. Sitzber kaiserl Akad Wiss, Wien, math.-nat. Cl. 12: 138-154.

Ettinghausen CR v (1861) Die Blatt-Skelette der Diktolyedonen. Mit bes. Rücksicht auf die Untersuchung und Bestimmung der fossilen Pflanzenreste. K.-K. Hof- u. Staatsdruckerei, Wien, 308 pp.

Europäisches Kolloquium für Ethnopharmakologie. European Society of Ethnopharmacology.

European Society of Ethnopharmacology.

Ewel JJ, Madriz A & Tosi A (1968/1976) Zonas de vida de Venezuela. MAC, Caracas.

Faber FC von (1912) Morphologisch-psychologische Untersuchungen an Blüten von *Coffea*.Arten. Ann Jard Bot Buitenzorg 1, Ser 10: 59-160.

Fagerlind F (1937) Embryologische, zytologische und bestäubungsexperimentelle Studien in der Familie der Rubiaceae nebst Bemerkungen über einige Polyploiditäts-Probleme. Acta Horti Bergiani 11: 195-470.

Fagerlind F (1939) Perisperm oder Endosperm bei *Coffea*-Svensk Botanisk Tidskrift 33.

Fahn A (1979) Secretory tissues in plants. Academic Press, London.

Fahn A & Leshem B (1963) Wood fibers with living protoplasts. New Phytol. 62: 91-98.

Fanshawe DB (1953) Akawaio Indian plant names. Caribbean Forester 14: 120-127.

Farmacopea Caribeña. Dr.M Delens, Av.2/CAlle 41, Apdo. 277, Urb. El Encanto, Oficentro El Encanto, Ofic. 503, Mérida, Venezuela.

Fernández del Valle A, Milano B, Vele G, Williams B, Rodríguez E & Michelangeli F (1999) Plantas medicinales de la región de Yutajé, Estado Amazonas. Mem Inst Biol Exp (MIBE) 2: 145-148.

Figueroa Marroquin H (1983) Enfermedades de los conquistadores. Edit. Universitario, Guatemala.

Finberg J (1960) Smoking compositions. ERD-Tobacco Products Co., Inc., USA.

Fischer F (1937) Beiträge zur Pharmakognosie der Plantaginalen und Rubialen. Anatomie des Laubblattes. Thesis, Basel, 82 pp.

Fischer F (1944) Praktikum der Pharmakognosie. Springer, Wien.

Foldats E (1969/1970) Ordchidaceae. In: T Lassner (Ed) 'Flora de Venezuela'. Vol XV, T. 1-5, Caracas.

Font Quer P (1963) Diccionario de Botánica. Ed. Labor, Barcelona, Madrid, Buenos Aires, Rio de Janeiro, Mexico, Montevideo.

Ford ES (1942) Anatomy and histology of the Eureka lemon. Bot Gaz 104: 288-305.

Franca F, Lago EL & Marsden PD (1996) Plants used in the treatment of leishmanian ulcer due to Leishmania (Viannia) braziliensis in an endemic area of Bahia, Brazil. Rev Soc Bras Mad Trop 29 (3): 229-232.

Francke G (1970) Nutzpflanzen der Tropen und Subtropen. Vol. I. Genussmittelliefernde Pflanzen; Kautschuk- und gummilief. Pfl. Öl- und fettlief. Pfl., Knollen- und Wurzelfrüchte, zuckerlief. Pfl. 3rd. rev. ed. Leipzig XIII. 1980. 460 pp.

Francke G (1984) Nutzpflanzen der Tropen und Subtropen. Vol II. Getreide, obstliefernde Pflanzen, Faserpflanzen. 4th. ed. Leipzig IX. 398 pp.

Freixa B, Cañigueral S, Adzet T & Vila R (1996) Antifungal activity of eleven Latin American plant drugs. Abstr. 44th. Ann Congr Soc Medicin Plt Res. Prague: 50.

Frischkorn CGB, Frischkorn HE & Carrazzoni E (1978) Cercaricidal activity of some essential oils of plants from Brazil. Naturwissenschaften 65: 480-483.

Frohne D & Jensen U (1979) Systematik des Pflanzenreichs. Unter besonderer Berücksichtigung chemischer Merkmale und pflanzlicher Drogen. 2. Aufl. G. Fischer Stuttgart New York.

Fryxell PA (1988) Systematic Botany Monographs, Vol 25. Malvaceae of Mexico. American Society of Plant Taxonomists.

Fuchs L (1932) Folia (Herba) *Plantaginis*. Pharmazeutische Presse. Wissenschaftlich praktisches Heft, 6 pp.

Fuentes E (1980) Los Yanomami y las plantas silvestres. Antropológica 54: 3-138.

Gallo P & Valeri H (1953) Investigación del poder antibiótico de algunos extractos de *Solanum mammosum* ('Manzanilla del diablo'), *Cassia occidentalis* ('Brusca macho') y *Cecropia* sp. ('Yagrumo'). REv Med Vet y Parasit 12 (1-4); 119-124, Caracas.

Gallo P & Valeri H (1953) Determinación del poder antibiótico de la semilla de aguacate (*Persea americana* C. Bauhin). Rev Med Vet y Parasit 12 (1-4): 125-129, Caracas.

Ganzera M & Stuppner H (1996) Phytochemical investigation of two Combretaceae: *Combretum glutinosum* and *Combretum quadrangulare*. Abstr. 44th. Ann Congr Soc Medicin Plt Res. Prague p. 137.

García EC de (1979) Estudio comparativo de la estructura y la ultraestructura de glándulas de especies arbóreas de manglar. Dr.-Thesis, Caracas.

García-Barriga H (1975) *Jatropha curcas*. In: Flora medicinal de Colombia.-Impr. Nac. Santafé de Bogotá, T. II: 102-104.

García-Barriga H (1975) Flora medicinal de Colombia. T. III. Inst. Cienc. Nat Univ Nac Colombia Impr Nac. Bogota.

García-Serrano R, Del Monte JP (1997) Uso ritual de las especies vegetales por las tribus Bribri y Cabecar de la zona Atlántica de Costa Rica. II Cingreso Internacional Etnobotánica '97. Mérida (Yucatán) Mexico.

Gassner G (1973) Mikroskopische Untersuchung pflanzlicher Lebensmittel (bearbeitet von F Bothe). G. Fischer, Stuttgart.

Gentry AH (1982) *Crescentia cujete*. In: Flora de Venezuela, p. 142-146.

Gentry A (1990) New nontimber forest products from Western South America. In: Suitable harvest and marketing of rain forest products. Proceed Meet Conserv Internat and Assoc Nac Conserv Nat. Panama City 1991 (M Plotkon & LM Famolare eds), Island Press.

Gentry AH (1992a) Bignoniaceae-Part II. Flora Neotropica 25 (II): 51-297. New York Botanical Garden.

Gentry AH (1992b) A synopsis of Bignoniaceae, Ethnobotany and Economic Botany. Ann Missouri Bot Gard 79: 53-64.

GEO 3, März 1990. Rezepte vom Schamanen (Mark Plotkin).

Gestetner B, Assa Y & Henis Y (1972) *Jatropha curcas*. In: Interaction of lucerne saponins with sterol. Biochem Biophys Acta 270: 181-187.

Ghosal S, Prasad BN & Lakshmi V (1996) Antiamoebic activity of Piper longum fruits against Entamoeba histolytica in vitro and in vivo. J. Ethnopharmacology 50: 167-170.

Gibbs RD (1974) Chemotaxnomoy of flowering plants. McGill-Queens University Press, Montreal London.

Girault L (1987) Kallawaya. Curanderos itinerantes de los Andes. Investigaciones sobre prácticas medicinales y mágicas. Servicio Gráfico Quipus, La Paz/Bolivia.

Girón LM, Freire V, Alonso A & Cáceres A (1991) Ethnobotanical survey of the medicinal flora used by the Caribs of Guatemala. J. Ethnopharmacol. 34: 173-187.

Glasby JS (1991) Dictionary of plants containing secondary metabolites. Taylor & Francis, London New York Philadelphia.

González F & Silva M (1987) A survey of plants with antifertility described in the South American folk medicine. Abstr. Princess Coongr. Bangkok, Thailand, 10-13. Dec., p. 20.

Gonzáles Esquinca AR, Luna Cazares L & Villatoro Vera RA (1997) El Quauhóloti (*Guazuma ulmifolia* Lam.) como recurso medicinal en enfermedades respirtorias y gastrointestinales. Il Congreso Internacional Etnobotánica '97. Mérida (Yucatán), Mexico.

Gonzales de Colmenares A, Vera V, Meza M (1999) Plantas aromáticas y medicinales en el Táchira. Mem Inst Biol Exper 2: 191-194.

Gonzáles de Colmenares N, Vera V, Meza M (1999) Plantas aromáticas y medicinales en el Táchira. Mem Inst Biol Exp (MIBE) 2: 191-194.

Gore UR & Taubenhaus JJ (1931): Anatomy of normal and acid injurd cotton roots. Bot Gaz 92: 436-441.

Gorecki P, Segiet-Kijawa E, Gaedcke F, Kurth H, Feistel B, Lehmann J & Roth-Ehrang R (1996) (6) - Gingerol as a standard for analytical evaluation of Ginger preparations - isolation from *Zingiber officinale* Roscoe and synthesis via Zingerone. Abstr. 44th. Ann Congr Soc Medicin Plt Res. Prague p. 134.

Gottwald H (1958) Handelshölzer. F. Holzmann, Hamburg.

Gragson TL & Tillet SS (1995) Aportes a la etnobotánica de Venezuela. 2. Etnobotánica de los Pumé. Ernista 5 (3): 89-120.

Grainge M & Ahmed S (1988) Handbook of plants with pest-control properties. J. Wiley & Sons, New York.

Granier Doyeux M (1965) Nativce hallucinogenic drugs: Piptadenias. Bull. on Narcotics 17 (2): 29-38.

Griggs B (1991) Green Pharmacy. Rochester, VI (Healing Arts Press).

Grosourdy R (1864) El médico botánico Criollo. Librería F. Brachet, Paris.

Grubert M (1974) Studies on the distribution of myxospermy among seeds and fruits of Angiospermae and its ecological importance. Acta Biol Venez 8: 315-551.

Grubert M (1981) Mucilage or gum in seeds and fruits of Angiosperms. A review.Minverva Publication, München, Fachserie Naturwissenschaften.

Guanchez F (1996) Plantas de uso medicinal, mágico y psicotrópico del Edo. Amazonas, Venezuela. Mimograf. 278 pp.

Guevara AP & col (1990) Antimutagens from *Momordica charantia* extract. Mutat Res 230: 121-126.

Gulati AS & Subba BC (1964) Drug analogs from the phenolic constituents of cashew nut shell liquid. Indian Journ Chem 2 (8): 337-338.

Gupta MP (1995) 270 plantas medicinales iberoamericanas. SECAB, Santiago de Bogota, Colombia, 'Ciencia y Tecnología No. 55.

Haarer AE (1963) Coffee growing. Oxford tropical Handbooks. Oxford University Press, London New York Toronto.

Hagen W von (1959) Südamerika ruft. Ullstein, Berlin Frankfurt Wien.

Hager's Handbuch der pharmazeutischen Praxis (1976). Springer, Berlin.

Handbook of South American Indians. Bureau of American Ethnological Bulletin.

Harborne JB (1977) Introduction to Ecological Biochemistry. Academic Press, London New York San Francisco.

Harborne (1998) Phytochemical Dictionary (bioactive compounds from plants). Taylor & Francis, UK.

Harborne JB, Boulter D & Turner BL (1971) Chemotaxonomy of the Leguminosae. Academic Press, London and New York.

Harborne B, Jeffery H, Baxter G, Moss (1998) A handbook of bioactive compounds from plants. 2nd. ed. Taylor & Francis, UK.

Hardy DE (1991) Tatoo time: Art from the heart. Honolulu, Hardy Marks Publication.

Harkin JM & Rowe JW (1971) Bark and its possible uses. U.S.D.A. Forest Service, Research Note, FPL-091.

Harnischfeger G & Cillien N (1996) Influence of *Cimicifuga racemosa* extract fractions on the proliferation of human carcinoma cells in vitro with regard to their estrogen receptor sensitivity. Abstr. 44th. Ann Congr Soc Medicin Plt Res, Prague p. 40.

Hartgerink HG (1962) Extract of *Bixa orellana* plant for use as a pharmaceutical. Appl. May 7, 1959. 2 pp.

Hartwell JL (1982) Plants used against cancer. Lawrence, MA (Quarterman Publications).

Hasegawa M, Rodríguez M, Bastidas E, Prieto F, López B, Orfila L, Arvelo F & Castillo A (1999) Fitoquímica y actividad biológica de los constituyentes alcaloidales del género Guatteria. Mem Inst Biol Exp 2 (MIBE): 195-198.

Hasegawa M, Rodríguez M, López B, Orfila L, Arvelo F & Castillo A (1999) Fitoquímica y actividad biológica de los constituyentes alcaloidales del género Guatteria. Mem Inst Biol Exper 2: 195-198.

Hausen BM & Vieluf IK (1988) Allergiepflanzen, Pflanzenallergene. Handbuch und Atlas der Allergie-induzierenden Wild- und Kulturpflanzen, EcoMed, Landsberg, München. 2. Aufl.

Hayword HE (1938) The structure of economic plants. New York.

Hegnauer R (1964) Chemotaxonomie der Pflanzen. Birkhäuser, Basel und Stuttgart. 743 pp.

Heinrich M (1989) Ethnobotanik der Tieflandmixe und phytochemische Untersuchungen von *Capraria biflora* L. Diss Bot 144, Berlin Stuttgart, Borntraeger.

Heinrich M, Velazco O & Ramos F (1990) Ethnobotanical report on the treatment of snake bites in Oaxaca, Mexico. Curare 13: 11-16.

Herbert JM et al. (1991) Verbascoside isolated from *Lantana camara*, an inhibitor of protein kinase C. Journal of Natural Products 54: 1595-1600.

Hernández L (1992) Gliederung, Struktur und floristische Zusammensetzung von Wäldern im Guayana Hochland, Venezuela. Gött. Beitr. Land- und Forstwirtsch. Tropen und Subtropen. Heft 70, 227 pp.

Hernández MM (1992) *Cecropia peltata*. In: Plantas Columbianas y su aplicación medicinal. Ed. Presencia, Santafé de Bogota, Colombia, p. 281-283.

Hernández Bermejo JE & León J (1992) Cultivos marginados, otra perspectiva de 1492. Colección FAO 'producción y protección vegetal' No, 26, Roma.

Hernández L, Williams P, Azuaje R, Rivas Y & Picón G (1994) Nombres indígenas y usos de algunas plantas de bosques de la Gran Sabana (Venezuela). Una introducción a la etnobotánica regional. Acta Bot Venez 17: 69-127.

Heywood VH (ed) (1971) The biology and chemistry of the Umbelliferae. Acad Press, London and New York.

Hiller K-O & Kato G (1996) Anxiolytic activity of psychotropic plant extracts. I. Test of ethanolic *Valeriana* extract STEI Val. Abstr. 44th. Ann Congr Soc Medicin Plt Res. Prague p. 65.

Hno Daniel (Brother Daniel) (1984) Utilización terapeútica de nuestras plantas medicinales. Publ Univ. La Salle. Ed. Tercer Mundo, Bogota.

Hobhouse H (1992) Fünf Pflanzen verändern die Welt. Klett-Cotta, Stuttgart (Deutsche Übersetzung).

Högermann CH (1987) Untersuchungen über die Blattnervatur einiger tropischer Baumarten von Venezolanisch-Guayana im Hinblick auf taxonomische Typisierbarkeit und Waldschichtung. Veröff. Naturf. Ges. Emden 1814: 9,3-D4.289 pp.

Holm T (1915) *Petiveria alliacea* L. Merck's Rep. 24: 266-270. See Bot Zbl. 131: 351, 1916.

Hostettmann K, Markston A, Maillard M & Hamburger M (1995) Phytochemistry of plants used in traditional medicine. Oxford, Clarendon Press.

Hostettmann K, Wolfender J-L & Rodriguez S (1996) Rapid detection and subsequent isolation of bioactive constituents of crude plant extracts. Abstr. 44th. Ann Congr Soc Medicin Plt Res. Prague P. 9.

Houk WG (1936) A note on the ovule and seed of *Coffea arabica*. Science 83: 464-465.

Houk W (1938) Endosperm and perisperm of *Coffea* with notes on the morphology of the ovule and seed development. Amer J Bot 25: 56-61.

Hoyos J (1976) Los árboles de Caracas. Soc Cienc Nat. La Salle, Caracas. Monografía No. 22.

Hoyos J (1978) Flora Tropical Ornamental. Soc Cienc Nat. La Salle, Monografía No. 24, Caracas.

Hoyos J (1985a) Flora emblemática de Venezuela. Publ Petrol Venez y sus empr filial. Gráficas Armitano, Caracas.

Hoyos FJ (1985b) *Cecropia peltata*. In: Flora de la isla Margarita. Soc Cienc Nat. La Salle, Caracas. Monografía 34. 643 pp.

Hoyos J (1989) Frutales en Venezuela. Soc Cienc Nat. La Salle, Caracas. Monografía No. 36.

Hoyos J (1992) Palma moriche, *Mauritia flexuosa* L. Natura 94: 18-23.

Hoyos J (1997a) El merey. Natura 107: 49-52.

Hoyos J (1997b) El Cacao o Cacaotero, *Theobroma cacao*. Natura 108: 14-19.

Huang Y, Cimanga K, Pieters L, van den Berghe D & Vlietinck AJ (1996) Antiviral and antibacterial activities of *Kyllinga brevifolia*. Abstr. 44th. Congr Soc Medicin Plt Res. Prague P. 28.

Hubert G (1921) Des Verbenacées utilisées en matieère medicale. Trav Lab Mat méd, Paris 13: 128 ff.

Humboldt AB v (1849) *Espeletia*. In: Geografía de las plantas o cuadro físico de los Andes equinocciales y de los países vecinos. In: Seminario de la Nueva Granada. Ed. Lessère. Paris. 546 pp.

Humboldt A & Bonpland A (1799-1800) Viaje a las regiones equinocciales del Nuevo Continente. Ed. Ministerio de Educación. Dir. Cultura, Caracas, 1956.

Hyde BB (1970) Mucilage-producing cells in the seed coat of *Plantago ovata*: developmental fine structure. Amer J Bot 57: 1197-1206.

Internat. Scient. Sympos. on natural phenols in plant resistance (1993) Freising, Germany.

International Society for the history of medicine, Glasgow G1 IQE UK.

International. Society for Horticultural Science. Internat. Science symposium on Natural Phenols in Plant Resistance. 1993. Fac Agr & Hortic. TU Munich, Freising - Weihenstephan.

Jackson BD (1965) A glossary of botanic terms (with their derivation and accent). 4th. ed. G Duckworth, London. Hafner Publ. New York.

Jackson DD (1989) Apotheke der Natur: der tropische Regenwald. Smithsonian Institution, Washington.

Jackson SJ, Houghton PJ, Photiou A & Retsas S (1996) *Kigelia pinnata* (Bignoniaceae) as a potential source of oncolytic drugs. Abstr. 44th. Ann Congr Soc Medicin Plt Res. Prague P. 21.

Jäger AK & van Staden J (1996) Zulu medicinal plants with antiinflammatory activity. Abstr. 44th. Ann Congr Soc Medicin Plt Res. Prague P. 29.

Jean S & Francis BJ (1969) Annatto tree (*Bixa orellana*) occurrence, cultivation, preparation and uses. Trop Sci 11: 97-102.

Jean de Lagardo B (1996) Plants used against malaria or fever by specialised traditional healers in the Dja biosphree Reserve (Cameroon). Abstr. 44th. Ann Congr Soc Medicin Plt Res. Prague P. 54.

Jensen W (1965) The chemical utilization of bark. (Unpubl. paper presented at Finland Symposium).

Jilka C & Col (1983) In vivo antitumor activity of the bitter melon, *Momordica charantia*. Cancer Res 43: 5151-5155.

Jiménez LC & Bernal HY (1992) El Inchi, Caryodendron orinocense Karsten. 2nd. ed. SECAB, Bogota.

Jolly SS (1966) Pharmacognostical aspect of the stem-bark of *Soymida febrifuga* A Juss - Ind For 9: 469-471.

Joshi AC (1938) A note on the morphology of the ovule of Rubiaceae with special reference to *Cinchona* and *Coffea*. Curr Sci 7: 236-237.

Joyce Ch (1994) Earthly goods: Medicine hunting in the rain forest. Little Brown. 304 pp. $ 22.95.

Juscafresa B (1875) Enciclopedia ilustrada. Flora medicinal, tóxica, aromática, condimenticia. Ed Aedos, España.

Karsten G & Weber U (1946) Lehrbuch der Pharmakognosie. Fischer, Jena.

Katzner K (1996) The languages of the world. New ed. Routledge & Kegan, London & New York.

Kayser O & Kolodzieg H (1996) Antibacterial activity of coumarins and extracts from *Pelargonium sidoides* and *Pelargonium reniforme*. Abstr. 44th. Ann Congr Soc Medicin Plt Res. Prague P. 48.

Killip EP (1938) The american species of Passifloraceae. Chicago Field Mus of Nat Hist.

Klucking EP (1986) Leaf venation patterns. Vol. I. Annonaceae. J. Cramer (Borntraeger) Berlin, Stuttgart. 256 pp.

Klucking EP (1987) Leaf venation patterns. Vol. II. Lauraceae. J. Cramer (Borntraeger) Berlin, Stuttgart. 216 pp.

Klucking EP (1988) Leaf venation patterns. Vol. 3. Myrtaceae. J. Cramer (Borntraeger) Berlin, Stuttgart. 279 pp.

Klucking EP (1989) Leaf venation patterns. Vol. 4. Melastomaceae. J. Cramer (Borntraeger) Berlin, Stuttgart. 283 pp.

Klucking EP (1991) Leaf venation patterns. Vol. 5. Combretaceae. J. Cramer (Borntraeger) Berlin, Stuttgart. 219 pp.

Klucking EP (1992) Leaf venation patterns. Vol. 6. Flacourtiaceae. J. Cramer (Borntraeger) Berlin, Stuttgart. 272 pp.

Klucking EP (1995) Leaf venation patterns. Vol. 7. The classification of leaf venation patterns. J. Cramer (Borntraeger) Berlin, Stuttgart. 96 pp.

Kotikova NM & Khaletskii AM (1969) Chemical composition of bog *Gnaphalium*. Tr. Leningrad. Khim Farm Inst No. 28: 95–102.

Kreher R (1989) Chemische und immunologische Untersuchungen der Drogen Dionaea muscipula, Tabebuia avellanedae, Euphorbia resinifera und Daphne mezereum, sowie ihrer Präparate. Diss Fak Chemie und Pharm. Ludwig-Maximilians-Universität München: 36–89.

Kučera A (1997) Wörterbuch der Chemie. Dictionary of Chemistry. Deutsch-Englisch, Englisch-Deutsch, Brandstetter, Wiesbaden.

Kupchan SM, Knox JR & Kelsey JE (1964) Calotropin, a citotoxic principle isolated from *Asclepias curassavica*. Science 146 (3652): 1685–1686.

Kupchan SM, Sigel WC et al. (1976) *Jatropha gossypifolia*. In: tumor inhibitors. Journ Am Chem Soc 98: 2295–2300.

Kupchan SM, Wanger CR & Bryan RF (1970) Jatrophone, a novel macrocyclic diterpenoid tumor inhibitor from *Jatropha gossypiifolia*. Journ Amer Chem Soc 92: 4476–4477.

Kusnick C, Jansen R, Gräfe U, Liberra K & Lindequist U (1996) Isolation and characterization of antibacterial active substances from *Kirschteiniothelia maritima*. Abstr. 44th. Ann Congr Soc Medicin Plt Res. Prague P. 21.

Laguna Hernández G, García Argáez A & Pérez Amador B MC (1997) Histoquímica de la raíz de 3 especies de Convolculaceae relacionada con la activitdad purgante. II Congreso Internacional Etnobotánica '97. Mérida (Yucatán), Mexico.

Lagunez Rivera L, Ortiz Hernández YD, Santiago Garciá P & Maldonado Bravo SE (1997) El achiote: fuente principal de ingresos de las mujeres mazatecas de la Sierra Norte de Oaxaca. II Congreso Internacional Etnobotánica '97. Mérida (Yucatán), Mexico.

Lamprecht H (1986) Waldbau in den Tropen. Paul Parey, Hamburg. 318 pp.

Lange D & Schippmann U (1997) Trade survey of medicinal plants in Germany. A contribution to international plant species conservation. Bundesamt für Naturschutz, Bonn.

Lasser T (1955) Nuestro destino frente a nuestra naturaleza. Ed. MAC (Bil. Cultura Rural 3). Col Rec Nat Renov. Caracas.

Lasser T (1956) El cultivo del ramio para la prducción de fibras y como forraje. Revista Pecuaria 256: p. 3.

Latharp DW (1973) The antiquity and importance of long-distance trade relationships in the moist tropics of pre-Colombian South America. World Archeology 5: 170–186.

Lawless J (1992) The encyclopedia of essential oils. The complete guide to the use of aromatica in aroma therapy, herbalism, health and well-being. Element books Ltd., Shaftesbury, Dorset, Great Britain.

Lawrence GHM (1951) Taxonomy of vascular plants. MacMillan, New York.

Lee-Huang S et al. (1990) Map-30. A new inhibitor of HIV-infection and replication. FEBS LETT 272: 12–18.

León E, Agostini G & Rodríguez P (1988) Morfología y anatomía foliar de especies leñosas Venezolanas de *Belencita, Capparis, Morisonia* y *Steriphoma*. Mem Soc Ci Nat, La Salle 48 (129): 169–193.

León de Pinto G, González de Troconis N, Rojas AC & Leal E (1989) Espectros de R.M.N. de la goma de *Albizia lebbeck* y de sus productos degradados. Aplicación a su elucidación estructural. Acta Cientif. Venez. 40: 335–340.

Liener IE (1969) *Jatropha curcas*. In: Toxic constituents of plant food stuff. Academic Press, Inc., London.

Liermann K (1926) Beiträge zur vergleichenden Anatomie der Wurzeln einiger pharmazeutisch verwendeter Umbelliferen. Thesis Basel, pp. 88 (See. Just's Jber., Pt. 2, 448).

Liese W & Parameswaran N (1972) On the variation of cell length within the bark of some tropical hardwood species. Res Trends in Plt Anat. KA Chowdhury Commem. Vol. ed. AKM Ghouse & Yunus. McGraw-Hill, New Delhi.

Lifchitz A (1981) Plantas Medicinales. Ed. Arneo, Costa Rica.

Lindley J (1951) Glosología o de los términos usados en Botánica. Transl and augment. Ed. Rothe Tucumán. Fund Miq Lillo, Misc 15.

Lindorf H (1978) Estudios morfológicos y anatómicos en las semillas de Zygophyllaceae de Venezuela. Mem Soc Cienc Nat. La Salle 110: 169–193.

Lindorf H (1980a) Estructura foliar de 15 monocotiledóneas de sombra del bosque nublado de Rancho Grande. I. Bifaciales. Araceae, Marantaceae y Musaceae. La Salle 113: 19–71. mem Soc Cienc Nat.

Lindorf H (1980b) Estructura foliar de 15 monocotiledóneas de sombra del bosque nublado de Rancho Grande. II. Eqifaciales: Arecaceae (Palmae) y Cyclanthaceae. Mem Soc Cienc Nat. La Salle 114: 9–57.

Lindorf H (1984) Estudio de los tejidos accesorios en la madera caulinar y radical de cinco árboles. Memoria. Soc Cienc Nat. La Salle 44 (122): 79–94.

Lindorf H (1987) Determinación del tipo estructural en la madera caulinar y radical de cinco árboles. Bol Soc Venez Cien Nat 41 (144): 113–129.

Lindorf H (1988) Contribución al establecimiento de diferencias anatómicas entre madera caulinar y radical. Bol Soc Venez Cien Nat 42 (145): 143–178.

Lindorf H (1992) Anatomía foliar de especies de un bosque húmedo en el Territorio Federal Amazonas, Venezuela. Memoria Soc Cien Na. La Salle 52 (137): 65–91.

Lindorf H (1993) Blattstruktur von Pflanzen aus einem feuchten Tropenwald in Venezuela. Bot Jb Syst 115: 45–61.

Lindorf H (1994) Eco-anatomical wood features of species from a very dry tropical forest. IAWA Journal 15 (4): 361–376.

Lindorf H (1997a) Wood and leaf anatomy in *Sessea corymbiflora* from an ecological perspective. IAWA Journal 18 (2): 157–168.

Lindorf H (1997b) Características antómicas de valor diagnóstico en la corteza de las especies suramericanas de *Uncaria* (Rubiaceae). II Congreso Internacional Etnobotánica '97. Mérida (Yucatán), Mexico.

Lindorf H (1998) Correlaciones eco-anatómicas entre la madera y la hoja. Mem Inst Biol Exper 2: (MIBE) 209–212.

Lindorf H (1999) Estudios anatómicos en plantas útiles de Venezuela. Mem Inst Biol Exp 1: 153–156.

Lindorf H (2000) Identificación de las uñas de gato (*Uncaria guianensis* y *Uncaria tomentosa*) mediante la anatomía de la corteza y de la madera. IX Congreso Italo-Latinoamericano de Etnomedicina. Urbino, Italia.

Lindorf H, de Parisca L & Rodriguez P (1985) Botánica – Clasificación, Estructura, Reproducción. EBUC, Caracas. 584 pp.

Lindorf H & Roth I (1992) Anatomía de algunas plantas útiles de Venezuela. Anales Congr Internat. Etnobotánica 92. Cordoba, España.

López-Palacios S (1984, 1987) Usos médicos de plantas comunes. ULA, Mérida, Venez. 3a. ed. 1987.

Lötschert W (1973) Baumborke als Anzeiger von Luftverschmutzungen. Umschau 73: 403–404.

Lozoya-Meckes M & Mellado-Campos V (1985) Is the *Tecoma stans* infusion an antidiabetic remedy? (Antidiabetic effect of tecomine and tecostanine). Journ Ethnopharmacology 14 (1985): 1–9. Elsevier Scient Publ. Ireland Ltd.

Luces de Febres Z (1963) Las gramíneas del Distrito Federal. Dir Rec Nat Ren. Ministerio de Agricultura y Cria, Caracas.

Maguire B & Steyermark J (1981) The botany of the Guayana highlands. Part XI. Mem. New York Bot. Garden 32: 1–21.

Maksuyitina NP, Nikitina NI, Lipkan GN, Gorin AG & Voitenko IN (1978) Chemical composition and hypocholesterolemic action of some drugs from *Plantago major* leaves. I. Polyphenolic compounds. Farm Zh 4: 56–61.

Manara B (1995) Guía ilustrada del jardín Botánico de Caracas. Fund Polar Fund Inst Bot. Venez.

Manara B (1996) Paria en el tiempo y en el corazón. Carúpano, Venezuela. Fund. Thomas Merle. 132 pp.

Mancera Aquayo M. El cacao ¿Moneda alimento? El Universal. 23.2.1992. Caracas, Venezuela.

Manfred L (1977) Siete mil recetas botánicas (12th. ed.). Ed. Kier SA, Buenos Aires.

Manfred L (1982) 7000 recetas botánicas a base de 1300 plantas medicinales. 13th ed.-Ed. Kier, Buenos Aires.

Mapes C (1993) El Amaranto. Etnobotánica 1: 3-4.

Maria de los Àngeles, Vielma: Una planta silvestre tiene más calcio que la leche. El Universal. 12.12.1991. Caracas, Venezuela.

Martin RE (1963) Thermal properties of bark. Forest Prod J 13: 419-426.

Martin RE (1967) Interim equilibrium moisture content values of bark. Forest Prod J 17: 30-32.

Martin RE (1968) Interim valumetric expansion values for bark. Forest Prod J 18: 52.

Martin RE (1969) Characterization of southern pine barks. Forest Prod J 19: 23-30.

Martin RE (1970) Directional thermal conductivity ratios of bark. Holzforschung 24: 26-30.

Martin RE & Crist JB (1968) Selected physico-mechanical properties of eastern tree barks. Forest Prod J 18: 54-60.

Martin RE & Gray GR (1971) pH of southern pine barks. Forest Prod J 21: 49-52.

Mater J (1967) Bark Utilization, a review and projections. Forest Prod Jour 17 (12) 15-20.

Mayolo KKA, DE (1989) Peruvian natural dye plants. Economic Botany 43: 181-191.

Mazzuca M, Klaiber I, Vogler B, Kraus W & Balzaretti V (1996) Bioassay guided isolation of biologically active constituents from *Prosopis* seeds. Anstr. 44th. Ann Congr Soc Medicin Plt Res. Prague P. 138.

Medicinal Plant Conservation (Medicinal Plant Specialist Group) Newsletter of the IUCN, Species survival Commission 1996 ff.

Mehta R, Arora OP & Metha M (1981) Chemical investigation of some Rajasthan desert plants. Indian Jour Chem Sect B 20 B (9): 834.

Melnik M (1995) Productos forestales comestibles p. 295-310. In: Carrillo A & Perera M. Amazonas modernidad en tradición. Venezuela GTZ, SADA Amazonas, CAIAH, ORPIA. 391 pp.

Metcalfe CR (1960) Anatomy of the Monocotyledons. I. Gramnieae. Clarendon Press, Oxford.

Metcalfe CR & Chalk L (1950) Anatomy of the Dicotyledons. Vol. I & II. Clarendon Press, Oxford.

Mibe (1999) Memorias del Instituto de Biología Experimental. I. Simposio Venezolano de Etnobotánica. Vol. 2 (1): 1-224.

Mila I, Smal A, Expert D & Scalbert A (1993) Tannin antimicrobial properties through iron deprivation. Abstr Internat Scient Sympos Nat. Phenols in plant resistance. Freising, Germany.

Miss VL, Flores JSI, Mena G & Rosado M (1997) Estudio de la actividad biológical del extracto acuoso y metanolico de hojas y semillas de *Persea americana* Mill (Aguacate). Congreso.

Molau V (1983) *Bixa orellana*. In: Flora of Ecuador. Nordic Journ Bot 20: 4-5.

Morell (1984) cited by Werner (1996) Fund. La Salle. Monogr. 41.

Morton J (1961) The cashew's brighter future. Economic Bottany 15: 57-78.

Morton J (1965) El Cundeamor (*Momordica charantia* L.) planta comestible, medicinal y tóxica. Rev Fac Farm. UCV 6 (14): 63-72, Caracas.

Morton J (1969) Folk-remedy plants and esophagal cancer in Coro, Venezuela. The Morris Arboretum Bulletin 25: 24-34.

Morton JF (1981) Atlas of medicinal plants of Middle America. Charles C. Thomas Publ., Springfield, III. USA: 1420 pp.

Morton JF (1994) Lantana or red sage (*Lantana camara* L., Verbenaceae), notorious weed and popular garden flower; some cases of poisoning in Florida. Economic Botany 48: 259-1994.

Moto DJ (1983) In vivo antitumor activity of the bitter melon, *Momordica charantia*. Cancer Res 43 (11): 5151-5155.

Muñoz M, Milano MB, Fernández A, Vele G, Williams B, Rodríguez E & Michelangeli F (1999) Reporte preliminar de actividad biológica de plantas de uso medicinal colectadas en Yutajé, Estado Amazonas. Mem Inst Biol Exp (MIBE) 2: 207-210.

Murphey WK, Beall FC, Cutter BE & Baldwin RC (1970) Selected chemical and physical properties of several bark species. Forest Prod Jour 20 (2): 58-59.

Murthy SS & Rao JVS (1978) *Sida acuta*. In: Ecological studies and chemical control of some terrestrial weeds. Comp Physiol Ecol 3: 61-64.

Mutis JC (1760-1790) Diario de observaciones (Real Expedición Botánica del Nueva Reino de Granada). Instituto Colombiano de Cultura Hispánica. Ed. Minerva. Bogotá. 1958.

Nagel GA (1993) 'Alternative' Methoden der Krebsbehandlung. Freiburger Universitätsblätter 119 (1): 89-93.

Nair MNB (1995) Some notes on gum and resin ducts and cavities in Angiosperms. Encyclopedia of Plant Anatomy. IX, 4. Ed. M. Iqbal 'The cambial derivatives': 322-342. Borntraeger, Berlin Stuttgart.

Narbaiza I (1999) La guama (Inga edulis) un recurso etnobotánico con potencial en la producción animal. Mem Inst Biol Exper 2: 69-72.

Narváez A & Stauffer F (1999) Products derived from palms at the Puerto Ayacucho markets in Amazonas State, Venezuela. Palms 43: 122-129.

Narváez Cordova A & Stauffer F (1999) Productos de palmas (Arecaceae) en los mercados de Puerto Ayacucho, Estado Amazonas, Venezuela. Mem Inst Biol Exp (MIBE) 2: 73-76.

Narváez A, Stauffer F & Gertsch J (2000) Contribución al estudio de la etnobotánica de las palmas del Estado Amazonas, Venezuela. In: Estudio de las palmas del Amazonas Venezolano (F. Stauffer ed.) Scientia Guianae.

Nash D & Moreno NP (1981) *Heliotropium indicum*. Flora de Veracruz 18: 86-89.

Nataraja K & Patil JS (1980) *Sida acuta*. In: Introduction of organogenesis in cultures of *Sida* and *Abutilon* in vitro. Proc Nat Symp Plt Tissue Cult Genet Manipul, Somatic hybrid plt. cells: 319-327.

Ndounga M & Ouama JM (1996) Antibacterial and antifungal studies on leaves and flowers essential oils of *Ocimim gratissimum* L. grown in Congo. Abstr. 44th. Ann Congr Soc Medicin Plt Res Prsague P. 50.

Ndounga M, Fliniaux MA, Carme B & Jacquin A (1996) In vitro antimalarial activity of *Triclisia patens* leaves from Congo. Abstr. 44th. Ann Congr Soc Medicin Plt Res Prague. P. 54.

Netolotzky F (1926) Anatomie der Angiospermen - Samen - Handb. Pfl. Anat. (ed. Linsbauer) 10 (14). Borntraeger, Berlin.

Newman LF (1985) Women's Medicine. A cross-cultural study of indigenous fertility regulation. New Brunswick, NJ Rutgers U. Pr.

Niedenzu F (1898) Punicaceae. In: Engler-Prantl. Die Natürlichen Pflanzenfamilien III. Teil, 7. Abt.: 22-25, Leipzig.

Niethammer A & Tietz N (1961) Samen und Früchte des Handels und der Industrie. Junk's-Gravenhage.

Norman J (1991) Das große Buch der Gewürze. Aarau, Schweiz, AT-Verlag.

Ocampo R: El empleo de la medicina tradicional en las comunidades de Cocles, Talamanca.

Oliva.Esteva F (1970) 12 árboles de Venezuela. Colección Científica Especial. Monte Avila Ed., Caracas.

Ossowski A (1929) Polish with English summary (see Metcalfe & Chalk 1950).

Pacheco JJ & Perez L (1989) Malezas de Venezuela. Tipogr. & Litogr. Central, San Cristobal, Venezuela.

Paez F (1998) Plantas útiles de Bajo Llano Apureno. Jardín Botánico UNELLEZ, Barinas, Venezuela.
Panapo edit. Caracas. 273 plantas medicinales de Venezuela (1988). (sin autor!).
Parameswaran N (1971) Über die Struktur der tropischen Baumrinde und ihre Verwertungsmöglichkeiten. Forstarchiv 10: 192–198.
Parameswaran N (1973) Untersuchungen über einige strukturelle chemische und physikalische Eigenschaften tropischer Baumrinden. Habil. Thesis, Univ. Hamburg.
Parameswaran N (1974) pH and buffering capacity of some tropical barks. J Indian Acad of Wood Sci 5: 28–31.
Parameswaran N & Liese W (1970) Mikroskopie der Rinde tropischer Holzarten. In: H. Freund (ed.). Handbuch der Mikroskopie in der Technik. Bd. V, Teil 1: 227–306.
Parques Nacionales de Venezuela. Creole Petrol Corp., Minist Agr y Cria, Caracas.
Paula JE & Hamburgo de L (1973) Anatomía de *Anacardium spruceanum* Bth. ex Engl. (Anacardiaceae de Amazonia). Acta Amazonica 3: 39–53.
Pennington TD & Sarukhan J (1968) Arboles tropicales de Mexico. UN, FAO & Sarukhan. Library of Congress Cat Card No 68–57357.
Penzig O (1887) Studi botanici sugli agrumi e sulle piante affini. Ann Agr 1887, Roma. Tipograf. Eredi Botta.
Perez-Arbeláez E (1956) Plantas útiles de Colombia. 3rd ed Rivadeneyra SA, Madrid.
Pérez-García F, Cartaño C, Muschietti L, Martino V, Marín E, Adzet T & Cañigueral S (1996) Nevadensin an anti-inflammatory and antioxidant principle from *Baccharis grisebachii*. Abstr. 44th. Ann Congr Soc Medicin Plt Res Prague P. 62.
Perkins KD & Payne WW (1978) *Momordica charantia*. In: Guide to the poisonous and irritant plants of Florida. Fla Coop Ext Serv. University of Florida, Gainsville.
Pertchik BH & White WC (1951) Flowering trees of the Caribben. Rinehart, New York & Toronto.
Peters Ch, Gentry A & Mendelsohn R (1989) Valuation of Amazonian rainforest. Nature 339: 655–656.
Pharmazie in unserer Zeit. 8. Jahrg. Nr. 4, 1979.
Phillipson JD, Hemingway RS & Ridsdale CE (1978) Alkaloids of *Uncaria*. Part V. Their occurrence and chemotaxonomy. Lloydia (41): 503–570.
Phytochemical Dictionary (Harborne, Baxter, Moss) 1999. Taylor & Francis.
Picaza S (1948) La medicina de las guerras de independencia cubana. Revista Bimestre Cubana 62 (4/5/6): 56–58.
Pigott CJ (1964) Coconut growing. Oxford Tropical Handbooks. Oxford University Press, London.
Pilger R (1937) Plantaginaceae. In: Engler's 'Das Pflanzenreich' IV, 269.
Pintão AM, Salomé Pais M, Coley H, Kelland LR & Judson IR (1996) the biotechnological production and anticancer activity of natural isothiocyanates. Abstr. 44th. Ann Congr Soc Medicin Plt Res Prague. P. 22.
Pittier H (1926) Manual de las plantas usuales de Venezuela (y su splemento). Fund. Eugenio Mendoza, New ed. 1970.
Pittier H, Lasser T, Schnee L, Luces de Febres Z & Badillo V (1947) Catálogo de la Flora de Venezuela, T. I & II, Caracas.
273 Plantas Medicinales de Venezuela (1988). Ed. Panapo, Caracas. Sin autor!
Planta Medica – Journal of medicinal plant research. Pharmacology, Toxicology and Botany. G. Thieme, Germany.
Plonczak M (1997) El uso de los bosques naturales de Venezuela. Quebracho 5: 63–69 (Argentina).
Pollak-Eltz A (1982) Folk-Medicine in Venezuela. Acta Ethnológica et linguística. Nr. 53. Wien-Föhrenau.
Pollak-Eltz A (1987) La medicina popular en Venezuela. Bibl Acad Nac Historia 86, Caracas.
Pompa G (1974) Medicamentos Indigenos. Ed. América, S.A. Miami, Florida and Panama. 313 pp.
Ponessa de Mercado G (1997) Plantas medicinales del valle de Tafi, Tucumán, Rep. Argentina II. Congreso Internacional Ethnobotánica '97. Mérida (Yucatán), Mexico.
Pons, Fachwörterbuch Medizin. PH Collin E Schick, R Livesey. Klett, Stuttgart Dresden.
Popovic MT, Gasic OS, Simmonds M & Malenčic (1996) Phytochemical and antimicrobial investigations of some Liliaceae and Cichoriaceae species. Abstr. 44th. Ann Congr Soc Medicin Plt Res Prague. P. 46.
Poulsen VA (1877) Pulpaens udvikling his *Citrus*.-Botan. Notis. 4: 97–103.
Prakash A, Varma RK & Ghosal S (1981) Chemical constituents of the Malvaceae. III. Alkaloid constituents of *Sida acuta, Sida humilis, Sida rhombifolia* and *Sida spinosa*. Planta Medica 43: 384–388.
Prance GT (1994) Ethnobotany and the search for new drugs. Ciba Foundation Symposium 185.
Prance GT & Kallunki AJ eds. (1984) Ethnobotany of the Neotropics. Advances in economical botany. Vol. I. New York Botanical Garden. Programa Interciencia de Recursos Biológicos (PIRB). *Bixa orellana*. In: II Seminario Interamericano de recursos biológicos nuevos Quirana, Colombia. Ed. Guadalupe Ltda: 173–174.
Programa Interciencia de recursos biológicos (PIRB) (1984) Bixa orellana Linneo. In: Seminario Interamericano de recursos biológicos nuevos. Quirana, Colombia. Edit. Guadalupe LTD: 173–174.
Puente M de la, Duran R & Peña L (1997) Analisis de variabilidad en el contenido de metabolitos con actividad biológica en frutos de *Solanum hirtum* Vahl. Congreso.
Purseglove JW (1968) Tropical Crops. Dicotyledons 1 and 2. Longmans Green, London.
Purseglove JW, Brown VG, Green CL & Robbins SRJ (1981) Spices. Vol. I and II. Longman, London.
Ramia M (1962) Datos etnobotánicos sobre los Indios Yaruros. Acta Biol Venez 3: 141–147.
Ramos E (1997) Tratamiento de términos botánicos indígenos en crónicas y otros documentos de los siglos XVI–XVIII. Incorpoiración de la lengua Española. Congreso.
Rauh W (1950) Morphologie der Nutzpflanzen. Quelle & Meyer, Heidelberg.
Record SJ & Hess RW (1943) Timbers of the New World. Yale University Press, New Haven.
Reynel C, Alban J, Leon J & Diaz J (1990) Etnobotánica Campa-Ashanica. Fac Ci Forest Univ Noe Agraria La Molina, Lima.
Rivera Lorca JA & Kuvera JC (1997) *Jatropha curcas* un recurso vegetal altamente promisorio para los sistemas agropecuarios tropicales. Congreso.
Robineau L Ed. (1991) Hacia una farmacopea caribeña: Investigacon científica y usos populares de plantas medicinales en el Caribe. ENDA-Caribe y Univ. Nac Autón Honduras, Reunión Tramil 4, Tela, Honduras, Santo Domingo.
Rodríguez P (1976) Estudio sobre frutos carnosos y sus semillas de las Rubiaceae de Venezuela. Acta Bot Venez 11: 283–383.
Rodríguez P (1983) Plantas medicinales utilizadas en el país de venta en herbolarios. Thesis, Caracas.
Rodríguez PM (1983) Plantas de la medicina popular Venezolana de venta en herbolarios. Publ Soc Venez Cienc Nat, Caracas.
Rodríguez P (1985) Aporte al inventario de las plantas medicinales de Venezuela. Parte I & II. Caracas.
Rodríguez Mesa P (1989) Notas de apoyo al estudio de las gramíneas. Ed. Americana, Caracas.
Römpp-Lexikon (1997) Naturstoffe. Ed. Fugman et al. Thieme Verlag.
Rogers GT (1997) Medical Dictionary (Diccionario medico). English–Spanish, Inglés–Espanol. McGrawHill, New York.
Roig FA & Mesa JT (1953) Diccionario botánico de nombres vulgares cubanos. La Habana.
Rollet B (1980) Intérèt de l'étude des écores dans la détermination des arbres tropicaux sur pied. Bois et forêts des Tropiques 194: 3–28.
Rollet B (1982) Intérêt de l'étude des écorces dans la déterminatión des arbres tropicaux sur pied. Bois et forêts des Tropiques 195: 31–50.

Rollet B, Högermann CH & Roth I (1990) Stratification of tropical forests, as seen in leaf structure. Part 2. Tasks veget. Sci. 21. Kluwer Acad. Publ., Dordrecht Boston Lancaster.

Roersch C (1994) Plantas medicinales en el Sur Andino del Peru. 2 vol., 1 map. 1188 pp. Doct Diss Cath Univ, Nijmegen, Holland.

Roosmalen MGM (1985) Fruits of the Guianan Flora. Inst Syst Bot Utrecht Univ Sylvicult. Dpt. Wageningen, Agric. Univ.

Roth I (1954) Entwicklung und histogenetischer Vergleich der Nektar- und Verdauungsdrüsen von *Nepenthes*. Planta 43: 361–378.

Roth I (1963a) Entwicklung der ausdauernden Epidermis sowie der primären Rinde des Stammes von *Cercidium torreyanum* im Laufe des Dickenwachstums. Österr Bot Ztschr 110: 1–19.

Roth I (1963b) Estudios fitomorfológicos en plantas tropicales. Acta Cient Venez 14: 41–44.

Roth I (1965) Histogenese der Lentizellen am Hypokotyl von *Rhizophora mangle* L. Österr Bot Ztschr 112: 640–653.

Roth I (1966) Anatomía de las plantas superiores. EBUC Col Ciencias Biológicas, Caracas.

Roth I (1968) Desarrollo de los nectarios extraflorales en *Passiflora foetida* L. Acta Biol Venez 6: 44–49.

Roth I (1969a) El crecimiento primario en anchura del eje de *Pedilanthus tithymaloides* (L.) Poir. Acta Biol Venez 6: 50–59.

Roth I (1969b) Estructure anatómica de la hoja de algunos xerófitos de la playa de Cumaná. Acta Biol Venez 6: 77–86.

Roth I (1969c) Características estructurales de la corteza de árboles tropicales en zonas húmedas. Darwiniana (Argentina) 15: 115–127.

Roth I (1969d) Estructura anatómica de la corteza de algunas especies Venezolanas de Anacardiaceae. Acta Biol Venez 6: 146–160.

Roth I (1969e) Estructura anaomica de la corteza de algunas especies arbóreas Venezolanas de Lecythidaceae. Acta Bot Venez 4: 89–117.

Roth I (1969f) Estructura anatómica de la corteza de algunas especies arbóreas Venezolanas de Sapotaceae. Acta Bot Venez 4: 119–150.

Roth I (1969g) Estructura anatómica de la corteza de algunas especies arbóreas Venezolanas de Bignoniaceae. Acta Bot Venez 4: 157–174.

Roth I (1969h) Estructura anatómica de la corteza de algunas especies arbóreas de Guttiferae. Acta Bot Venez 4: 175–203.

Roth I (1969i) Estructura anatómica de la corteza de algunas especies arbóreas de Rutaceae. Acta Bot Venez 4: 205–226.

Roth I (1969j) Estructura anatómica de la corteza de algunas especies arbóreas de Combretaceae. Acta Bot Venez 4: 227–239.

Roth I (1969k) Estructura anatómica de la corteza de algunas especies arbóreas de Myristicaceae. Acta Bot Venez 4: 241–257.

Roth I (1971) Estructura anatómica de la corteza de algunas especies arbóreas de Meliaceae. Acta Bot Venez 6: 239–259.

Roth I (1972a) Estructura anatómica de la corteza de algunas especies arbóreas Venezolanas de Capparidaceae. Acta Bot Venez 7: 33–46.

Roth I (1972b) Estructura anatómica de la corteza de algunas especies arbóreas Venezolanas de Vochysiaceae. Acta Bot Venez 7: 47–65.

Roth I (1972c) Estructura anatómica de la corteza de algunas especies arbóreas Venezolanas de Bombacaceae. Acta Bot Venez 7: 67–82.

Roth I (1972d) Estructura anatómica de la corteza de algunas especies arbóreas Venezolanas de Celastraceae. Acta Bot Venez 7: 83–100.

Roth I (1972e) Estructura anatómica de la corteza de algunas especies arbóreas Venezolanas pertenecientes a 3 familias diferentes (Rhamnaceae, Rhizophoraceae y Sabiaceae). Acta Bot Venez 7: 101–119.

Roth I (1973) Vida y uso de los frutos. Monte Avila Ed., Caracas.

Roth I (1973a) Estructura anatómica de la corteza de 6 especies arbóreas de las familias: Araliaceae, Dichipetalaceae, Lacistemaceae, Olacaceae, Opiliaceae y Quiinaceae. Acta Biol Venez 8: 103–129.

Roth I (1973b) Estructura anatómica de la corteza de algunas especies arbóreas Venezolanas de Boraginaceae. Acta Biol Venez 8: 131–153.

Roth I (1973c) Estructura anatómica de la corteza de algunas especies arbóreas Venezolanas de Rosaceae. Acta Bot Venez 8: 121–161.

Roth I (1973d) Estructura anatómica de la corteza de algunas especies arbóreas Venezolanos de Lauraceae. Acta Bot Venez 8: 255–279.

Roth I (1973e) Anatomía de las hojos de plantas de los páramos Venezolanos. 2. *Espeletia* (Compositae). Acta Bot Venez 8: 281–310.

Roth I (1974a) Estructura anatómica de la corteza de algunas especies arbóreas Venezolanas de Annonaceae. Acta Bot Venez 9: 177–195.

Roth I (1974b) Desarrollo y estructura anatómica del merey: *Anacardium occidentale* L. Acta Bot Venez 9: 197–223.

Roth I (1974c) Morfología, anatomía y desarrollo de la hoja pinada y de las glándulas laminales en *Passiflora* (Passifloraceae). Acta Bot Venez 9: 363–380.

Roth I (1974d) Anatomía de las hojas de los páramos Venezolanos. 1. *Hinterhubera imbricata* (Compositae). Acta Bot Venez 9:381–398.

Roth I (1974e) Passiflora: La variabilidad de la hoja pinada. Anatomía y desarrollo de la glándulas en la lamina de la hoja. Acta Bot Venez 9: 363–380.

Roth I (1976) Estructura interna de los domacios foliares de *Tococa* (Melastomaceae). Acta Biol Venez 9: 221–258.

Roth I (1977a) Anatomía y textura foliar de plantas de la Guayana Venezolana. Acta Bot Venez 12: 79–146.

Roth I (1977b) *Arracacia xanthorrhiza*: el tubérculo del apio criollo y su estructura interna. Acta Bot Venez 12: 147–170.

Roth I (1977c) Estructura anatómica de la corteza de algunas especies arbóreas Venezolanas de Mimosaceae. Acta Bot Venez 12: 293–366.

Roth I (1977c'd) Estructura anatómica de la corteza de 2 especies arbóreas de Melastomaceae encontradas en la Guayana Venezolanas. Acta Bot Venez 12: 367–384.

Roth I (1977e) Estructura anatómica de la corteza de *Caryocar nuciferum* L., Caryocaraceae. Acta Bot Venez 12: 385–394.

Roth I (1977f) Über die anisokitylen Keimlinge von *Gyranthera caribensis*. Bot Jahrb Syst 97: 515–547.

Roth I (1977g) Fruits of Angiosperms. Encycl. Plt Anat X ptl Borntraeger, Berlin Stuttgart

Roth I (1980) Blattstruktur von Pflanzen aus feuchten Tropenwäldern. Bot Jb Syst 101: 489–525.

Roth I (1981) Structural patterns of tropical barks. Borntraeger, Berlin Stuttgart

Roth I (1983a) Estructure anatómica de la corteza de algunas especies arbóreas Venezolanas de Sapindaceae. Acta Bot Venez 14: 7–27.

Roth I (1983b) Estructure anatómica de la corteza de unas especies arbóreas Venezolanas de Araliaceae. Acta Bot Venez 14: 29–39.

Roth I (1983c) Estructure anatómica de la corteza de unas especies arbóreas Venezolanas de Hernandiaceae. Acta Bot Venez 14: 41–47.

Roth I (1983d) Estructure anatómica de la corteza de unas especies arbóreas Venezolanas de Humiriaceae. Acta Bot Venez 14: 49–59.

Roth I (1983e) Estructure anatómica de la corteza de unas especies arbóreas Venezolanas de Elaeocarpaceae. Acta Bot Venez 14: 61–69.

Roth I (1983f) Estructure anatómica de la corteza de unas especies arbóreas Venezolanas de Rubiaceae. Acta Bot Venez 14: 71–95.

Roth I (1983g) Estructure anatómica de la corteza de 2 especies arbóreas Venezolanas de Simaroubaceae. Acta Bot Venez 14: 97–107.

Roth I (1983h) Estructure anatómica de la corteza de algunas especies arbóreas Venezolanas de Flacourtiaceae. Acta Bot Venez 14: 109–127.

Roth I (1983i) Estructure anatómica de la corteza de 2 especies arbóreas Venezolanas de Sterculiaceae. Acta Bot Venez 14: 129–139.

Roth I (1983j) Estructure anatómica de la corteza de 2 especies arbóreas Venezolanas de Polygonaceae. Acta Bot Venez 14: 141–155.

Roth I (1983k) Estructure anatómica de la corteza de algunas especies arbóreas Venezolanas de Verbenanceae. Acta Bot Venez 14: 157–171.

Roth I (1983l) Estructure anatómica de la corteza de algunas especies Venezolanas de Tiliaceae. Acta Bot Venez 14: 173–185.

Roth I (1983m) Estructure anatómica de la corteza de una especie arbórea Venezolana de Moraceae. Acta Bot Venez 14: 237–245.

Roth I (1983n) Estructure anatómica de la corteza de 2 especies arbóreas Venezolanas de Nyctaginaceae. Acta Bot Venez 14: 247–257.

Roth I (1984) Stratification of tropical forests as seen in leaf structure. Junk Publ, The Hague, Boston Lancaster.

Roth I (1987a) Estructure anatómica de la corteza de algunas especies arbóreas Venezolanas de Papilionaceae. Acta Bot Venez 15: 13–22.

Roth I (1987b) Estructure anatómica de la corteza de algunas especies arbóreas Venezolanas de Burseraceae. Acta Bot Venez 15: 23–41.

Roth I (1987c) Estructure anatómica de la corteza de algunas especies arbóreas Venezolanas de Apocynaceae. Acta Bot Venez 15: 43–57.

Roth I (1987d) Estructure anatómica de la corteza de algunas especies arbóreas Venezolanas de Caesalpiniaceae. Acta Bot Venez 15: 59–86.

Roth I (1987e) Estructure anatómica de la corteza de algunas especies arbóreas Venezolanas de Myrtaceae. Acta Bot Venez 15: 49–64.

Roth I In press. Estructure anatómica de la corteza de algunas especies arbóreas Venezolanas de Euphorbiaceae.

Roth I (1987f) Stratification of a tropical forest, as seen in dispersal types. Tasks for veget sci 17 (ed H Lieth, HA Mooney). Dr. W. Junk Publ., Dordrecht Boston Lancaster.

Roth I (1990) Leaf structure of a Venezuela cloud forest (in relation to the microclimate). Encyclop Plt Anat XIV, 1. Borntrager, Berlin Stuttgart.

Roth I (1992) Leaf structure: Coastal vegetation and mangroves of Venezuela. Borntraeger, Berlin Stuttgart.

Roth I (1994) Strukturelle Anpassungsmerkmale sowie taxonomische Struktureigenheiten der Hochgebirgspflanze *Krameria lappacea*. Bot Jb Syst 116: 145–155.

Roth I (1995) Leaf structure: Montane regions of Venezuela (with an excursion to Argentina). Borntraeger, Berlin Stuttgart.

Roth I (1996) Microscopic venation patterns of leaves (and their importance in the distinction of (tropical) species). Encyclopedia of Plt Anat Vol XIV, P. 4, Borntraeger, Berlin Stuttgart.

Roth I & Ascensio J (1977) Crecimiento anómalo del eje de varias especies de *Bauhinia* y su relación con la filotaxis. Acta Bot Venez 12: 23–77.

Roth I & Carrasquel N (1969) The peeling off of the epidermis in fruits of *Jacquinia loeflingii* Carrasquel. Acta Biol Venez 6: 139–145.

Roth I & Carvajal Sandoval A (1983) Anatomía de las hojas de plantas de los páramos Venezolanos. 3. *Chaetolepsis*. Acta Bot Venez 14: 187–211.

Roth I & Clausnitzer I (1969a) Los tipos de desarrollo estomático en las Angiospermas. Acta Bot Venez 4: 259–292.

Roth I & Clausnitzer I (1969b) El descolgamiento del follaje y su explicación fisiológica en *Brownea*. Phyton (Argentina) 26: 87–105.

Roth I & Clausnitzer I (1972) Desarrollo y anatomía del fruto y de la semilla de *Carica papaya* L. (Lechosa). Acta Bot Venez 7: 187–206.

Roth I & Cova O M (1969) Estructura anatómica de la corteza de algunas especies de Moráceas Venezolanas. Acta Biol Venez 6: 60–76.

Roth I & Delgado M (1968a) Desarrollo de la epidermis pluriestratificada y de la corteza primaria en el tronco de *Parkinsonia aculeate* L. Acta Bot Venez 3: 265–277.

Roth I & Delgado M (1968b) Sobre la formación del ritidoma en el tronco de *Parkinsonia aculeata* L. Acta Bot Venez 3: 279–291.

Roth I & Gimémez del Bolzón A-M (1997) Argentine Chaco forests (Dendrology, tree structure and economic use). I. The semi-arid Chaco. Encycl Plt Anat XIV, 5. Borntraeger, Berlin Stuttgart.

Roth I & Lindorf H (1971a) Anatomía y desarrollo del fruto y de la semilla del café. Acta Bot Venez 6: 197–238.

Roth I & Lindorf H (1971b) Desarrollo y anatomía del fruto y de la semilla de *theobroma cacao*. Acta Bot Venez 6: 261–295.

Roth I & Lindorf H (1972a) Anatomía y desarrollo del fruto y de la semilla de *Achras zapota* L. Acta Bot Venez 7: 121–141.

Roth I & Lindorf H (1972b) Desarrollo y anatomía del fruto y de la semilla de la granada (*Punica granatum* L.). Acta Bot Venez 7: 143–162.

Roth I & Lindorf H (1972c) Desarrollo y anatomía del fruto y de la semilla de *Citrus*. Acta Bot Venez 7: 163–186.

Roth I & Lindorf H (1974) Desarrollo y anatomía del fruto y de la semilla de *Myristica fragrans* van Houtt. Acta Bot Venez 9: 149–176.

Roth I & Lindorf H (1990) Blatt- und Rindenstruktur von *Tepuianthus auyantepuiensis*, Tepuianthaceae, einer neuen Familie aus Venezuela. Bot Jb Syst 111: 403–421.

Roth I & Lindorf H (1991) Leaf structure of 2 species of *Pachira* indigenous of Venezuela from different habitats. Bot Jb 113: 203–219.

Roth I & Lindorf H (1993) Structure and use of economically important Venezuelan wild plants. Consol Rep activ. EC Andean Pact Countries: 29–32.

Roth I & Lindorf H (1997) Structure and use of the neotropical wild plant *Chenopodium ambrosioides* L. Plant structure and morphogenesis: 19–24, India.

Roth I & Mérida de Bifano T (1971) Morpholocal and anatomical studies of leaves of the plants of a Venezuelan cloud forest. I. Shape and size of the leaves. Acta Biol Venez 7: 127–155.

Roth I & Mérida de Bifano T (1979) Morphological and anatomical studies of leaves of the plants of a Venezuelan cloud forest. II. Stomata density and stomatal patterns. Acta Biol Venez 10: 69–107.

Roth I, Mérida T & Lindorf H (1986) Morfología y anatomía foliar de plantas de la selva nublada. La Selva Nublada De Rancho Grande, Parque Nacional 'Henri Pittier', Ed. Otto Huber. Acta Ci Venez Fondo Edit, Caracas.

Roth I & Mérola G (1969) Sobre la consistencia de la hoja: La consistencia de los catáfilos de *Brownea grandiceps*. Acta Bot Venez 4: 295–301.

Roth I & Yee S (1991) Ökologisch-taxonomische Untersuchungen über Leitbündeldichte und Leitbündelmuster bei Angiospermen-Laubblättern. Bot Jb Syst 113: 7–71.

Roth L & Kornmann K (1997) Duftpflanzen-Pflanzendüfte (Ätherische Öle und Riechstoffe). Ecomed, Landsberg.

Royero R, Narbaiza I & Gil E (1999) El papel de la mujer Piaroa en la conservación de diversidad genética de los cultivos. Mem Inst Biol Exper (MIBE) UVV Caracas, Vol. 2: 81–84.

Ruiz Zapata T (1999) Capparidaceae venezolanas y sus usos. Mem Inst Biol Exp (MIBE) 2: 157–160.

Ruppolt W & Högermann Ch (1986) Schulexperimente mit Gewürzen. Praxis Schriftenreihe Biologie: Bd. 33. Aulis Verlag, Deubner, Köln.

Rutter RA (1990) Catáloga de plantas útiles de la Amazonia peruana. Serie Comuniades y cultivos Peruanas 22. Centro amazónico de lenguas autóctonas Peruanas 'Hugo Pesce' Yarina cocna Pucallpa, Peru, 238 pp.

Sachsse H (1965) Bau und Eigenschaften der Rinde und ihre derzeitigen Nutzungs- und Verwendungsarten. Forstarch 36: 25–32.

Saksena SK (1971) Study of antifertility activity of the leaves of *Momordica* (Karela). Indian J Physiol Pharmacol 15: 79–80.

Samson JA (1980) Tropical fruits. 250 pp. London. Reprint 1982.

Santos JK (1925) A pharmacognostical study of *Chenopodium ambrosioides* Linn. from the Philippines. Philipp J Sci 28: 529–542.

Sauer CO (1959) Age and area of American cultivated plants. Actas 33rd Int Congr Am 1: 215–229.

Scannone TH (1948) Contribución al estudio de algunas quinas Venezolanas. Rev Ven San y Asis Soc 13 (5–6): 495–518, Caracas.

Schauerte A, Alban S & Franz G (1996) Anticoagulant and fibrinolytic activity of new heparinoids sythesised using natural polysaccharides as starting polymers. Abstr. 44th Ann Congr Soc Medicin Plt Res Prague. P. 70.

Schery RW (1956) Plantas utiles al hombre (Botánica Económica). Salvat Ed, S.A. Barcelona. 335 pp.

Schneider A & Baums A (1970) Wohin mit der Rinde? DRW-Verlag, Stuttgart.

Schmidt E (1951) Überseehölzer. F. Haller, Berlin-Dahlem.

Schmithüsen F (1982) Venezuela. Goldstadt-Reiseführer, Pforzheim.

Schmook B & Serralta L (1997) El sikilte o piñón (*Jatropha curcas* L.) en la península de Yucatán-Usos y Potencial. II. Congreso Internacional Etnobotánica '97. Mérida (Yucatán), Mexico.

Schnee L (1960) Plantas comnes de Venezuela. Rev Fac Agron Univ Central, Venez. Maracay.

Schneider A & Baums A (1970) Wohin mit der Rinde? DRW-Verlag, Stuttgart.

Schultes RE (1990) The place of ethnobotany in the ethnopharmacologic search for psychotomimetic drugs. Curare 13: 31–48.

Schultes RE & Hoffmann A (1979) Plants of the Gods. Origins of hallucinogenic use. McGraw Hill, New York.

Schultes RE & Raffauf RF (1990) The healing forest (Medicinal and toxic plants of the Northwest Amazonia). Dioscorides Press. Portland, Oregon (USA).

Schwanitz F (1967) Die Evolution der Kulturpflanzen. Bayr Landwirtschafts-Verlag, München Basel Wien.

Schweppe H (1992) Handbuch der Naturfarbstoffe (Vorkommen, Verwendung, Nachweis). Ecomed, Landsberg/Lech.

Scott FM & Baker KC (1947) Anatomy of Washington navel orange rind in relation to waterspot. Bot Gaz 108: 459–475.

Seaforth CE (1991) Natural products in Caribbean folk medicine. Univ West Indies, St. Augustine, Trinidad.

Seaforth CE, Adams CD & Silvester Y (1983) A guide to medicinal plants of Trinidad & Tobago – Common Wealth Secretariat Marlborough House, Pall Mall, London SW7Y 5HX.

SECAB: Especies vegetales promisorias. Ed. JE Correa & HY Bernal. Bogota, Colombia.

Seidenmann J (1966) Stärke-Atlas. Parey, Berlin Hamburg.

Serra Gay A (1999) Estudio etnobotánico de plantas medicinales en Cuyagua y Cata (Edo. Aragua), Venezuela. Mem Inst Biol Exper 2: 161–164, Caracas.

Seul H (1989) Leben im Wald. Der Kampf der Kautschukzapfer für die Erhaltung des Regenwaldes. In: B. Wilcek & E. Pabst (eds.), Amazonien, ein Lebensraum wird zerstört: 77–86. Raben Verlag München (Germany).

Silva A (1999) Vegetation der halbimmergrünen submontanen Wälder, landwirtschaftlich genutzten Flächen und ihrer Brachestadien auf der Halbinsel Paria, Venezuela. Dr. Thesis, Freiburg.

Simmonds NW (1962) The evolution of the bananas. Tropical Science Series. Longmans, London.

Sinnott EW & Bloch R (1943) Development of the fibrous net in the fruit of various races of *Luffa cylindrica*. Bot Gaz 105: 90–99.

Small J (1913) The identification value of hairs. Pharm J Ser 4, 36: 587–591.

Smet PAGM de (1991) Is there any danger in using traditional medicine? Journ Ethnopharmacol 31: 181–192.

Solereder H (1908) Systematic Anatomy of the Dicotyledons. Vol. I & II. Printed in India, New Delhi.

Sosa S, Kastner U, Glas S, Jurenitsch J, Tubaro A & della Loggia R (1996) Anti-inflammatory activity of sesquiterpene derivatives from the genus *Achillea*. Abstr. 44th. Ann Congr Soc Medicin Plt Res Prague. P. 53.

Souleles C, Chinou I & Loukis A (1996) Antibacterial activity of the essential oils from cultivated plants of *Origanum vulgare* L. Abstr. 44th. Ann Congr Soc Medicin Plt Res Prague. P. 47.

Souza Cardoso de J, Resende EMPT & Prous A (1997) 'Silos' prehistóricos – Valle del Rio Peruacú (Brasil central). II. Congreso Internacional Etnobotanico '97. Mérida (Yucatán), Mexico.

Sprecher v. Bernegg A (1929) Tropische und subtropische Weltwirtschaftspflanzen, ihre Geschichte, Kultur und volkswirtschaftliche Bedeutung.

Sprecher v. Bernegg A (1929) Tropische und subtropische Weltwirtschaftsplanzen. I. Stärke- und Zuckerpflanzen. Enke, Stuttgart.

Sprecher v. Bernegg A (1962) Tropische und subtropische Weltwirtschaftspflanzen. II. Ölpflanzen. 2. Aufl. Enke, Stuttgart.

Sprecher v. Bernegg A (1960) Tropische und subtropische Weltwirtschaftspflanzen. III. Genusspflanzen. 1. Kakao und Kola (1934). 2. Kaffee (1960). Tee und Mate (1936). 2. Aufl. Enke, Stuttgart.

Srivastava LM (1964) Anatomy, chemistry and physiology of barks. Int Rev For Res 1: 204–277. Academic Press, NY.

Staub JE, Frederik L & Marty TL (1986) Electrophoretic variation in cross-compatible wild diploid species of *Cucumis*. Canad Journ Bot 65: 792–798.

Stauffer F (in press) Estudio de las palmas (Arecaceae) del Estado Amazonas, Venezuela.

Steinberg P (1994) *Uncaria tomentosa* (uña de gato). Una hierba maravillosa de la selva peruana. Townsend letter for Doctors. s/n.

Stephens EL (1910) The development of the seed coat of *Carica papaya*. Ann Bot London 24: 607–610.

Steyermark J (1967) Flora del Auyantepui. Acta Bot Venez. 2.

Steyermark JA & Huber O (1978) Flora del Avila. Soc Venez Cienc Nat 'Vollmer' y Minist. Ambiente Y Recursos Nat Renov, Caracas.

Strauss LC (1952) The use of wild plants in tropical South America. Economic Botany 6: 252–270.

Strnad M (1996) Anticancer compounds derived from plant hormones. Abstr. 44th. Ann Congr Soc Medicin Plt Res Prague. P. 8.

Swarnalakshmi T et al. (1981) Antiinflammatory activity of (–)-epcatechin, a bioflavonoid isolated from *Anacardium occidentale*. Indian J Phram Sci 43: 205–208.

Tamayo F (1961) Exploraciones botánicas en el Edo. Bolívar Bol Soc Venez Ci Nat. 22: 25–180.

Taylor KI (1974) Sanuma Fauna: prohibitions and classifications. Fund. La – Salle de Ciencias Naturales, Caracas.

Tejerina Oller JL & Arenas Martizen R (2001) Coimw o Amaranto – CAB, Ciencia Y Tecnologia No 86.

Teppner H (1993) Die Tomate. Kat. OÖ. Landesmus. Neue Folge 61: 189–212.

Thomas OO (1989) Re-examination of the antimicrobial activities of *Xylopia aethiopica, Carica papaya, Ocimum gratissimum* and *Jatropha curcas*. Fitoterapia 60: 147–166.

Thuiller Y & Gionobarber O (1971) Antihypertensive effect of Anacardiaceae bark. Patent S. Africa 7004, 836: 17 pp.

Tierra Yanomami (1984) (K. Weidmann, A. Perez, J. Lizot). 2a ed. O. Todtmann, Caracas.

Tillett S (1995) Guía introductoria de Etnobotánica. Ed. UCV Caracas, Fac. Farmacia. ISBN 980-00-0859-4.

Tillett S & Fuentes E (1983) Los Yanomami y las plantas silvestres. Antropológica 54: 3-138. E. Fuentes, Caracas.

Tillett S & Haiek NG (1999) Veinticinco años de investicagiones en el herbario Dr. Víctor Manuel Ovalles (MYF). Mem Inst Biol Exp (MIBE) 2: 165-168.

Tobler F (1938) Deutsche Faserpflanzen und Pflanzenfasern. München Berlin.

Tomlinson PB (1966) Anatomical data in the classification of Commelinaceae. J Linn Soc (Bot) 59: 371-395.

Tomlinson PB (1969) Anatomy of the Monocotyledons. III. Commelinales, Zingiberales. Ed. CR Metcalfe. Clarendon Press, Oxford.

Toro de Vieira N (1999) Aspectos odontológicos de la medicina tradicional en el Distrito Monagas (Edo. Guárico), Venezuela. Mem Inst Biol Exper (MIBE) Vol. 2: 93-96.

Torres AM (1968) Algunas plantas leñosas de Cumaná. Ed. Universitario de Oriente, Cumaná.

Trapp G (1933) A study of the foliar endodermis in the Plantaginaceae. Trans Roy Soc Edinb 57: 523-546.

Trockenbrodt M (1990) Survey and discussion of the terminology used in bark anatomy. IAWA Bull n.s. 11: 146-166.

Troll W (1964/69) Die Infloreszenzen. Vol. 1 and 2, Fischer, Stuttgart.

Tung C-L (1935) Development and vascular anatomy of the flower of *Punica granatum* L. Bull Chinese Bot Soc 1: 108-128.

Ugent D, Pozorski S & Pozorski T (1986) Archaeological manioc (*Manihot*) from coastal Peru. Economic Botany 40: 78-102.

Unger W (1926) Beitrag zur anatomischen Kenntnis der Kräuterdrogen (Folia *Plantaginis*) Arch Pharm Berl 264: 754-762.

Uphof JCTH (1968) Dictionary of economic plants. J Cramer, Lehre.

Usubillaga A, Khouri N, Abad A & Rojas L (1997) Estudio de los componentes de la resina del *Protium crenatum* Sandwith. II. Congreso Internacional Etnobotánica '97. Merida (Yucatán), Mexico.

Valery H & Gimeno F (1953) Estudio fito-quimico-toxicologico de los frutos de aguacate (*Persea americana*, C. Bauhin, Pinax 441, 1623). Rev Med Vet y Paras 12 (1-4): 131-165, Caracas.

Valery H & Narváez P (1950) Investigaciones sobre el yagrumo. Rev Med Vet y Paras 9 (1-4): 103-116, Caracas.

Vaněk T & Valterová I: Society for medicinal plant research. Abstracts. Sept. 1996, Prague.

Vareschi V (1970) Flora de los páramos. Univ. Los Andes. Ed. Rectorado. Mérida.

Vareschi V (1979) Plantas entre el mar y la tierra - una florula de las playas del Caribe. Ed. E. Armitano, Caracas.

Vareschi V (1980) Vegetationsökologie der Tropen. Ulmer, Stuttgart.

Vareschi V (1986) Cinco breves ensayos ecológicos acerca de la selva virgen Rancho Grande. In: La Selva Nublada de Rancho Grande. Ed. O. Huber. Acta Cientif Venez, Caracas.

Vargas L, Vila R, Lozano N, Cañigueral S & Adzet T (1996) Screening of some Peruvian medicinal plants for antifungal activity. Abstr. 44th. Ann Congr Soc Medicin Plt Res Prague. P. 51.

Vaughan JG (1970) The structure and utilization of oil seeds. Chapman & Hall, London.

Vaughan JG & Geissler CA (1997) The new Oxford book of food plants. Oxford Univ Press, Oxford New York Tokyo.

Velásquez J (1971) Contribución al conocimiento de las especies del género *Cecropia* L. (Moraceae) 'Yagrumos' de Venezuela. Acta Bot Venez 6: 25-64.

Vele G, Milano B, Fernández A, Williams B & Michelangeli F (1999) Plantas medicinales recopiladas de la etnobotánica nacional y el uso herbal por la población Venezolana. Mem Inst Biol Exper 2: 169-172.

Vélez-Boza F (1943) Cuatro observaciones clínicas de los efecttos producidos por la ingestión de semillas del coco de mono. Mem Soc Cienc Nat, La Salle 3 (8): 28. Caracas.

Vélez-Boza F, Velez de GV (1990) Plantas alimenticias de Venezuela. Fund Bigott Soc Cienc Nat, La Salle, Monogr. No. 37. Caracas.

Vélez-Salaz F (1959) Plantas medicinales de Venezuela. Ed. Las Novedades, Caracas.

Venanzi de F (1952) Acción de la túa-túa (*Jatropha gossypifolia*) sobre la diabetes aloxánica de la rata. Acta Cient Venez 3 (2): 61-52, Caracas.

Venezuela, fascinante (1987). K. Weidmann (fotografía). O. Todtmann, Caracas.

Vera A (1994) Inventario de plantas medicinales del Táchira. Rev Cient Univ Nac Exper, Táchira 8: 55-70.

Vera VA & Pinto Pabon D (1999) Inventario de las plantas medicinales del Táchira. Mem Inst Biol Exper (MIBE) 2: 97-100.

Vila M-A (1970) Conceptos de geografía histórica de Venezuela. Monte Avila, Ed. Caracas.

Volz K-R (1971) Über den pH-Wert einiger Baumrinden. Holz-Zentralblatt 123. Oktober 1971.

Volz K-R (1972) Verwertung von Baumrinde. Umschau 72: 796-797.

Volz K-R (1973) Herstellung und Eigenschaften von Fichten-, Kiefern- und Buchenrindenplatten. Holz als Roh- und Werkstoff 31: 221-229.

Wagner H (1993) Pharmazeutische Biologie, Bd. 2. Drogen und ihre Inhaltsstoffe. 5. Aufl. G. Fischer. 522 pp. Stuttgart New York

Waldron KW & Selvendran RR (1993) Bioactive cell wall and related components from herbal products and edible plant organs as protective factors. In: Waldron KW, Johnson IT & Fenwick GR (eds.). Food and cancer prevention: chemical and biological aspects. The Royal Soc Chemistry, Cambridge: 307-326.

Waltzl B & Leitzmann C (1995) Bioaktive Substanzen in Lebensmitteln. Hippokrates-Verlag, Stuttgart.

Weber D (1907) Beiträge zur Anatomie einiger pharmakognostisch wichtiger Samen und Früchte. Diss Univ Bern (Druck Budapest).

Weber D (1908) Beiträge zur Anatomie der Samen und Früchte einiger wichtiger Pflanzenfamilien. Növ Közl. 7: 228-233. (Ref Just's Bot Jb 36, I: (40)-(42).

Welhinda H et al. (1986) Extrapancreatic effects of *Momordica charantia* in rats. J Ethnopharmacol 17: 247-255.

Wellmann FL (1961) World Crops Books (ed. N. Polunin). Coffee (Botany, Cultivation and Utilization). L. Hill, London New York.

Weniger B et al. (1984) Seminario TRAMIL I. Port au Prince.

Werker E (1997) Seed Anatomy. Encycl Plt Anat X, 3. Borntraeger, Berlin Stuttgart.

Wettstein R (1935) Handb. syst. Botanik. Deuticke, Leipzig und Wien.

Wiesner J v (1962-1968) 5. Aufl. ed. C. v. Regel. 1. Tanning materials. 2. Antibiotics. 3. Organic acids. 4. Insecticieds. 5. Glycosides. 6. Starch. 7. Ethereal oils.

Wilbert W (1995) Conceptos etnológicos Warao. Scientia Guianae 5: 335-370.

Wilbert W (1996) Fitoterapia Warao. Monogr 41. Fund. La Salle de Cienc Nat, Caracas.

Wilbert W & Haiek G (1991) Phytochemical screening of a Warao pharmacopoeia employed to treat gastrointestinal disorders. Journ Ethnopharmacology 34: 7-11.

Willemann JJ Alkaloids in foods. (Aristeguieta 1976).

Williams KA (1966) Oils and fatty foods. 4th. ed. J&A Churchill.

Williams RO (1951) The useful and ornamental plants in Trinidad and Tobago. Guardian Commerc Print. Port-of-Spain.

Wolters B (1992) Jahrtausende vor Kolumbus. Indianische Kulturpflanzen und Arzneidrogen. Deutsch Apotheker Ztg. 40: 2067-2075.

Wolters B (1994) Arzneipflanzen und Volksmedizin Chiles. Dtsch Apotheker Ztg 134 (39), Sept. 1994: 25–40. Dtsch Apotheker-Verlag, Stuttgart.

Wolters B (1994/95). Drogen, Pfeilgift und Indianermedizin (Arzneipflanzen aus Südamerika). URS Freund Verlag, 86926 Greifenberg 22.

Wong CM et al (1985) Screening of *Trichosanthes kirlowii, Momordica charantia* and *Cucurbita maxima* (Cucurbitaceae) for compounds with antilipolytic activity. Ethnopharmacol 13: 313–321.

Wong W (1976) Some folk medicinal plants from Trinidad. Economic Botany 30: 103–142.

Woodson RE jr (1951) *Aspidosperma quebracho-blanco*. Flora of Panama. Ann Missouri Bot Gard. 38: 180–183.

Woodson RE jr & Schery RW et col (1965) Flora of Panama: P. VI. Malvaceae. A. Robyns. Ann Missouri Bot Gard. 52.

Wyatt-Smith J (1954) Suggested definitions of field characters (for use in the identification of tropical forest trees in Malaya). Malayan For 17: 170–183.

Yanes A, Hasegawa M & Finol HJ (2000) Actividad de extractos y fracciones de cuatro plantas contra Tripanosomiasis americana. Revista Latinoamericana de Química. Vol. 28, Supl Esp 2000. X. Congr. Italo-Latinoamericano de Etnomédica, Urbino, Sept. 2000.

Yajia ME, Ama AG & Aguilera I (2000) Estudios anatómicos y etnobotánicos de especies medicinales comercializados para las afecciones cardíacas, respiratorias y neoplásticas, en el Dpto. Capital, Prov. Misiones, Argentina. II Parte: Monocotiledoneas. X Congr. Italo-Latinoamericana, Urbino, Sept. 2000. Revista Latinoamericana de Química, Vol. 28, Supl Esp 2000.

Zin J (1922) La salud por medio de las plantas medicinales. 3rd ed Santiago de Chile, 685 pp.

Subject index

* Indicates a figure

B

C

D

E

F

G

H

I

J

K

L

M

N

O

P

Q

R

S

T

U

V

W

X

Y

Z

GPSR Compliance
The European Union's (EU) General Product Safety Regulation (GPSR) is a set of rules that requires consumer products to be safe and our obligations to ensure this.

If you have any concerns about our products, you can contact us on

ProductSafety@springernature.com

In case Publisher is established outside the EU, the EU authorized representative is:

Springer Nature Customer Service Center GmbH
Europaplatz 3
69115 Heidelberg, Germany

www.ingramcontent.com/pod-product-compliance
Ingram Content Group UK Ltd.
Pitfield, Milton Keynes, MK11 3LW, UK
UKHW061705190726
13853UKWH00008B/2413
9783662046999